CAMBRIDGE LIBRARY COLLECTION

Books of enduring scholarly value

History

The books reissued in this series include accounts of historical events and movements by eye-witnesses and contemporaries, as well as landmark studies that assembled significant source materials or developed new historiographical methods. The series includes work in social, political and military history on a wide range of periods and regions, giving modern scholars ready access to influential publications of the past.

The Book of the Farm

Henry Stephens (1795–1874) was a farmer and later a writer on agriculture. After attending lectures on chemistry and agriculture at the University of Edinburgh he boarded with a Berwickshire farmer, George Brown, and gained practical experience of agricultural work. In 1820 Stephens acquired his own farm, on which he used modern and experimental farming methods. In 1837 he sold the farm, and devoted the rest of his life to writing guides to farming for the use of inexperienced farmers. These influential volumes, first published in 1842, and reprinted many times (the 1844 edition is reissued here) contain Stephens' detailed descriptions of contemporary farming practices. He describes in meticulous detail all aspects of farming, including livestock care and slaughter, dairying, irrigation practices and crop culture. Arranged by season and including copious high-quality illustrations of farming equipment, these extremely popular and fascinating volumes were considered the standard work on practical Victorian agriculture, and were used as an 'instructional manual' on the recent BBC television series, 'The Victorian Farm'. Volume 3 describes farming tasks performed in summer and autumn.

Cambridge University Press has long been a pioneer in the reissuing of out-of-print titles from its own backlist, producing digital reprints of books that are still sought after by scholars and students but could not be reprinted economically using traditional technology. The Cambridge Library Collection extends this activity to a wider range of books which are still of importance to researchers and professionals, either for the source material they contain, or as landmarks in the history of their academic discipline.

Drawing from the world-renowned collections in the Cambridge University Library, and guided by the advice of experts in each subject area, Cambridge University Press is using state-of-the-art scanning machines in its own Printing House to capture the content of each book selected for inclusion. The files are processed to give a consistently clear, crisp image, and the books finished to the high quality standard for which the Press is recognised around the world. The latest print-on-demand technology ensures that the books will remain available indefinitely, and that orders for single or multiple copies can quickly be supplied.

The Cambridge Library Collection will bring back to life books of enduring scholarly value (including out-of-copyright works originally issued by other publishers) across a wide range of disciplines in the humanities and social sciences and in science and technology.

The Book of the Farm

Detailing the Labours of the Farmer, Farm-steward, Ploughman, Shepherd, Hedger, Cattle-man, Field-worker, and Dairy-maid

Volume 3

Henry Stephens

Cambridge University Press

CAMBRIDGE UNIVERSITY PRESS

Cambridge, New York, Melbourne, Madrid, Cape Town, Singapore,
São Paolo, Delhi, Dubai, Tokyo, Mexico City

Published in the United States of America by Cambridge University Press, New York

www.cambridge.org
Information on this title: www.cambridge.org/9781108024969

This edition first published 1844
This digitally printed version 2010

ISBN 978-1-108-02496-9 Paperback

THE

BOOK OF THE FARM,

DETAILING THE LABOURS OF THE

FARMER, FARM-STEWARD, PLOUGHMAN, SHEPHERD, HEDGER, CATTLE-MAN, FIELD-WORKER, AND DAIRY-MAID.

BY

HENRY STEPHENS, F.R.S.E.

IN THREE VOLUMES.

WITH NUMEROUS ILLUSTRATIONS.

VOL. III.

"Wherefore, come on, O young husbandman!
Learn the culture proper to each kind."
VIRGIL.

WILLIAM BLACKWOOD AND SONS,
EDINBURGH AND LONDON.
MDCCCXLIV.

CONTENTS OF VOLUME III.

SUMMER.

" The season now is all delight,
  Sweet smile the passing hours,
And SUMMER's pleasures, at their height,
  Are sweet as are her flowers ;
The purple morning waken'd soon,
  The mid-day's gleaming din,
Grey evening with her silver moon,—
  Are sweet to mingle in.

" How sweet the fanning breeze is felt
  Breath'd through the dancing boughs ;
How sweet the rural voices melt
  From distant sheep and cows.
The lovely green of wood and hill,
  The hummings in the air,
Serenely in my breast instil,
  The rapture reigning there."

CLARE.

I have represented Winter, in the agricultural sense, as the season of dormancy, in which every thing remains in a state of quiescence. In the same sense I have said that Spring is the season of restoration to life, in which every thing again stirs and becomes active. SUMMER, on the same principle, is the season of *progress*, in which nothing is begun or ended—none of the greater operations of the field are either commenced or terminated, but only advanced a step towards the consummation of all things in Autumn ; and, therefore, the mere advancement of the greater operations involves no change of principle, whilst the smaller ones present so varied an aspect as to excite considerable interest.

The astronomical cause of Summer, as one of the seasons, and the safety in which every living thing—whether animal or vegetable— progresses or grows under its influence, are thus succinctly and well described by Mr Mudie. " Summer is the bloom of the year—the period during which all the growing and living children of nature, which wax and wane with the revolving seasons, are in the spring-tide of their activity, and when all those general agencies by which they are stimulated are working to the very top of their bent. Summer is, both in the literal and the metaphorical sense, the season of blossoms ; and as the blossoms make the fruit, the time of them is really the most important of the whole. In our middle latitudes, there is a very beautiful instance of design and adaptation in this. The grand stimulating agent in all terrestrial action, at least in a natural view of it on the surface of the earth, and the intensity of this action is made up of two elements—the position of the twenty-four hours during which the sun is above the horizon, and the altitude of the sun above that horizon. Both of them, in either hemisphere, increase as the sun declines toward that hemisphere, or rather as the hemisphere inclines to the sun; though, as the first of these is the apparent result of the second as a reality, our using an expression on the other does not affect the result. The increase or decrease of altitude is the same, with the same change of declination, in all latitudes; but the variation in time above the horizon increases with the latitude: consequently the higher the latitude the greater the change of solar action with the same change of declination. The change in declination increases from the solstice to the equinox, and diminishes from the equinox to the solstice. Thus the increase of solar action begins to slacken at the vernal equinox in March, and gradually diminishes till it becomes 0 at midsummer; after this the decrease commences. This, however, only in so far as depends on the altitude of the sun; for the other element, the time which the sun is above the horizon, goes on increasing till the longest day, or day of the solstice. Thus, in the advanced part of the summer, there is a diminished increase of the momentary intensity of the solar action, and a lengthening of its daily duration. What is given to the presence of the sun above the horizon is taken from its absence below it; and thus, as the summer advances towards the longest day, all that works by the action of the sun works with his increase of intensity and for a longer time. After the longest day is past, both elements of the solar action diminish, slowly at first, and more rapidly afterwards, until the summer merges in the autumn. Near the equator the changes are comparatively small, and they increase with the latitude; and the differences in this respect are what may be called the celestial differences of the character of summer in different latitudes; but terrestrial causes modify them so much, that the practical results as observed are very different from what the celestial theory would give. Still, any one who thinks but for a moment will not fail to discover how beautifully the season of bloom is secured from violent action either in the one way or the other. This is enough to convince us that the action which goes on in the production of nature during the summer is really the most important of the whole year; for it is performed with the maximum of power in the agents, and the minimum of disturbance in their operation. That resistance of winter, which but too often shrivels the young leaf and blights the early blossom in the spring, is vanquished and completely staid from making any inroad till the seasonal purposes of nature are accomplished; and the ardour of the stimulating causes which have vanquished the destructive one, are slackened, so that they may not injure that which during the struggle of the early part of the year they have preserved. All this, too, is accomplished by means so very simple, that their simplicity proves the most wonderful part of the whole, for it is nothing more than the planes of the annual and daily motions of the earth intersecting each other at an angle of about 23° 28′; and the line of intersection passing through the equinoctial points of the diurnal orbit."*

The atmospherical phenomena of summer are of the most varied and complicated nature. At one time the air is highly elastic, and feels balmy and bracing, indicated by the high position of the mercury in the barometer; at another the mercury descends, and almost always suddenly, to the lowest point, accompanied with gusts of wind and a deluge of rain, forming what is called a tornado in the tropics. The heat of the air at one time is so scorching as to cause us to seek the shade, and the thermometer marks its intensity; at another, a chilling gust, accompanied with a heavy shower of hail, suddenly brings down the thermometer many degrees. The air to-day is so calm and breathless, that not a ripple is visible even on the bosom of the great ocean; to-morrow a hurricane raises the waves of the sea to a state of agitation dangerous to the safety of the mariner. Not a cloud is seen at times to stain the purity of the blue vault of heaven; at others the thunder-cloud hovers over the earth, and blackens its surface with a portentous shadow. These changes, in summer, are usually sudden and of short duration, and they are requisite to preserve the healthy state of the air. Did rain not fall in quantities, the vapour absorbed by the great capacity of heated air for moisture would accumulate beyond due bounds. Were there no colder strata of air moving about to condense the warmer, the warmer portion containing a large quantity of vapour in solution would always be elevated beyond the reach of the earth, and there waste its latent heat. The thunder-storm of sheet-lightning passing from cloud to cloud, the most common display of electric action in summer, restores the electric equilibrium of the air, and the forked lightning relieves both earth and air. Did not the hurricane at times force its way through the calm and inert air, the same portion of the atmosphere would always remain over the same locality, and become vitiated by the breath of animals and the exhalations of vegetables. If the dews failed to descend upon the grass, the pasture would soon become parched by the meridian fervour of the summer sun. Thus, all the agencies

* Mudie's Summer, p. 1-6.

of nature are required to keep the air in a healthy state for animals and vegetables, and these operate in the most beneficial manner in summer,—the season of the intensest action of the solar rays.

As I have already entered very fully into the state of the weather in winter, in section 19, p. 229 to 301 of the 1st volume, I have less occasion to revert to the subject here; but the consideration of a few of the atmospherical phenomena which only occur in summer was purposely reserved until the arrival at that season; and these phenomena were, thunder and lightning, dew, hail, and easterly winds. The origin of the electricity of the air was indicated at sufficient length from (303.) to (316.). The presence of electric matter being accounted for, the *rationale* of a *thunder-storm* is simple, and it is this:—The electric fluid, as the agency of electricity is called, for want of a better name, derived from sources enumerated in the above paragraphs, accumulates in the clouds of vapour. When two clouds, thus provided with electric matter beyond their usual state, are not far from each other, the electricity of the one becomes *positive*, that is, in an active state; and that of the other, as a necessary consequence, becomes *negative*, that is, in a passive state. Why these opposite states of electricity always co-exist when near each other, it is impossible to say; but the fact is established as an unvarying law of electric matter. When two clouds thus in these opposite states are near each other, they attract and approach; and when the approach comes within the distance in which the force of the positive electricity is able to overcome the resistance of the air between the positive and negative clouds, the fluid leaves the positive and enters into the negative cloud in such quantity as to restore the equilibrium of both. The forcible passage of the electric fluid causes a concussion in the air between the two clouds, and the vibrations occasioned by the concussion striking against the earth and mountains, cause the noise which is heard in thunder. The time taken by the electric fluid in passing from one cloud to another is inappreciable, but the velocity of sound is calculable. For every 4½ seconds of time which elapse after seeing the lightning to hearing the thunder, the clouds are situate as many miles from the auditor. Far at sea, where there are no objects for sound to be reflected from, thunder is never heard; whereas, in a mountainous country, it inspires terror, though it is obvious that thunder, a mere sound, can do no harm, and lightning, which can do harm, does all the mischief it can do before we are made aware of its danger.

The phenomenon of *dew* is familiar to every person who lives in the country. In the hottest day of summer the shoes become wet in walking over a grass-field at sunset, and they may then become as effectually wetted as in wading through water. The late Dr Wells investigated the phenomena of dew more closely than any other person; and his experiments, as detailed in his instructive essay on that subject, appear to have been very satisfactorily conducted; and the theory which he established on these experiments is the one now embraced by philosophers. From that essay I shall first relate a few of the circumstances which influence the production of dew; and then give Dr Wells' theory of its formation, in contradistinction to those of other philosophers. "Aristotle and many other writers," says Dr Wells, "have remarked, that dew appears only in calm and serene nights. This remark of Aristotle, however, is not to be received in its strictest sense, as I have frequently found a small quantity of dew on grass, both in windy nights, if the sky was clear or nearly so, and in cloudy nights, if there was no wind. If, indeed, the clouds were high and the weather calm, I have sometimes seen on grass, though the sky was entirely hidden, no very inconsiderable quantity of dew. Again, according to my observation, entire stillness of the atmosphere is so far from being necessary for the formation of this fluid, that its quantity has seemed to me to be increased by a very gentle motion of the air. Dew, however, has never been seen by me on nights both cloudy and windy. If, in the course of the night, the weather, from being calm and serene, should become windy and cloudy, not only will dew cease to form, but that which was formed will either disappear or diminish considerably. In calm weather, if the sky be partially covered with clouds, more dew will appear than if it were entirely covered, but less than if it were entirely clear. Dew probably begins in the country to appear upon grass, in places shaded from the sun during calm and clear weather, soon after the heat of the atmosphere has declined; and I have frequently felt grass moist in dry weather several hours before sunset. On the other hand, I have scarcely ever known dew to be present in such quantity upon grass as to exhibit visible drops before the sun was very near the horizon, or to be very copious till some time after sunset. It also continues to form in shaded places after sunrise; and if the weather be favourable, more dew forms a little before, and in shaded places, a little after sunrise, than at any other time. The formation of dew, after it has once commenced, continues during the whole night, if the weather remain still and serene. During nights that are equally clear and calm, dew often appears in very unequal quantities, even after allowance has been made for any difference in their lengths. One great source of their difference is very obvious, for, it being manifest, whatever theory be adopted concerning the immediate cause of dew, that the more replete the atmosphere is with moisture, previously to the operation of that cause, the more copious will be the precipitation of moisture in the atmosphere, must likewise tend to increase the production of dew. Thus dew, in equally calm and clear nights, is more abundant shortly after rain than during a long tract of dry weather. It is more abundant during S. and W. winds, than during those which blow from the N. and the E. Dew is commonly more plentiful in spring and autumn than in summer; the reason is, that a greater difference is generally found between the temperature of the day and the night in the former seasons of the year than in the latter. Dew is always very copious on those clear and calm nights which are followed by misty or foggy mornings; the turbidness of the air in the morning shewing that it must have contained, during the preceding night, a considerable quantity of moisture. I have observed dew to be unusually plentiful on a clear morning, which had succeeded a cloudy night. For the air having, in the course of the night, lost little or no moisture, was in the morning charged with more watery vapour than it would have been if the night had also been clear. Heat of the atmosphere, if other circumstances are favourable, which, according to my experience, they seldom are in this country, occasions a great formation of dew. For, as the power of the air to retain watery vapour in a pellucid state, increases considerably faster while its temperature is rising than in proportion to the heat acquired, a decrease of its heat in any small given quantity during the night must bring it, if the temperature be high, much nearer to the point of repletion before it be acted upon by the immediate cause of dew, than if the temperature were low. I always found when the clearness and stillness of the atmosphere were the same, that more dew was found between midnight and sunrise, than between sunset and midnight, though the positive quantity of moisture in the air must have been less in the former than in the latter time, in consequence of a previous precipitation of part of it. The reason, no doubt, is the cold of the atmosphere being greater in the latter than in the prior part of the night."

Theories of the formation of dew have been proffered by many philosophers, from the days of Aristotle to the time of Dr Wells. "Dew, according to Aristotle," remarks Dr Wells, "is a species of rain formed in the lower atmosphere, in consequence of its moisture being condensed by the cold of the night into minute drops. Opinions of this kind, respecting the cause of dew, are still entertained by many persons, among whom is the very ingenious Mr Leslie of Edinburgh." This view is erroneous, because "bodies a little elevated in the air become moist with dew, while similar bodies, lying on the ground, remain dry, though necessarily from their position as liable to be wetted by whatever falls from the heavens, as the former." Dufay concluded, that dew is an electric phenomenon, but it leaves untouched bodies which conduct electricity, while it appears upon those which cannot transmit that influence. All the theories on dew, to the time of Dr Wells, omitted the important part, that the production of dew is attended with cold, and this is a very important omission, since no explanation of a natural phenomenon can be well founded which has been built without a knowledge of one of its principal circumstances. "It may seem strange to many," continues Dr Wells, "that neither Mr Wilson nor Mr Six applied the fact of the existence of cold to its production, to the improvement of the theory of dew. But, according to their view of the subject, no such use could have been made of it by them, as they held *the formation of that fluid* to be *the cause* of the cold observed with it. I had many years held the same opinion, but I began to see reason, not long after the regular course of my experiments commenced, to doubt its truth, as I found that bodies would sometimes become colder than the air without being dewed; and that when dew was found, if different times were compared, its quantity, and the degree of cold which appeared with it, were very far from being always in the same proportion to each other. The frequent recurrence of such observations at length corrected the doubt of the justness of my ancient opinion into a conviction of its error, and at the same time occasioned me to conclude, that dew is the production of *a preceding cold in the substance* upon which it appears." Dr Wells' theory, therefore, is, "that the cold observed with dew is the previous occurrence, and, consequently, that the formation of this fluid has precisely the same immediate cause as the presence of moisture upon the outside of a glass or metallic vessel, where a liquid considerably colder than the air has been poured into it shortly before." As an obvious application of this theory, the experiments of Dr Wells, which led to its establishment, evince, that of all natural substances grass is peculiarly adapted to the exhibition of dew, inasmuch as it becomes, under ordinary circumstances, colder than the air above it, by the radiation of more heat towards the heavens than it receives in any way, and, accordingly, whenever the air is calm and serene, dew may be seen on grass, when it may not be observed on other substances.

But it has been alleged by Dufay that dew is the condensation of vapour *rising* out of the earth upon the grass on it, because objects removed higher from the surface of the earth, as trees, are exempt from dew; and this is a very popular opinion; but it is an erroneous one, and the phenomenon can be explained on other principles, because the lower air in a clear and calm evening is colder than the upper; it is less liable to agitation than the upper, and it contains more moisture than the upper; and hence, on all these considerations, it will sooner deposite a part of its moisture. At the same time, it is true that vapour does rise from the earth, and it may be condensed as dew; for we find the grass first becoming moist with dew, then the substances raised above it, while both indicate an equal degree of cold; but all the quantity of dew from this cause can never be great, because until the air be cooled by the substances attractive of dew with which it comes in contact below its point of repletion with moisture, it will always be in a condition to take up that which has been deposited upon grass, or other low bodies, by warm vapour emitted by the earth, just as the moisture formed on a mirror by our breath is, in temperate weather, almost immediately carried away by the surrounding air. Agreeably to another opinion, the dew found on growing vegetables is the condensed vapour of the very plants on which it appears; but this also is erroneous, because dew forms as copiously upon dead as upon living vegetable substances; and "if a plant," as Dr Wells observes, "has become, by radiating its heat to the heavens, so cold as to be enabled to bring the air in contact with it below the point of repletion with moisture, that which forms upon it from its own transpiration will not then indeed evaporate. But although moisture will at the same time be communicated to it by the atmosphere, and when the difference in the copiousness of these two sources is considered, it may, I think, be safely concluded, that almost the whole of the dew which will afterwards form upon the plant must be derived from the air; more especially when the coldness of a clear night, and the general inactivity of plants in the absence of light, both lessening their transpiration, are taken into account." Hoarfrost is just frozen dew, but as it only appears when the surface of the earth is sealed with frost, the vapour of which *it* is formed cannot, of course, at the time, perspire from the earth.*

Another remarkable phenomenon in summer is *hail*. "The difficulty for accounting for the retention of masses of ice in the free atmosphere," observes Professor Forbes, "is certainly very great. Perhaps no hypothesis more satisfactory, certainly none more ingenious, has followed that of Volta, who conceived, from the highly electric condition of the atmosphere, almost univer blly attending the production of hail, that the frozen masses were kept in a state of reciprocating motion between two clouds oppositely charged with electricity, until the increase of the mass

* Wells on Dew, p. 1-116, second edition, 1815.

rendered the force of gravity predominant, or the electric tension of the clouds was exhausted by mutual reaction." * As hail is a very curious and highly interesting phenomenon, a few facts regarding its occurrence and the form of hailstones may prove instructive. " Hail generally falls in the hottest hours of the day in Spain, Italy, and France. It falls in Europe generally in the day, and seldom in the night. It seldom falls in winter, though at Plymouth, according to Mr Giddy, there seems to be an exception. Thus, in the course of 21 years, the recurrence in each month was, in

January,	23 times.	July, . .	1 times.
February,	25 ...	August, .	0 ...
March, .	25 ...	September,	5 ...
April, . .	27 ...	October,	17 ...
May, . .	7 ...	November,	22 ...
June, . .	5 ...	December,	43 ...

In August is absolute zero, and in December is the maximum. Rain falls in all seasons, snow in winter, hail principally in summer. The appearance of hail-clouds seem to be distinguished from other stormy clouds by a remarkable shading; their edges present a multitude of indentations, and their surfaces disclose here and there immense irregular protuberances. Hail seldom falls on mountains, which indicates lowness of clouds, which Arago has seen cover the valley with fog, while the mountains were clear above. The form of hailstones varies. They are nearly uniform when they fall on the same level; and in the same storm they have fallen smaller on the tops of mountains than on the plains. Change of temperature or wind changes the form of hail. On 7th July 1769, M. Adanson observed 6-sided pyramids fall, but the wind changing to N.E., changed them to convex lenses, and so transparent, as to transfer objects without distortion. Hail is sometimes attended with spongy snow, which may have formed the interior of the hailstone, while its exterior was transparent ice. It has been supposed from this, that the different portions have been formed under different circumstances. Leslie imagines the spongy texture to result from an atom of water having been suddenly frozen, and particles of perhaps rarified air suddenly driven into the centre. The pyramidal form fell at Aberdeen on 29th November 1823. The usual form of hail is a concentric lamellar structure, with a stellular fibrous arrangement. There is great difficulty in accounting for large masses of ice in the atmosphere, as hail cannot take above 1 minute in falling from low clouds."†

* Forbes's Report on Meteorology, vol. i. p. 253.

† Encyclopedia Metropolitana, art. *Meteorology*, p. 129.

The form of the *clouds* in summer are very distinctly marked. When a deposition of vapour is taking place in the highest part of the atmosphere, the *cirrus*, or curl-cloud, appears (287.), and it sometimes soon disappears, which is a sign of fine weather; but instead of disappearing, it may descend a little lower, and be converted into the *cirro-cumulus* (291.), which is the form of that elegant, light, flocculent cloud so often seen in a fine summer day. A farther deposition changes the small cloud into the larger *cumulus*, called the day-cloud in summer, because it disappears in another form in the evening. The *cumulus* or heap may be seen distinctly represented in the plate of the Leicester tup, both above the animal, and above the horizon, where it frequently takes up its position for the greater part of the day, resting on the vapour plane. When a large cumulus rises from the horizon in the day-time, and shews white towering heads, it is a sign of a storm or fall of rain from that quarter; and the wind will change to that direction in the course of the next 24 hours. This threatening cloud, called *cumulo-stratus*, is characteristically given in Plate XVII., beyond the portrait of the Draught Mare. In calm serene evenings in summer, the day-cloud or cumulus descends and spreads itself along the bottom of valleys, or in hollows of the open country, covering the ground like a lake as seen in moonlight, or with a partial sheet of snow. This is the true *stratus* cloud. Tall objects, such as trees, steeples, and even elevated ground, jut through it like rocks and islands in a lake. The air is then perfectly calm, the temperature delightfully warm, and the intenseness of the silence is broken only by the snipe drumming in its curious somersets in the air—by the harsh ventriloquous cry of the corn-craik amongst the grass—or by the occasional barking of the watch-dog at some distant homestead. The morning after such a night is sure to usher in a bright and peerless sun, whose steady heat will soon evaporate the sheet-like stratus-cloud from the valley and hollows, and elevate it, in the form of the beautiful day-cloud, above the mountain top or the horizon.

The direction of the *wind* forms an important item in the consideration of the weather in summer; for in no other season does the slightest variation of the wind make so decided changes on the weather. If we assent to the general conclusion in regard to the cause of winds, that it is the partial changes of temperature which are their chief general cause, we may expect the winds to be most variable in summer, since changes of temperature are most likely to occur at that season; and the change of the wind is the most ready indication of the change of temperature, especially in the upper portion of the atmosphere. In the torrid zone, whilst the barometer seldom varies but in a trifling degree, in the temperate zone it is not less fickle than the wind. This indication of a loss of weight in the atmosphere, can arise only from a local diminution of its elasticity, most likely from changes of temperature. But the most remarkable effect in relation to the wind in summer, is the constancy with which it blows from the east in the early part of that season; and so invariable is this phenomenon, that every person who dwells on the east coast of Great Britain is quite familiar with it, as every one feels the keenness of the east wind, and every one knows the aptitude with which catarrhal affections are produced by it. An explanation of this remarkable phenomenon cannot fail to prove interesting; and as a very rational solution of its recurrence has been given by Mr Samuel Marshall, I shall copy his succinct but satisfactory account of it. After stating that the east winds, in early summer, usually prevail from the middle of April to the 7th or 8th May, or even to the 18th May, and I may add, that it continued all June in 1843, Mr Marshall then proceeds to account for the cause of the phenomenon. "In Sweden and Norway, the face of the country is covered with snow to the middle of May or longer. This frozen covering, which has been formed during winter, grows gradually shallower to the 15th or 16th of May, or until the sun has acquired 17° or 18° N. declination; while, on the other hand, the valleys and mountains of England have received an accession of 24° or 25°. On this account, when the temperature of Sweden and Norway is cooled down by snow to 32°, that of Britain is 24° or 25° higher than that of the preceding countries. Because, while the ground is covered with snow, the rays of the sun are incapable of heating the air above 32°, the freezing point. For this reason, the air of England is 24° or 25° more heated than that of the before-mentioned countries. The air of Sweden and Norway will then, of course, by the laws of comparative specific gravity, displace that of England, and, from the relative situation of those countries with this country, will produce a N.E. wind. The current is in common stronger by day than by night, because the variation of temperature is at that time the greatest, being frequently from 50° to 60° about noon, and sinking to 32° in the night."* Some of the most obvious prognostics of the wind in summer are, when the wind is variable, rain is not far distant; when it blows low, and raises the dust much, it has the same effect; but when there are small whirlwinds raising the dust along the road or corn-fields, it is a sign of dry weather. I remember of seeing a beautiful whirlwind, in a calm hot day, in the neighbourhood of Berlin, raise the sand of a field, the soil being mostly composed of that substance, in a perpendicular direction to a great

* Brewster's Journal of Science, vol. viii. p. 39.

height in the air. When currents of air are seen to move in different directions, that which the upper current takes will ultimately prevail. At times in summer, and particularly in the evening, it is hard to say from which quarter the current comes, when a wetted finger held up will tell the quarter, from the cold produced on it by consequent evaporation of the moisture. The pleasant phenomena of the land and sea breezes are distinctly marked in fine warm weather in summer (370.). When winds blow strongly from any quarter, even from the warm west, for two or three days in succession, the temperature of the air is much diminished, sometimes as much as 20°, and seldom less than 10°.

So much for the aërial speculations of summer. Let us now notice the humbler toils of the field. These, as I have already observed, require to be advanced a stage toward their completion in autumn. The first operation which calls for the ploughman's attention in summer is the turnip-land, which is now drilled up, dunged, and sown. The culture of the turnip is a most important and stirring operation, affording much interesting work in singling and hoeing the plants for the greater part of the season. In the height of summer, young stock luxuriate on the riches of the pasture-field, while forage plants, consisting of vetches, rape, or broad clover, are allowed to grow until the general season of want, between the failure of pasture and the premature consumption of turnip. Before stock take possession of the pasture-fields, the hedger makes it a point to put the fences in a complete state of repair, and to second his exertions, the carpenter and smith make the field-gates secure for the season. Fattened stock are seldom allowed to taste the pasture, they being disposed of off the turnips to the butcher or dealer. The fat cattle are almost always then sold. Young cattle and cows are sent to the grazing field of the farm, though turnip sheep are not unfrequently retained on grass until the fleece is clipped from their backs (the season being nigh at hand), and after that they also are disposed of. The separation of ewe and lamb is now effected; and the respective marks of age, sex, and ownership, are put on each. Horses now live entirely another sort of life, being transferred from the confinement of the collar in the stable to the perfect liberty of the field, and heartily do they enjoy themselves there. The brood mare now brings forth her foal, and receives immunity from labour for a time. Hay-making is represented by poets as a scene of unalloyed pleasure. No doubt lads and lasses are then as merry and chirping as grasshoppers, but, nevertheless, in spite of buoyant spirits, haymaking, in sober truth, is a labour of much heat and great toil—the constant use of the hay-rake and pitch-fork, in hot weather, being no sinecure. Early as the season is, preparations are made in summer for the next year's crop. The bare fallow is worked and dunged, and it may be limed too, in readiness for the seed in autumn. Summer is of all others the season in which the farmer most seriously makes his attacks on those spoilers of his clean fields, and contaminators of the samples of his grain—the weeds. Whether in stocked pastures, upon tilled ground, along drills of green crops, amongst growing corn, or in hedges, young and old, weeds are daily exterminated, and the extermination is most effectually accomplished by the minute and painstaking exertions of female field-workers. For these purposes they are provided with appropriate cleaning instruments. This is the season, too, in which his stock and crops are sometimes seriously affected by the attacks of insects. Where building-stones are plentiful, and the risk great from the overflowings of rivulets in winter, summer is also the season for the erection of stone dykes as fences between fields, and of embankments along the margins of rivers. The former afford a substantial fence at once, the latter form insuperable barriers against an element powerful alike whether exerted for or against man's operations.

Every operation requires constant attention in summer, for the season being active in its influences, farmers must then put forth their energies to meet its rapid effects, whether these tend to forward or retard his efforts. The long hours of a summer day, of which at least ten are spent in the fields—the ordinary high temperature of the air, which suffuses the body of the working man in constant perspiration—and the fatiguing nature of all field-work in summer, bear hard as well on the mental as the physical energies of the labourer, and cause him to seek for rest at a comparatively early hour of the evening. None but those who have experienced the fatigue of working in the fields in hot weather, and for long hours, can truly appreciate the luxury of rest—a feeling truly described in these simple lines :—

"Night is the time for rest.
How sweet when labours close,
To gather round the aching breast
The curtain of repose—
Stretch the tir'd limbs, and lay the head
Upon one's own delightful bed !"

MONTGOMERY.

The hours devoted to field-work vary in summer in different parts of the country. On the Borders it is the practice to go very early to the morning yoke, as early as 4 o'clock, that the forenoon's work may be over by 9, and that there may be time to rest in the heat of the day; the afternoon's yoking commences at 1 o'clock, and continues till 6 o'clock. There are thus 10 hours spent in the fields. But in most parts of the country, the morning yoking does not commence till 6 o'clock, and, on terminating at 11, there are only 2 hours for rest and dinner till 1 o'clock, when the afternoon's yoking begins. In most places the afternoon yoking does not commence till 2 o'clock, and, finishing at 6, only 9 hours are spent in the fields. In other parts, only 4 hours are spent in the morning yoking, when the horses loose at 10 o'clock, and, yoking again from 2 to 6 in the afternoon, only 8 hours are spent in the fields, and the men are employed elsewhere by themselves for 2 hours. This is practised where the ploughmen are made to do the work of field-workers, and where a large number of draughts are kept. Perhaps the best division of time is to begin the yoking at 5 o'clock in the morning, loose at 10, yoke again at 1, and loose at 6 in the evening, affording 3 hours for rest to man and horse at the height of the day, and 10 hours in the field. Day-labourers, when not dependent on the horses, as well as field-workers, usually work from 7 to 12, and from 1 to 6 o'clock in the evening, having 1 hour for rest and dinner. When labourers take their dinner with them to the field of their operations, this may be a good enough division of time; but when they have to go home to dinner, 1 hour is too little for the purpose, and affords no time for rest between yokings, which is to be deprecated, as neither men nor women are able to work 10 hours without an interval of quiet rest. It would, therefore, be a better arrangement for field-workers to go to work at 6 instead of 7, and loose at 11 instead of 12, when they have to go home to dinner; but if they took their dinners with them to the field, then 1 hour is sufficient for rest and dinner at the same time.

Summer is the only season in which the farmer has liberty to leave home without incurring the blame of neglecting his business, and even then the time which he has to spare is very limited. There is only about a fortnight between finishing the fallow, the turnip and potato culture, and hay-making, and the commencement of harvest, in which the farmer has leisure to travel. This limitation of time is to be regretted, because it is proper that he should take a journey every year, and see how farm operations are conducted in other parts of the kingdom. An excursion of this nature is seldom undertaken by a farmer, who is generally a man capable of observation, without acquiring some hints which may induce the adoption of a practice that seems good, or the rejection of one which is bad. Such a journey exhibits mankind in various aspects, and elevates the mind above local prejudices; and as husbandry is a progressive art, a ramble of a week or two, through different parts of the country, cannot fail to enlighten the mind of the most experienced farmer much beyond any thing he can observe by always remaining at home.

61. OF THE SOWING OF TURNIPS, MANGEL-WURZEL, RAPE, CARROTS, AND PARSNIPS.

"They require the land to be well broken by frequent ploughings and harrowings, and also to be well dunged; this is of great importance, not only as the crop of these roots is thereby rendered larger, but as the land is thereby prepared for carrying good crops of corn."

COLUMELLA.

(2499.) The first great field-operation in summer is the completion of the preparation of the soil for the sowing of the turnip-crop. This crop commences the rotation of crops, is a substitute of bare-fallowing, and is of the same nature, as regards the amelioration and working of the soil, as the potato-crop, and therefore admits of the soil being manured; and, indeed, on account of all these properties, it is regarded and denominated a *fallow*-crop. Being thus a renovater of the condition of the soil, the turnip-crop necessarily succeeds the crop which terminates the rotation, and beyond which the exhaustion of the soil is not permitted; and, being a fallow-crop, the preparation of the soil for it requires much labour, and should therefore be begun as early as the breaking up of the stubble in the beginning of winter. This fact is indicated in (1135.); and the different modes of ploughing stubble for fallow, according to the nature of the soil, are mentioned in (1136.), and the due precautions to be used to keep the soil in a dry state all winter are stated in (1137.). These constitute the winter preparations for the turnip-crop; and those in spring are begun by cross-ploughing, fig. 368. Should potatoes have been planted in the same field intended for turnips, the cross-ploughing for the potato-land should be extended across the turnip-land, if there is time for it; but should the time required for this extended cross-ploughing encroach upon that which is devoted to potato-planting, the turnip-land should be let alone, until there is leisure. From the cross-ploughing to the drilling of the land for the reception of the dung, the turnip-culture is exactly the same as for potatoes, with perhaps the exception, that as there is more time for working and cleaning turnip-land, it receives one or more ploughings or stirrings with the grubber than the potato-land; and in this cleansing process the grubber will be found a most efficient implement, and will save a ploughing, while it keeps the upper soil uppermost and in a fine loose state. When turnip land is manured with farm-yard dung, the drilling is best and most expeditiously done in the single mode, fig. 369; but as the drills have to be kept in exact proportions, for the sake of the better operation of the sowing-machine that is to follow, the best ploughman should be desired to make them. The position of the plough which makes the single drills is shewn at *a*, fig. 411; the ground occupied by one feering in drilling is shewn from *a* to *e*; and the process of dunging, both in carting out and spreading the dung, and in splitting the drills in the double way, are conducted in the manner described for that figure, with the exception of the potato-planters at and *s*. So far, then, the culture of potatoes and turnips correspond, but after this point a considerable difference ensues, which arises from the difference in the nature of the seed. After the soil on the top of the drills has become a little browned with the sun, or *rizzared*, as it is technically phrased in some, or with a proper *tid* in other places, the turnip sowing-machine, which sows 2 rows at a time, such as the one described below by Mr Slight, is then used, with one horse, for sowing the seed. The soil should be dry at the top of the drills before the seed is sown, because damp soil clings to the rollers of the machine, and causes them to make bad work. One of these machines could sow a great breadth of land in a day, but it is seldom that it can be employed throughout a whole day, for two reasons: One is, that the soil is seldom in a dry enough state in the morning to be thus sown; and the other reason is, that a sufficient quantity of land will not be dunged and split in the course of a day to keep a machine going constantly, because one plough can only split ⅓ more land in a day than it can plough, so that 3 ploughs will only split 5 acres at most of drills over dung in a day, and thus 1 machine could hold 4 ploughs splitting drills; and as the dunging is carried on at the same time, there are few farms so large as to employ 4 ploughs splitting drills.

(2500.) The quantity of seed sown need not exceed 3 lb. to the English acre, nor should the quantity be much less, as thick sowing ensures a quick braird of the turnip-plant, and the seed is not a costly article, being usually from 9d. to 1s. per lb. Fortunately, that the land may receive its due labour, the different kinds of turnips cultivated require to be sown at different times. Swedes, for instance, should be sown by the 15th of May at latest, and if the land is ready to receive the seed by the 10th, so much the better. Swedes will grow on any kind of soil, except perhaps what is *in a state of pure peat;* but they grow best in rich alluvial sandy loam—best, because largest, and in that state this particular turnip is firmest too. The yellow turnip follows the Swedes, and then the white, which may be sown any time in June. In England, white turnips are sown as late as July, because, if sown earlier, they would come too soon to maturity. The seeds of Swedish turnips are much larger than those of yellow or white, and require to be sown through a larger hole in the cannister of the machine. They will retain their freshness for several years, and may be confidently sown though kept for 3 years; but it would be hazardous to sow yellow or white turnip-seed after the first year; they somehow lose their vitality after that time. The reasons why I give the preference to the purple top Swedish, the Aberdeenshire yellow bullock, and the white globe turnips, will be found in paragraphs from (1241.) to (1250.) inclusive, and these turnips are figured in fig. 214, at the same place. At the same time, you should partly be guided by the practice of the district in which your farm is situate, what sorts you should sow on your first coming into the district; but you should hear very satisfactory reasons why those kinds will not thrive in that particular district, and should even experience their failure before you determine on preferring the culture of other kinds. When I began to farm in Forfarshire, I was told that the Swedish turnip would not thrive on such soil as my farm, a light turnip-soil, and that I would find the red-topped white the best; but a very short trial shewed every one that Swedes and the white or green-topped globe throve much better than the favourite red-topped, which is an early turnip, no doubt, but a very spongy one on that soil. For my part, I would have no fears of raising a good crop of Swedes on any ordinary soil by proper culture, that is, by giving them plenty of old well made dung, and sowing them early. Give Swedes 15 tons, or 20 good cart-loads, of dung in that state to the imperial acre, and sow them *before* the 15th of May, and little fear need be entertained of the crop on the most ordinary soil; 12 tons of well made dung will suffice for yellow turnips; and 10 tons of the same for white globes; but, of course, the more dung each of the kinds receive, the larger crop may be expected.

(2501.) It is not an unusual practice in England to sow turnips *broad-cast on the flat ground*, instead of in rows, and on ridglets or drills, as in Scotland; and the reason I have heard stated in vindication of the broadcast method is, that it resisted the bad effects of drought on the land in summer. No doubt excessive drought in summer is inimical to the full development of the turnip, and it is on this account that the turnip-crop fails so frequently in Germany; but, for my part, I cannot see how a broad-cast crop can escape drought more cerainly than one in rows, since the plants have to grow and be thinned out to proper distances in both cases; and should the ground be foul with weeds, it must be stirred to get rid of them in both cases; and as the work of weeding is done by hand instruments in the case of the broad-cast crop, it is, of course, not so *effectually* done as with horse-hoes, in the case with the crop in rows; while I can see, that as the dung must be spread broad-cast for a broad-cast crop, it will not have the same opportunity of promoting the growth of the crop at its early stage as when it is deposited in rows, so that the means employed defeats its own purpose; because I think it cannot admit of doubt, that the same quantity of manure placed in bulk immediately under the seed should promote the growth of the plant more rapidly than when it is spread over a larger surface of ground, and there can be no doubt, also, in regard to the turnip-plant itself, that the more rapidly it grows in its early stage, the more certainly it will be free of danger from drought and from some insects, at least, for it is well known, in regard to the habits of some of those which attack the turnip, that they become innocuous to its leaves after the development of the rough ones. I do not aver that the sowing of turnips in drills will render them invulnerable to the attacks of insects or to the effects of drought, but these evils being merely seasonal, the drill system places the crop more immediately under the management of the cultivator, inasmuch as it enables him to apply the whole powers of the manure at once, and of cleaning the soil quickly with the assistance of horse power. Besides all this, the period of sowing the crop should be suited to the climate of the locality. If drought is too great in July, or the insects too powerful, the crop should be sown earlier, and though it should reach maturity sooner than desired, it can be drawn and stored until the season of its consumption arrives; or it should be sown later, such as in August, when the genial climate of the south of England—where the nights are warm as well as the day, thereby affording every day the growth of two ordinary days in Scotland, where the nights are always cold—is sufficient to mature the crop before the end of October, which is as early as the turnip-crop is required for consumption in that part of the country, and where till then the grass continues good. And were the soil properly cleaned before the crop is sown, comparatively little labour would be required to keep it so in the height of summer, and of course the drought would not then much affect it. Or, a part of the turnip-crop could be taken after winter vetches, which, on being cleared off the ground in time by feeding sheep, or by cutting, would enable the land to receive a short fallowing for turnips before the end of July.

(2502.) Nor is a much earlier fallowing and cleaning of the turnip-land an impracticable thing in England, since the corn-crop is frequently cleared from the fields by the end of August, when the stubble could be broken up, harrowed, cross-ploughed, cleaned, drilled, and even dunged, before the arrival of winter, as has been proved in Scotland, by the practice of Mr James Scougall, at Balgone, East Lothian, the seat of Sir George Grant Suttie, Bart., in the autumn of 1841, when he drilled and dunged good turnip-land, at 32 inches apart in the drill, and otherwise finished its tillage. Purple-top Swedes were sown on the 10th of May 1842, the plants thinned to 15 inches apart, and the matured crop was drawn and stored by the middle of September following, when the ground was sown with wheat. On comparing the produce of this mode of culture with the usual one of labouring the turnip-land in spring, and at the usual distance of 28 inches between the drills and 12 inches between the plants, the ground prepared in autumn yielded, in 429 links measured along one drill, 82 stones of turnips, whereas that worked in spring yielded only 58 stones, though the number of turnips in the former weight was only 238, while that in the latter was 276, thus giving to each turnip a weight of 4 lb. 13 oz. in the former, and of only 2 lb. 15 oz. in the latter case. A single horse-load of turnips selected from the ground prepared in autumn only numbered 141 roots, and weighed as much as 109 stones, shewing the weight of each root to be 10 lb. 13 oz.* This instance proves that land for Swedish turnips may be prepared in autumn, and a heavy crop obtained on drills as wide as 32 inches, and from plants 15 inches apart. A somewhat similar success attended the trial of raising turnips on strong clay-land, at ordinary distances, by Mr Peter Thomson, Peffermill, near Edinburgh. So soon as the ground was cleared of tares, he cross-ploughed it with a deep furrow; well worked it with harrows and grubber early in October; drilled it in the single way at 28 inches asunder, but, owing to the unfavourable state of the weather, was prevented applying the dung until December and January, which consisted of 32 single horse-loads of street-manure per imperial acre, and which was covered in with the common plough. In March, the soil was found in a fine state, but, as it was rather foul with weeds, the drills were harrowed a double tine, the ground stirred between them with the single-horse grubber, and set up with the double mould-board plough, and again set up just before sowing the seed of the green-top yellow turnip on the 15th, and of the white globe on the 25th and 26th of May. The yield was 32 tons of turnips per imperial acre, including tops, and quite free of mildew. The field had been dunged 5 years before, and twice since top-dressed with 8 bags of 4 bushels each of soot to the imperial acre.† Were such modes of culture adopted in the south of England, I have no doubt certain and abundant crops of turnips would be raised in spite of drought

* Mark-Lane Express, 17th October 1842. † Ibid, 21st November 1842.

and insects, and the slovenly practice of broad-cast culture would then give way to the more scientific mode of the drill system.

(2503.) Besides farm-yard dung, a great breadth of turnips is raised every year with *bone-dust*. This manure is not only cheap, but insures a good crop of turnips in ordinary circumstances, and its use expedites field-labour very much. When the land is ready to be drilled up from the flat, the drills are made in the double way (2178.); and that is all the preparation the land requires for bone-dust, which is then applied by the bone-dust sowing-machine, represented in Plate XXXI, and described below by Mr Slight. This machine deposites the bone-dust and turnip-seed at the same time, and finishes the sowing of the turnip-crop. The quantity of bone-dust used is commonly 16 bushels or 2 quarters to the imperial acre. There is something in the action of bone-dust on the soil, and its consequent power to produce a turnip-crop, which I do not understand, the means being apparently so inadequate to produce the results obtained. What I mean is, that, up to a certain quantity used, this manure has evidently a beneficial effect, but, beyond that quantity, there is derived from its use no apparent benefit, in as far, at least, as the crop is concerned. I have tried to raise turnips with different quantities of bone-dust, varying from 12, 16, 20, and 24 bushels to the imperial acre, and have found the crop improved up to 16 bushels, but any quantity beyond that, even to 24 bushels, produced no greater effect on the turnips in the same field, and on the same sort of soil, than 16 bushels. Nay, more than this, my late agricultural preceptor, Mr George Brown, when he farmed Hetton Steads, in Northumberland, raised as good crops of turnips, as 16 bushels of bone-dust, with only 8 bushels of bone-dust, combined with an indefinite quantity of sifted dry coal-ashes; and yet 8 bushels of bone-dust, or an indefinite quantity of coal-ashes applied separately, produced a very poor crop of turnips. It is, therefore, unnecessary, in so far as the crop of turnips is concerned, to sow more than 16 bushels of bone-dust alone, or 8 bushels with coal-ashes, or perhaps street-manure. Both coal-ashes and street-manure, when proposed to be used with bone-dust, should be kept dry under cover and sifted free of large lumps. It is truly surprising what an effect on the soil so small a quantity of bone-dust produces. I have raised a portion of the manured soil of a drill with my hand before the turnip-seed had germinated, and found it agglutinated together in a lump with a greasy matter, and the lump interspersed partly with white mouldiness, and partly with minute fibres of plants. When the turnip-seed germinates, which it will do in 8 or 10 days, according to the state of the weather, its radicle strikes into the greasy mass of earth, and sends out an immense number of white fibres around and through it. Its cotyledons then expand upwards into two rudimentary smooth leaves, and immediately thereafter two true or rough leaves appear, and these last are called *rough* leaves, because they feel rough by reason of the small spiculæ which may be observed to occupy the surface of every leaf of the common turnip. The rudimentary leaves of the Swedish turnip are not rough but smooth, because the plant is not a true *turnip*, but a species of cabbage, which are all smooth-leaved. The smaller bone-dust is ground the more effective it is as a manure, because it then mixes more intimately with the soil, though its action upon it continues for a shorter time ; and, on the other hand, large or *drilled*, or *inch-bones*, as they are called, remain longer in the soil undecomposed, but produce less immediate effect. On these accounts, bone-dust is the more valuable manure for turnips, and inch-bones for wheat. But even *bone-dust* has effect beyond the turnip-crop ; it extends its influence over all the crops of the rotation. I pulled off 4 acres of turnips raised with bone-dust, and 4 acres adjoining these in the same field that had been raised with 15 loads of farm-yard dung ; and the crop of turnips was not only better after the bone-dust than after the dung, but all the crops that followed in the rotation, namely, barley, hay, pasture, and oats, were also better in their respective years. *Bone-dust*, when purchased in that state on ship-board, is almost always adulterated with old plaster, brick-dust, ashes, &c. ; and should, therefore, be purchased direct from the grinder, one who is on the spot. There are now many bone-mills scattered over the country. To avoid adulteration, an association of farmers was formed a few years ago in Perthshire to import bones, rape-cake, &c., and grind them at their own mills under the superintendence of a manager in whom they have confidence, and in this way they continue to supply themselves with genuine manures. Bone-dust is best conveyed in sacks, and 40 bushels will fill a double-horse cart and sow 2½ acres imperial. It weighs 47 lbs. per bushel. It should be immediately emptied out of the sacks and kept in small heaps in a cool shed until it is used, as it is very apt to heat, and one consequence of *recent* heating is to become lumpy and troublesome to sow by the machine, though otherwise it is not injured but rather improved by it. In some parts of the country, particularly on the Borders, bone-dust is sown by hand either along drills made up in the single way, which are then split in the double way, while this plan imposes the trouble of a second drilling, or it is sown on the flat ground and covered by drilling in the double way. In both cases, the seed is sown afterwards by itself with the common turnip-sowing machine. The only reason I have heard in favour of sowing bone-dust by hand instead of machinery, is saving the cost of the machine; but whatever advantage is gained by this saving, it is, I think, evident that the machine must deposite the bone-dust much more regularly than the hand; and as to the cost of a machine the saving must be trifling, as hoppers for containing bone-dust can be attached and made to remove at pleasure from an ordinary sowing-machine. I always used a machine of this form myself. But in a case of this kind accuracy of work is a more potent consideration than the cost of a small machine, even though it should be used but for a few weeks every year. There is another consideration, too, of even greater import, that the nearer bone-dust is placed to the turnip-seed the quicker will the seed vegetate, and experience has proved that turnip-seed may safely be placed amongst bone-dust. In sowing by hand, the manure is not placed near the seed in as far as the sower knows, and when the seed is sown by itself after the bone-dust has been covered up by the drill, the sowing-machine is as likely to deposite it away from, as near to, the manure, and hence a regular braird cannot be insured by the practice. The ordinary bone-dust sowing-machine, as well as sowing by the hand, deposites the bone-dust in a continuous line along the drill; but a sowing-machine has been lately invented, and is partially in use in Perthshire, which deposites the bone-dust in small portions at regular intervals. The argument in favour of this mode of depositing manure is the comparatively small quantity required by it to produce a full crop; and certainly when bone-dust is either very scarce or very dear, this is a fair argument to use in its favour. If by depositing as much bone-dust at given distances of 12 inches as would be deposited at those distances were 16 bushels per acre sown in a continuous line, and if the quantities so deposited at intervals of 12 inches are found to produce as well-grown turnips as a continuous sowing, then the saving of manure must be as great as from $\frac{1}{2}$ to $\frac{2}{3}$, because the intervals between the distances are not manured at all; and even if a greater quantity than usual were sown on the spots at intervals, still a saving would be effected upon the whole; and if these greater quantities are found to produce a greater crop than the usual mode, then the *plumping* mode, as this method of sowing by intervals is termed, may be regarded as a valuable discovery. Still more experience of its results are required before implicit faith can be placed in it as a practice fit for general adoption; although it must be owned, that the anomalous circumstance regarding the action of bone-dust already noticed, namely, of its maximum effect being produced by a given quantity, and not by indefinite quantities—favours the pretensions of the plumping mode, and renders it deserving of experiment. The very best mode of using bone-dust in small quantity, both for increasing the fertility of the soil and rearing a good crop, is to sow the seed along with it in drills already manured with farm-yard dung. The bone-dust secures a good and quick braird of the plant, and the dung supports it powerfully afterwards. This plan I would recommend to be pursued, particularly in England, on the land prepared for turnips in autumn, and were it practised we need not despair of raising heavy crops of turnips, especially Swedes, on the strongest soils, and most certainly they would be obtained after thorough-draining. On using bone-dust, it should be measured from the heaps in the shed in bushels, and put into the bodies of close carts, which should be left at stated distances upon the headridge; it being exceedingly inconvenient to take bone-dust out of sacks. A field-worker should take the charge of replenishing the hoppers of the machine as the steward returns with it to the headridge at every bout, with a rusky, fig. 371, filled by the frying-pan shovel, fig. 176. As bone-dust is apt to heat in heap, and although it is improbable that you will keep bone-dust over the year, it being more profitable to put it into the ground, yet in case you should have any left over, or in case you should purchase a lot cheap out of season, it may be proper to let you know how to keep it in the best state until it is used for turnips or for any other purpose. The mode of keeping it is this: Whenever you get it, you should put it on a dry pavement floor, as a damp one rots it fast away. It will heat again, but not so violently as at first after being made, and would heat again every time it is turned, which it should not be. It should not be kept in bags, as it will soon rot them, as I have experienced, nor should it be kept upon or under a wooden floor, as it will rot them both; nor should it be kept near horses and cattle, as they evince a strong dislike to its smell, on feeling which, horses actually become restive and troublesome. If new bone-dust obtained from the mill is desired to be heated at once, which it should be when it is to be kept, the process is much accelerated by the addition of a little sifted coal-ashes or earth, and as much water as will make the whole mass only damp, and turning it over several times until the entire mass is incorporated; and in 48 hours the heat will be so great as that you cannot hold your hand in it. The heat will gradually subside, the mass become dry and in a good state for passing through the sowing-machine. Heating has the effect of increasing the weight of bone-dust from 47 lb. to 49 lb. per bushel, and this increase is no doubt affected by the heating having taken off the angles of the particles, and allowing them to come closer together in the bushel. The colour is also changed to blue and yellow, and on examination by the microscope, the mass will be found full of mites.*

(2504.) The established manures for raising turnips are farm-yard dung, street-manure in the neighbourhood of towns, and bone-dust. There are many other substances which have been recommended for the same purpose, such as guano, animalized carbon, &c.; but as they are only of comparatively recent introduction, and cannot be said to have yet established their characters, I shall decline entertaining their pretensions here, and shall rather notice what promise they hold out when I come to mention the subject of making experiments in agriculture.

(2505.) The young turnip-plants may be expected to make their appearance above ground in the course of 8 or 10 days at soonest, and later if the weather is unfavourable to vegetation. When the plants have attained about 3 inches in length, it is time to prepare for their being *singled*, that is, thinned out singly to determinate distances. The first operation in preparation of the process of singling, is passing the horse-hoe between the intervals of the rows of the turnip-plants. This implement, assuming different forms, is drawn by a single horse, and guided by a ploughman, and is figured and described below. The object of using this implement at this time, is partly the removal of any young weeds that may have shewn themselves between the drills, but chiefly to pare away a little of the earth from each side of the drills, in order to afford facility to the hand-hoe in singling out the plants. The implement used for singling turnips is represented in fig. 423, and consists of an iron

Fig. 423.

THE HAND-DRAW-HOE.

plate faced with steel, *a*, 7 inches in length and 4 inches in breadth: this has an eye *b* attached to its upper edge to receive the shaft *c*, usually made of fir, to make the implement as light as possible. The shaft should not exceed 3 feet in length, though in some parts of the country it is 4½ feet, whilst in others it is as short as 33 inches. The shorter it is the better for the work, as it enables the field-worker to bow closer to the ground; but this position is very severe upon the back, and on this account the shaft is in some places made so long as to allow the

* Quarterly Journal of Agriculture, vol. xiii. p. 347.

worker to stand nearly upright, in which position, however, the eye and hand being removed a considerable distance from so small an object as a turnip-plant, she cannot command the implement so effectually in the thinning of the plants, as when the hands are placed nearer the working part of the hoe. The attitude of the workers, the best mode of using the hoe, and of arranging the field-workers at singling, is endeavoured to be represented by fig. 424. This work is performed by the field-workers of the farm, and they are placed or *stented* to the work, as it is termed, at

Fig. 424.

THE SINGLING OF TURNIPS.

every 2 rows, that is, beginning at one side of a field, the first worker is stented to the first and second drills *l m* and *i k*, the second worker to the third and fourth drills *g h* and *e f*, the third worker to the fifth and sixth drils *c d* and *a b*, and so on with every other worker. The reason for this particular arrangement, instead of giving 1 drill to every worker, is, that each may have sufficient room to work, and, having 2 drills each, the whole body of workers have to shift their ground the seldomer. It is not easy to give a short account of the mode of using the hoe in singling turnips; but the following directions may serve to shew the leading requisites to perform that operation in the best manner. On commencing to single the first drill *l m*, the feet are placed on each side of the second drill *i k*, so that the side of the worker is presented to the drill to be singled, the shaft of the hoe is held near its end with one hand, while the other hand, being that of the side in front, is placed a little in advance. The foremost hand indicates whether the person is right or left handed, as it is rare to find a worker that can single turnips with either hand. The foremost hand is steadied by being rested partially on the bend of the leg of the same side. The hoe is pushed chiefly by the weight of the body of the worker against the row of plants, when as many plants are removed by the forward push as the length of the face of the hoe covers, and the body is then brought back to its former position, so there is an oscillation of the body forwards and backwards in the act of singling. In pushing away another portion of the plants, the side of the hoe covers the plant next to the one intended to be left growing, which is, of course, made to stand *single*. This singling constitutes the difficulty of the operation, for, if attention and dexterity are not exercised in it, the plant will be dragged up by the roots with the slightest hold of a portion of its leaf; and, although the leaves may be quite free, the stem or roots may intertwine with those of the adjoining plants. It is found, that the best mode of avoiding these difficulties is to single the plants before the leaves become so enlarged as to be confounded with those of other plants, or the stems to become so drawn up as to intertwine amongst each other. It is also found, that in pushing the hoe is a much safer mode of leaving the plants single than by drawing it towards the worker. The plant, on being left single, falls away from the worker upon its side, partly from want of the support of the others, and partly from the taking away of a portion of the soil from its roots. The plants are represented on their sides in the row *n o*, fig. 424. They receive no injury in falling over, for, if the weather is at all favourable, they will have nearly recovered their upright position by the following day; and it is found, that all that portion of the root which was drawn a little farther out of the earth, being all the part left exposed above the ground, is converted into bulb. When the first of the two rows are thus singled, the second row is then singled, on the completion of the bout. Thus, singling the turnips on going up the drill *l m*, fig. 424, the worker returns down *k i*; on going up the drill *g h*, she returns down *f e*; and on going up *c d*, she returns down *b a*. All the figures in the cut are represented going up singling the first drills of the stent. There are only 3 figures introduced, but the number of course can be increased to any amount that there is room for in the field. On shifting the workers from one stent to another, the worker nearest that part of the field which has yet to be singled forms the pivot upon which the rest turn. Thus the figure on the drill *l m*, on finishing the drill *n o*, shifts to the next drill *l m*, forming the pivot for the rest to turn, who, before the shift, had worked in the drills *beyond n o*, and, in like manner, after the figure in the drill *c d* shall have returned by the drill *b a*, she will take her position at the drill on this side of *a b*, and form the pivot upon which the others will take up their position on this side of her. This alternate shifting, whilst it keeps every worker in her own relative place, and prevents confusion, serves to mark the point from which each successive stent is measured. In some parts of the country, especially in England, a triangular form of hoe, of 3 faces, is used instead of the common hoe represented above. It is preferred, I believe, on the idea that its 3 faces will last longer than the one face of the common hoe, and is therefore considered a cheaper instrument; but this notion, I suspect, is not well founded, as the common hoe costs only 2d. per inch, while the triangular costs 4d. per inch along the working face. There is this insuperable objection against the triangular hoe, that, in pushing away the superfluous plants with it, the very acute angle of its corner is so small as to be apt to miss a hold of the plant that should be taken away from the one to be left single; whereas, the straight side of the common hoe covers that plant completely from leaf to root, and takes it away decidedly with the rest. In using the hoe, it is not an uncommon practice, both in England and Ireland, to make regular gaps with it amongst the plants, and to leave their *singling* to boys or girls by the use of the fingers, and hence they act as assistants to those, usually men, who form the gaps with the hoe. No doubt turnips may be singled in this way, but at greater expense, and it entrusts the selection of the plants to be singled to the judgment of mere boys and girls, who cannot be supposed to know so well the properties which render plants valuable, as experienced adult persons. Indeed, I consider the employment of boys and girls in singling turnips a questionable policy, and certainly they ought not to work in company with an experienced band of field-workers, because they work much slower, as may be expected of all inexperienced persons, and, of course, they restrain the progress of the working band, for these will not at all times leave their young companions far behind; but their want of skill causing them to destroy many plants that should be left, is even a greater evil than slow work. Two young people might be put on one drill, but this deranges the form of the whole band, and rather than do that, they should be put by themselves at another part of the same or another field. It is true that boys and girls must acquire a knowledge of this work, but to effect the acquisition they should at first be placed in a part of a field which offers peculiar facilities for singling, such as a smooth state of the ground, and the most proper age of the plants; and so circumstanced they will learn to single at leisure, and understand instructions more readily from an experienced worker, who should be set to superintend them. Singling of turnips should only be prosecuted when the ground is dry, and the plants themselves also dry, as they then separate from one another more readily. Whenever the ground becomes cloggy on the hoes, even with a shower, the work should be desisted from until drought returns. The quantity of work done in singling turnips varies in different parts of the country. In the midland counties of Scotland it is estimated to take 3 women to single 1 Scotch acre in a day of 10 hours, and there the long-shafted hoes are in general use, and the women work with upright backs. In other parts 1 woman will single ½ an English acre in a day of 10 hours, and there short-shafted hoes are used, and the women work with bent backs. One summer I superintended 16 field-workers, and they singled about 90 acres imperial in 8 days of 10 hours each, which was equivalent to nearly ¾ of an acre a day to each worker. This is above the average rate of work, but then the weather was exceedingly fine all the time, the land mellow and dry, the plants were of a proper age in the progress through the Swedes to the white globes, and the women were all first-rate workers. I set an excellent steady hand as leader to the band, whom she carried on like clock-work. She herself preferred to work with a hoe of only 33 inches in length, which allowed her to bow down to her work, and which she performed, in consequence, in the most perfect manner; the hoes of the rest were 36 inches in length. A rest of 20 minutes was given them at each mid-yoking, and, instead of fatiguing themselves walking home, they brought their dinners to the field, consisting of barley and pease bread, and a bottle of milk. I took my own bottle of milk and loaf of home-made bread to the field, and enjoyed my repast at the hedge-root with a genuine relish, in as light-hearted a company as ever undertook laborious work.

(2506.) After the plants that have been removed by the singling into the middle of the drills are partly decayed, and the leaves of the single plants have attained about 6 inches in length, the horse-hoe or scuffler is again set to work to harrow the removed plants to the surface, and to destroy any surface weeds that may have made their appearance. After the turnips are all singled and scuffled, the field workers *hoe* the turnips in the order they were singled, that is, remove the weeds from *between* the plants, and loosen the soil immediately *around* them. This is done by setting one foot on each side of a drill, and on grasping the hoe short, the earth is loosened with it around every plant, all double plants removed by the hand, and every weed pulled by the hand that is found growing too near a single plant for the hoe to remove, which, if attempted by the hoe, might cut the plant through; and care should also be taken that no plants are cut through by the root under ground with the hoe. A second hoeing of a similar kind generally finishes the hand-culture of the turnip-crop. In stenting the drills for hoeing, a drill is given to each worker, because when they hoe beside each other they can see whether the ground is entirely hoed over, which it should be, wherever there is a weed seen growing; and as each worker embraces a drill between her feet, and, as it were, covers her own drill, there is no chance of her jostling her neighbour, which she would do in singling were each to take a drill, for the worker has to stand over the drill next the one she singles to find room for the action of the hoe. Before sowing turnips, the land should generally be as clean as it can be made; but when it has been allowed to run excessively foul to weeds, rather than be too late in sowing the crop by cleaning the land, and rendering it too fine, that is, *deaf, sow the crop in due time;* and provided the weeds can be kept down by the horse and hand hoe, until the luxuriance of the leaves cover the drills, the weeds will be smothered under them and rot, and serve partly to support the turnips; and after the turnips have been eaten off with sheep in winter, not a single weed will be seen in the land in spring when it is ploughed up, either for spring wheat or barley. I had a favourable opportunity of witnessing, one season, an instance of what I am now describing. I had a field of 25 acres so foul with common couch-grass, that it was utterly impossible to clean it in time. After removing the roughest part of the wrack to the compost stance, the land was dunged and sown with white globe turnips, and the result was as I have stated. And here I may mention, that the turnip sowing-machine, which has movable coulters and handles, is the best for sowing foul land.

(2507.) With regard to the *setting up* of turnips, I am of opinion that it is an unnecessary piece of work in most cases, and certainly on dry turnip-soil, and on thorough-drained land, I would not set them up, for two reasons, namely, that smooth ground is the best for sheep to lie on on turnips in winter; and that on ground with a considerable inclination hollow drills serve only as channels for surface-water to carry away the best of the soil to the bottom of the inclination. There is, besides, the objection of setting up or working in any way amongst turnips after the leaves have nearly met in the drills; and this objection is founded on a physiological fact connected with the growth of the plant which I shall mention. Early one summer morning, many years ago, I observed a whiteness, like hoar-frost, in the bottom of the drills between the rows of young turnips before their leaves had met across the drills. Knowing from the temperature of the air that the phenomenon could not be hoar-frost, I examined it particularly, and found that it arose from the deposition of dew on innumerable minute fibres proceeding from the roots of the turnips, which had not yet attained the form of a bulb, and interlacing one another across the drills from either row of turnips. A similar appearance may be observed at any time on a spider's web bespangled with dew or mist on a furze bush. Having traced these delicately minute vegetable fibres to the roots of the turnips, it occurred to me that it was surely an error in practice to work amongst turnips after the development of such fibres, for there cannot be a doubt that these are sent out for some special purpose connected with the growth of the plants. Either, then, the setting up should be concluded before the exhibition of these fibres, or, if there is not time to do that, the setting up should be abandoned altogether, for I apprehend that more injury will accrue to the crop from disruption of those fibres than advantage from the formation of a channel in the bottom of the drills, the only benefit of which to the land can be to carry off superfluous surface-water, when the land is undrained; but when it is drained, these channels are of no use, for then there will be no surface-water to *run* off, as it will be absorbed by the soil as fast as it falls in the shape of rain. These fibres are not to be seen after the dew is evaporated, except on the most careful examination of the surface of the soil. The same phenomenon may be partially observed amongst drills of potatoes; but the fibres from the roots of the potato spread more under the immediate surface of the ground than those from the turnip. Such is the anxiety of most farmers to set up drills, that I have seen the double mould-board plough smashing amongst the leaves of both potatoes and turnips after they had met across the drills.

(2508.) In regard to setting up the *potato* crop, it is perhaps necessary to do it to a partial extent, inasmuch as those potatoes which grow in a cluster around the stem of the plant, near the surface of the ground, and even above it, would be too much exposed to the air, and, on becoming entirely green, be bitter to the taste. An interesting experiment on the depth to which potatoes should be earthed, was made by Mr Peter Mackenzie, Plean, near Stirling. "On well-drained land," he says, on describing the results of the experiment, "three modes of cultivating the potato were tried; the dung used being what is called well-made farm-yard manure. The first was similar to the plan usually adopted, namely, earthing up the crop, until the interval between the rows was 2 or 3 inches deeper than the roots and dung. The second plan differed only from the first in being less earthed, or what may be called by some a half setting up. The third mode had no earth drawn to the stems of the plants, and the earth was only hoed between the rows. The dunging of the crop and the distance between the rows were equal. When the potatoes were dug, the advantage of the second mode of culture over the first was fully more than ⅓ part of the increase, and better in quality; for the potatoes grown by the first plan would not bring the same price in the market which the second would. The produce of the third plan was nearly equal in bulk with the second, but rather inferior in quality, many of the potatoes having their sides *greened* by being exposed to light. While growing, the second and third lots had a much more healthy appearance than the first; and when dug, what remained of the dung that was used was well mixed with the soil; while the dung of the first lot was dry and little decomposed, clinging in clusters to the roots of the potato plants when they were dug. It would be little trouble for farmers and others who grow the potato," advises Mr Mackenzie "to try the experiment on land that has been well drained; for it would certainly be a great advantage to themselves if they could increase their potato-crop ⅓ more than is usually grown, and with less labour to themselves and horses. When we bear in mind the number of acres which are planted with potatoes, it must greatly increase the food of our country, both for man and beast, if 20 or 30 bolls were added to every acre in the produce of that essential article of food. When we remember the number of square feet there are in an acre, and if upon every square foot an extra potato were raised weighing only ¼ lb., more than 4 tons would be added to the crop of every acre." * It thus appears that a partial earthing up is beneficial to the potato crop; but the same reason does not apply to the turnip.

(2509.) From what I have said on the effect of 1 or 2 inches between the turnips decreasing the weight of a crop of turnips several tons per acre, in (1258.), the singling and hoeing of the turnip-crop ought to be regarded as one of the most important particulars that demand your attention. If you wish to preserve a distance of 9 inehes between the plants of white globe turnips, the first consideration is, whether the plants have brairded so equally over the field as to allow you to preserve an equal distance between the plants. Being satisfied that the germination of the seed is pretty equal over the field, then it ought to be your endeavour to single off the plants at the stated distances desired, say 9 inches of white globes and 12 inches of Swedes. The hoe commonly in use is 7 inches long in the mouth, so there is plenty of room for it to have a play in both sides, to allow plants to stand 9 inches asunder. If *care were bestowed* on this process of singling out the plants, there can-

* Quarterly Journal of Agriculture, vol. xiii. p. 363.

not be a doubt but that they would grow at those stated distances, and produce such a weight of crop, as the average weight of bulb from an average portion of the crop would indicate. In like manner, were the after hoeings of the crop to receive due attention, there would be no cutting up of single plants, and, of course, no blanks. Were both these operations conducted with proper care, the estimated weight of the turnip crop could not so much exceed the realized, as is too frequently the case. But instead of care being bestowed on either of these operations, field-workers are often left to themselves, and then talking, the proverbial failing of women, occupies their attention much more earnestly than the work they are engaged in. Plants are knocked off without regard to the space left between them; and even when women are superintended, many are so disingenuous, that when they clear a space too bare, they will stick in a plant into the place, which, of course, dies off. Gravelly soil is, with the best work-people, difficult to work in with a hand-hoe, as the corner of the hoe will sometimes start off a small stone, and cut through or remove a plant against the will of the worker. Such soils, therefore, require particular attention in singling or hoeing the crop. Instead, therefore, of stewards being anxious merely to gather a large number of field-workers to the singling of the turnips, their anxiety should be evinced to obtain a number of practised hands, in whom they can place confidence for attention and skill.

(2510.) Swedish turnips transplant very well, like the common cabbage; but the true turnip, the white globe or yellow, do not transplant, and any attempt to fill up blanks with them ends in disappointment. When first introduced into this country, the seeds of the Swedes were sown in the garden, and the plants transplanted in the field; but now by far the most common way of procuring plants is sowing the seed at once in the field. I have tried several times to fill up blanks in drills with carefully removed plants, but though most of them grew, they never attained the size of the rest of the field. Mr Howison of Crossburn House, Lanarkshire, adopted a mode of transplanting Swedes on strong land, which deserves attention, and may be even extensively practised in a season unfavourable to the working of strong land in time for Swedes. There is first the method by which plants are procured to be described, and then the transplanting of them. "In an open piece of ground," says Mr Howison, "I form raised drills at the distance of 12 inches from centre to centre, in each of which I place a layer of short dung, closely laid on, on which the turnip-seed is pretty thickly sown, and afterwards covered with ½ inch of fine mould. The breadth of the drills at top should not be more than 4 inches, so that the earth and plants may be more completely lifted together by the spade, when to be transplanted. The time of sowing should be regulated according to the purposes intended; and as a fall of ground converted into drills will produce plants sufficient for transplanting 3 or 4 acres, it is better always to have a superabundance. As to the best age for taking up the plants, I have found no difference in their success from the time they have got their proper leaves, until they are 3 or 4 months old; however, those intended to remain long should be thinned out in the rows. I need scarcely mention, that in dry weather they should be carefully watered, which in so small a space can easily be done. Having provided a wheelbarrow, a garden spade, and a couple of flower-pot saucers of a large size, I, with the spade, lift up its breadth of one of the drills, taking care that the spade enters below the roots of the plants. This spadeful is then carefully placed in the barrow, and the same operation is repeated until the quantity of plants wanted is taken up. When that is done, they are then removed in the barrow to the field, where they are transplanted with the implements above mentioned. It is necessary for expedition that 2 persons be employed in the transplanting—one to prepare the plants, the other to transplant them. One spadeful is then taken from the barrow, and with a knife divided into 3 or 4 pieces, 1 of which is taken into the hand and carefully drawn asunder so as to lay open the roots of the plants with as little injury to them as possible, and, taking hold of the leaves of the one that appears uppermost, draw it gently out with as many of the little balls of earth and dung adhering to its tender roots as practicable, and place it in one of the saucers. In this way when the saucers have been carefully filled with the plants laid in regular rows, the transplanter may commence his operations. He should then, with a short dibble not thicker than his finger, make a hole which should only reach to the dung; and then lifting up a plant by its leaves drop it into the hole, and with the fingers of both hands press the earth gently around it. In this way 2 drills may be planted at the same time. His follower with the barrow will be able to supply him with prepared plants; and from my experience the 2 men should be able to finish 1 rood of ground in the course of the day if the plants are 8 inches distant from each other. I may here warn the transplanter against using plants that have not one or more balls of dung or earth adhering to their roots; for if he does, the chance of their growing will be very small." The advantages attending this mode of transplanting Swedes, which I have no doubt is most effectual, are stated by Mr Howison to be three:—it enables the farmer to fill up the blanks of a turnip or potato field with good plants of Swedes; it is more certain of procuring a crop than sowing the turnips over again. This was exemplified on the farm of Green Burnside, Lanarkshire, in 1839, tenanted by Mr James Cassie, when in October of that year "the crop that had escaped the ravages of the fly was excellent; the transplanted very good, but the turnips not so large in general as the first. The third, although sown on the same day as the others were transplanted, were none of them the size of an egg, which I believe is always the case with late sown turnips, although the result is difficult to be accounted for. I may, in conclusion, mention another advantage, that it is *best* performed in *wet* weather when all other field-labour is at a stand. I consider rainy and cloudy weather as of the utmost importance to successful transplanting; and so much so, that I prefer waiting weeks to attempting it in sunny and dry weather."*

(2511.) The culture of *mangel-würzel* is exactly similar to that of Swedish turnips, up to the point of sowing the seed, which is, however, done in a very different manner, owing to its peculiar structure; that of Swedish turnip is smooth, and can be sown with a machine, but the seed of mangel-wurzel is contained within a persistent capsule which is so rough as to be impracticable for any machine yet invented to sow it. I have had very little experience myself in raising mangel-würzel; it is but little cultivated in Scotland, the climate being too cold for it, but is extensively cultivated in England, and is now to be found in Ireland, where it grows in the greatest luxuriance, as I have had opportunities of observing. There are 3 kinds recommended for field-culture, the common long red, the sugar-beet, and the orange-globe-beet. The sugar-beet is too delicate for this climate, and of the other two sorts, the orange-globe, though of late introduction, is the favourite for yielding a sweet root and large crop. The way I sowed the common long red or marbled mangel-würzel, after the drills had been dunged and split in the double mode (2178.), on good hazel loam, in the end of April, was this:—A rut was made by a field-worker along the top of the drill with a common draw-hoe, fig. 422, another field-worker followed and dropped in the seed along the bottom of the rut with the hand, and a third followed last and covered up the seed with the earth raised by making the rut, with the back of an iron garden rake trailed along the rut. The crop was singled by hand, hand-hoed, and horse-hoed. The crop proved a good one; was taken up before the arrival of frost, stored amongst dry sand like carrots, and given to milch-cows in winter, who were very fond of it. Another season I tried a little on a rich alluvial

* Prize Essays of the Highland and Agricultural Society, vol. xiii. p. 513.

clay, which was recovered from the bottom of a marsh in the middle of a field, with exactly the same culture, but the crop was not worth the trouble bestowed upon it. "The mode of culture I adopt," says Mr Miles, M.P., "up to depositing the seed in the ground, is the same as that adopted in Northumberland for ridging the Swede; great care, however, must be taken that the seed of the mangel-würzel is not buried too deep, or it will not vegetate. Dibbling, as you can never ensure an equal depth, does not answer; nor does the seed drill well, if properly prepared by steeping, which I should recommend, for at least 24 hours before planting. To ensure, therefore, a proper depth, I have been in the habit of using an iron wheel, round the circumference of which, 18 inches apart, iron points project, broad at the base and tapering towards the point, about 2½ inches long; this is wheeled upon the top of the ridge, the men walking in the furrow, and thus holes are formed which can never run into the excess of too great depth, and into which the seeds are deposited by women and boys following the wheel, and generally covering the seed by drawing the foot as they advance at right angles with the ridge over the holes; the roller follows, and thus the sowing terminates."* The wheel should be, I presume, at least 3 feet in diameter. I tried, on one occasion, to form holes for the seed by short pins attached to a long roller of wood which was pressed by the foot, and the seed deposited by the hand in the holes and covered up with the rake, but the plants did not braird so equally as when the seed was sown in ruts. The earlier in April mangel-würzel is sown the better, and the deeper the tilth the greater probability of a heavy crop; and although the common long red sort requires a deeper and stronger soil than Swedes, the orange globe beet will thrive where Swedes succeed. The quantity of seed required is from 5 lb. to 7 lb. per imperial acre.

(2512.) *Rape* is sown at two seasons, in autumn to be ready to be eaten as spring food, and in summer, to be consumed in autumn before the consumption of turnip commences. Its culture in summer is precisely that given to Swedes, and as its seeds are smooth, though larger than turnip seed, they are sown by the same machine. The plants are not thinned out in the drills like turnips, because it is the leaves, and not the root, which are used; but the ground is horse and hand-hoed until the leaves are so far advanced as to be able to keep down weeds. Rape is cultivated in this country solely for the use of sheep, for ewes before they are tupped in autumn, to bring them into season, and perhaps for hoggs too in preparation for turnips, and for great-ewes and hoggs in

* Journal of the Royal English Agricultural Society, vol. ii. p. 299.

spring, to bring the milk on the ewes. It is extensively cultivated on the continent, in Belgium, Holland, Germany, for its seed, out of which is expressed the rape-oil that is much used in manufactures. It is the compressed husks of the seeds of this plant, after the oil has been expressed from them, which constitute the well-known manure *rape-cake*.

(2513.) *Carrots* are only occasionally raised in the field for farm use in this country, not but that the climate is quite congenial to their growth, but their growing best in light soils, the culture is necessarily confined to certain localities. Carrots will not succeed with dung applied directly under them, as they are then very apt to become much affected with worms, and to fork into a number of roots. The dung should be applied, and in large quantity, as for Swedes upon the soil in autumn. Such a treatment as that described for Swedes by Mr Scougall, or Mr Thomson (2502.), would answer well for carrots. The seed being confined in a capsule covered with small hooked spines, cannot be sown in a satisfactory manner with a machine, though attempts have been made to effect the purpose. It must therefore be sown with the hand, and as the capsules are apt to adhere together, they will separate more freely if mixed with a little dry earth, or still better with sand. A slight rut should be made along the top of the drills with a hoe, the seed strewed along, and covered with earth. The plants should be singled out at the distance of 4 inches by the hand, as their roots strike down too deep at once for the hoe to remove, without at the same time taking away too much of the soil from the top of the drill. The ground may be horse and hand-hoed, according as weeds are found to be troublesome. The land being of slight texture, requires no setting up. There is a mode of raising carrots practised by the feuars of the parish of Barrie, Forfarshire, on sandy soil, which deserves attention in similar circumstances. They begin a trench of 2 feet deep in the sand in autumn, after a white crop, and collecting sea-weed as it is washed ashore after storms in the course of winter, they half-fill trench after trench with it, till the break of soil allotted to the carrot crop is manured. Of course, other manure would answer the same purpose as sea-ware, but when it can be found in sufficient quantity for the gathering and carriage, it affords a cheap manure for the purpose. Of farm-yard manure, horse-dung is found to be best when treated in the manner I am describing. Ruts are formed with the hoe in rows of about 14 inches apart, and old dung is crumbled along the bottom of these to ensure the brairding of the seed on so poor a soil as sand generally is, the seed is sown on the dung, and the soil returned over it. The plants are thinned out by the hand, and the ground kept clean by hand and hoe together. The carrots grown in this manure, and in this soil, are excellent, as may be witnessed in the Dundee market in autumn. The carrot that has hitherto been mostly cultivated in the field is the Altringham, which is long and tapering in the root, pretty thick at the top, of a dark red outside and yellow with core. Recently the Belgian white carrot has been introduced into field culture, and bids fair to become a favourite, being large of size and sweet to the taste. From 8 lb. to 12 lb. of seed are sown on the acre, when good, and as much as 20 lb. when inferior.

(2514.) The *parsnip* requires a milder climate than this country affords to be successfully raised in the field. Where it is desired to be raised, it may be cultivated as the carrot, either on drills, or in rows upon the flat ground, or broadcast, but it affects a much stronger soil than the carrot, though it may be raised upon sand, and even peat, if an additional quantity of manure is applied. The seed is contained in a broad thin capsule, is very light, and 10 lb. the acre are required when good. It should be sown in April. It should be new, and steeped before being sown, but if sown in a very dry soil, when soaked, it is apt to be destroyed by the drought. The thinning and weeding of the crop may be effected in the same manner as the carrot. Colonel le Couteur, of Bellevue, in the island of Jersey, where the parsnip is cultivated to the highest perfection, describes both the broadcast and drilled modes of cultivating this root, but the account is not very explicit; for instance, he does not mention the season when the crop should be sown. Of the mode of conducting the broadcast made in Jersey, he says, "An old grass lea is broken up by some persons in September, by others just before the parsnip-land is sown. The former I consider to be the best mode. When the turf is well rotted, 20 tons of stable manure per acre are spread over the land. A trench is then opened through the centre of the field, between 2 and 3 feet wide, and, where the soil will admit of it, from 1 foot to 18 inches deep. A 2-horse plough then turns the manure and about 3 inches of soil into the trench, and is immediately followed by a large trench-plough with 3 or 4, and in many cases here, with 8 or 10 horses, which turns 1 foot or more of clean soil upon the manure and scurf, when the land has been recently skim-ploughed. The soil is then harrowed, and the parsnip-seed, which should be *new*, is sown at the rate of 3 or 4 lb. to the acre, and lightly harrowed. When the plants are 1 inch high they are weeded. The plants, from the first, should be thinned out to 6 inches apart, and, according to the soil, should be again thinned out to 9 inches or more at the second hoeing." In regard to the drill method, Colonel le Couteur says, that "The land may be prepared as above described. In one case I found the plants to answer well by spreading a portion of the manure on the surface of the ploughed land, and then earthing it up into small ridges, 1 foot apart, with a double mould-board plough. The seed is then sown on the top of the ridge and rolled in, which succeeded extremely well. The hoeing was performed with a horse-hoe in the drills, and the plants were cross-hoed with a hand-hoe. This mode does not appear so neat as the following: when the land is well harrowed and levelled, sow the seed broad-cast, harrow and roll it; then, when the plants appear, hoe it into drills, either with a horse-hoe or hand-hoe. A drill-machine will be the best method if one could be found to sow parsnip-seed regularly; mine sows it much too profusely. The parsnips require hoeing and thinning as in the broad-cast husbandry. In a dry season it is well to observe, that moistening the seed with wet sand and earth, and stirring it daily, to be sown in the first moist weather, or after a shower, will forward its growth a fortnight." In regard to taking up the crop, Colonel le Couteur conceives this to be the best method. "A plough with a blunt or worn-out share, and without a coulter, being drawn by a pair of horses along or across the ridge as is most convenient, the pressure of the plough and earth forces the plants out of the ground; and although a small portion of the long tapering root is sometimes broken off, yet the time saved by thus raising them, and afterwards throwing them out of the loosened soil, as is done with potatoes, is an enormous gain of time over the ancient practice of forking out each parsnip from the solid ground." It may be here observed, that the above modes of cultivating this or any other green crop, does no credit to Jersey agriculture. It thus appears that a green crop is taken and dunged after lea—8 or 10 horses used in a trench plough—the land dunged before it is trenched—and the plants are placed in rows by sending an implement through a broad-cast braird. Notwithstanding these slovenly modes of culture, such is the fineness of the climate, that 27 tons, 8 cwt. of parsnips have been raised on an imperial acre, "a quantity," as Colonel le Couteur remarks, "nearly sufficient for 10 cows during the 6 winter months, according to the calculation of the Flemings."*

(2515.) Both *turnip* and *rape* are comprised in the natural order of *Cruciferæ*, a name derived from *crux*, *crucis*, a cross, and *fero*, to hear, in allusion to the 4 petals being disposed crosswise; and in the order *Tetradynamia*, *Siliquosa* of the Linnæan system; and in both systems under the genus *Brassica*. The *Swedish turnip* is placed under *Brassica campestris*, whose specific characters are, "leaves rather fleshy, covered with glaucous bloom; first ones rather hispid or ciliated, lyrate, toothed; the rest cordate, stem-clasping, acuminated,

* Journal of the Royal English Agricultural Society, vol. i. p. 419.

partly pinnatifid. Is a native of Britain, Lapland, Spain, Transylvania, and in the Crimea in fields," and it is immediately classed in the variety *rutabaga.* The *common turnip* is the *Brassica rapa*, the specific characteristics of which are, "radical leaves lyrate, destitute of glaucous bloom, green, covered with bristley hairs, middle cauline ones cut, upper ones quite entire, smooth. Native throughout Europe in cultivated fields, and their borders." The seed of the turnip has been found in remarkable situations. On cutting through a gravelly knoll of about 30 feet in height to the eastward of the town of Lisburn, in making the railway betwixt that town and Belfast, I remember of observing a great number of plants of the wild turnip growing on the side of the embankment amongst the gravel that had been wheeled out in levelling the knoll. Their leaves were beautifully expanded, and the plants seemed quite healthy, but there was no appearance of bulb at the root. The *Rape* is the *Brassica napus*, whose botanical characters are, "leaves smooth, of a greyish-glaucous hue, radical ones lyrate, stem ones pinnatifid and crenated, uppermost ones cordate-lanceolate, clasping the stem; siliques devaricate-spreading. Native country not known. Cultivated in fields. It is to be found almost naturalized in waste ground, and on ditch banks in Britain." The British rape or colsat is *Brassica napus oleifera.* "It is distinguished from the colsat or colsai of the continent by the smoothness of its leaves, the other being hispid. It would be desirable, De Candolle observes, if all cultivators would examine whether the plant they cultivate is *Brassica campestris oleifera*, or the *Brassica napus oleifera*, which can easily be ascertained by the roughness or smoothness of the leaves. Experiments made by Gaujac shew the produce of the first compared to that of the second, to be 955 to 700. . . . . The soils best suited for the culture of rape are the deep, rich, dry, and kindly soils. Arthur Young says, that on open fen and peat soils and bogs it thrives well, and especially on pared and burned land, which is the best preparation for it; but it may be grown with success on fenny, marshy, and other coarse waste-lands that have been long under grass, after being broken and reduced to a proper state of preparation. . . . Wheat is considered the best crop to follow rape, by its being eaten off early in autumn, there is sufficient time allowed for getting the land in order for sowing wheat. The time for sowing rape is the same as that for the turnip, and the manner either broad-cast or in rows. Where the object is to keep sheep in autumn or winter, by eating it down, the broad-cast method and thick sowing is evidently the best, and is that generally resorted to in Lincolnshire and the fenny districts. The quantity of seed, when sown thick, may be 1 peck per imperial acre, but when drilled or sown thin, from 2 lb. to 3 lb. will suffice. Vacancies may always be filled up by transplanting. No hoeing or thinning is necessary. *Rape-cake* is the adhering masses of seed-husks after the oil has been expressed, and *rape-dust* is the dry loose husks. In harvesting rape for seed, great care is necessary not to lose the seed by shaking, or by exposing it to high winds or rain. It is reaped with the hook, and the principal point is to make good use of fine weather, for it must be thrashed as fast as reaped, or at least without being stacked like other crops." * The quantity of *rape-seed* "imported for home consumption in 1836, was 561,457 bushels; in 1837, 937,526 bushels." †

* Don's General System of Botany and Gardening, vol. i. p. 242-5.

† Ure's Dictionary of the Arts and Manufactures, art. *Rape-Seed.*

" Rape-cakes were worth, in December 1833, from L. 5 to L. 6 a ton. In 1830 we imported about 330,000 cwt. of rape and other oil-cake."* The duty charged in the new tariff for rape-seed is 1d. per quarter. I can find no mention made of rape-cake or of oil-cake in the new tariff, and suppose the duty on their import must be abolished; it was formerly on rape-cake 2d. per cwt. Rape-cake, when subjected to chemical analysis, yields the following constitution according to Dr Henry R. Madden :—

	Water,	10.5
Organic matter,	Soluble in cold water,	24.7
	... in hot water,	4.8
	... in weak potass,	31.5
	... in strong potass,	10.2
	Destroyed by heat,	14.3
	Earthy phosphates,	3.0
	Silicate of potass,	1.0
		100.0†

The culture of *turnip-seed* is profitable to the farmer. It may be raised either by transplanting the bulb, or by pulling up the faulty bulbs in the field in which the crop is growing, and allowing the remainder to run to seed. There is a difference of opinion amongst the growers of turnip-seed which is the best mode; but other considerations enter into the question before the matter is determined. When the bulbs are removed and transplanted, any spot the farmer chooses, such as an otherwise useless piece of ground, may be selected for the purpose, without interfering with the ordinary culture of the turnip-field, and this of itself is a strong argument in favour of transplantation; but besides this, when a crop of turnips is allowed to stand on the ground it grows until next year when the crop of seed is fit for reaping, the crop of seed must be regarded as a substitute for the white crop which usually follows the turnip-crop on the farm; or if it is not so regarded, serious disputes may arise between landlord and tenant regarding the nature of the crop, and the landlord has certainly a very rational plea for declaring the turnip-seed a scourging crop, and one at least equally so to any white crop; and if turnip seed is to be regarded as equivalent to a white crop for the land, the profit derived from this mode of culture must be very much reduced. For my part, I would always prefer raising turnip-seed by transplantation of the bulbs, because a complete choice of each bulb can then be made, and because, also, the practice does not interfere with the ordinary rotation of crops; and a concentrated crop, such as a transplanted one always is, is more easily watched from the depredations of small birds, which are remarkably fond of turnip and rape seed as articles of food. Suppose, then, that the bulbs are to be transplanted; this is best effected any time between November and January, and let them be selected of the finest form and the largest size of bulb, and let them be taken to a piece of ground out of the way of any other piece of turnip or cabbage seed in the neighbourhood; and to render the plants as strong as possible, let the ground be dunged as the bulbs are set into the trench formed by the spade. The trench should be as deep as to cover the bulbs, to protect them from frost. When the plants bloom, every stem bearing

* Macculloch's Commercial Dictionary, art. *Rape*.

† Prize Essays of the Highland and Agricultural Society, vol. xiv. p. 529.

spurious blossom, however well formed the bulb may be, should be removed; not but that an individual variety may be better than the general stock, but its existence there until the seeds have been formed may be the means of producing a hybrid, which may possess worse properties than either. Bees being the active agents in carrying the pollen of flowers from one place to another, and thereby forming hybrid impregnations injurious to the variety of plant wished to be preserved pure, care should be taken to have your plot of turnip seed as far as practicable, from any other, or from that of cabbage, broccoli, or cauliflower, raised by nurserymen or gardeners. Long before the seed is ripe, small birds will be busy in shelling it out of the husk; and were they to destroy merely what they consumed, the loss, perhaps, might not be grudged, but as they spill a great deal more than they consume, a whole pod is destroyed for the sake of one seed. In this depredation none are so active as the *Linaria cannabina*, variously denominated grey, brown, or rose linnet, or rose lintie, one of the sweetest warblers of our woods. There is no way of evading the attacks of these active marauders but by constant watching from dawn to eve; and to render the watching more effectual, alleys should be left crossing each other when the bulbs are transplanted, to allow the watchers to pass at pleasure in various directions through the plot. The crop should be cut down with the sickle before it is ripe, as the seed is very apt to shake out; and the best mode of preserving and winning the seed is to place the stems in frames of wood of 20 feet in length, 12 feet square, and perhaps 10 feet in height. The frame in its length is left hollow, to allow the air to pass along, and the stems are placed in it as upright as that the butt-ends of the stems shall project over the outer laths of the frame, and above one another, so as to form a thatching of stems. The upper part of the frame is filled up and rounded with the smaller stems of seed, cut off from the larger, and the whole is thatched with straw, and bound down with straw-ropes. The seed is thrashed out when wished to be disposed of or used. A crop of Swedish turnip-seed, when grown in the fields where the turnips grew, is considered good when it yields 28 bushels per imperial acre; of yellow turnips, 20 bushels; and of globes, 24 bushels. When transplanted, I should suppose the yield will at least double these quantities. A bushel of Swedish turnip-seed weighs 47 lb.; of yellow, 51 lb.; and of globes, 51½ lb. The price of Swedish varies from 1s. to as high as 2s. per lb.; of yellow from 1s. to 1s. 3d.; and of globe, from 9d. to 1s. There is a remark of Mr S. Trewecke of Breage, Cornwall, regarding the remarkable facility of raising spurious turnip-seed, compared with good, which is very true. "I have found," he says, "from several years' experience, that the better the quality of the turnip, the less is the quantity of the seed, and the worse it is to bring it to maturity; while with the stringy, spongy, long necks and bushy roots, a large quantity of seed may be saved with little trouble. I think farmers in general pay too little attention to the quality of the seed, if they can get a cheap article." *

(2516.) It is surprising what a waste of turnip-seed is occasioned by the time the turnip-crop is thinned out. About 3 lb. of seed is sown per imperial acre, and I have already stated in (1258.), that 25,813 plants of globes at 9 inches apart, 23,232 of yellow at 10 inches apart, and 19,360 of Swedes at 12 inches apart, are left on an acre in drills at 27 inches apart. Now there are 1387 Swedish turnip-seeds in 1 drachm, so that 1 oz. 6 drachms of seed

* Mark Lane Express, July 1841.

are sufficient to produce plants for 1 acre; consequently there is a waste of seed of 27 to 1. There are 1645 of purple-top yellow turnip-seed in 1 drachm, and 1800 of white globe; and what is remarkable, that in both these turnips the weight of the seed bears the same proportion to the distance left between the plants in the drills of the respective kinds, as to make the waste in each the same, namely, 27 to 1. It must not be supposed, from these results, that 1 oz. 6 drachms of seed of either of these kinds of turnips would suffice to sow an acre, for many of the seeds may want vitality, and many others in sowing, are, no doubt, buried too deep to vegetate with the rest. Abundance of turnip-seed not only secures a full braird, and tends to draw the plant quickly to a state for being singled, but quickness, combined with abundance, of growth, is a great safeguard against the injurious effects of the attacks of insects.

(2517.) The turnip-plant is subject to be preyed upon by many insects, but though its enemies are very numerous, there are but few which inflict grievous injury, the principal being the turnip flea-beetle and the turnip saw-fly. It would occupy much too large a space to give a characteristic description of every insect that infests the turnip, so that I shall enumerate very little more than a catalogue of them, with a reference to the authorities by whom they are particularly described, and whom you may consult, with a view to the recognition of the insects and their destruction. The insect which first infests the turnip-plant, and attacks its seed-leaves, is the turnip flea-beetle, *Haltica nemorum*, usually, though improperly, designated the turnip-fly, which is a very different sort of insect. The flea-*beetle*, as its name denotes, is a coleopterous or hard-shelled insect, capable of either penetrating the ground or of bearing a considerable pressure. "It is a small insect," says Mr Duncan, "scarcely $\frac{1}{8}$ of an inch in length. It is smooth, shining, and of a brassy black colour, with a slight tinge of green, particularly on the wing-cases; the antennæ black, with the second and third joints, and the apex of the first, of a pale colour. The thorax is convex above, and pretty deeply punctured; the wing-cases are much wider than the thorax, likewise thickly and irregularly punctured, each of them with a pale yellow or slightly sulphur-coloured stripe running along the middle, curved inwards posteriorly, and not reaching quite to the extremity; the under side of the body and thighs black; all the tibia and tarsi of a pale hue. Several kinds occur, presenting the above characters in a somewhat modified form, many of which have received different names from entomologists; such as *Haltica flexuosa*, *Haltica sinuata*, *Haltica ochripes* of Curtis, *Haltica intermedia* of Westwood, &c., which differ from each other chiefly in the colour of the legs. and in the form of the longitudinal yellow stripe on the wing-cases. It is very unlikely that they are more than varieties of the same species; and however important it may be to discriminate closely allied kinds of insects in most other instances, it is of little consequence, in a practical sense, as all those mentioned, whether varieties or species, are alike prejudicial to the turnip, and are doubtless precisely similar in their habits and economy. This little insect feeds on the turnip, which it attacks both in its perfect and larva state. When the plants have acquired some degree of strength, and the foliage is considerably developed, the injury done by it is insignificant, but unfortunately its favourite food is the young plant, just as it is beginning to unfold its cotyledon leaves. These it consumes with the utmost avidity, both as a larva and full-grown insect; and when it abounds, the field is often wholly stripped of its crop in a very short time. Indeed their powers of mastication are surprising for creatures of such small size. An individual who confined a few, for the purpose of observing their habits, found that they consumed 10 young turnip-plants every day. This may serve to give an idea of the extent of their devastations when their numbers become excessive. They are found to attack the turnip-plants as soon as the latter make their appearance; and one of the difficult points to determine is, how they are produced so speedily and so opportunely. In regard to the turnip saw-fly and lepidopterous insects, the process is obvious, the eggs being laid upon the plant by the parent fly, and the larvæ evolved more or less speedily, but after the lapse of some considerable time. The appearance of the plant and insect being in the present case almost simultaneous, it has been thought difficult to conceive how the same process should be gone through." * Various conjectures have been formed to account for the early appearance of this insect on the turnip-plant, such as, that their eggs are deposited in the ground, and hatched by exposure to sun and air, which is not in accordance with the usual economy of such insects as feed on the leaves of plants;—that the eggs are laid *within* the seed, a notion that could only be entertained by one utterly unacquainted with the structure of the insect, for it is unprovided with any thing analogous to a boring-instrument or ovipositor, which forms so indispensable an appendage to insects of a different economy;—that the dots or specks observable on a large proportion of turnip-seed, in the proportion of about 3 out of 5, are the insects' eggs, but experience proves the groundlessness of such a supposition;—that the eggs are deposited in the manure, which is utterly inconsistent with the provident care and economy of insects, to place them in a situation where they would be exposed to continual risk, or where they would be buried in the earth, for the larvæ, on first appearing, are unable to provide themselves, unless their food be immediately before them. The only other conjecture left to account for the early appearance of the insect is the laying of the eggs upon the seedling plants by those beetles which have passed the winter in a dormant state concealed in the furrow till they have access to their favourite food, for although they can fly, they are seldom observed to use their wings, and it may be confidently affirmed that they never migrate from one field to another; but the short time that elapses before the grubs and perfect insects appear, certainly implies that their development and metamorphoses are more than usually rapid. The egg seems to be hatched shortly after its exposure, the larva speedily to go through the various stages of its growth, and the state of the pupa to be of similarly brief duration. " In the case of those insects," continues Mr Duncan truly, " which feed on the foliage of plants in their larva state, and afterwards derive their aliment from other substances, the general law seems to be, that a much longer duration is assigned to the larva than to the perfect insect; and it may be that this is not observed in regard to such as always consume vegetables, because in either of these conditions they serve the same purposes in the economy of nature, to which the prolonged existence of the larvæ bears reference in the other instance. Parallel examples are of frequent occurrence amongst insects. Unless the eggs of the common flesh-fly were hatched with extreme rapidity, the larvæ, when they appear, would neither obtain their food in perfection, nor fulfil the useful purposes for which they are now subservient." I have dwelt the longer on the nature of this insect, in order to make these few

* Quarterly Journal of Agriculture, vol. viii. p. 353.

observations which apply to the habits of all insects, and which will not have again to be repeated.

(2518.) The remedies against the attacks of this insect are, I fear, of a hopeless character; at least, it is better to prevent their appearance, than wage war against them when they appear, for even in the efforts to effect their destruction, the farmer is the chief sufferer. The *preventive* measures seem to be to keep the land in as clean a state as possible of all weeds, and especially of those of the cruciferous kind, such as wild mustard, and charlock, which are the special favourites of this beetle;—to sow the turnips in drills instead of broad-cast; whether it is this difference in the culture of the crop which makes it less vulnerable, I do not know, but the attacks of the turnip insects are far less frequent in Scotland than in England;—to sow the seed thick and of the same age, for it is found that the more rapidly the plants grow at first, they are the less often attacked;—to put the seeds for some time before they are sown amongst flour of sulphur, and sow the sulphur amongst them. The late Mr Airth informed me, that when he farmed the Mains of Dun, Forfarshire, his young turnip crops were often very much affected, and even destroyed, by these insects, but that after he used the sulphur, he never suffered loss, though his neighbours did who would not use the same precaution, and that for as long as he possessed the farm afterwards, namely, 15 years. It may be that the juices of the plant are tainted by absorption of the sulphur, so that the insect had no relish for the plant, while the disagreeable odour arising from the sulphur strewed in the soil would help still farther to drive the insect away. The sulphur was found in no degree to injure the vegetative powers either of seed or of plant. Being a simple preventive, it is worth trying by those whose crops are usually affected by insects. The *remedial* measures recommended are, to sow turnip-seed on the top of the drills as usual, and to sow it also thick in the bottom, in order to give the insects plenty of food, and to induce them to leave as many plants as will still furnish a crop; but the plan is both troublesome and expensive, and the same result might be obtained by strewing lime on the rows that are to be preserved; for if the insects are not partial to the thin rows when their attractions are not greatly inferior to those of the others, they will surely avoid them altogether when they thus become positively distasteful. The clumsy expedient of brushing a newly-painted board along the plants, so as to entrap the insects on the paint, no one would think of adopting on a large scale; nor is the bag-net to catch them in a similar manner a much better expedient, although it is alleged that it has succeeded in destroying many. A long-haired hearth-brush switched along the drills by field-workers would cause the insects to fall off from the plants better than any board or net; and if quick-lime were strewed immediately upon the plants, as recommended from the experience of 102 practical farmers of the Doncaster Agricultural Association, their destruction would likely be more certain.*

(2519.) The insect which commits the greatest ravages on the turnip crop in the order of time to the one just described, is the turnip saw-fly, *Athalia spinarum*, first described by Fabricius under the name of *Tenthredo spinarum*. It belongs to the order Hymenoptera, having 4 *membranous* wings, and it is denominated a *saw*-fly from the use and appearance of the instrument with

* Report of the Doncaster Agricultural Association on the Turnip-fly, 1834. For a fuller account of the *Haltica nemorum*, see Journal of the Royal English Agricultural Society, vol. ii p. 193-213, and Kirby and Spence's Introduction to Entomology, vol. i. p. 186.

which it deposites its eggs. It is placed at the extremity of the abdomen of the female on the under side, and is so constructed, that it combines the properties of a saw and auger. This insect is represented magnified in fig. 425, at *a*, its natural dimensions being shewn by *b*. It is distinguished by the following

Fig. 425.

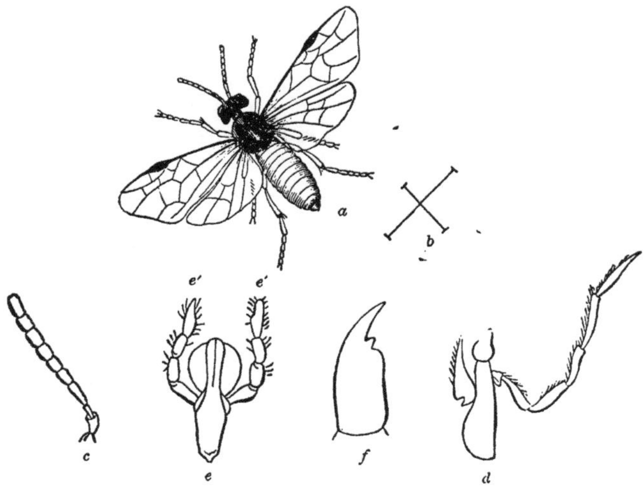

TURNIP SAW-FLY—ATHALIA SPINARUM.

characters:—"Head wider than long; deep black, with 3 ocelli in the centre; eyes oval; antennæ black above, and for the most part dull yellow beneath; labrum and palpi light yellow; thorax black above, with a triangular space in front; the scutellum and a spot behind it reddish orange; the collar, which is rather long and slender, black on the sides, and yellow in the middle; abdomen rather short, entirely orange-yellow, inclining to red, with a small black spot on each side of the first segment; legs likewise orange-yellow, the tarsi paler, approaching to whitish; the tip of the tibiæ, and of each of the tarsal joints, black; the tibiæ with two spines at the apex, and the joints of the tarsus each with a very slender lobe beneath; extremity of the ovipositor black; wings yellowish at the base; the corta and stigma black. Length 3 to 3½ lines, exclusive of the antennæ: *Antennæ*, *c*, short, and somewhat club-shaped, 9 or 10-jointed in the male, but generally with the appearance of 11 joints in the female, the radical joint slightly thickened at the extremity, the second shorter and ovate, the third as long, or longer, than any two of the other joints taken together, the remainder decreasing somewhat in length to the terminal one, which is large and ovate: *Maxillary palpi*, *d*, long 6-jointed, the radical joint shortest, the sixth more slender than the rest, and somewhat spindle-shaped: *Labial palpi*, *e*, 4-jointed, the joints nearly of equal length, the terminal one having the apex, *e′ e′*, curvated in the inner side: *Mandibles*, *f*, thick at the base, terminating in a curved claw, with a tooth or slight notch on the inner edge: *Maxillæ*, with 2 lobes, one of them of an ovate shape at the extremity, the other long and pointed, projecting from the inner side: *Superior wings*, with 2 marginal and 4 submarginal cells. Several slight varieties occur. In many specimens the thorax is entirely black above, and the scutellum is frequently of the same colour. The flies which appear in the early part of summer, and deposite their eggs on the young turnip plants, have probably survived the winter under ground in the pupal state, enveloped in their cocoon. Emerging from them as soon as the milder weather is confirmed, in their winged state, the females immediately lay their eggs, after which they very soon die. The eggs appear for the most part to be placed round the outer margin of the true leaves. In favourable weather they are hatched in a short time, and the young larvæ immediately commence their attack on the plant. At first these larvæ are of a deep black colour, and, of course, small size; but they grow rapidly, and in the course of a few weeks attain their full dimensions. In the course of their growth, they change their skin several times, and most of these moultings are attended with a slight change in the colour. After casting their last skin, they are of a dark lead or slate-grey colour, paler beneath. Mr Curtis states that they are sometimes *green*, a colour which we never saw them assume, for in general they are not liable to such variations in this respect. Like most of the other larvæ of their tribe, when touched, or in any way disturbed, they coil themselves up, and remain motionless. . . . When full grown, the larvæ cease to eat, and allow themselves to drop from the plant that nourished them to the ground, in which they usually bury themselves, or they take shelter among rotten leaves, moss, &c. When examined after a short time, they are found to be completely enclosed in a cocoon, composed of 2 distinct layers of silk. The inner layer is of a fine satiny lustre, and when the cocoon is opened, it appears as if it had been washed with a solution of silver. When the fly is fully matured, it makes its exit by gnawing with its mandibles, *f*, fig. 425, a hole in one end. The larvæ are known in different parts of the country by the names of *black caterpillar*, *blacks*, *nigger*, *canker*, &c. . . . The loss they occasion to farmers is very considerable, but data cannot be easily obtained to form an estimate of its amount. In some instances the crop is wholly destroyed, and where the caterpillars are most numerous, the injury they occasion to the plant appear in the diminished size of the bulb, its vegetative functions being impeded by the partial consumption of the leaves. A belief at one time pretty generally prevailed that they did not attack the Swedish turnip, and they certainly seem less partial to that plant; for examining a field in which Swedes alternated with the ordinary kind, the caterpillar was found much less plentiful in the former, and in many places did not appear at all. The late Earl of Leicester, however, lost in 1836 upwards of 200 acres of Swedes by them, and that plant by no means escaped in other parts of the country. They are said to have attacked the mangel-würzel, but this seems not at all probable." My own turnips never suffered from insects but in one season, when the Swedes were partially attacked by this saw-fly after they were singled. The leaves were eaten to the stem, and the crop seemed as if it would be lost, but it recovered and produced a fair crop after all. "Unfortunately," adds Mr Duncan, "it is more easy to describe their depredations than to suggest an efficient remedy of easy application. A distinguished delineator of insects, who has published a beautiful figure of the fly, expresses his belief that it is not difficult to destroy them, 'for if they were brushed off the leaves, it seems they are unable to crawl upon the ground and recover their station; they consequently perish unless they are full grown at the time; but as there is a constant succession from August till near November, the operation of drawing a hurdle or board over the turnips ought to be repeated at intervals during that period.'* Of course, this method can be effectual only on the supposition that the caterpillar is unable to crawl; it might have been presumed, however, that it did not receive such a complement of legs merely as a matter of form, and accordingly the slightest observation shews that it can move about with facility,"† Rain will destroy the caterpillars; ducks are very fond of them: I have often thought that women and boys might be employed to whisk them off the plants, and though they would not thereby be destroyed, they would be much annoyed, and in the time of such repeated annoyances, the crop might advance so far as to escape serious injury from their ravages. The most effectual destruction would be that of the fly itself before it lays its eggs. It is very sluggish in its flight, does not fly far at a time, and may be easily caught by hand, or rather with an entomologist's net. Swallows are useful creatures in capturing the flies. Each fly so destroyed would prevent the coming into life of from 250 to 300 caterpillars. It has been ascertained that a boy of 10 years of age can gather 180 caterpillars in an hour; and a troop of young ducks, 100 preceded by a boy or girl to switch off the caterpillars to the ground, would destroy them in great numbers. Young fowls are also efficacious, but old ducks and fowls do not work well.

(2520.) As the term caterpillar is so frequently used in reference to insects, it is right to be able to distinguish between *true* and *false caterpillars*. The larvæ of the saw-fly are called *false* caterpillars, from the general resemblance they bear to the larvæ of butterflies and moths, to which the name of caterpillar is properly applied. A very slight examination, however, soon shews decided marks of distinction. Caterpillars, *properly so called*, have never more than 16 feet, while the larvæ of saw-flies have generally from 18 to 22; a few have only 6, a circumstance which again distinguishes them from true caterpillars, in which the number of these organs is never below 10. Another mark of distinction is afforded by the structure of the feet. In lepidopterous larvæ the abdominal legs are surmounted by a coronet of small hooks, which is never found to be the case in those of false caterpillars, as they are simple mammiform protuberances. This minute difference, which can only be detected by the microscope, has, however, a material influence over their habits, and often enables us to distinguish between the respective kinds at first sight. The coronet of hooks converts the membranous or abdominal legs of caterpillars into efficient instruments of prehension, and they accordingly fix their body by means of them to the place of position, while the head and anterior part remain free. The abdominal legs of the others, on the contrary, are mere points of support, incapable of clinging to an object, and the larvæ consequently fixes itself by its pectoral or fore-legs, which are much developed for the purpose. The whole of the abdominal portion of the body is thus left at liberty, and it is either borne curved inwards, as in the gooseberry saw-fly, *Nematus grossulariæ*, or projects into the air in variously-contorted and singular postures, as is remarkably the case with

* Curtis's British Entomology, vol. xiii. folio 617.

† Quarterly Journal of Agriculture, vol. vii. p. 562-6. Other accounts of this insect will be found in Kirby and Spence's Introduction to Entomology, vol. i. p. 186; Journal of the Royal English Agricultural Society, vol. ii. p. 364-89; and Newport's Observations on the Turnip Saw-fly.

the willow saw-fly, *Nematus capreæ*, and the larvæ of *Hylotoma rosæ*, the rose saw-fly, which has the extremity of its body almost raised and curved in the form of the letter S. Additional distinctive characters might be mentioned; such as the form of the eyes, which are pretty large in the false caterpillars, and placed one in each side of the head, while in true caterpillars they are small, almost invisible points, disposed in a circle, which will suffice to distinguish the two tribes, and it is of importance to be able to do so, as they are often associated together in the work of destruction. The body of false caterpillars is generally composed of 12 segments, but the incisures are distinctly defined, and liable to be confounded with the transverse wrinkles which thickly cover the whole surface. Many of them are marked with bright and varied colours, but the majority are of one colour. In this respect they often undergo a remarkable change after they have cast their last skin, the colour becoming entirely unlike what it was before, so that it is impossible to recognise the same individual. This change, indeed, extends even farther than to colour, for such kinds as are furnished with tubercles or spines in their earlier stages, lose them at their last moult, and become smooth; that of the gooseberry species, *Nematus grossulariæ*, for example, loses the black tubercles, which made the surface appear as if chagreened. Like the flies which they produce, their larvæ are sluggish and inactive, seldom moving from the place where they fix themselves, unless when requiring an additional supply of food. When not engaged in feeding, or when apprehensive of danger, most of them roll themselves into a circle, sometimes with the tail elevated in the centre. The greater number lie exposed on the foliage of plants, but others take up their abode in the interior of the slender shoots, and feed on the immature pith; others lodge in the interior of fruit, and cause it speedily to decay.

(2521.) Among the other insects which infest the turnip-crop, I believe there is no doubt but that the *wire-worm* frequently destroys the young plants, after they have been some time singled, by dividing the root from the stem. The leaves in this case are observed to become flaccid, and the plant either dies on the spot, is pushed out by hoeing, blown away by the wind, or pulled up by crows. I remember of seeing one season at least 3 acres of young turnip-plants pulled up by crows, no doubt to get at the insects at the root, on the farm of Haughhead, near Brechin, at that time farmed by Lord Panmure. "When full grown it is about 7 lines long, and rather less than 1 in breadth. The shape of the body would be perfectly cylindrical were the back not a little depressed. It is entirely of an ochre yellow colour, except the anterior part of the head, which is brown. Owing to the rigid consistency of the skin or crust, the larvæ can scarcely contract its body, but as it is composed of wings or segments, it is flexible enough from side to side. These rings are 12 in number, the 3 nearest the head each provided with a pair of conical legs, and the caudal segment having a fleshy tubercle beneath, which serves the purpose of a seventh foot. The last named segment is cervical, terminating in a point, and is remarkable for having 2 circular holes like 2 brown points on the surface. The whole body is smooth, with a few scattered hairs. The perfect insect is of a brown colour, thickly covered with short pubescence. The antennæ are about the length of the head and thorax, the colour obscure yellow. The insect is called *Cataphagus lineatus*. Another species occurs in similar places, and generally in much greater abundance, at least in Scotland, namely *Cataphagus obscurus*; and it is so closely allied to the above, that it may readily be taken for a variety of it, and there is no doubt that the description of it will apply to this. From the great abundance of the *obscurus*, it may be presumed to be the species which commits most injury in this country. There is another species even more plentiful in ploughed land. It is the *Hypnoidus riparius*, a small insect of a brassy-black colour, with pale-reddish legs."* A small ichneumon searches out the retreats of wire-worms, and deposites its eggs in their bodies, which are consumed by the parasitical larvæ as soon as hatched. I may mention, that the wire-worm is the larva of those beetles named *Elateridæ* or *click-beetles*, which are readily known by having the sternum produced behind into a strong spine fitted to enter a groove in the abdomen, situated between the intermediate pair of legs. By bringing these parts suddenly into contact, the insects are enabled to spring to some height into the air, and thus recover their natural position when they happen to fall on their backs, which they frequently do when dropping from plants to the ground. A special provision of this kind is rendered necessary, in consequence of the shortness and weakness of their legs. Upwards of 60 species of these insects occur in Britain, and it is probable that a considerable portion of them feed upon our most valuable cultivated plants. A very small weevil, scarcely 1 line in length, of a uniform black colour, slightly tinted with metallic blue on the elytra, the latter with punctured lines, the *Ceuterhynchus contractus*, is found occasionally associated with the turnip flea-beetle, *Haltica nemorum*, feeding on the young leaves of that plant. This species does also good by attacking that pestilent weed, the wild mustard, *Sinapis arvensis*. Another species, *Ceuterhynchus pollinarius*, attacks the nettle, *Urtica dioica*, a troublesome weed near buildings; and a third, *Ceuterhynchus assimilis*, occasions the knobs on the roots of the wild radish, *Raphanus raphanistrum*, one of the most troublesome weeds which infest cultivated fields, particularly those in which the turnip is cultivated. That multitudinous tribe of insects, named the *Aphis* or plant louse, sends a few of their number to attack the turnip crop, and their attacks are chiefly directed against the Swedish turnip. Mr Curtis says, that there are 3 if not 4 species of aphides which live upon the turnip; one he has found under the rough leaves of the English varieties, as well as one which he believes to be distinct; another appears to be attached to the Swedes, and the last is secreted amongst the flower stalks; but numerous as these minute creatures are, their enemies are no less numerous and active. I cannot enter upon so extensive a subject, and must refer you to Mr Curtis' paper.† I shall only add a few words descriptive ot the nature and peculiarities of the curious aphis. "As almost every animal has its peculiar louse, so has almost every plant its peculiar *plant-louse;* and next to locusts these are the greatest enemies of the vegetable world, and like them are so numerous as to darken the air. The multiplication of these little creatures is infinite and almost incredible. Providence has endued them with privileges for promoting fecundity, which no other insects possess: at one time of the year they are viviparous, at another oviparous; and what is most remarkable and without parallel, the sexual intercourse of one original pair serves for all the generations which proceed from the female for a whole succeeding year. Reaumur has proved, that in 5 generations 1 aphis may be the progenitor of 5,904,000,000 descendants; and it is supposed that in 1 year there may be

* Quarterly Journal of Agriculture, vol. viii. p. 102.

† Journal of the Royal English Agricultural Society, vol. iii. p. 49-68.

20 generations." Bennet says there may be 30. " This astonishing fecundity exceeds that of any known animal, and we cannot wonder that a creature so prolific should be proportionably injurious; some species, however, seem to be more so than others. Those that attack wheat, oats, and barley, of which there are more kinds than one, seldom multiply so fast as to be very noxious to those plants; while those which attack pulse spread so rapidly, and take such entire possession, that the crop is greatly injured, and sometimes destroyed by them."* There is a class of insects called *turnip-leaf miners*, the larvæ of which destroy the energy of the leaves of turnips, by boring galleries between their upper and under surfaces. One is the *Drosophila flava*, which form their dwellings so carefully under the *upper* cuticle of the larva, that not a trace of them can be seen on the under side Its length is 1 line. The larva of this insect is destroyed by two little parasitic hymenopterous flies, the *Ceraphron niger*, and the *Microgaster viridis*. The other leaf miner is named *Phytomyza nigricornis*, which is bred from the *under* sides of the turnip leaves, where the maggots form long irregular galleries inside of the lower cuticle, and these miners are not visible on the upper side of the leaf. The fly is 1 line in length. Lepidopterous insects, that is, those of the butterfly and moth kind, afford several species injurious to the turnip-crop. One is the turnip diamond-back moth, *Cerostoma xylostella*. When at rest the wings are closed and deflexed, and the horns are projected forward in a straight line. It is more or less brown. The upper wings are long and narrow, and when closed form 2 or 3 diamonds upon the back; the inferior wings are lance-shaped, and of an ash-colour, with a very long fringe. Its length is 2½ lines. The caterpillar is green, about ½ an inch in length, slender and tapering to both ends. They are exceedingly active, and on the slightest touch wriggle themselves off the leaf they are feeding, and let themselves down by a silken thread, and remain suspended until the cause of alarm subsides. As many as 240 have been counted on one leaf, and such is their avidity, that not the smallest vestige of a green leaf is left by them. This larva is destroyed by a black ichneumon, named *Campoplex paniscus*.† The cabbage-moth, *Mamestra brassicæ*, will live on the Swedish turnip. It is of sombre colours, and most of them have a mark resembling the letter W on the disk of the anterior wings, which expand 1½ inch. The caterpillar is greenish or brownish, with a dark stripe along the back, on which a pale line is visible; the sides are marked with an obscure yellowish stripe, having a tendency to become reddish on the upper side. The spiracles are white surrounded with black. The White line brown-eyed moth, *Mamestra oleracea*, is of a rusty-brown colour, the fore-wings at times slightly clouded, the cross-lines obliterated. The caterpillar reddish or yellowish-brown, with a dark line along the back, and another on each side, and beneath the latter a white line; under side and feet light-brown. The Gamma or Y-moth, *Plusia gamma*, is easily recognised by the silvery character on its fore-wings resembling the letter Y, or rather the Greek γ, upon a variegated dusky-brown ground. Head and thorax ash-grey; the abdomen of a lighter hue. It expands from 1⅓ to 1½ inch. The eggs of this moth are very beautiful, being of an orbicular shape, with elevated ribs, and slender, transverse, raised lines. They are chiefly laid on the under side of leaves, sometimes singly, but more commonly in small

* Kirby and Spence's Introduction to Entomology, vol. i. p. 174.

† Journal of the Royal English Agricultural Society, vol. iii. p. 68-72.

clusters. When the caterpillars reach maturity, they are green, with 6 white or black lines along the back, and a faint yellow streak on each side; the breathing pores black. They possess only 4 abdominal feet and 2 anal ones, thus indicating an approach to the geometrine or looper caterpillars. The chrysalis, which is pitchy-brown, is inclosed in a white woolly cocoon, spun between the folds of a leaf, or among herbage. The moth flies about in the day with much rapidity, keeping the wings, when feeding, in constant vibration. They may be seen, often in great numbers, hovering in this manner about a turnip-field, over the yellow blossom of the charlock and field-mustard, or the blue heads of the scabious and devil's-bit.* There is no moth more shy and difficult to catch by day, for it will seldom allow any one to come near it, but whether it detects the approach of man by its eyes, which sparkle like living rubies, or by its hearing, is not known: it darts off, however, in an instant, when disturbed, and stops again a few yards off, or entirely vanishes. There is, perhaps, nothing but hand-picking, switching the turnip-plant, to get quit of these insects. The White cabbage butterfly, *Pontia brassicæ*, lays eggs in clusters, of 20 or 30 on the under side of the turnip leaves. The caterpillars are green before and yellow behind, when young, but when matured to full growth, are 1½ inch long, and as thick as a small goose quill. They chiefly attack the Swedish turnip. Their larva is destroyed by the ichneumons, *Microgaster glomeratus*, and *Pimpla instigator*, and by the cynips, *Pteromatus brassicæ*, and *pontiæ*. The Small White or turnip butterfly, *Pontia rapæ*, as its name implies, is another enemy to the turnip. The superior wings of the *male* are tipped with black, and the inferior have a black spot on the upper edge; the *female* is similar, but has 2 large black spots likewise beyond the centre of the superior wings; underside of the same white, apex yellow, and 2 black spots beyond the middle, the lower one sometimes obliterated; inferior wings yellow, freckled with black. Length of male 8 lines, and expanse about 2 inches. The eggs are not unlike those of the cabbage-butterfly; but the caterpillar is totally different, being entirely green, and so densely clothed with minute hairs, as to be velvety. They have a yellowish stripe down the back, and another along each side. They are more than 1 inch long, and about as thick as a crow quill. Another species is the Green veined white butterfly, *Pontia napi*, so named, because it feeds chiefly upon the rape. The superior wings, with tips of powdery black, and the nervures greyish, inferior wings with a black spot on the upper margin, and the dark nervures shining through. Caterpillars are delicately green, clothed with velvety hairs. Male flies are nearly ¾ of an inch long. The larva of this species is destroyed by the ichneumon, named *Hemiteles melanarius*. Hand-picking and young fowls to pick up the larvæ as they are switched off the plants, is the most easy mode of getting quit of them. The small oval eggs so deeply embedded in the pulpy substance of the bark of some turnip leaves, are laid by the *Chrysomela betulæ*, a brilliant shining blue or green oval beetle, with the under sides, horns, and legs black, and about 1½ line long. The *Cetonia aurata*, Green rosechafer, is found on the flowers of the turnip plant, and renders them abortive. It is one of the most beautiful of our insects, having a brilliant metallic green, often with a golden or copper hue. Its length is ¾ of an inch. It is found in numbers in the south of England, and in Morayshire in Scotland, but not in any intermediate locality. The larva commits a good deal of damage,

* Quarterly Journal of Agriculture, vol. xiii. p. 167-70.

where it prevails, by feeding in the same way as other chafers. The perfect insect destroys the petals of roses by gnawing them. The turnip-seed, when growing, is seriously injured by a weevil named the *Ceuterhynchus assimilis*, the Turnip-seed weevil. It is 1¾ line in length, including the rostrum; is black, clothed with short white depressed hairs above, and scales beneath, which give the insect a grey tinge. As weevils are so sensitive as to fall down, if suddenly approached, they may be easily collected when they abound in the turnip-flowers left for seed, by shaking the stalks over a bag-net or cloth; and being hard they cannot be destroyed by stamping upon them, they must be killed, when collected, with boiling water in a pail.* Besides the leaves, flowers, and seeds, the turnip is attacked by insects in the bulb, among which noxious insects are many large caterpillars, called by farmers and gardeners *surface grubs*, which commit very extensive depredations upon turnips. Among these is the *Triphæena pronuba*, the Great yellow underwing. This moth is sometimes very plentiful in hay-fields, where they will rise from the swathes when disturbed, and alight again in another swathe. The *Noctua* or *Agrostis exclamationis*, the Heart and Dart moth, so named from the markings of the wings resembling the note of exclamation (!), and a heart and a dart, affords a caterpillar of a dull lilac colour, with a lurid space down the back of a more ochreous hue, which is a most destructive animal to crops of turnips at every stage of their growth, and especially in separating the crown from the root. They also attack the potato-tuber, and so insensible are they to cold, that, so late as the 20th November, they have been taken out of the tubers of potatoes quite alive. An equally if not more destructive insect is the Common dart-moth, *Noctua*, or *Agrotis segetum*, which is generally of a redddish-brown, but varies so greatly in the tint of the upper wings as to be sometimes of a clay colour. It is in length from 8 lines to ¾ of an inch, and its expanse is from 1¾ to 2 inches. The moth is sometimes seen flying in multitudes about the tops of hedges soon after sunset in June and July. The larva is smooth and shining, and of a pale lurid ochreous colour, faintly freckled, with a broad space down the back, often rosy, and a few short hairs scattered over the body. It does great mischief to young mangel-würzel plants, the roots of which it cuts through just below the crown, and it attacks the potatoes as they are pushing out of the ground, and is exceedingly voracious. One cause of the great mischief arising from the attacks of the caterpillar of this and of the preceding species, is their capability of travelling at a very rapid rate from one spot to another; and in this way, as soon as a caterpillar has eaten through the root of a young plant, it marches off in quest of another, and thus the evil is greatly multiplied. This grub is a very formidable assailant in the more advanced state of the turnip plant, near to which it forms a round hole in a vertical direction about 2 or 3 inches deep in the earth. At the bottom of this it remains during the day, unless it be dark and moist, and at night emerges from its burrow, and commences an attack upon a plant by eating round the neck of it, and eventually detaching the upper part from the root. In this way singled Swedish turnip-plants may be destroyed one after another until very few are left. The rook is useful in searching for this grub, and in quest of it will also tear up the plant. Young pigs are fond of this grub, and would dig for them in a turnip-field, were it not that they would dig up the plants at the same time. There seems no ready means of getting quit of this pest. The excrescences which

* Journal of the Royal English Agricultural Society, vol. iii. p. 406-19.

frequently disfigure the turnip bulb, and are not confined to any particular variety of turnip, on being opened will be found to contain a small maggot. This is produced by the *Curculio plurostigma*, the Turnip-gall weevil. It is very similar to the turnip-seed weevil, but is black instead of grey; the wing-cases are not so rough or strongly tuberculated at their extremities, and *all* the thighs have a small tooth beneath. It is not uncommon in hedges from May to August, and, closely contracting all its members when alarmed, it looks like a black seed. The female pierces a hole in the rind of the turnip with her proboscis and deposites an egg in it, and the young maggot, which is fat and whitish, often of a bright flesh-colour, lives on the substance of the bulb. Except in affecting the beauty and symmetry of the bulb, this insect does no great harm to the turnip.* These are all the insects known to be injurious to the turnip crop which requires mention here. The diseases of the turnip called Fingers and Toes, and the Anbury, have been ascribed to the agency of insects; but as I am not of that opinion, it appears to me irrelevant to mention the names of those to which blame has been imputed for producing those diseases. Nevertheless, a few words are required to be said on the cause of those malformations in the turnip.

(2522.) A plan of destroying moths in vineyards has been tried by M. Audouin with success, and which consists of lighting lamps covered with bell-glasses smeared with oil, when the light attracts the insects, and they are captured on the glass. Such lights are much more effective than open fires of brushwood. In this way 200 lamps the first night, and 180 in the following nights, placed at 25 feet apart, in 4 nights in August 1842, in a vineyard of 4 acres extent, and lighted for 2 hours each night, destroyed these numbers of moths in the respective nights, 30,000, 14,400, and the two last nights 9260; in all, 53,660. Of these it was reckoned that $\frac{3}{4}$ were females, which, supposing they would have laid 150 eggs each, caused a destruction of 6,000,000 of eggs. Many moths, I apprehend, might be caught by the employment of an entomologist's net while flickering around the lamps.†

(2523.) *The Anbury in Turnips.* Mr Marshall, in allusion to the anbury, says that it is a large excrescence produced below the bulb; and when this was just forming and not larger than a green walnut, the anburies were as large as a goose's egg, awkward and irregular in form, with excrescences below, not unlike races of ginger depending from them. After arriving at maturity they exhibit a putrid fermentation, and emit a most offensive smell. When the anburies are divided they are hard; but with the assistance of a lens, veins or string-like vessels may be seen dispersed through the tumour. When turnips are affected with this disease, the tops become yellow, and flag in the heat of the sun, and they are thus readily distinguished. He says it has been attributed to the land being too long continued under this green crop; but it is certain that the anbury appears on land where turnips had never before been grown. He, however, considers that it proceeds from the formation of an insect in the vessels of the tap-root, by which the course of the sap is divided, and instead of the natural bulb an excrescence is produced. He recommends that the dis-

* Journal of the Royal English Agricultural Society, vol. iv. p. 103 21.

† Gardener's Gazette for 1843.

eased plants should be removed as soon as possible, and the earth stirred about those that remain; and he adds, that it may be wholly avoided by well preparing and richly manuring lands subject to produce anbury.* If the disease were occasioned by the puncture of insects, better cultivation would not abate its virulence, but rather increase it, as the turnip would thereby be rendered even more palatable to insects; but the truth is, all such diseases arise from poverty in the soil arising either from want of manure when the soil is naturally poor, or when it is rendered effete by overcropping, or by too much wet, or by too much drought. Labour, clean, and manure the soil fully, according to the condition it presents, and no anbury will appear, unless it may happen in certain seasons whose inflrence counteracts the effects of culture, and affects plants in a manner similar to want of nourishment. This disease is not now so prevalent as it was 30 years ago, because the culture of the turnip is now better understood, and the ground is manured with greater liberality.

(2524.) *Fingers and Toes in Turnips.* Of this disease Mr Dickson says, that "it occasionally happens that turnip-plants, instead of swelling and forming bulbs, send off numerous stringy roots, which soon decay and come to no account. It occurs most generally when the crop is sown on *fresh* land, and no remedy is said yet to have been discovered to prevent it. More perfect tillage, and the use of such measures as have a tendency to render such lands more mellow and friable, may perhaps be beneficial."† No doubt the disease has been observed on fresh virgin soil that had never before borne a crop of turnips; but it has been remarked in a long experience, that land which had often carried turnips was most affected by the disease. The county of Roxburgh long experienced this disease, and it continued to increase for 30 years; but latterly it has decreased, and may now be said to have disappeared, in consequence of the superior manuring of all the crops and the better tillage of the soil, and this will be the result everywhere else. I may here remark, however, that spurious seed will have the same effect on the turnip, in unfavourable seasons, as want of manure, or the injurious effects of weather. As a matter of history, then, more than as an instructive narrative, the symptoms of this disease may thus be described. The disease affects the turnip plant from the period of singling to the first hoeing of the crop. The plant becomes flaccid, and the leaves assume a yellowish hue, but it does not die, nor does it bear the slightest mark of insects; and when it is once affected it never gets free of the influence of the disease, though it continues to live and grow in size. The disease never affects a whole field at once, nor does it run along the drills, but invariably begins in spots which increase in diameter, and spread out into large patches. The patches, however, never come in contact, but, on finding interruptions, assume irregular forms. Those interruptions are the ends of drills and the hollows of fields. The patches never commence in hollows or drill-ends where water may lodge, but on the driest knolls where sheep would take to rest for the night. In conformity with this circumstance, light loamy turnip-soil, on an open bottom, is much more apt to be affected than clay land, or any soil resting on a retentive bottom; and it prevails more in dry than in wet seasons. Hence in the wet seasons of 1816 and 1817 it was but little felt in Roxburghshire; whereas in the dry summers

* Marshall's Rural Economy of Norfolk, vol. ii. p. 33.

† Dickson's Practical Agriculture, vol. ii. p. 665.

of 1818 and 1819, and particularly 1818, it was both extensive and destructive. Hence also Roxburgh, with its light soil, was always more affected than the neighbouring county of Berwick with its heavy soil. The ultimate effects of the disease are to produce a distorted bulb like a boxer's glove, with fingers and thumb, which are longer or shorter, smaller or larger, in proportion to the bulb. The leaves are unhealthy in colour, and the top has a tendency to shoot. Inequalities occur in the bulb which collect water, the freezing of which in winter causes premature decay of the bulb. The texture of the bulb becomes fibrous, and the juice tastes acrid like the skin, and the smell is somewhat putrescent. Hence the disease affects the weight of the crop, as also its nutritive properties as an article of food for live stock.*

(2525.) *Rape-cake and Rape-dust.* This substance, as a manure, I shall have occasion to notice afterwards; meantime I mention the mode of keeping it in the best state. Rape-cakes should be of a yellowish-green colour when new, but they become dark on being long kept. They should be put past in the apartment allotted for their accommodation in a dry day and on a dry clay or wooden floor, for except by their weight they do no injury to wood; and they should be neatly built up free of the walls in case they should draw damp therefrom. The air, and especially damp air, should be excluded from them, as it is the cause of their becoming mouldy, and of losing their light colour. New made cakes will heat a little after they are built up, but after the sweat they have had in the ship this will not be great; and to prevent heating as much as possible the small dust of the cargo should be kept by itself. It is not likely that you will have occasion to keep rape-dust, that is, rape-cake after it has been crushed to powder; but in case you should wish to keep it in that state, the mode of doing so is as follows:—It should not be put together in a thick heap, for however dry it may seem, and however dry it may be kept, sooner or later it will heat. When put by dry on a dry floor, it may be 3 weeks or a month before it will heat; but if damp it will heat soon, and the heat will become so as to be insufferable to the hand. Whenever it begins to heat it should be carefully and slowly turned over in shallow bins to cool it. When heated to excess it becomes burnt as black as soot, and gets into lumps like coal-cinders from which it is not easily distinguished, and in which state its efficacy is impaired. Many throw rape-dust into a corner of a cart or turnip-shed where pigs may dig and fowls scratch amongst it, and where it is constantly exposed to the moisture of the air; and which is a practice that cannot be too much reprobated. You should purchase none but new crushed rape-dust.† Rape-seed is larger and heavier than that of the Swedish turnip, so that only 820 are required to weigh 1 drachm.

(2526.) *Mangel-würzel* belongs to the family of beet. It is of the class and order *Pentandria digynia* of Linnæus, and of the order *Chenopodiaceæ* of the natural system; its chief botanic character is—"calyx, when green, soft, and fleshy, when ripe, hard, and somewhat woody in texture, and into which the rough kidney-shaped seeds are deeply imbedded." The seeds which have a per-

* See paper by me on this subject in vol. i. p. 429–39 of the Quarterly Journal of Agriculture, in which I have endeavoured to shew that the insects which are found in diseased bulbs of turnips have come there *in consequence* of the diseased state of the plant, although I *then* believed more in the agency of insects than I do *now*.

† Quarterly Journal of Agriculture, vol. xiii. p. 349.

sistent capsule are large and heavy, only requiring 184 to weigh 1 drachm. The field beet is commonly called mangel-würzel, two German words which have been retained in consequence of the root having been long and is still extensively cultivated in Germany, where it thrives better than any of the turnip tribe, and they mean literally *root of want.* The marbled variety, *Beta vulgaris campestris,* has been longest in cultivation, but the orange-globe variety is daily becoming a greater favourite, both on account of its heavier crops and greater feeding properties. The white field-beet, *Beta vulgaris campestris alba,* has long been cultivated in France for the sugar which it affords. Many mills are at work in France for the manufacture of this sugar.* It was the policy of Bonaparte to encourage the culture of this plant, in order to render the French people independent of the supply of sugar from our colonies. Both carrot and mangel-würzel are very apt to run to seed when drought sets in early after they have been thinned out in the rows; and seeds used from plants that have shewn a tendency to premature shooting encourages that tendency still more than drought. It is not an unfrequent practice to strip off the under leaves of the mangel-würzel plant in summer as fodder for cows and pigs; but the practice, as may be supposed, is injurious, as may be seen from an experiment made on this subject by Mr R. Rand, Hadleigh, Suffolk, in 1842. He selected 3 portions of mangel-würzel, containing each 7 square yards, and from the 1st lot he stripped 4 or 5 of the under leaves on 8th of July, 6th of August, and in the 1st week of September; from the 2d lot he stripped the same number at each period of the 6th of August and the first week of September; and from the 3d lot there was none stript off at all. The roots were carefully cleaned and weighed, and the produce was as follows:—

From 1st lot,	47½ lb. net weight, or	14 tons,	13 cwt.	0 qrs.	27 lb.	per acre.
... 2d ...	52	16 ...	1 ...	0 ...	2	...
... 3d ...	61	16 ...	16 ...	2 ...	9	...†

(2527.) *Carrot.*—This plant is of the natural order *Umbelliferæ,* from *umbella,* an umbel, and *fero,* to bear, from the shape of its flowering: this order comprehends a very extensive tribe of plants, including parsnip, celery, parsley, carrot, coriander, hemlock, &c. In the Linnæan system it stands in the class and order *Pentandria digynia.* Its botanical cognomen is *Daucus carota sativa,* the first from a Greek word signifying to make hot from its supposed medicinal effects, and *car,* a Celtic word meaning red, the colour of the root; and its botanical characters are, "prickles of secondary ribs slender, separate even to the base; rays of umbels nearly equal or gradually shorter to the centre; seeds small, protected by the incurvation of all the flower-stalks, by which all the umbels are rendered hollow, like a bird's nest." To save its seed, plant some of the largest and best roots in October, November, or last fortnight of February, 2 feet apart, and insert them a few inches over the crowns. They will yield ripe seed in autumn, of which gather only from the principal umbel, which is likely not only to afford the ripest and largest seed, but the most vigorous plants. A considerable quantity of carrot-seed, for the supply of the London seedsmen, is raised near Weatherfield, in Essex; and much is imported from Holland, on

* Quarterly Journal of Agriculture, vol. ii. p. 922.

† Henslow's Letters to the Farmers of Suffolk, Letter XIV.

which the duty by the new tariff is 10s. per cwt., and 5s. on that from British possessions. The entire culture of the carrot is well described by Mr Burrows, a Norfolk farmer.*

(2528.) *Parsnip.*—This root is also in the natural order *Umbelliferæ*, and Linnean system *Pentandria digynia*. It is the *Pastinaca sativa edulis*, from *pastinum* a dibble, in reference to the form of the root; leaves glabrous on both sides, shining above; root thick, fleshy. This root has long been an inmate of the garden, and was formerly much used. In the times of Popery it was a farmer's Lent root, being eaten with salted fish, to which it is an excellent accompaniment. "In the north of Scotland," as Dr Neill observes, "parsnips are often beat up with potatoes and a little butter; of this excellent mess the children of the peasantry are very fond, and they do not fail to thrive upon it. In the north of Ireland, a pleasant table beverage is prepared from the roots, brewed along with hops. Parsnip wine is also made in some places; and they afford an excellent ardent spirit, when distilled after a similar preparatory process to that bestowed on potatoes destined for that purpose." To save the seed, transplant some of the best roots in February, 2 feet apart, inserted over the crowns; they will shoot up in strong stalks, and produce large umbels of seed, ripening in autumn.†

(2529.) *Bone-dust.*—The specific gravity of bone, as determined by Dr Thomson, is as follows:—

Os femoris of a sheep,	2.0345
Tibia of a sheep,	2.0329
Ilium of an ox,	1.8353
Human os humeri,	1.7479
Vertebræ of haddock,	1.6350

The nature of this remarkable substance, as a manure, demands attention. Bone consists of organic and inorganic matter, and these are found in the following proportion in the bones of those animals which supply the largest quantity for manure, namely, the ox and sheep:—

		Ilium of a Sheep.	Ilium of an Ox.	Tibia of a Sheep.	Vertebræ of Haddock.	Snout of Saw-fish, deprived of teeth.
Organic matter or cartilage,		43.30	48.05	51.97	39.49	46.310
Inorganic	Phosphate of lime,	50.58	45.02	40.42	56.08	42.550
	Carbonate of lime,	4.49	6.10	7 03	3.57	2.638
	Magnesia,	0.86	0.24	0.22	0.79	0.101
	Soda,	0.31	0.20	0.19	0.79	0.141
	Potash,	0.19	0.11	trace.	...	water 6.920
		99.73	100.35	99.83	100.72	98.66‡

* Communications to the Board of Agriculture, vol. vii. p. 72.

† Don's General System of Gardening and Botany, vol. iii., p. 338 and 354.

‡ Thomson's Animal Chemistry, p. 234 and 241-2.

In an analysis made by Dr Henry R. Madden, to ascertain the relative component parts of bones and farm-yard dung, the following results were obtained, the bone being used in the state it is applied to the soil, and the dung consisting of a proportion of both stable and byre manure :—

	Bone-dust.	
	Water, ...	11.5
Organic matter, with some saline materials,	Soluble in cold water,	4.7
	... in hot water, .	5.5
	... in weak potass,	26.0
	... in strong potass,	15.5
	Destroyed by heat, .	6.0
	Phosphate of lime, . .	28.0
	Carbonate of lime, . .	2.8
		100.0

	Farm-yard dung.		
	Water, ...		45.535
Organic matter,	Soluble in water, .		10.750
	... in potass, .		14.250
	Destroyed by heat,		18.565
	Salts of potass,		
	... of soda, .		7.900
	... of lime, .		
	Silica, ...		
	Earthy phosphates,		3.000
			100.00

Another analysis gave of azote in bone-dust 1.77 per cent., in farm-yard dung 45; so that in this respect these substances stand comparatively thus :—

	Bone-dust.	Farm-yard dung.
Soluble matter, ...	10.20	10.750
Easily rendered soluble, ...	41.50	14.250
Azote, ...	1.77	.450
Total organic matter, ...	38.50	33.565
Saline matter, ...	55.00	10.900

Hence 1 ton bone-dust equals, as regards	organic matter, .	1	ton farm-yard dung.
... ...	soluble matter, .	1	... ...
... ...	easily dissolved do.,	2.9	... ...
... ...	azote, . .	3.9	... ...
... ...	saline matter, .	5	... ...
... ...	earthy phosphates,	18.3	... ...

If all the various degrees in which bone-dust is superior to farm-yard dung be added together, 1 ton of it equals 30 tons of dung, but as only 16 bushels of bone-dust are applied to the acre, which, at 47 lb. per bushel, weigh 7 cwt., this quantity is equal to 10½ tons of dung.* It is still a question whether the organic or inorganic portion of bone-manure bestows the greatest benefit on the land, and, from its effects extending over more than one crop, some are of opinion, among whom are Leibig and Sprengel, that its principal efficacy is in all cases to be ascribed to the earthy ingredients, and especially to the phosphate of lime; but it has been brought to light, that bones do not lose their organic matter entirely, though they should have been buried for years under ground. Thus the bones of a bear and a stag, after being long buried, were found by Marchand to consist of—

* Prize Essays of the Highland and Agricultural Society, vol. xiv., p. 528 31.

	Bones of the Bear buried		Femur of a Stag.
	Deep.	Shallow.	
Animal matter,	16.2	4.2	7.3
Phosphate of lime,	56.0	62.1	54.1
Carbonate of lime,	13.1	13.3	19.3
Sulphate of lime,	7.1	12.3	12.2
Phosphate of magnesia,	0.3	0.5	2.1
Fluoride of calcium,	2.0	2.1	2.1
Oxides of iron and manganese,	2.0	2.1	2.9
Soda,	1.1	1.3	...
Silica,	2.2	2.1	...
	100.	100.	100.

"The most striking change," as Professor Johnston observes, "undergone by these bones, was the large loss of organic or animal matter they had suffered. The relative proportions of the phosphate and carbonate of lime had been comparatively little altered. The *main* effect, therefore, produced by bones, when buried at the roots of trees, and their first effect, in all cases, must be owing to the animal matter they contain—the elements of the animal matter, as it decomposes, being absorbed by the roots with which the bones are in contact." Still, it is found that bones, after having been boiled, and of course deprived of the greater part of their animal matter, make excellent manure; but as they at the same time take up a considerable quantity of water, which will cause them to decompose more rapidly when mixed with the soil, they will appear to act as beneficially, Professor Johnston conjectures, as unboiled bones. "He who candidly weighs the considerations above presented, will, I think, conclude," says the Professor, "that the whole effect of bones cannot in any case be ascribed exclusively either to the one or the other of the principal constituents. He will believe, indeed, that in the turnip husbandry, the organic part performs the most permanent and most immediately useful office, but that the earthy part, nevertheless, affords a ready supply of certain inorganic kinds of food, which in many soils the plants could not otherwise easily obtain. He will assign to each constituent its separate and important function, being constrained at the same time to confess—that while in very many cases the *earthy part of bones applied alone* would fail to benefit the land, there are few cultivated fields in which the *organic part applied alone* would not materially promote the growth of most of our artificial crops."* When bones are heated to redness in the open air, the organic part burns away, and leaves the white earthy matter, in the form and nearly of the bulk of the original bone. It is very brittle, and consists chiefly of phosphate of lime, and is sold at the chemical works at 9s. per cwt. Bones of cattle and other animals, and of fish, except whale fins, whether burnt or not, or as animal charcoal, pay a duty by the new tariff of 6d. per ton. The declared value of bones imported into this country in 1832 was L.91,755 : 5 : 5; and the duty paid thereon was L.940 : 5 : 9.† I should suppose there has been

* Johnston's Lectures on Agricultural Chemistry, p. 660.

† Macculloch's Dictionary of Commerce, art. *Bones*.

a considerable increase in the import since that period. The price of bones at Hull, the great mart of their import, in 1843, was from L.3, 15s. to L.4 per ton.

(2530.) [*Turnip-drills*, like the other members of the class of implements employed under the drill-system of cultivation, mark an improved state of the art, for in the early history of the turnip culture, we find the broadcast method of sowing generally adopted, and in some parts of England at the present day, where farming is otherwise well understood, we yet see the broadcast system of sowing turnips not only practised, but advocated as the most productive. That great aggregate weight of turnip may be produced in broadcast in particular soils and seasons, may be true, but doubtless a greater certainty of success is to be obtained from the drill practice; and it appears now, from the latest English authorities,* that the former practice will soon be rooted out by the unflinching hand of experience. In the early stages of the drill practice of turnip culture, the breadth of land sown being but small, a single row hand-drill or barrow seems to have been generally used, but a simpler implement than even this has been employed—the hand-flask sower. As the practice extended, machines of two rows were introduced, drawn by one horse, and this in various forms continues to be the chief instrument employed in sowing the seed of this important root.

(2531.) The varieties of the turnip-drill are too numerous to be detailed here in full, nor would it be profitable to follow all the fancies of machine-makers, some of which have produced but cumbrous and inconvenient vehicles, which, when their purpose is considered, are cumbrous not only in their bulk, but in the multiplicity of their parts, and hence are complicated and tender, liable to derangement and failure, and are marked by an absence of that simplicity of construction so desirable in agricultural *mechanics*. In many of our more modern turnip drills, however, there is to be seen a marked simplicity of construction, that accords well with the objects in view, and this holds especially with those machines which are employed for sowing the seed alone; and in none more so than the machine now very generally known as the East Lothian drill, which I shall have occasion more particularly to notice. The recent introduction of numerous granulated manures has called forth a new class of machines, whose object is to deposite the manure along with the seed either in immediate contact, or in close contiguity, and these compound machines have again involved a degree of complexity of construction; for whenever a machine is required to perform compound functions, a necessarily increased complication of structure is entailed upon it to a greater or less extent, proportionate, perhaps, to the mechanical talents of the fabricator. An additional cause of complication in the compound-drills has arisen within the last few years, from a desire to economize the distribution of the modern expensive manures, by depositing small portions of it at the points only where the future root is intended to grow, leaving the intervals destitute of manure. The propriety of thus dealing so niggardly with the soil in withholding those substances in abundance by which it is enabled to continue its fertility, is at least questionable; but the experiment has effected what I am at present endeavouring to establish—the further complication of structure in the machines employed. The drill-sowing machines adapted to this purpose are designated in Scotland *plumpers*, from their dropping their gifts on one point. In England they are better known by the name of dibbling-machines, or drop-drills.

* Journal of the Royal English Agricultural Society, vol. iv. p. 49.

(2532.) The whole tribe of turnip-drills may be conveniently divided into two classes, from the manner in which their movements are produced, namely, those that move on rollers which press the ridge before the seed is deposited, and those which move on wheels, having rollers generally attached as an appendage rather than as a mover. The first class embraces most of those machines that deposite the seed alone; the second, chiefly those which deposite manure along with the seed; but the one-row hand-machines approach to this class, from the necessity of their being furnished with wheels to render their movements more light upon the hand. Without entering fully into the numerous varieties of the first class, I shall proceed to give illustrations of two of its forms.

(2533.) The East-Lothian *Turnip-drill*, represented in fig. 427, is one of the most efficient and simple in construction of its class. It is uncertain

Fig. 427.

THE EAST-LOTHIAN TURNIP-DRILL.

whether its origin can be traced to East-Lothian. More probably it is of Berwickshire; but it must be observed that John Wightman, an industrious mechanic of Upper Keith, East-Lothian, obtained the award of a premium from the Highland and Agricultural Society of Scotland in the year 1827 for his invention or improvement of this machine, and the form adopted by him still continues to be very generally employed, though in some districts with slight variations; but from the simplicity of its construction, and the ease with which it is managed and kept in order, it is not likely to undergo any very important change for the sowing of seed alone. Referring to the figure annexed, the bed-frame *a a* is 4 feet 6 inches long, and 2 feet wide over all, having 2 transverse and 3 longitudinal bars, each 2¾ inches square. The 3 pendant or upright bars *b*, *b*, *b*, are of the same scantling, and are mortised into the longitudinal bars of the bed-frame; and for additional strength to these pendants, a slot-bar of $2\frac{1}{2}$ by 1 inch, is frequently passed through them immediately above the rollers, though not exhibited in the figure. Stay-braces *c c* are attached to the bed-frame on the bottom of the pendant by bolts; but in many of the inferior sorts of machines these are omitted, greatly to the deterioration of their strength and durability. The horse-shafts *d d*, are bolted upon the two transverse bars; their scantling is usually about 3 by $2\frac{1}{2}$ inches at the front bar; they are also supported by stay-braces *c*, and furnished with the usual horse mounting. The two rollers *f f*, which are both motive and compressing, were in the early forms of the machine made of wood, but are now made almost always of cast-iron; they are 16 inches in length, 14 inches diameter at the ends, and from 6 to 8 inches in the middle. These rollers are cast as light as possible, and with cross arms at each end, through the eyes of which, and through the lower end of the three pendants *b*, an iron rod of 1 inch diameter is passed, forming the axle of the rollers, and upon which the machine rests and moves. From the varying diameter of these rollers, their surfaces at different points move with different velocities, producing friction and disturbance on the surface of the ridges. To obviate this, it has been proposed to cut the rollers transversely into three sections, thus giving each section freedom to move with its own individual velocity. Two seed-boxes *g g*, one of which is here shewn with its cover thrown open, are attached to an iron-frame or bow *h*, through the ends of which the axle also passes, whereby the whole becomes movable upon the axle. The seed-boxes are in this figure 12 inches long, and 9 inches wide, outside measure, and including the cover are 8 inches high to the apex. The bottom of the boxes is formed of tin-plate funnel-shaped, and terminating in a short nozzle to enter the sheath of the coulter *i*, the latter being also firmly attached to the iron bow *h*. For the convenience of regulating the depth of the rut made by the coulter, and for moving it out of or into the ground, iron connecting rods *k k* are attached to the seed-box frame, and again at top to the levers *l l*, the latter serving as handles to the machine, are jointed at *l'*, and when lifted up by the sower and rested upon the iron brackets or guards, by which each lever is embraced, the coulter is raised out of the ground by the ascent of the seed-box frame; and, when again set in work, the sower, by holding down the handles till they rest upon the hind-bar of the bed-frame, keeps the coulter at a uniform depth in the soil; the rod *k* being inflexible when once set, but being furnished at top with a plate palm, having several perforations, it can be adjusted to any depth of rut that may be judged requisite; hence, in this form of the machine the connecting rod *k* is the regulator of depth. The seed-box *g* is only the cover or shield; for the true seed-box, which is usually made of tin-plate, is in form of a small barrel, and is hence called the seed-barrel. The length of the seed-barrel is 6 inches; and its diameter 6 inches, its ends being of hard-wood, and 4 inches diameter, a small axle is passed through the barrel, having bearings on the two ends of the outer box, and upon the outward extremity of each axle, there is placed the pulleys *m m*, having an acute groove formed in their edges. A pulley corresponding to each of these is fixed upon the outer ends of the rollers *f f*, and being concentric to the rollers, the main axle passes also through them. A band of strong jack-chain *m* is now passed round each pair of pulleys, whereby the locomotion of the rollers gives a revolving motion also to the seed-barrels, and the consequent distribution of the seed through the small orifices in the middle zone of the seed-barrel. An important function of this machine has yet to be noticed, its self-adjustment to the width of the ridges. This is accomplished by the width between the pendants *b b* being greater by 4 or 5 inches than the length of the rollers, together with their attached pulleys and iron bows *h*, which admits of a ready lateral motion of the roller, with its accompaniments of bow, coulter, and seed-box, so soon as the machine is put in progressive motion, and the curved rollers feel any unequal resistance right or left. Any such unequal resistance, on either end of the rollers, draws it immediately to that side where the resistance is felt, until it is fairly adjusted to the slope of the ridge or drill; the effect in this case being produced entirely by the action of the sloping sides of the drill against the conoidal sides of the rollers.

(2534.) Fig. 428 is a longitudinal section of this turnip-drill, for the purpose of shewing more in detail the exact relation of the parts referred to in fig. 427,

Fig. 428.

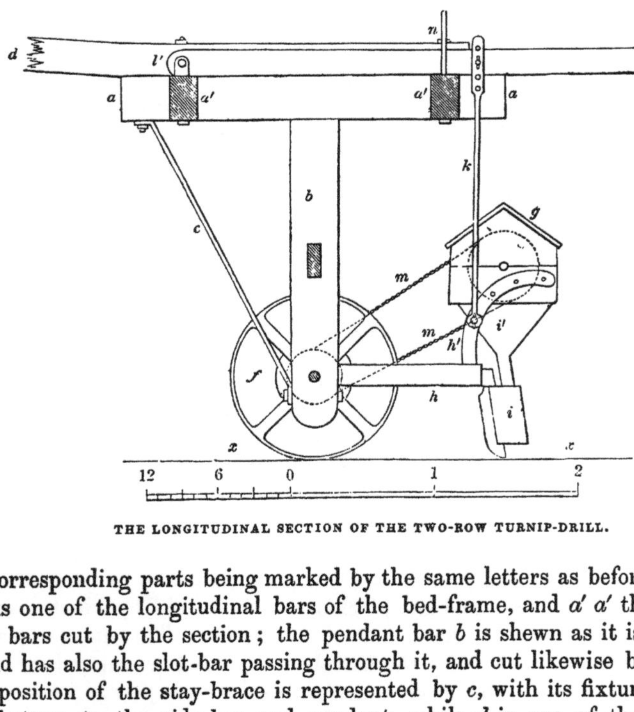

THE LONGITUDINAL SECTION OF THE TWO-ROW TURNIP-DRILL.

the corresponding parts being marked by the same letters as before. Thus *a a* marks one of the longitudinal bars of the bed-frame, and *a' a'* the two transverse bars cut by the section; the pendant bar *b* is shewn as it is connected to *a*, and has also the slot-bar passing through it, and cut likewise by the section. The position of the stay-brace is represented by *c*, with its fixture-bolts at top and bottom to the side-bar and pendant, while *d* is one of the horse-shafts broken off. The end of the roller *f* is seen in its position beyond the pendant bar, and *g* gives also an end view of the seed-box, which is fixed between two light bent iron standards *h'*; these are forged in pairs, and are first bolted to the coulter-frame or bow *h*, the seed-box being afterwards affixed between the standards. The coulter and its shield *i* is fixed upon the back of the bow, its length is about 8 inches, the cheeks of the sheath being 5 inches by 3 inches, and standing about 1½ inch wide at the tail. The hopper *i'*, of tin plate, is fixed in the seed-box as a bottom, the funnel terminating in the spout that conveys the seed between the cheeks of the sheath, and thence into the rut prepared by the coulter in the soil. The connecting rod *k* is jointed to one of the seed-box standards at bottom, and at top to the lever l' l, as before described; l' shews the joint of the lever, and n is the bracket in which the lever is embraced, and on which it is supported when the coulters are raised out of the ground. The chain m m which communicates motion from the roller to the seed-barrel, by passing over their respective pulleys, shewn in the figure by the dotted circles, is kept always at uniform tension, whatever be the position of the seed-box, by reason of the bow h and its apparatus moving round the main axle of the rollers, when acted upon by the lever l' l. The machine, as here represented, is to be supposed in the act of sowing, the lever l being close down upon the transverse bar; the line x x therefore represents the line to which the point of the coulter penetrates the soil, which, from the curvature of the roller, will be 3 inches below the surface, the sheath reaching only a depth of 2 inches.

(2535.) Fig. 429 is a perspective view of the seed-barrel, detached from its seat; a a is the axle or spindle in which it revolves, and on the longer end of which the pulley is placed. The barrel is formed of tin-plate, in two conical frustæ, joined base to base, with a cylindrical band b of 1 inch broad interposed between the two, and the truncated ends are closed with discs of hard wood. The band b is usually divided into six equal parts, and at each point of division three small apertures are punched out, each three varying in size from $\frac{1}{16}$ to $\frac{1}{8}$ inch diameter, but all in the same order from more to less. A separate band is then fitted to the first, closed with a clasp-joint, and capable of being slid round, to a small extent, upon the interior band, and is, besides, provided with a pinching-screw, by which it can be fixed at any point within its range of motion, which does not necessarily exceed one inch. The movable band is likewise divided into six equal parts, and at each division a perforation is made larger than any of those in the interior. By these arrangements the movable band can be placed so as to expose any of the three sets of the six perforations of the inner band, whereby a greater or lesser quantity of seed can be sown, according to circumstances. In the figure, the perforations are seen on the outer band; the clasp-joint also is seen near the upper side b of the figure; and the pinching-screw and slit by which it can be fixed or moved, are seen in the middle of the figure. The slider d covers a hole of $\frac{3}{4}$ inch diameter, by which the barrel can be filled or emptied.

Fig. 429.

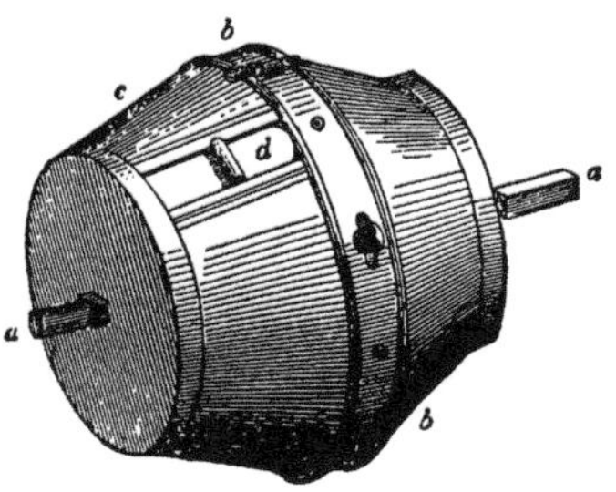

THE SEED-BARREL.

(2536.) In some cases this machine is furnished with a pair of small covering rollers, of 18 inches in length, and 4 inches in diameter, made of any hard wood. When these are adopted, which is but seldom, the rollers are mounted in light iron-frames or sheers, which are hooked on to a bolt in the coulter-frame, and are thus dragged behind the machine. These appendages are not, however, considered as forming an essential part of this drill. In conclusion, I may state that the figures given here are from the machines as manufactured by James Slight and Co., Edinburgh, who have also been successful in applying to it an apparatus by which some of the richer granulated manures, such as guano, can be applied along with the seed. The price of this drill in the ordinary state for sowing seed alone, ranges from L.5, to L.6, 10s.

(2537.) Several varieties of this drill are to be met with, in which the chief difference lies in the mode of communicating motion to the seed-barrel. In some of these bevelled-geer is employed, in others spur-geer, and in some, examples of both, means are applied for throwing the seed-barrel out of motion, when it is desirable that the sowing should cease, such as at the turning at the land-ends, and the like. The want of this in the machines above described may appear as a defect, and, viewing the article as a piece of machinery, it no doubt is a defect; but the advantage to be gained from the adoption of that which may be considered to constitute the more perfect machine is so trifling a saving of seed, perhaps to the amount of from ½ to 1 per cent. of the whole, that few farmers seem to consider the saving to be an object of such importance as to compensate for the additional expense and complication induced by the adoption of the disengaging principle. But among these varieties, there appeared, some years ago, at the Highland and Agricultural Society's Show in Dumfries, a drill of this class, which, to the adoption of spur-wheel geering, added a mode of distributing the seed which gave it a claim to notice, and for which a small premium was awarded to Mr Geddes of Cargen Bridge, the originator, as it appeared, of this improvement.

(2538.) The drill above referred to is here represented in perspective in fig. 430, and a glance will satisfy the reader that the general principle of its construction

Fig. 430.

GEDDES' TWO-ROWED TURNIP-DRILL.

is the same as that already described; but with a depression of the parts forming the frame-work of the machine, which gives to it an appearance of compactness and strength. A solid plank *a a* takes the place of the bed-frame of the former, and a pendant *b* at each end is secured to the plank by means of tenon and bolts; the pendant being no longer than just to admit the free motion of the rollers under the bed-plank. The horse-shafts *c* are bolted upon the back of the plank, and the handles or levers *d d* jointed to it, and embraced by the open guard, as in the former figure. The rollers *e e* are in like manner mounted upon a rod passing through them and the pendant, from side to side, and likewise through a V-form brace of iron appended to the middle of the bed-plank, supporting thus the middle of the rod, which forms the common axle of the machine. The coulter-frames *ff* are applied also on the same principle, but differently constructed; for instead of being a continuous bow, as in the former, they here consist of single bars appended on the axle, and connected transversely by a bed-plank of hard-wood which forms the sole or bottom of the seed-boxes *g g*. The coulter is also attached to this plank, and a perforation formed for the discharge of the seed, which is conducted into the sheath of the coulter by a tube as usual. The arms *ff* of the coulter-frame are prolonged and turned downward to form the bearings of the small covering-rollers *i i*, which, in this case, become permanent members of the machine; and as they must necessarily be used when the land is in different states of dryness, they are each furnished with a permanent scraper, to prevent the accumulation of soil upon them, when it is in a damp state. The motion of the main rollers, as before stated, is communicated to the seed-distributor by means of spur-geering in a series of small-toothed wheels, the first of which is attached to the end of its corresponding roller, and the last fixed upon the axle of the distributor, and these, together with the intermediate wheels, are inclosed in a case, one of which is partly seen at *h* with the cover of the case removed. These cases, with their wheels, are attached to and movable with the coulter-frames and all their appendages, and the coulter-frames being supported behind upon the covering rollers, these last become the regulators of the depth to which the coulters are to penetrate; and, for the purposes of this regulation, the rollers are capable of being shifted up or down in their bearings in the end of the bent arms of *ff*, to suit any required depth. As a consequence of this also, the connection between the coulter-frame and the levers *d d* is formed with a chain in place of an inflexible rod, the levers being required only for lifting the coulter from the ground; while the weight of the frame, with its load of coulter, seed-box, wheels, and covering rollers, possess sufficient weight to keep the coulters in the ground, and the covering-rollers prevent their going beyond the pre-arranged depth.

(2539.) The distributing apparatus in this machine has been considered to contain its principal merit, and has been supposed to afford a more correct means of graduating the quantity to be sown than the common seed-barrel. This apparatus is very simple; the interior of the seed-box is formed into a semi-cylinder of 7 inches diameter and 5 inches in length, which may be of wood or of tin-plate. Fig. 431 is a transverse section of this, *d d* being the interior surface of the box, in the bottom of which an opening is made to receive a brass roller *b*, $1\frac{1}{2}$ inch diameter, having a groove of $\frac{1}{10}$ inch wide running round it, as in the figure. The roller is mounted on an axle *a* which is prolonged to a sufficient distance beyond the box for receiving the last wheel of the series already described, the connection with which gives motion to the roller *b*. A slider *c* is attached to the interior of the box, and capable of nice adjustment by a screw or otherwise. The lower end of the slider, which comes in contact with the roller, is formed with a tongue that enters into the groove, and the adjustment of the opening between the point of

Fig. 431.

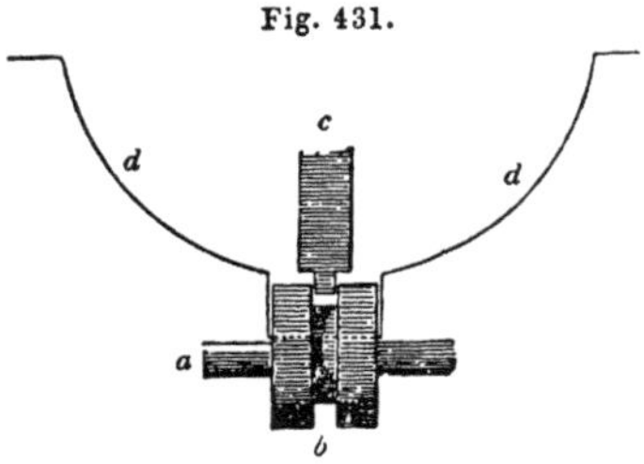

THE VERTICAL SECTION OF THE SEED-DISTRIBUTOR.

the tongue and the bottom of the groove determines the quantity of seed to be delivered.

(2540.) The dimensions given of the former machine will apply generally to this, taking 10 to 12 inches for the breadth of the bed plank, while its length is the same as the frame of the former. The price, under similar circumstances, will, in consequence of the greater expense of the wheel-geering as compared with that of chains, be somewhat higher than the first described machine; but the peculiar points here detailed merit the attention of machine makers.

(2541.) Passing over the many and varied forms of the two-row drill that are in extensive use, and many of which are very effective, but much too numerous to be detailed here, I proceed to the description of one of those compound machines already alluded to as forming the second class of drills, and which are employed for depositing manure along with the seed in *continuous lines*.

(2542.) *The Two-row Turnip and Bone-dust Drill.*—The machines of this class have not yet received that general sanction from agriculturists which the first class has long maintained, and this probably arises from the consideration, that turnip will, and must be, sown more or less in all localities, but the kinds of manure to be resorted to will vary considerably, according to the circumstances of those different localities. The simple seed-sowing machine will be employed in all places, but the manure-sower will be resorted to principally in those localities where the granulated manures are more extensively employed; hence the demand for the machines of this class will always be greatly under that of the former. The difficulty, also, of producing an efficient compound machine is great, and the expense considerably higher than of those of simple construction, and these circumstances combine to check the extension of the former. The machine now to be described is the result of some experience and numerous experiments, in which a variety of arrangements have been experimented upon, and the result has produced an implement that yields satisfaction, and possesses all the requisites of the operation of depositing seed and manure together continuously and in any required quantity.

(2543.) Plate XXXI. fig. 432, exhibits a view in perspective of this turnip and bone-dust drill, but which, from its combination, scan only be described in a general way from this figure. The main or bed-frame *a a*, is constructed in a similar manner to those of the corn-drills. This frame is 3 feet 9 inches wide over all, and measures 4 feet 2 inches longitudinally over the front and back rails; the axle of the carriage, which passes across and under the frame, is supported in pillow-blocks bolted to the side-bars. The wheels *b b*, are 3 feet 6 inches diameter; one of them is fixed dead upon the axle, carrying the latter round with it, and thus forming the mover of the acting parts; the other wheel being left at liberty to revolve on the axle, for the convenience of turning the machine round. The horse shafts *c c*, are bolted to the two foremost transverse bars, and are made of bent timber to bring them to the proper height for the horse. The seed-barrels, of the form already described (2535.), are enclosed in the boxes *d d*, through which axles pass to, and are supported on longitudinal bearers, each axle carrying a pulley, one of which is seen at *d'*. The two manure hoppers *e e*, are 2 feet 2 inches square at the top, 1 foot 8 inches deep, and 8 inches square at bottom. They are constructed with a cast-iron bottom, having an opening of 1½ inch wide, and the length equal to the breadth of the entire bottom, adapted for the reception of the manure distributing-wheel. Flanges *e′e′* rise from the bottom, by which it and the boarded sides of the hoppers are connected, and four ears project laterally also from the bottom, by which it is bolted to the bed-frame. The latest improvement in this machine has been the addition of a pair of pressing rollers *ff*, of the same form and dimensions as those of the common drill. Each roller is also furnished with a coulter-frame *g g*, which carries the coulters *h h*; the coulter being the only appendage attached in this case to the frame or bow. The rollers and bows are threaded upon a rod as usual, the ends of which are supported in the malleable-iron brackets *i*, one being attached to each side of the bed-frame, with adaptation for being raised or lowered at pleasure. The pressing rollers have also the usual extent of lateral play, whereby they possess the property of adjusting themselves to the ridges,—of carrying the coulter-frame and coulter along with them, and of securing the object of the seed being always sown directly in the middle of the drills. Two lever handles *k k*, are jointed to the front bar of the bed-frame, though not seen in the figure; and to these are attached the connecting rods *f′*, whose lower ends are jointed to the coulter-frame, thus bringing the operation of the coulters under the control of the person who takes charge of the levers *k k*. The connecting rods are capable of adjustment, by means of a series of holes in their upper ends; and when sowing, the adjustment is such, that, when the coulter is at the proper depth, the lever rests upon the hind-bar of the bed-frame, and when thrown out of working, the coulters are raised by lifting up the levers, till they rest upon the bridge *n*. An iron-lever is also joined upon the front bar; its handle, extending backwards to *k′*, serves to disengage the action of the manure distributors from the motion of the main axle; and as the motion of the seed-barrels is taken from that of the manure-distributors, all the secondary motion ceases on the movement of the lever *k′*, and is again brought into action by moving it in the opposite direction. The motion of the manure-distributor is conveyed by small spur-wheels, which are not seen in the figure, and the seed-barrels are driven by separate chains from the shaft of the former.

(2544.) Such is a general outline of this drill; but in order to make the illustration complete, I have here added a geometrical plan and elevation drawn to the annexed scale, which thus exhibits the true relation of the parts. Fig. 433 is the elevation in which *a a* is one of the side bars of the bed-frame, and *a′ a′ a′* the 3 transverse bars; the former 4½ by 2¼ inches, the latter 4 by 2¼. The carriage-wheels *b*, as before noticed, have the axle supported on pillow-blocks bolted to the lower edge of the side-bar *a*, and the horse-shafts seen broken off at *c* are bolted upon the upper edge of the two foremost transverse bars. The position of the seed-box is shewn at *d*, and *e* marks the position of the manure-hoppers *o*, lower *e′* marking also the junction of the hopper with the cast-iron bottom, and the form of the same. While the carriage-wheel rests on the line *x x* representing the bottom of the drill, the relative position of the pressing roller is shewn at *f* with its coulter-frame *g* and coulter *h*, *g′* being the connecting-rod between the coulter-frame and the lever-handle, and *h′* a funnel inserted within the sheath of the coulter to retain the manure and seed directors. The pressing-roller axle is supported in the malleable-iron pendant and stay *i i*, which is bolted on at each side of the bed-frame. The lever-handles *k* are jointed at the fore-end in iron sheaves placed under the front transverse bar as at *a′k*, the handle being at liberty to rise and fall as required. A kneed tube *l*, the director of the manure after it passes the distributor, is secured by flanges that lie under the ears of the hopper bottom; it is jointed at *l* so as to have a

Fig. 433.

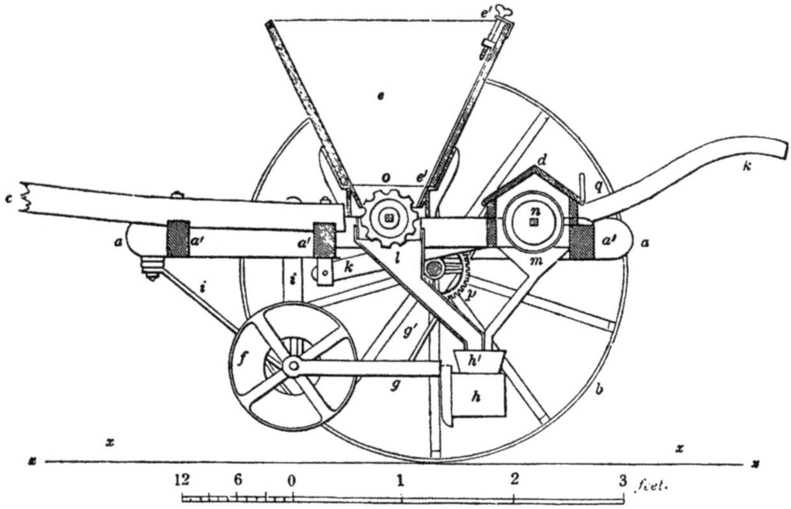

THE LONGITUDINAL SECTION OF THE TURNIP AND MANURE-DRILL.

lateral swinging motion to adapt it to the movements of the roller *f* with its appendage, and it terminates with the funnel *h'*. The seed-director *m* is formed in a similar manner, and terminates with the other in the funnel *h'*. It follows from this arrangement, that the seed and manure are deposited in a mixed state, and at the same instant. A seed-barrel of the form of fig. 429, is placed within the box *d*, *n* being the spindle on which it revolves, and *o l* is one of the manure-distributors, being simply an indented wheel of 6 inches diameter, 1½ inch thick, and having 10 indentations or teeth; they are mounted on an axle which passes quite across the bed-frame. This wheel is so placed in relation to the opening in the hopper, as to be quite close to the fore-end of the opening, while an aperture is left at the opposite side sufficiently large to pass the largest allowance of manure that is to be given out; and in order to graduate this quantity, a sliding sluice *e' e'* is attached to that side of the hopper, and is adjusted by means of the screw at top. By these means the area of the discharging orifice can be regulated to any desired quantity per acre. The motions for the discharge of the seed and manure are produced from the wheel *p*, which is placed on the main axle, and gives motion to a similar wheel not seen in the figure, placed upon the manure axle.

(2545.) In the plan, fig. 434, which exhibits the bed-frame and what is immediately attached to it above, but leaving out the upper works, as well as those below, the parts not seen, or imperfectly seen in the former figures, are here further brought into view. The longitudinal and transverse bars of the bed-frame are again marked *a a*, &c., but here four secondary bars *a' a'*, &c. are exhibited; these are introduced to form the bearings of the seed-boxes and manure-hoppers with their shafts. The bars *a' a'*, &c. are mortised into, and lipped over the two transverse bars, their upper surfaces being all upon a level with those of the main side-bars. The carriage-wheels are again marked *b*, the

Fig. 434.

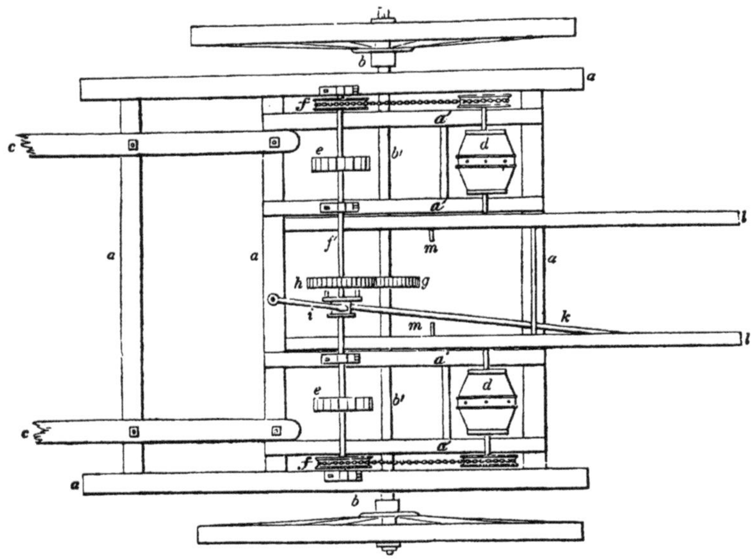

THE PLAN OF THE TURNIP AND MANURE DRILL.

main axle *b′ b′*, and the horse-shafts *c c*, broken off as before. The seed-barrels, with their separate spindles and pulleys, are seen in their position at *d d*, the boxes being removed. The seed-barrel pulleys are driven by means of the chains which pass round the pulleys *f f*, mounted on the auxiliary shaft *f*, which carries also the manure-wheels *e e*, and the clutch-wheel *h*. This last wheel, which carries round the auxiliary shaft, runs loose upon the shaft, but can at pleasure be put in connection with the clutch-fork *i*, which slides upon a square on the shaft, and moves at all times with it. The loose wheel *h* is also constantly in geer with the driving-wheel *g*, which is fixed upon the main axle, and at all times, when the machine moves, keeps the wheel *h* in motion, while the sowing-geer is all at rest. The lever *k* is jointed to the front-bar, and has hold of the clutch-fork; this, in the figure, is shewn disengaged; but when the whole machinery is required to move, the lever is shifted to the right, which brings the fork of the clutch into contact with the loose wheel, and immediately carries round the clutch, and with it the shaft *f′* and all its dependencies. The lever handles *l l* are, as already noticed in fig. 433, jointed below the front-bar, and extend backward to a convenient distance; their chief duty is to lift the coulter, and keep it at a uniform depth in the ground, both of which are accomplished through the medium of the connecting-rods *g′*, fig. 433, which are hooked to the levers on the studs *m m*, fig. 434.

(2546.) Fig. 435 is a horizontal section of the pressing-roller, with the coulter and its frame; *a a* is a portion of the roller-axle, *b* the roller in section, its contour being arcs of a circle. This form, though very generally adopted, is inferior in point of usefulness (especially in giving shelter to the young plants) to that form of roller that has its central part nearly cylindrical, and its termi-

Fig. 435.

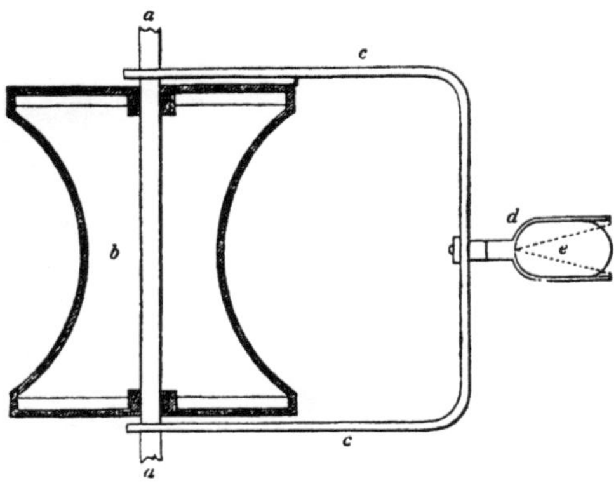

THE PRESSING-ROLLER AND COULTER-FRAME.

nations two opposite conical frustæ, giving to the top of the ridge a surface nearly level to the breadth of 8 inches. The bow or frame *c c* is made of malleable iron, 2 inches in depth, and $\frac{3}{8}$ inch in thickness; *d* is the sheath of the coulter, which for this machine requires to be made wider than usual to receive the manure-director; the dotted lines *e e* shew the width and form of the cheeks of the sheath at bottom. In this case also the opening behind in the sheath is closed up by the oval-shaped funnel, the body only of which is here shewn, the spreading funnel being cut off in the figure, but is seen in fig. 432, Plate XXXI. The present figure shews the bow *c c* as applicable to the seed and manure machine, but the only difference between this and that of fig. 427 is, that in it the bow is made so much wider as to admit the chain pulling between the end of the roller and the cheek of the bow.

(2547.) The machine thus described may be considered as one of the best of its kind, and though apparently complicated, it is yet as free of that fault as it is perhaps possible to attain, while the essential objects are kept under command. The graduation of the discharge of manure is attainable by it to any desired limit, and the discharge is also regular and uniform. The means of engaging and disengaging both the seed and manure geering is perfectly efficient and simple. From the materials and labour necessarily expended in the construction of a machine of this kind, the price is consequently higher than the common two-row drill, being L.11, 10s. It is believed that machines of this construction have only been made by myself, of the house that I have had frequent occasion to mention, namely, James Slight and Co., Edinburgh.

(2548.) Though the cultivation of the turnip can only be carried on to its most productive extent on the large scale, there are yet many who cultivate it successfully, in proportion to their extent, on a moderate or even small scale. I deem it therefore proper to give an illustration of a hand sowing-machine or one-row turnip-barrow, which is suited to farms of small extent.

(2549.) The *turnip-barrow* in common use is represented in perspective by fig. 436 : it consists of a frame of timber, formed of the two handles *a a*, framed upon a broad transverse bar which carries the seed-box. The stilts are 5 feet in length, 2½ by 1½ inch, the width of the frame at the handles is 2 feet, and

Fig. 436.

THE TURNIP-BARROW.

at the point 14 inches; besides the broad bar, a round stretcher is introduced near the point of the stilts, chiefly intended for the attachment of a drag rope; an iron axle is placed below the frame, running in bushes or small pillow-blocks, and the two wheels *b b* are fitted to it, one of them fixed the other running free. These wheels are 2 feet diameter, with cast-iron naves, and of very light construction. Two iron legs *c c* are bolted to the stilts, with stretcher and braces to render them steady. A toothed spur-wheel *d*, of 7 inches diameter, is fixed upon the axle, and this acts upon another *e* of equal size fitted upon the spindle of the seed-barrel, which last is of the same construction as fig. 429. The seed-barrel is mounted in the case *f*, and, for the purpose of disengaging the wheel *e* to set the seed-barrel at rest, the slide-bar *g* is applied, which having its upper edge worked into an inclined plane, the drawing out of the slider *g* raises the spindle and wheel *e*, and disengages it from the driving-wheel. The bottom of the seed-box is formed into a funnel, terminating in a director-pipe, as seen at *d*, which descends into the sheath of the coulter *h*. The coulter is simply a bar of hardwood, 2½ by 1½ inches, set in a mortise in the transverse bar of the frame; and fixed at the proper position by means of a wedge, and shod at the bottom with a strong sheet-iron sheath. J. S.]

(2550.) I would say a few words on the effects of the general disuse of the hind rollers of the turnip-drill. There is no doubt that when rollers are so placed, the best work is made when the surface of the ground is dry, and as little doubt they make bad work when the surface of the ground is damp, and when that is wet they cannot be used at all; and the finer the surface of the ground has been pulverized these effects are the more strongly observed. This being the case, when the ground is damp, and rollers used, the sowing of the turnip-seed will be delayed until the ground become sufficiently dry, and in this delay a material effect for the worse may be produced on the future crop. But, on the other hand, the disuse of the rollers inflicts a positive injury of another kind, so that it is worth considering whether the injury alluded to is of a nature to induce the employment of rollers on all occasions. The mechanical effect of the hind rollers is to fill up the rut made by the coulter, and to smoothen the top of the drills. Now, the utility of this smoothening and compressing of the top of the drills is, not only to prevent drought reaching beyond the surface of the ground; which, in both light and strong soils, is an advantage, but to render the singling of turnips more easy and certain to the field-worker. Hence, after the sides of the drills have been pared by the scuffler, it will be found that the turnip-plants are much more easily singled when the top of the drills have been smoothened than when left rough with a rut; because the hoe can be made to seize any individual plant more certainly on a smooth surface than at the bottom of a rut where the plant is comparatively out of reach, and partially out of sight. A greater portion, too, of the drill is required to be pushed away with the hoe when singling is performed in a rutted drill; the dung is more apt to be torn up along with the plant; the plants cannot be singled at so early a stage of their growth, for until they have reached some way *above* the edges of the rut, it is not safe to touch them with the hoe at its bottom; and the drill, after singling, is of course pushed away into a flatter state. The advantages of a smooth surface are not imaginary, for I have experienced these opposite inconveniences when I have been induced to remove the hind roller in damp weather, under the desire to proceed with the sowing before the ground was perfectly dry on the surface. A scraper is of some use on the hind rollers, but still they cannot make the rollers work *well* when the ground is damp. On carefully weighing the disadvantages of both modes, I have long been convinced of the superiority of the smooth drills, inasmuch as I am of opinion that the most proper singling of the plants is of far greater importance to the future turnip crop than any mischief that can arise from waiting 2 or 3 hours in the morning until the ground becomes dry with the sun and air, and I would much rather wait that time in the morning and work the longer in the evening, than sow turnip-land in a dampish state without hind-rollers.

(2551.) The concluding portion of the description of the turnip-harrow, which should have been continued on page 800, having been by some oversight omitted, is here given in this separate paragraph.—[The distance between the barrow-wheels is made to correspond with the prevailing width given to the ridges, which is generally 27 inches, but in the best constructed barrows the arms of the axle are made cylindrical and elongated; the wheels can then be accommodated to any width. Ten years ago I engrafted an apparatus upon the turnip-barrow for depositing manures along with the seed, but, as might be expected, the combination has not been of much importance. The barrow for seed only, requires the labour of a man, and to load it with additional apparatus, and a supply of manure, is to place it beyond his powers; the aid of a horse is then called in, which is spending the labour of man and horse, while they perform only half the work that they could, with equal ease, produce from a two-row machine, hence the combination in this form can never be of extensive application. As a general rule, indeed, it may be said that little economy is derived from the attempts to make machines perform operations that do not properly come within the range of their intended objects. The price of the common turnip-barrow, as here figured, is from L.2, 5s. to L.2, 15s.

(2552.) The *drop-drill* or *dibbling* machine for sowing turnip and manure together, not continuously, but in small deposites, has been the object of numerous and varied experiments. The propriety of this mode, and the advantages to be derived from its achievement, seems yet doubtful, but the accomplishment of the mechanical part of the operation is now tolerably well performed by machines of various makers. In the earlier forms of the drop-drill they were found to perform tolerably well when moving at a slow rate, but failed when impelled at the usual rate of sowing. This defect seems to have arisen from the peculiar form of the manure-depositor, which, in these machines, was usually a short cylinder chambered in the periphery, and half encircled by a metal cover, whereby the manure was allowed to drop from the chambers as these came successively from under the cover. One of the first successful machines of this kind is believed to have been that patented by Mr William Grounsell, Louth, Lincolnshire, in 1838, and soon after they were brought to an equal degree of perfection in Scotland by Mr Sidey, Pitcairn Green, Perth, who still holds by the chambered cylinder, but more correctly encased, so as to drop distinctly with a moderate speed.

(2553.) The latest improvement that has yet appeared on the drop-drill has been effected by Messrs Smith of St Ninians, near Stirling, who, adopting a combination of the best parts of previous machines, have produced one that seems more likely to attain the desired object than any of its predecessors. In this the metal trunk is introduced for receiving the manure from the ordinary distributing wheel, and being provided with a *valve* capable of being opened and shut at certain intervals, the manure is retarded in its descent within the trunk by means of the valve, until the requisite quantity is collected, when the valve, being suddenly opened, makes the deposite and is again shut; but the arrangement of the whole machine will be better understood by reference to the accompanying figure, which represents a single row machine in this improved form.

(2554.) This improved drop-drill is represented in perspective by fig. 437. It is constructed entirely of cast and malleable iron, though these materials are by no means essential to its formation. The bed-frame *a a* is of cast-iron, 24 inches in length, and 18 inches in breadth, and consists of an outer frame, with a broad longitudinal central bar, to which are attached the seed and manure hoppers and trunks. The bed-frame is mounted on a carriage axle, supported on wooden plummer blocks bolted to the frame, and bearing the two carriage wheels *b b*, which are likewise made of iron, and are 3 feet diameter. The axle carries a spur-wheel *c* (better seen in the succeeding figure), of 7½ inches diameter, which acts upon the pinions *d* and *e*, the first being upon the shaft of the manure distributing-wheel, the second upon that of the seed-distributor. The shafts of these two pinions are supported at the outward ends in the attached brackets seen in the left of the figure, their inward ends being supported in their respective trunks or conductors. The shaft of the pinion *d* carries the manure distributing-wheel placed within the manure trunk, and it extends beyond the trunk to carry the wiper wheel *f*, while the shaft of the pinion *e* passes through the seed-trunk, and carries the seed-distributor, as afterwards described. The manure-trunk, with its sheath *g*, is fixed to the middle bar of the bed-frame, which is perforated for the reception of the trunk, the sheath being of stout sheet-iron, and attached to the trunk in front by means of the shank-plate *h*, with two bolts, on which it can be slid up or down to suit the depth of the furrow. A discharging-valve is placed within the trunk, hinged upon an axis carrying at one end the lever *k*, with a similar, but shorter, lever at the opposite end, which is acted upon by a spring, as seen in the figure at *i*,

Fig. 437.

SMITH'S DROP-DRILL.

keeping the valve shut. The seed-trunk is fixed in the middle bar, in the same manner; its lower extremity is seen at *l*, while its hopper or seed-box, which is a tin-plate funnel shipt on the head of the cast-iron trunk, is seen at *m* between the large hopper and the wheel. The manure hopper *n* is 22 inches square at top, 4 by 5 inches at bottom, and 15 inches deep; it is held in its place by a knee-plate before and behind, and is furnished with the usual slide-gauge, the handle of which is seen at *o*, for regulating the discharge. For the purpose of covering in the seed and manure, and for consolidating the ridge, the concave roller *p* is attached to the hind part of the machine, in a permanent position, by two pendant bars *p'*; this roller is 18 inches in length, 14 inches diameter at the ends, and 9 inches in the middle, and is suspended with its centre at 9 inches above the sole of the carriage wheels. Though this machine sows but one row, it requires, like others of its class, to be drawn by a horse, and that the horse may walk in the furrow, as well as the man who guides it, the bar to which the horse is yoked, together with the handles for the man, are placed at one side of the bed-frame and bolted thereto; of these parts *q* is the yoke-bar or beam, supported by the stay-rod *r*, and *s s* the handles, supported in like manner by the stay *t*. The point of the beam *q* is 20 inches above the sole-line of the wheels, and the handles the usual height of 3 feet.

(2555.) For the clearer illustration of this machine, especially of those parts which are imperfectly seen in the above figure, a longitudinal section of it is given in fig. 438, the beam and handles being left out, and, in this, the same letters are applied to the parts alluded to in the former figure. Thus *a a* is the

Fig. 438.

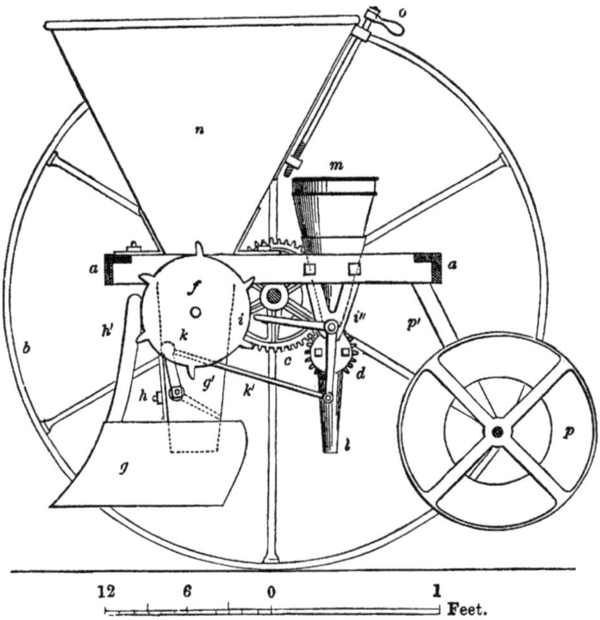

SECTION OF THE DROP-DRILL.

bed-frame, *b* the carriage wheels, *c* the main spur-wheels, *d* the pinion of the seed-axle, that of the manure-axle being hid behind the wiper-wheel *f*. The manure-trunk *g'* dotted is seen coming from behind the wiper-wheel and entering into the coulter *g*, while the dotted parallel lines below *g'* show the position of the valve; and *h* is the shank-plate attached to the sheath by which the latter is fixed upon the trunk, *h'* being a coulter-bar or feather fixed in the front of the sheath and rising above it, to throw off any impediment that may rise upon the coulter. The lever *k* is fixed upon the axis or spindle of the valve, and acts in opposition to the spring already noticed, by opening the valve in the following manner. The wipers of the wheel *f*, as they revolve, touch in succession the horizontal arm of the bent lever *i'*, whose fulcrum is attached to the bracket *i'*, and its vertical arm being jointed to the connecting-rod *k'*, which again is jointed at its opposite end to the lever *k* of the valve; the combination thus effected opens the valve once for every wiper of the wheel *f* that passes the lever *i*, and is quite unconnected with the seed-distributor except through the wheel and pinion *c d*. The lower extremity of the seed-depositer is seen at *l*, and its seed-box at *m*, the manure-hopper *n*, with its slide-gauge *o*, is seen as fixed upon the bed-frame by the knee-plates, and the covering-roller *p*, with its pendant bars *p′*, are seen also in position.

(2556.) Fig. 439 is a detached section of the manure-trunk; it is a box of iron with one side removable; *a* is a portion of the middle-bar of the bed-frame to which the trunk is bolted, the manure-hopper, when in its place, being inserted in the top *b* of the trunk; *c* is the manure-distributor, a wheel of 4 inches diameter and 2 inches broad, with twelve thin boxed teeth, and, as before described, the axle of this wheel carries also the wiper-wheel. The valve *d* is here shewn shut, its usual position, being held so by the spring and lever, except when opened for an instant by the contending levers to drop the charge of manure. That portion of the side of the trunk against which the valve shuts at *e* is cut off below the valve, which admits of its closing upon the edge, and gives also a free egress to the charge the moment the valve opens.

Fig. 439.

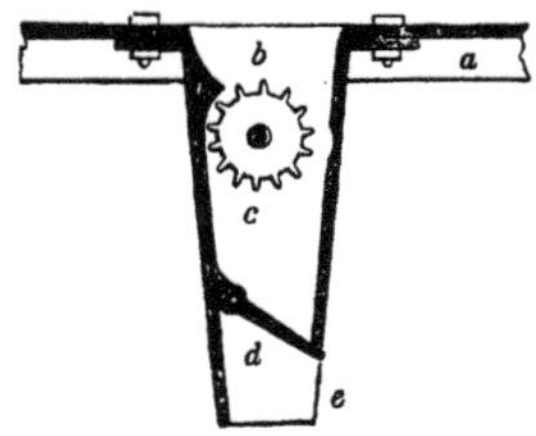

SECTION OF THE MANURE-TRUNK.

(2557.) Fig. 440, *a*, is a similar section of the seed-trunk; *c c* are the flanges by which it is bolted to the bed-frame, *d* that part on which the tin-plate funnel is shipt, and these, together with the lower space *e*, forms the seed-receptacle, which diminishes downward to an aperture of one-fourth inch, opening into the nicely turned chamber in which the distributor revolves. The chamber is 1½ inch diameter and 1½ inch deep, and the outle tfrom it to the seed-depositor is again ¼ inch. The distributor is shewn in the detached fig. *b*, in which *g* is the distributing cylinder, exactly fitted to the chamber *f*, and is in one piece with its axle *h*. The part *i* is a separate flange fitted upon the axle, the smaller portion of it follows the cylinder into the chamber, and the thin flange fixes it to the trunk by two screws, as seen at *d*, fig. 438, leaving the cylinder freedom to revolve. The cylinder has six little chambers perforated in its circumference, which in its revolution coincide successively with the upper and lower openings of the trunk, receiving from the receptacle a charge of seed amounting to six or eight grains, and delivering the same, when that chamber has gone half a revolution of the cylinder, through the lower orifice into the seed-depositor.

Fig. 440.

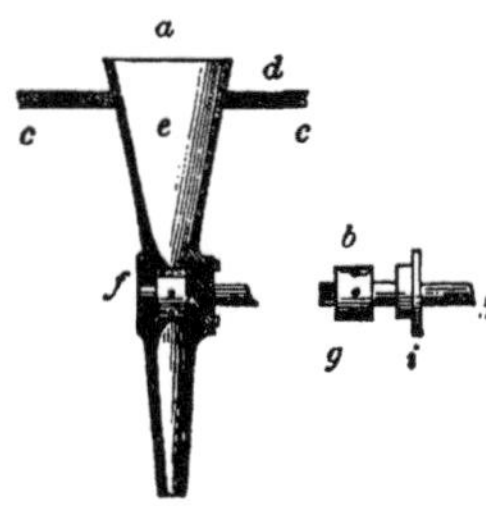

SECTION OF THE SEED-TRUNK.

(2558.) In sowing with this machine, the effects of the combination of its machinery is to be thus understood. The carriage-wheels being 3 feet in diameter, or 113 inches in circumference, the wheels will turn once round while the machine passes over that space. The main spur-wheel will also make one revolution in the same space; but, as the pinions which are upon the axle of the seed and the manure distributors are just half the diameter of the other, they will each make two revolutions while the machine is moving over 113 inches; and as the wiper-wheel has six wipers, each revolution of it will lift the valve six times, or twelve times in the space of 113 inches, making twelve depositions of manure in that space, which is 9½ inches to each nearly. In the same manner the seed-distributor, which also makes two revolutions in the same time or space, and as it has six little cups or perforations, it will discharge twelve times, as before, in 113 inches. Then, the seed-depositor being placed 9½ inches distant and behind the manure; and as they are arranged to drop at the same instant, the manure-deposite will be always one space in advance of the seed, and the seed will be dropped over the immediately preceding deposite of manure.

(2559.) There are apparently some advantages derived from this successive mode of deposite of the seed and manure, especially with those manures that are held the most active, such as bones, guano, and the like, for here, the manure is laid into the rut, the earth of which partially falls in and mixes with it, thereby reducing its intensity; and the seed is dropped upon this mixture, instead of falling directly amongst the pure matter, as is generally the case with machines that sow both seed and manure continuously. It is said that a more speedy and vigorous vegetation is produced by this than by the continuous mode of sowing; but it may yet be deserving of observation, whether the more speedy development of the young plant does not arise from the circumstance of the seed, under this mode of treatment, being deposited nearer to the surface of the soil, than it is when put in immediately behind a coulter; and, viewing the subject in this light, it may suggest the question, whether deep sowing alone may not be the cause of the protracted vegetation so often and so seriously experienced in the turnip crop? It is well known that the vegetation of all seeds is decidedly affected by the depth at which they are planted in the soil, so much so, that at or beyond certain depths, the seeds lie perfectly dormant; the depth, however, requisite for producing this effect varying considerably with the nature and qualities of the seeds. Thus, a potato-tuber will vegetate if within 2 feet of the surface, but the process will be very much retarded; the seeds of some Cruciferæ, again, to which family the turnip belongs, are supposed to become dormant, but not dead, at the depth of ordinary ploughing. There need be no surprise, then, though we should find the vegetation of the turnip retarded to the extent of days, or even weeks, from the seed being deposited at 2, 3, or even 4 inches, as is sometimes done, below the surface. The subject, as regards the turnip crop alone, appears to me deserving of careful experiment, and, if determined in the affirmative, much disappointment and loss may be prevented by adopting due precautions to insure sowing at proper depths.

(2560.) The saving of manure, in the first instance, by the use of the drop-drill, appears to be considerable, since it has been frequently asserted that 10 or 12 bushels of bone-dust per acre will produce a braird equal if not superior to 16 or 18 bushels put in by the continuous mode. The subject is, therefore, of great importance, and calls for close observation, for if the drop system is really so important, it cannot be too widely adopted. In the view of its more general adoption, the form of the machine must be changed from the single to the two-row drill, a change of which it is quite capable, and which may be done at less than double the expense of the single machine. In its present form, the process is too slow for large farms; and on any such, the additional expense of the double machine is not to be put in comparison with the advantages of dispatch in sowing. The price of the single machine is about L.6; if extended to two rows, the price would not exceed L.10.—J. S.]

62. OF REPAIRING THE FENCES OF GRASS-FIELDS, AND OF THE PROPER CONSTRUCTION OF FIELD-GATES.

> " At neglected gaps
> Burst scrambling through, and widen every breach.
> A stake put timely in, or whinny bush,
> Until the season come when living plants
> May fill the vacant space, much harm prevents."
>
> GRAHAM.

(2561.) The season being almost arrived when the grass is able to support stock, and, of course, when the cattle are permitted to leave their winter quarters in the steading, it is necessary to ascertain, in the first place, whether the fences of the grass-fields are in such a state of repair as will offer no temptation to stock to scramble through neglected gaps, much to the injury not only of the fence, but perhaps of themselves, or at least much to their disquietude ; and, in the next place, to watch the period when the grass is in a fit state to receive them. Sometimes a good deal of work is required to put grass-fields in a proper state for the reception of stock, owing principally to the nature of the soil, and partly to the state of the weather. On every kind of land the small stones lying on its surface should be gathered by the field-workers and carted off for the use of drains, or be broken into metal for roads. It may happen that the throng of other work may prevent the assistance of horses and carts being given for this purpose, in which case the stones should be gathered together in small heaps on the furrow-brow of every other single ridge ; but in doing this, it should be remembered that these heaps occupy so much of the ground, and, of course, prevent the growth of so much grass, that, on this account, it is a much better practice to cart them away at once if practicable. When carts are used the stones are thrown directly into them ; whereas in making heaps, the stones require some care to be put together, and, of course, waste time, and they have to be removed after all. Some farmers are regardless of gathering the stones from any of their fields, even from grass-fields which are to be in pasture ; while all acknowledge that fields of grass which are to be made into hay ought to be cleared of stones to save the scythes at hay-time. On clay soils there are very few, or perhaps no stones to clear off, and in wet weather no cart should be allowed to go on new grass. As every field, whether of new or of old grass, should be rolled some time before the stock enter them, it is clear that the ground cannot receive all the benefits of rolling as long as stones are allowed to remain on its surface. The best time for rolling is when the surface is *dry*—mark you, not when *hard* and dry—for when grass, especially young grass, is rolled in a wet state, it is very apt to become bruised and blackened. When dry, grass is elastic and able to bear the pressure of the roller without injury. Light land will bear rolling at any time when the surface is dry; but plants are very liable to be bruised by the roller against the hard clods of clay-land, and in a soft state, on the other hand, clay-land is apt to become hardened or encrusted by rolling. The rolling of heavy land is thus a ticklish matter; but a good criterion to judge of its being in a fit state for the roller, is when clods crumble down easily with the pressure of the foot, and not press flat, or enter whole into the soil. The rolling is always given *across* the ridges. The stones should be gathered, and the land rolled at least a fortnight before the stock are put on grass, to allow the grass time to grow after these operations, when it will be found to grow rapidly, if the weather is at all favourable.

(2562.) While the surface of the field is thus preparing for the reception of stock, the hedger should be engaged in *repairing the fences*. In this he is frequently assisted by the shepherd, and in cases where there is no professed hedger on a farm, the shepherd himself undertakes the duty. The repairing chiefly consists in filling up gaps, and these are rendered fencible by driving stabs on the face of the hedge-bank behind the gap, and nailing 2 or 3 short rails on them as in fig. 393, or by wattling them with branches of trees or thorn, or by setting a dead hedge like fig. 391. There should nothing be put *into* the gap, as is often done, to the prevention of the lateral extension of the plants on either side of it, and which of themselves in time will fill up a narrow gap. A wide gap should be kept clear, and filled up in due season with living plants, as described in (2333.), or with layers as shewn at *i*, fig. 389, and described in (2334.). Every gateway in a field not required for the season should be filled up with a dead hedge, fig. 391. Stone-fences should be repaired by replacing the cope-stones, and rebuilding the few stones that may have been driven down by violence.

(2563.) In making repairs of fences, it should be borne in mind to keep an easy passage for the shepherd from field to field, when looking after his flock. Facilities should be afforded him, by leaving openings at the corners of fields, or setting stiles across the fence; because it is better that these should be formed for him at once, than that he should have to make them for himself. He is the best judge of where they should be placed, in the short cuts he must necessarily take by the fields.

(2564.) Besides the fences, the *gates* of grass-fields require inspection and repairs, so as they may be put in a useable state for the season. When any of the timbers, posts, or bars, are broken or wanting, or the fastenings loose, the carpenter or smith should be made to repair them; and the posts on which the gates hang should be made firm in the ground when loose, or renewed when decayed. In putting up new gate-posts, the firmest mode I have found, is to dig as narrow a hole as practicable 3 feet deep for the hanging-post, and then to ram the earth, by little and little, firmly around the post without any stones. Charring or pitching the part under ground is a pretty good prevention from rot for some time. The simplest mode I have seen of fastening field-gates is with a small chain attached to the fore stile of the gate, to link on to a hook on the receiving-post. The most convenient position for field-gates is at the ends of head-ridges, which may be regarded as the roads of fields. Field-gates should be made to fold back upon a fence; to open beyond the square; and not to shut of themselves. When they shut of themselves, and are not properly set when opened, and which requires greater care than is usually bestowed on these matters, they are apt to catch a wheel of the cart which is passing, and, of course, to be shivered to atoms, or the post to be snapped asunder; and more than this, self-shutting gates are apt to be left unfastened by most people who pass through them, and are therefore unavailing as a fence to stock, especially to horses when idle, which seem to take delight to loiter about gates, and they not unfrequently find out the mode of opening them. One reason, perhaps, for horses loitering about gates is to rub themselves, to prevent which thorns are wattled into the bars; but independent of this casual safety, every pasture-field should be provided with one good *rubbing-post* at least, standing 6 feet in height. The proper construction of field-gates is generally very little attended to, if one may judge by those usually to be seen in the country. Some judicious remarks by Mr Slight, with appropriate figures, in reference to this much neglected subject, will be found below, and which I hope will tend to the diffusion of an improved form in this necessary portion of farm furniture.

(2565.) [The importance of field-gates, whether viewed as the means of security to crops and live stock, or as a portion of the perishable stock in trade of the farm, is an object deserving of grave consideration. Hitherto, it may safely be averred, that very little attention has been paid to principle in the construction of field-gates; and for the truth of this, we have only to look around us, where the eye of the constructive student will detect probably not more than one in a hundred of our field-gates that are not glaringly defective as pieces of extremeiy simple constructive carpentry. Defects in point of construction are not even confined to the field-gate; we find them in many of those gates of much higher pretensions, and where the hand of a master having been at work, we might be led to expect something like an approach to the true and simple principle; yet how seldom does the eye, experienced in directing those geometrical and dynamial principles, on which alone a just and permanent system of construction depends, light upon a form that satisfies its discriminating glance. We do, indeed, occasionally meet with examples wherein are evinced a correct knowledge of those arrangements of the parts, in a rectangular frame, whether of wood or iron, that at once secures permanency of form, and stability in the entire structure. In such cases we are sure to find, that the proprietor possesses not only an educated mind, but one that has not disdained to stoop to the drudgery of acquiring a competent knowledge of practical mechanics, leading him also to see the advantage of securing the assistance of mechanics, whose education qualifies them to perform their duties in accordance with the laws of that science. It may seem trivial to apply remarks of this kind to such a simple matter as the construction of a field-gate; but assuredly there is no part of our agricultural economy that points out so broadly the ignorance of our artizans in those important branches of their education, the elementary truths of geometry and mechanics.

(2566.) A gate, generally speaking, may be described as a rectangular frame; there are exceptions to this definition applicable to gates as a whole, but to *field-gates* there are none. A gate to be permanent, should be immutable or unchangeable in its form, a simple rectangular frame without upfillings, or even with upfillings, if they are placed at right angles to each other, is the most liable to change of any connected structure of frame-work. The triangle, on the other hand, is the most immutable or least liable to change; it is, in short, so long as the materials remain unchanged, perfectly immutable, but a gate in the form of a triangle would, in most cases, be very unserviceable, though a combination of triangles may produce the requisite figure for a serviceable gate. If then we take the rectangular frame so essential to a field-gate, and apply a bar in the position of the diagonal of the parallelogram, we immediately convert the original rectangular figure into two triangles, applied to each other by their hypothenuse, and this gives us the true elements of a properly constructed gate, all the other parts being subordinate to these, and adapted solely to the practical purposes of the gate as a defence or for ornament. In many cases depending upon the material employed, an opposite diagonal may be applied, dividing the gate into four triangles; but, in general, this is only necessary where flexible rods of iron are applied as the diagonals.

(2567.) In looking at the construction generally of field-gates, we observe traces of an incipient knowledge of the usefulness of a diagonal bar, but in very many cases it is applied with that uncertainty of purpose, that marks a doubtful and hesitating knowledge of the subject; we see it, in short, applied in all the possible positions that may be conceived to deviate from the one, true, and simple position,—extending from an angle to its opposite,—which is the simple and universal rule applicable to gates. Let it then be borne in mind, that the essentials of a field-gate, whether of wood or iron, are a rectangular frame, consisting of the heel and head-posts, and a top and bottom bar or rail, which four parts properly connected at the angles are rendered of an unchangeable figure, by the application of one or more diagonal bars, and these diagonals should in *no case* be applied short of the whole length between any two of the opposite angles. The upfilling, whether of rails or otherwise, as may be desired to attain any particular object, are mere accessories, and not in any way tending to the stability or durability of the fabric.

(2568.) The choice of the material for the diagonal is of some importance, and here the maxim of universal application in mechanical construction,—"*Tie with iron and strut with wood*," is especially applicable. In field-gates constructed entirely of wood, the diagonal should invariably be applied as a strut; that is to say, it should *rise* from the *foot* of the heel-post, and terminate at the *top* of the head-post. Placed in this position, the diagonal supports the head or swinging end of the gate, by its resistance to compression a duty which, from the *area* of its cross section being considerable, and hence capable of resisting lateral flexure, it is well adapted to perform; while, at the same time, the above sectional property gives it a broad terminal resistance, where it abuts upon the angles of the external frame. The same diagonal bar, if applied in the opposite position, and performing the duty of a tie or stay, its great sectional area would avail but little, for, though woody fibre is capable of resisting very considerable tension, larch-wood having, with equal sectional areas, a power of resistance to tension equal to about $\frac{1}{4}$ that of malleable iron of medium quality: and though this wooden tie might be found to possess in its aggregate section a cohesive force greatly beyond that of an iron bar applied in the same position, for the wooden bar would have a sectional area at least twelve times greater than could be requisite for an iron tie, thus yielding an aggregate force double of the iron; yet as the wooden *tie* must depend for its connection in the structure, upon nails or bolts only, its ultimate power of resistance to tension depends not on its own sectional area, but on that of the nails or bolts by which it is fastened, and these, again, may be very greatly reduced by the rending of the extremities of the wooden tie.

(2569.) The advantages of iron as a tie, and its disadvantages as a strut, are just the converse of the foregoing. From the smallness of sectional area requisite in an iron-bar applied in this construction, as compared with its length, it is not capable of withstanding compression even to the smallest extent without suffering lateral flexure, and this defect unfits it entirely for the purpose of a strut. Applied as a tie the iron bar is perfect; the cohesion is such that a very small sectional area is sufficient for the purpose under consideration; thus, a rod of $\frac{1}{2}$ inch square even of inferior iron will bear a tension with safety, of two or three tons, while the best quality will bear six tons; a rod of half this sectional area may, therefore, be held as sufficient for the diagonal *tie* of a gate, and as it can be fixed by bolts without risk of fracturing its ends, as in the case of wood, its application in principle becomes as perfect as it is possible to approach. But though this construction is in principle completely supported by the *tie* from any force tending to depress the head-post of the gate, there is yet a defect in practice; for a gate is liable, from various causes, to be forced upward at the head-post. We have seen that a slender and flexible iron-rod cannot resist compression without flexure, and, therefore, a gate with only one iron diagonal tie will still be practically imperfect, and it becomes necessary to apply an antagonist placed in the position of a strut, but virtually performing the duties of a *tie* arising from the antagonist effects of the two, the tendency of the one being to hold the head of the gate up, while the other exerts an equal force to keep it down, whereby a perfect equilibrium is preserved within the structure.

(2570.) In treating of the *practical* construction of field-gates, it is, perhaps, unnecessary to dwell upon the strains that occur in the individual horizontal bars, because, if the principles inculcated above are attended to, all cross strains in the principal joinings are avoided; and, except when any extraneous force is applied, the strains are resolved by construction into those of direct compression or of tension. Thus, in fig. 441, which may represent the elements of a fly-gate, applicable to a drive or thoroughfare, and opening either way; if we take

Fig. 441.

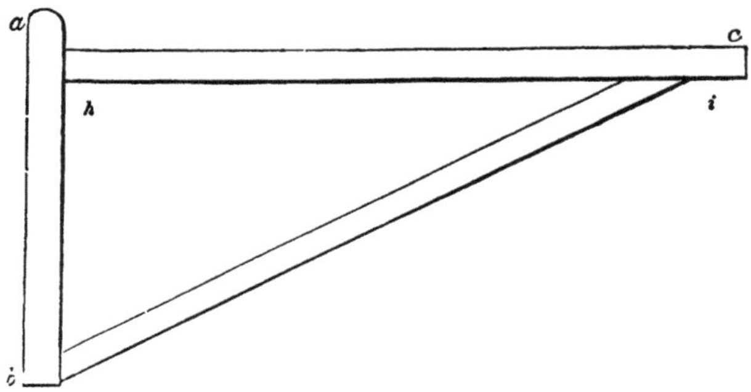

THE ELEMENTS OF THE TRUSS IN A WOODEN FIELD-GATE.

the heel-post *a b* and the bar *a c* alone, and hinged in any manner at *a* and *b*; and if the bar *a c* is 10 feet long, the breadth of the heel post at *a* 5 inches, and the parts being joined by mortise and tenon: suppose then a load applied to the bar at *c*, the mechanical effect of the load would be a cross strain at *h* tending to break the bar directly across with a force equal to 24 times the load, exclusive of the effect of the bar itself, the bar *a c* forming here a lever of the first order, whose arm *h c* is to the arm *h a*, or the breadth of the heel-post, as 24 to 1. By increasing the number of bars we do not alter the *total* effect of the load, but simply divide it equally over the superinduced bars, supposing them to be connected with a head-post at *i*. But reverting again to the single bar *a c*, and applying to it the diagonal *b c*, we have now the triangular figure *a b c*, the parts of which being firmly connected at their points of junction, the form becomes unchangeable, and the effect of the load at *c* is instantly altered. The effect of the load at *c* is now resolvable by the parallelogram of forces into two others, the one of tension on the bar *a c*, the other of compression on the diagonal *b c*, and taking the length *a b* as a representative of the gravitating or direct effect of that load, then the tension on *a c* will be to the absolute load as the length of *a b* is to that of *a c*, and the compression on the diagonal will be as *b c* to *b a*. In this example the hinges of the gate are not shewn, but in actual practice the tails of the upper hinge stretch along part of the top bar, binding it securely to the heel-post, while the bottom hinge may be made to form an abutment to the foot of the diagonal.

(2571.) The practical insertion of the diagonal is also of some importance. Fig. 442 exhibits one mode of performing this, where *a b* is the strut half lapped upon the heel-post at bottom and upon the head-post at top; it is not an elegant mode of insertion, but it is efficient, in so far that while it acts as a strut, it binds the top of the head-post to the top-bar, and prevents it from flying off by the thrust of the diagonal.

(2572.) Fig. 443 exhibits another mode of insertion of the diagonal, which is perhaps preferable to the former; here the head of the diagonal is attached to the top bar, half lapped upon it at *b*, and notched into the lower edge as an abutment, forming a very efficient strut, the tension strain being brought entirely upon the top-rail, which is always securely tied to the heel-post by the tails of

Fig. 442.

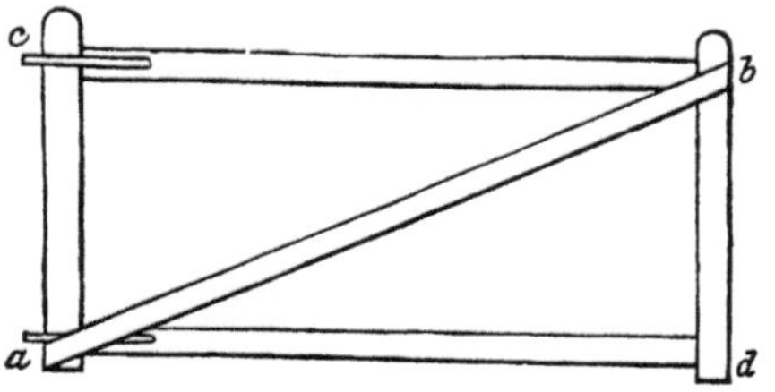

Fig. 443.

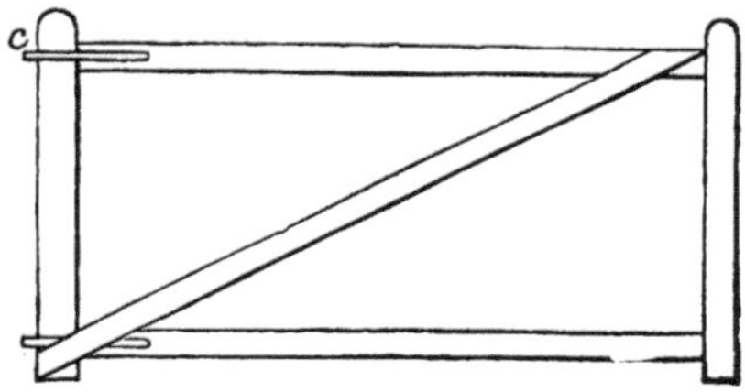

EXAMPLES OF THE APPLICATION OF THE DIAGONAL.

the hinge at *c*. These examples are applicable to field-gates where the bars are always much thinner than the heel and head-posts, the latter being generally about 2½ to 3 inches thick, and the former only 1½ inch. In such cases the diagonal is slightly notched upon all the bars, and deeper upon the heel and head-posts, where the method, fig. 442, is adopted; but in no case should the bars be notched except in the edge of the top bar, where it receives the head of the diagonal.

(2573.) I have hitherto alluded only to the simplest efficient form of wooden field-gates, in the construction of which cheapness is always an object, and I have, in order to avoid confusion of ideas, restricted the description to the essential parts, the number of bars, or other means of rendering the gate a sufficient fence, is left to be filled up at discretion. In making these upfillings the maker should studiously keep in mind that no curved bars or timbers, of any description, should enter into the construction either for ornament or ostensibly for use.

(2574.) In the construction of wooden gates for drives or approaches, where utility is still the chief object, strict attention should still be paid to the principles of construction, but a little more latitude may be admissible in point of finish and expense. For such purposes, the rails and posts of the gate should be all of one thickness, or at most, the only difference should be a gradual diminution in thickness towards the head, to lessen the effect of gravity on the hinges and gate-post; and for pleasant effect, there should not be more than three horizontal rails, with two diagonals; and if it is necessary to have a closer upfilling, it should be of an upright light ballustrade form. Fig. 444 is an example of this form of gate, which I adopted 26 years ago, and the originals then constructed are still good and serviceable, but it is considerably more expensive than the common field-gate. The heel-post *a b* is 5 inches broad and 3 inches thick, while the head-post *c d* may be reduced to 2¼ inches if thought advisable. The top and bottom bars are formed with abutment pieces at both ends, which are 5 inches broad, the intermediate parts being reduced to 3½ inches; the middle bar has the same breadth, but is made up in the middle with corresponding abutments, and the two diagonals, of 2 inches in breadth, are inserted in four pieces, exactly fitted between the abutments of the bars. The hinges are of the common double-tailed form, binding the top and bottom rails firmly to the heel-post, and the gate may be hung upon pillars of stone, or of wood well secured. The best ballustrade for a gate of this kind is rods of iron ½ inch diameter, as in the figure, passed up through the bottom and middle bars and the diagonals, the holes for these being easily bored with an auger after the gate is formed; but a simple and

Fig. 444.

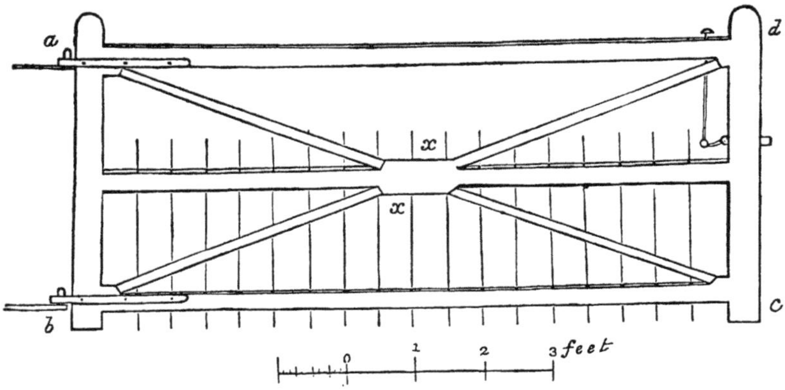

WOODEN GATE SUITED FOR THE APPROACH TO A VILLA.

cheaper ballustrade is formed of light wooden spars sunk into the bars and diagonals.

(2575.) For all wooden gates, the method of bracing with light iron diagonals is to be preferred to wooden struts, but to be effective two diagonals must always be applied. In some cases, they may pass from one angle to its opposite in one length, but in others it is necessary to apply them in four pieces, the connection at the centre of the gate being effected either by a ring of iron, in which the four ends are screwed, or by bolting the palmated ends of the four parts, two and two together, through the middle bar, as applied at *x x* fig. 444, one bolt securing the four ends, and in either case the rods pass through the top and bottom of the heel and head-posts of the gate, and are there secured by screw-nuts. It is obvious that iron diagonals would apply in this manner to the gate, fig. 444, instead of wooden braces.

(2576.) Field-gates of wood, from the cheapness with which they may be furnished when imperfectly constructed, have been long in use, and in many localities continue to be more extensively used than any other; but *malleable iron* is gradually taking the place of wood, and, from its greater durability, may be expected to supersede wood entirely. In the construction of malleable iron gates, we as frequently find malformations as in those of wood, such as placing all the bars on edge except the heel and head-post, misplacing the diagonal, if single, and not unfrequently applying bars variously formed in curves and fanciful figures, to serve the purpose of the diagonals. The field-gate maker should be instructed to hold steadily in view, that there is but *one position* and form for that member of the structure that can be fully efficient, and these are, the straight bar extending from the upper angle at the heel to its opposite angle at the head post; and, if the materials of the gate are light, to apply an antagonist diagonal crossing the first. In framing the gate, also, the top and bottom bars should be set flat-ways, to enable the structure to resist lateral strain from animals rubbing or pushing against it.

(2577.) One of the latest improvements in iron field-gates is the introduction of *angle-iron*, now so extensively used in boiler-making, ship-building, and other purposes. In the application of the angle-iron to the construction of gates, the fabric acquires the rigidity of a massive wooden gate, with all the tenacity and strength of the iron, while its weight is little more than that of wood. Fig. 445 is a form of gate of this construction, which I have lately adopted, with some alteration, from a design of Mr William Dunlop, Edinburgh, and is manufactured by James Slight & Co., Edinburgh. The external form is composed of four bars of angle-iron, measuring 1½ inch on each side, and to give security to the joinings at the four angles of the truss, the ends of the bars are rivetted upon cast-iron corner-plates, those of the heel-post *a b* being formed with strong projecting pivots, by which the gate is hinged. Any number of interior bars may be applied to suit the objects of the gate. The figure exhibits the arrangement adapted to retain sheep and lambs. The diagonal *b d* is contrary to the general rule, for it is apparently a strut, but being a bar of angle-iron, of the same breadth as before, it possesses the stiffness of wood, to resist lateral strains—and is hence properly adapted for a strut; to render the bracing complete, the antagonist diagonal *a c* is applied, and this, acting as a tie, is only a flat bar 1 inch by ⅛ inch. The external frame is thus rendered unchangeable in figure by any force that may be applied to the head-post in a vertical direction, either upward or downward, short of what will fracture the gate; and the point *e*, where the diagonals cross each other at the centre of the gate, becomes also immovable in the plane of the truss; hence the perpendicular bar *f e g*, being rivetted to the diagonal at *e*, acquires the same property, and by attaching all the horizontal bars to *f e g* at their several crossings, each of them is rendered permanent in its position at that point, and no force short of break-

Fig. 445.

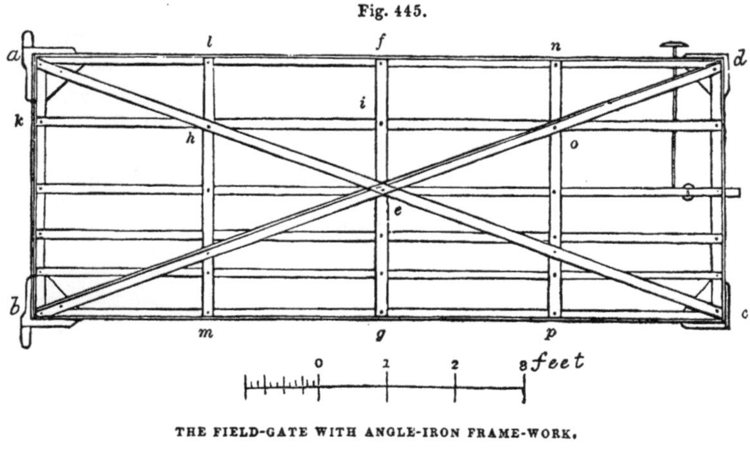

THE FIELD-GATE WITH ANGLE-IRON FRAME-WORK.

ing down the gate can bend any of the parts upward or downward in the direction of the bar *f e g*, so long as this last remains attached to the crossing of the diagonals at *e*. In order to give farther support to the horizontal bars by the *principle* of construction, we have only to take a point where a diagonal crosses a bar, as at *h*, forming the opposite triangles *e h i* and *a h k*, which, when the bar and diagonal have been connected, become also immutable, and the perpendicular bar *l h m* being secured to the point *h*, and again to the different bars at their intersection with *l h m*, the whole are again rendered immovable as in the middle. The support given to the horizontal bars in the line *l h m*, would have been still more complete if there had been only one intermediate bar below the middle one, as the three parts would then have met in one point, as they do at *h*, but two bars are introduced to render the gate fencible for sheep of all ages. The perpendicular bar *n o p* is applied on the same principle as laid down for *l h m*, the point of support in this case being *o*.

(2578.) In this construction of gates, the greatest possible amount of mutual support among the parts is obtained with a given quantity of materials; hence gates of this construction may be made lighter than any other form, where iron is the material employed, and yet have a greater amount of strength. In this example, the dimensions of the angle-iron are $1\frac{1}{2}$ inch each way, and about $\frac{1}{4}$ inch thick; all the other parts are $1\frac{1}{4}$ inch broad by $\frac{1}{8}$ inch thick, the cast-iron corner plates being, of course, stouter, and the entire weight of the gate is 112 lb. It may be of use to those who make iron-gates, but who have not taken time to study the first principles of their construction, to notice this further remark. Any number whatever of additional upright bars to those shewn in fig. 445 would add strength or support to the horizontal bars only on the principle of superposition, or adding bar to bar, without the advantages which arise from the principle of unchangeableness in the triangle when applied in the construction of frame-work, whether in a simple field-gate, or in the highest branches of constructive carpentry in wood or in iron. In the one case, the stress on the parts continue to act at right angles to the bars, the direction in which they are weakest, while in the other,—the principle of throwing the frame into triangles,—the whole stress is thrown upon one or more parts in the direction of their length, in which position all bars and beams are strongest.

(2579.) An example of an ingenious construction of iron-gate is given in Parnell's work on road-making, which has been improved on by Mr Buist, now of Bombay, and described in the Prize Essays of the Highland and Agricultural Society These gates consist of a wrought iron external frame, which is supported by a very perfect system of bracing, with diagonal ties of iron wire, and filled up in a variety of forms with the same material. Fig. 446 represents Mr Buist's gate with the fundamental braces and ties, which he thus describes:

Fig. 446.

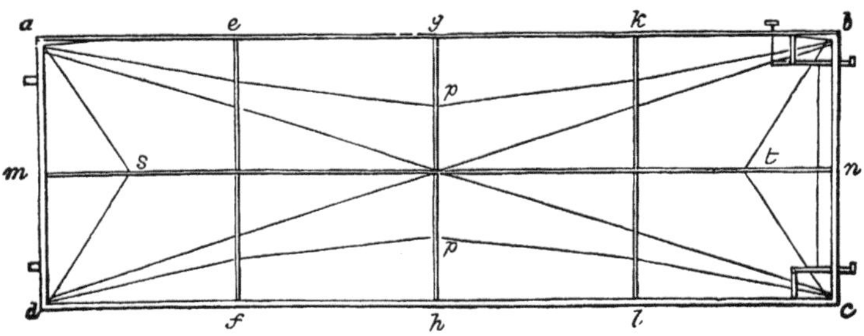

THE ELEMENTS OF THE BRACING OF A WIRE FIELD-GATE.

" The framing *a b c d* is fashioned like that of an ordinary gate; *e f*, *g h*, and *k l* are three light slips of iron parallel to the ends of the gate, and rivetted to the upper and lower rails; *a p b* is a wire about the thickness of a goose quill, fastened by a rivet at one end, and a screw and nut at the other; it passes through holes in the slips *e f*, *g h*, and *k l*, and serves as a brace to support the bar *a b*. In the same manner *d p c* serves as a brace to *d c*, while the two sides of the gate being coupled together by the slips *e f*, *g h*, and *k l*, the lower and upper rails have severally the benefit of both braces. The diagonals *a c* and *d b* keep the frame in shape, while *a s d* and *b t c* are braces to *a d* and *b c*, by means of the light bar *m n*. It will be seen that all the wires and straps which act as fills-up, are either braces or supports, so that nothing can be more stiff than the gate thus completed. It weighs about 80 lb., and costs L.1, 8s. Its dimensions are 9 feet by 3½ feet, but may be made of any size, the price varying in proportion. It may be observed, that a gate with one bolt, when shut suddenly, vibrates for some time at the fore-foot; this is obviated by two bolts coupled together, as in the figure, near *b* and *c*, and acting simultaneously. It is also convenient for gates opening into policy grounds, getting bolted when thrown back by means of a short stump driven into the ground, with a catch at the height of the lower bolt *c*."

(2580.) Fig. 447 represents a gate of this construction filled up with wires in the form of rays from a centre. "The horizontal bar *m n*, and the braces *a s d*, and *b t c*, of fig. 446, are here omitted as superfluous. The rays consist of wires of the same thickness as before, and about 9 inches from each other.

Fig. 447.

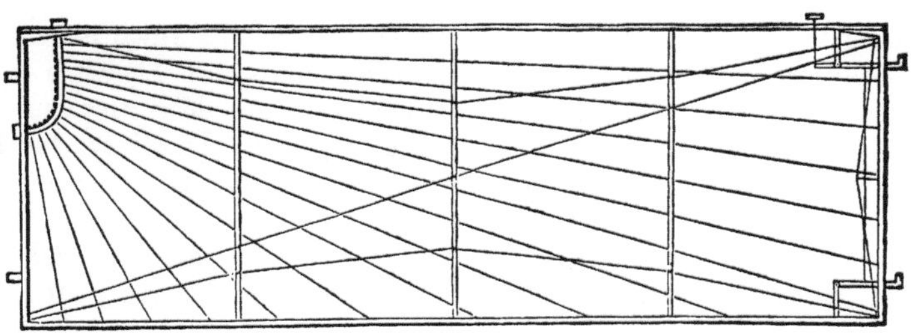

THE WIRE FIELD-GATE WITH RAYED UPFILLINGS.

Their lower extremities are upset as nail-heads or rivets, and their upper ends are fastened with a nut and screw. For the admission of the screws a strong iron arch is placed in the corner of the gate, and fastened at each end with screw nuts; its range down the heel post, as compared with that along the upper rail, should be as the length of the gate is to its height, which, in this example, is 18 inches down the heel-post, and 6 inches along the upper rail. A gate of this form, and 9 feet by 3½ feet, costs L.1, 15s."*

(2581.) The wire-gates above described are admirable examples of the principles of trussed frames, and for gates. So far as that principle gives them firmness and support, they can hardly be excelled; but there is one defect attendant upon the wire upfilling, its too great tenuity, which renders the wires liable to derangement on being loaded with any cross-strain, such as a person attempting to climb over the gate, and setting foot on the wires. A diagonal wire un-

* Prize Essays of the Highland and Agricultural Society, vol. xiv. p. 612-3.

dergoing such treatment will be liable to stretch, and thereby lose its effect. Could such accidents be effectually guarded against, these gates might be regarded as almost perfect. In regard to the expense of the gates first described, the common wooden field-gates may be considered to range from 15s. to 25s. The gate fig. 444 about L.1, 15s.; and the selling price of the angle-iron-gate, fig. 445, is about L.1, 5s.

(2582.) *Gate-pillars and fastenings.* For field-gates where hedges form the enclosures, the simplest kind of gate-posts are those of larch or common Scots fir. They may be set in their natural state, or peeled or dressed in any form to the taste and means of the farmer; but in whatever form the body of the post may be, the top should be cut off either in form of a cone or pyramid, or a hemisphere, to prevent the lodgment of water on it. The simplest means of fastening wooden gate-posts is by beating or *pinning* in the earth around them as described (2564.), but the most effectual and permanent mode is to form a pit of at least 2½ feet square, and of the same depth, and the post being set in it, the pit is filled up with rubble masonry in mortar packed firmly, and grouted round the post. This is an expensive mode, but is more permanent, and a greater durability of the timber is effected by contact with the lime.

(2583.) In the Isle of Man, though its agriculturists cannot boast more than ourselves of the perfection of their gates, nature has supplied them with an admirable material for gate-posts. The transition-slate with which the island abounds is found in one locality, Spanish-head, of a nature that serves all the purposes of beams and planks of wood, and of this many indeed, most of the gate-posts are formed. In Scotland we have now in abundance a material equally suitable for the same purpose, in the Caithness pavement, which is now applied for gate-pillars, and for which no material can surpass it. They are built into the ground in the same manner as wooden posts, and look very much like a plank of wood, but possessing strength, and especially durability, far beyond wood. The crook-hinges are either fixed into them by passing through them as bolts through a jumper-hole, or they may be put in in the form of a hoop, and secured with wedges. For field-gates built pillars of stone are seldom used, and only when the enclosures are formed of stone-dykes, and these forms are so various, that it is out of place to describe them here.

(2584.) The *hanging of gates* has already been slightly adverted to; the crook and band hinges are those chiefly employed, and of them it is only necessary to say, that the hinge should be always double tailed. The fastening of gates are still more various, and many of them so simple and efficient, that it were difficult to point out which is the superior.—J. S.]

(2585.) While treating of fences and iron gates, I may mention a simple plan by which grass lawns, or the small plats usually formed in front of farm-houses, may be kept short without the use of scythes, or incurring much trouble and expense. Let a bught be made of wire hurdles of the form of the gate fig. 446, and, that it may be movable, let the hurdles be fixed together. If the plat is very small, one enclosure composed of 4 hurdles, forming a square of 9 feet, will suffice; if it is larger, any number of such bughts may be employed. The use of these enclosures is to fence 1 or 2 sheep within each, so that when they have eaten down the enclosed grass, the bughts may be moved forward, or backward, or sideways to a fresh division of grass, until the whole grass plat is eaten bare. The sheep may then be removed for a time till the grass grows, and again brought back. Or, since such bughts would be rather ornamental as otherwise, only as many sheep may be enclosed within them as will just suffice to keep the grass in good order, and where they may constantly grace the lawn. A few Leicester tups or tup-lambs may thus be kept separate all the summer from the rest of the flock, and be pretty objects on the lawn, while, at the same time, they may be doing good in keeping the grass in order, and top-dressing it.

---

63. OF THE WEANING OF CALVES, OF BULLS AND OF THE GRAZING OF CATTLE TILL WINTER.

" Dissolv'd in pleasure, crown'd with buds of May,
They, for a time, in their fat pastures play."

FLAVEL.

(2586.) The cattle which were accommodated in their respective places in the steading at the beginning of winter, continue to be treated through the spring months in the same manner as is pointed out in the 34th section, page 109 of this volume. In fact, their treatment is throughout the same until turned out to grass, which is usually some time in May; unless variety of food may be regarded as difference of treatment. It is found that cattle in a state of confinement in a steading, thrive better on a variety than on the same food; and yet when on grass they require no variety of food, and thrive the better the longer they are kept upon it. except that a change of pasture is desirable when it becomes bare. Grass is thus evidently the natural food of the ox, and his anatomical structure is peculiarly suited for it. Whatever kind of food he is supplied with in winter partakes of an artificial character, and that being only a succedaneum for grass when it cannot be obtained, the food he receives should be made as palatable to him as circumstances will allow, whether by variety or in superior quality.

(2587.) The grass should be ready to afford a bite for cattle whenever the turnips are all exhausted; at which time the cattle will be found to be in this state:—The 2-year-olds, now 3-year-olds, will be full fat, and ready to be disposed of to the butcher or dealer. The 1-year-olds and calves will have grown much to bone, and their condition will have improved in proportion to the quantity of nourishing food they have received during the feeding season. The cows will all have calved, or should all have calved, for a May calf is too late to bring up and form a part of the herd; they will be in milk, and in fresh condition to put to grass. The early calves will have just been weaned, and in excellent order to put on grass; and the remainder will be in rapid progress towards weaning. These cattle, in their respective states are treated in summer in a different manner from each other, and different from the way they were treated in winter, and they are treated thus.

(2588.) The *fat* cattle are seldom put on grass, being disposed of to the butcher or dealer, either at home or in markets held for the purpose. Breeders of their own stock seldom dispose of their fat cattle until the turnips are nearly consumed, being anxious to keep them as long as possible, for the sake of the excellent manure which the turnips afford. Those who purchase oxen to fatten usually buy a larger lot than can be maintained on full turnips till the grass is ready, in order to dispose of them before the season arrives when fat oxen are usually sold. Such fatteners of cattle dispose of their lots in whole, or in part, from the end of the year to April, whenever the market offers them the most profit. As you are supposed to breed and feed your own stock, you are supposed to retain your fat oxen to the end of May, when Swedish turnips are still fresh, sweet, and nutritious, though perhaps a little shrivelled and even sprouted. There are then plenty of opportunities to dispose of them, such as the weekly markets of large towns, fairs for fat stock, butchers from the neighbouring towns, dealers from a distance, and shipment by sea to London, under consignment to a respectable salesman. Before disposing of your oxen, however, you should be acquainted with their weight, otherwise you will not know what to ask for them at the current market prices. Experience will teach you to estimate the weight of an ox by the eye, not the actual live-weight of the animal, but the weight of beef it will yield after deducting the weight of what is called the *offal*, that is, the weight of every part but the beef and bones, such as the skin, head, entrails, and loose tallow. It is understood that the farmer is entitled to the value of the entire weight of the beef at the current prices, and these are always understood to relate the prices of the meat alone; and that the profit of the purchaser consists of the value of the offal, subject to the deduction of incidental expenses before the animals are slaughtered. The available parts of the offals are supposed to be equal to the value of $\frac{1}{5}$ of that of the animal, so that an ox of L.20 value should leave the purchaser L.4 for profit, subject to incidental expenses. But if the buyer is a good judge of cattle, and knows their exact available weight of beef, which he will know to within a very small fraction, he may make such a bargain with the farmer as will not only cover his expenses, but leave a profit on the carcass over and above the value of the offals; and as dealers are constantly in practice, they generally contrive to make pretty good bargains for themselves, in as far as their judgment is concerned, at the time the bargain is made, but their purchases are subject to fluctuation of price which may fall without an adequate cause, and occasion them serious loss; or they may rise and leave them a handsomer profit than was anticipated, so that the profession of dealers resolves itself into a speculation, which, if prolonged in its issue, is involved in uncertainty. Dealers formerly made large fortunes when breeders and feeders of stock could estimate the value of even their own stock but indifferently; but now-a-days I believe few dealers make fortunes, owing partly to competition, and partly to greater skill on the part of the farmer, and partly to the great expense incurred in driving their purchases to suitable markets, and in leading a roving sort of life.

(2589.) But assistance may be afforded you in ascertaining the weight of your cattle until you are better taught by experience, and this consists in *measuring their bulk, or in weighing their live-weight.* The live-weight of cattle is easily ascertained by placing the animals upon a steel-yard, of which convenient forms have been recommended for the special purpose; but I am not aware that any of those which have been specially recommended are as accurate or convenient as might be. The rule to determine the quantity of beef by the live-weight is to multiply the gross weight by .605 of a decimal, if the ox is ripe fat, but if not so, by .5 of a decimal; that is to say, that the offals of an ox in ordinary condition weigh about as much as its beef and bones. An ox should not be weighed immediately after it has fed, as it will weigh too heavy, but after it has chewed the cud, and is ready again to feed. Ascertaining the weight by measurement is a more convenient method than by weighing; and when the measurement is properly taken, and the ox of an ordinary size, it is about as accurate, though every person cannot measure an ox, that process requiring judgment to do it properly and accurately. Suppose fig. 448 to represent an ox whose weight, sinking offals, is desired to be ascertained by measurement. The mode is, measure with a tape line from the top of the shoulder *a* to the tail-head *c*, and mark this for the *length*, then measure round the body at *f*, immediately behind the shoulder, and mark this for the *girth*, and on consulting the tables calculated for the purpose of affording the results, the weight of beef will be found. Upon what principle this rule

Fig. 448.

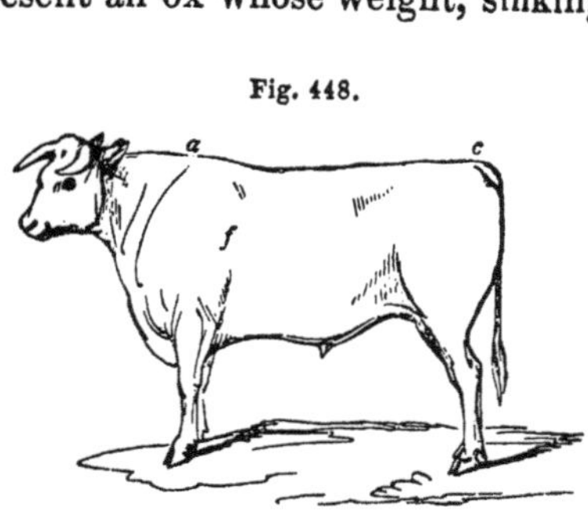

HOW TO MEASURE A FAT OX IN ORDER TO ASCERTAIN HIS WEIGHT, SINKING THE OFFALS

for measurement is founded I cannot say, and suspect that it is entirely empirical. The rules by which the tables are calculated seem to be these two, namely ;—multiply the *square* of the *girth* in *inches* by the *length* in *inches*, and divide the product by 7344, and the quotient is the weight in imperial stones. Or, square the girth in *feet* and multiply it by the length in *feet*, and multiply again by the decimal .238, and the sum is the weight in imperial stones. For example : Suppose the girth is 7 feet, or 84 inches, and the length 5 feet, or 60 inches, the weight of beef in imperial stones, according to Strachan's tables, which are the most recent, is 57 st. 10 lb. by the former rule, and by the latter it is 56 st. 4 lb. These results shew that there is no fixed principle upon which either rule is founded.

(2590.) This seems a very simple process, and so it would always be, were the form of the ox always perfect, which is very seldom the case, the fore and hind quarters, instead of being equal, being more frequently otherwise ; and were the condition always the same. Therefore, the judgment is called into exercise to make allowance for those differences. When the fore-quarter seems heavier of the two, the line should be stretched nearer the head than *a*, the exact top of the shoulder ; and so, in like manner, when the hind-quarter seems heavier than the fore, the line should be stretched beyond *c*. In regard to the girth, it is no uncommon fault in an ox to have the carcass gripped small behind the shoulder, so that the exact girth of such a shoulder would be below the truth. It is very rare to find the girth too much filled out. You thus see that judgment is required to apply the tape-line to an ox ; and as an illustration of the practical effects of misapplying the tape-line, I may state, that 1 inch only added to the girth and length given in the above example, makes a difference in the above weight of upwards of 2 st. The addition of 1 inch in the length is a mistake which may easily be made by the ox merely standing with his head down and back up ; and a similar position may as easily cause an error in the girth. Experience alone can make you proficient in measuring. I knew a steward in Berwickshire who had so much practice in measuring, and so frequent opportunities of verifying his measurements, that he measured any ordinary sized ox, whether fat or half lean, from 1 st. to ½ st. of its true weight.

(2591.) To an ox from 40 to 70 stones the rule of measurement applies pretty near, if it is applied with judgment, and it is but fair to the principle that it should be so applied ; but in weights both below and above these figures the rule is usually at fault. I have no instances of error in very small weights to adduce, but many of large ones, and shall adduce only two. A short-horn white ox belonging to Mr Boswell of Kingcausie was exhibited at the Highland and Agricultural Society's Show at Aberdeen in October 1834. Its measurement was 9 feet 3 inches in girth, and 6 feet 2 inches in length. According to Renton's tables, the farmer of Dykegatehead, in Berwickshire, the weight should be 126 st. 9 lb.; Strachan's tables, 124 st. 2 lb.; Ainslie's tables, 122 st.; and Stewart's tables, 117 st. The actual weight of beef yielded by it was 136 st. 10 lb., after being slaughtered by Deacon Sparks of Aberdeen, being 10 st. 1 lb. more than the heaviest weight indicated by all these tables. This difference of weight, at 6s. per stone, would make a loss to the feeder of L.3 upon a single animal! A 2-year-old light red ox, also belonging to Mr Boswell, and exhibited at the same time, measured 7 feet 10 inches in girth, and 5 feet in length. Stewart's tables give a weight of 80 st. 3 lb.; Renton's, 73 st. 8 lb.; Ainslie's, 73 st.; and Strachan's, 72 st. The actual weight obtained by Deacon Sparks was 89 st. 10 lb., shewing the enormous difference of 17 st. 10 lb. between the real weight obtained and the lowest weight indicated by Strachan's tables, and which, estimated at 6s. per stone, would cause a loss to the feeder of L.5 : 6 : 3. The rule derived from live-weight is also liable to err, or when applied to oxen of inordinate weights. Thus, Lord Kintore's black ox, 7 years off, exhibited on the above occasion at Aberdeen, weighed 28 cwt., or 224 stones. By the rule of multiplying the live-weight by the decimal .605, the dead weight should have been 135 st. 7 lb.; but when the ox was slaughtered by Mr Roger, Crown Street, Aberdeen, his 4 quarters weighed 173 st. 4 lb., or 37 st. 11 lb. more than anticipated, which, at 6s. per stone, made him worth more by L.11 : 6 : 8! On the other hand, a small spayed heifer, belonging to Mr Boswell, weighed 88 st. live-weight, and should have yielded 53 st. 3 lb., but her 4 quarters only weighed 49 stones.* In consequence of such discrepancies, it would certainly be desirable were means used to render the rules of measurement, as well as of weight, more correct than they are; and I see no way of effecting this desirable end than by collecting data, by instituting a set of experiments in different parts of the country, to measure and weigh every animal before they are slaughtered, be it ox, sheep, or pig, be it large or small, for a given period, say twelve months, and see how far the results coincide or differ from the actual weight of meat afforded by the 4 quarters.

(2592.) The fat cattle being disposed of, the pasture should be judiciously distributed amongst the remaining stock; and first, as to the *cows*,

* See paper by me on this subject in the Quarterly Journal of Agriculture, vol. v. p. 612-17.

they should be supplied with the best pasture, the object of keeping them being to breed calves and to afford plenty of milk to bring them up, the more milk they yield the better will the calves prove, and the more profitable they will themselves prove after the calves are weaned. Cows in summer are treated in a different manner by different people, some taking them into the byre at night, and even at all times of milking, whereas others allow them to lie out all night, and milk them in the field. Whichever mode is adopted, it should be kept in mind that cows are peculiarly susceptible of sudden changes of temperature, especially from heat to cold and from drought to rain, so that whenever cold or rain, or both together, which is the most common circumstance, occur, they should be brought into the byre. For some time after they are put out to grass they should be brought into the byre at night, where they are milked, and again in the morning before they are let out to the field, and milked in the field at mid-day. After the nights become warm, I have found it conducive to health, and it is both a rational and a natural plan, to allow them to lie out in the field all night, and to milk them there at stated ours, three times every day, the shepherd or cattle-man taking it as a part of his duty to bring them to a certain spot of the field to be milked, and which is usually named the *milking loan*. This mode of allowing them to lie out always in a sheltered field, no doubt imposes a good deal of labour on the dairy-maid and her assistant in carrying the milk to the dairy after the calves have been weaned; but I am persuaded it is an excellent system for the health of the cows. Under it, cows rise from their lair at day-break and feed while the dew is still on the grass, and by the time of milking arrives, say 6 o'clock, they are already partially filled with food, and stand contented, chewing the cud, while the milking proceeds. They then roam and fill themselves, and by 9 o'clock, lie down in a shady part of the field and chew their cud until milking time arrives at mid-day, when they are again brought to the loan and milked. Again they roam for food, and, when the afternoon is hot, will stand in the coolest part of the field whisking away the flies with their tail and ears. The evening milking takes place about 7, after which they feed industriously, and take up their lair about sunset, and from which they rouse themselves in the morning before being milked. Some people are apprehensive that cows must injure themselves by eating grass which is wet with dew in the morning; but it is a fact, which I believe is not so sufficiently known as it should be, that bedewed grass before sunrise, and grass after it is dried by the sun, are alike uninjurious to animals, and it is only when the dew is in the act of being evaporated, immediately after sunrise, that grass proves injurious to animals. Why it should be injurious at that particular state I do not precisely know, but imagine it to be so, because the grass then becomes suddenly cold by evaporation of the dew. When cows lie out, they have nearly filled themselves by the time the dew is evaporated, and therefore feel less impelled to eat much of the grass in the dangerous state; whereas, cows that are housed in the byre all night, are usually milked about sunrise, and put out to grass just at the very time the dew on it is being evaporated, and is, of course, in the most dangerous state; and it is, moreover, just then that cows feel the greatest hunger, and eat most grass. The effect of this state of grass is to produce *hoven* (1514.), and especially on aftermath in autumn. Be the cause of injury what it may, it is certain that cows let out to grass from the byre are far more liable to be affected with hoven on wet grass than those which lie out all night, and especially on new grass. The lying out at night, too, saves the trouble of providing supper for the cows, which they must have when housed in the byre. But *whenever* the weather becomes cold and wet, cows should be brought into the byre at night, and supplied with supper, such as cut aftermath or tares.

(2593.) The *weaning of calves* should not exceed one month after the cows have been on grass, that is, by the end of June, for a calf later weaned than that period, has been too late brought into the world to be worthy of belonging to the standing stock of a farm. As cows increase in gift of milk after the grass has fairly passed through them, the late calves should have as large an allowance of new milk, three times a day, as the quantity obtained will allow, reserving a little for the use of the house. The elder calves are off sweet-milk by the time the cows go to grass, and have received lythax (2112.) with a little skimmed milk amongst it, cut Swedish turnips and hay, until the grass is ready. These calves having been in the court *k*, fig. 3, Plate III. of the steading, for some time, may be put on grass when the cows are let out. The most convenient first grass-field for calves is a contiguous paddock, from which they should be brought into the court for a few nights and receive turnips and hay, until the grass has safely passed through them, and the weather prove mild and dry for them to lie out all night in the paddock. The youngest calves should now leave their cribs R, and pass a few days in the court *k* until they become accustomed to the air and sun, when they also may be put into the paddock during the day, and there supplied with their diets of milk, and brought into the same court at night until they are able to lie out all night. In weaning the youngest calves, the milk should be gradually taken from them until they take with the grass, upon which they must then entirely depend. A little after all this has happened, say by the middle of July, the pasture in the paddock will become rather bare, and the whole lot of calves should then be taken to good pasture, where they will have a full bite, for nothing can be more injurious for calves than to place them on bare pasture to fall away in condition immediately after weaning, and which they will assuredly rapidly do, and from which it will be very difficult to recover them all summer. Calves may be grazed amongst cows, or in a field by themselves. In their peregrinations through the field, it will be found that those which have been brought up and weaned together, will keep together for the greater part of the season. The older calves may be 4 months old before they are weaned; but as the season of grass approaches, the younger ones may be weaned at an earlier age, being seldom indulged with milk for more than 13 weeks, but it should never be forgotten, that the first month's nourishment to a calf is of much greater importance to its future growth and health, than any period beyond 13 weeks, supported on a stinted allowance of inferior milk. There are parts of Ireland where calves are brought up on butter-milk and gruel, after the first 8 days they have received sweet milk, and it is boasted that they thrive well on that beverage. This is possible, but they will thrive far better on sweet-milk.

(2594.) *Bull*-calves should have good milk every day until the grass is able to support them, in order to strengthen their bone, and maintain their condition. When a number are brought up together, they should be grazed by themselves on the best grass the farm affords, or they may go along with the cows, or they may go along with the ox-calves while the quey-calves are with the cows. Under any arrangement, they should not be allowed to accompany the quey-calves. I knew an instance of a quey calf being stinted at so early an age, as to bear a calf at 15 months old; and I knew another quey-calf, one of my own, that was so injured by a young bull-calf, that she was thereafter rendered incapable of impregnation, though her season recurred periodically. To avoid every such vexatious casualty, it is far better to keep apart young animals of different sexes that are capable of breeding. When there is only a single bull-calf, he may go with the cows, or with the young oxen.

(2595.) The young 1-year-old bulls should now be furnished with a *ring* in their nose. This instrument is useful not only in leading the animal, but, being constantly in use, in keeping his temper in subjection. I have no doubt whatever that such a ring affords the most complete command over the most furious bull. In case of a bull becoming more irritative and troublesome as he advances in years, which many bulls are inclined to be, the ring furnishes the means of curbing him at once, when it would otherwise be impossible to get hold of his nose. It affords also an easy means of suspending a light chain from the nose to the ground, upon which the fore-feet are ready to catch the chain in walking, when the nose receives so sudden a check that whenever the bull attempts to run at any one in the field, he pains himself. Even a young bull in a field may follow you at first in sport, and run at you afterwards in earnest. I remember of being encountered by a 2-year-old bull in the midst of a field. Instead of recognising me, which he used to do, he advanced towards me, bellowing, scraping the ground first with one fore-foot, then another, and casting the earth over his back. Feeling it to be vain to reach a fence or gate before he could overtake me, and knowing it to be dangerous to indicate any motion like retreat, I determined on standing still, and on doing one or both of two things, should he approach near enough; namely, to hit him on the forehead with a large stone, and should that fail to scare him, to seize him by the tail. With this intent I picked up a large stone in each hand, and watching the moment when he came near enough, about 5 yards, when his head was at the ground, while he was bellowing and preparing to make a rush at me (which singular enough, bulls always do at people with their eyes closed), I deliberately aimed and struck him a blow with all my might between the horns with a stone, the shock of which so terrified him, that he turned round, shook his head, and ran away from me at a round trot. Had this manœuvre not succeeded in scaring him with the second stone, there was no alternative for me but that of laying hold of his tail, and there is no great difficulty in effecting this with a bull, by boldly going up, and slipping instantly behind him, seize the tail. When the tail is held firmly the bull has no power to kick, or throw you off; and your policy is to kick his shins if you have no stick, and if you have a stout one, to belabour his shanks until he lies down exhausted, which he will soon do under such a punishment, and then you effect your escape. A number of sharp strokes on the bony or bare part of his legs will deprive him of courage much more quickly than severer punishment upon the fleshy part of his rump and flanks. Not content with defeating him in the manner described, I had him immediately brought home, and putting a strong rein-rope in the ring of his nose, led him out to the highway, and gave him such a punishment, by pulling at, and checking his speed by, the nose, while trying to run away from me, that he became subdued at the sight of a man ever after. To keep him constantly in check, however, a chain of such a length as to trail on the ground was suspended from the ring. This was the first and only instance of bad temper he ever shewed. The ring is put into a bull's nose in this way:—Let a ring of iron be provided, of perhaps 2½ inches in diameter over all, and ¼ inch diameter in the rod, when finished. It should have a joint in it, to let the ring open wide enough to pass one end through the nose, and the two sides of the ring, on being closed again after the operation, are kept together with two countersunk screws. An iron rod tapering to the point, and stouter than the rod of the ring, should be provided. Let a cart-rope have a noose cast firm at its middle, and put the noose over the bull's head, and slip it down his neck, with the knot undermost, till it rests upon the breast. Any mortared wall sufficiently low to allow the bull's head to reach over it will answer to put him against; or what is safer for his knees, any gate-way with a stout bar of wood placed across it ashigh as his breast. Place the bull's breast against the wall or bar, and pass the rope from the lowest part of the neck along each side round the buttock like a breeching, and bring one end of the rope over the wall or bar on each side of the bull's head, where a stout man holds on at each end, and it is the duty of both these men to prevent the bull from retreating backwards from the wall or bar. A man also stands on each side of the bull's buttock to prevent him shifting his position. The operator having the iron rod given him heated in the fire, just red enough to see the heated part in daylight, he takes the bull by the nose with his left hand, and feeling inwards with his fingers, past all the soft part of the nostrils, until he reaches the cartilage or septum of the nose, he keeps open the nostrils, so as on passing the hot iron through the septum, it may pass clear through without touching the outer skin of the nostrils, taking care to pass the iron parallel to the front skin of the nose, otherwise the hole will be oblique. Immediately after the rod has been passed so far as to make the hole sufficiently large, and the wound has been sufficiently seared, the operator takes the ring, opened, and still keeping hold of the bull's nose with the left hand, passes it through the hole, and on bringing the two ends together, puts in the screws, and secures them firmly with a screw-driver. On being satisfied that the ring turns easily round in the hole, and hangs or projects evenly, the bull is then released. The ring, as it should appear in the nose, is represented in the portrait of the Short-horn bull. It cannot be too highly finished or polished, and its cost in this state may be 2s. The ring should not be used until the wound of the nose is completely healed, though it is not uncommon to see the poor animal tormented, in the attempt of being led about by the ring immediately after the operation, when every part of the nose is still tender and sensitive. So alarmed do some bulls become from the operation, that they hang back from the leading rope of the ring with such force as to pull the ring through the nose ; but such a use is an abuse of the rope, and rather than such a result shall occur the rope should be slackened, and the animal relieved from pain as often and until he learns to yield to the rope. The readiest and neatest way to attach a rope to a bull's ring is with a swivelled hook, retained in its place by a spring, and a rope should be kept for the purpose. On first trying to lead a bull by the ring, the drover should not endeavour to *pull* the animal along after himself, but allow him to step on while he walks by his side, or even behind him, with the rope in his hand. While so following, to relieve the animal as much as practicable of the weight of the rope upon the nose, the drover should throw the middle of the rope upon the bull's back, and retain a hold of its end. Should he offer to step backwards, a tap on the shank with a stick will prevent him ; and should he attempt to run forward, a *momentary check* of the rope will slacken his pace. On no account should the drover attempt to struggle with the bull on the first occasion ; on the contrary, he should soothe and pacify him, and endeavour to inspire him with confidence in himself and the rope, and to shew him that he will receive no hurt if he will but walk quietly along. The animal, in the circumstances, will soon learn the nature of the tuition he is undergoing if he is properly dealt with, but if tormented merely that the drover may shew his power over a powerful animal, it may be a long time, if ever, before he will learn to behave quietly when led. A useful instrument for leading a bull by occasionally, when he has not been ringed, or for leading a cow to the bull at some distance, or for taking away any single beast, and at the same time retaining a power over it, is what is named the *bullock-holder*. It consists of iron in two parts jointed, which are brought together or separated by a thumb-screw passing through them. The ends furthest from the joint terminate in a ring having an opening at its extreme side, each end of which opening is protected by a small ball. The arms of this ring embrace the septum of the nose gently between them ; and the shank of the instrument being screwed close together, the balls approach no closer than just to embrace the septum, and the nose of the animal prevents them slipping out. The leading rope is attached to the jointed end of the instrument, which is formed for the purpose into a small ring.

(2596.) A bull that is serving cows is never in a better position than when attending them in the field. He knows much better than either the shepherd or the cattle-man when a cow is coming into heat, and he will attend her faithfully until the proper time of service ; and I believe it is a fact, that a bull which is constantly amongst cows in a field never teases or abuses them so much as one that is taken to them for the occasion out of his own house. At any rate, I have seen very striking instances of what I have now stated. But a bull can only be left in the field where he is intended to serve all the cows. It may be necessary, however, in the course adopted by you for the improvement of your stock, that different bulls shall serve the cows, in which case no single bull can have access to all of them alike, and therefore cannot be grazed in the same field with them. When a bull goes amongst cows he is usually quite safe to approach, and is quiet within the fence; but a bull is always troublesome by himself in a paddock or field, or even amongst oxen. He is restless, often bellows, and few fences will retain him when he is resolved on breaking through, especially where he can snuff the cows at a distance. In such circumstances, it is much better to confine him in his hammel or byre, and support him on cut forage of some kind, such as clover, lucerne, tares, and the like. When so confined, bulls, like watch-dogs always kept on the chain, dislike the approach of any one but their keeper, and even a keeper has been known to fall a victim to resentment. Some bulls become so prone to mischief when constantly confined, that they will attempt to run on any one, when brought out of the house to serve a cow, the presence, or smell of the cow in heat having so maddening an effect upon them as to render them reckless. The air and daylight together have an intoxicating effect upon them. Besides the rope or chain in the ring, a safe precaution for the keeper, in such a case, is to have a stout stick about 6 feet long, with a swivelled hook on its end to fasten into the ring, and with this he will not only have a better command over the ring than by the rope, but it will enable him to keep the bull off to a certain distance, the animal not being able to run at him without first giving a warning of his intention by pushing the stick. Bulls display a natural fondness for calves. Bulls that have served cows should never be allowed to herd together, as they will inevitably fight, and a serious bull fight is a terrific thing, seldom terminating before the infliction of serious injury to both parties.

(2597.) The young cattle—the 1-year-old and 2-years-old—may be put on the inferior pasture, that is, on the second or third year's grass, —inferior, not because it is insufficient to maintain the stock in full condition, but because new grass is usually considered the most nourishing on a farm for the youngest stock, and most profitable for cows when bringing up calves. Certainly it is the most succulent grass, and stock are fondest of it, and it springs the earliest of any on a farm; but then all the stock cannot have new grass at the same time, and that part of the stock which are not employed in reproduction, should receive the least valuable grazing in summer, though the entire pasture may be very good.

(2598.) Grass-land requires peculiar management to render it the most available as pasture in every variety of season. The circumstances which most injure grass are *overstocking* and *continual stocking*. The most obvious plan, of course, of avoiding overstocking is to have no more stock upon the farm than its grass will in summer maintain in good condition, and to avoid continual stocking, the stock should not be allowed to remain too long on the same field. I believe the safest principle upon which each grazing field can be treated, is to stock it at once as fully, as that it shall be eaten bare enough in a short time, say in a few weeks, and then it should be left unstocked altogether, *hained*, as it is technically called, for perhaps a fortnight, in order to allow the grass to grow sufficiently to afford a bite for cattle. One obvious advantage of this plan is, that the stock at periodic times during the grazing season will enjoy fresh grown grass; and another is, that the same growth of grass will not be so long depastured as either to cloy the appetite of the animals, or become foul by being constantly trodden upon. That this is a rational and natural mode of managing grass-land is evinced by the fact, that all stock delight in consuming fresh grown grass, and all loathe grass which has been long trampled and dunged upon, and the breath passed over it times out of number. There is another important consideration to be borne in mind in conducting the grazing of grass-land, which is, the modes in which different kinds of stock crop the grass, one kind biting it close to the ground, while another bites it high. The ox gathers each mouthful of grass with his tongue before he cuts it with his teeth, and therefore requires the grass to be some inches in length before he can obtain a full bite. Grass, which is not sufficiently long for this purpose, either before it is grown, or after it has been eaten down, is in an unfit state for cattle; they may live, but cannot gain condition upon it, and as long as they are confined upon it will lose so much time. Horses and sheep, on the other hand, crop grass in quite a different way; their lips being very mobile and muscular, seize the grass firmly, while the teeth cut it over, and in doing this they bite very near the ground. This is a remarkable peculiarity in regard to the sheep, inasmuch as the arrangement of their teeth being the same as that of the ox, both wanting them in the upper jaw, one would have expected them to employ a similar mode of cropping; but the difference in the form of their lips explains the peculiarity, those of sheep being decidedly prehensile, like those of the horse, while those of the ox are thick and inactive, but the ox's tongue is prehensile, and thus we see that it is the prehensile organs which are the active agents in the collection of food. And this is a wise distinction in the formation of these two classes of ruminants, in as far as it is suited to their respective natures,—sheep being suited to mountainous regions, where pasture is always short, and where they are able to crop it with the assistance of their prehensile lips, notwithstanding the want of teeth in the upper jaw; whereas the ox is better suited to the plains and valleys, where grass grows long, and is therefore better suited to be cropped by the scythe-like operation of the ox's tongue and teeth. The practical conclusion to be drawn from these different distinctions in the formation of these animals is, that the horse or sheep should follow the ox in grazing, or graze in company with him, but not precede him. When they follow, the pasture will be eaten barer by the horse or sheep than when the ox left it; and, when in company, it will be eaten barer by the horse or sheep where the ox has eaten before, and it may first be topped by them before the ox has touched it. Of the two modes of treating the horse or sheep, the latter is the preferable, because choice is offered them to take the long grass as well as the short. The same reason, however, that should graze the ox *with* the horse or sheep should cause the separation of the horse *from* the sheep, and especially in the later part of the pasture-season, for both kinds of animals biting close, will render the grass too bare for the maintenance of both. Horses too, work-horses especially, seem to have a greater dislike to sheep than to cattle.

(2599.) There is a difficulty attending the grazing of all kinds of stock on a farm of mixed husbandry which I must mention, that as there is every summer the same number of stock, there may not be the same quantity of grass to support them; for the same number of acres of grass, secured by following a regular rotation of crops, may produce different quantities in different seasons—one year being scanty, another superabundant. The number of stock, therefore, should correspond with the produce of an average year. In such a case, a bad year very much stints the condition, while a good year supplies perhaps even more than the same stock can consume. The stinted condition cannot be alleviated by the sale of any of the stock, as it is impolitic to disturb the equilibrum of the stock of several years produce which exist on the farm at the same time; and if the stock suffer hunger, as it did in the severe drought of the summer of 1826, there is no alternative but to make up the deficiency by hay, even though it should be purchased for the occasion. On the other hand, superabundance of pasture does no harm, for, independent of its maintaining the entire stock in high condition, the rough aftermath will be of essential use to sheep in winter. Of farms whose stock are purchased every year, the number can be regulated by the likelihood of the crop of grass, but even in this case the season may turn out worse than was expected. Seeing, therefore, that no one can foretel the future state of any crop, the prudent plan is, in every case, to keep the number of stock under the mark which a farm can support.

(2600.) One essential requisite in all pasture-fields is an abundant supply of *water* for stock to drink. Both cattle and horses drink largely, and sheep, grazing early on the dewy grass, do not require so much water to keep them in a healthy condition, still, when there is no dew, they do drink water. The proper construction of a watering-pool is sadly misunderstood in this country. The entrance to it generally consists of poached mud of at least half a foot in depth, and to avoid this, the animals go into the water before they drink, when, of course, it is at once rendered muddy. Not unfrequently there is scarcity of water, and if there is just a sufficient supply to prevent the pool being evaporated to dryness, the water is rendered almost stagnant. So obvious are the objections to this mode of administering so necessary and wholesome a beverage to the brute creation as water, that all that seems necessary to an amendment of the system is to point out its inconveniences. True, in some cases, tanks of wood or stone are provided in fields, which are supplied from some adjoining spring, or even pump-well, and, as far as the quality of the water is concerned, this is a much better mode of supplying it than in pools; but this mode, good as it is as far as it goes, does not provide all the requisites of a good watering-pool. In hot days, a walk through a pool is very wholesome to the feet of cattle, and in dry weather, a stand for some time amongst water is an excellent preventative of that troublesome complaint the foot-sore. The external application of water in this manner allays inflammation, and prevents irritation, and permits animals to take their food in peace even in scorching drought. Besides, the tank of a pump-well is not unfrequently neglected to be filled—because in cold and rainy weather it is scarcely visited by the cattle, in hot weather it is supposed to be viewed with the same indifference; and, even where tanks are duly attended to for cattle, there are none set down at a lower level for sheep. A watering-pool should be securely fenced, as cattle are very apt to push one another about while in it, and for that reason it should also be roomy. It should be of considerable length and narrow, to allow access to a number of animals at the same time, if they choose to avail themselves of it; and I have often observed cattle delight to go to the water in company. Pools are usually made too small and too confined. The access to them should be made firm with broken stones in lieu of earth, and gravel placed on its bottom keeps the water clean and sweet, while the water should flow gently through the pool.

(2601.) The want of *shade* in pasture-fields is also a sad reflection on our farmers. Observe, in summer, where the shade of a tree casts itself over the grass, how gratefully cattle resort to it, and where a spreading tree grows in a pasture-field, its stem is sure to be surrounded by cattle. The stirring breeze under such a tree is highly grateful to these creatures; and such a place affords them an excellent refuge from the attacks of flies. In cold weather, also, observe how much shelter is afforded to cattle by a single tree, and how they will crowd to the most wooded corner of a field in a rainy day even in summer. Ought not such indications of animals teach us to afford them the treatment most congenial to their feelings? I am no advocate for hedge-row trees, even though they should cast a grateful shade into a pasture-field and still less do I admire an umbrageous plane in the middle of a field that is occupied in course with a crop of grain or turnips; but similar effects as good as theirs may be obtained from different agencies. A shed, erected at a suitable part in the line of the fence of a field, would not only afford shade in the brightest day in summer, but comfortable shelter in a rainy day, or in a cold night in autumn. Such an erection would cost little where stone and wood are plentiful on an estate, and they could be erected in places to answer the purpose of a field on either side of the fence when it was in grass. But no matter what it may cost, when the health and comfort of stock are to be maintained unimpaired by its means. The cost of a shed may, perhaps, in this way be repaid in the first year of its existence, and it would stand, with slight occasional repairs, during the currency of a long lease. When such an erection is properly constructed at first, it is surprising how many years it will continue to be useful with a little care. Let it be roomy, and its structure light, as it may be roofed at a moderate cost with zinc, or composition of some sort, or tiles, when they are manufactured in the neighbourhood. It may be troublesome to carry straw for litter from the steading to a shed situate at a distance, but there is no occasion for straw for litter in summer; the rough grass from an adjoining plantation or ditch will supply litter, and the dung at any rate should be shovelled up and carried away before it becomes uncomfortable to the animals. I should like to see a farm with such a shed erected in it for every 2 fields.

(2602.) Young cattle are not grazed on *carse*-farms, nor on farms *in the neighbourhood of towns*. On the former there is little land appropriated to grass, and that little being kept for only 1 year, it is used chiefly as forage grass; and on the latter class of farms it is more profitable to let the grass for forage and hay. All the grazing on these farms is confined to the farm milch-cows and work-horses. On *dairy* farms the pasture for cows is always preferred of old lea abounding in natural grasses, and, no doubt, these afford by far the finest flavoured butter and cheese. On such farms, cows being kept for the express purpose of yielding dairy produce, they form the principal stock, and receive the largest share of attention. They are most commonly put into the byre at night, where they are milked in the morning and evening, but are milked in the field at mid-day. The milk being appropriated to other purposes than bringing up calves, it is passed through the milk-sieve, or *sey-dish*, as it is named, on the spot. This sieve consists of a circular basin of wood, with a circular opening in the bottom, covered with fine brass wire-cloth, or of a square hopper of wood, and its use is to strain all hairs and other impurities from the milk that may accidentally have fallen into the pails during the act of milking. It is most profitable for the dairy-farmer to give his cows at all times as much food as they can consume; and with this view, beside the fine pasture they receive, they are supplied with an ample suppering of clover-grass or tares, according to the season, the clover coming first into use, and then the tares, and where lucerne is grown it is used for the same purpose. I have no doubt but that it is the most profitable, and certainly it is the most convenient plan, for dairymen in towns to keep their cows constantly in the byre even in summer, and to feed them on cut forage, and purchase such litter for them as can be found cheapest at the season. In pursuance of this system, the irrigated meadows in the neighbourhood of Edinburgh are taken by dairymen at auction every year, and their produce cut and carted daily from April to November. The number of cuttings may be 4 or 5 in that period, according to the growing state of the weather. The usual rent is from L.16 to L.20 an acre imperial; and in the excessively dry year of 1826, such was the demand for this grass for cows, when none other could be obtained, that L.40 per acre was given for it. But such is the conviction of the healthy nature of pasturage for milch-cows, that many dairymen, even in towns, take pasture for them at high rents in the neighbourhood, where they allow the cows to lie out all night, and incur the trouble and expense to send people to milk them three times a day.* On *pastoral* farms devoted exclusively to the rearing of *cattle*, the calves are allowed to go with their dams, from whom they draw as much milk as they can get, and support themselves besides on grass. The calves thrive well in this way, and attain to a large size, and could they be maintained in the same condition until they are 3 years old, they would become very fine beasts; but unfortunately they are apt to fall off in condition in autumn, and it is difficult to bring them into a sleek condition in winter; and it is this circumstance which raises a doubt in my mind whether it would not, upon the whole, be better for the calves to wean them from their mothers at 4 months old, and place them entirely dependent on grass thereafter, during the best season of pasture, and there would then be no difficulty of maintaining, in the ensuing winter, the growing and healthy condition thus attained. In such a plan, no doubt, the trouble of milking the cows would be incurred, for it would be impracticable to let them run dry in the height of the grass season, and otherwise it would be impolitic, beside the loss of the milk, as they would get in too high condition before winter, through which it would be imprudent to

* Although there is nothing peculiar in the management of cows on a dairy-farm to other milch-cows in summer, you may peruse with advantage Aiton's Treatise on the Ayrshire Dairy Husbandry, 1825, and Harley's Account of the Willowbank Dairy in Glasgow.

maintain them at that pitch till their next calving. And besides this, whatever may be the condition which stock attain in the beginning of winter, it never fails to prove injurious to them if it is in any degree lessened during winter, for when the animal system falls into a declining state it is very difficult to rally it. The young stock on pastoral farms graze on the lower and more sheltered portions of the grazings till the weather becomes less stormy and cold in the upper parts, when they stretch their walks upwards by degrees until the highest points are at length attained. This procedure, of course, cannot be strictly adhered to where there are no fences to mark the different points of elevation; but there is the natural check of the state of vegetation to deter them from proceeding higher up until the grass grows. It would be better, were fences so laid off as to divide the pasture into portions which could be occupied in their proper season at proper intervals of time; and it has often occurred to me, that this object would be best attained, were fences to run in horizontal parallels along the face of a height, or round a hill, than up and down the slope of the ground. As it is, pastoral farmers seem contented with only a ring fence round their farms.

(2603.) Young cattle are purchased for farms on which none are bred immediately before the grass is ready to receive them; and not unfrequently this class of farmers hire grass-parks for the season and stock them with young cattle on speculation. Such cattle are to be obtained at public markets, or from breeders who do not keep the stock they have bred beyond two years. Young cattle when purchased for grazing should have all the symptoms of health, a clear eye, dewy nose, and glossy hair. To attain full size they should have a strong bone, and their appearance should be what is commonly termed *raw-boned;* that is, the bones or skeleton should seem large in comparison with the condition of the flesh, and which appearance implies a growing state. To be a good thriver and attain condition the hair should feel mossy, and the touch of the skin mellow. The skin should not be too thin, nor feel hard and tight, and it should be covered with abundance of hair. A thick tail with plenty of hair on it and at its point, indicates strength of back and constitution. The hams should not be too full of flesh, that is, *lyary*, which in a young animal indicates that the carcass will soon *set* from growing. A very deep body, and short carcass between the shoulder-blade and the hook-bones across the ribs, indicate a disposition to fatten at a small size. Flat ribs and a high back-bone indicate difficulty of fattening, and though the bone may be large enough, it will nevertheless be coarse and round, instead of flat and fine; and the sinews of the legs will be indistinctly marked and seem heavy, instead of being small and clean. On selecting a lot of young cattle, they should be nearly all of the same size and appearance, and they should be what is called *level*, that is, presenting an even surface over all their backs when placed together, or walking in a drove, with but the sharp horns and ears of horned cattle, and the crown of the head and ears of polled ones, rising above the common level. This last is a very enticing property in every lot of cattle, and the individual properties enumerated above are common to every breed of cattle. Where a number of cattle of the same breed have to be compared, the properties which distinguish the particular breed must form the standard of comparison, and what these standards are you shall know by and by, when we come to treat of the properties of the different breeds.

(2604.) As to the state of cattle most profitable for the farmer to purchase, I believe that 2-year old queys have been found the most kindly feeders in every breed. They are, no doubt, subject to periodic disturbance when in season in summer, and to avoid these the operation of spaying (2117.) was recommended, and which was at one time pretty extensively practised, and so long as the general lot of heifers possessed properties that unfitted them for becoming good cows, the practice was unobjectionable. Now, however, that every farmer possesses a large proportion of his heifers well formed, spaying has fallen into desuetude, and it is found more profitable to dispose of them as breeding stock than to fatten them, and for this reason the heifer market is not so well supplied as it used to be; and it is seldom that a lot of good 2-year old queys can be obtained without being in calf, or such a price demanded for them as to render a profit from feeding them, hopeless. Of the different breeds of cattle, I may here anticipate, so far as to say, that the Short-horns have proved themselves the surest feeders; and of this breed, perhaps, more heifers may be obtained for feeding than of any other. The reason for this is, that this kind of stock being now so universally improved, every animal possesses good properties, and, of course, as every heifer cannot be transferred into the cow stock, they are brought into the market for other purposes. I have known, even lately, handsome profits returned from such heifers which never could be obtained from steers of any kind. Still, as every farmer cannot be supplied with heifers, oxen must be purchased in lieu, and of these the age at which they should be purchased must be regulated by the nature of the soil. Some soils bear pasture which will fatten oxen of large size, and on which aged oxen of 3 years will return most profit; on weaker soils, it would be folly to attempt to feed heavy beasts, and on light soils, young beasts must be chosen for them. Of the black breeds, the hornless Angus and Galloways are most safe to one another in large lots in courts in winter; but on pasture in summer, this consideration is of less importance, and the West Highlanders and Aberdeenshires, or a cross of all these with Short-horn bulls, would be found kindly feeders. Those who possess pasture that will feed a heavy ox, have the advantage, as they can purchase light stock if they choose, but those who possess weak land, can only purchase small, and, probably, for most profit, young stock, and must let the heavier and more aged alone.

(2605.) Cattle are subject to *very few diseases while upon grass in summer.* Sometimes they may receive a chill in a sudden change of the weather to wet and cold; but were sheds provided in every field, there would most probably be no chills felt, as cattle never suffer from cold when they have shelter at will. The immediate effects of such a chill is a staring coat and hide-bound skin, which may be removed by a cordial drink, composed of 1 quart of gruel and 1 bottle of ale in a lukewarm state, in which has been dissolved some treacle, and spiced with 1 oz. of ginger and 1 oz. of caraway seeds ground fine. The drink is administered with the drinking-horn (1633.), and the animal kept in a shed for a night or two. The ultimate danger from such a chill is inflammation of the lungs, which is in most cases a fatal disease.—In reference to the mode of extracting chaff from the eyes of cattle in (1521.), I may mention another mode which has been practised many years by Mr Sadler, Norton Mains, near Edinburgh, with most complete success. It is this:—"Let the animal be held firmly by the head, and take a beard or awn of barley and draw it slowly over the chaff, and the rough edge of the awn will cause the chaff instantly to adhere to it, and thus easily remove it." This, I have no doubt, is an efficacious plan when the chaff can be seen; but, as it cannot be observed in the eye on all occasions, another mode, such as the one referred to above, seems indispensable—Besides lice, cattle are not unfrequently troubled, towards the latter end of the feeding season, with what are named *warbles* or *wommals*, that is, small swelled protuberances along the chine, caused by the larvæ of the *Œstrus bovis*, or cattle-bot. Fig. 449, *a*, gives a representation of the fly which originates these larvæ. This is the female, which has the abdomen attenuated behind, and terminating in a black coloured style, composed of cylinders which slide into each other like the tubes of a telescope, as seen at *b*, but greatly magnified. It is not well ascertained whether the fly merely lays her eggs on the hair or skin, and the larva, when disclosed, is left to force its own way beneath it, or a perforation is made by the fly and the egg deposited with it. That the latter is the case seems most probable, as the ovipositer *b* seems constructed for the express purpose. Cattle feel great pain, and become almost furious, when attacked by this fly. The larvæ *c* are of an oblong shape; the body divided into 11 segments by transverse bands, which again are crossed at the sides by longitudinal lines, and on each side of all the segments there is a distinct spiracle or breathing-hole. The young larva is found to occupy a small cyst or cell within the substance of the skin, which gradually enlarges with its growth; while the pus which is abundantly secreted by

Fig. 449.

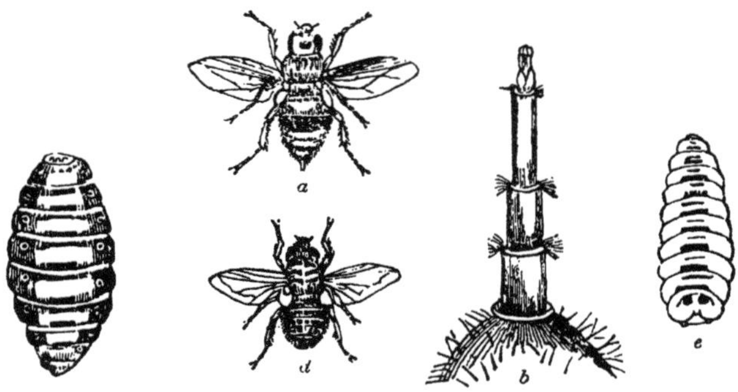

THE CATTLE-BOT—ŒSTRUS BOVIS, AND SHEEP-BOT—ŒSTRUS OVIS.

the irritation, serves for its subsistence." The tumour is never entirely closed around it," says Mr Duncan, " there being always a small aperture on the upper side. On attaining its full growth, the larva makes its exit by the aperture just mentioned, and the wound speedily closes up and is healed; but the hide never recovers its original strength, as afterwards appears when it comes under the operations of the tanner. It is remarked by Reamur as a singular circumstance, that the larva commonly issues from the tumour, to assume the pupa state, at a very early hour in the morning, and thereby avoids many of the dangers to which it would be otherwise exposed. The warbles are so conspicuous on cattle, that if searched for scarcely any could be overlooked, and they may be killed with the utmost ease by simple pressure of the finger and thumb, or by pouring some corrosive liquid into the aperture of the tumour. They are seldom so numerous but that a short time would suffice to inspect a whole herd, and if this were done simultaneously at different places, a whole district might in a short time be nearly or altogether freed from this pest."* A very tormenting insect to cattle on the Highland moors, though it is scarce in the lowlands of Scotland, and far from being rare in England, is the Cattle cleg, *Tabanus bovinus*. It is about one inch in length, being the most bulky of our native Diptera. When the proboscis is fixed in the skin, and employed in pumping the blood, the insect can in general be easily got at, and killed with a stroke of the hand. The instrument by which the skin is pierced, and the blood extracted, is of curious and complicated structure. The larva *a*, fig. 450, is long and cylindrical, narrowing at the head into an elongated cone. The body is divided into 12 rings, the anal one being very minute, and resembling a spiracle; colour dirty white, the head brown and shining. The pupa *b* is nearly cylindrical, of a greyish brown colour, the segments fringed on the posterior margins with grey hairs. The anal segment is small, and armed with 6 sharp scaly points *c*, which seem to enable the pupa to push its head above the surface of the soil.†

Fig. 450.

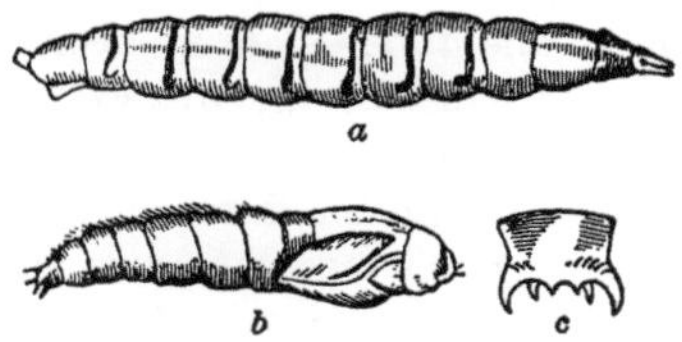

LARVA AND PUPA OF THE CATTLE CLEG—TABANUS BOVINUS.

(2606.) The teats and udder of cows are at times subject to certain complaints in summer; and these are *chapped teats*, *sore teats*, *warty teats*, and *cow-pox*. Chapped teats consist of cracks occurring across the teats, which, when drawn downwards, the cracks are forcibly opened, and, of course, inflict pain on the animal; and the cows then become troublesome to milk. The easiest mode of milking them while under this complaint, is that described by nievling in (2094.), while stripping aggravates the complaint. I do not know the certain cause of this complaint, but suppose it to arise from leaving the teats in a wet state after milking; and perhaps cows lying upon wet ground may have the same effect. Sore teats are, when blotches of skin come off the teats, and their fleshy substance becomes sores by exposure to the air. This complaint may arise from the milker who strips seizing a particular part of the teat too hard, where an inflammation being set up, terminates in a sloughing of the skin, and consequent exposure of the fleshy substance to the air. Warty teats, I conceive, may originate in the skin of the teats being ruffled by too much force in stripping, or by too long a nail upon the thumbs; and the warts produced in consequence may be exuberances of the skin in covering the injured parts. I am not sure that these conjectures, for they are nothing more, will explain the causes of these complaints, but I believe when care is used not to abrade the skin or pinch the substance of the teat, but to keep it clean and dry, these complaints seldom or never occur. As to the cow-pox, it is a constitutional disease, and cannot be either induced or retarded. The pock makes its appearance both on the udder and the teats of the cow; and as milking must be performed frequently by all the teats, the operation feels very painful to the cow, and she, of course, becomes very troublesome to milk. Very soon the pustules are rubbed off by the operation, and their sites become skinless sores

* Quarterly Journal of Agriculture, vol. x. p. 544.

† The British species of the Horse flies, *Tabani*, are described at length in the Magazine of Zoology and Botany, vol. i. p. 359.

Nothing but the utmost gentleness will prevent the cow becoming distracted under the torture. After having run its course, the disease declines, the sores become less acute, and heal up by degrees. The disease fortunately is not of frequent occurrence; I have only seen it once in the course of a fifteen years experience, and it affected all the cows I had at one time—9 in number. I acquainted several of my medical friends of the existence of the disease, imagining that they would have been glad of so favourable an opportunity of obtaining fresh vaccinating matter, but they treated the offer with indifference. There was an ointment I found very efficacious in affording relief to the cows when afflicted with the cow-pox, sore teats, or hacked teats. It consisted of fresh butter melted and burnt in a frying-pan, and mixed with half its quantity of tar. While hot it was poured into a gallipot, and applied cold to the affected parts. The tar had the effect of keeping off the flies, while the burnt butter never became dry. The ointment was washed off with warm water before milking commenced, and the udder and teats dried with a soft linen cloth; they were again bathed with warm water after milking, again dried with the soft linen cloth, and the ointment again applied. Calves, after being weaned, are subject, towards the end of summer, to a disease commonly called the *joint-fellon*, which, when oxen take it upon the loins, is named the *chine-fellon*. It is nothing else than acute rheumatism, ending in a resolution to low fever, and so severe is it at times upon calves, that they cannot bear to be moved when lying stretched out all their length upon the ground. Had cattle sheds to retire to whenever a cold dash of rain came in the evening of a cold day, even in summer, this disease would perhaps never occur. Its treatment is removal to the courts and sheds of the steading amongst straw, bleeding, moderate purging, with fomentation, and embrocations of liquid blister, forcibly and long rubbed in, of the swelled joints. Another effect of the same febrile affection on calves in autumn is the *quarter ill or evil*. "Its characteristic symptoms are general disturbance of the circulation, and feeble, rapid pulse, weakness, prostration of strength, determination of blood to particular, but in different instances and epidemics, very different, parts, producing pain, and manifesting a tendency to inflammation, but of a degenerate kind, so that the very texture of the tissue becomes disorganized. The progress of the disease is often rapid, and the result very fatal. In some cases the lungs or heart are attacked, in others the liver, bowels, or even some external part of the body." Its immediate cause is plethora, or fulness of blood in the system, which shews its effects in this manner:—"When the supply of food is greater than the exigencies of the system requires," as Professor Dick observes, "an animal usually becomes fat, but still may be tolerably healthy. When, however, a sudden change is made from poor to rich feeding, not fatness but plethora may be the consequence; more blood is formed than the system can easily dispose of, and it becomes oppressed. The effect is often witnessed in cattle and sheep, which, after indulging for a time in luxuriant pasture, take what is called a *shot of blood*. All at once they become very ill; some part of the body swells, becomes puffy, as if containing air, and in 2 or 3 hours the animal is dead, from the *quarter evil* already described. Upon dissection a large quantity of black and decomposed blood is found in the cellular membrane, which during life was distended."* This disease is of frequent occurrence on farms where fine stock are bred, and from the above description of its nature, there

* Dick's Manual of Veterinary Science, p. 11 and 88.

is no wonder that the best calves first fall victims to it. As its name implies, the disease attacks the hind-quarter, and its effects are as sudden as described. Since its cause is known, calves should not be put at once on strong rank foggage from a comparatively bare pasture, nor, for the same reason, should calves in low condition be put on rank foggage; the transition, both as regards the pasture and the state of the calves, should be gradual. As a preventative, some farmers introduce a seton into the dewlap of all their calves before putting them on foggage in autumn. The use of the seton is to produce counter-irritation. The seton consists of a piece of tape or soft cord passed under a portion of the skin by a seton-needle; the ends may be tied together, and the cord may be moved every other day from side to side, being previously lubricated with oil of turpentine or blister-plaster, and in this way the amount of irritation may be regulated. As to the cure, I believe every one is unavailing after the disease *has been observed to exist;* but as a remedial measure applied by anticipation, large blood-letting with purging of repeated doses will reduce the plethoric tendency of the animal system. Perhaps a cribful of hay, with some salt, placed in a foggage field, would not be a bad alternative for calves to resort to at times, in order to modify the effects of the succulence of rank aftermath.

---

64. OF MARES FOALING, OF STALLIONS, AND OF HORSES AT GRASS.

" Yet when from plough or lumbering cart set free,
They taste awhile the sweets of liberty:
E'en sober DOBBIN lifts his clumsy heel
And kicks, disdainful of the dirty wheel."

BLOOMFIELD.

(2607.) Whenever field-labour commences in spring, whether with bean or oat-seed, draught-horses are indulged with hay instead of straw, and their allowance of corn is increased. The hay is supplied to them at will, and a strong work-horse is supposed to eat about 1½ stone, or 33 lb. a-day, and some will eat even more. The oats are increased to 3 feeds a-day, which, at 40 lb. the bushel, will weigh 7½ lb., but if the measure is heaped, which some farmers allow to be done with horse-corn, the feeds will weigh 8 lb. The quantities of hay and corn here specified are supposed to be given in their ordinary state, but when given in a cooked state, I have already described their treatment in (1574.) to (1576.) inclusive.

(2608.) The *hay-stack* is seldom broken upon until the horses get hay in spring. Its site in the stack-yard is marked by *n* in Plate IV., fig. 4, of the steading, and the hay-house is contiguous to it at H, the apartment which also contains the corn-chest *y*, and the whole adjoins the work-horse stable O. When the hay is to be used, as much of the stack is brought in as will fill the hay-house, and thence distributed to the horses. Every portion cut off the length of the stack should be 4 or 5 feet, and they are cut off with a *hay-knife*, fig. 451, which represents the usual form of that instrument. It will be observed in the figure, that the line of the back of the blade is not so much as at right angles to the handle, a position which gives the cutting edge of the knife an inclination to the line of section, which, when the knife is used alternately up and down in a perpendicular motion with both hands, causes it to cut a longer space than the breadth of the blade. The person who uses the knife, usually the steward, while it is the field-workers who carry the hay and fill the hay-house, sets himself on his knees upon the part he is cutting off, with his face to the body of the stack. This form of knife requires considerable strength in its use, and unless the edge is kept remarkably keen with a whet-stone, and the hay is firm, it does not make good work. The hay knife I prefer is of the form of the dung-spade in fig. 364, which, being used when standing, can be wielded with more force, and make a deeper cut; and it cuts equally well in both directions, whereas the common knife cuts only in one direction—to the right.

Fig. 451,

THE HAY-KNIFE.

(2609.) There is considerable waste of hay as it is commonly used in hay-racks *i*, fig. 7. The ploughmen stuff the rack not only full, but squeeze the hay firm, from a mistaken notion that they cannot give too much at a time to their horses; but when horses find it in this compressed state, and are unable to select the morsels they want, they toss out a considerable quantity amongst the litter, with which, after being trampled upon, and refused to be eaten, it is thrown into the dung-yard. True, the cattle there will eat it, and prefer it much to straw, because of its being hay and having a saline taste, so that it cannot be said to be entirely lost, yet the object of supplying the horses only upon hay is frustrated. If it is desired to give hay to cattle also, it should be given them in a direct manner. To avoid waste, therefore, small quantities should be put into the racks at a time, and these frequently; but the best way to prevent waste is to chop the hay and bruise the corn, and the mixture will also go a greater length in feeding the horses. Young horses should also receive hay after the stack has been broken up, straw becoming too hard and dry after March; and hay, besides, serves to improve their condition, and prepare them for grass. They should also get a little corn every day, until grass time, as nothing tends so much to enlarge and strengthen the bone, and confirm the spirit of young horses as corn. So much am I convinced of the truth of such a result, that I entirely agree with the observation of the late celebrated Nimrod (Mr Apperley), that the belly of the young horse should be the measure of his corn. It is to the niggardly disposition, in regard to corn, which farmers evince towards their young horses, that is to be ascribed the number of weak horses that are met with on farms; for the greatest abundance of corn administered to the adult state, will never compensate for its want when the animal was in the state of adolescence.

(2610.) May is the particular month in which *draught-mares* usually *drop their foals*. They continue to work until the immediate symptoms of foaling are observed. These are great loosening of the ligatures on each side of the root of the tail, and the appearance of a waxy-like matter projecting from the point of the teats. As it is impossible to predict a mare's foaling within a few hours, it is proper to put her into a loose-house or box by herself, and to watch her every night. Too many farmers contemn such precautions, and allow their mares to foal in their stall in the stable, to the risk of having both them and their foals kicked by the other horses. That watching is necessary at night, even in a loose-house, is evinced by a fact mentioned to me by the late Mr Airth, Mains of Dun, Forfarshire, of a mare of his which, having been neglected to be watched at night, or even removed from the stable, was found in the morning lying on the floor with her womb protruded, and the foal smothered in it. The mare shortly after died in great agony. It is a remarkable fact, that few people have observed mares to foal, even though watching for the purpose, for somehow they contrive to foal when left by themselves for even a few moments. I have endeavoured for successive years to witness the foaling both of blood and draught mares, and was always disappointed. A mare will eat with heartiness until the pains of labour seize her, when she suddenly lies down, foals easily, quickly, forcibly, requiring no assistance, and starts to her feet almost immediately after parturition, takes up with, though licks but little at, her foal, and soon begins to eat again. The foal is not long of gaining its feet after a few staggering attempts on its long legs, but some time elapses before it can steady itself, so as to lay hold of the teat. It should be assisted in this its first attempt, in order to get filled with milk, after which it may lie and sleep amongst straw until it becomes dry. The placenta soon drops from the mare, and should be immediately removed. The thin pellicle which covers the foal, is, when dried, very like the finest gut-skin used by gold-beaters, and it forms a very good protection from the air when applied as a plaster over the surface of a green wound. The mare should have a drink of lukewarm water and oatmeal, and a few handfuls of corn, after parturition, and, on partaking of which, she may be left with the foal. For the sake of increasing her milk, she should be put to grass immediately after foaling, if she is not on grass, and should remain there without being put to work for a month; by which time she will have recovered her tone of body sufficiently to bear again the fatigue of labour. The sort of work for a mare having a foal, should be of a nature to interfere but slightly with the more important operations of the farm; because having to suckle the foal every half yoking, she should be employed singly, such as at sowing and scuffling turnips, scuffling potatoes, and leading grass or other forage. When the mare is again made to work, the foal should be left by itself to rest in a well-littered apartment of the steading until it become accustomed to be alone, rather than in a court or hammel, out of which it may attempt to escape and injure itself. Should there be two mares with foals, both should be worked together; the treatment of both being alike, little inconvenience will arise to work, from being taken to their foals out of yoke together (1562.). A stout mare will be able to perform her own share of summer work, and bring up a foal at the same time; but should she be in a weakly state, which she will be when becoming old, or is overworked, she should be put either to very gentle work, or to work only in half yokings, or one yoking a-day, or be idle altogether and constantly with her foal. She should not be neglected of corn, however, though idle at grass with her foal. It is seldom that any illness attacks a draught-mare while bringing up a foal, and it is as seldom that any thing is the matter with a foal. I had one work-foal, however, which, though safely and easily foaled, and seemed lively enough, could never stand upon its feet, or lay hold of a teat, and it died in the course of the day after it was foaled, though fed on cows' milk. A mare when thus deprived of her foal, should be occasionally milked, and kept on dry food for a few days, until the tendency of the milk to secrete subsides. The mare will be ready to receive the horse in 3 weeks or a month at farthest after parturition, and the first symptom of heat is frequent twitching of the vulva, and emission of a clear fluid. When a mare is touched in this state, she immediately presses towards the object that touches her.

(2611.) In presenting a mare to the stallion, caution should be used to prevent her striking him with her heels when she refuses his attentions; and this consists simply in holding her by the head with a bridle across the outside of the stable door, while the stallion is kept within, and allowed to snuff and pinch her flank. If she takes the teasing kindly, presses closer towards him, twitches the vulva, and emits, she is in proper heat, but if she squeal and kick whenever he touches her, she is in an unfit state for him; but the tickling of the horse for a time not unfrequently confirms the season of the mare when it at first evinces a doubtful issue. If in season, she should be taken to an open piece of level ground, and held by the head as long as the horse covers her, and the time occupied by a stallion in covering is much longer than by the bull or tup, which is instantaneous. A horse which is safe to use requires no encouragement from his leader, but many need some assistance from him. Making a mare stand to the horse by a twitch on her nose is an unnecessary act of cruelty; for if she will not voluntarily receive him, she will not become impregnated by any *force* that can be used against her. One cover is quite sufficient at a time. In about 3 weeks it will be seen whether the mare has held to the horse; and should she again exhibit symptoms of season, simple and safe expedients may be used to secure her holding, such as throwing a bucket of cold water upon her rump the moment the horse leaves her, or drawing blood from her neck vein while the horse is covering, or, what is better than all, unless the season is going rapidly off her, retaining the horse all night and offering her a fresh cover in the morning, or, to adopt a different plan altogether, covering her with another horse, or another kind of horse—one or other of which expedients generally secures the holding, unless the mare is past bearing, which casualty befalls mares at very different ages. I was told by a man who led stallions for many years, that the expression of a sigh, from both horse and mare, immediately after an embrace, is an infallible sign of the mare proving in foal. The circumstances which militate against a mare's holding in foal is too high and too low condition. Whenever a mare is seen to eject semen as soon as the horse has left her, she will certainly not hold. Sometimes the fault is as much that of the horse as the mare, for when subjected to much travelling, and is, moreover, not a good traveller, and has undertaken more service than he can easily overtake, he is often so much fatigued when brought to a mare, especially towards evening, as to be quite unfit for effective service. When a horse is observed to be in a state of lassitude, the best policy for the farmer is to give the horse and his leader a night's quarters, and let him cover the mare in the morning when he is comparatively fresh. Many farmers grudge maintaining a horse and man all night, but much better incur that small expense than run the risk of a mare proving barren. When a mare has been covered 3 separate periods without success, it is needless to persevere with her, as the foal will come too late next season, and a late foal is as objectionable stock to bring up as a late calf.

(2612.) No farmer rears a stallion for the use of his own mares only, the cost of maintaining one for the purpose would soon exceed the profit. The usual practice of the owner of a stallion is, to travel him within a certain extent of country for the season, which commences at the beginning of April and terminates at the end of June. He commonly enters his horse in competition at a local agricultural show, where if it happen to obtain a prize, a district is allotted him. Besides the value of the premium, and gaining a district, the prize obtains for the horse a good name and a profitable season, for most farmers are desirous to employ a prize-horse. The number of mares limited to a horse which obtains a prize is commonly 60, at 1 guinea a mare, but the number is seldom adhered to, because many of the farmers, instead of paying the guinea, make a bargain, offering less money, or only agreeing to pay even the lesser sum, should the mare prove in foal; and to secure a good season, the leader of the horse agrees to the terms, and makes up the gross sum he considered by the rules of competition he was entitled to, by either taking more mares than the stipulated number, or by travelling beyond the district, or by doing both; and who can blame him in the circumstances? In this, as in many similar matters, farmers are very shortsighted, for in thus attempting to save a few shillings they run the risk of losing a foal, by *compelling the horse to be overworked.* The horses which fail of obtaining a prize take up districts for themselves, and receive any amount of fee they can bargain for; and these extra horses usually prove the means by which farmers beat down the fee of the prize-horse. In these various ways the country generally is supplied with the service of stallions. Sometimes a man purchases a stallion from the breeder and travels him on his own account, and if he occupies a small farm on which he can support the horse out of season, he may derive a handsome profit from the horse every year. Of course, such a man's interest is to purchase a first-rate horse, and, when trammelled with the conditions of a premium, to exact moderate fees.

(2613.) With regard to the treatment of a *stallion* on a farm, he should be placed under the care of a man who will work him when at home, and lead him for service when from home. While the stallion is young, say 1 year old, he should be bridled in spring and taught to handle and lead, and in summer get a run on good grass, in company with colts. The next winter he should have a loose-box to himself, such as *u*, in the work-horse stable O, fig. 4, Plate IV., or a loose-house, and supported on the best food, prepared for him according to the directions given in (1574.) to (1576.) inclusive. Next spring he should be regularly broke in by an experienced horse-breaker, and taught to work as described in section 58, page 691 of this volume. His work, however, should never exceed his ability, in case his shape should be injured; but a little work, even at so early an age, encourages the growth of bone and muscle, and renders the horse more easily handled and commanded. In spring, when in hands, he may be exhibited at a show as an entire colt, to help to make him known, if his figure is good; if not, he had better be castrated at once. A good run at grass, for a couple of months, after this discipline, in a securely fenced field, in company with colts, is of great service; and if the fence is not trustworthy, he might be supplied with cut grass in one of the hammels or courts. If thus confined, every thing loose should be removed from the court or hammel, that he may not blemish his legs upon them. In the following winter he should be supported on the best food, in a loose-box; and towards spring should be well kept, and groomed, and clothed, in order to keep him clean from dust, and should be regularly exercised, to put him in high order and condition by April, in time to be exhibited for a prize. When young, not exceeding 3 years, he may appear leggy, and want middle in comparison with older horses, and may therefore be defeated in competition; but if he have good shapes and promising points, he may, nevertheless, get a few mares to serve; and should he obtain a district by a premium, the number of mares allotted to him should be very limited. Many a farmer objects to a 3-year-old stallion serving at all; and, in ordinary circumstances, I dare say it is better to refrain from service, and then next year, when a 4-year-old, he will be in great vigour, and possess plenty of substance. Such matters must be regulated by the state of the horse. When a stallion undertakes to travel a district he should be provided with a sheet and roller, with a light wallet strapped across his back containing corn and beans, a few cleansing instruments, such as curry-comb and brush, water-brush, foot-pick, and mane-comb. Besides a bridle with curb-bar, to keep him in check, he should be provided with a stall-collar and water-chain, to fasten him at night in a stall when a loose-box cannot be had. His shoes should be light, and to be durable, they should be *steeled* in the fore-bits and heels, the former being only a thickening, and the latter a little turning up of the shoes. The shoes usually worn by stallions are very clumsy, and, in case of excited action in the horse, are apt to cause him tramp himself. It is perhaps too much fatigue for a man to walk with a stallion in all his journeys during a season, and the only remedy for the inconvenience is to give him a pony gelding to ride upon; but should possession of a pony induce him at any time to trot the stallion along the hard road, to make up for time spent in his own indulgence, or to overwalk him in too long journeys, it would be much better to cause the man to walk who would so far forget his duty; and, at the same time, to forbid him, with strict injunctions, to mount the horse's back. If the leader has a proper idea of his work, he will divide the district so as to go over it all in regular order in the time a mare would come again into heat should she prove not in foal. He should keep a book and enter the services of the horse day by day, not merely as a memorandum, but as a detailed document, by which to make up his accounts correctly, and in case of dispute arising from alleged negligence of service, to prove the regularity of his attendance. I have witnessed disagreeable disputes arise from the leader neglecting to keep an account of the services of the horse. It is customary, when the farmer affords a night's quarters, to do so gratuitously, and even supply the corn; but unless otherwise arranged, the understanding is, that the man supplies the corn and beans from his own store. A stallion in his travels requires at least 5 feeds or 1 stone of corn a-day, with a proportionate quantity of beans, at 5 separate times. He should always rest at noon. He should be supplied frequently with water during the day. Whenever he halts, his skin should be wisped and brushed, and his tail and mane combed. Every night he should have his feet searched with the foot-pick, and washed clean with the water-brush; and should they feel hot and hard when travelling in dry weather on dusty roads, a stuffing of cow-dung and clay forms a cooling poultice. A bran-mash at night, twice, or, at least, once a week, on Saturday night, with 1 oz. of nitre, proves an excellent alterative. His litter should be ample to encourage him to lie down and rest at night. It should always be borne in mind to give his food at stated hours every day, along with the conviction, that it is his food alone which enables the horse to maintain his condition and consequent spirit on his very exhausting travels. A stallion that loses condition and spirit to a considerable degree on his travels, one, in short, that wants bottom, is unworthy of serving draught-mares, for his progeny will assuredly prove as soft as himself.

(2614.) The usual treatment of *draught-horses* in summer is to allow them to lie out in the pasture-field all night, and give them cut grass between the yokings in the stable. Forage is supplied them, because the time between the yokings is too short to fill themselves with grass on pasture; but on the Borders, where the first yoking is over by 9 or 10 o'clock in the forenoon, the horses are put into the pasture-field until the afternoon yoking at 1 o'clock; and this plan saves all the trouble of cutting grass for the horses. The piece of grass allotted to horses is cut by the ploughmen, who each take the duty for a week by turns, and he quits the yoking in time to allow him to cut the requisite quantity and cart it to the stable. It is not his duty to supply the racks in the stable, except his own horse's, but to empty the load of grass on some convenient spot near the stable-door—a dirty and slovenly practice. No doubt it is better to keep the fresh cut grass in the open air than to put it into a house; but still a crib or enclosure could be made in a shady place conveniently near the stable, to keep the grass fresh, and place it off the dirty ground.

(2615.) Grass is cut with the common scythe, which is so well known an implement, that a particular description of it seems unnecessary. A few words, however, on the *choice of scythes* and of the *manner of mounting them*, may be useful to you. The handle, or *sned* or *sneath*, fig. 452, *a b*. is made either curved to suit the sweep of the instrument by the hands round the body, as in the figure, or straight, and the suitableness of both for work you shall learn when we come to consider harvest-work. The curved sned is usually made of willow, which, being shaped in hot water, and restrained, on being released when the wood becomes cold, retains the shape. Of scythes there are various kinds: the *common* kind keeps its edge but a short time, and in the long run is, I believe, more expensive than the *patent* kind, which consists of a steel plate rivetted between 2 small rods of iron, and which plate will continue to cut keenly until it is worn to the back-bone. The length of the blade of scythes varies from 28 inches to 46 inches, and the price of the common kind varies between these lengths from 2s. 4d. to 3s. 3d. each, and the patent from 3s. 2d. to 4s. 8d. each. There are besides these other kinds termed *crown*, *labelled*, and *extra-warranted* scythes. Bent sneds cost from 1s. 3d. to 1s. 6d. each, straight ones 1s. each. The straight are made of any sort of wood; I have seen good ones of larch. Scythes are sharped with *strickles* and *stones*. The strickles are made of fine sand embedded in an adhesive medium laid over the surface of a piece of wood of square or flat shape, and cost 6d. each. They are used to smoothen the edge after the stone, and serve of themselves, for a time, to keep the edge keen; and they are always attached to and carried along the upper end of the sned at *d*, by a T headed nail and spike. Scythe-stones are either square or round, are formed of the same sandstone as grindstones, and cost 4d. each. They are only occasionally used to set a new edge on the blade. Ragstones cost 4d. per lb. The blade of a scythe is mounted

Fig. 452.

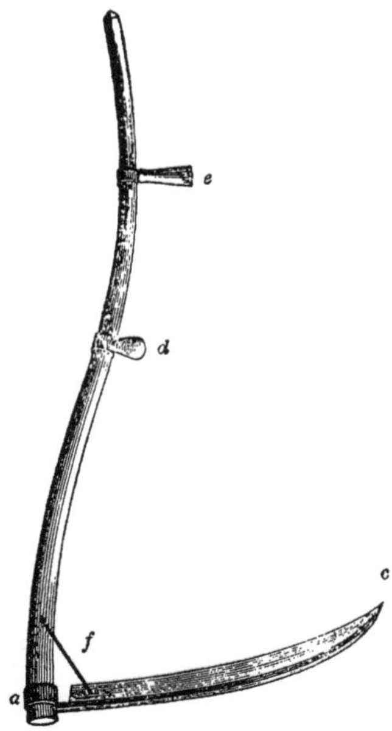

THE PATENT SCYTHE WITH BENT SNED.

in this manner:—the sned is furnished with an iron ring at the end, to which the blade is attached; the projecting hook at the butt-end of the blade is embedded flush into the sned by taking away a portion of the wood, and the ring is then slipped over the embedded hook, and is held tight in its position by an iron wedge as at *a*. The peculiar position which the blade bears to the sned is determined by measuring the length of the blade *a c* straight along the sned from *a* to *d*, where is fastened the handle for the right hand, and the same length from *d* to *c* fixes the point of the scythe, so that *a d c* forms an equilateral triangle; and of course the blade stands at an angle of 60° with the sned. Theory would advise the placing of the *plane of the blade* parallel with the ground, when the scythe is held as intended for cutting; but practice requires the cutting edge to be a little elevated above the ground, and above the back of the scythe which sweeps along the surface of the ground; and the reason for keeping the edge elevated is, that it would not only be apt to run into the ground if swung parallel with it, but the scythe would be worked with greater labour, as the stems of the plants to be cut would present an obstacle directly at right angles against the blade, whereas the edge set upwards cuts the stems in an oblique direction. The blade is still farther secured in its position by the grass-nail *f*, which is hooked by one end into a hole in the blade, and is nailed through an eye by the other to the sned; and a great use of the grass-nail is to prevent any plants cut from being entangled between the blade and sned. The left-hand handle *e* is placed to suit the convenience of the workman. I shall reserve remarks on the mode of using and sharping the scythe until we treat of harvest-work.

(2616.) Objections have been made to pasturing grass at all by any species of stock, inasmuch as the same extent of land will maintain a greater number of animals when cut and given them in houses or yards. To express this proposition shortly, *soiling* is a more profitable and less wasteful mode of using grass than pasturing. Although there is much truth in the observation, it is too generally expressed to be true in all cases. In the case of mountain-grass it is evidently an impracticable proposition to use it by soiling. Much cattle and sheep must, therefore, be allowed to pasture; and with regard to much of the old grass of the low country a great part of the summer would elapse before it would be fit for the operation of the scythe. What is to become of stock in the meantime? The only other grass left are the cultivated kinds, such as clover and ryegrass, and that from irrigated meadows. Of these two kinds it is quite possible to procure a supply of cut grass from water meadows by the time the Swedish turnips are exhausted in the beginning of June; but the cultivated grasses are not fit for cutting by that time except in the neighbourhood of large towns. What, again, is to become of stock in the meantime? Some other plants than clover and ryegrass must be cultivated to support the stock till that period; perhaps lucerne and Italian ryegrass might be cultivated for the purpose, but both cannot be cultivated everywhere, for lucerne will not thrive in Scotland, and before either could be cultivated any where for an extensive system of soiling, a different system of husbandry will have to be contrived, and a system to produce early forage in a late climate will not be easily discovered. In regard to the comparative extent of ground required for soiling and pasturage, it has been alleged to be 3 to 1 in favour of soiling; that is, 33 head of cattle were soiled from 20th May to the 1st of October 1815 on 17½ English acres, and which number of cattle, it was said, would have required 50 acres to pasture them.* On the face of this statement I would say, that any 33 head of cattle that could be maintained on 17½ acres of cutting grass, would as easily be maintained on the same land on 33 acres of pasture—1 acre of pasture to an ordinary-sized ox being quite sufficient to maintain it from May to October. So that the proportion is reduced from 2 to 1, which I believe is near the truth in regard to the maintenance of oxen. In regard to work-horses, it is different, for they will certainly require a much greater extent of ground in cutting grass than in pasture. To cut grass, however, for all the cattle on a large farm, to lead it to the steading, and to supply them with sufficient litter in summer, is what I consider an impracticable thing, were it for no other reason than that the crop of grain on any farm that admits of its grass being pastured cannot afford sufficient straw to litter stock the whole year; and if sheep are to be included in the soiling system, where is the steading that could afford them accommodation? Nor is the objection against grazing of the manure of animals being entirely lost, altogether valid, because land that is constantly grazed will support stock to an indefinite time; whereas where grass is cut and carried off every year, the time will soon arrive when the grass can no longer be cut, until manure be applied to the ground. Does not this circumstance of itself shew, that the dung dropped on pasture is not entirely lost, and that the land derives considerable advantage from being pastured? I have often thought that all the work-horses on a farm might be supported in the steading night and day upon cut grass. I have tried the experiment twice myself, but failed in both cases, at one time for want of cutting grass, the second cutting having entirely failed that year, and the other for want of straw for litter until the arrival of the new crop. The straw might have been

* See Sinclair's Code of Agriculture, p. 424, and Notes, p. 73, paragraph 354.

economised in the stable, but the stable in summer, even with open doors, ventilators, and no hay-loft, is insufferable; and the horses required far more straw to keep them dry in the hammel on cut grass than they did on straw and corn in the stable in winter. Taking all these untoward circumstances into consideration, they lead to the conviction that soiling on a large scale is impracticable; and until early growth of grass, as well as a late growth of aftermath, and plenty of straw, are assured to the farmer every year, I cannot see how soiling can be established as a regular practice in husbandry on a farm of even moderate extent. On a small scale where only a few animals of every kind are kept, I conceive that soiling might be practised with advantage, and it behoves all small farmers to make their grass land go as far as possible.

(2617.) Work-horses, when on grass, are subject to few distempers, the principal being annoyance from a host of insects, and amongst these the common Horse-fly or Cleg, and the Bot-fly, are the most troublesome. The *cleg* or *gleg*, a term derived from the Danish *klaeg*, *Hæmatopota pluvialis*, represented by *a* in fig. 453, is so well known, that a particular description of it seems unnecessary. It may be said generally of the tribe of *Tabanidæ*, of which this is one, that they appear in June, and come into full force in autumn. They are more plentiful in the southern than in the northern parts of the country. They delight in warm and sultry weather; are most active on the wing during the day, and therefore most troublesome to horses and cattle when they stand most in need of repose. They are particularly excited and eager for blood when the atmosphere is in a warm and humid state, such as it usually is after a thunder shower; and it is this circumstance which has obtained the specific name of *pluvialis* for the cleg. A remarkable fact in reference to this species is, that the males are seldom seen, their numbers seem to be remarkably few in proportion to those of the other sex, and they appear to subsist entirely on the juices of flowers, and, in conformity to their innoxious habits, the organs of the mouth are much less developed than in the female. The wings *b* are speckled; the eyes *c* are green, with transverse undulating purple-brown bands; and *d* is an antenna. Another pest to the horse is the Great Spotted Horse-bot, *Gasterophilus equi*, seen at *a*, fig. 454. It is about 7 lines in length; general colour clear yellowish-brown; thorax inclining to grey; abdomen rust-brown, with a tinge of yellow; wings whitish; and legs yellowish. The antennæ are inserted in the cavity of the face, as seen at *b*, the second joint large and kidney-shaped, the remaining three forming a naked bristle as at *c*. The eyes are equally distant in both sexes; mouth either entirely wanting, or consisting merely of an indistinct line or opening. This insect takes no nourish-

Fig. 453.

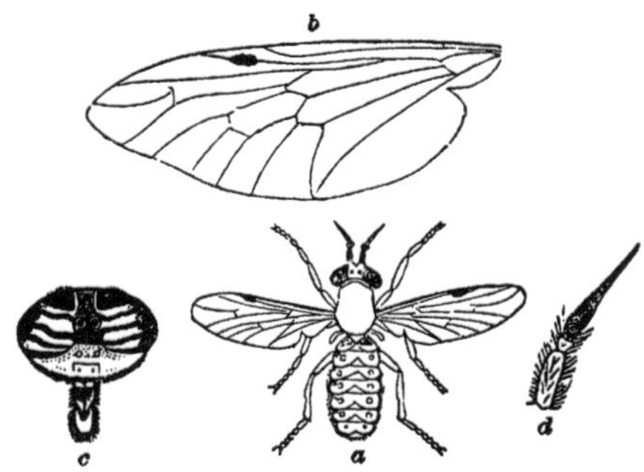

THE CLEG OR GLEG—HÆMATOPOTA PLUVIALIS.

ment of any kind; in fact, the alimentary canal has no opening at its anterior extremity. It flies in company, producing a humming sound. "The female

Fig. 454.

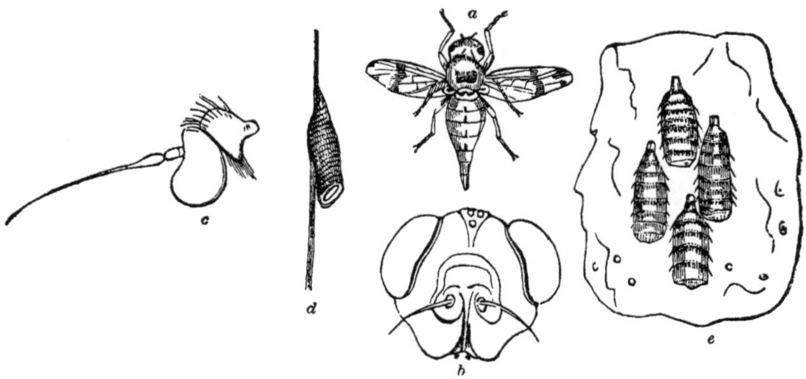

THE HORSE-BOT—GASTEROPHILUS EQUI.

having selected the individual to which her treasure is to be entrusted," says Mr Duncan, "she continues to hover about for a short time till the egg be propelled through the oviduct, and placed in the pincers at the extremity of the anal tube. Thus prepared, she makes a sudden descent upon the horse—her body carried nearly in a perpendicular direction, and the ovipositor curved forwards—and deposites the egg upon a hair, to which it instantly adheres by means of a glutinous matter secreted along with it. This process, which is performed with such expedition that the fly can scarcely be said to alight on the horse, is repeated at intervals till the whole of the mature eggs are discharged." These eggs, which are very numerous, 400 or 500 being sometimes placed on a single hair, are somewhat pouch-shaped, and chagreened with transverse and longitudinal striæ, as seen at *d*. Under the guidance of an instinct which cannot be sufficiently admired, the fly almost invariably attaches her eggs to some part of the fore-quarter of the horse, the inside of the knee and the shoulder being the spots most commonly selected, so as to be within the reach of his mouth, for he is himself to be made the unconscious instrument of conveying them into his stomach, where alone they can be brought to maturity, the temperature of a horse's stomach being as high as 102° Fahrenheit. Even when beyond the reach of the mouth, the eggs are not necessarily lost, for horses are in the habit of licking each other, and a horse free from bots may thus receive them from another. "When the eggs are mature," continues Mr Duncan, "it would seem that the larvæ make their appearance very soon after they are touched by the tongue, the warmth and the moisture both contributing to their immediate development. Indeed, if the larvæ were not disclosed before reaching the stomach, or very shortly after, the eggs would very soon pass into the alimentary canal. The larvæ fix themselves by hooks to the inner tissue of the stomach, where they remain in security, uninjured by the powerful action of the gastric juice, and enjoying the warmth of a tropical climate." A small group of these larvæ adhering to the coat of the stomach are represented by *e*. Their colour is pale reddish-yellow. Their only food seems to be the humour secreted by the internal membrane of the stomach, or it may be the chyme, the latter undergoing a farther elaboration to adapt it to their system. Bots take up their quarters in the stomach in the end of summer or autumn, and pass the whole winter and spring months there, without undergoing any change, save gradually enlarging and advancing to maturity. When that is complete, they cease to retain their hold, pass into the intestinal canal, and are ejected by the anus. On account of the many ordeals which this insect has to pass in its transformation, perhaps not 1 in 100 of the eggs ever arrive at the perfect state of fly. The Red-tailed Horse-bot, *Gasterophilus hæmorrhoidalis*, though only half the size of the preceding, is nevertheless a greater torment to the horse. The female parent fly deposites her eggs on the lips of the horse, and this operation is attended with so much pain, that no sooner does it make him aware of the presence of the fly, than he tosses his head and gallops off to a different part of the field, or, if he has the opportunity, betakes himself to the water, where his tormentor generally leaves him, having a peculiar dislike to that element. Indeed all the tribe of gad-flies have, and, to avoid them, it is not uncommon to see numbers of cattle lying on the sea-shore until the approach of the tide alone compels them to retire. When this fly succeeds in fixing an egg, the horse rubs his mouth against the ground or upon his fore-legs in great agitation, frequently striking out with his fore-foot, which occasionally comes in contact with the jaw, and serves but to increase his irritation. The larvæ are taken into the stomach, and fix themselves there, exactly in the same manner as the greater bot. When they reach the intestines, they remain a long time, casting anchor again in the rectum, where they cause great uneasiness to the horse, causing him to kick frequently, and even rendering his movements awkward. These bots should occasionally be looked for in horses that have been out at grass the *preceding* year, at the extremity of the anus. The only speedy remedy for getting quit of them is in back-raking by the anus. The more rare species are the *Gasterophilus nasalis*, *salutiferus*, and *Clarkii*. Mr Bracey Clark was of opinion that the presence of bots in no way injured the horse, but on the contrary, by stimulating the stomach, they tend to prevent cholic, gripes, and other indigestions which affect the head of the horse and produce staggers. "The appearance of exanthemous eruptions on the skin," he says, "and the formation of local abscesses, from the same cause of partial irritation, often relieve a general disorder of the system. The mucous membranes of the skin possess this power, when irritated, in the most eminent degree, and to these the larvæ of the *Œstri* are applied. Irritating the membranes of the stomach in other animals would excite nausea and vomiting; but the horse not possessing this power, his stomach is peculiarly fitted for the stimulus of such animals."* An annoying insect of less importance, is the *Chrysops cæcutiens*, *a*, fig. 455, which is of a bright colour, and though frequent in some parts of England, is not so great a pest there as to horses on the Continent: *b* are the organs of the mouth, and *c* is an antenna.

Fig. 455.

A HORSE-FLY—CHRYSOPS CÆCUTIENS.

* Clark's Essay on Bots, p. 40.

Another fly, the *Stomoxydæ calcitirans*, is in size and markings not unlike the common house-fly, *Musca domestica*. This insect attacks various animals, as well as man himself, and becomes very troublesome in certain localities. It attacks the legs, and its punctures are attended with great pain, especially in damp moist weather.* Another annoyance to horses is the forest-fly, *Hippobosca equina*, seen at *a*, fig. 456; its antennæ *b*, consist of a single tubercular articulation, with 3 setæ at the extremity. "This insect," says Mr

Fig. 456.

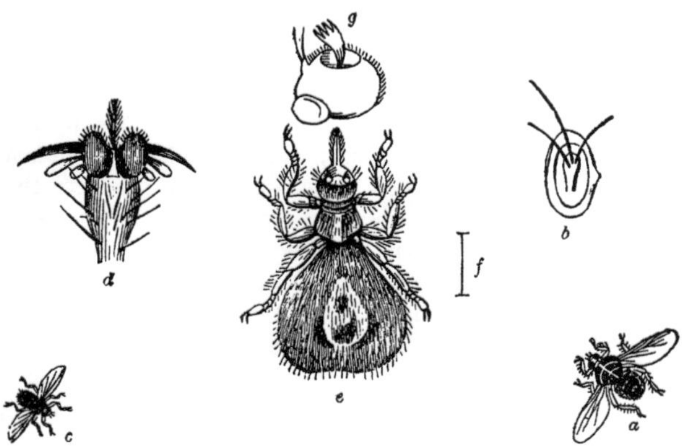

THE HORSE-FOREST-FLY—HIPPOBOSCA EQUINA; BIRD SPIDER-FLY—ORNITHOMYIA AVICULARIA; AND SHEEP KED—MELOPHAGUS OVINUS.

Duncan, "is generally distributed throughout the country, but it is scarce in all the northern quarters, and does not seem to become abundant till we reach the wooded districts of the central counties of England. It flies with facility, but seldom appears on the wing except during bright sunshine. June, July, and August are the months it is in force. Its attacks are principally confined to horses. It occasions no other harm than an extreme degree of irritation, and if the flies are numerous, the animal is apt to become unmanageable. It insinuates itself by a sideling crab-like motion beneath the hair, and anchors itself to the skin by means of its largely-toothed claws. It also runs about among the roots of the hairs with great ease, creating an insufferable titillation, which is still more increased by the frequent insertion of its proboscis into the pores of the skin. The places to which it prefers attaching itself are the under side of the belly, beneath the tail, and on the under side of the jaws. The insect is so flat, tough, and unyielding, that it is by no means easy to kill it by pressure, and it is, moreover, tenacious of life. It is said that horses long accustomed to its attacks, become, in some measure, indifferent to them; those which have never experienced this plague, which is enough to render some animals almost frantic, may be saved from it, according to M. Köllar, by the following application. Take of mineral earth 8 oz., and of lard 1 lb., and make them into a salve. Some of this salve is to be rubbed on here and there upon

* Quarterly Journal of Agriculture, vol. x. p. 529-46.

the hair, and worked in with a wisp of straw. After 24 hours the salve is to be washed off with warm water, in which brown soap has been dissolved. Care must be taken that the horse does not catch cold.* This insect and its allies are neither oviparous nor viviparous. The egg, when fecundated, descends from the ovarium into a kind of matrix, consisting of a large musculo-membranous bag, expressly designed for its reception, and analogous to the uterus of mammiferous animals. The egg is here hatched; and the larva passes its life and is converted into the pupa in the same receptacle. When the pupa is completely formed, it is then extruded from the body of the mother in the form of a soft, white, oviform body. It soon changes its colour to brown, then to black, and at the same time the skin becomes hard and strong, so much so as almost to resist the edge of a knife. Out of this body the insect makes its exit by a kind of lid." The smallest insect *c* on the same cut is the Bird-spider fly, *Ornithomyia avicularia*, of the same nature as the insect just described, but scarcely half the size. Its claws are very strong, with two lobed appendages near the base of each, as seen at *d*. It infests various birds, particularly the grouse and black-cock. *Anthomyia meteorica* is often troublesome to horses, by collecting in clouds round their heads in warm weather, and flitting about with a kind of jerking flight, and occasionally alighting on the lips and nostrils of the animal, to his no small annoyance.†

(2618.) There is so curious a fact connected with the manufacture of scythes, that I cannot refrain mentioning it. "In the manufacture of scythes, the length of the blade renders it necessary that the workman should move readily, so as to bring every part on the anvil in quick succession. This is effected by placing him in a seat suspended by ropes from the ceiling, so that he is enabled, with little bodily exercise, by pressing his feet against the block which supports the anvil, to vary his distance to any required extent." ‡

---

65. OF SHEEP-WASHING, SHEEP-SHEARING, AND WEANING OF LAMBS.

"Gay shearing time approaches. First, howe'er,
Drive to the double fold, upon the brim
Of a clear river—gently drive the flock—
And plunge them one by one into the flood."

DYER.

(2619.) I have said, that as lambs become strong enough to be put to pasture, they always get new grass, in order to increase the milk of the ewes (2371.). The new grass, to be pastured by ewes and lambs, should be selected with judgment, and that intended for hay should first be stocked, because it is found that new grass, if moderately

* Köllar's Treatise on Insects injurious to Gardeners, Foresters, and Farmers, translated by Misses J. and M. Loudon.

† Quarterly Journal of Agriculture, vol. xi. p. 51-7.

‡ Babbage on the Economy of Machinery and Manufactures, p. 27.

eaten down in spring, stools out, and affords a thicker cutting at hay time, than if it had not been so pastured. For the same reason, the new grass intended to cut for horses' forage, should also be earlier pastured, than what is to be pastured all the season, not only to give both it and the hay-grass time to attain their growth when they shall be wanted, but to give the pasture-grass time to become so strong as to support being pastured. None of the new grass should be eaten too bare, even the part which is to be pastured by the ewes; and rather than commit suchia mistake, even in a late season, the ewes should have a hasty run over the best of the older grass for a fortnight or so, till the hained new grass has revived. After the castration of the lambs (2377.), there is nothing to do to them until the ewes are washed preparatory to shearing the wool from them, and which is done about the beginning of June (2379.).

(2620.) When the turnips are all consumed, and the time has arrived for the last of the turnip land to be ploughed up for barley (2388.), the *hoggs* receive a change of treatment. The *wether*-hoggs are either sold to the dealer off the turnips, or put to grass till shorn of their wool, and then disposed of; but the *ewe*-hoggs are always retained on the farm, because it is they which supply the waste of ewes, and are, of course, clipped of their wool in due time. The circumstance that determines which of these ways the *wether*-hoggs should be treated, is the state of the wool and mutton markets. If you find, on examination, that the hoggs are so forward in condition as to realize as much money off the turnips, with current prices, as they probably would, after being kept for a month on grass, and washed and clipped, it is, of course, more profitable to dispose of them at once; and, independent of this circumstance, should you fear the grass to prove insufficient to support them well till they are clipped, there is a necessity for parting with them immediately off the turnips; but should you find that you have grass to maintain their condition, and that the wool market is likely to be brisk, it would, of course, be advisable to clip them. It is of less importance to *increase* the condition of the *ewe*-hoggs off the turnips by putting them on the best grass, nevertheless they should not be allowed to *fall off* in condition, for fear of injuring the quality of their wool. If you determine on selling the wether-hoggs, you should first ascertain their value; and in attempting this you will at once perceive, that a sheep wearing its coat of wool cannot be subjected to the ordinary rules of measurement; nor can its true weight be found by weighing it alive, because the weight of the wool enters as a disturbing element into the calculation, and the value of that material depends on very different circumstances from that of mutton. A new clipped sheep, however, may either be measured or weighed, and its value ascertained pretty nearly. The eye and the hand must be employed to judge of the weight of a rough sheep, and there is no more certain way of acquiring a correct judgment of its weight in that state, than by first handling the sheep in the way I have described in (1387.), and weighing the 4 quarters after it is slaughtered. An average sized sheep of a large flock thus treated will enable you to form a pretty correct idea of the average weight of the whole, and the market price of mutton per stone is quoted in all the newspapers of the day.

(2621.) The season for *washing* sheep having arrived, a fit place should be selected for the purpose. It should consist of a natural rivulet, or, where that is wanting, of a large ditch, having both its banks clad with clean sward. The next step is to form a damming across the rivulet, if it is not naturally sufficiently deep of water to conduct the operation of washing. The bottom of the river or ditch should be hard and gravelly, and the water in it pure, or it will not answer the purpose, as a soft and muddy bottom, and dirty water, will injure the wool more than do it good. A damming may be made either entirely of turf-wall built across the stream, though that imposes considerable labour and waste of grass, or with an old door or two or other boarding placed across the bed of the stream, supported by stabs against the weight of water, and the chinks at the bottom and sides filled up with turf; and over which, when the water accumulates, the water falls. In constructing this dam, the overflowing should be as great as to cause such a current in the pool as to carry away quickly all impurities, such as earthy matter, greasy matter, small locks of wool, and scum. One side of the pool is occupied by the unwashed, and the opposite by the washed sheep. They are confined in their respective places by flakes, fig. 216, or nets, fig. 217. To prevent any sheep taking the water of themselves, which they are apt to do when they see others in before them, the fence should be returned along the sides of the pool as far as the men who wash the sheep are stationed. Fig. 457 brings out all these particulars pretty well. The damming *a a*, by means of doors and stabs, and turfing, retain the water until it overflows. The net on each side of the pool is returned so far down both its sides. The depth of the water is seen to take the men to the haunches—the proper depth.

(2622.) Everything being thus prepared at the pool, the sheep are also prepared for the washing. The lambs not being washed, and to save trouble with them at the washing pool. they are separated from their mothers, and left in a court of the steading until the washing is over. The ewes, hoggs, and dinmonts are all taken to the pool in a lot. They should be driven gently, and not allowed to be at all heated. The

Fig. 457.

SHEEP-WASHING.

ewes will be troublesome to drive, being always in search of their lambs; but notwithstanding this annoyance, they should not be dogged, and rather give them plenty of time upon the road. They should be driven along the road most free of dust or mud. The men who are to wash also prepare themselves by casting their coats, rolling up the sleeves of their shirts, and putting on old trowsers and shoes to stand in the water with. The shepherd and other 2 men are quite enough to wash a large number of sheep thoroughly, but if the stream is very broad, another may be required to save time in handing the sheep across. These 3 men are represented in the figure, *e* being the shepherd, and the last man to handle the sheep, and *d* and *c* his assistants. At least 2 men are required to catch the sheep for the washers, of whom 1 is seen at *b*. On an occasion of this kind the men receive a gratuity of bread and cheese, and ale, and also a dram of spirits as a stimulus, as at *h*, where the dog is seen to keep watch. Indeed, some stimulant of this kind is requisite for men who stand for hours in the water, the lower half of their body being chilled by the water, and the upper half being heated by the work. If they were provided with large fishermen's boots they would less require such a fillip.

(2623.) The *washing* is performed in this way:—While the 3 men are taking up their positions in the water, the other 2 are catching a sheep; and to render this fatiguing work more easy, the fold should not be made larger than to contain the sheep easily. A sheep is caught, and is being presented by *b* to the first washer *c*, who takes the sheep into the water, and allowing its wool to become saturated with it, turns it over upon its back, keeping up its head, and taking a hold of the arm of a fore-leg with either hand, and of the wool on the opposite side of the head with the other. He then dips the sheep up and down, to and fro, and from one of its sides to the other, slowly, causing all the wool to wave backwards and forwards, as if rubbing it against the water. In doing this the water becomes very turbid about the sheep, and he continues to do it till it clears itself, when he hands the sheep to the next washer *d*, standing in the middle of the stream. Whenever *c* gets quit of one sheep, another should be ready by the catchers for him to receive into the water. The second washer *d* holds and manages the sheep in the same manner; and then hands it to the shepherd *e*, and is immediately ready to take another sheep from the first man. It is the duty of the shepherd to feel if the *skin* of the sheep is clean, and every impurity removed from the wool. The position of the sheep on its back is favourable for the rapid descent of earthy matter from the wool. Wherever he feels a roughness upon the skin, he washes it off with his hand, and wherever any clots are felt in the wool which have escaped the other washers, he rubs them out. The belly, breast, and round the head, he scrubs with the hand. Being satisfied that the sheep is clean, he dips it over the head while turning it to its natural position, when it swims ashore, and gains the bank at *g*. Its first attempts at walking on coming out of the water are very feeble, its legs staggering under the weight of the dripping fleece; in a little after it frees itself from the water entirely by making its fleece whirl like a large mop. It will be observed in the figure that the 3 men do not stand abreast across the water, but in eschelon, and they stand so for this reason, namely, the sheep being in the dirtiest state when in the hands of the first man *c*, his standing farthest down the stream allows the dirtiest water to flow away at once, and not come near the stations of the other men. The sheep being in a comparatively clean state when it reaches the second man *d*, the water from it cannot render that more dirty which runs past the first man *c*. In like manner, the same sheep, when with the shepherd *e*, can but very slightly render the water impure to those below it, and being washed highest up the stream, gets an immediate supply of the cleanest water for finishing its washing. This is the way in which sheep are washed in the Lowlands, and from 2 to 3 scores may be washed in 1 hour, according to the size of the sheep, the activity of the washers, and the supply of water. Tups are washed 2 or 3 weeks before the rest of the flock, and clipped as long before to allow time to their wool to grow; and as they are usually in high condition, their new wool rises soonest, and, of course, soonest allows the old fleece to be taken off. After washing, sheep should not be driven along a dirty or dusty road, nor should they be put into any grass-field having bare earthy banks, against which they might rub themselves. In fact, they should be kept perfectly clean until their fleeces are taken off. How long they should be kept depends on the state of the weather, for the wool must not only be thoroughly *dry*, but the *yolk*, as the natural oil of the wool is called, must return again into it. Perhaps 8 or 10 days may suffice to produce both these effects. But another circumstance more important than either should receive attention before the wool is clipped, which is, that the new wool be risen from the skin before the old be attempted to be taken off. Disregard to this state of the fleece will make good clipping very difficult to accomplish, and it will certainly deteriorate the appearance of the fleece. You need be under no apprehension of the wool falling off when the new growth commences, for wool will remain for years upon the sheep's back if not clipped off, and the sheep be in a clean and healthy state. How many years the fleece may continue to grow I do not know, but I have seen a 3-years' fleece. Lord Western shewed Anglo-merinos, at the Show of the Royal English Agricultural Society at Oxford in 1839, the fleeces of which, when clipped, weighed, I think, 20 lb. each.

(2624.) A place under cover should be selected for clipping the fleeces. The straw-barn L, fig. 4, Plate IV., of the steading, is a very suitable place for the purpose. The end next the chaff-house *r*, between the two doors, will answer for the clipping process, and the remainder will contain the sheep under cover until they are returned clipped to the fields again. The clipping floor is prepared in this way: Let clean wheat straw be spread equally over the floor 2 or 3 inches thick, and spread over it the large canvass barn-sheet, the edges of which should be nailed down tight to the floor. The use of the straw is to convert the floor into a sort of soft cushion for the knees of the clippers, as well as for ease to the sheep. A broom should be provided to sweep the cloth clean. A board 2½ feet above the floor should be provided to wind the fleeces upon, and it may be placed along the wall of the barn opposite to that occupied by the clippers. A space near this should be cleaned, and a sheet spread upon it, for putting the rolled fleeces upon. The remainder of the barn should be cleared of dust both from the floor and walls as high as the sheep can reach, and a little clean straw strewn upon the floor for them to lie upon.

(2625.) The instrument by which the wool is clipped off sheep is named *wool-shears*, as seen in fig. 458. They require no particular description farther than to explain that the bend or *bowl a*, which connects the two blades, acts as a spring to keep them separate, while the pressure of the hand on each side of the handle *b* overcomes and brings the blades together. There are wool-shears which have additional springs placed between the handles *b* to separate the blades more forcibly, but are not so agreeable to the hand as the simple bent spring *a*. Sometimes the spring of *a* is so strong as soon to tire the hand, to relieve which a piece of cord is wound slackly round the handles. Strong shears are more easily worked if held near the blades; but if held upon their sharp edges they will soon hurt the hand. When not in use, the blades are held together by their points being passed through a ring of leather. A rag-stone is used to sharpen wool-shears. The cost of the shears is from 2s. 6d. to 3s. 6d. each. A shepherd requires 2 or 3 pairs to do other jobs, and he makes it a rule to use clipping-shears on no other occasion than at regular shearing.

Fig. 458.

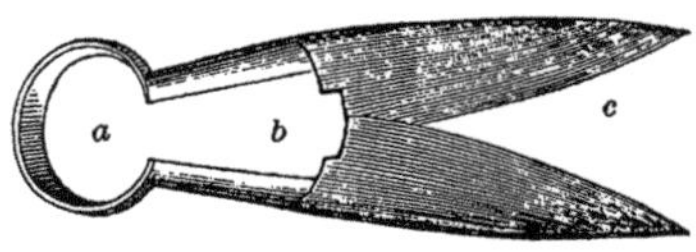

THE WOOL-SHEARS.

(2626.) In case of dew or rain in the morning, it is customary to bring into the barn as many dry sheep on the previous evening as the clippers will shear on the ensuing day. About 1 score of Leicester sheep to each clipper is considered a very good day's work. It is a frequent custom for neighbouring shepherds to assist each other; and though the plan does not perhaps expedite the entire sheep-shearing of the country, yet a number of men clipping at the same time makes work seem lighter, and it gets the clipping of any individual flock the sooner through. It is seldom that a steward can clip sheep, but is an accomplishment not unfrequently possessed by a hedger, and if the cattle-man has been a herd, he can lend a hand as far as he is able, and in this way a party of clippers is mustered upon the farm itself. Clipping being both a dirty and heating work, the coat should be stripped, and the oldest clothes worn; and the hat and the vest are commonly thrown aside. Garters or tight knee-breeches will be found very irksome pieces of dress in clipping. There is an order followed in the classes of sheep as they are clipped; the tups are taken first, then the hoggs, and lastly the ewes; and where there are dinmonts or wethers, they follow the tups. The reason for adopting this order is, that as tups are highest in condition, and having been earliest clipped the season before, their fleece rises earliest in this; and if they are to be let in autumn on hire, the sooner they are washed and clipped they will then shew in better wool. Hoggs, too, being forward in condition, should be ready before ewes, as the latter lose their flesh by suckling, or require longer time to bring the yolk again into their old wool, and for the new to rise. The object of washing the sheep perfectly clean, will be apparent at shearing; for if the shears grate upon anything, they will make bad work; and if any dirt is found on clipping, either upon the skin or fleece, the shepherd is alone to blame who had the sheep last through hands, and had the charge of them afterwards on pasture.

(2627.) *Clipping* is done in this way: Whenever a sheep is caught in the barn, every straw or dirt on the wool or hoofs should be taken away before it is laid on the canvass carpeting. Clipping consists of 3 stages, the *first* of which is represented in fig. 459. After setting the sheep on

Fig. 459.

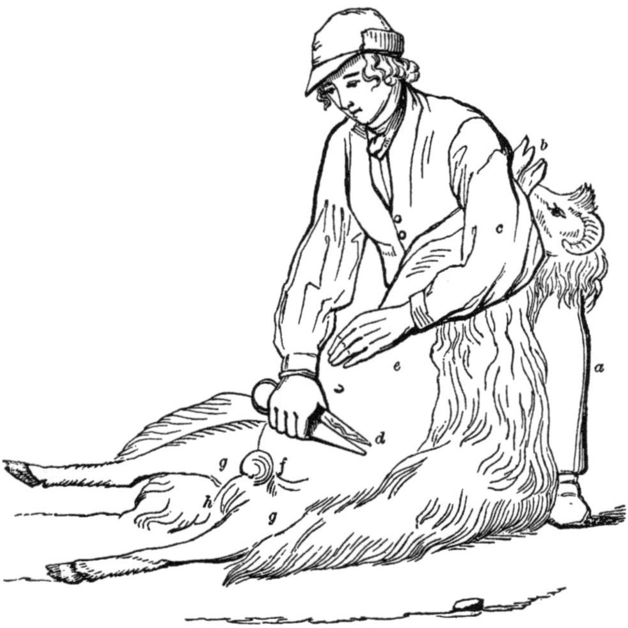

THE FIRST STAGE OF CLIPPING A SHEEP.

its rump, and, on the supposition that the clipper is a right-handed man, he goes down on his right knee, and leans the back of the sheep against his left leg *a*. Taking the shears in his right hand, and holding up the sheep's face with his left, he first clips the short wool on the neck, and passes down the throat and breast between the fore-legs to the belly. Then placing the fore-legs under his left arm, as seen at *b* under *c*, the belly is left exposed to be next shorn across from side to side down to the groins. In passing down here, while the shears *d* are at work, the left hand *e* is engaged keeping the skin tight where it is naturally loose. The scrotum *f* is then bared, then the inside of the thighs *g g*, and then the under side of the tail *h*. These complete all the parts of this position. For clipping this part of the sheep, small shears, as in the figure, will suffice ; and as the wool is short and of a detached character on the under side of a sheep, it is best clipped away by the *points* of the shears, as by *d*.

(2628.) The shears are used in a particular manner, to be safe alike to the fleece and the skin of the animal. The essential thing is to keep the *points always clear of the skin*, for if held downwards, they will inevitably run into it, and should such a prick be made when the hand is about to close, the consequence will be, that a large piece of the skin will almost be clipt out before the clipper is aware of what he is about. This is a common error committed by new clippers, and it is a great offence in any clipper's hands. The only way to avoid this serious injury to sheep, is to rest the *broad* part of the shears only and always upon the skin. In this position, with the skin drawn tight by the left hand, the shears are made to move forwards with a hold of the wool not exceeding 1 inch in breadth, in very short and frequent clips, taking care only to bring together the broad parts of the blades from where they are seen to separate in fig. 458 to as far as *c*, keeping the points always apart. The form of the carcass of a sheep being round, it is clear that the shears cannot make a long clip by bringing the points of the blades together at every stroke, without cutting the wool with the points at a considerable elevation above the skin. *Very* short clips, no doubt, make slow work, but rather work slow with short cuts than injure the wool with long. Experience will teach you to make longer clips effective when you know how to manage the shears dexterously.

(2629.) Fig. 460 represents the *second* stage of clipping. Its position is gained by first relieving the fore-legs from the first position, fig. 459, and, gently turning the sheep on its far side, the fore legs *c* are put under the right or clipping arm *f*, while the clipper, going on both knees, supports the shoulder of the sheep upon them, thus giving the animal an easy reclining posture. You may rely upon this fact, that the more at ease the animal feels, the more readily will it lie quiet to be clipped. Supporting the head of the sheep with his left hand, the clipper first removes the wool from behind the head, then around the entire back of the neck to the shoulder top. He then slips its head under his left arm, as *a* under *g*. Having the left hand thus at liberty, he keeps the skin tight with it, while he clips the wool with the right, from where the clipping in the first position was left off to the back bone. In the figure the fleece appears to have been removed about half way down the carcass, the left hand *b* is laid flat, keeping the skin tight,

Fig. 460.

THE SECOND STAGE OF CLIPPING A SHEEP.

while the right hand *e* holds the shears at the right part and in the proper position. The clipper thus proceeds along the thigh and the rump to the tail *d*, which is entirely bared at this time.

(2630.) Clearing the cloth of the loose parts of the fleece, the clipper, holding by the head, lays over the sheep on its clipped side, and still continuing on his knees, slips his left knee *a*, fig. 461, over its neck to the ground, while his left foot *b*, resting on the toe, supports the left leg *c* over the neck of the sheep, and keeps its head *d* down on the ground. This is the *third* position in clipping. The wool having been bared to the shoulder in the second position, the clipper has now nothing to do but to commence where it was left off in the first position, and clear the fleece entirely to the back bone, meeting the clips where they were left off in the second position, the left hand *e* being still at liberty to keep the skin tight, while the right hand *f* uses the shears along the whole side to the tail. The fleece *g* is now quite freed from the sheep. In

Fig. 461.

THE THIRD AND LAST STAGE OF CLIPPING A SHEEP.

allowing the sheep to rise, care should be taken that its feet do not become entangled in the fleece, for in its eagerness to escape from the unusual treatment it has just received, its feet will tear the fleece to pieces. Immediately that the lot of sheep in the barn is clipped, it is taken to the field, and another brought in its place.

(2631.) A new clipped sheep should have the appearance of fig. 462., where the shear-marks are seen to run in parallel bands round the carcass, from the neck and counter *a*, along the ribs *b*, to the rump, and down the hind leg *c*. When pains are taken to round the shearmarks on the back of the neck *d*; to fill up the space in the change of the rings between the counter and the body at *e*; to bring the marks down to *f* to the shape of the leg; and to make them run straight down the tail, a sheep in good condition so clipped forms a beautiful object. A sheep clipped to perfection should have no marks at all, for they are formed of small ridglets of wool left between each course of the shears; but such nicety in clipping with shears is scarcely possible, and, at any rate, the

Fig. 462.

A NEW CLIPPED SHEEP—THE PARALLELOGRAM SHEWING ITS SYMMETRY.

time occupied in doing it would be of more value than all the wool that would be gained. It should be borne in mind, however, that the closer a sheep is clipped, it is the better clipped, and is in a better state for the growth of the next year's fleece.

(2632.) I have introduced so many illustrations on this subject, in order to shew the most easy positions that can be assumed in clipping, both to man and animal, that these positions may easily be compared with the common ones. On making this comparison, it is necessary to look again at the *first* stage of the process, fig. 459, the common practice being to place the sheep upright on its tail, and the clipper to stand on his feet, supporting its back against his legs—an insecure and painful position for the sheep, and an irksome one for the man when he bows down to clip the lower part of the animal. Again, in the *second* stage, fig. 460, it is customary for the man still to remain on his feet, and the sheep upon its rump, while he secures its head *sideways* between his legs, in order to tighten the skin of the entire side which is now bent outward. The skin is thus certainly tightened, but at the expense of the personal ease of the animal; for the hand can tighten the skin as well, as shewn in figs. 460 and 461, by *b* and *e*, whilst the bowing down so low, and as long, until he clips the entire side, cannot fail to pain the man's back. The *third* position is nearly the same in both plans, with this difference in the old one, which keeps the right leg bent, resting on its foot—a far more irksome position than kneeling on both knees. The treatment of the fleece I shall consider in a separate section; and the mode of washing and clipping Highland sheep will be found below. I may here remark, that sheep-shearing is usually held as a merry season, a sort of harvest, in which an allowance of good victuals—beef and broth and beer—is usually awarded to clippers engaged at their fatiguing task.

(2633.) Clipping makes such a change on the appearance of sheep, that many lambs have difficulty at first in recognising their mothers, whilst a few forget them altogether, and wean themselves, however desirous their mothers may be to suckle them; but as the ewe is content with one lamb, many a twin which does not follow her is weaned on this occasion. It should be the shepherd's particular care to mother the lambs frequently after clipping; but I am aware of the difficulty of bringing an old lamb and ewe together, without much disturbance to the rest of the flock; and besides, the shepherd cannot attend constantly on the clipped portion of his flock while engaged with clipping the rest, and this being the case, constitutes another reason, besides those given above (2626.), for clipping the ewes last. Leicester lambs are *weaned* at the end of June or beginning of July; and the process is simple and safe, as most of them by that time chiefly depend upon grass for support. All that is required is to separate the lambs from the ewes, in fields so far lying asunder, as to be beyond the hearing of the bleatings of each other. Where there is the convenience, lambs should be put on hilly pasture for some weeks at this time, the astringent quality of which gives an excellent tone to their system, and renders them more hardy for winter. Some farmers even hire rough hill pasture for their lambs, but where such cannot be had, they are put on the oldest, though good pasture, for a few weeks before the aftermath is ready to receive them.

(2634.) The ewes, when separated from their lambs, should be kept in a field of rather bare pasture, near at hand, until their milk be dried up. They must be *milked* by the hand, for a few times, till the secretion ceases—once, 24 hours after the lambs are taken away—again, 36 hours thereafter—and the third time perhaps 2 days after that. Even beyond that time a few may feel distressed by milk, which the shepherd should relieve at intervals until the udders become dry. Indeed, milking after weaning of lambs, is essential to the safety of ewes, and I fear it is not so effectually performed as it should be until the udders go dry. The danger to be apprehended from its neglect is the plugging up of the teats with caseous matter, deposited therein by the milk which should have been drawn away ; and which plugging, in the next lambing-season, will probably prevent the natural flow of the new milk ; and the consequence will be, that inflammation will be set up in the udder, and the ewe either take puerperal fever, that is, *udderclap*, or *garget* (2383.), and die, or her lambs be so restricted of milk as to be half starved. Ewes are milked in a very different manner from cows. A long narrow bught formed of hurdles on both sides is erected along a fence close to the gate of a field near the steading or dairy, and it should be no larger than to contain all the ewes in a crowded state. The ewes being driven into the bught head inwards, women proceed with the milking, which is accomplished by holding a small handy, that is, a small vessel formed like a milking-pail, in the left hand, and squatting down *behind* the ewe, the milk is stripped clean from the teats alternately, with the right hand into the handy. The milker requires to be always on her guard, and remove the handy the instant she sees the ewe shewing the least symptom of voiding either water or fœces, and a ewe is very apt to void the former whenever she is touched behind for the first time. Every ewe is turned out of the bught by the shepherd as it is milked, to prevent its coming in hand again. Time was when ewe-milking created a great stir in the farm-house for the making of ewe-milk cheese, and so much anxiety did housewives evince to make it, that the ewes were milked till they were perfectly lean to supply a sufficiency of this sort of milk. Better ideas now prevail, and farmers very properly will not allow their ewes to be milked oftener than is requisite to render them completely dry. It was misplaced economy to reduce the condition of the entire ewe-flock for the poor boast of making a few strong-tasted cheeses.

(2635.) When lambs cease to bleat for their mothers, they should be *marked* and *buisted*, not only to identify them with the flock of the farm on which they are bred, but as a record of the particular blood from which they are descended. The *markings* are confined to the *ears*, and consist of small pieces being cut out, or slits made in the tips, with a knife, from the fore or back margin, or of holes made with punching-nippers, or of a combination of both sorts of marks. The female stock are always marked on the *near* ear, and the male on the *far* one. Thus, a single round hole is punched through the near ear of all the ewe-lambs, and a similar hole through the far ear of the wether lambs ; and should any of the ewe-lambs be considered fit for breeding tups, they either receive an additional hole through the near ear, or a bit cut out from a margin of the same ear, corresponding to a similar mark on their dams or sires, to distinguish their particular blood from the rest of the flock. Twin ewe-lambs receive a hole through both ears. Tup-lambs receive no *ear* marks, their long tail serving the purpose till they are weaned, when they are at once transferred to the tups. Individual tups are so easily identified, and their descent so well known by the shepherd, that they require no marking. Fig. 463 are the *punching-nippers*, of which the inverted hollow cone *a*, having its small end sharpened, is employed to cut the hole out of the ears; and, to save the ears from being unduly pinched, a pad of horn *b* is inserted into the straight under-arm of the nippers, the pieces nipped out rising out of the orifice *c*. The figure at once shews how the instrument is used, being similar to the one used by shoemakers to punch holes into the lappets of shoes, through which the shoe-strings are passed, and it costs 2s. 9d. *Buisting* consists of stamping a letter or letters, expressive of the initials of the name of the owner or of that of the farm, or of both. The buist or mark is effected by a simple instrument *a*, fig. 464, which carries the capital letter S. This mark is made on the same principle as the ear marks, the *near side* indicating the female, and the *far side* the male sheep. The material of which the buist is made is boiled tar, made viscid by a little pitch. The sheep to be buisted are put into a convenient apartment of the steading, and handed out one by one, and kept steady by holding the head and rump by the hands, and a knee being placed against the opposite side, causes the side to be marked to project. The buist is then dipped lightly into the melted tar, to prevent its dripping; and to make the mark vivid, it should be applied with some force, and with the entire surface at once, to compress the wool equally, and then quickly withdrawn. The wool must be quite dry, or the tar will not adhere to it. All new clipped sheep are buisted in this manner, and though but a temporary mark, being in time obliterated,

Fig. 463.

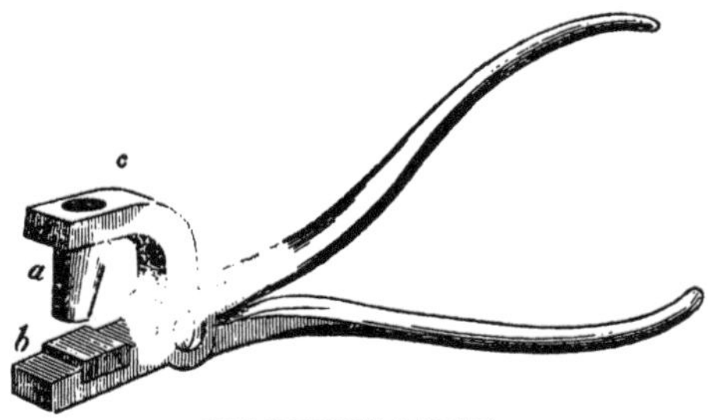

THE PUNCHING-NIPPERS.

Fig. 464.

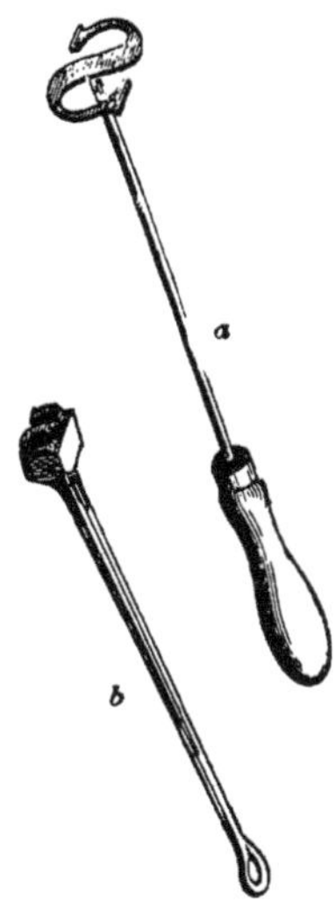

THE BUISTING-IRON AND BRANDING-IRON.

though not on short wool, it serves the present purpose well. To my taste, the buist looks best on that point which is the roundest part of the rib, but others prefer it on the shoulder, the rump, or the loins. In fig. 464 is another instrument *b*, the *branding-iron*, also bearing the capital letter S. It is sometimes used, and on being heated in the fire, is applied to brand the letter on one side or other of the nose, or on one of the horns, of Blackfaced sheep, or on the horns of Highland cattle, in lieu of the punching-nippers or knife, and it makes an indelible mark. To save twice handling of the lambs, they should be marked and buisted at the same time, one person making the marks, another applying the buisting-iron. The buist costs 3s. and the brand 4s., though a smaller instrument, as the letter has to be cut out of the brand like that of a die, and it must be made of the best iron to stand frequcnt heating. These are all the operations to which sheep are subjected in summer.

(2636.) Some breeders of even Leicester sheep annually dispose of all their lambs, immediately after being weaned, except a few ewe-lambs to freshen the ewe-flock, and such breeders keep a larger flock of ewes. This difference of management does not arise from caprice, for it is found that some soils will maintain ewes in good keeping condition which will not put hoggs in the condition they should be ; and there are large breeding farms which cannot raise a sufficiency of turnips to maintain hoggs in winter. When lambs are sold they are not marked or buisted, those operations being left to be done by those who become their future owners. There are many markets convenient for the sale of lambs in summer enumerated in the almanacks. It is not uncommon to see from 50,000 to 70,000 lambs at a fair.

(2637.) Neither *carse*, *dairy*, nor farms in the *neighbourhood of towns*, support a breeding flock. and therefore have nothing to do with washing and clipping sheep. The *washing* of sheep on *pastoral* farms is conducted on a somewhat different manner from what has been described. A natural deep pool in a river, or the edge of a lake, is selected for the purpose, and failing these, a damming is made in the gully of a rivulet, or even a pool is dug in the plain ground near a supply of water. A bught of flakes or hurdles, such as in fig. 216, is made near the edge of the pool, a narrow passage fit to contain at most 2 sheep and 2 men in breadth, is made also of flakes to a jetty, which projects into the pool and is raised 5 or 6 feet above the water, and from which the sheep is made to leap into the water one by one. On leaping down from this height, the sheep go over the head, and on swimming across the pool reach the dry land at the opposite side, where another bught of flakes is ready to receive them. They are thus treated several times till they are clean. There are store-masters who prefer hand-washing before leaping and swimming them across the pool. Merely with hand-washing, I cannot see how any sheep can be thoroughly washed without inverting their backs, and especially those which have been smeared. I have seen it somewhere stated in print, that the more greasy the water becomes the cleaner will the sheep be washed, and, therefore, the water in which sheep are washed should not be changed. I suppose that this opinion prevails pretty generally, as in any pool in pastoral districts that I have seen, no means are used to let the used water flow off, except where it is at liberty to flow away from the side of a lake. Theoretically, this opinion may be correct, for M. Raspail observes, that "when the wool is washed, this soap (the yolk) is dissolved, and takes the salts along with it. Hence it follows, that the water that has been used in this process becomes, at each repetition, better adapted for the purpose." Practically, however, the notion of greasy and dirtied water washing wool better than clean will gain no converts from those who use clean water; and, besides, every fleece has, or ought to have, as much natural soap in it as will wash it clean in clean water. No doubt *soft* water will wash wool better than *hard*, but all river water, when exposed to the air for some time, becomes soft, unless it contain an inordinate proportion of lime or tannin; but be the state of the water what it may in particular localities, there can be no doubt that wool, like every thing else, is best washed with clean water. There is more probability in what M. Raspail states immediately after delivering the above opinion, that, "it has been calculated that the grease obtained from the washing of wool in France might be sufficient to manure about 370,000 acres of ground."* In small lots of sheep, I have seen the wool clipped before it was washed. The clipping on a dirty skin, as may be supposed, was performed very roughly, but independently of this, wool washed off the sheep's back can have no yolk in it, and, consequently, when dry, will feel harsh, and be in an unfit state for certain processes of manufacture.

(2638.) A mode of washing sheep, differing still more widely, is practised in Wurtemberg Advantage is taken of a fall of water at a sluice in a river or mill-lead, to place a number of spouts to convey the water in small broad rills, and let them fall from a height of 5 or 6 feet into a pool below. Men hold the sheep, floating in the pool, in different postures, under the spouts from which the water falls upon different parts of their body. The water first falls upon the head and shoulders, the sheep being held upon its rump; then it falls upon the belly, the sheep being floated on its back; it falls on one side and then on the other, the sheep being placed on its side; and lastly it falls on its back, the sheep swimming in the water. The washers all the time shed the wool this way and that with the bare arm, to let the water reach every part of the body, and not with the hand, in case its manipulation should break the wool. The sheep are first rubbed with soap in a trough in which they are made to stand, and the cost of the process is about 1d. a-head.† I am not sure that the constant beating thus inflicted by the fall of the water is beneficial to the sheep.

(2639.) The *clipping* of sheep in most *pastoral* districts is conducted in a most slovenly manner. The old-fashioned practice of tying the legs of the sheep together, on the grass sward in the open air, is still practised in most parts of the Highlands, and after the creature is thus placed in a helpless state between the legs of the clipper, who sits on the grass with the head of the sheep towards him, the shears are made to ply, from the neck to the tail, in irregular long slashes to their very points, that the fleece may be snatched off in the

* Raspail's Organic Chemistry, p. 457.

† In Zeller's Landwirthschaftlichen Maschinen, Apparate, und Geräthe, zweite lieferung p. 15-21, is given a detailed description and plates of this singular process.

shortest time. The legs are then loosened and the sheep set at liberty. Women are frequently employed at this work, to which there is no objection, provided they do it properly; but the result is, that a considerable proportion of the wool is left on the sheep, and their appearance, as a piece of work, is most disgraceful. The entire proceeding is the less justifiable, that the proper way can be done as expeditiously in the hands of dexterous people as this slovenly method. If herds cannot clip any other way, let them go and learn it in places where it is better done.

(2640.) An improvement has been effected on this mode of clipping by Mr Colin Munro, Dingwall. It consists of erecting an awning of canvass, tied to stabs driven into the ground at stated distances, and of appropriate heights, along the side of a wall. This tent, as it may be called, is sufficient to hold as many sheep as is desired, by making it large enough; it protects them from rain, allows the clipping to proceed uninterruptedly, and it screens the workers from the heat of the sun. If the part of the ground occupied by the clippers were covered with canvass, the wool would be kept clean. A tent of this sort can be erected at little cost, and it would last many seasons.

(2641.) *Scalded Heads.*—Sheep are much infested in summer with flies. As a protection to the *head* against them, the simple *cap*, or *hood*, as seen in fig. 465, is effectual. It may be made of stout linen, and fastened with 4 tapes tied crosswise under the chin, or of leather, and buckled at the same place. Leicester tups should not be without these caps in summer, especially when grazing near

Fig. 465.

THE HEAD-CAP, OR HOOD, FITTED ON THE SHEEP.

woods; and as tups are occasionally apt to box each other, any little portion of skin which may thereby be abraded on the head will receive immediate protection from the cap.

(2642.) *Bots, Keds,* and *Fly-blows.*—Sheep are troubled with bots as well as cattle. The fly is called *Œstrus ovis*, and will be found at *d*, fig. 449. It is a smaller species than the cattle-bot, being about 5 lines in length. It is supposed to deposite its eggs on the margin of the nostrils, which, whenever it does, the sheep lies down upon dusty bare spots, holding its head close to the ground, or, when a number are attacked at the same time, they form a dense phalanx, with their noses pushed towards each other. The warmth and humidity of the nostrils very soon bring the eggs to maturity, and the larvæ find no difficulty in gaining their way into the frontal maxillary, and other sinuses and cavities of the face. There they adhere with 2 hooks, the secretions of the cavities constituting their food. In time they wriggle down the nose and fall on the ground, in which they undergo their future transformation. The larva *e* is flat on the under side, and convex above, of a delicate white colour, without spines of any kind, save the terminal hooks already mentioned. A series of black transverse spots are visible on the under side, covered with rough points. The *ked* or *keb*, the Sheep spider-fly, *Melophagus ovinus*, is an insect so well known in its nature and habits, that a particular description here seems unnecessary. It is magnified at *e*, fig. 456, the line *f* shewing its natural size. Its antennæ *g* are nearly concealed in a cavity before the eyes, consisting of a large subglobose articulation, from which a flat membranous style arises, having 4 teeth at the apex. The remarkable circumstance attending the tribe of *keds* or *ticks*—which belong to the family of Hippoboscidæ, and are included among the dipterous or 2-winged insects, though they are wingless—is, that the young is retained in the body of the mother until it becomes a pupa, there being no other instance amongst other 2-winged flies of the period of gestation extending beyond the state of larva. This peculiarity has caused the Hippoboscidæ to be termed *nymphiparous* or *pupiparous* insects. Much more dangerous tormentors of sheep are *Blow-flies*. When sheep are *struck by the fly*, the symptoms of disease cannot be easily misunderstood. They almost constantly hang down their heads, sometimes turning them on one side as if listening; shake the tail with a quick jerking motion; run rapidly from one place to another, and in doing so, at times stop suddenly and stamp with the fore-feet. The flies deposite their eggs on any bare skin they can find, and failing that, on the wool on the rump, below the tail, and about the groins. If the larvæ are left undisturbed, when in large numbers, in 2 days they will destroy the sheep, having in that short time eaten the flesh into the very bones, and sometimes exposing the entrails! Warm moist weather, in fields inclosed by woods, and in the bottom of dells, are the circumstances and places most favourable to their attacks. The smell arising from excrementitious discharges, the glutinous matter left after milking ewes, and long wool, are all attractive objects to blow-flies. A shepherd ought to be able to detect sheep that have been struck by the fly the moment he enters the field. Dogs have been known to point them, as truly stated by Mr Price. "A looker's dog," he says, "when properly trained, the moment he enters a field in which are any sheep struck by the fly, instantly singles out the diseased animals, and runs up to them, as much as to say they ought to be caught." * Dogs require little training to do this, partly because the symptoms which struck sheep exhibit are unequivocal, but more probably from the peculiar smell which maggots no doubt emit; or the sheep themselves may emit a peculiar odour after being struck. The Ettrick Shepherd is of opinion (and it is a probable one) that flies give a preference to one sheep over another, probably on account of the selected sheep being either actually subjected to diarrhœa, or emitting such a peculiar flavour along with its perspiration, as to be attractive to flies, and which may be indicative of a predisposition to disease.† It is culpable in a shepherd to allow any sheep to be dangerously injured by the fly. He cannot prevent their attack, but he should be able to detect it before it proves

* Price on Sheep, p. 472, Note.

† Hogg's Shepherds' Guide, p. 106.

serious in its consequences. The sheep should be carefully observed one by one when the flies are active, and being gathered in a convenient part of the field, the suspected ones should be caught with the crook, fig. 397, and examined, and every maggot removed by the hand. As maggots are not killed by being thrown on the ground, they should be collected in some vessel, and carried away and buried. I have seen a shepherd fill his hat with maggots, in the course of 1 hour's search amongst a small flock of Leicester hoggs. Should the maggots have broken into the skin, rubbing the part with a strong solution of corrosive sublimate, or a strong decoction of tobacco-liquor and spirit of tar, will check a farther attack on that part; and should the part affected be larger than is seen between the sheds of the fleece, the wool should be removed with the shears, and the corrosive sublimate applied upon, and around, and rubbed into, the wound. Should the wound, on healing, indicate a dryness of the skin, in consequence of the application of the corrosive sublimate, an ointment of tar and lard will soften it, and keep off the flies. Mr George Mather, shepherd, New Scone, recommends a wash containing arsenic, which I have no doubt would prove effectual; but I have so great an aversion to using arsenic in any shape on a farm, that I cannot recommend it to be used in this case.* The most dangerous, perhaps, of all the flies is the Chequered blow-fly, *Sarcophaga carnaria.* It is somewhat larger and more elongated in shape than the common blue-bottle fly; general colour changeable grey; thorax with black longitudinal lines; abdomen covered with black quadrate spots, which give it a tesselated appearance; body pretty thickly beset with strong hairs. It *produces its young alive;* hence the unaccountable appearance of *maggots* in a short time after the sheep have been examined. "The larva, when full grown," says Mr Duncan, "is scarcely ½ an inch in length, as at *x*, fig. 466; the head *t* is small and membranous, having 2 fleshy prominences above, with a small nipple-shaped knob *m m*, so that they bear a perfect resemblance to small mammæ. Beneath these mammiferous protuberances are 2 strong black movable hooks *c* placed by the side of each other; *a a* is the first segment of the body, and *r* one of the anterior stigmata. The principal use of these hooks is to tear off and separate the fibres of the flesh on which the creature feeds. The last segment of the body *d* is, as it were, cut across. Two large air-vessels may be seen running along each side of the body, terminating at both ends in breathing-holes, *p p* and *s s*. The description of the maggot of this fly will serve for that of the other flies about to be mentioned.—*Musca Cæsar* is readily known by its brilliant green hue, which has a silvery play of colours when seen in certain lights.—*Musca vomotaria*, common Blue bottle-fly or blow-fly. This fly is well known in our houses, and may easily be iden-

Fig. 466.

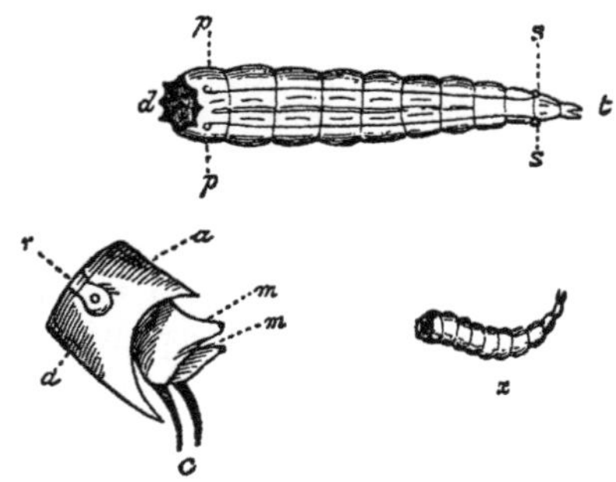

MAGGOT OF THE CHEQUERED BLOW-FLY—SARCOPHAGA CARNARIA.

* See some excellent practical remarks on the sheep-maggot by Mr Mather, in vol. x. p. 221, of the Prize Essays of the Highland and Agricultural Society.

tified by its buzzing noise when on the wing. "This is the species," observes Mr Duncan, "of which Linnæus affirmed that 3 individuals could devour the carcass of a horse as soon as a lion. There is no doubt that we must impute to it a large share of the injury our flocks sustain from this tribe of insects."—*Anthomyia lardaria* is rather more than half the size of the blue-bottle-fly, and of a bluish-black colour; thorax with longitudinal lines; and abdomen slightly tesselated. It is very common in low sheltered woods, and is doubtless often associated with the other in preying on living subjects.* It is said that the green-fly first attacks the sheep, and is succeeded by the more greedy blue-bottle-fly, which, having made a suitable place, is, in its turn, succeeded by the chequered blow-fly.

(2643.) *Doddering.*—One great means of warding off the attack of the fly on hoggs is *doddering*, especially in localities obnoxious to flies. This operation consists of clipping away all the wool off the tail and between the hind-legs. Should hoggs scour, which they are not unapt to do when put on foggage immediately after being weaned, and when there is no rough moory or hilly pasture to put them upon previously for some weeks, the removal of the wool will prevent the discharge remaining about the animal, and, of course, deprive the fly of one object of attraction. The use of docking the tails of sheep is now made obvious (2377.).

(2644.) Young sheep are subject to serious, and even fatal, injury on farms situate on the rocky cliffs of the ocean, from the attacks of the Raven, *Corvus corax.* This formidable bird has been seen, on approaching lambs asleep, and left alone by their dams while grazing at a distance, to peck a hole in their abdomen, and draw out the entrails; or, should the lambs awake on its approach, to pick out their eyes. Even hoggs, when fallen on their back, have been known to have their eyes picked and their entrails pulled out. The raven frequently builds its nest on cliffs overhanging the sea, and, when feeding its brood, becomes very bold, and will take advantage of many animals which they would not dare to approach at other times. Many a hare falls a victim to them. There is no way of destroying this bird but by shooting after long watchings, or by setting baited steel traps near their haunts. The Carrion-crow, *Corvus corone*, will perch itself on the rump of a sheep that is infested with maggots, and while devouring them, will aggravate their attack, by picking the flesh to the bones. I have saved many a young leveret from being destroyed by the carrion-crow.

(2645.) *Lying awkward.*—Leicester hoggs, before being clipped, are so loaded with wool, that, when annoyed by the ked, often roll upon their backs, and should this happen in the hollow of a furrow, cannot get up again. They are then said to be *awkward* or *awald.* Should they lie for some time with their head down the hill, with the stomach full of food, they may die on the spot by apoplexy. It is a disgrace to a shepherd to allow sheep to die when lying awkward. He cannot prevent them falling awkward, but he should not be so long in visiting a field, where rough sheep are grazing, as to give them time to die Sheep are not easily discovered lying awkward in a furrow, and, on this account, when the field is examined, the ridges should be crossed, and every furrow viewed in length. An accustomed eye, however, can detect the hind-hoofs elevated in the air. Many dogs are quick in observing sheep in this state, and

* Quarterly Journal of Agriculture, vol. xi. p. 58-63.

some I have seen run and take a hold of the wool on one side near the ground, and pull the sheep so far over on its other side as to enable it to regain its feet. Sheep that have lain awkward but a short time will run instantly away on getting up, but those which have lain a good while will void much dung, and thump the ground with their head and rump in their struggles to get up, and at length become so fatigued, that, on regaining their feet, will appear stupified, and walk away as if light-headed.

(2646.) *The Sturdy.*—There is a disease in sheep called *sturdy* or *turnsick*, which produces so much light-headedness as to cause its victims to turn round, and seem stupified. When affected by it the sheep separates from the flock, wanders into a corner of the field, feeling an aversion to put its head down to the ground. It seldom eats, and thus loses condition, and at length, becoming emaciated, falls into a ditch or burn, and is killed or dies, if it has been so long neglected as to be allowed time to wander until such a catastrophe overtake it. Hoggs are most affected by this disease, and it seldom attacks large numbers, one or two now and then. What induces the disease I do not know, for it may be observed when hoggs use little exertion in walking to obtain their food on turnip-land, and it also occurs when they wander over a large range of pasture. Accurate dissection of the head has proved that the disease is caused by a living animal in the brain, the Many-headed hydatid, *Hydatis polycephalus cerebrali.* "Instead of a single head, there are a great number spread over the surface of the parasite, and opening into the same general cavity. When the sac is distended, they appear only as opaque spots upon it; but a lens of no great power will give a distinct view of their heads, or rather necks, with the tentacula or barbs projecting from the apparent opening or mouth which forms the extremity of them. These hydatids vary in size from that of a pigeon's to a hen's egg. The wall of the cyst appears to be composed of 2 or 3 layers, the centre one of which seems to possess a muscular character. On examining them with lenses of a high magnifying power, their coats resemble paper made upon a wire frame, their muscular film so plainly and regularly interlace each other."* The complaint may be cured, though it is seldom attempted, the sturdied hogg being killed whenever it is seen to be affected. I was once tempted to try the experiment of a cure on a very fine Leicester gimmer, which was evidently with twin lambs, and would lamb in the course of a fortnight. The creature became so bewildered under the disease, that she was obliged to be brought into the lambing paddock, and fed by hand on cabbages and kail. On examining the head minutely, I discovered a soft spot near where one of the horns would be. Experimenting for life or death, I cut an incision into the skull round the margin of the soft spot, leaving only a small attachment which acted as a hinge to the piece cut out, and admitted of its being so far raised up as to permit me to see the hydatid distinctly, and to extract it entirely with a small cobbler's awl, the only small instrument the shepherd or I had that would answer the purpose. Shutting the lid, covering it with a plaster of tar on a piece of linen, and putting a cap, such as in fig. 465, on the head, the gimmer was allowed to remain in the paddock and was fed, and in the course of a few days I had the satisfaction of seeing her perfectly recover, and in due time bring forth twin lambs, and she continued to breed for several years after. The success attending this experiment,

* Youatt on Sheep, p. 379.

was probably owing to the proximity of the hydatid to the skull, where its presence had produced, and was indicated, by the softening of the skull, and in all such cases, I have no doubt a similar cure may be easily effected; but when hydatids exist deeply in the brain, it is questionable that a cure can be effected. The Ettrick Shepherd recommends the insertion of a wire through the nose and brain into the hydatid, by which puncture it would be destroyed, and the wire may be resorted to before the softness appears, or when, from the situation of the hydatid, there cannot be a softness, and he says, that "several years passed before he failed in this operation in any one instance;" * but it is clear, that no hydatid can be touched by the wire, but such as may happen to be situate in the determinate line of its passage. A trocar is now used for the purpose, like fig. 272, but, of course, much smaller in diameter. The sturdy is not an infectious disease, and I agree with Mr Parkinson, that it is not hereditary, for although I have seen several sturdied hoggs, I could never trace any connection of the disease with their parents.†

(2647.) *Blinding.*—This affection is sometimes produced in the eyes of sheep by the wind blowing into them the pollen of the grasses. The eyes become almost clogged up with pollen-dust, and such a degree of inflammation is sometimes set up in them as to cause the effusion of pus. In many cases the eyelids become glued together with the pus. No serious injury is produced by this accidental annoyance, though it is disagreeable to the sheep for the time it exists. Wiping the eyes with a wet spunge affords great relief, and this relief should be cheerfully bestowed by the shepherd once a-day. The annoyance can only last till the largest portion of the pollen is blown off at the termination of the period of flowering of the plants, and it may not be of annual occurrence, for there may be no wind, or there may be rain during the time of the efflorescence of the plants.

(2648.) *Sheep worried by dogs.*—Sheep are sometimes destroyed by being worried by dogs, and the vexatious catastrophe happens, perhaps, most frequently early in a summer morning. Experienced dogs mostly go singly to do this mischief, and take care not to bark while engaged in it; and their only object seems to be to obtain a feed of mutton. Dogs most addicted to this mischief are mastiffs, bull-dogs, bull-dog terriers, and lurchers having a cross of the colley in them, and they are most prone to it when escaped from the chain, on which they are mostly kept as watch-dogs. An old colley addicted to this vice practises it with consummate art, and obtains mutton with the least trouble to himself, whilst he commits the least extent of mischief. Pointers are very apt to run sheep when hunting, and especially when self-hunting. The part of the sheep chiefly attempted to be seized by the dog is the throat, which he tears open, and eats the flesh to the neck-bone; and were he content with this morsel, or his appetite satisfied, the loss sustained by the depredation would not prove of a serious nature; but his propensity to destroy seems only bounded by the numbers of the flock; he worries a few to death and bites a great many more. The extent of destruction is probably aggravated by the conduct of the sheep themselves, who run away in a body from the dog; and in fear of losing the rest whilst he is bringing down one, he leaves that one lying on the ground and pursues the others, seizing the nearest him at the time, one

* Hogg's Shepherds' Guide, p. 50.

† See Parkinson on Live Stock, vol. i. p. 412.

by the back, another by the throat, and a third by the haunch, until a great number are rendered lame by bites. It is rare that a dog tears more than one sheep, as he is probably scared by some noise, or other circumstance, before he has time to break into another. When a lamb is run after, it is so easily overcome that the dog tears its neck open at once and satisfies himself upon it. I am not aware that a dog which worries sheep can be deterred from the practice by any means, certainly an old dog cannot, and the only fate that should await so hardened an offender is the rope and tree; but a young dog, especially a pointer, may be deterred; and the most effectual way, I believe, of deterring him for the future, is to couple him for a few days to the carcass which he has worried, and cause him to drag it about with him; or, in a hill country, to couple him to a Black-faced ram for some days, with a sufficient length of chain to allow the ram to turn round and butt him severely with its horns. An inexperienced dog attempting to seize a Black-faced wether by the neck incurs the hazard of having an eye poked out with the point of a horn by a sudden movement of the head. In every attack of sheep by dogs there are comparatively few deaths to the number injured, and were time afforded most of the bitten sheep would most probably recover; but the usual custom, in the excitement which such an occurrence creates, is to kill the sheep with the view of preventing a total loss of the mutton; for, *were* they to die in the blood, the mutton would be rendered unfit for use. There were once 9 of my Leicester ewes worried by a dog, 3 of whom were bled on the spot, and the shepherd would have bled other 2, but I would not let him; and from the recovery of these bad cases by the means used, I was persuaded that the 3 which were killed would have recovered had they been permitted to live. There is no doubt, however, that sheep which have been run and worried ever so little are a long time, if ever, of recovering their customary composure, and on this account alone, the owner of a dog that runs other people's sheep should be severely fined over and above the value of the sheep actually injured.

(2649.) *Lamb for the table.*—I think that every farmer who possesses sheep should supply his own table with lamb in summer. I am aware that breeders of the valuable kinds of sheep grudge to use their lambs at table, and no doubt the table would thus be supplied at considerable cost; but there are other modes of obtaining the end without making any sacrifice, such as purchasing a few Black-faced ewes and crossing them with a Leicester tup. Such a cross would supply the requisite number of lambs of the finest quality for the table, and the ewes, before becoming old, could be fed off and sold. There is no sort of meat so readily obtained and suitable for the table in summer, on account of small joints, and being fit for use when killed, as lamb. The jigot boiled, and the fore-quarter or loin roasted, and relished with a thin slice of toasted ham, makes an excellent dinner in a summer's day; and who does not enjoy a cold roast rib and shoulder of lamb, with dressed salad, and cucumber, even in the hottest period of the dog-days, when the very sight of any other sort of meat would cloy the appetite? A sheep's haggis has been lauded by poet and peasant, and though classed amongst the "puddin' race," must be regarded as too substantial a dish for summer, and, in fact, possesses at all times too *grateful* a sense of the honour conferred upon it; but a lamb-haggis is a delicacy even in summer; and as it is seldom seen at table, I will give you a recipe by which to judge of the merits of so neglected a delicacy:—Take the lights and heart of a lamb and parboil them for 20 minutes, and allow them to stand till cold.

Then mince them pretty small, and mix with the mince 2 table-spoonfuls of flour, and the same of oatmeal. Season the whole with a small onion shred fine, with pepper and salt, and a small sprig of common thyme. Clean the bag thoroughly, and turn it outside in; that is, let the minced meat be contained within the smooth outside surface, and the rough interior be left on the outside. After the mince has been put into the bag, pour an English pint of water into it, and then sew up the bag tightly with a needle and thread. In careless hands the bag should be tied loosely in a pudding-cloth to prevent its bursting if too hastily boiled; but if boiled for 1 hour over a gentle fire the precaution of the cloth is unnecessary. Then serve the haggis on the table in a pretty large dish. To prevent the meat running out into the dish in carving, take a firm hold of the uppermost part of the bag with the fork, and raise up the skin before making an incision in it with the knife large enough to admit a table-spoon, by which serve the meat direct from the bag. Green pease is the best vegetable to accompany lamb-haggis, and cauliflower, kidney-beans, and vegetable marrow, are all excellent accompaniments.

(2650.) Those who have many horse-chestnut trees, *Æsculus hippocastanum*, may be pleased to learn that their fruit, which is considered worthless, and even poisonous in this country, though both opinions are erroneous, may be usefully employed in feeding sheep. "Whilst I was at Geneva in the autumn of 1837," says a correspondent, "I observed every one collecting carefully the fruit of the horse-chestnut, and on inquiry I learnt that the butchers and holders of grazing stock bought it readily at a certain price per bushel. I inquired of my butcher, who himself kept a very extensive grazing farm, and he told me it was given to those sheep in particular that were fattening. The horse-chestnuts were well crushed, something in the way, so I understood, that apples are, previous to cider being made. In Switzerland, they are crushed or cut up in a machine kept solely for that purpose, then about 2 lb. weight is given to each sheep morning and evening. Sheep eat it greedily; it must be portioned out to them, as too much would disagree with them, it being of a very heating nature. The butcher told me that it gave an excellent rich flavour to the meat. The Geneva mutton is noted for being as highly flavoured as any in England or Wales."*

(2651.) There are various modes of regulating *hill-sheep stock*. Some land is suited to the breeding, whilst others are adapted to the bringing up of sheep. On what is called *bare* or *hard* land, ewes are preferred, and to preserve room for them their lambs are sold off every year, with perhaps the exception of a few ewe-lambs to maintain the vigour of the ewe stock, and where there is no such exception, *great* ewes, that is, ewes in lamb, are bought in lieu of the draft ewes sold. On the other hand, *soft* land is best suited to lambs—to wether-lambs to be reared until they become wethers, when they are sold in autumn to farmers who raise turnips to feed the sheep they do not breed themselves, or to English graziers, who fatten them upon grass. The ewe-lambs are sold to others who rear them until they are tupped, and then sell them as *great* ewes to breeders, who purchase them to the extent of the old ewes they get quit of. Others purchase ewe-lambs to convert into ewes, and after taking a few crops of lambs from them, sell them while yet young to be fed off on turnips in winter. It is found dangerous to change the ewe stock on some lands, because new ewes

* The Gardener's Chronicle, 21st October 1843.

become diseased on new ground, and the fear of disease is felt so strongly, that the ewes on such lands are not permitted to leave them, the incoming tenant being bound to take them at a valuation. Hard land bears scanty pasture, which is found sufficient and wholesome for breeding ewes, though insufficient for rearing young sheep to size and condition. Soft land bears sufficient pasture to support young sheep, and to put even old sheep into good condition, but it is considered improper to rear ewes upon it, as they would not only be maintained beyond the pitch of profitable condition, but probably become diseased, especially in a wet season.

(2652.) These various modes of regulating the breeding of hill-sheep have probably originated from local circumstances, which cannot now perhaps be traced; but there is one circumstance, namely, the *rot*, whose consequences have frequently had so fearful an effect upon hill-sheep, and especially upon ewes, that every means have been devised to avert its recurrence; and a store-master is justified in trying every means to prevent such a calamity. Other circumstances besides the dread of rot may account for the many varied modes adopted for managing hill-sheep, which would otherwise appear questionable practice. For example,—Land may support ewes in keeping condition, which could not fatten wethers; and land may support lambs well, though not wethers. Young sheep may pine on land that supports wethers, because its elevation and steepness may fatigue young sheep too much to travel over it; and its herbage may be too hard for young sheep. Such circumstances as these may have affected the practice of different grazings, but still the dread of the rot has most probably caused most of the varied practices which exist amongst store-farmers, for, without a paramount reason, it would be difficult to account for so great a variety of practice which exists in situations that present no marked difference of feature. If this view be at all near the truth, we may imagine when draining is more generally practised by hill-farmers, it will render pasture more uniform in similar localities, and management thereby become more uniform. Whether uniformity of practice would ultimately be serviceable to store-farms, I cannot pretend to foretel; but the general draining of hill-pasture would undoubtedly have this beneficial effect, namely, that it would give farmers liberty to follow their own plans, whereas, at present, all are obliged to act as if placed under the control of an overpowering influence. Let the physical effects of wet pasturage be overcome, that is, let the rot be subdued, and the mind, emancipated from its dread, will originate a general system of managing hill flocks more in accordance with sound principle.

(2653.) Sheep on hill-pasture delight in summer to spread themselves over, and to go to the highest point of their range. Ewes are restricted in their range by the lambs, which, when young, shew little inclination to wander afar, but rather to lie down and sleep after being satisfied with milk. Hoggs keep much together, and on that account do not wander far from their morning lair, whereever that may be. Wethers, on the other hand, attain the height of their pasturage at an early period of the day, and remain till dusk. You thus see, when sheep of different ages are brought up together, how usefully they distribute themselves over their entire pasture; and where only one class of sheep are reared, they extend their range according as their age increases, or their food becomes bare. On contiguous estates, where there is no march dyke to define their common boundary, it is quite possible that the flock of one property may occasionally trespass on the pasture of another. Should this happen in the early part of the day, the shepherds should not dog off the strange sheep, as that may render them restless even for days, but to wait till nightfall, and then point them gently over the march to their own ground, where they will take to their usual lair. Sheep usually select a spot for resting at night, and it will mostly be that which is safest for them, especially if the sheep are aged, and well acquainted with the ground. In fine weather they should not be disturbed in thus selecting their lairs; but in case of threatening storm, they had better be directed to the sheltered side of the pasture, or even near the stells, if need be. Within enclosed fields in the Lowlands, sheep can hardly go wrong in summer in selecting their lairs for the night.

(2654.) All the uppermost part of our mountain pastures, as well as many portions of lower elevation, consist of a soil very different in its nature from what is met with in valleys; I mean *peat-earth*, not the soft peat of bogs, but the hard peat-earth which covers our mountains. The usual and natural produce of this earth is heath, consisting usually of 3 kinds, the *Calluna vulgaris*, common ling, the *Erica tetralix*, cross-leaved heath, and the *Erica cenerea*, fine-leaved heath. Peat-earth is only found in the colder portions of the temperate zone, and it is formed from the partial decomposition of several cryptogamic plants. Professor Jameson's theory of the formation of peat-earth is in these terms:—"The woody or other vegetable matters, being slowly deprived of a portion of their hydrogen by the putrefactive process of heat and moisture, become brown by the precipitation of carbon, and at the same time somewhat soluble in water and spirit of wine, thus forming a kind of bituminous matter. By a further decomposition more hydrogen is separated, when the vegetable matter becomes insoluble in water, but still soluble in alkali. Lastly, nearly the whole of the hydrogen is separated, when a black substance is left, what is called *peat-earth*. This last, by exposure to the air, combines with a portion of oxygen, forming suberic acid." Peat-earth might be greatly improved by surface-draining and top-dressing, and in many situations would yield a superior produce than heath for the support of sheep, at least in summer. It has long been known, that lime, when applied to peat-earth, causes it to produce an abundance of white clover, *Trifolium repens*, and as the rationale of this remarkable effect, as I have reason to know, was first propounded by Professor Jameson, I give it in his own words:—"We have already observed, that peat contains the suberic acid, or one nearly allied to it, which appears to be formed in greater quantity the longer the peat is exposed to the action of the air; thus assisting in retarding the decomposition of the peat, of course preventing its being useful in vegetation. Marl, shells, and limestone, are useful, in a triple capacity: 1*st*, by removing the acid, the vegetable matter is allowed to decompose more rapidly; 2*d*, the combination of the acid with the lime forms a compound, which may assist in vegetation; and, 3*d*, this acid having a stronger attraction for lime than the carbonic, will disengage it in considerable quantity, when it will assist vegetation. I shall conclude, with remarking, that there is a considerable prejudice in favour of quick-lime: if the explanation now given have any plausibility, it is plain that carbonate will answer as well, if not better, in the improvement of moss-lands."*

(2655.) *Pounded Limestone.*—If limestone would answer the purpose of improving peat-earth, as suggested by Professor Jameson, it would be worth

* Jameson's Mineralogy of the Scottish Isles, vol. ii. p. 120-57.

while to erect mills for pounding it in those Highland districts where it abounds, and to apply it as top-dressing after the soil is surface-drained, in the manner attempted about the beginning of the century on the estate of Struan in Rannoch, Perthshire. After being pounded, the limestone was carried away by a run of water to 3 different ponds, one above the other. The upper pond contained the grossest particles, and the lowest the finest part of the limestone, which there resembled clay or marl from its smoothness. On being put on the land at Struan, its effects were visible and much approved of. I do not see the use of putting pounded limestone into water, as it might be used in the manner of limestone gravel in Ireland, which proves of much advantage in reclaiming bog as well as a general manure. In its effects limestone is slower than burnt lime slaked, but more lasting, and on that account might prove a more judicious application to hill-pasture, where rain is much more abundant than on the plains.* Were a large proportion of the peat-earth of our mountains to bear white clover, and other succulent plants, instead of heath, there is no calculating the additional number of sheep that might be reared in this country.

(2656.) The subject of peat-earth is suggestive of that of *muir-burning*. Store-farmers have long been in the habit of burning the heath on their farms every year, with the view of allowing it to grow again, that its young shoots may support the sheep in parts of grazings where there is little or no grass. The injudicious mannner in which the burning is conducted, and the late period of the season at which it is done, destroys not only the heath plant itself by the root, but also the eggs of grouse. The destruction of game by this means naturally determined the owners of grazings to prohibit muir-burning altogether, and the consequences were what might have been anticipated, namely, the heath grew so tall and became so old, that the young grouse could not reach the young shoots from the ground, nor did a sufficiency of these grow every year to support both sheep and grouse,—the sheep nibbling off the short shoots which the old plants produced, food became scarce for the grouse, and they died. Hence, burning causing an abundant growth of young heath—thereby raising the common food of sheep and grouse—it is clearly the interest of both landlord and tenant, that heath shall be *so* burned as to produce the greatest growth of young shoots. The question of burning betwixt landlord and tenant being thus established in principle, the only difficulty is to discover a mode of burning which will produce the desired effect. Now, there is, in fact, no difficulty of pursuing a plan of burning which will answer the purpose ; and it is this :—Let that part of a hill-farm which bears heath be supposed to be divided into 8 or 10 equal parts, if the whole farm forms one hirsel or haift ; and, if it contains more hirsels, let each hirsel be divided into 8 or 10 equal parts according as it is known the number of years the heath takes of growing on the particular farm till it is too old for use, which may be so considered when it reaches 1 foot in height. Suppose it takes 8 years ; then let the farm be divided in this way :—

1844.	1845.	1846.	1847.	1848.	1849.	1850.	1851.	1844.	1845.	1846.	1847.	1848.	1849.	1850.	1851.
1	2	3	4	5	6	7	8	1	2	3	4	5	6	7	8
First Hirsel.								Second Hirsel.							

* See Robertson's Agricultural Report for Perthshire, p. 62 and 309.

Burn No. 1 of the first and second hirsel in the year 1844, and No. 1 of all the hirsels on the farm. No. 1 may contain 2 acres, or 10 acres, as the extent of the farm will afford. Since No. 1 of all the hirsels are at a considerable distance from each other, it is clear that the burning of any one number cannot commingle with that of its corresponding number of the same year. When No. 1 is burnt in 1844, the heath on it will have become so old as to be ready to be burned the year after No. 8 is burned in 1851; and as Nos. 1 and 8 will be in the least useful state as to herbage at the same time, the rest of the lots between these numbers will be in a progressively available state both for sheep and grouse. For, Nos. 4, 5, and 6 being but recently burnt and grown, the heath on them will not be so tall as to be beyond the reach of young grouse, and at the same time there will be sufficient old heath in Nos. 3, 2, and 1, not only to afford food and shelter to lambs and young grouse in summer, but food above the snow to old grouse in winter, while the snow will cover the young heath in Nos. 7 and 8, and protect it from the frost. In addition to these particulars, I may state it as a fact, that young heath plants withstand much severer frost than very old plants, as the spring of 1837 proved. This arrangement is unexceptionable, but still how are the lots to be burned? The usual practice is to set fire to the heath on the *windy* side, when the blaze soon towers to a great height, and is seen at a great distance, and the plants crackle in fine style amidst a scorching heat; but the heat which produces the crackling destroys the plants by the roots, and the flame fanned by the following gale runs along the ground, catching every bush of heath that presents itself the most readily, until a much larger space of ground is on fire than was desired, and the conflagration becomes so extensive that the shepherd and all his family cannot extinguish it. Indeed they don't mind it, and retire to bed, and the flame goes wherever the wind lists till it has no more heath to consume, or till the wind lulls, or the rain falls. The proper way is to begin the fire along the entire *lee*-side of each lot, when the flame will not mount high, but eat, as it were, its way among the heath against the wind, and if any part of the lot does not take fire from the rest, it is easy to take a burning brand and set it on fire. Let the burning be watched all day, and if the extent of the lot has not been burnt by the evening, let it be watched all night, and whenever it has reached its prescribed limit, let it be put out by going to windward and by beating and pushing a board nailed upon a long and limber pole against the burning plants. Where the fire has not reached the bounds of the lot, let the flame burn until it reaches the limit. There is none more easy than this plan, and if it were followed on every heath-farm every year, there would every year be grown a certain quantity of young heath well suited to support both sheep and grouse in the very best condition that the plant is capable of sustaining them.* In whatever position the farm is situated, it should be considered from what direction the wind prevails at the period most favourable for burning; and No. 1. should be fixed on as the lot farthest from the wind, so that if the wind prevails from the E., or right hand, No. 8 of the above table would be converted into No. 1. When the wind prevails from either the E. or W., the burning should be be begun on the *side* of the lot, but when it prevails from the N. or S., it should be begun at the *end*.

(2657.) *Law of Muirburn.*—" Persons making muir-burn, or setting fire to any heath or muir in Scotland, from the 11th April to 1st November, will be fined 40s. for the first offence, L.5 for the second, and L.10 for every other, or

* See Prize Essays of the Highland and Agricultural Society, vol. xiv. p. 640-8.

suffer imprisonment for 6 weeks for the first, 2 months for the second, and 3 months for every subsequent offence (13th Geo. III. 54. 4.). The tenant or possessor of the ground will be deemed guilty of the offence, unless he prove that the fire was communicated from other grounds, or raised by some one not belonging to his family (*ib.* 5.). Proprietors of high or wet muir may burn the heather thereon, between the 11th and 25th April, or if the ground be let, he may give permission to do so in writing (*ib.* 6.). Which permission, however, must be recorded in the Sheriff's books (*ib.* 7.). Prosecutions for the above offence prescribe in 6 months (*ib.* 14.)." *

(2658.) The figures of the bull's ring and bullock-holder not having reached me when their use was described in (2595.), I here give them before concluding the account of the treatment of stock in summer. Fig. 467 represents the *bull's ring* in an open state, ready to be inserted into the bull's nose, the joint

Fig. 467.

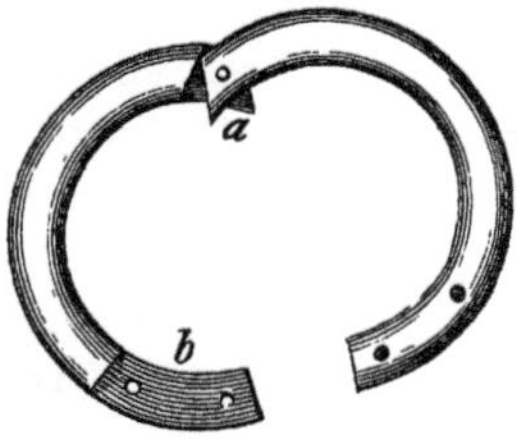

THE BULL'S RING IN A STATE TO BE INSERTED INTO THE BULL'S NOSE.

Fig. 468.

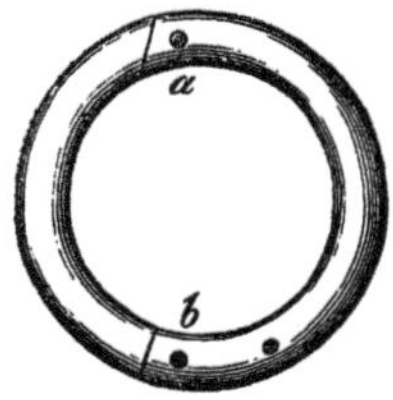

THE BULL'S RING AS FASTENED IN THE BULL'S NOSE.

a allowing the two sides of the ring to open as wide as the end *b* may be inserted into the hole burned into the septum of the nose. Fig. 468 shews the ring closed, after it has been fastened into the nose, *a* being the joint, and *b* the two ends of the ring lapped over and secured together by 2 countersunk screws. The ring is formed of $\frac{1}{4}$ inch of rod-iron, and its diameter over all is $2\frac{1}{2}$ inches, and it should be highly finished.

(2659.) The *Bullock-holder* is well represented by fig. 469, where *a* is the joint which allows its two parts to open so far as each to enter a nostril of the animal. The lever nut *c* brings the 2 knobbed points *b* as close as to embrace and hold firm the septum of the nose. The leading rein is fastened to the under ring *e*. This form of bullock-holder would allow the points *b* to be screwed to any degree of tightness, until they meet each other; and it is, in my opinion, so far objectionable, as the screwing may be carried as far by a rash hand as to hurt the animal when the instrument was moved in the least degree to either side. Another form which I have seen, and approve of, never allows the 2 knobs *b* of the instrument to be screwed closer than just to embrace the septum of the nose, allowing the holder to swing from the nose at ease whilst it holds the nose firmly whenever the animal attempts to move away. The cost of a bullock-holder of either of these forms is 4s. each.

Fig. 469.

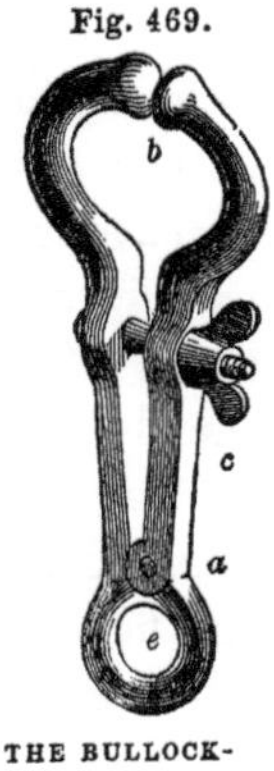

THE BULLOCK-HOLDER.

* The Farmer's Lawyer, p. 193-4.

66. OF ROLLING THE FLEECE, AND OF THE QUALITIES OF WOOL.

" Come, gentle swains, the bright unsullied locks
Collect: * * * *
Be faithful; and the genuine locks alone
Wrap round: nor alien flake, nor pitch enfold:
Stain not your stores with base desire to add
Fallacious weight."

DYER.

(2660.) Whenever a fleece is clipped from the sheep, a field-worker should be ready to roll it up. I have already said that a board should be provided to roll the fleeces upon, and it may be erected either in the clipping-barn, or any other adjoining apartment, such as the corn-barn; but the most convenient place for it is beside the clippers, as is also the stance for the rolled up fleeces to lie upon till the end of the day's work, when they are removed to the wool-room. Suppose the board to be placed in the clipping-barn, the fleece, whenever separated from the sheep, is lifted unbroken from the cloth, and spread upon the board on its clipped side. The winder examines the fleece that it is quite free of extraneous substances, such as straws, bits of thorn, of whin, or burs, and removes them, and she also removes, by pulling off, all locks that have lumps of dung adhering to them, and which have escaped the notice of the washers. The farmer should be very particular in giving instructions on the purity of the fleece, as the purchaser cannot unloose every fleece he buys; and should he find as much filth in the fleeces after purchase as to warrant the belief that it had been purposely left in them, he may either relinquish his bargain, or make a large deduction from the price —in the former case implying fraud on the part of the farmer, and in the latter diminishing his profits. And besides the disgrace, any person who attempts to commit fraud in the case of wool is liable to be merced in heavy fines when informed against, which he puts in the power of the purchaser of the wool to do. The winder being satisfied that there are no impurities in the fleece, folds in both its sides, putting any loose locks of wool into the middle, and making the breadth of the folded fleece from about 24 to 30 inches, according to its size. She then begins to roll the fleece from the tail towards the neck as tightly and neatly as she can, and when arrived at the neck, draws the wool there as far out, twisting it into a sort of rope, as will go round the fleece, hold its own end firm, and make the entire fleece a tight bundle. The fleece is then easily carried about, having the clipped surface outside, which, being composed of wool saturated with yolk, exhibits a shining silvery lustre.

Fleeces are by no means all alike either in structure or colour. Those of ewes, for instance, will be found thin and open in the locks, of pale colour, and feel light in hand; those, again, of hoggs will be close and long in the pile, of a rich colour, and the rolled fleece will feel heavy and be bulky in hand. Neither will all the fleeces be alike entire and in proper condition. An occasional fleece may want some part, having been shed off; and another may be *coated*, that is, have its wool felted together like a piece of very thick cloth; while another may be a black fleece. Whenever any such difference is observed in a fleece, it should be laid aside. The inferior stray locks, and those clotted with dirt, should be put into a basket by themselves, to be afterwards washed, dried, and used at home for various purposes, such as in repairing of saddlery. Every day's clipping is carried into the wool-room, at W in the steading, fig. 4, Plate IV., which enters by a stair from the straw-barn. Previous to being occupied, the room should be completely swept of all dust from its floor and plastered walls, and washed clean and dried. The fleeces are built up on the floor at a little distance from the wall, putting the hogg and ewe fleeces in separate divisions. The whole are covered with cloths, and the shutters of the window closed. The reason for these precautions, which are too seldom attended to by farmers, is, that the cloths keep off dust, prevent too quick evaporation of the yolk of the wool, which, if allowed, will diminish its weight, and the window-shutters being closed excludes the light, which obscures the bright lustre of clipped wool. The odd fleeces and locks should not be brought into the wool-room at all, but the former sold, and the latter prepared for use immediately, as their unclean state creates an effluvium which induces the wool-moth, the *Tinea sarcitella* of Linnæus, to come into the wool-room.

(2661.) Wool is a ticklish article for a farmer to keep long. If the wool-room is too dry, the natural moisture, occasioned by the yolk, evaporates, and the fibres become curled, and feel harsh; and on the other hand, if it is too damp, which is its usual condition, the fleeces become compressed, feel clammy, and affected with green and yellow mould. The wool-moth then takes up its residence, in summer, amongst such fleeces, and breeds numerous larvæ, which subsist on the very fibres of the wool, and of course entirely destroy their character. There are farmers who have no wool-room, but keep their wool in the granary or an out-house, where, of course, either of these effects are aggravated. The best means that I know of preserving wool for a length of time in the fleece, is to keep it in a wool-room with a wooden floor, packed in the pack-sheet, in which it will be out of reach of dust, light, and moths, and where no more air and light need be admitted than is desirable. If it is meant to keep wool only for a few weeks, the means spoken of in the last paragraph are all that are requisite. The safest plan for the wool-grower is to sell it every year at the current prices, which are determined at the great wool fairs that take place in summer in many parts of the country, where wool-dealers attend, and whose proceedings are duly reported in the newspapers.

(2662.) When a wool-dealer purchases wool from a farmer, he sends his own people to pack it in his own pack-sheets. Wool is weighed and packed in this way:—Wool is sold in Scotland by the wool-stone of 24 lb. avoirdupois, and it is weighed out in double stones of 48 lb., each being called a *weigh*. Usually 7 Leicester hogg and 11 ewe fleeces make 1 weigh. In England, wool is sold by the lb., and weighed out by the *tod* of 2 stones of 14 lb. each, or 28 lb. In weighing out, the above number of fleeces may not exactly weigh the double stone; and, as fleeces are never broken to equalize the scales, a few small weights are used to balance the scale either on the side of the wool or of the weights, at each weighing. In this way the weight of the number of scalefuls required to fill each pack are correctly ascertained, a memorandum being taken of the number of weighings. While a simple and large beam and scales for weighing the wool is erecting, the pack for containing the fleeces is also making ready, so that they may be packed immediately from the scales, and save much handling. Pack-sheets are made of thin canvass, of the shape of an oblong rectangle, about 8 feet long when empty, and open along one side. A small stone being placed in each end of the opening of the sheet, a rope for each end being suspended from the ceiling, the stones form knobs which prevent the corners of the sheet slipping through the ropes, as at *a a*, fig. 470. The sheet *d* is suspended just to swim above the floor. Two men, *b* and *c*, then get into the sheet, and placing the fleeces, as handed to them by the woman *e*, in regular order lengthways *across* its bottom, trample them down with considerable force, especially at the corners, where they are pushed down with both feet set together, while both hands are holding firm by the outside of the corner of the sheet immediately under the tying, as shewn in action by the man *b*. The second layer of fleeces is laid contrary to the first, that is, in length along the sheet, placing 2 or 3 fleeces parallel in the breadth of the sheet; but the fleeces at the ends are always placed across the length of the sheet, in the same position as before, and pressed down in the same manner. The sheet is thus filled with alternate layers of fleeces to the top, when the packers leave it, and then loosening the ropes, and reserving the small stones for the next sheet, immediately close the mouth of the pack; for if left open, the elasticity of the wool will cause the fleeces to rise so far as to render the closing afterwards impracticable, and a pack is difficult to close at anytime. With the aid of hand-cramps, inserted into the opposite sides of its mouth, the edges are brought togteher,

Fig. 470.

THE WEIGHING AND PACKING OF WOOL.

and are so held by iron skewers being passed through both edges. When a farmer is packing wool on his own account, it is as well to know that common table-forks answer as well as hand-cramps for pulling the edges of the pack-sheet together, and for keeping them as close as skewers. The edges being thus brought together, they are permanently secured by sewing with packing-needle and stout twine, and the skewers are removed as the sewing proceeds. Thus, one pack is filled after another. A pack of wool *f* contains 10 stones, that is, 240 lb. Any wool I have seen packed in the Highlands, was not put so regularly into the sheet as I have described; the fleeces were crammed in and trampled down in the most irregular and promiscuous manner. I once had an opportunity of seeing Merino wool packed on a large scale at Leipzig. The sheets were made of horse-hair, and, during the packing, were occasionally subjected to the pressure of a long pole of wood acting upon them as a lever. The pole was fastened at one end by a ring to the ground, and heavy weights were suspended, and a rope passed from the other through a ring in the ground, to keep good what the lever had gained. The wool was packing for Great Britain, and was to be conveyed in large waggons, each drawn by 8 stallions, to be shipped at Rotterdam. In the rest of the cut the wool is seen piled up at *g*, and the man *i* is in the act of weighing a scaleful with the large beam and scales *h*.

(2663.) Even on the slightest inspection of a fleece on the sheep's back, one can perceive that it contains wool of different qualities; the coarser is evidently below and the finer above; but none but wool-staplers would discover 10 different qualities of wool in the same fleece. As a general description of a fleece, I may mention, that the finest wool is upon the shoulder and along the top of the back to the rump; the next best is below the shoulders, along the ribs to the rump—the coarsest being on the haunches—and that below the belly is short and detached, and cannot be classed with the rest. Each of these parts have their respective qualities, which wool-staplers classify, in order to satisfy the wants of their customers, the manufacturers. The subdivision of the fleece by wool staplers, after they have purchased the wool from the farmers, is technically in these terms:—Prime—choice—super—head—downrights—seconds—fine abb—coarse abb—livery—short coarse or breech-wool. It would be well for wool-growers to receive lessons from wool-staplers on the essential properties which constitute good wool, that they may be able to judge whether the wool which they grow be intrinsically good or bad, comparatively improving or deteriorating, or to what species of manufacture it is best suited. According to present practice, wool-growers, I believe, demand prices for their wool on the faith of markets, without knowing whether their wool is really worth a high or low price.

(2664.) Good wool should have these properties:—The *fibre* of the staple—a staple being any lock that naturally sheds itself from the rest—should be of uniform thickness from root to point, it should be *true*, as the phrase has it; the finer the wool, the smaller is the diameter of the fibre; the fibre should be elastic, and not easily broken; its surface should have a shining silvery lustre; and it should be of great density or specific gravity. Of a *staple* all the fibres should be of the same length, otherwise the staple will have a pointed character; the end of the staple should be as bright as its bottom, and not seem as if composed of dead wool; the entire staple should be strong, and its strength is tested in this manner:—take the bottom of the staple between the finger and thumb of the left hand, and its top between those of the right, and, on holding the wool tight between the hands, make the third finger of the right hand play firmly upon the fibres, as if in staccato on the strings of a violin, and if the sound produced be firm and sharp, and somewhat musical, the wool is sound; if the fibres do not break on repeatedly jerking the hands asunder with considerable force, the staple is sound; if they break, the wool is unsound, and, what is remarkable, it will break at those places which issued from the felt of the sheep when the sheep was stinted of meat or had an ailment; though it will not break at every place simultaneously, because the weaker part, occasioned by the greater illness, will first give way. A *good fleece* should have the points of all its staples of equal length, otherwise it will be a pointy one; the staples should be set close together; and it should be *clean*. One essential good property of wool is *softness* to the feel like silk, which does not depend on *fineness* of fibre, but on a peculiar property of yielding to the touch at once, and readily returning to the hand. There should be no *hairs* in wool, neither long ones, which are easily distinguishable from wool, and give the name of *bearded* to the fleece; nor short ones, soft and fine, like *cat's* hair, which are not easily distinguishable from wool, and are denominated *kemps*. The long hairs are frequently of a different colour from the wool; but the kemp hairs are of the same colour, and of the two, the latter are much the more objectionable, as being less easily detected.*

(2665.) Keeping all these properties in view, it is clear that a farmer who breeds sheep having fleeces with pointy staples, thinly set on, and of unequal lengths; who stints his sheep of food at one time, and overfeeds them at another, thereby producing wool of unequal size, and therefore *untrue;* who, moreover, does not wash his sheep clean, or having so washed them, allows their wool to be again dirtied before being clipped, thereby creating much waste to the manufacturer to bring the wool again to a clean state; the farmer who manages his sheep and wool so as to produce these effects, injures himself to an incalculable extent.

(2666.) It is remarked by Mr Culley, that "the Herefordshire sheep that have the finest wool are kept *lean*, and produce $1\frac{1}{2}$ lb. each; if better kept they grow larger, and produce more wool, but inferior in quality."† This is true of every breed of sheep, and particularly of the Merino, whose *propensity to leanness* caused their culture to be abandoned in Great Britain as being unprofitable. Their wool did not so much deteriorate in this cold climate, as there was no possibility of getting mutton upon their carcass. But though leanness produces wool of finer quality than high condition, yet the remark is only strictly true when applied to breeds which yield fine wool in every state of condition, for no degree of leanness will cause a coarse-woolled breed of sheep, such as the Black-faced, to produce *fine* wool. To obtain any given quality of wool, therefore, it is necessary to possess the breed that produces it, and then the wool will be finer or coarser in comparison as the sheep are kept in low or high condition. There is, no doubt, that the general quality of wool in this country has become coarser than it was years ago; not because the breeds of sheep have deteriorated, for, on the contrary, they have all improved, but because the animals are now kept throughout the year in much higher condition; and this result might have been anticipated, for if there is any analogy between the vegetable and animal economy, we know that well manured soil will produce flax of thicker and longer fibre than the same soil in poor condition; so, in like manner, sheep when in high condition produce wool of thicker and longer fibre than when lean. During the improvement that has taken place in the breeds of sheep, a counteracting influence, as I conceive, has been at work to retain the wool of finer quality than the

* See Luccock on Wool, from p. 168 to 179; Ed. 1805.

high condition would produce, I mean the influence of shelter; and though it may only be of a negative character, preventing unequal evaporation of the yolk of wool, by warding off cold and drying winds; yet, in preventing these, its effects are positively beneficial, inasmuch as Mr Luccock observes, that "the silky softness, like most other good qualities of the fleece, depends very much upon the breed of the sheep, *and the quality of yolk which they constantly afford.*" It is difficult to say whether the density of the fibre of wool, that most desirable property, depends on some general law connected with the breed, or the circumstances in which the fibre is produced, for the mere coarseness or fineness of the fibre does not affect its specific gravity; as, for instance, the close full-grown wool off the shoulder of a sheep, does not differ materially in density from that from the thin and hairy breech; and hence, perhaps, the density does not depend on the breed. I am inclined to believe, that soil and climate very much affect the general condition of fleeces, for we find wool grown in the chalky districts of England much drier and coarser than that which is produced on fine soft hazel loam; and wool grown upon turnips appears to me coarser than when grown on grass in the same soil in similar condition. This fact is undeniable, that fleeces from the same breed, reared even in similar circumstances, differ much in density. The conclusion to be inferred from all these considerations seems to be, that whatever induces the greatest secretion of yolk, whether it be breed, condition of animal, nature of soil, or climate, will produce fibre of the greatest specific gravity; and hence on grass on a deep mellow soil in good heart, and in a sheltered situation, a breed of sheep, capable of continuing in good condition throughout the year, should produce the densest and the finest quality of fibre of wool. If these views be at all correct, you can easily perceive how much depends on the judgment of the farmer himself, to produce wool that will possess the greatest number of good qualities.

(2667.) Chemically, "wool has not yet been subjected to a rigid examination," says Dr Thomson, "but from the experiments made on it by Berthollet, there is reason to conclude, that its chemical qualities do not differ much from those of hair. When growing upon the sheep, it is enveloped in a kind of soapy matter, which protects it from the attacks of insects, and which is afterwards removed by scouring. Vauquelin has examined this matter, and found it to consist of the following ingredients:—A soap of potash,—carbonate of potash,—a little acetate of potash,—lime,—a very little muriate of potash,—and an animal matter."*

(2668.) Mr Youatt has examined the external structure of wool with the microscope, and has ascertained that the surface of the fibre is covered with a sort of scale which forms a series of serrations along the entire length. The general outline of the woolly fibre consists of a central stem or stalk, probably hollow, or at least porous, and possessing a semi-transparency not found in the fibres of hair. From this central stalk there springs at different distances, in different breeds of sheep, a circlet of leaf-shaped projections. It is thus ascertained that wool possesses a property common to all independent horny fibres which issue from the felt of animals, namely, an irregularity which constitutes a certain degree of roughness upon their surface from the root to the point. Hence, both physically and chemically, wool and hair are analogous substances. Some of the results of Mr Youatt's investigations with the microscope, in conjunction with the micrometer, were these:—

* Thomson's Animal Chemistry, p. 305.

	Diameter.		Serrations.	
Merino wool, ...	$\frac{1}{750}$	an inch, and	2400	in an inch.
——— Picklock, . . .	$\frac{1}{750}$	... ...	2560	... ...
——— Saxony, . . .	$\frac{1}{840}$	... ...	2720	... ...
Leicester, ...	$\frac{1}{500}$	... ...	1860	... ...
Deccan, black, . . .	$\frac{1}{1000}$	... ...	1280	... ...
Odessa, ...	$\frac{1}{750}$	... ...	2080	... ...
Wallachian, ...	$\frac{1}{750}$	... ...	2080	... ...
Australian, ...	$\frac{1}{750}$	... ...	1920	... ...
New South Wales, . .	$\frac{1}{750}$	... ...	2080	... ...
——— Mr Arthur's, .	$\frac{1}{780}$	... ...	2400	... ...
Van Dieman's Land, .	$\frac{1}{750}$	... ...	...	... ...
South Down, ...	$\frac{1}{660}$	... ...	2080	... ...
Wiltshire, ...	$\frac{1}{500}$	... ...	1860	... ...
Ryeland, ...	$\frac{1}{750}$	... ...	2420	... ...
Cheviot, hill-fed, . . .	$\frac{1}{500}$	... ...	1860	... ...
——— good pasture, .	...	... ...	1440	... ...
Norfolk, ...	$\frac{1}{580}$	... ...	1600	... ...
Lincoln, ...	$\frac{1}{480}$	... ...	1280	... ...
Irish, ...	$\frac{1}{560}$	... ...	1920	... ...*

Another instrument beside the micrometer, named the eirometer, the invention of the late celebrated optician, Dolland of London, and which reads off diameters to a very minute fraction of an inch, is used for measuring the diameters of wool, and it may be employed by an inexperienced hand with less chance of error than a micrometer. It produces a double image of the fibre which are brought in contact, and the result is then read off from a circular index.

(2669.) Wool is well known to have a felting power, but it is unknown in what that power consists. A *coated* fleece is a natural instance of the felting tendency of wool. Mr Youatt seems to believe that the discovery of the serrations on the fibre of wool accounts for its felting property. " It is a curious and interesting point that has been established," he says, " the existence of an irregularity of form in the wool, accounting for and necessarily giving it a felting power—is there a variation in this structure corresponding with the degree of felting power?" Mr Boyd of Innerleithen, Peeblesshire, is much disposed to question Mr Youatt's views. " Mr Youatt asserts with much confidence," he remarks, and in this remark Mr Boyd is supported by Mr Luccock, " that the felting properties depend entirely on the structure of the wool. During an experience of many years I have found this not to be the fact, and therefore state, without fear of contradiction, that in many instances it is impossible to estimate the extent of the felting properties in a variety of wools, until they have been submitted to the actual test of experiment; and I am decidedly of opinion," he adds, " that however perfect the structure of wool may be, if produced in the absence of an oily or saponaceous substance, it cannot possess the requisite properties of a clothing material."† Mr Boyd is again supported by Mr Luccock, when the latter says, " If the wool-grower be anxious to promote the growth of fleeces,

* Youatt on Sheep, p. 87 to 91.

† Prize Essay of the Highland and Agricultural Society, vol. xiv. p. 669.

in which the felting quality greatly prevails, I should recommend, from the little knowledge at present possessed, that he attend closely to the supply of natural, rich, and nutritious yolk, which the pile receives while growing." I confess that Mr Youatt's theory to account for the felting property of wool, in reference to the action of the serrations on the surface of the fibre upon one another, appears to me unsatisfactory. On the authority of Mr Luccock, the application of *moisture, warmth,* and *pressure* are necessary to bring the felting property of wool *into action.* "*Without* the aid of moisture," he affirms, "it remains perfectly *dormant;* the warmth and pressure are required to *quicken* the process." And he adds, "the degree of heat required to make the felting property act with the utmost force, is considerably below the boiling point of water," and that "a higher temperature loosens the texture of the thread, and increases the elasticity of the hair, thus giving it a disposition to start from the substance of the cloth and spoil its surface."* If the action of the serrations on the fibres is the principal means of felting wool, it must be proved that they change their structure on being immersed in water of a temperature near the boiling point, which has not yet been done. It is also known, that wool, after being combed with heated iron combs, will not felt, and yet the figures given by Mr Youatt of combed and uncombed wool of different variaties indicate no such decided change by combing on the structure of the serrations, as to warrant us in believing that in the one state wool shall felt, and in another it shall not.

(2670.) Having mentioned the application of the microscope to the structure of wool, it may prove instructive to shew by the same instrument in what manner wool grows. "If the fœtus of a sheep," says M. Raspail, "taken when it is of the length of about 4½ inches, and preserved in alcohol, be examined, it will be found studded with globules of uniform size, elegantly arranged, and almost at equal distances, round certain white spots disposed in quincunxes, which seem, even at this early period, to indicate the places where the hairs are to grow. On the epidermis of the temple, instead of thin white spots, we find vesicles projecting in the form of bottles, or rather of urns, whose sides are granulated in the same manner as the epidermis. These vesicles are the rudiments of hairs." †

(2671.) The number of sheep in the empire estimated by Mr Macculloch, from different sources of information, is the following:—

In England, ...	26,148,463
In Scotland, ...	3,500,000
In Ireland, ...	2,000,000
	31,648,463

By a statement made by the late Mr Hubbard, an eminent wool-stapler in Leeds, the number of packs of wool grown every year in England is 463,169 of 240 lb. each, or of 111,160,560 lb., which gives an average weight to each fleece, including those of lambs, of 4¼ lb. ‡

(2672.) The term *merino*, applied to a particular breed of sheep and variety of

* Luccock on Wool, p. 163-8.

† Raspail's Organic Chemistry, p. 283.

‡ Macculloch's Dictionary of Commerce, art. *Wool.*

wool, is of obscure origin. Mr Southey informs us, that " Merino is an old Leonese title, still preserved in Portugal, though long since obsolete in the other kingdoms of Spain. Perhaps it is a mongrel diminutive of the Arabic title *mir* or *emir*, likely enough to have been formed when the two languages, Spanish and Moorish, were, as it were, running into each other. *Mirquebir*, the augmented title, was in use at Ormuz. *Merino* would be sufficiently explained by supposing it a diminutive grade. The old laws of Spain define it thus: 'He is a man who has authority to administer justice within a certain district.' The first mention of this office is in the reign of Bermudo II. The merinos then commanded the troops of their respective provinces in war; but, before the time of Henrique II., it was become wholly a civil office, and the title was gradually giving place to that of Alguacil, mayor. Most probably the judge of the shepherds was called the merino, and hence the appellation extended to the flocks under his care." It is the general opinion that merino sheep came to this country from Spain, and so they did at the end of the last century, but it appears that *fine-woolled* sheep were sent from England to Spain a very long time ago. That *sheep* were sent from England to Spain at a known period is certain, for Mr Youatt quotes from the chronicles of Stowe, that " this yere (1464) King Edward IV. gave a license to pass over certain *Cotteswolde* sheep into Spain." And he quotes Baker also, who says, " King Edward IV. enters into a league with John, King of Arragon, to whom he sent a score of *Costal* ewes and four rams—a small present in show, but great in the event, for it proved of more benefit to Spain, and more detrimental to England, than could at first have been imagined." The wool of the Cotteswolde sheep of the present day is long, and not remarkable for fineness. Perhaps the old Cotteswolde wool was finer than the present, because the latter has been much crossed with the Leicester. But if the old Cotteswolde conferred so much benefit on Spain, it may be fairly inferred that the wool of Spain was not so fine as that of England at the time. But sheep were exported from England to Spain prior to the reign of Edward IV., as Mr Southey intimates, that " Fernan Gomez de Cibdareal, in one of his letters (Epist. 73.), mentions a dispute between two Spaniards concerning rank in the presence of Juan II. 1437. It was objected tauntingly to one of them, that he was descended from a judge of the shepherds. The reply was, that this office had always been held by hidalgos of great honour, and that 'King D. Alfonso had instituted it in the person of Inigo Lopez de Mendoza, *when the English sheep were first brought over to Spain.*'" This dispute occurring in 1437, and referring to an ancient title of honour which had been conferred as far back as the time of the introduction of English sheep into Spain, and a taunt being given in 1437 to a *descendant* of a judge of the shepherds, that is, of a merino, thereby implying that he, the descendant, was at least of the second or third generation, it is clear that the English sheep referred to could have no reference to the Cotteswoldes exported in 1464, as mentioned by Stowe. " How long was it before the merino fleece became finer than that of the original stock?" asks Mr Southey; and he replies, " Brito, who wrote towards the close of the 16th century, says in praise of the wool grown about Santarem, it is so fine that *it may vie with that of England* (*Monarchia Lusitania*, f. i. p. 93.) If the Spanish wool had been as fine then as it is now, he would hardly have drawn his comparison from the English."* While these facts are recorded

* Southey's Omniana, vol. ii. p. 103-8.

in Spanish literature regarding the origin and ancient quality of the wool of Spain, the opinion of Mr Youatt seems much too strongly expressed, when he says, that " Europe and the world are *originally indebted to Spain* for the most valuable material in the manufacture of cloth." And, again, " The chunahs, therefore, may be descendants of the English sheep," namely, those sent to Spain in 1464, " mixed with the common breed of the country ; but farther than this England cannot, with any degree of justice, urge the claim which some have done, of being instrumental in producing the invaluable Spanish wool."* And yet, as we have seen, sheep were probably sent from England to Spain long before that date, or even long before 1437, for if the King Alphonso mentioned above as having instituted the order of judge of the shepherds, be Alfonso the Wise, king of Leon and Castile, who is stated to " have digested a code of excellent laws, and rendered his name famous in history by his patronage of the arts and sciences,"† he reigned at the early period of from 1252 to 1284. And another fact mentioned by Mr Southey, when Catherine, daughter of John of Gaunt, was espoused to Henrique III., she took English sheep with her as her dowry, fixes another exportation of sheep to Spain about 1390, a considerable while prior to the Cotteswolde exportation of 1464, and if the English sheep had been of an inferior description to those of Spain, it is not likely that the future Queen of Castile would have taken them with her as her own dowry. In concluding this subject, Mr Southey puts this doubting query ;—" Can there possibly be any truth in the remark of Yepes (t. 7, § 134.), who says, ' *Daily experience shews us*, that if a lamb is sucked by a goat, the wool becomes hard and hairy ; and, on the contrary, if a kid is suckled by a ewe, the hair becomes soft ?' "

---

67. OF THE MAKING OF BUTTER AND CHEESE.

" And now the DAIRY claims her choicest care,
And half her household find employment there :
Slow rolls the churn, its load of clogging cream
At once forgoes its quality and name :
From knotty particles first floating wide,
Congealing butter's dash'd from side to side ;
Streams of new milk through flowing coolers stray,
And snow-white curds abound, and wholesome whey."

BLOOMFIELD.

(2673.) THE dairy operations on a farm of mixed husbandry are limited, both in regard to the season in which, and the quantity of materials by which, they can be prosecuted. Until the calves are all weaned, which can scarcely be before the end of June, there is no milk

* Youatt on Sheep, p. 146-7.

† Bigland's View of the World, vol. ii. p. 589.

to spare to make butter or cheese, but what of the former may suffice for the inmates of the farm-house; and as some of the cows, at least, will have calved 4 months before all the cows are free to yield milk for the dairy, a full yield of milk cannot be expected from them even when entirely supported on grass. But though thus limited, both in regard to length of time and amount of milk, there is ample opportunity for performing every dairy operation, according to the taste and skill of the dairy-maid. For example: Butter may be made from cream, or from the entire sweet milk. It may be made up fresh for market, or salted in kits for families or dealers. Cheese may also be made from sweet or skimmed milk, for the market; and any variety of fancy cheese may be made at a time; such as cream-cheese, imitation Stilton, Gloucester, or Wiltshire. With all these means at command, to a moderate, indeed, but available extent, it is quite possible for the dairy-maid to display as much skill and taste in her art on a mixed as on a dairy farm, not only in these respects, but in the endless forms in which milk may be served on the table of the farmer. In short, the only advantage a dairy-farm possesses over one of mixed husbandry, is, that all its dairy operations are conducted on a larger scale.

(2674.) The *milk-house*, large enough and otherwise convenient and suitable for a farm of mixed husbandry, may be seen represented in plan at *m*, fig. 32; and the cheese-room adapted to the same is seen in plan at *h*, fig. 33. The mode in which a milk-house should be fitted up, is described in (231.), as well as a cheese-room in (237.).

(2675.) The *utensils* with which a dairy should be supplied, comprise a large number of articles, though all of simple construction. The *milk-dishes* are composed of stoneware, wood, metal, or stone. The stoneware consists of Wedgwood and common ware; the wooden of cooper-work, of oak-staves bound with hoops of iron; the metal of block-tin, or of zinc; and the stone are hewn out of the block, and polished. Besides these, utensils formed of a combination of materials are used, such as, wooden vessels lined with block-tin or zinc, and German cast-iron dishes lined with porcelain. Of the whole variety, the stone, and wooden ones lined with metal, are stationary, and the rest movable. All milk-dishes should be of a broad and shallow form, for the purpose of exposing a large surface with a shallow depth of milk, in order to facilitate the disengagement of the several parts of the milk. There seems to be a difference of opinion, which of those substances have the greatest influence in disengaging the greatest quantity of cream from the milk. I have heard it maintained at one place, that stone is much the best; at another, that wood is the best; at another, that stoneware is preferable to all others, and of the two kinds of stoneware, that the common ware is better than Wedgwood's; and at another, that zinc is by far the best; from all which difference of opinion I would infer, that the subject has not been satisfactorily ascertained by comparative experiment, and which I believe to be the case; but I am sure that different management, even in the same circumstances, will produce very different results in this product of milk. Independently of the consideration of cream, however, other circumstances should be regarded in making choice of milk-dishes. Wooden ones require much labour to keep them thoroughly clean, though they are the least liable to injury in the use. Metal ones also require much cleaning, and are liable to be bruised; and it is not questionable that they are unwholesome for milk, if not kept thoroughly clean. Tin produces no injurious salt with the acid of milk; but the salt produced by the action of lactic acid upon zinc, is believed to be in a slight degree poisonous to the human stomach. At all events, the thought of *keeping* milk in metallic dishes is unpleasant to the mind. Stoneware is easily frangible, but is, nevertheless, so cheap, so easily cleaned, and so safe in use, that it forms the most convenient material for milk-dishes to every class of country people. The advantage of Wedgwood ware consists in its hardness, and the durability of its glazing. The price of dishes of common ware, of 15 inches in diameter, is 9d. each; of Wedgwood's ware, from 12 inches to 24 inches, from 2s. to 8s. each; of wood, 16 inches diameter and 4 inches deep, 2s. each; and of zinc, 18 inches in diameter, 3s. 9d. each, and others are 3d. per inch more or less, as the diameter increases or diminishes from this size.

(2676.) The other utensils are: Creaming scallop, for taking the cream off milk. A jar for containing the cream until it is churned; a Wedgwood one, with top and opening in it to be covered with muslin, to keep out dust and let in air, costs from 6s. to 7s. 6d. A churn, of which there are many forms, all of which will be found described below by Mr Slight. A flat wooden kit, to wash butter in. Scales and weights for weighing butter, whether in pounds, fractional parts of a pound, or in the lump. Jars or firkins for packing salted butter. Moulds for forming prints of butter for the table. Covered dishes for holding fresh butter in pounds. A tub for earning the milk in, when about to make cheese. A curd-cutter, and a curd-breaker. A drainer to lay across the cheese-tub while the whey is straining from the curd. Cheese-vats for giving the form to cheese. A cheese-press, a figure of the most convenient and powerful form of which is given below. A furnace and pot for heating water and milk. And a supply of spring-water is an essential concomitant to a dairy.

(2677.) A word or two on *churns.* The old-fashioned upright hand plunge-churn is now confined chiefly to the use of small farmers and cottars; but, when inanimate power is applied to the making of butter, the upright churn is yet used by many dairy-farmers who conduct their operations on a large scale. About 20 years ago the barrel-churn was much in vogue, and it was very well suited for making butter, as it kept all the cream in constant agitation; but the trouble required to keep it clean, from its inconvenient form to get within it, has tended to bring it into disrepute. It is now superseded by the box-churn, whose construction admits of its being easily taken to pieces and cleaned. Of the two forms of box-churn, I prefer the one in which the agitators move horizontally, because it can be filled to any degree of fulness, without danger of the cream or milk working out of the journal-holes of the axle. Long ago, small churns were in repute; though they fell into desuetude, but lately have again come into use, and for making small quantities of butter they are admirably adapted. Various constructions of small churns, intended to regulate the temperature of the cream, have of late been proposed, some of which are noticed below.

(2678.) A word or two also on *cleanliness.* Unless the milk-house is kept *thoroughly* clean, in its walls, floors, and shelves, the milk will become tainted; and in order to keep them clean, the floor and shelves should be of materials that will bear cleansing easily and quickly. In most farm-houses the shelving is of wood, and the floor of pavement or brick. Wooden shelves can be kept clean, but are too warm in summer. Stone shelving is better, but must be *polished,* otherwise cannot be sufficiently cleaned; and to be kept clean, requires at times to be rubbed with sandstone. Marble shelving is the best of all for coolness and cleanliness, and they are not so expensive as many imagine. Polished pavement makes a more durable, easier cleaned, and cooler floor than brick. There should be ample means of ventilation in the dairy when required; the principal object, however, not being so much a constant change or a larger quantity of air, as an equality of temperature through summer and winter. To obtain this desideratum, the windows, which should face the N. or E., should not be opened when the temperature of the air is above or below the proper one, which, on an average, may be stated at 50° Fahrenheit. The milk-house should be thoroughly *dry,* the least natural damp in the walls and floor will emanate a heavy fungus-like odour, very detrimental to the flavour of milk and its products. The utensils should all be kept thoroughly clean, and exposed to and dried in the open air. Some dairy-maids are so careless in this respect, that I have seen seams of green and yellow rancid butter left in the corners and angles of churns, and a heavy smell of dirty woollen rags pervading wooden utensils. However effectual woollen scrubbers may be in removing greasiness left by milk and butter on wooden articles, they should never be employed in a dairy, but only coarse linens, which should always be washed clean in hot water without soap, and dried in the air. All the vessels should be quickly dried with linen cloths, that no feeling of clamminess be left on them, and then exposed to the air. In washing stoneware dishes, they should not be dried at that time, but set past singly to drip and dry; and they should be rubbed bright with a linen cloth when about to be used. If dried and set into one another after being washed, they will become quite clammy. The great objection to using stone milk-coolers is the difficulty of drying them thoroughly before being again used. No milk-house should be so situate as to admit the steam rising from the boiler which supplies hot water for washing the various utensils; nor should the ground before its windows contain receptacles for filth and dust, but be laid out in grass, or furnished with evergreens. It is said that the odour from the blossom of the common elder, *Sambucus nigra*, in summer, is a complete muscafuge, if I may be allowed to coin a word.

(2679.) The articles which engage the dairy-maid's attention within the dairy are milk, butter, and cheese; and, first, as to *milk*. I have already said that the milk is drawn from the cow into a pail (2097.), the most convenient form of which is given in fig. 471, and the size of which may vary to suit the pleasure of the dairy-maid. It is made light of thin oak staves bound with iron hoops, and costs from 2s. 6d. to 5s., according to the size. The milk, in being drawn from the cows, is put into a tub, and left to cool; but not to become so cold or stand so long as to separate the cream. The tub should be placed in the air, and out of reach of animals, such as cats and dogs. After it has cooled, the milk is passed through the milk-sieve into the milk-dishes, and as much only is put into each dish as not to exceed 2 inches in depth. To know at once the age of milk in the dishes, *one* mark or score should be made with chalk on the dishes just filled, to shew that they contain the last drawn milk, or freshest meal; a *second* mark is made, at the same time, on the dishes containing the meal before this; and a *third* is put on the dishes containing the milk drawn before the second meal, and which constitutes the third meal, or oldest milk. If the cows

Fig. 471.

THE MILKING-PAIL.

are milked three times a-day, when the first mark is put on the dishes of the evening meal, those of the morning meal of the same day will have 3 marks, to indicate its being the third meal previous, and the dishes of the mid-day or second meal will have 2 marks. At every meal all the utensils that have been used should be thoroughly cleaned, and set past dry, ready for use when required.

(2680.) The next care of the dairy-maid is taking the *cream* off the milk. In ordinary weather in summer, the cream should not be allowed to remain longer on the milk than 3 meals; that is, when a fresh meal is brought in, the cream should be taken off the dishes which have 3 marks, when the milk will be 20 or 22 hours old; but should the weather be unusually warm, the milk should not be allowed to be more than 18 hours old, or that having 2 marks, before the cream is taken off it. For example, in ordinary routine, the cream of the previous mid-day's milk should be taken off in the morning, and at mid-day the milk of the previous evening would be creamed, and so on; but when the weather is very warm, it would be well to anticipate the creaming of one meal, and take the cream of the two oldest meals at one time, and in this way take off all the cream that can be got every 18 hours. The reason for using this precaution in taking off cream is, that the milk should on no account be allowed to become sour before the cream is taken off, because the cream of sour milk makes bad butter. Let sweet cream become ever so sour after being taken off the milk, and no harm will accrue to the butter. Not that sour cream off sour milk is useless, or really deleterious, for it may be eaten with relish by itself, as a dessert, or with porridge. The cream is skimmed off milk with a thin shallow dish, called a *skimmer* or *creamer*. It may be made of wood or of stone-ware, and of the two substances the ware is preferable for cleanliness, and of ware Wedgwoood's or porcelain is the best, being light, thin, hard, highly glazed, and smooth. There is no other way of taking cream off dishes but with a skimmer; but in stationary coolers of metal or of stone, a spigot is drawn cautiously from a hole in the bottom, through which the milk runs slowly down into a vessel, and the cream is left on the bottom of the vessel; and this manner of separating the cream from the milk is said to be the most effectual; but, of course, the skimmer can be used for creaming the milk in coolers, as well as other dishes. The cream when taken off the milk is put into a *cream-jar*, in which it accumulates until churned into butter. Every time a new portion of cream is put into the jar, the cream should be stirred in order to mix the different portions of cream into a uniform mass. The stirring is usually done with a stick kept for the purpose, but spoons of Wedgwood ware

are made for doing it. The cream soon becomes sour in the jar, and it should not be kept too long, as it is apt to contract a bitter taste. Twice a-week it should be made into butter, however little the quantity may be at a time. The skimmed milk is put into a tub and made into cheese; but if a cheese is only made every other day, the milk kept for the following day should be scalded before it is put into the tub.

(2681.) On converting cream into *butter*, the first act is to put the churn into a proper state. It is assumed that the churn when last used was put aside in a thoroughly clean and dry state. This being the case, a little hot-water, about 2 quarts, should be poured into it to scald and rinse it. In summer it should be rinsed with cold water after the hot, but not in winter. Some people sprinkle a little salt into the churn before the cream is put into it, but whether it does any good or not I cannot say. The churn being thus prepared, the cream is strained into it through a bag of coarse open sort of linen cloth, well known under the name of cheese-cloth. This cloth is always kept in a sweet state, no soap ever being employed to wash it. It is dipped in water, and then held over the churn; and on the cream being slowly poured into it from the jar, the greater proportion will run through into the churn; but the clotted part which will contain in it dust, drowned flies, moths, and spiders, and other impurities, which it is impossible to keep out of an open cream-jar, the cloth will keep back, on being gently pressed. In the small hand plunge-churn, and in a barrel-churn, a churn-cloth is required to be put round the mouth of the former, and under the bung of the latter, to prevent the cream being thrown out in churning. The temperature at which *cream* is put into the churn has a considerable influence on the time which the butter will take to make, and also on the weight of butter obtained from a given quantity of cream. It has been found that 55° Fahrenheit is the temperature which best attains these ends, and it is one easily attained in a cool apartment early of a summer's morning. The churning should be done slowly at first, until the cream has been completely broken, that is, rendered a uniform mass, when it becomes thinner, and the churning is felt to be easier. During the breaking of the cream a good deal of gas is evolved, which is usually let off by a small spigot hole if the churn be tight, such as a barrel-churn; but in other churns, which have a cover, the air escapes of itself. When the motion of churning is rotatory, it should be continued in the same direction, and not changed backward and forward. I am not sure that a satisfactory reason can be given for continuing uniform motion; but the opinion is, that the butter is formed more simultaneously; and that the backward and forward motions are apt to make the butter soft.

It is certain, at all events, when the motion is uniform and rather slow, the butter, whenever it is formed, is felt to stop the motion of the agitators at once. After the cream has been broken, the motion may be a little increased, and continued so until a change is heard in the sound within the churn, from a smooth to a harsh tone, and until an unequal resistance is felt to be given to the agitators. The butter may soon be expected to form after this, and by increasing the motion a little more, it will form the sooner, and, the moment it is formed, the motion should cease. The rate of motion in churning butter is of some importance, for when performed too slowly, a longer time will be spent in churning than is necessary, and the butter will be strong-tasted; and, on the other hand, when the motion is too rapid, the butter will be soft and frothy, when the churning is said to have *burst*. In very warm weather, and when the cream is put in too warm, the churning is liable to burst with any degree of fast motion, and hence the judgment is required to be exercised in the circumstances. I suppose that the most proper motion in churning has never been ascertained by experiment, and to determine which would probably be tedious, but it would be worth while being tried. When butter forms from cream in $\frac{3}{4}$ of an hour to 1 hour churning, it is satisfactory work, when it comes, much sooner it will be soft, and when much later it will be strong-tasted. The temperature, by agitation during churning, rises 3° or 4°.

(2682.) Immediately on being formed, butter should be taken out of the churn and put into the butter tub, one of a broad and shallow form, to be worked up. A little cold water being first put into the tub, and the tub set in an inclined position, the butter is spread out, rolled up round the edges, and pressed out by the palm of the hand, in order to deprive it of all the butter-milk, for the least portion of that ingredient retained in it would soon render it rancid. The milky water is poured off and fresh poured in, and the butter is again washed and rubbed as often as the water becomes milky. If intended to be kept or disposed of in a fresh state, the large lump is divided and weighed in scales in 1 lb. or $\frac{1}{2}$ lb. lumps each, and placed separately in the tub amongst water. Each lump is then clapped firmly by the hand and formed into the usual form in which pounds and half-pounds of butter are disposed of in the part of the country in which your farm is situate. For the table any requisite number of the pounds should be moulded from the lump into prints according to taste, or rolled into forms with small wooden spades, figured or plain. The made-up butter is then floated in jars with covers, in a clear strong brine of salt and water fit to float an egg, made ready for the purpose.

(2683.) Objections have been urged against the use of the hand in making up butter, and small wooden spades recommended to be employed for the purpose; and the use of water has also been objected to making up of butter, as it is said to deprive the butter of its pleasing aroma. A woman who has hot clammy hands should never become a dairy-maid, for butter being very susceptible of taint, its flavour will, no doubt, be injured by the heavy smell of sweaty hands; but clean cool hands—rendered so by washing in warm water and oat-meal, *not soap*, and then rinsed and steeped in cold water—hands so prepared will make up butter far cleaner and more solid than any instrument, whether of wood or of any other material; and as to cold water injuring butter, there can be no such strong affinity between a fatty matter like butter and cold water, as that the latter shall dissolve any essential ingredient out of the former; at all events, water will more effectually take away any milky substance from butter than any dry instrument that can be used, or even all the art the hand alone can accomplish. Let the trial be made both ways, and their comparative efficacy be tested by keeping the butter fresh, and seeing which will keep the longest *sweet*.

(2684.) If the butter is intended to be salted, it is somewhat differently treated. After being washed clean as above described, it is weighed in the scales, and salt is immediately applied. Practice varies much in the quantity of salt given to butter, so much as from 1 oz. of salt to 1 lb. of butter, to ½ oz. of salt to 1½ lb. of butter: 1 oz. to 1 lb. is too much, it is like curing butter with as little art as a salt herring; ½ oz. of fine pure salt being quite sufficient; and this quantity is intended for keeping-butter, for as to powdered butter for immediate use, ½ oz. to 2 lb. is quite sufficient. In the process of salting, the butter is spread out in the tub, and the salt, ground fine, is sprinkled over it by little and little, and the butter rolled up and rubbed down with the side of the hand until the whole mass appears uniform, and is considered to be incorporated with the salt. To insure uniform salting, only half the salt should be applied at once, and the butter lumped and set aside until next day, when the other half of the salt should be rubbed in. Whatever of brine or milk may have subsided from the lump in the mean time should be poured off. The salted lump is then put into the jar or firkin on the second day. One great advantage of deferring the making up of butter until the second day is, that, without it, the butter will not assume that firm, smooth, waxy texture, which is so characteristic a property of good butter. Butter when kitted is pressed firmly down in all points, and great care taken that it be particularly pressed with the side of the finger round the circumference of the jar, and its substance made solid, and its surface flat and smooth. If a former churning of butter had been put into the jar, its surface should be raised up into regular furrows, that the new lump of butter may be commingled with what was put into the kit before. The compressing of butter, then, into the kit, is of great importance, inasmuch as if the least cell of air be left in its mass, or get access by the side of the kit, it will *wind* the butter, that is, impart to it a rancid taste. After the kit has been filled within an inch of the top, the butter is made smooth, and covered with a new piece of wetted white linen or cotton cloth. To secure its goodness, butter should be salted immediately on being made.

(2685.) The state of the kit should be particularly examined before it is used for packing butter. If composed of stoneware, it is easily cleansed and rendered sweet. A wooden kit that has been used before should be filled with water for some time, to render it water-tight by the swelling of the edges of the staves. It should then be repeatedly scalded with hot water, and exposed to the air; and, just before being used, should be rinsed with cold water, and a slight sprinkling of salt scattered over its bottom. A new wooden kit requires somewhat different treatment, because the odour from the new wood will impart a disagreeable flavour to the butter. It should be filled with water mixed with garden mould, or with limeshells and water, for some days, and the mixture occasionally stirred; after which it should be thoroughly scrubbed and cleansed with hot water, and rinsed with cold water, and salted before being used.

(2686.) The quality of the *salt* has a material effect on the taste of the butter that has been salted with it. Ordinary sea-salt contains a considerable proportion of other salts which are bitter, and, of course, they will have effect upon the butter as well as the true salt. "It is easy, however, to purify the common salt of the shops from these impurities," says Professor Johnston, "by pouring 2 quarts of boiling-water upon 1 stone or 2 of salt, stirring the whole well about, now and then, for a couple of hours, and afterwards straining it through a clean cloth. The water which runs through is a saturated solution of salt, and contains all the impurities, but may be used for common culinary purposes, or may be mixed with the food of the cattle. The salt which remains in the cloth is free from soluble salts of lime and magnesia, and may be hung up in the cloth till it is dry enough to be used for mixing with the butter, or with cheese."*

(2687.) Butter assumes a *texture* according as it has been treated. When burst in the churning, it is not only soft but frothy, and on being cut with the knife, seems as if it could be compressed into much less bulk. When churned too rapidly, especially in warm weather, the

* Johnston's Lectures on Agricultural Chemistry, p. 828.

churning may not advance to the degree of bursting, but the butter will always continue soft, and never assume a firmness, though worked up with ever so much care, and in the coolest manner, and when one piece is separated from another, they are drawn asunder with a jagged surface, and stick to the knife that cuts them. Soft butter will not keep long, whether salted or fresh. When over-churned, that is, when the churning has been continued after the butter had been formed, it becomes soft, not unlike the state when it is too rapidly churned. When properly churned, both in regard to time and temperature, butter becomes firm with very little working, and is tenacious; but its most desirable state is that of waxy, when it is easily moulded into any shape, and may be drawn out a considerable length before breaking. It is only in this state that butter has the rich nutty flavour and smell, which impart so high a degree of pleasure in partaking it. To judge of butter, it is not necessary to taste it; the smooth unctuous feel on rubbing a small piece between the finger and thumb expresses at once its richness of quality; the nutty smell indicates a similar taste; and the bright, glistening, cream-coloured surface, shews it to be in a clean state.

(2688.) What I have stated in reference to the making of butter, applies especially to that obtained from cream alone, and from cream in the usual state for butter, namely, after it has become sour by keeping; but butter can be obtained from sweet cream as well, though churning renders its butter-milk sour, as well as that always is from sour cream. To have butter in perfection from sweet cream, it should be churned every day, and, as the supply of cream daily is usually very limited, a smaller churn than usual is most convenient to be used; and for this purpose, there is perhaps none better than the Table-churn figured below; by the peculiar construction of which, being placed in a vessel, the temperature of the cream can be regulated in all seasons. I see it alleged in advertisements, of churns of similar construction to this, that butter may be made from cream in 10 or 12 minutes. I have made several experiments with such a table-churn, in churning cream at different temperatures, and with different velocities, but never obtained good butter in less than 30 minutes; and when formed so quickly as in 15 minutes the butter was as soft as froth. I have heard it alleged that butter of the finest quality cannot be obtained from sweet cream; but the allegation, I suspect, is made by persons who have little experience of butter from sweet cream. So far am I from coinciding in this opinion, that I know butter of the richest quality, and of the finest flavour and appearance, can be made from sweet cream. Were this butter not necessarily good, would the nobles of the land have it upon their tables every morning? It is true, that sweet cream requires longer churning than sour; still butter is obtained from it in from 30 to 40 minutes, and if it is an unprofitable mode of using cream, that is a different question. For my own taste, I would never desire better butter, all the year round, than that churned every morning in a small churn from sweet cream. Such butter, on new baked oat-cake, cooled, with a little virgin flower-honey, and a cup of hot strong coffee, mollified with crystallized sugar and cream, such as the butter had been made from, is a breakfast worth partaking of, but can seldom be obtained.

(2689.) Besides cream, butter is made from *sweet milk;* but to obtain which, a large churn is required, and the churning continued for a long time, seldom less than 3 hours. When the quantity of milk is large, power other than human, whether of steam, water, or horse, is employed to move the churn. The butter obtained from this method is very good. The obvious objection to this method is, the labour imposed in agitating a great quantity of milk, and in consequently having a large quantity of butter-milk, which, however, may easily be disposed of in towns, and may even be converted into very good cheese. The method has its advantages in the uniform character of the butter which it affords in all seasons, from the certainty of obtaining a churning at the proper temperature, which is required to be higher than that of cream, being 65° Fahrenheit, and easily obtained, in winter and summer, by the addition of hotter or colder water amongst the milk. Churning from sweet-milk is thus a comparatively simple process. The milk is poured into coolers at first, and from them " it is drawn off by siphons into vats sufficiently large to contain both the evening and morning meals; and the vats are then put by, to stand totally undisturbed, till the whole acquires a sufficient degree of acidity. The time required for this purpose varies a little according to the heat of the weather, and the temperature of the milk-house. The point is ascertained by the formation of a strong thick *brat* or scum on the surface, when this becomes uneven." All the milk is not of the same age, but this does not affect the quality of the whole. The times of churning are these :—" The milk of Sunday and Monday is churned on the Thursday morning; that of Tuesday, Wednesday, and Thursday morning, on the Saturday evening; and that of Thursday evening, Friday, and Saturday, on the Monday morning."*

(2690.) In churning either ingredient the residuum is *butter-milk*, which, when obtained in large quantity from milk, may be disposed of in towns, or converted into cheese; and when derived in small quantity from cream, a part may be used for domestic purposes, and the remainder mixed with the food given to the brood-sow.

* Transactions of the Highland and Agricultural Society, for July 1843, p. 245.

(2691.) It is now time to say something on the making of *cheese*. On a farm of mixed husbandry, as much skimmed milk cannot be procured every day as to make a cheese of ordinary size, but there may be one made every other day. To save skimmed milk from souring in warm weather till the next day, it is necessary to scald it, that is, to put it into a furnace pot, and heat it sufficiently, and then let it cool. The fire should be a gentle one, and the milk should be so carefully attended to as neither to burn nor boil, nor be made warmer than the finger can bear. After being thus heated in the morning, the milk should be poured into a cheese-tub, to await the cheese-making of the following day. The skimmed milk of next morning is poured into the same tub, except about $\frac{1}{3}$ of it, which is put into the furnace or another pot, and made warm for the purpose of rendering the entire milk of the tub sufficiently warm. The heat is applied slowly to the pot, the milk occasionally stirred with a stick, and made as warm as the finger can hardly bear. This warm portion is then poured into the tub, the contents of which is thereby made as warm as new milk, that is, about 110° or 112°. The various degrees of heat here referred to are in rather vague terms, but I believe no specific degree of heat is required in scalding, provided the milk is not allowed to burn at the bottom of the pot, and is prevented from boiling, for if it boil, the milk will coagulate instantly, and be useless for cheese-making, because the cheese will then be hard or flinty; and so far as these two points are concerned, the ready and practical test of the finger is sufficiently accurate. On the contents of the tub being mixed by stirring, the rennet or earning is added to the milk, which is allowed to stand some time to coagulate, with a cloth thrown over it, to keep the proper heat.

(2692.) In the meanwhile, I shall describe the method of preparing the *rennet*, or *reed*, or *earning*. A calf's stomach is usually recommended for this purpose; but as calves' stomachs are not easily obtained in districts where calves are reared, a pig's stomach, which can be easily obtained on every farm, will answer the purpose equally well, and, indeed, many believe that it makes the stronger earning of the two. When the pigs are killed for hams in winter, their stomachs should be preserved for rennet, and they are preserved in this manner: Let the inside skin of the stomachs be taken out; the operation is somewhat troublesome, but may easily be done by an experienced dairy-maid. Any curdling in it is thrown away, as being unnecessary, and tending to filthiness, and the skin is then wiped clean with a cloth, not washed. It is then laid flat on a table, and rubbed thickly over with salt on both sides, and placed on a dish for 4 days, by which time it has imbibed sufficient salt to preserve it. It is then hung stretched over a stick near the fire to dry and won, and in the dried state is kept for use as rennet by the next season. Some people place a layer of clean wheat-straw on the skin, after it is salted, and roll the skin over it to keep it open, tie a piece of paper around it, and then hang it up near the kitchen-fire to dry and won. This plan is good enough, but not better than the other. When the rennet is to be used, a strong brine of salt and boiling-water, sufficient to float an egg, is made and sieved through a cloth, and allowed to cool, to the amount of 3 imperial pints to each skin. One skin is allowed to remain in that quantity of brine in a jar, with its mouth covered with bladder, for 3 or 4 days, when the coagulating strength of the brine is tested by pouring a drop or two into a tea-cupful of luke-warm milk; and when considered sufficiently strong, the brine is freed of the skin, bottled and tightly corked for use. The skin is again salted as before, and spread over a stick to dry and won, and is again ready for use when required. Half a tea-cupful of this rennet will coagulate as much milk as will make a 15 lb. cheese.

(2693.) When the milk is sufficiently coagulated, which it will be in half an hour, the curd is cut in the tub with the *curd-cutter*, fig. 472, which consists of an oval hoop of copper *a b*, 9 inches long and 6 inches wide, and 1½ inch deep, embracing a slip of copper, of the same depth, along its longitudinal axis *a b*. The stem *c* of round copper rod rising from each side of the oval hoop unites at *c*, and after attaining in all 18 inches in length, is surmounted by a wooden handle *d*, 9 inches in length, but 6 inches would be enough, by which it is held either by one or both hands, and on the instrument being used in a perpendicular direction, cuts the curd into pieces in the tub. Some people break the curd at first with the hand, but this instrument cuts it more effectually. On being cut, the curd lets out its whey, which is drained off by means of a flat dish being pressed against the curd-cloth, linen of open fabric, spread upon the curd. As much of the whey is removed in this way as practicable, and the curd will be left comparatively dry, when it receives another cutting with the cutter, and the whey again expressed from it. The curd is then lifted out of the tub, and wrapped into the curd-cloth, which, in the form of a bundle, is placed upon a drainer lying across the mouth of the tub, and the whey is pressed out of it by main force. This is the laborious part of the operation, and to save both time and labour in large cheese dairies, the bundle of curd is placed in a large cheese-vat, and subjected to pressure in the cheese-press to get quit of the whey.

Fig. 472.

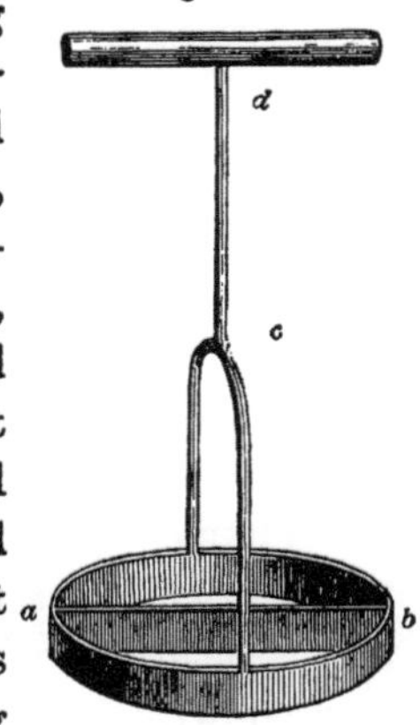

THE CURD-CUTTER.

The curd becomes very firm after this pressing, and must be cut into small pieces by some instrument before it can be put into the cheese-vat. In most small dairies, I believe, that the hand alone, or a table-knife, is employed to divide the curd, but a simple implement, such as represented and described below, effects the purpose with comparative ease and sufficient minuteness. The curd, being made small enough, is salted to taste with salt ground fine. In some parts of the country, such as in Cheshire, and in Holland, cheeses are salted by being floated in a strong solution of salt in water, which no doubt penetrates the new formed cheese, but it seems an uncertain mode of giving any desired degree of saltness. After being salted, the curd is put into a cheese-cloth, spread over a cheese-vat, and firmly packed into the vat higher than its edge, and on the curd being covered with the cloth, the vat is placed in the cheese-press and subjected to pressure, upon which a quantity of whey will probably exude by the holes in the bottom of the vat. In a short lapse of time, 2 hours or more, the cheese is turned out of the vat, a clean and dry cheese-cloth put in, the cheese replaced into it upside down, and again subjected to increased pressure in the press. Should whey continue to exude, the cheese must again be taken out of the vat, and a clean cloth substituted; in short, a clean cloth should be substituted, and the pressure increased, as long as any whey is seen to exude; but if the prior operations have been properly performed, the exudation should cease in about 12 hours, after which the pressure is continued until the press is wanted for a new cheese on the second day. Fig. 473 shews the common cheese-vat or *chessart*, as it is called, the form being varied according to that adopted for the cheese. This vat is built in elm staves, as being least liable to burst with pressure, and strongly hooped, and is furnished with a substantial bottom, pierced with holes, to allow the whey expressed to flow away, and a strong wooden cover cross doubled. It is of advantage that the cover fit the vat exactly, and that the vat have as little taper interiorly as possible. In some parts, as in Cheshire, cheese-vats are made of tin, with holes in the bottom. The old and inconvenient form of cheese-press, as well as the new and convenient one, are both found figured and described below by Mr Slight. In some dairy districts large cheeses are subject to immense pressure. In Cheshire the cheese is subject to three presses, the first giving a pressure of ¼, a second of a ½, and a third of 1 ton weight. There the cheeses being

Fig. 472.

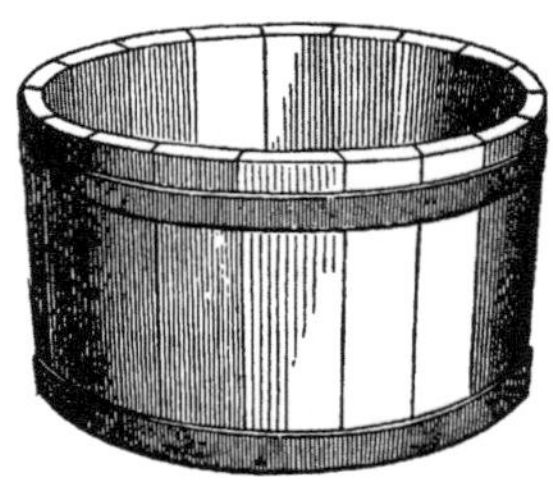

THE CHEESE-VAT.

unusually large, are bound round with fillets of linen until their form attains consistency.

(2694.) After the cheeses have been sufficiently pressed, they are put into the cheese-room *h*, fig. 33, which should not be exposed to too much heat, drought, or damp, as heat makes cheeses sweat; drought dries them too quickly, and causes them to crack; and damp prevents them hardening and wonning, and causes them to contract a bitter taste. Cheeses being exposed to a cool, dry, and calm air upon the shelves *m*, will dry by degrees, and obtain a firm skin. The skin becomes harder by being dipped in hot water, but I see no benefit to be derived from such a practice. They should be wiped with a dry cloth, to remove any moisture that may have exuded from them, and turned daily. To expedite the process of turning and rubbing, a cheese-rack, such as is figured below, has been in use in England, and found to answer the purpose. New cheeses, as they are made, are set in the rack, while as many of the oldest are removed from it and placed upon the shelves. Some cheeses burst, and throw out a serous-like fluid, which accident happens in consequence of the whey which was left in it fermenting, and which should have been pressed out. Any cheese that changes the shape which the cheese-vat gave it, should be suspected of some organic change taking place within it; but if such a cheese does not crack, so as to admit the air into it, it may soon become ripe and mould, and prove of fine flavour. The inconvenience of cracks in cheese, is the facility afforded to the cheese-fly to enter and deposite its eggs; and to prevent their egress, the cracks should he filled up every day with a mixture of butter, salt, and pepper, made to a proper consistency with oat-meal.

(2695.) But the casualties which I have mentioned are less likely to befall skimmed-milk cheese, the making of which I have been describing, than new or sweet-milk cheese. These are made exactly in the same manner with the milk as it comes from the cow. One day's milk being insufficient for a cheese, the fresh morning's meal is mixed with the meals of the previous day, the oldest part of which will have thrown up a covering of cream, which is mixed through the milk, and the entire gatherings are heated with a portion of the morning's milk. The rennet is applied in the same manner, but in rather larger quantity. Greater difficulty will be found to squeeze the whey entirely from the curd than with skimmed milk, and it is the difficulty of expressing all the whey out of them which renders sweet-milk cheeses more liable to ferment, and burst and loose their shape.

(2696.) I have said nothing of employing *annotto* or *arnotto* for dyeing cheese, because I think it a very useless piece of trouble which cheese-farmers impose upon themselves. It is employed in Glocestershire to the extent of 1 oz. of arnotto to 1 cwt. of cheese; and in Cheshire, 8 dwts. to 60 lb. of cheese, and it costs from 1s. to 1s. 6d. per lb. Arnotto is a precipitate from maceration of the pulp which covers the seeds of the *Bixa orellana* of Linnæus. It is manufactured in two forms, one in *flags* or *cakes* of 2 lb. or 3 lb. each, of a bright-yellow colour, soft to the touch, of good consistence, and comes from Cayenne wrapped in banana leaves, and is much used in giving an orange tint to silk and cotton goods, but which is not permanent; and the other kind is called *roll* arnotto, which is small, not exceeding 2 oz. or 3 oz. each, hard, dry, and compact, of a brownish-colour outside and red within, is brought from Brazil, and is the kind used in the dairies. The duty on the roll kind used to be L.5, 12s. per cwt and is now 4s.* When employed it is put into the milk before the earning, and it is prepared by rubbing down the requisite quantity in a bowl of warm milk. All the quantity employed is said to impart no peculiar flavour to the cheese, and this being the acknowledged case, of what utility is it? for as to the appearance of cheese, I suppose it will be allowed that Stilton and Dunlop cheese look as well on a table as that of Glocester and Cheshire. Marigold flowers, saffron, and carrots, are also employed to give colour to cheese.

(2697.) The dairy operations of a *carse*-farm are confined to supplying the milk and butter required for the support of its inhabitants; and the same remark applies to a *pastoral* farm. Dairymen *in* towns derive their chief profit from the sale of sweet-milk and cream, and what skimmed-milk they have is readily purchased by the working-classes. From dairy-farms in the *immediate vicinity* of towns sweet-milk and cream are also disposed of; and the butter-milk sold to the work-people in the town. On farms *farther off*, all the milk is converted into butter, and the butter-milk disposed of to work-people in town. True *dairy*-farmers conduct their operations irrespective of the immediate wants of town; and butter is made from cream, and also from new milk, skimmed milk is made into cheese, new milk is made into cheese, and butter-milk is employed to fatten pigs.

(2698.) The yield of dairy-cows varies so much, even in the same circumstances, that no average result can be depended upon. I have had cows of the same breed give from 8 to 17 Scotch pints a-day, that is, from 16 to 34 quarts; and I have known a cow of a cross-breed give 60 quarts a-day. So is the variation in respect of the quality of the milk. I have had cows whose milk would only yield a thin film of cream in the course of 24 hours, whilst that of others would almost admit at its cream of the same age being lifted off with the fingers. But though I can say nothing satisfactory on this head, I may state a few of the circumstances which are supposed to have a controlling influence on the quantity and quality of the milk of cows.

(2699.) The circumstances which affect the *quantity* of milk are, the *breed;* the smaller breeds yield the smaller quantity, and the yearly quantity of the large and small breeds may vary from 4900† to 2400‡ quarts; *the kind of*

* Macculloch's Dictionary of Commerce, and Ure's Dictionary of the Arts, art. *Annotto.*

† Dickson on Live Stock, vol. i. p. 226. ‡ Aiton's Treatise on Dairy Husbandry, p. 43.

food, cows fed on succulent food, such as moist meadow-grass, brewers' and distillers' refuse, and new sown grass, will yield a larger quantity of milk than when fed on hay and roots, and old pasture: and the *time from calving*, which, according to a statement of Mr Aiton's, causes a variation from 1200 quarts in the first 50 days, to 300 quarts in the sixth 50 days after calving.*

(2700.) The circumstances which affect the *quality* of milk are more various. The *breed* has an effect; the small ones yield richer milk than the large, in which respect the small Kerry cow is superior to the large Yorkshire. *The kind of food*, hay, corn, and oil-cake, produce richer milk than turnips and straw, and yield more butter; and bean-meal and tares afford more cheese than oil-cake, corn, potatoes, and turnips. In the *time from calving*, it is well known that the first milk of a cow, called the *biestings* (2104.), is much richer than the ordinary milk which the cow afterwards gives. It has a different composition from milk, and acts as a purgative to the new-born calf, which is of essential use to it at that period of life, in removing the sticky sort of dung called the *meconium*, from its bowels. You thus see how erroneous is the practice of those breeders of calves who throw away the biestings as injurious. In *wet and cold* weather the milk is less rich than in dry and warm; and on this account more cheese should be obtained in cold, and butter in warm weather. The *season* has its effect; the milk in spring is supposed to be best for drinking, and hence it is then best suited for calves; in summer it is best for cheese, and in autumn for butter; and hence, perhaps, autumn butter keeps better than that of summer, because it contains less of the caseous principle. Cows *less frequently* milked than others will give richer milk, and consequently more butter; *morning's* milk is richer than the evening's; and the *last drawn milk* of each milking, at all times and seasons, that is, the afterings or strokings, are well known to be the richest part of the milk, and the first drawn the poorest. A cow, *before she is again in calf*, gives richer milk than when she is pregnant, a portion of the secretion which supplies the richer milk being, no doubt, withdrawn to support the fœtus. A *well-formed* cow will give more and better milk than an ill-formed one. *Old pasture* will produce richer butter than new. Cows *kept constantly in the byre* are said to give richer milk than those allowed to go at large at pasture, but the latter are supposed to yield more cheese;—the exercise, perhaps, preventing the due proportion of the richer secretion. Many other circumstances may be known in different localities to affect the quantity and quality of the milk of cows; but a sufficient number are here related to shew how perplexing a thing it is to conduct a dairy in the most profitable way.

(2701.) The phenomena accompanying the changes in milk are well known to every dairy-maid, but few of them know that the constituent parts of milk are only mechanically commixed; and this must be their state even in the udder of the cow, otherwise the afterings, which had occupied the upper part of the udder, would not be the richest portion of the milk, nor the first drawn the poorest. All, therefore, that is required to separate the different parts of milk is rest and time. The cream or fatty part floats to the surface in the course of a few hours; in a little longer time, according to the state of the temperature, the caseous portion becomes sour, and in a still greater length of time the acidity becomes so powerful as to coagulate the milk in one mass, and in a still greater lapse of time

* Aiton's Treatise on Dairy Husbandry, p. 45.

the coagulated mass separates into two parts, one becoming firmer, or cheese, the other again fluid, or whey. The rationale of this natural process is thus well given by M. Raspail :—" Milk, when viewed by the microscope with a power of only 100 diameters, exhibits spherical globules, the largest of which are not more than .0004 of an inch in diameter, and which, from their smallness, appear of a deep black at the edges. These globules disappear on the addition of an alkali, such as ammonia, and the milk then becomes transparent. If the proportional quantity of milk be more considerable, it forms a coagulum of a beautiful white colour, on the addition of concentrated sulphuric acid. This coagulum does not arise simply from the adhesion of the globules to each other, but it may be plainly seen by the microscope, that the globules are evolved in a transparent albuminous membrane, which has no appearance of a granular structure. Milk, then, is a watery fluid, holding in solution *albumen* and *oil*, by the agency of an alkaline salt or a pure alkali, and having suspended in it an immense number of globules which are in part albuminous and in part oily. The albuminous globules must tend to subside slowly to the bottom of the vessel by their specific gravity, while the oily globules must have a tendency to rise to the surface. But the oily globules being dispersed in myriads amidst equally numerous albuminous globules, they cannot rise to the surface, without taking with them a greater or less number of the globules of albumen. Hence, at the end of 24 hours, we find on the surface of the milk a crust composed of 2 layers, the upper one of which contains more butter than milk, while the lower contains more milk than butter. This separation will take place equally with or without the contact of the air. The liquid part which lies under the crust contains the dissolved albumen and oil, with a portion of the sugar, the soluble salts, and a certain quantity of the albumen and oily globules."* Cream cannot rise through a great depth of milk. If milk is, therefore, desired to retain its cream for a time, it should be put in a deep narrow dish; and, on the other hand, if it is desired to free it most completely of its cream, it should be poured into a broad flat dish, not much exceeding 1 inch in depth. The quantity of cream which any given milk contains can be easily measured by the *Galactometer*, which consists of a narrow tube of glass not more than 5 inches in length, 3 of which is divided into 100 parts, and on being filled with milk to the top of the graduated scale, whatever number of degrees the thickness of the cream embraces, will be the per-centage of cream yielded by the milk. For example, if the cream covers 4 lines of the scale, it is 4 per cent., if 8 lines, 8 per cent. The evolution of cream is facilitated by a rise, and retarded by a depression, of temperature. At the usual temperature of the dairy at 50° Fahrenheit, all the cream will probably rise in 36 hours, and at 70° it will, perhaps, all rise in half that time; and if the milk is kept near the freezing point, the cream will rise very slowly, because it becomes partially solidified. Milk boils and freezes about the same temperature as water. Milk may be prevented becoming sour by being kept in a low temperature; in a high temperature, on the other hand, it rapidly becomes sour, and, at the boiling point, it curdles immediately. The acid of milk is called the *lactic acid*, and, in its nature, resembles acetic acid, the acid of vinegar.

(2702.) *Milk* contains many very different substances in its composition, to each of which may easily be traced the origin of its various properties. I shall

* Raspail's Organic Chemistry, p. 380-2.

give the composition of biesting, along with that of milk, from the analyses of MM. Henri and Chevalier, in order to shew the great difference between them. It appears that biesting contains nearly 3 times more casein than milk, and only a trace of sugar of milk, no salts, and a large proportion of mucus; and more than 9 times more casein than mare's milk.

	According to Henri and Chevalier.						According to Luiescius and Bondt.
	Cow.						
	Biestings.	Milk.	Ass.	Goat.	Ewe.	Woman.	Mare.
Casein, . .	150.7	44.8	18.2	40.2	45.0	15.2	16.2
Mucus, . .	20.0	...	...	...	...	...	...
Butter, . .	26.0	31.3	1.1	33.2	42.0	35.5	trace.
Sugar of Milk,	trace.	47.7	60.8	52.8	50.0	65.0	87.5
Salts, . . .	...	6.0	3.4	5.8	6.8	4.5	} 896.3
Water, . .	803.3	870.2	916.5	868.0	856.2	879.8	
	1000.	1000.	1000.	1000.	1000.	1000.	1000.

Butter gives its richness to milk, sugar its sweetness, casein its thickness, water its refreshing property as a drink, and salts its peculiar flavour. Of the different kinds of milk enumerated above, the superior sweetness and thinness of mares' milk are accounted for by the large proportion of sugar and the small quantity of casein it contains. " The change which takes place when milk becomes sour, is, therefore, easily understood," as well observed by Professor Johnston. " Under the influence of the casein the elements of a portion of the milk-sugar are made to assume a new arrangement, and the sour lactic acid is the result. There is no loss of matter, no new elements are called into play, nothing is absorbed from the air, or given off into it, but a simple transposition of the elements of the sugar takes place, and the new acid compound is produced. These changes appear very simple, and yet how difficult it is to conceive by what mysterious influence the mere contact of this decaying membrane, or of the casein of the milk, can cause the elements of the sugar to break up their old connection, and to arrange themselves anew in another *prescribed* order, so as to form a compound endowed with properties so very different as those of lactic acid."

(2703.) " *Cream* does not consist wholly of fatty matter (butter)," observes Professor Johnston, " but the globules of fat, as they rise, bring up with them a variable proportion of the casein or curd of the milk, and also some of the milk-sugar. It is owing to the presence of sugar that cream is capable of becoming sour, while the casein gives it the property of curdling when mixed with acid liquids, or with acid fruits. The proportion of cheesy matter in cream depends upon the richness of the milk, and upon the temperature at which the milk is kept during the rising of the cream. In cool weather the fatty matter will bring up with it a larger quantity of the curd, and form a thicker cream, containing a greater proportion of cheesy matter. The composition of cream, therefore, is very variable—much more so than that of milk—and depends very much upon the mode in which it is collected." In warm weather, therefore, the cream should be rich though thin. Cream, at a specific gravity of 1.0244, according to the analysis of Berzelius, consists of

Butter, separated by agitation,	4.5
Curd, separated by coagulating the butter-milk,	3.5
Whey,	92.0
	100.

(2704.) "*Butter*," says Professor Johnston, "prepared by any of the usual methods, contains more or less of all the ingredients which exist in milk. It consists, however, essentially of the fat of milk, intimately mixed with a more or less considerable proportion of casein and water, and with a small quantity of sugar of milk. Fresh-butter is said to contain about $\frac{1}{6}$ of its weight (16 per cent.) of these latter substances, and $\frac{5}{6}$ of pure fat, according to Chevreul. How much of the 16 per cent. usually consists of cheesy matter," may be seen by this statement:—"two samples of fresh-butter from *cream*, examined in my laboratory, have yielded only 0.5 and 0.7 per cent. respectively. This is certainly a much smaller quantity than I had expected. Does butter from the *whole milk* contain more?" * The proportions of butter yielded by milk varies considerably, from 1 lb. of butter to 15 quarts of milk, as in Holstein, to 8 quarts of milk of the Kerry cow.†

(2705.) *Curd* in a state of purity is named *casein*. "Casein has many properties in common with the albumen of blood," says Dr Thomson, "and, like albumen, may be obtained in two states, namely, uncoagulated, when it is soluble in water, and coagulated, when it is insoluble in that liquid. It is precipitated from its aqueous solution by acetic acid, which is not the case with albumen. It is coagulated by a boiling heat, but slowly; separately in films, which collect upon the surface of the liquid. Coagulated casein subjected to pressure to free it from the whey constitutes cheese. If cheese consist of nothing but casein, it has a bluish-white colour, is very hard, almost like horn, and is quite insipid. Good cheese is always made from milk still retaining its cream. It is impossible to state the proportion of casein which exists in milk, because it varies so much, not only in the milk of different animals, but also in that of the same animal at different times. According to Berzelius, 100 parts of skimmed milk, which he analysed, contained 2.8 of casein." The average proportion of curd to milk is thus given by Dr Dickson,—"Exact and repeated trials have shewn that about 15 gallons of milk are necessary for making about 11 lb. of 2-meal cheese, and that 1 lb. of curd is produced from 1 gallon of new milk. And as the food afforded by the extent of from $2\frac{1}{2}$ to $3\frac{1}{2}$ acres of land is commonly supposed sufficient for the support of 1 cow the year round, by taking the medium of 355 lb. of cheese for each cow, the quantity of cheese produced by 1 acre will be 118 lb., which is supported by the authority of many statements. But during the summer season, cows will afford from 14 lb. to 20 lb. of cheese, or more, in the week, when no butter is made."‡

(2706.) *Whey* has a yellowish-green colour, and an agreeable and sweetish

* Johnston's Lectures on Agricultural Chemistry, p. 797, 802, and 807.

† Journal of the Royal English Agricultural Society, vol. i. p. 386.

‡ Dickson on Live Stock, vol. i. p. 237.

taste, in which the flavour of milk may be distinguished. Almost the whole curd may be separated by keeping the whey for some time at a boiling temperature. "It still retains its sweet taste," says Dr Thomson, "but much of the milky flavour is dissipated. If it be now evaporated over the steam-bath, it deposites a number of crystals of sugar of milk. Towards the end of the evaporation, some crystals of chloride of potassium, and some of common salt, make their appearance. According to Scheele, it contains also a little phosphate of lime, which may be precipitated by ammonia." "The sugar of milk constitutes, at an average, about 3.5 per cent., while the saline ingredients do not exceed 0.22, or $\frac{2}{9}$ of a per cent. of whey. The water, of course, constitutes about 93.3 in the 100 parts."

(2707.) Schwartz found that 1000 parts of cows' milk left 3.697 of ashes, and of woman's milk 4.407, composed of

	Cows'.	Woman's.
Phosphate of lime,	1.805	2.5
Phosphate of magnesia,	0.170	0.5
Phosphate of iron,	0.032	0.007
Phosphate of soda,	0.225	0.4
Chloride of potassium,	1.350	0.7
Soda, combined with lactic acid,	0.115	0.3
	3.697	4.407

(2708.) There are a number of other facts given by Dr Thomson connected with milk in various states, worth mentioning. The constituents of *skimmed* milk, for example, according to the analysis of Berzelius in 1808, are—

Water,	92.875
Curd, not free from butter,	2·800
Sugar of milk,	3.500
Lactic acid, and the lactate of potash,	0.600
Chloride of potassium,	0.170
Phosphate of potash,	0.025
Phosphate of lime and magnesia, with a trace of iron,	0.030
	100.000

Brisson states the specific gravity of various milks: but it is important to remark, that it varies so much even in the milk from the same animal, that it is impossible to give a correct mean. The specific gravity of cows' milk is the lowest, being 1.0324; its whey is, of course, still lower, 1.0193; and that of ewes' milk is the highest, being 1.0409. Lassaigne examined the specific gravity of cows' milk at various distances of time before and after parturition, at a temperature of 46° Fahrenheit, and the results were generally, that at 21 days before parturition it was highest, being 1.064; and lowest at 6 days after parturition, being 1.033. The cow, from 42 days before to 30 days after parturition, was fed on the same kind of food, namely, beet-root, hay, and straw. Lassaigne also ascertained the quantity of cream yielded by the same animals at different times. There was no difference in the ratio between the bulks of cream and whey from the same cow, fed on beet-root, hay, and straw, from 42 days before to 4 days after parturition, when the quantities were 200 volumes of cream to 800 of whey; but by 30 days after parturition, the volume of cream had *decreased* to 64, and that of whey had *increased* to 936, and by that time the water in 100 parts of milk had also increased to 90.0. Lassaigne made a curious remark respecting the milk of a cow, which he examined at 10 different periods, 4 of these before and 6 after parturition. The milk examined during the first three of the former periods, namely, 42 days, 32 days, and 21 days *before* parturition, contained no casein at all, but in place of it albumen; no sugar of milk and no lactic acid, but a sensible quantity of uncombined soda. The milks examined 11 days before and just after parturition, contained both albumen and casein, while milk 11 days before parturition, and always after it, contained free lactic acid and sugar of milk, but no free soda. The milks examined 4 days, 6 days, 20 days, 21 days, and 30 days *after* parturition, contained casein and no albumen. It would appear from these observations that the milk of the cow is at first very similar to the serum of blood, and that the casein, sugar of milk, and lactic acid, to which it owes much of its distinguishing characters, begins first to make their appearance in it about 11 days before parturition.

(2709.) M. Raspail alleges, that the dairymen in the neighbourhood of Paris take off the cream from their milk, and supply its place with raw sugar, and an emulsion either of sweet almonds or hemp-seed. Milk is sometimes adulterated by the addition of starch, and sometimes a portion of carbonate of potash is added to it to prevent it from curdling.* I have detected magnesia in cream in Holland put in to thicken it. In London, the milk is so adulterated with water, that some dairymen have adopted the practice of driving their cows along the streets, and supplying their customers direct from the cow. Both the milk and cream obtained from the dairies in the neighbourhood of Edinburgh are generally free from adulteration, and the worst material put into it in the public dairies in town is water. It has been observed that the *Equisetum fluviatile*, the great water horse-tail, gives the milk a leaden or bluish colour, and deprives it of its cream. It is believed that the leaves of the bulbous buttercup, *Ranunculus bulbosus*, is pernicious to the milk, but there is no sufficient foundation for the assertion, as cows will not eat the plant; but it is well known that the broad-leaved wild garlic, *Allium ursinum*, is eaten by cows, and gives a most offensive flavour of garlic to milk. This is an annoyance to which the settlers in some parts of Canada are peculiarly subjected.

(2710.) A great deal more might be given in detail of experiments made in different parts of the country on churning butter at different temperatures, and of the various recipes for making the innumerable variety of cheeses to be found in this country; but as those experiments are merely so many isolated attempts to illustrate particular points, without reference to general principles, their results are unsatisfactory. In like manner the numerous experiments in the laboratory on milk and its constituents have as yet led to no improvement in practice. The field, therefore, to the practical and chemical experimenter is still open, and it might be usefully explored by their combined labours, with a view to the establishment of general principles on the best mode of keeping dairy-stock, and of evolving their products, in a superior degree than hitherto attained. As one instance of a want of common understanding on this subject, I may adduce the numerous forms in which the cheeses of Scotland are made, whether in the homely *kebbuck*, or in the finished new-milk cheese intended for city epicures.

* Raspail's Organic Chemistry, p. 385.

Everywhere else but in Scotland the place of origin of a cheese may be told by its form. A Parmasan, a Gruyère, a Gouda, a Kanter, an Edam, a Cheshire, a Glocester, whether double or single, a Wiltshire, a Cheddar, or a Stilton, can be named at sight, but who can tell where a Scotch cheese, whether of skimmed milk or new, whether poor or rich, comes from, until examined by a connoisseur? for its Stilton or Wiltshire form would not necessarily imply an imitation of those celebrated substances. There should be some settled conventional forms adopted for the identification of the respective classes of cheese of this country, that they may pass current in commerce without suspicion and challenge. I refer you to authorities who have treated of milk practically and chemically,* to consult at leisure, and shall finish this interesting subject with recipes for making a number of pleasant dishes from milk, all of which are in the power of every farmer to have in summer as a cool diet. *Curds* are obtained by simply earning a dishful of new milk. When served up with grated loaf-sugar, sprinkled over the curd deprived of whey, it has the appearance of a prepared dish, and is eaten with sugar and cream. A *sour cog* is a dish of milk allowed to stand with its cream until the milk becomes thoroughly coagulated by sourness, and the sour cream and milk are eaten together. *Plain cream*, whether sweet or sour, is an excellent accompaniment to oat-meal or barley-meal porridge, or to sowens. The *clouted cream* of Devonshire is prepared by straining the new-milk into a shallow dish, into which a little warm water has previously been put; and after allowing it to stand from 6 to 12 hours, it is carefully heated over a slow fire or hot plate till the milk approaches to the boiling point, but it must not actually boil, or the skin of cream be broken. The dish is then removed to the dairy, and the cream allowed to cool, when it may be used as cream or made into butter. *Milk oat-meal porridge* is more agreeable to the palate than water porridge, and when eaten with cream forms a rich diet. *Half churned cream* is better with oat-meal porridge than plain cream, having a slight taste of acidity. *New unwashed butter* is a great treat to breakfast. *Hatted kit* is one of the pleasantest preparations of milk. Make 2 quarts of new milk scalding hot, and pour upon it quickly 4 quarts of fresh butter-milk, and let it stand, without stirring, till it becomes cold and firm. Then take off the hat or upper part, drain it in a hair-sieve, and put it into a shape for half an hour, and serve with cream. The slight acidity of this dish, with the rich sweetness of the cream, and the addition of a little sugar, combine to make this a very delicious dessert. *Float-whey* is another preparation equally good as hatted kit, and more delicate. Pour in all the whey drained from the new-milk cheese that has just been made, into a small furnace pot; apply a slow fire, and raise the whey near the boiling point, but not to let it boil, else the curd will fall to the bottom. During the heating, a scum of curd forms upon the surface of the whey. Take then 1 quart of fresh butter-milk,

* Aiton's Treatise on Dairy Husbandry gives an account of the Ayrshire breed of cattle, which is reared in the dairy district of Scotland exclusively for dairy purposes.—In vol. vii. p. 155, and vol. ix. p. 228, of the Prize Essays of the Highland and Agricultural Society, will be found recipes for making imitation Glocester and Wiltshire cheeses; and in the Transactions of the same Society for July 1843, p. 14-27, are remarks, experimental and practical, on the making of butter.—Dr Johnston's Lectures on Agricultural Chemistry, Lecture xx. p. 779, contains the rationale of all the phenomena exhibited by milk in its various stages.—And in the Mémoires de l'Academie Royal des Sciences de l'Institut de France, tome xvii. p. 201, will be found some very curious microscopic researches made by M. Turpin upon different milks obtained from cows affected by the disease vulgarly called *Cocote* in France.

and pour it gently over the scum, and pour as much more butter-milk till the scum has attained some thickness and consistency. After pouring in some cold water to lower the temperature of the whey, thereby rendering the scum more consistent, skim off the scum upon a hair-sieve, put it into a mould, and on turning it out a short time after, serve with sugar and cream. A *treacle posset* is made by boiling a little milk in a pan, and putting a table-spoonful of treacle or molasses into it. On removing the curd, the whey is ready for use for a sore throat or cold. *White-wine whey* is made exactly in a similar manner, but is sweetened in addition. To make *Irish two-milk whey*, put $\frac{2}{3}$ of sweet-milk into a sauce-pan, and make it boiling hot, then pour in $\frac{1}{3}$ of butter-milk, gently stirring it round the edges of the pan. Let the whole come to a boil; take it off the fire, let it settle, and strain off the whey, which is an excellent drink in fever. *Cream* may be used as an emulsion with all sorts of preserved fruits, and it enhances the flavour of every kind of fruit used with it; and there is, perhaps, no form of cream more agreeable, or more generally admired, than *blancmange* flavoured with almonds. *Iced-cream* flavoured with pine-apple is a delightful cooler in warm weather. I conclude with a recipe for making *cream-cheese*. "One pint of cream being mixed with 12 pints of noon-day milk, warm from the cow, a little rennet is added, and when the curd is come, the whey is poured out gently, so as to break the curd as little as possible. It is then laid in a cloth, and put into a small sieve; the cloth is changed every hour during the day, and in 24 hours it will be fit for use. It may be served on a breakfast plate with vine leaves under it, and it will keep perfectly good only one day."*

(2711.) Notwithstanding the immense quantity of cheese made in this country, a large importation of foreign takes place every year, not less than 132,000 cwt., chiefly from Holland. The old duty of 20s. per cwt. is retained by the New Tariff.†

(2712.) Cheese is liable to many casualties besides those already alluded

Fig. 474.

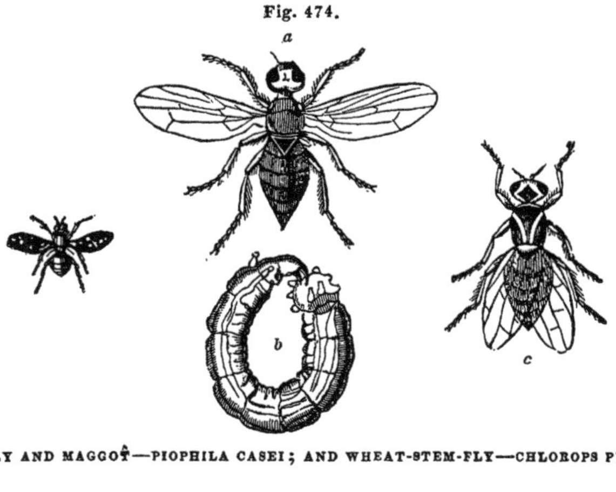

CHEESE-FLY AND MAGGOT—PIOPHILA CASEI; AND WHEAT-STEM-FLY—CHLOROPS PUMILIONIS.

to as arising from fermentation, chiefly from the attacks of animals. When yet quite fresh, the cheese-fly, *Piophila casei*, *a*, fig. 474, is ready to deposite

* Dalgairns' Practice of Cookery, p. 467.

† Macculloch's Dictionary of Commerce, art. *Cheese*.

its eggs in the deepest crack it can find, by means of an extensile abdominal tube. The specific distinguishing characters of this insect are, in the words of Mr Duncan,—" About 2 lines in length, the whole body of a greenish-black colour, smooth and shining; front of the head reddish yellow, paler yellow on the under side. Thighs ochre yellow at the base and apex; tibiæ deep ochre, the first and last pair black at the apex; anterior tarsi black, the others ochrey, with the 2 last joints and the claws black; wings clear and iridescent, slightly tinged with rust colour at the base; halteres ochrey." * The cheese-maggots produced from this fly are commonly called *jumpers*, and it prepares itself for jumping as seen at *b*. " When it prepares to leap, our larva first erects itself upon its anus, and then bending itself into a circle, by bringing its head to its tail, it pushes forth its unguiform mandibles, and fixes them in 2 cavities in its anal tubercles. All being thus prepared, it next contracts its body into an oblong, so that the 2-halves are parallel to each other. This done, it lets go its hold with so violent a jerk, that the sound produced by its mandibles can be easily heard, and the leap takes place. Swammerdam saw one, whose length did not exceed the fourth part of an inch, jump in this manner out of a box 6 inches deep, which is as if a man 6 feet high should raise himself in the air by jumping 144 feet! He had seen others leap a great deal higher."† When cheese passes its stage of ripeness, it is liable to become mouldy, if kept in a damp situation, which all ripe cheeses should be to retain their moisture and flavour, and where the latter is much enhanced by the production of blue mould. It is possible to inoculate new cheese with the blue mould of old, and thereby at once impart the flavour of ripeness. This process is easiest done by inserting rolls of moulded cheese, extracted by the scoop or *spyler*, into holes previously made in the new cheese by the same scoop, an instrument usually employed by cheesemongers to taste cheese.‡ Towards a still farther period of decay, cheese is attacked by the well-known, and by some, the highly prized cheese-mite, *Acarus siro* of Linnæus. " We often wonder how the cheese-mite is at hand to attack a cheese wherever deposited; but when we learn from Leewenhock that one lived 11 weeks gummed on its back to the point of a needle without food, our wonder is diminished," says Kirby and Spence. Both cheese-maggots and mites, when numerous, destroy cheese rapidly, by crumbling it into small pieces, and by emitting a liquid substance, which causes the decayed parts to spread speedily. They may easily be killed, however, by exposure to strong heat, or by plunging the cheese in some liquid capable of destroying the larvæ, without communicating any disagreeable flavour, such as whisky. Rats and mice are remarkably fond of, and commit sad havock amongst, old cheese. Nothing but a cat can deter them from a cheese cellar, where poison cannot be employed with impunity.

(2713.) [The process of churning, or agitating milk and cream for the production of butter, is performed in such a variety of machines, and in vessels so variously formed, each of which has its advocates, and, probably, with but little difference in the principal results, that it becomes a question of no small difficulty to answer, which of these numerous forms is the best? That there will be slight differences, is at the same time sufficiently apparent, but these will arise

* Quarterly Journal of Agriculture, vol. xii. p. 126.

† Kirby and Spence's Introduction to Entomology, vol. ii. p. 283.

‡ Prize Essays of the Highland and Agricultural Society, vol. ix. p. 232.

more from the capacity of the vessel, as affecting temperature, than from peculiarity of structure, and probably, also, from other conditions, affecting the chemical changes that take place during the process. It will also be obvious, on a full consideration of the subject, that causes of difference will arise to some extent from the circumstances attending the more or less perfect nature of thè agitation produced, as arising from the peculiar formation, and the motions, whether of the containing vessel or of the agitators.

(2714.) The peculiarities of form may be viewed under four distinct classes. 1. Those in which both the fluid and the containing vessel, with its agitators, are in rotative motion; 2. In which the containing vessel is at rest, and the agitators in rotative motion horizontally; 3. In which the containing vessel is at rest, and the agitators in rotative motion vertically; and 4. Wherein the containing vessel is still at rest, and the agitator having a rectilineal vertical motion.

(2715.) In the first class is to be noticed those machines acting by their gyration on a centre, such as the old barrel-churn, wherein the contained fluid will have a tendency to acquire a motion of rotation approaching to that of the vessel; and if this rotation were continued always in the same direction, the process would be very much retarded from a deficiency of agitation; hence, in such machines, very frequent changes in the direction of the rotative motion is absolutely necessary to the completion of the process.

(2716.) The second class embraces those in which the vessel is permanent, and an agitator of two or four arms revolves horizontally within it. If the vessel is cylindrical, the agitation will be imperfect, for the fluid will very soon acquire a motion *en masse*, and will be carried forward by the arms of the agitator, undergoing such agitation only as will arise from the central portion, acting by the centrifugal force, having always a tendency to fly towards the circumference of the revolving mass. This will no doubt ultimately, though slowly, produce the requisite effect, but it will be more speedily accomplished if the vessel is of a square form; for in that case the rotation of the mass will be interrupted at every angle, and eddies formed therein of such force as will not only cause disturbance of the fluid in these angles, but in doing so will produce corresponding counter-motions in the entire mass. From such causes, churns of this class and of the square form are found, especially on the large scale, to be very efficient, though, from the inconvenience of adopting manual power to a horizontal motion, it is seldom resorted to in the small scale.

(2717.) In the third class of machines, where the vessel is also permanent, but having the agitators revolving in a vertical direction, the agitation is regular, and pervades at all times the entire mass of fluid; and in this respect there is little if any difference in effect, whether the bottom of the vessel be flat or cylindrical, nor is there any necessity for changing the direction of motion. We accordingly find, that, whether on the large or small scale, churns of this class find favour in almost all localities.

(2718.) The fourth class of this useful machine is the upright or plunge churn. The vessel in this is also permanent, and its height, unlike all the others, is considerable, as compared with its breadth. As a matter of convenience, it is always made cylindrical, or rather slightly conical, which last property is given to it for the convenience of hooping, for in every other respect its tapering form is a defect. The plunger must necessarily always move vertically, without any rotative motion, and, when extended throughout the entire column of fluid, the agitation must be as complete as it is possible to attain. Plunge or vertical churns, therefore, are by many held as the most perfect for the production of butter, as well as for saving time in the operation; but there are objections to it as regards fitness for the adaptation of manual power. It is obvious to even a careless observer, that the human arm, if applied directly to the upright staff of this churn, the body being also upright, will be employing that power under the greatest possible disadvantages. The muscles of the arm acting, as they do, in all cases under great natural disadvantages in respect of their leverage, will, in this peculiar position, be deprived of nearly all aid which the muscles of the trunk in many positions are calculated to give out to assist those of the arm; hence it is that the labour of working these churns by hand, is found so oppressive, that the operator is unable to keep up a constant action; and the principle on which the churn is based is blamed for that delay in the operation, that ought rather to be attributed to the defect in the medium through which the power is applied. In proof of this, we find that when animal, or any of the inanimate powers, are applied to the vertical churn, it attains a character superior to all others, both as to time and production, and this character is sustained throughout some of the best dairy districts in Scotland. The advantages of applying hand-power through a proper medium are also observable in the case of this churn, when such power is applied through a winch-handle with a fly-wheel. Examples of this arrangement have been attended with the best effects, and with much ease, as compared with the direct application of the power to the plunger-rod.

(2719.) In proceeding to the details of the churning machinery, the first class embraces those machines that act by their gyration round a centre, the fluid and the containing vessel revolving together, or partially so; of which the common barrel-churn may be taken as the type. The barrel, which is of capacity suited to the dairy, is sometimes provided only with a large square bung-hole, secured by a clasped cover, by which it is charged and emptied; while in other cases, one of its ends is movable, and made tight by screwing it down on a packing of canvass cloth. In all cases, it is necessary that the interior of the barrel should be armed with three or more longitudinal ribs of wood standing as radii towards the centre, and perforated in various forms; these constitute the agitators of the churn. Each end of the barrel is furnished with an iron gudgeon or journal strongly fixed to it, and to one of them is applied the winch-handle by which the machine is turned; while it is supported on a wooden stand, having bearings for the two journals. More than one imperfection attends this construction of churn; from the circumstance of its rotatory motion, it will always have less or more of a tendency to carry the fluid round with the barrel and the agitators, more especially if a rapid velocity of rotation is given to it; and to counteract this tendency, it becomes necessary to reverse the motion at every few turns, which is of itself an inconvenience. There is, besides, the great inconvenience of getting access, either to remove the butter that may adhere to the agitators, or to cleanse the interior of the barrel. This is especially the case where there is no movable end; and even with this convenience for cleaning, the trouble of opening and closing the end is considerable. To those imperfections it is, no doubt, chiefly owing that this churn, once in high repute in East Lothian and Berwickshire, has gradually fallen into comparative disrepute: and from these circumstances, it has been deemed unnecessary to give figures of this class of churn.

(2720.) The machines to be here noticed under the second class, are of a less varied character than any of the others, are seldom, if ever, used with hand-labour, but are generally worked by horse-power, and chiefly in town-dairies, or in those around Edinburgh. From the comparatively small importance of this churn, it has been deemed unnecessary to give a figure of it; but its structure is so simple, that the following description will convey a tolerably correct knowledge of its construction to the mechanical reader. The horse-path, especially in towns, where space is valuable, seldom exceeds 16 feet; the horse-beam, with yoke, is fixed to an upright central shaft, which carries a spur-wheel placed at a height of about 3½ feet from the floor, the wheel being about 4 feet diameter. One, or sometimes two, churning vessels are attached to the machine; their dimensions vary with the extent of the dairy, but for one of 25 or 30 cows, where much of the produce is sold as sweet-milk, the vessel may be 2 feet 8 inches square and 2 feet 8 inches deep; they are placed securely on the floor at a proper distance from the central shaft, to suit the spur-wheel. A foot-step of metal is placed in the centre of the bottom, and a cross-head is attached to two uprights fixed on opposite sides of the vessel. The foot-step and cross-head serve as the two bearings for an upright spindle, which carries the four-armed agitator, and a pinion of 3½ inches diameter adapted to the spur-wheel. These parts, and a close cover for the vessel, completes the machine, which is, therefore, extremely simple, and in a dairy which is too extensive for manual power being applied effectually to churning, while want of space may preclude the adoption of more bulky machines, the one here described may be resorted to with advantage.

(2721.) In the same class stands the *table churn*, remarkable for its elegance and cleanliness, and which, though adapted only for the lighter purposes of the butter-dairy, I have considered as deserving a place here. This utensil is represented in fig. 475 in perspective, exhibiting it in the most recently improved form, with outer case to contain hot or cold water. This churn has for the last few years been brought prominently before the public as a recent invention; but on close inquiry, I find that the merit of its original introduction is due to the late celebrated Mr Wedgewood, and that it has been in use for a period of 30 or 40 years. The containing vessel is still formed of the Wedgewood stoneware; and, as I am informed by Mr Child of Edinburgh, the indefatigable purveyor of elegance, utility, and comfort, in the china, stoneware, and glass trade, it was in great request about 30 years ago, its lately extended appearance being merely a revival. The chief part of this utensil is the Wedgwood receptacle *a*, formed of the finest and strongest white glazed ware of that manufacture; they are of various sizes, from 1 to 4 gallons capacity; it is furnished with a varnished wooden cover *b*. The outer case *c* is made of sheet zinc or of tin plate; it is 2 inches wider than the churn, furnished with handles *d*, and two ears to which the iron cross-bar *e* is attached by two thumb-screws *e* and *d*, serving to secure the cover to the top of the vessel. A brass socket *f* is fixed in the cover, and an iron spindle armed with three vanes is fitted to turn in the socket; a wooden pulley is usually placed in the position of the wheel

Fig. 475.

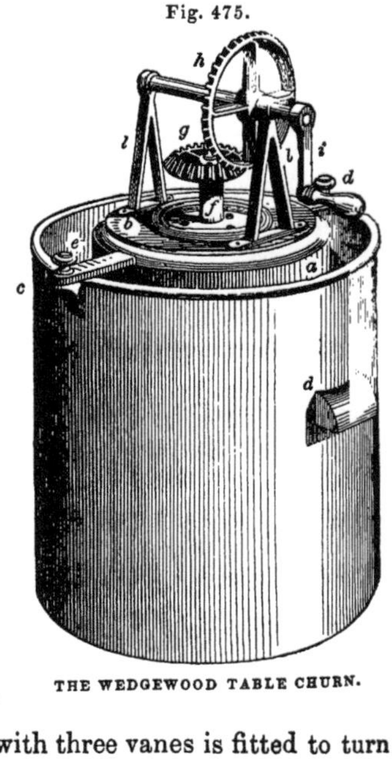

THE WEDGEWOOD TABLE CHURN.

g on the top of the spindle, and when secured there it holds the spindle and cover in constant connection. The common drill-bow is the usual medium of power, the string of which being held in tension by the elasticity of the steel-back or bow, any movement backward or forward of the instrument will cause the pulley and spindle to revolve, and the movements are effected by applying the hand to the handle of the drill-bow. When the cover and spindle has been secured by the screws *e d*, and the bow-string applied as above, which is effected by bending the bow until the string is sufficiently relaxed to allow of its being laid once round the pulley, the bow is then allowed to expand, and the operation proceeds. It has been stated (2716.) that vessels in this class, if cylindrical, do not produce the effects of churning so speedily as if the form were square, to obviate which the motion of the agitator must be frequently reversed in the cylindrical vessel. The drill-bow motion is admirably adapted to the reversing process; for in pushing the bow forward by the hand, the agitator will be made to revolve 2 or 3 times, the number being in the proportion of the length of the string to the circumference of the pulley; and in drawing them back the same number of revolutions will be performed in the opposite direction, and so on till the process is completed. Fig. 476 is a view of the agitator, *a a a* are the vanes of strong tin-plate with perforations, *b* is that part of the spindle that falls within the socket, and *c* that on which the pulley or the wheel is fastened.

Fig. 476.

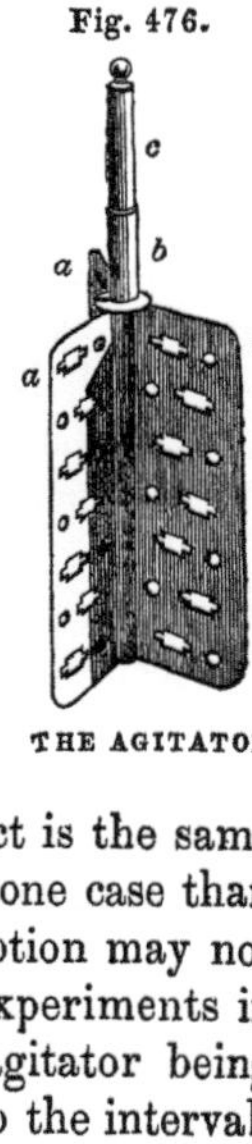

THE AGITATOR.

(2722.) The drill-bow being rather an awkward medium of power, especially in non-mechanical hands, an attempt has lately been made to substitute for it, in this case, the common winch-handle turning horizontally. This arrangement is exhibited also in fig. 475, where *h* is a toothed-bevelled wheel of 4 inches diameter, on the axle of which the handle *i* is fixed, and it works into the wheel *g* fixed on the top of the agitator spindle in the place formerly occupied by the pulley, the two standards *l l* being fixed on the cover *b*, to carry the axle of the wheel *h*. By this arrangement two turns of the handle *i* produces the same result in the agitator as was done by one stroke of the bow, and the motion of the handle being reversed at every second revolution, the ultimate effect is the same as before, and the manual operation is more easily effected in the one case than in the other. With a view to determine whether the alternate motion may not be dispensed with even in a cylindrical vessel, I have instituted experiments in which the interior of the vessel is armed with combs, and the agitator being converted to a like form, but having its prongs or teeth adjusted to the intervals of those of the vessel; with this apparatus I have found that like quantities of milk and cream, and with the same temperature, yield their butter in the following times:—

In the plane cylindrical vessel, with the agitator always moving in one direction, butter was not obtained in 55 minutes. In the same vessel, with the agitator moving alternately right and left, at every two revolutions butter was obtained in 25 minutes. In this same vessel, fitted as above described, with counter-agitators, but with the agitator moving in one direction only, butter was obtained in 20 minutes. From these results it appears, that, even with a cylindrical vessel, if properly armed, the process is performed in a shorter time than with the inconvenient reversed motions.

(2723.) It is well known, that a certain elevation of temperature is acquired by the fluid in the process of butter-making, and that the process is accelerated by producing this temperature artificially, from the application of heated water. For this purpose also, a practice is becoming prevalent to apply the water externally to the vessels containing the milk and cream, and not in mixture with them. Fig. 475. exhibits the application of this process to the utensil now under consideration, where *c* is the water-case formed of tin-plate, zinc, or of wood, at the bottom of which is fixed a circular stand to place the receptacle *a* upon, that the water may be under as well as around the receptacle. Wood, from its nonconducting quality, is perhaps better adapted than any metal for a water-case. Water brought to the proper temperature is poured into the space between the case and the retaining vessel, and if found necessary to increase or diminish the temperature, part of the contained water is drawn off below by a spigot, and hot or cold added to restore the requisite degree of heat. Experience seems to point out, that, in operating on the large scale *in wooden-vessels*, no extraneous heat is required, the naturally acquired heat appears to be sufficient, especially if aided in winter by the admixture of a small quantity of moderately heated water, and the non-conducting quality of the wooden vessel retains it; whereas the stone-ware vessel will be continually abstracting heat, and giving it off by radiation, if not surrounded by a medium of equal temperature.

(2724.) *Churns* of the third class are much more numerous than the two former, and though in themselves not differing much in the essential points, they yet exhibit a variety of structure in their details. They are usually distinguished by the name of *box-churns*, though the class embraces not only the cubical and the oblong box, but also the cylinder or barrel, the distinctive character being an agitator revolving in the vertical direction within a stationary case of any form.

(2725.) To illustrate this class, I have selected, first, the common *box-hand churn*, in very general use. The present example, fig. 477, is 18 inches in length, 11 inches in width, and 20 inches in depth, inside measure. Birch or plane-tree are the best material for the purpose, and it requires to be very carefully joined so as to be water-tight. As before noticed (2717.), it is of very small moment whether the bottom is formed to the circle of the agitator, or remains flat in as far as the production of butter is to be considered; but for the process of cleansing, the curved bottom will present some little advantages, a cover of the same material is fitted close in the top of the box, with convenient handles. The agitator, fig. 478, is of the usual form, the dimensions of its parts are unimportant, except that they have sufficient strength, and present sufficient surface to produce the requisite degree of agitation in the fluid. The two pairs of arms are half lapped at

Fig. 477.

THE BOX-HAND CHURN.

the centre, and the cross-bars mortised into them; the dimensions in length and breadth being such as to allow it to move with freedom within the box. At the centre a perforation is made through the sides to admit the iron spindle, which, at this part, is a square bar, fitting neatly into socket-plates of iron let into the agitator on each side, as seen in the figure at *a*. The further end of the spindle projects about an inch beyond the agitator, and is rounded to form a journal, which has its bearing in a close brass plate-bush or socket, which is sunk into the side of the box, and fixed with screws. The outward end of the spindle is furnished with a conical journal, the smallest diameter of which is equal to the diagonal of the square part of the spindle, and is furnished with a raised collar or ruff; the conical journal lying in a thorough brass plate-bush, leaving the collar outside, but embraced by a cup of the bush, upon which a coupling-ring is screwed, covering the collar, and, pressing home the conical journal, preventing thus the spindle from being withdrawn until the coupling-ring is removed. The spindle extends beyond the collar about 2½ inches, and is here formed into a square stud, upon which the eye of the winch-handle *f* is shipt when in work. In rigging this apparatus, the agitator is placed within the box, and the spindle is pushed through the outer bush and the agitator, until its two journals rest in the bushes; the coupling-ring is then screwed on to the outer bush, until the spindle with the agitator just turns round with freedom in the conical bush. To prevent the ring from turning round by the motion of the spindle, a smooth ring or washer of steel may be interposed between the collar and the brass ring. Various other modes of securing the spindle are employed, but in all, the object is to prevent leakage at the bush. To prevent taint from galvanic influence also, it is not uncommon to apply bone or other animal substance for the bushes.

Fig. 478.

THE AGITATOR OF THE BOX-CHURN.

(2726.) Churns of this kind are made of all sizes, without any change in the principle of their construction. They are applied to all kinds of power, and the capacity is proportioned to the extent of the dairy; but the *entire* capacity of the churn must be in general about double the quantity in gallons, of the fluid intended to be acted upon. In operating with the box-churn, it may be filled to the height of the spindle, or, if that member is well fitted, the milk may rise considerably above it, though in general practice it rises but little above. To adapt the churn to power, it is only necessary to put a pulley for a strap, or a chain, upon the end of the spindle, in place of the handle; or the spindle may be attached to another shaft, having the proper velocity, by means of a slip-coupling. The velocity of the agitator must depend upon the size of the churn; one of 24 inches diameter may make 60 revolutions per minute on the average, but may be higher or lower in the different stages of the process.

(2727.) A *new churn* of this third class has been lately introduced, and meets with much favour in the north of Ireland. It is believed that its introduction from France is due to Mr Blacker of Armagh, who is ever zealous in the advancement of objects tending to the improvement of his country; but the manufacturing of it is conducted by Mr R. Robinson, Lisburn. This churn is represented in figs. 479 and 480; the first a perspective of the machine as it appears when in work, the second a cross section, shewing the chief peculiarity of its construction. In the principle of its operation, this churn is the same as the one last described, but, like that alluded to under our second class, it is furnished with a case for containing water to regulate the temperature. It differs also from the common box-churn, in having the spindle or axis of the agitator passing through the length, instead of the breadth, of the containing vessel.

(2728.) In construction, fig. 479, represents a *hand-churn* of ordinary size, the containing vessel is of an oblong form, with a semicylindrical bottom, its length *a b* is $22\frac{1}{2}$ inches, the breadth *b c* 14 inches, and the depth to the

Fig. 479.

THE IRISH BOX-CHURN.

axis of the semicylinder $11\frac{1}{2}$ inches, the entire depth being 21 inches, all inside measure. The sides and ends are formed of birch or plane-tree, $1\frac{1}{4}$ inch in thickness, securely joined at the angles; the ends, of the vessel thus constructed, descend to the full depth of the semicylinder, while the sides are only 12 inches in depth. The bottom is formed in two plates of zinc, the edges of which are secured to the outer and inner faces of the sides, by means of close nailing, and also to the ends making water-tight joinings all round, in the manner seen in section at *d* and *e*, fig. 480, where the space *d f e*, between the outer and inner plates form the receptacle for the water. In the edge of one side there is inserted a funnel *g*, by which the water is introduced into the receptacle, by passing down through a small bore in the side of the vessel, and the spigot *h* serves to draw off the water when required. A wooden cover is fitted into the opening in the top, with nobs for the convenience of lifting it out. The chamber thus constructed is set in the iron frames *i i i*, which are attached to each end of the box, with screw-nails, and are further secured by the stretcher bolts *k*, and furnished with stud handles *l l*, for the convenience of lifting the churn.

(2729.) In fig. 480, the agitator is also seen in section, whereof *m m*, &c. are the four arms, the three transverse bars in each arm being in this case round rods, which are here shewn cut across by the section, as is likewise the square spindle passing through the centre. The spindle is formed and inserted in the same manner as described for the box-churn (2725.), and differs only from it in the mode of securing it in place. In this machine the spindle is pressed home by a perforated oblong plate, as seen in fig. 479, which passing upon the outward end of the spindle till it rest against the collar, it is then passed upon two stud-bolts fixed in the end of the vessel, having a screw-nut on their point, which presses the plate and the spindle, by means of the collar, home to the conical bush with any degree of tightness; and the winch-handle is then applied to the extreme end. It will be observed, that both this and the last described churn require being placed upon a stand or table about 2½ feet high when worked by the hand, in order that the handle may be brought to a proper height for the full effect of the hand. When either of them are worked by power, they are then most conveniently set on the floor of the dairy.

Fig. 480.

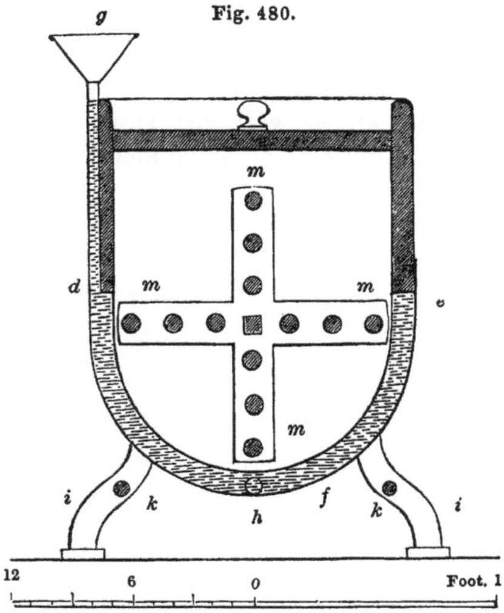

THE TRANSVERSE SECTION OF THE IRISH CHURN-BOX.

(2730.) In the fourth class of our subdivision, there is only to be considered the *plunger churn;* for in this class there is no variety, except as to the dimensions. The means of working the plunger, and the different media through which the power is applied, whether of man or of animals, are of a character much more varied than in all the others put together; and all these varieties seem to have been suggested, with a view to overcome the very unfavourable position in which a man applies his force directly to the plunger-rod of this churn. Amongst these varieties of construction in the mechanical media through which power, of whatever kind, may be applied, we find all possible varieties of the lever and its combinations, the loaded pendulum, with combinations of the lever and of rack-geering, crank and lever, and crank with connecting-rods; but it would be profitless to enumerate all the forms that have been devised for the improvement of this particular kind, and much more so to attempt to describe them. I will, therefore, rest satisfied with *one* that has long been in extensive use, and is equally well adapted to the hand, or to *power* of any kind.

(2731.) The upright or *plunge churning-vessel*, is always a piece of cooper-work varying in capacity, according to the extent of the dairy in which it is to be employed, from 10 to 130 gallons. It is built slightly tapering upward for the convenience of being tightly hooped, having a strong bottom and a movable cover, which may be fastened down with an iron clasp hoop; a perforation is formed in the centre of the cover for the passage of the churn-staff or plunger-rod, and this is surrounded by a wooden cup, to prevent the lashing over of the fluid in the act of churning. The plunger or agitator, which is attached to the bottom of the plunger-rod, is a circular reticulated frame of wood; the meshes may be from $1\frac{1}{2}$ to 2 inches square, and the bars forming the reticulation about $\frac{5}{8}$ inch thick and $1\frac{1}{2}$ inch deep, the whole embraced by a wooden hoop, whose diameter will just enter the top of the churn. The structure of the plunger is by no means important, the only essential point being, that it shall not present too much resistance to the passage of the fluid through its interstices, nor too little, to give it a too easy passage, which might render the agitation so slight as to be ineffectual; a good medium is to make the horizontal area of all the interstices 1.5 times the area of the solid parts, and this rule will apply to the agitators of all churns. Such is the simple construction of this churning vessel, which is capable of adaptation to any kind of power, and to any extent of dairy.

(2732.) As an example of the application of power to the plunge-churn, the accompanying illustrations are taken from those extensively used in Lanarkshire and the neighbouring counties, where they are usually worked by one horse. It will suffice at present to shew the internal mechanism, leaving the horse-wheel and intermediate geering for another occasion. Fig. 481 is a

Fig. 481.

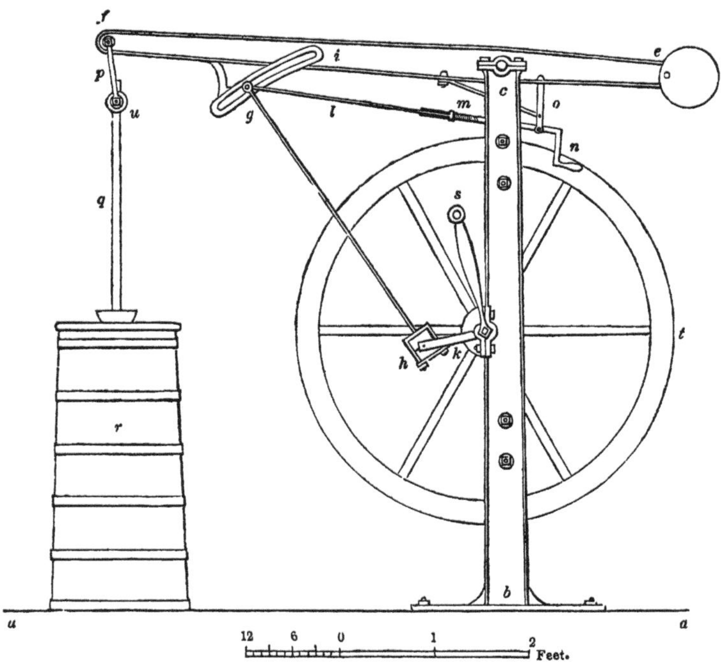

THE PLUNGE-CHURN WITH POWER-MACHINE.

side elevation of the churn, and the mechanism required within the churn-room of the dairy. It is here represented as being for hand-power, but the arrangements of the machine, so far as here exhibited, are the same whether for man or horse. Fig. 482 is a back view of the same, and the same letters apply to the corresponding parts in each figure. The floor-line of the churning-room is represented by the line *a a*, and *b c* is an upright frame of cast-iron, which is bolted to the floor. The frame consists of two cheeks, which are bolted together on the flanged bars *d*, *d*; it is 5 feet 9 inches high, and 16 inches wide. The reciprocating cast-iron lever *e f* is supported by its fulcra on the top of the frame, by means of a centre-shaft passing through the lever. The length from the fulcrum to the head *f* is 4 feet 2 inches, and from the same point backward to the extremity is 2 feet 3 inches; a counterpoise is here placed upon the lever to bring it nearly to an equilibrium. The connecting-rod *g h* is jointed upon a bolt that is fitted to move along the oblique groove *g i* formed in the lever, and the crank *k* is formed on a shaft that turns in bearings in the upright frame; its throw is 8 inches, less or more, proportioned to the height of the churning vessel, and the foot *h* of the connecting-rod is fitted to the crank-pin. *l* is a light sheers of malleable iron, whose forked ends embrace the head of the connecting-rod, which is also forked at the head; and the pin of the groove passes through all the three. The adjusting-screw *m*, with its winch-handle *n*, is supported in a swivel-socket in the bracket *o*; while its screwed end *m* works in the bend of the sheers *l*, which is screwed as a nut to receive it, for the purpose of shortening or lengthening the distance *m g*, and so changing the position of the head of the connecting-rod. The head *f* of the lever is mounted with a pair of side-links *p p*, jointed on a cross-head in *f*; the links are again jointed on a wooden cross-head *u*, into which the head of the plunger-rod *q* is inserted, and fixed by a pin or key; these side-links, jointed as they are at top and bottom, produce a very imperfect parallel motion, but sufficient to answer the rise and fall of the plunger-rod. The handle *s* is fixed on one end of the crank-shaft, and on the other is placed the fly-wheel *t*, to equalize the motion of the lever, and compensate the inequality of the resistance to the plunger in its ascent and descent. The churning-vessel *r* is 3 feet high, with an average diameter of 15 inches, and its capacity about 23 gallons. In extensive dairies, it is common to have two churning-vessels attached to the machine, in which case the wooden cross-head *u* is elongated so as to receive the head of the plunger-rods of both vessels, the vessels standing side by side during the operation.

Fig. 482.

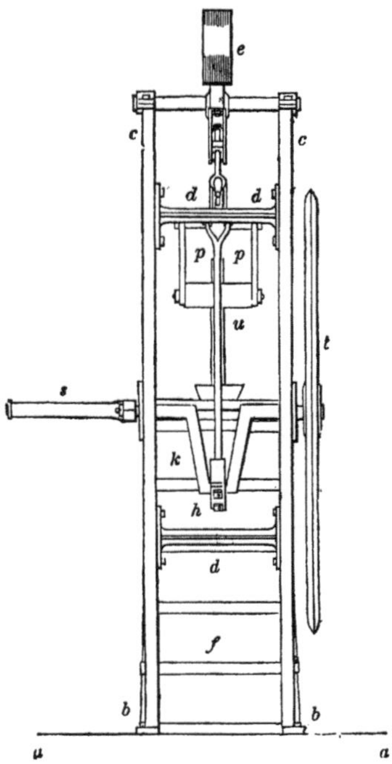

BACK VIEW OF THE PLUNGE-CHURN WITH POWER-MACHINE.

(2733.) When this machine is in operation, the revolutions of the crank produce a reciprocating action in the connecting-rod, which is communicated to the lever, and thence to the plungers; and it will be seen that, by moving the head *g* of the connecting-rod in the oblique groove of the lever, the strokes or reciprocations of the plunger will be long or short, as the joint at *g* is moved upward or downward in the groove *g i*. It is found from experience, that there are advantages to the process derivable from this; hence, at the commencement of the operation, the head of the rod *g h* is kept at the lower extremity of the slit, producing the shortest stroke; as the fluid becomes heated, and from the consequent effervescence its bulk is increased, the stroke is gradually lengthened by turning the handle *n* of the screw, and by thus shortening the distance *g m*, the pin at *g* is brought to the head of the slit, producing a stroke of the greatest length; and when the effervescence ceases and the butter has begun to form, this state is reversed, the stroke is gradually shortened, till the process is finished with the shortest stroke.

(2734.) The application of power to this machine is easy and commodious. It may be effected by a pulley placed on the crank-shaft instead of the handle *s*, and so driving with a strap or a chain; or it may be driven by a spur-wheel placed on the same point, calculated to the speed that may be afforded by the power, whether horse, water, or steam. The usual rate of the plunger in these churns is about 50 to 55 double strokes per minute, subject to the usual variation that is required in the different stages of the process. The price of the hand-machine, as in the figure, is from L.6 to L.8; and, when completed with horse-wheel and geering, it is from L.15 to L.18.

(2735.) Besides the four established classes of churns now described, there are a few other anomalous cases which have of late years been brought forward by inventors, and though they ultimately may be found to possess merit, they cannot yet be received as of established character. The machines to which I here allude may be named " oscillating churns." They possess no distinct agitator, but produce their effects by a species of oscillation produced in the fluid. Amongst these may be mentioned a vessel of an oblong form, placed upon skids, curved in a manner that, when the vessel is touched with even a gentle force, it will assume a motion resembling the well-known swing of a child's cradle, which will thereby produce a constant succession of irregular oscillatory motions in the contained fluid, which will produce the effect of churning.

(2736.) Another of these forms is an upright cylindrical vessel containing the fluid, and which is attached by its bottom to a crank revolving horizontally. The vessel is restrained from revolving on its own axis, but is made to describe a small circle, and every point in its base will virtually describe that circle which will be equal to that described by the crank that gives it motion. The effect of this oscillating motion in the vessel is to produce oscillation in the fluid, which will ultimately make gyrations within the vessel, combined with an undulatory motion, that will again produce the effect of the churn. In the first of these proposed machines, the motion would be so easy and equable, that there appears no mechanical difficulty in the scheme; but in the second, when we consider the mass of fluid that may be contained in the vessel, and then the whole put in that kind of motion that may be described as reciprocating-circular, which, when combined with the unequal effects of the undulatory motion on the fluid within, would, it is feared, require either a much stronger fabric than is usually allotted to such operations, or otherwise the machine would be very liable to derangement and to fracture. A churning-machine, on this principle, was exhibited at the Show of the Royal Agricultural Improvement

Society of Ireland, held at Belfast in August 1843, and excited considerable interest from its apparent novelty.

(2737.) The utensils and machines yet remaining to be described in the cheese-making department of the dairy, consist principally of the curd-breaker, cheese-press, and cheese-turner.

(2738.) The *curd-breaker*,* of which fig. 483 is a plan, and fig. 484 a transverse section, consists of a hopper-shaped vessel *a b*, 17 by 14 inches at top, and 10 inches deep. It is fixed upon two bearers *e e*, which are set upon a stand or tub when in operation. A wooden cylinder *f*, having an iron axle, which passes

Fig. 483.

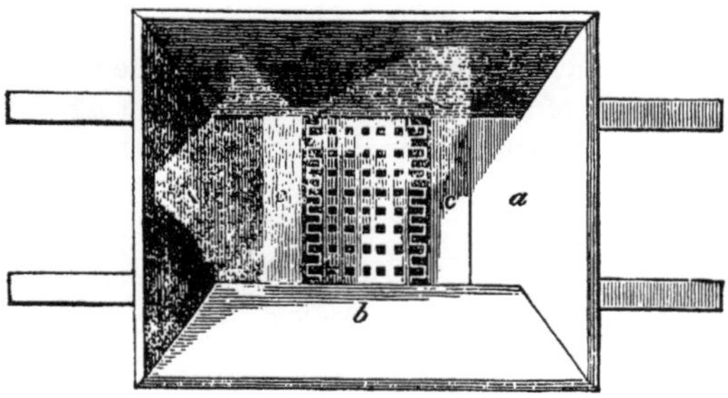

THE CURD-BREAKER.

through the bearers *e*, and is turned by a winch-handle *d*. The axle is kept in its bearings in *c* by means of slots *h* on each side, which slide in grooves in the bearers, and are held in their place by a slider at *h*. The cylinder is 7 inches long and 3½ inches diameter, studded all over with pegs of hardwood ¼ inch square, and projecting ⅜ inch. These pegs or teeth are set in eight regular zones round the cylinder, each zone containing 18 teeth. Two wedge-shaped pieces *c c* are attached to the sides of the hopper below, serving at the same time to reduce the opening between the cylinder and the hopper, and to carry a row of pegs similar to those of the cylinder, but falling into the spaces between the zones of pegs, as seen in fig. 484. In using this machine, it is placed over the tub or

Fig. 484.

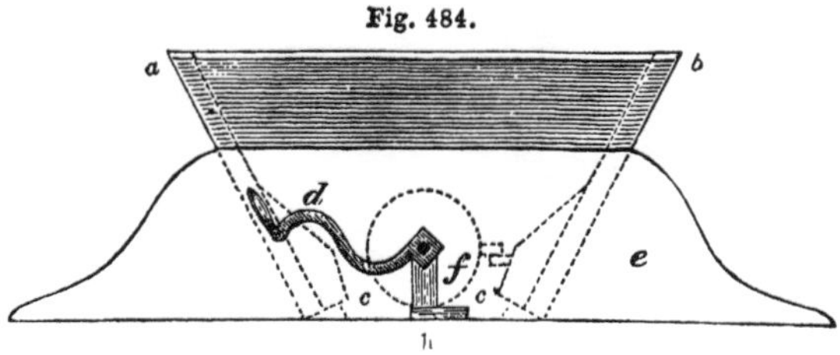

THE SECTION OF THE CURD-BREAKER.

other recipient for the broken curd; the hopper is filled with the curd that requires to be broken; and while one hand is turning the winch-handle, which may

* Quarterly Journal of Agriculture, vol. iv. p. 384.

be turned either way, the other hand may occasionally be required to press the curd down upon the cylinder. To prevent the curd being passed beyond the ends of the cylinder, it is covered at each end, to the extent of ½ inch, with an overlap of the sides of the hopper.

(2739.) The *cheese-press* comes next in the order of business, and of that machine the varieties are very numerous, though they may be all resolved into three kinds, namely, the common old stone press; the combined lever-press, of which the varieties are the most numerous, embracing from the single lever, through the various combinations of simple levers, to the more elaborate one of the rack and levers; and the atmospheric or pneumatic press. An essential characteristic of these presses must always be, that the load, in whatever way produced, shall, when left to itself, have the power to descend after the object being pressed, as that may sink by the expression of the whey from the curd.

(2740.) An example of the first of these is shewn in fig. 485; it consists of a strong frame of wood, of which *a* is the sill, 2 feet long, 18 inches broad, and

Fig. 485.

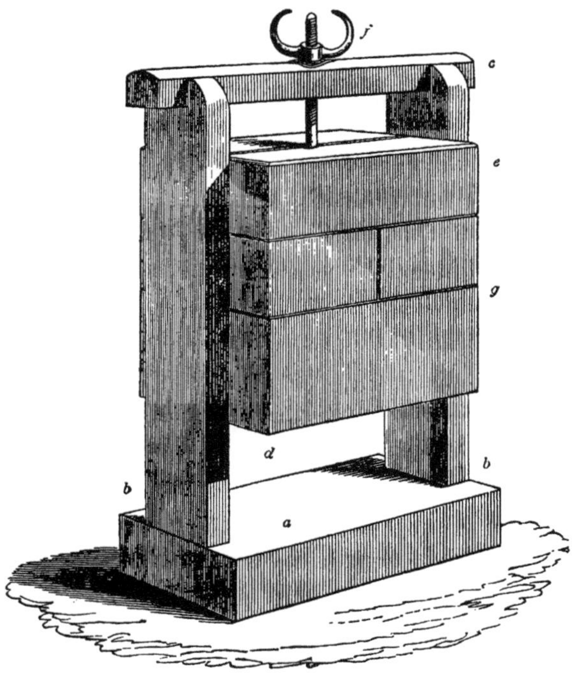

THE STONE CHEESE-PRESS.

4 inches thick; two uprights *b b* are mortised or dovetailed into it; these are each 6 inches broad, by 2½ inches thick, and 3 feet high, and are connected at top by the cross-head *c* mortised upon the posts. A cubical block of stone *d e* is squared to pass freely between the posts; an iron stem of 1 inch diameter is fixed into the upper surface of the block, and the upper end of it being screwed, is passed through the centre of the top-bar, and the lever-nut *f* is applied to it for raising or lowering the block. In each end of the block a verti-

cal groove is cut corresponding to the middle of the posts; and a baton of wood is nailed upon the latter, in such form and position as will admit the block to rise and fall freely, while it is prevented falling to either side. When put in operation, the block is raised by means of the screw until the cheese-mould, with its contents, can be placed upon the sill *a* under the block. This being done, the nut is screwed backward till the block rests lightly on the cover of the mould; it is let down by small additions, as the curd consolidates, until it is thought safe to let the entire weight press upon the mould, which is done by withdrawing the nut *f*. Instead of the solid block of stone *d e*, which, when left to itself, will always produce the same pressure, it is better to have one block *d g* into which the suspending bolt is fixed, and the remainder of the mass made up of smaller pieces, as shewn in the figure, by which means the amount of free pressure can be regulated to the particular size and state of the cheese; or blocks of cast-iron are sometimes used in the form last described, which are more commodious, and less liable to be broken.

(2741.) The next class of this machine, *the lever-press*, to speak in general terms, possesses the sill of the former in some shape or other, but generally of wood, with two uprights as before; but instead of a heavy block of stone, a simple movable sill of plank is employed, having an iron or a wooden stem attached to it, upon which simple or compound levers are made to act in producing the pressure. An improved form of this lever press was brought out by the Shotts Iron Company,* made entirely of iron, of a more commodious and compact construction than had hitherto been attempted, and, with slight alterations, is here presented as the type of the class.

(2742.) The *combined lever cheese-press* of iron is represented in the perspective view, fig. 486, and is constructed in the following manner: *a a* are a pair of cast-iron feet, on which the machine is supported; they have a socket formed at the crown to receive the malleable iron pillars *b b*. The sill-plate *c* is 18 inches in diameter, cast with two perforated ears, through which the feet of the pillars *b b* also pass, and secure the sill to the feet,—the cross lines in the sill indicate channels for the escape of the expressed whey. The movable sill *d* is of the same size as the one below, with corresponding ears perforated and fitted to slide on the pillars, and having the rack-bar *f* fixed in its centre. A top frame *e*, 18 inches by 9 inches, and 3 inches deep, is seated upon the top of the pillars, where it is fixed by two screw nuts, and adapted to carry the geering of the machine. The action of the rack and its sill is effected in the following manner:—The ratchet wheel *g* is fixed upon an axle that has its bearings in the top frame; on the same axle is fixed a pinion of eight teeth, not seen in the figure, which works in the wheel *i* of twenty-four teeth, fixed upon an axle which has its bearing also in the top frame; and this axle carries also a pinion of eight teeth, which acts upon the rack, but is also hid from view in the figure. The ratchet wheel *g* stands clear of the top frame about an inch, and its axle is prolonged beyond the wheel double that extent. The lever *k* is forked at the extremity *g*, and the terminations of the furcation are received upon the axle of the wheel *g*,—the wheel being embraced by the fork of the lever, but the lever moving freely upon the axle. The furcation is also lengthened towards *k* to an extent that receives a pall at *h*; between and in the throat of the fork, the pall is jointed upon a pin that passes through both,—the edge of the pall

* Prize Essays of the Highland and Agricultural Society, vol. x. p. 52.

pointing downwards to catch the ratchet, while it has a nob-handle standing upward, by which it can be conveniently disengaged from its wheel. A small

Fig. 486.

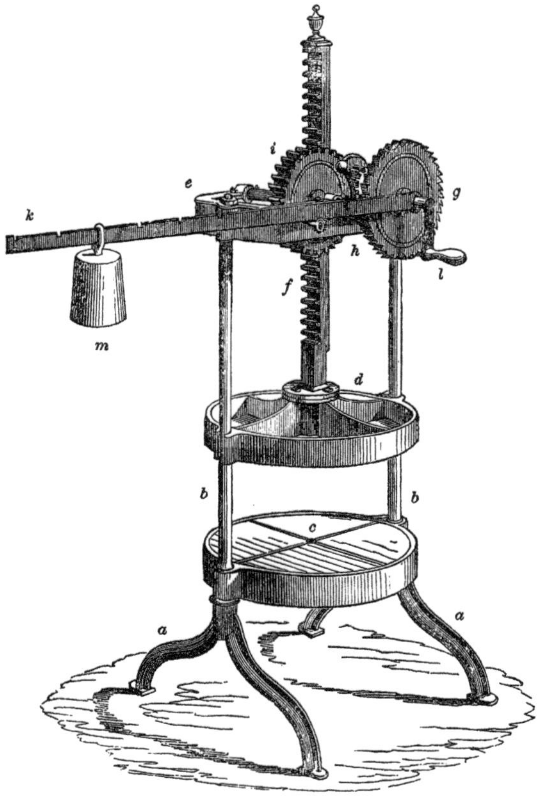

THE COMBINED LEVER CHEESE-PRESS.

winch-handle *l* is also fitted upon the axle of the ratchet wheel, and a pin seen near *h* is adapted to a perforation in the top frame, by the insertion of which, the descent of the lever is checked, when such is required, and this completes the mechanism of the press.

(2743.) In *pressing* with this machine, the cheese-mould is placed upon the lower sill, and the lever being supported on the pin at *h*, the winch-handle is turned to the left, depressing the rack and its sill till the sill presses upon the cover of the mould. The lever is now lifted by the hand, and the pall allowed to take into the ratchet; while the lever, being loaded by the weight, will cause the ratchet to turn, and produce the descent of the rack. If necessary, this is repeated again and again, till a considerable pressure is produced; and if it is wished that a continued pressure is to go on, the lever is again raised considerably above the horizontal line, and left to descend gradually, following the consolidation of the cheese. If it is wished that the load shall not follow the shrinking of the cheese, the pin *h* is inserted, which, when the lever comes to rest upon it, checks further descent. The amount of pressure is also regulated by the disposal of the weight *m* in the different notches of the lever. The usual selling price of this machine is L.4, when constructed of iron, as in the figure; but with wooden frame-work, and the rack and other geering of cast-iron, the price is L.3, 5s.

(2744.) In the third division of the cheese-presses, there is not such a variety as in the others, and what does exist is of recent introduction. I have to notice only one,—the *pneumatic cheese-press* of the late Sir John Robison of Edinburgh,—as an elegant application of science to a homely though important domestic purpose. The principle here applied, is the pressure of the atmosphere, brought to bear, with any degree of force compatible with that natural pressure, upon the curd when placed in the mould, by exhausting a vessel placed below and in communication with the mould, by means of a pipe descending from the bottom of the mould to a receiving vessel. To this pipe of communication a small air-pump is attached, and the mould having, besides its true bottom, a movable one of wire or of wicker-work, the curd, wrapt in a cloth, is laid upon the pervious bottom, and gently pressed with the hand till it fill the vessel all round. The pump is worked a few strokes to produce a partial vacuum in the receiver and below the curd, when the atmosphere, by its pressure on the surface of the curd, causes the whey to separate and descend into the receiver. The pressure may be continued or increased at pleasure by a few more strokes of the pump, until the cheese has acquired sufficient consistency to allow of its being handled, when, as recommended by Sir John, it is to be removed from the mould and placed within another of close wire-work, with a weight placed over it to complete the consolidation.* The whey that is discharged from the curd while under the atmospheric pressure, being collected in the receiver, is drawn off by means of a stop-cock.

(2745.) There falls yet to be taken notice of an important article in the cheese department of the dairy machinery—the *tumbling cheese-rack*, or *cheese-turner*. This machine is the invention of Mr William Blurton of Fieldhall, Uttoxeter, and its merits are believed to be sufficient to warrant its adoption on dairy farms. The object of the machine is to save much of the labour required in the daily turning of a large number of cheeses in the drying-room, and this it does very effectually, for with a rack containing 50 cheeses, they are turned over in very little more time than would be required to turn a single one.

(2746.) Fig. 487, is a view of the cheese-turner, as constructed to stand alone, and on its own feet; though this is not the best mode of constructing the machine. It consists, first, of an external frame *a b c d*, of which the two parts *a b* and *c d* are 6½ feet high, and 7 inches by 2½ inches; they are here represented each with a cross foot *a* and *c*, and connected at top by a top-rail *b d*. If constructed in a cheese-room, the posts should be at once fixed to the floor at bottom, and to the joisting or tie-beams overhead, becoming thus a fixture in so far as regards the external frame. The second part of the machine is a movable frame or rack, formed by the two interior posts *e* and *f*, which are framed upon the 12 shelves *g h i*, &c.; the posts are 6 feet high, and are again 7 by 2½ inches, the shelves being 7 feet long and 14 inches broad, or more, according to the size of the cheeses manufactured, by 1 inch thick. The shelves extend to 5½ feet in height over all, and are tenoned into and lipped over the posts, and each shelf is finished on both sides with a knife-edged lath, nailed along the back edge; these laths are 2 inches broad, and ¼ inch thick at the back,

* Prize Essays of the Highland and Agricultural Society, vol. x. p. 198.

thus increasing the thickness of that edge of the shelves to 1½ inch. As the figure represents a rack that will contain 5 cheeses on each shelf, a correspond-

Fig. 487.

THE CHEESE-TURNER.

ing number of pairs of vertical laths *k*, *k*, &c. are nailed upon the back edge of the shelves. These laths are 1½ inch broad and 1 inch thick, chamfered off to one side to the thickness of ⅜ inch or thereby at the edges, against which the cheeses are laid, and are checked upon the shelves, and securely nailed. The shelf-frame thus formed is provided with two strong iron gudgeons or pivots fixed in the side-posts at mid-height, and these are received into corresponding holes in the outer or bearing posts, so that the shelf-frame swings poised upon the two pivots; and it is further provided with an iron latch at top and bottom on one end, by which it may be tilted and secured with either the shelf *g* or *e f* uppermost. The catches of the two latches are both placed at top on the external post at one side, suited to the motion of the shelf-frame, and to prevent its being turned with the back edge of the shelves upward.

(2747.) When cheeses are placed upon the shelves, it will be found that the knife-edge laths keep them free of the body of the shelf, and thus permit air to pass under them, while the pair of vertical laths keep the cheese in its proper position on the shelf. The height between the shelves is such as to leave a free space of 1 inch between the cheese and the shelf above it; and whatever number of cheeses may be lying upon the shelves, the simple act of tilting the frame will place every cheese which *was* resting on a sholf, on its opposite side, *upon* that shelf which immediately before was *above* the cheese, but by the tilting is now *below* it. It will be observed, that the vertical laths serve to prevent the cheeses from falling out while the frame is tilting, and each cheese has only to fall one inch in that operation, or from the one shelf to the other, in a reversed position.

(2748.) It will be also observed, that the fixed external frame is best adapted for an extensive cheese-room, where the racks may be placed in rows extending the length of the room, leaving free passage between the rows. The width of the passages requires to be equal to half the height of the shelf-frame, or 3 feet; a room, therefore, 20 feet wide would contain 4 rows of such *racks;* and if the length were equal to 10 diameters of the cheeses, or containing that number in the length, the room would contain in all 440 cheeses in the best possible condition for their being prepared for market, having free ventilation, and access for the dairy-maids to handle and wipe any cheese at any time. We have no experience of this cheese-rack in Scotland; but, judging from its apparent capabilities, there is much reason to think that it might be employed with excellent effect in the extensive cheese-dairies of Ayrshire and Galloway. The price of a portable rack, as here figured, capable of holding 55 cheeses, is L.4.—J. S.]

---

68. OF WEEDING CORN, GREEN CROPS, PASTURES, AND HEDGES, AND OF CASUALTIES TO PLANTS.

> "Now is the time before the thistle blow,
> While gule is in the flower, and charlock breathes
> Its cloying scent around, the weeding task
> To urge."
>
> GRAHAM.

(2749.) From page 345 to 352 of the first volume, I gave a list of common plants which indicate to a certainty the nature of the soil upon which they grow. Where any of them occur as a constituent of natural pasture, they are useful, and should be cherished; but wherever they occur in cultivated soil, they assume the character of *weeds*, and should be eradicated. Viewing the occurrence of plants in this light, we may arrive at a correct definition of the circumstance which renders any plant a weed; and the circumstance is, that wherever a plant is found to grow where it should not, it is a *weed*. For example, a stalk of wheat in a bed of tulips in a garden is a weed; and, in like manner, a tulip in a field of wheat is a weed. When you look at the number of the plants noticed in the place referred to above, and the list is not given as comprehending the entire number found in fields, but only as decidedly indicative of the soil upon which they respectively grow; and when you consider that they are all *weeds* in cultivated fields, in the sense of the definition given above, you may imagine how actively the farmer should be employed in destroying them in the season they present themselves in the greatest vitality. No doubt, the entire number of these weeds will not make their appearance simultaneously, because each has its season and soil in which it is in its most vital state, and were they then allowed to grow, such is the vigour of growth in all wild plants, that they would soon occupy the ground to the exclusion of the entire cultivated species. It is the duty of the farmer to refuse them this freedom, and to check at all times their intrusive tendencies. It is partly for this purpose that the land is ploughed before winter, that the roots and seeds of weeds may be exposed and directly destroyed by frost and rain, and indirectly destroyed by those natural agents comminuting the soil so much as to render the removal of plants by the roots comparatively easy. By perseverance, it is quite possible for the farmer to get quit of all weeds which propagate themselves chiefly by the agency of roots, though it is impossible for him to prevent amongst crops the appearance of weeds which originate from seed; yet it is quite possible, even in regard to this latter class of weeds, to prevent their ultimate development to leaves and seed. Summer is the season in which weeds thrive in the greatest luxuriance, and, of course, summer ought to be the season also of the farmer's greatest activity in employing his plans of destruction against them. I shall first describe the mode of extirpating the most troublesome weeds amongst growing crops, pasture, and fences, and then notice the casualties to which those crops are respectively subject. There is a principle affecting the vitality of weeds which should never be lost sight of by every farmer attempting their destruction, and if he would be guided by it whenever weeds presented themselves, they would infallibly be prevented from coming to maturity. It is a well-known law in vegetation, that elaboration of sap by its leaves is the chief means by which a plant is enabled to develop its component parts, so that by destroying its leaves the life of the plant will be destroyed. The simple plan, therefore, of destroying all sorts of weeds, is to destroy their leaves as soon as they appear.

(2750.) The first crop which requires weeding is the *oat*, and the weed which most infests it is the creeping plume thistle, *Cnicus arvensis*. It is perennial, and its roots very creeping, by which it easily spreads itself, and, when allowed to stand, will attain the height of from 1 to 3 feet, and waft its plumed seeds over many adjoining fields. When allowed to flourish amongst corn, it is extremely troublesome to reapers at harvest. The plant should not be cut down before it has attained about 9 inches in height, otherwise it will soon spring again from the root, and require another weeding, and by the time it has attained that height, the oats will be about 1 foot high. This weed is destroyed amongst corn by cutting with the weed-hook. A troublesome weed among oats in the neighbourhood of large towns where street-manure is employed, is wild mustard or charlock, *Sinapis arvensis;* and it so completely covers the soil in moist seasons, that its bright yellow blossom obscures the crop altogether. Where the corn is sown broadcast, I have seen the wild mustard pulled by hand, and, indeed, there is no other way of extirpating it in such circumstances; but where its seed is expected to exist, the oats are sown in drills of 9 inches in width at least, by a drill-machine, Plate XXVII. fig. 382 and fig. 396, to give room to the hand-hoe, fig. 423, to be used between them.

(2751.) The only implements used in weeding corn are the hand draw-hoe already alluded to, and shewn in fig. 423, and the *weed-hook.* The hand-hoe is used amongst drilled corn, turnips, potatoes, and pasture. In using it among drilled corn, the field-workers each occupy a row, and, on walking abreast in the form of a double eschelon, the best worker taking the lead in the centre, the entire ground is cleared of weeds at once, as far as the workers extend across the drills. The weed-hook, fig. 488, is used with one hand, the field-worker walking upright and holding the hook before her near the ground in the position represented in the cut. The stem of the weed is embraced between the clefts of the iron hook *a*, which being sharpened to a cutting edge on both sides, the stem is severed by the worker pulling the handle towards her in a slanting direction upwards. The handle is of wood, and about 4 feet in length, and the weed-hook costs 6d., or 4d. for the head alone. I have seen a weed-hook with its outer edge also sharpened, with the view of cutting a weed with a push forwards; but such a weed-hook cannot be used amongst standing corn, as its sharp outer edge would inadvertently cut the stems of corn The usual way for field-workers to arrange themselves when using the weed-hook, and, indeed, when weeding broadcast corn in any way, is for 2 workers to occupy every single ridge of ordinary breadth, or 4 upon the double ridge, each taking charge of the half-ridge from the open

Fig. 488.

THE WEED-HOOK.

furrow to the crown. The weeds, when cut over with the weed-hook, are left upon the ground, and, being young, soon decay.

(2752.) *Barley* is weeded after oats. Besides being troubled with the creeping plume thistle, *Cnicus arvensis*, though not to the same degree as oats, and the charlock, *Sinapis arvensis*, which is pulled when the crop is sown broadcast, and hoed when sown in drills, barley is much infested with the common red poppy, *Papaver rhœas*, especially in England, and the long smooth-headed poppy, *Papaver dubium*. The deep scarlet flower of the former may be seen at a great distance, and produces a striking effect. The seed of the sticking-grass, or cleavers, *Galium aparine*, is naturally so hard, that even after having undergone the process of boiling, along with pot-barley, still retains its hardness.

(2753.) *Wheat*, though next in order, is not weeded until the stalks have reached the knee, when the principal weed which infests it, the corn-cockle or popple, *Agrostemma githago*, is in bloom. This plant having a woody stem, is not cut with the weed-hook, but pulled by hand. Its seed is black, rough, and round, and is much objected to in samples by bakers and corn-dealers, though the kernel being white, its flour is not distinguishable from wheat-flour, and is otherwise innocuous. In dry ground is found the long prickly-headed poppy, *Papaver argemone*, and in strong ground, the hairy tare, *Ervum hirsutum*, though a low insignificant-looking plant, yields a hard heavy seed, difficult to winnow from wheat. The blue-bottle, *Centaurea cyanus;* smooth rye-brome grass, *Bromus secalinus;* as well as the corn and common sow-thistles, *Sonchus arvensis* and *S. oleraceus;* the corn gromwell, *Lithospermum arvense*, the seeds of which contain a large proportion of siliceous matter; the bearded darnel, *Lolium temulentum*, which is now rare, are all found among wheat. The cleavers, or sticking grass, *Galium aparine;* and the common hemp-nettle, *Galeopsis tetrahit*, and *G. versicolor*, large-flowered hemp-nettle, are also found among wheat. Any of the natural grasses will do no harm amongst corn; their seeds being light, are easily blown away in the process of winnowing, or even of thrashing; but all the plants formerly named bear so heavy seeds as cannot be easily got quit of in thrashing and winnowing, and are therefore troublesome in the barn, and injurious to the sample. The common reed *Arundo phragmites*, is not uncommon in corn-fields on carse-land, and is not disliked, as it serves to keep the corn open both in the stooks and stack.

(2754.) In *pastures*, the biennial spear plume-thistle, *Cnicus lanceolatus*, is prevalent, and not unfrequently the welted thistle, *Carduus acanthoides*, both in dry ground. In marshy pastures, the marsh plume-thistle, *Cnicus palustris*, is the most prevalent plant of the kind. The ragwort, yellow weed, or weebo, *Senecio Jacobæa*, is often seen in pastures, in deep dry loam; and wherever seen, no sheep have been pastured there, for they never fail to keep the plant down. The tribe of docks are also numerous: the broad-leaved *Rumex obtusifolius*, and *acetosella*, sheep sorrel, are found on dry pastures, and the common sorrel, *R. acetosa*, and sharp dock, *R. acutus*, in moist and damp pastures. All these weeds are best destroyed by the hand-hoe, cutting them over by the ground just before the flower-buds are formed. In moist pastures in low situations, the soft rush, *Juncus effusus*, and common rush, *J. conglomeratus*, are most frequent. All the tribe of rush is best destroyed by being frequently cut over with the scythe. The great white ox-eye, or nowt-gowan, *Chrysanthemum leucanthemum*, and the common daisy, *Bellis perennis*, disfigure the pastures on low grounds, and the yellow nowt-gowan, *Chrysanthemum segetum*, disfigures as much upland pastures, and all indicate the soil in a state of poverty. The common brake, or fern, *Pteris aquilina*, prevails in upland pastures, where the soil is deep and dry. When growing together so thickly as to injure the grass, it is injurious, and should be removed as a weed, which may be done either with the constant use of the scythe whenever it makes its appearance, or by irrigation with water. When it is as thin on the ground as to permit grass to grow, it is rather an advantage, inasmuch as it affords shelter to sheep in the first instance, and protects the grass from frost, and thereby cherishes it into early verdure in spring. Circumstances alone should guide you in the destruction of the braken in upland pastures.* What constitutes the windlestraw of rich pasture is the stems of the crested dogs'-tail, *Cynosurus cristatus*, one of its most valuable constituents.

(2755.) Besides pastures, weeds are found among the *sown grasses* and *hay fields*, such as occasionally the corn and stinking chamomile, *Anthemis arvensis* and *cotula*. The soft brome-grass, *Bromus mollis*, deteriorates the value of hay. The lesser dodder, *Cuscuta epithymum*, is a parasite which fastens itself on various plants, and lately the great dodder, *C. europæa*, has been imported from the Continent, and in the two last seasons has much infested the clover-plant in the south of England. The only way of getting rid of its seeds is by sifting, as they are not above $\frac{1}{6}$ the size of clover seeds. Perhaps the sieve No. 17 may answer the purpose. There are other parasites which fasten themselves on the clover plant. These are *Orobanche major* and *minor*, the greater and lesser broom-rape. Their seeds are like those of the clover, only not so

* See Journal of Agriculture for October 1843, p. 143.

plump; neither so large nor so glossy, the skin being rough and of greyish colour. They cannot, therefore, be easily sifted out. The seeds, too, of the common self-heal, *Prunella vulgaris*, mix themselves with those of clover, but being smaller are easily riddled out.* The seeds of the *Festuca bromoides* and *F. myurus*, constitute the hair-seed among rye-grass seed. In upland hay fields, the common yellow-rattle, *Rhinanthus crista-galli*, occurs. Of this weed Dr Hooker says—"When the fruit is ripe, the *seeds* rattle in the husky capsule, and indicate to the Swedish peasantry the season for gathering in their hay. In England, Mr Curtis well observes, the hay-making begins when this plant is in full flower."†

(2756.) The weeds most troublesome to *fallow land* are couch-grass, *Triticum repens*, and knot-grass, *Arrhenatherum avenaceum*, var. *bulbosum*. The annual meadow-grass, *Poa annua*, is a very common weed in cultivated soils. In damp subsoils in strong land, the common colt's-foot, *Tussilago farfara*, is a weed which cannot be eradicated; the only way of destroying it, after draining the land, is to cut off the leaves as they rise, and the plant will die out. On light soils, the rest-harrow, *Ononis arvensis*, is a great pest.

(2757.) Next to corn and grass, the *hedges* require weeding, and if they are cleaned at an early period of the summer, before many of the smaller weeds have dropped their seeds, they will be the more easily kept clean during the remainder of the season; but this is a department of farm-work which I am ashamed to say is very much neglected by most farmers. One would suppose that time were thrown away in weeding a hedge; whereas the work is not only proper in itself on account of the hedge, but useful in saving the adjoining cultivated fields being overstrewn with the seeds of weeds. There are a few implements requisite for weeding hedges different from those described for weeding corn and pasture fields. These are seen in fig. 489, and consist of a *hedge-spade a*, which, when viewed horizontally, is in its proper position, and costs 4s. or 5s. with its short handle; a common *Dutch-hoe b*, for removing the weeds on the hedge-bank, and costs 3d. per inch along its face. A narrow faced one is most convenient for this purpose, for getting at the weeds near the root of the hedge; and *c* is a *hooked stick* for drawing the weeds, cut over by the Dutch-hoe, through between the stems of the thorns into the ditch. The manner of using these implements is this:—The hedger steps into the bottom of the hedge-ditch, with his

* See Journal of the Royal English Agricultural Society, vol. i. p. 173–76.

† Hooker's British Flora, p. 283, edition of 1831.

face towards the hedge, having in his right hand the cross-head of the hedge-spade *a*, and resting its handle at *a* in his left hand, works the spade in a horizontal direction, removing all the grassy and other plants

Fig. 489.

THE HEDGE-SPADE AND DUTCH-HOE FOR WEEDING HEDGES.

growing on the face of the hedge-bank below the line of thorns along the entire side of the ditch, and pushing the refuse into the bottom of the ditch. If the thorn-bed was formed of turf, the grass below the line of a young hedge may be expected to be pretty strong; but if made on fallowed ground, the weeds will be few. Whatever be the state of the weeds, they should be removed from the young hedge, to prevent them scattering their seed. While the hedger takes the lead of weeding the *face* of the hedge-ditch below the hedge, a field-worker follows him, and removes the weeds with the Dutch-hoe along the *top and face* of the hedge-bank behind the hedge, stirring the whole surface, but not deeper than necessary to remove the weeds. Such of the weeds as fall on the top of the hedge-bank get leave to lie and rot, whilst those on its face, immediately behind the hedge, are drawn through between the hedge-stems into the ditch, with a crooked stick *c*, by each of the field-workers, who both follow the hedger in the ditch, and the field-worker on the top of the hedge-bank. The women take their turn at the Dutch-hoe, as it is more severe to work than using the crooked stick, which merely saves the hands and fingers of the workers being scratched by the thorns in reaching between the stems. In this way the party proceed in their relative positions. I never could ascertain what the weeding a young hedge in this way costs, as it depends on the state of the weeds; but I remember of an old man taking on the job one summer at 1d. per rood of 6 yards. He first used the Dutch-hoe along the bank for one yoking, then the hedge-spade for another, and completed the clearing the space previously gone over of weeds with the crooked stick. When the hedge has attained 3 or 4 years old, it so overshadows the face of the hedge-bank behind it, that few, if any, weeds appear upon it, so that the hedge-spade in front, with only a skimming of the Dutch-hoe on the top of the hedge-bank, is all the work required. When the hedge covers the hedge-bank entirely, the Dutch-hoe is dispensed with, and only the hedge-spade used. To accomplish this latter work, the same old man undertook it at ½d. per rood of 6 yards, so that the spade work is about half the amount of the whole, which, I dare say, may be near the cost, in hedges not allowed to be overrun with weeds.

(2758.) The weeds which infest hedges are numerous, many of which are common to them and corn and pasture, but others are most numerous in hedges, such as the sticking-grass, *Galium aparine;* great bindweed, *Convolvulus sepium*, a creeper abundant in England; Bishops' or goutweed, *Ægopodium podagraria*, which cannot be eradicated; ground-ivy, *Glechoma hederacea*, a creeper; hog-weed, *Heracleum sphondylium*, a coarse rank plant growing from 4 to 5 feet in height; meadow vetchling, *Lathyrus pratensis*, a strong creeper with yellow flowers; common hedge-mustard, *Sisymbrium officinale*, flowers very small, pale yellow; woody nightshade, *Solanum dulcamara*, in moist situations; hedge-woundwort, *Stachys sylvatica*, flowers purple, whorls of about 6 flowers; upright hedge-parsley, *Torilis anthriscus*, fruit densely clothed with incurved bristles; zigzag trefoil, *Trifolium medium;* tufted vetch, *Vicia cracca*, with fine bluish-purple flowers, and hedge-vetch, *Vicia sepium*, with large leafets. Numerously as these crowd together to take shelter in hedges, there are other intruders which insert themselves between the thorn-plants, and, acquiring strength from the protection afforded them, at length overcome their protector. Of these, the most destructive to hedge-plants, are the crab-apple, *Pyrus malus*, which easily makes room for itself with its stiff and elbowing branches; the sloe, *Prunus spinosa;* common dog-rose, *Rosa canina*, and even the favourite sweet-briar, *Rosa rubiginosa*, placed as it is in a hedge for the sake of its odour, both these never fail to injure thorn-plants, as far as their spreading arms can reach; the broom, *Cytisus scoparius*, and whin, *Ulex Europæa*, both displace the thorn, and on themselves dying out by old age or severe frost, leave unseemly gaps; the common bramble or blackberry, *Rubus fruticosus*, and the hazel-leaved bramble, *Rubus corylifolius*, being powerful climbers, completely overrun thorns in many parts of England. The common barberry, *Berberis vulgaris*, is a well-known intruder in hedges in England.

(2759.) Weeds growing in the *bottom* and *sides of ditches*, cause the water to fill up the bottom, and break down the sides. Among these are the beautiful yellow corn-flag, *Iris pseud-acoris*, which takes up its station on the sides of ditches, and directs the water to the opposite side; the water-cress, *Nasturtium officinale*, which grows in the bottom of ditches, and arrests mud in its progress down the ditch, but will only grow where spring-water flows; the common butter-bur, *Petasites vulgaris*, which, with large expanded leaves, occupies a prominent position at a ditch side. Dr Hooker relates, that "the early flowering of the plant induces the Swedish farmers to *plant* it near their bee-hives." The red canary-grass, *Phalaris arundinacea*, which, with its creeping roots, tufting here and there, proves troublesome in unduly hardening the parts of the sides of ditches where it grows; the reed meadow-grass, *Poa aquatica*, which acts a similar part to the canary-grass; the floating meadow-grass, *P. fluitans*; which floats its long narrow leaves down the stream of water in the ditch, retarding and stagnating it; the curl-leaved dock, *Rumex crispus*, which strikes its roots down the side of the ditch; while the brooklime, *Veronica beccabunga*, is found in the direct water-course. The only mode of destroying weeds in ditches is scouring the bottom, and paring the edges with the common spade, and extracting the roots of the obnoxious plants growing in both. In connection with ditches, I may mention, that, when the amphibious persicaria, *Polygonum amphibium*, finds its way into a drain, whether of stone or tiles, it chokes it up, and, being a perennial, there is no chance of its dying out. Where this weed is suspected to exist, the drain should be carefully taken up from the bottom, and every vestige of the plant removed.

(2760.) Numerous weeds lurk about the *margins of fields* and *dyke-sides*, rendering the cultivated ground near them foul. Most farmers allow weeds to grow without molestation in ground not touched by the plough, though within the fences of cultivated fields. Around the fences of every field is a narrow space, which cannot conveniently be ploughed, and in the four corners of every field there is a triangular space of waste ground, all of which are usually occupied with weeds, especially when the fields are under grain crops. This waste ground being well sheltered, and its soil being as good as that of the field in which it is situate, and unexhausted by cropping, weeds grow on it easily and luxuriantly. Instead of allowing it to be *waste* ground, the plough should turn over the soil *towards* the very foot of a stone fence-dyke, and even when the soil is turned *from* such a fence, the distance from it and the plough need not exceed 9 inches in breadth, by putting the horses *atrip*, that is, one before the other in the plough, and giving the bridle of the plough more land. And the corners should be dug with the spade by the hedger. But independent of considerations of absolute economy, in lessening the extent of waste land in any field by such simple means, the weeds should be cut down by the field-workers going round the margins of all the fields bearing green and grain crops, at intervals of time during summer and autumn. Besides the slovenliness exhibited in the neglecting to weed such places, there is positive loss incurred elsewhere, by allowing the seeds of syngenesious plants to be carried about by the wind. Besides thistles, ragwort, dock, whin, and broom, other weeds are found in waste places, such as the common burdock, *Arctium lappa*, which is not the least formidable; the hemlock, *Conium maculatum*, a well-known poisonous plant; as well as the purple fox-glove, *Digitalis purpurea;* the annoying dandelion, *Leontodon taraxacum;* and the great nettle, *Urtica dioica*. In damp situations, *Ænanthe crocata*, water sap-wort grows. What is remarkable in this plant, is the fact of its being poisonous in England, but inocuous in Scotland.

(2761.) I shall now proceed to consider the cleansing of the turnip and potato crops from the particular weeds which infest them. After the ample details given on the culture of the *turnip*, which terminated in (2506.) and (2507.), all that remains to be said, is on the use of the implements employed in cleaning the ground of surface-weeds, and of putting a finish on the turnip culture. The most useful implement for this purpose is the *scuffler*, which assumes various forms in different parts of the country, all of which are noticed, and some described below, by Mr Slight. In the intervals of manual labour in first singling and then hoeing turnips, the scuffler is employed to remove the weeds that may have sprung up between the rows of turnips, and to stir the soil, for the purpose of loosening it for the roots of the plants. It was once a prevalent practice to pare the soil from both sides of each turnip-drill with the paring or small plough, fig. 370, because it was considered that strong land could not be pulverized between the drills without such an operation; but the better practice now is, to pulverize the land well before the crop is sown; and it is desirable to dispense with every tedious process which requires a bout in each drill, while an implement, such as the scuffler, can be had to cover the same space of ground in a landing. When the scuffler has cleaned and stirred the land, after the last *hoeing*, the next operation, performed with horse-labour, is to set up the drills with the double mould-board plough. I have given reasons for dispensing with the setting up of turnip-drills on light soils in (2507.); but on strong soils, where water must remain a while on the surface, however thoroughly they may have been drained, it is necessary to set up the drills, in order to provide channels for surface water to run off freely. The double mould-board plough finishes the setting up of every drill at a landing, and this is the last operation the turnip culture receives; but care is requisite to put the mould-boards as close together as that no earth shall roll upon the leaves of the young plants, which, at this period of their growth, spread themselves just above the ground.

(2762.) The *weeds* most troublesome amongst turnips is the wild mustard or charlock, *Sinapis arvensis*, and the wild radish, *Raphanus raphanistrum*, and a few others which spring from seed, for as to those which spread from roots, they should have been eradicated pretty well in the preparation of the soil for the sowing of the crop, and although they should not all have been subdued in the preparatory process, they receive a severe check from the scufflers; and rather than employ that implement too long, and prevent the spreading of the fibrous roots of the young turnip plants in search of food, it is better to allow a few weeds to remain, and be overcome by the future luxuriance of the crop, as is shewn in (2506.).

(2763.) The implements required for cleaning the potato crop are the same as for turnips, and the periods and modes of their use are also similar. There is a difference, however, in setting up the drills, which, in the case of potatoes, is done with the common plough, instead of the double mould-board plough, and, in doing which, the plough must go a bout in every drill. A peculiar sort of double mould-board plough, described below under fig. 494, should be substituted for this purpose, for the common plough. Potatoes require to be set up in all soils, because, being tubers occupying the ground below the surface, the earth should be loosened and heaped about them; but it is possible to set up potatoes so as to injure them, which is always the case when the drills are earthed up too narrow, by the plough going too deep in the furrows, because on light land such a practice is useless, and on heavy land, when it is damp, it has the effect of inclosing the tubers in encrusted ridglets. This subject has been treated experimentally, and is noticed in full in (2508.).

(2764.) The *potato crop* is not so much infested with the wild mustard and radish as the turnip, but in consequence of the ground being obliged to be early prepared in spring, there is not sufficient time to destroy couch-grass, and the oat-like grass, commonly called knot-grass. The weeds that frequent the potato ground are pretty numerous, because the stems of the potato plant permitting the air to pass between them, and the ground being usually in good heart, and there being plenty of room between the rows of plants, weeds get up after the operations of the plough have ceased, and they are usually these :—white goose-foot, *Chenopodium album;* common fumitory, *Fumaria officinalis;* mugwort, *Artemisia vulgaris;* chicken-weed, *Stellaria media;* nipple-wort, *Lapsana communis;* shepherd's purse, *Capsella bursa-pastoris;* ivy-leaved speedwell, *Veronica hederifolia;* small annual nettle, *Urtica urens;* all on their respective soils.

(2765.) The first writer of practice on the "Weeds of Agriculture," was Mr Benjamin Holdich, who classified weeds into 5 very natural divisions, in as far as the experience of a farmer could suggest, viz.:—1. Weeds which infest samples of corn, such as the corn-cockle, *Agrostemma githago.* 2. Fallow weeds, such as couch-grass, *Triticum repens.* 3. Weeds which are principally objectionable, as they incumber the soil, or whose roots are annual, and whose seeds pass the corn-sieve, such as the charlock, *Sinapis arvensis.* 4. Weeds which never rise into the crop, nor come into the sickle, such as spurry, *Spergula arvensis.* 5. Weeds of pastures, such as yellow goat's beard, *Trogapogon pratensis.** Other authors divide weeds into fibrous and fusiform rooted, annual and perennial; but it is of little moment to the farmer whether the weed that annoys at the present moment has a bushy or long root, or is an annual, biennial, or perennial, as ordinary ploughing will eradicate almost any weed that infests cultivated fields. These are botanical distinctions, and their study, with the botanical names, may afford you exercise in botany, but will give you little insight into the weeds you may expect to encounter, when cultivating a particular kind of soil, or raising a particular crop.

(2766.) The casualties which befall the growing crops are numerous, and most of them cannot be averted. I shall begin with the *oat.* This plant is liable to attack at a very early period of its growth by the oat-grub, the larva of the meadow crane-fly, *Tipula oleracea,* the effects of which have been described in (2356.), and the insect figured in fig. 394. Another complaint of the oat plant is *segging* or sedging, so named, in consequence of the plant being short, and the leaves becoming broad, and the roots thickened like those of the sedge. What the true cause of this complaint is I do not know, for any insect that may then be detected in the plant, I regard as the effect, not the cause; but I have cured a piece of land of its constant tendency to grow sedged oats, simply by draining. Oats are liable, in damp seasons, to have blackened ears when the entire panicles are clustered together, and covered with a black powder, which is named *Uredo segetum,* a parasitic fungus. The blackened heads do not occur to a serious extent, but whether it would be entirely avoided by pickling the seed-oat as the seed-wheat is, I cannot say. The dart-moth, *Agrostis segetum,* whose wings expand from 1½ inch to 1¾ inch, produces larvæ, which attain the length of 1½ inch, and live upon the portion of the corn-plant below the surface. Rolling heavily at night, when larvæ generally come to the surface, may effect the destruction of some; but all caterpillars which live in the earth have a tough elastic skin capable of considerable resistance, and such of them as have not a stone or other hard substance under them, would be merely pressed into the soil, without sustaining any material injury.†

(2767.) *Barley* is extensively attacked in Sweden by the *Chlorops frit,* which destroys the principal stem; and Mr Chrisp, Rugley, near Alnwick, mentions the effects of an insect on barley, its "larva living in the sheath and feeding upon the ear previous to its appearance from the shot blade. Every year I have observed more or less injury to the crop by the same cause, but it appears most serious in a late or wet cold year upon stiff soils."‡ Barley, as well as oats, have blacked heads, occasioned by the *Uredo segetum.*

(2768.) But of all species of grain, *wheat* is most affected by casualties. It is attacked by the larva of the wheat dart-moth, *Agrostis tritici,* under ground,

* See Holdich's Weeds of Agriculture. Edited by G. Sinclair, 1825.

† See Quarterly Journal of Agriculture, vol. xiii. p. 162-7, for an account of this insect.

‡ Ibid. vol. xiii. p. 208.

feeding on the radical portion of the plant, in autumn, and never appearing above the surface except in the night. The wheat stem-fly, *Chlorops pumilionis*, *c*, fig. 474, derives its specific name, which signifies a dwarf, in consequence of the effects it produces on the plants it attacks, which, having their central shoots destroyed, throw out many lateral ones, and seldom attain any considerable height. " The colour of the fly *c* is black," says Mr Duncan, " the under side of the head and 2 narrow longitudinal lines in the thorax yellow; under side of the body pale yellow, with 2 black spots on the mesosterum, halteres white; the legs ash grey, and black at the tips." The maggot is small and white, the pupa yellow, smooth, and shining, rather more than the $\frac{1}{12}$th part of an inch in length.* The wheat-fly, *Cecidomyia tritici*, has been much talked of, for the last dozen years, and its habits are now pretty well known. It may be seen at *d*, fig. 394, greatly magnified, its natural size being represented by the cross lines beside the figure, and which does not exceed a line in length, that is, about the $\frac{1}{12}$th of an inch, resembling a small gnat or midge. Mr Duncan describes it as having " the colour of the whole body reddish yellow, deepest on the thorax; antennæ dusky, eyes black; legs pale reddish yellow; wings longer than the body, rounded at the tip, of a whitish hue, with iridescent reflections. The antennæ of the female *b* consists of 13 sub-ovate joints, those of the male *c* of 25 globose remote joints."—" The fly makes its appearance in wheat fields just about the time the ear is beginning to emerge from its leafy envelop, most commonly in the early part of June. In a calm evening they fly about in little undulatory clouds. An ear just emerging from the sheath is generally preferred as a receptacle for the eggs. They are introduced, by means of the ovipositor, into the floret, being usually placed upon the interior valvule of the corolla, just above the stigmata. The eggs are placed in clusters from 2 to 20, and are hatched in 10 or 14 days, and after the larvæ begin to feed on the pollen, they acquire a saffron-yellow hue *e*, fig. 394. Minute as these creatures are, when it is considered that 47 have been counted in one floret, it is easy to perceive how they must interfere with its impregnation. It is not improbable, as they increase in size, that they likewise attack the milky pulp destined for the nutriment of the grain. At all events, the grain shrivels and decays, to the grievous disappointment of the husbandman. The loss sustained by the farmers of the Carse of Gowrie in 1829, was estimated by Mr Archibald Gorrie, Annat Cottage, from 6½ to 16 bushels per imperial acre."—" Nature herself has provided a considerable check in the multiplication of these flies, by making them the prey of no fewer than 3 kinds of ichneumons, viz.:—*Encyrtus inserens*, about half the length of the wheat-fly; another, *Platygaster tipulæ*, which commits its eggs to the larva of the wheat-fly; and the third, *Eurytoma penetrans*. Some of these ichneumons appear in great numbers where the fly abounds, and multitudes must become their victims."† Mr Westwood describes a species of *Chlorops*, *glabra*, which was found in great profusion amongst wheat, whilst removing it from the stack in which it had stood the winter.‡ The young wheat plant is not unfrequently destroyed by the wire-worm, which cuts over the plants by the ground; and it is apt to be eaten down by a slug, especially when sown in autumn after clover or beans. Many remedies have been proposed, but I believe there is none effective against the wire-

* Quarterly Journal of Agriculture, vol. xii. p. 121.

† See Mr Duncan's entire account of this insect, of which the above is but a small portion in vol. xi. p. 372-8, of the Quarterly Journal of Agriculture.

‡ Loudon's Gardener's Magazine, vol. xiii. p. 289.

worm except hand-picking. Young ducks would be effectual against the slug, could they be provided in sufficient numbers, and sowing quicklime over them has been tried. Wheat is destroyed in the grain by insects as well as in the plant. The *Calandra granaria*, grain-weevil, has obtained a bad eminence amongst its compeers, of which this is the only species native of this country. It is about 2 lines in length, the body being of a brown pitchy colour when mature, legs rusty red, antenna scarcely longer than the head and rostrum. The female frequents granaries, and on burying herself in the heap, deposites her eggs in the grain, which are usually hatched in a few days. The larva is small and vermiform, and of a white colour, furnished with a pair of strong jaws, with which it erodes the farinaceous substance of the grain, until it is entirely consumed, when it assumes the pupa state, which is also white, and then lies like a kernel in a nut within the interior shell of the grain. When perfect, the insect gnaws a hole through the shell and makes its escape. So numerous are the eggs laid by a single female, that her descendants, in a single season, have been estimated to amount to 23,600 individuals. The grains they inhabit *appear* sound, but will be found to float if put in water; if so light, they may be winnowed out, but still there is no certainty of them all being disposed of in this way. The best period for destroying the beetle is before she lays her eggs, and to effect this, a plan has been recommended, which is, to place a heap of barley, the grain they are fondest of, in the granary, and then shovel over the wheat repeatedly, so as to disturb and annoy the creatures, when they will take refuge in the heap of barley, in which they can be destroyed with boiling water.* Besides insects, the wheat-plant is made the prey of parasitic fungi. Mildew and rust constitute the two diseases of this nature which affect the straw of the wheat-plant. Mildew or blight, *Puccinia graminis*, forming blackish-brown parallel lines upon the straw, seems to affect the entire plant, so that when generally diffused over it, to deprive its sap of the power to form seed in a healthy state, and hence the grain is either very much shrivelled when it is formed, or no grain is formed at all. So generally did this disease affect the wheat crop of Berwickshire in three successive years, 1810, 1811, and 1812, I think, when the price of wheat was at the highest pitch during the year, that in many instances it was not considered worth while to thrash it. There is a prevailing opinion in England, where the berberry bush is common in hedges, that it is the cause of mildew in wheat. That this is too broad an assertion is obvious from the fact, that wheat suffers from mildew in places where not a berberry bush is within many miles of it. I question if there was a berberry bush in the entire district of Berwickshire affected by the blight in the years above mentioned; and it seems to me preposterous to suppose that the mildew of the berbery, *Erysipha berberidia*, can be converted into the *Puccinia graminis* of the wheat, by any possible mode of transference from the one plant to the other. It is probable, however, that similar external circumstances may superinduce both plants to be affected with mildew simultaneously; and if, as Professor Hunslow plainly states, "one of the berberry blights, *Æcidium berberides*, can only be a different form of development belonging to the same species of fungus which produces the mildew," there is no absurdity in the supposition, that it is only this species, and not both, of the berberry mildew which infects wheat; nor is it absurd to suppose, that the berberry is more liable to be affected by mildew than wheat, and,

* See Quarterly Journal of Agriculture, vol. ix. p. 3–6, for an interesting account of this insect by Mr Duncan.

when violently attacked itself, the disease may infect the wheat in its immediate vicinity. Rust is chiefly found scattered over the inner surface of the outer chaff scales, the skin of which is raised into blisters, which are most frequently of an oval form. It is occasioned by the parasitic fungus *Uredo rubigo*, and is of an orange-yellow colour.* Smut is so well understood in its origin and effect on the wheat crop, as to require no description. It is occasioned by the parasitic fungus *Uredo caries*, a brownish-black dust, filling the kernel itself of the wheat with a fetid greasy powder. Though not externally conspicuous, it causes the seed to swell, and then to look diseased; and when broken in thrashing affects the whole grain. Of yellow rust, brown mildew, and black smut, mildew is the worst. Rust affects the straw, and, on disappearing, the straw is not materially injured, and it is only the rusted grains which suffer. Smut destroys the grains it attacks, but leaves the straw untouched. Mildew, on the other hand, affects both straw and grain, depriving the grain of its subsistence, and the straw assumes a pale sickly hue, and is easily cut to pieces on being thrashed. I have seen a field of wheat so much affected with yellow rust at the time of weeding, before the ear had shot out, that the clothes of the field-workers who weeded the crop were covered with orange dust as far as the knee, but the crop was nevertheless good. Mr John Lawson junior, of Elgin, examined the structure of smut-balls in comparison with sound grain, and the

Fig. 490.

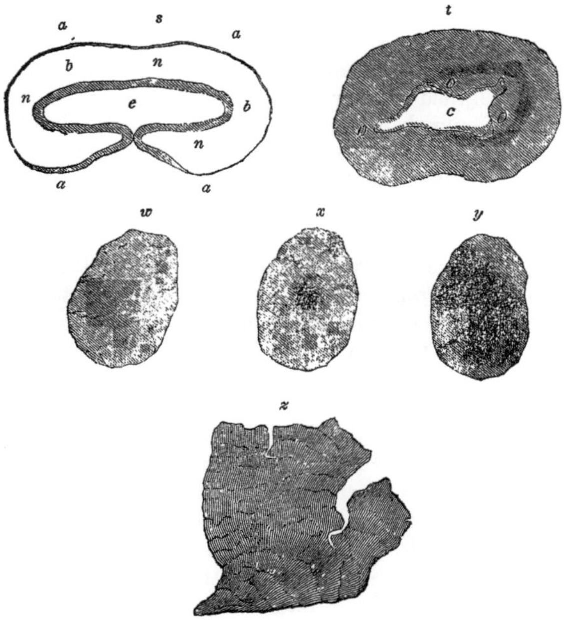

THE STRUCTURE OF SOUND WHEAT AND SMUT-BALL COMPARED.

different appearances are shewn in fig. 490, where *s* is a longitudinal section of a sound grain of wheat in its stage of growth, when the anthers have just pro-

* See Journal of the Royal English Agricultural Society, vol. ii. p. 9 and 220, in which Professor Henslow endeavours to shew that the *Uredo rubigo* and *Uredo linearis*, and the *Puccinia graminis*, are all one species of parasitic fungus. See also Banks on Blight in Wheat.

truded beyond the corolla; *e* is an empty space lined internally with a greenish border *n n n*. As the grain advances, the space *e* contracts, and its substance *b b* lying between the green border *n* and the outer green cuticle *a*, becomes filled with milky juice; *t* is a longitudinal section of a smut-ball taken when the stamens are fully formed within the corolla, for the anthers never protrude beyond the corolla when the grain is affected with smut; *c* is an empty space surrounded by a dark green substance, which extends to the outer cuticle. In a very short time the whole interior of the smut-ball changes from green to white, as at *w*, the outer cuticle continuing green. The white substance soon has a black speck in its centre, as at *x*, which gradually spreads through the ball, as at *y*; and as the ball still advances to maturity, the dark green cuticle changes to brown, as at *z*. The green substance occupying the place of the milky juice, at once explains the difference betwixt a smut-ball and a sound grain.* How do smut and rust originate? It is easy to conceive the spongioles of the roots of the wheat-plant taking up the sporules of the parasitic fungus into the circulation of the sap, but it is not so easy to explain how mildew or its sporules attach themselves to the exterior of straw. Wheat sometimes exhibits blacked heads, like oats and barley, occasioned by the same *Uredo segetum*.

(2769.) *Pease and beans* are destroyed in store by the *Bruchus granarius*, pulse beetle. The *Bruchus pisi*, rather more than 2 lines in length, lays an egg in each pea, when the young pods are forming, till the contents of her ovarium are exhausted, but fortunately this insect is rare in this country. Worm-eaten pease are occasioned by the larvæ of a saw-fly. It is no uncommon occurrence for the tops of bean-stalks to be covered with a black *aphis* vulgarly named the *collier*. The quickest way of getting rid of the collier is to cut off the tops of the bean-stalks, which field-workers can easily do with a sickle.

(2770.) *Clover* is injured by insects. The red clover, *Trifolium pratense*, is affected by a weevil named *Apion apricans*, in length about 1½ line, and the colour of the body is black. By the time the heads of the red clover are ready to flower, the apion deposites her eggs on the calyx of the florets. As soon as they are hatched, the larva, an extremely minute whitish worm, with a black head, eats its way through the base of the floret, and consumes the rudiments of the future seed. So extensive is the injury occasioned at times by this creature, that, in 1798, a piece of ground measuring 4½ acres of red clover produced 16¾ bushels of seed, which was worth L.41 : 17 : 6; when the like extent of ground in 1800, produced only 7½ bushels of seed, worth L.18, 15s.; thus this little insect occasioned a loss of L.23 : 2 : 6 on the produce of 4½ acres of land. Another species, *Apion flavipes*, about 1¼ line in length, with a black shining body, attacks the white Dutch clover, *Trifolium repens*, and as the insect is very common, the cultivators of white clover would require to be on their guard against it.† Slugs devour the broad leaves of red clover.

(2771.) *Pasture grass* is injured by a host of insects. The grub of the meadow crane-fly, *Tipula oleracea*, destroys the roots of grass as well as of oats. The year 1762 was called the *wormy* year, in consequence of the devastations occasioned by this grub in Selkirkshire. It again appeared in 1802, 1812, 1824, and 1826, and in Peeblesshire in 1830.‡ The ear-beaked weevil, *Otiorhyncus sulcatus*, about 5 lines in length, of a brownish-black colour, and in-

* See Quarterly Journal of Agriculture, vol. ix. p. 269-72.

† Ibid. p. 15-18.

‡ New Statistical Account of Scotland, *Selkirkshire*, *Yarrow*, p. 41.

capable of flight, in consequence of the junction of the wing-cases, produces a larva nearly half an inch long of a whitish colour, thick and fleshy, and thinly beset with long bristles, which is, for the most part, subterranean, and lives indiscriminately on the roots of all gramineous plants. The common kinds of ants, *Formica fusca* and *rufa*, sometimes almost usurp the entire of dry pastures. The caterpillars of several butterflies also destroy pasture plants. The meadow brown-butterfly, *Hipparchia janira*, whose wings expand nearly 2 inches, produces a light-green caterpillar, with a white line along each side, which prefers for its food the smooth-stacked meadow-grass, *Poa pratensis*, one of the most nutritious grasses for cattle; and the caterpillar of the large heath-butterfly, *Hipparchia tithonus*, considerably less than the preceding, is of a green colour, with a reddish line on each side, and a brown head, and feeds on the annual meadow-grass, *Poa annua*, which forms the chief covering of our meadows and pasture lands. The caterpillar of the antler-moth, *Charæas graminis*, is brown or blackish, with light-yellowish stripes along the back and sides, and attain about 1 inch in length. It lives under ground, and feeds on the roots of grass, and the injury which it does to pasture-land in hills is sometimes very great, but it avoids low-lying damp meadows. The effects of its ravages are very similar to those produced by the burning of heath. But the greatest devastation of this kind is committed by the larvæ of the common cock-chafer, *Melolontha vulgaris*, an insect but little known in Scotland, though sometimes it occurs in countless myriads in England and Ireland. There the beetle "wheels its droning flight" in the summer twilight. Fig. 491 represents 2 females *a* and *b*, about 1 inch in length, oblong and convex, the head reddish in front, the hinder-part

Fig. 491.

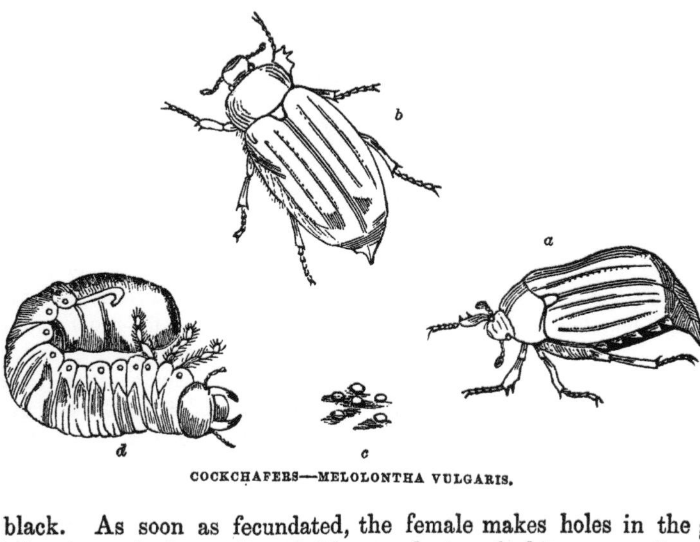

COCKCHAFERS—MELOLONTHA VULGARIS.

pitchy black. As soon as fecundated, the female makes holes in the ground about ½ foot beneath the surface, and lays a cluster of white eggs *c*, tinged with dirty yellow at the bottom of each hole, and she lays a great many in all. Thel arvæ *d*, which proceed from them, are of a dingy-white or yellowish colour, and soft and inactive. They are 1½ inch in length, and furnished with a pair of powerful mandibles. The perfect insect is short-lived, but the larvæ survive for 3 years, passing into a state of perfect repose in winter, and recovering their activity in spring. So completely do they sever the pasture from the soil that the turf may be rolled up in large sheets. Another chafer, named *Amphimalla solstitialis*, produces a grub very similar in its habits to the one described, though smaller and more restricted in its distribution.*

(2772.) The *thorn hedge* is subject to the attack of many insects, though the effects are different. One of these is the black-veined white, or hawthorn butterfly, *Pieris cratægi*, *a*, fig. 492, an insect unknown in Scotland, but exists

Fig. 492.

THE HAWTHORN BUTTERFLY—PIERIS CRATÆGI.

in England, and is at times so numerous on the Continent, that its flight has been mistaken for a shower of snow. The caterpillar is seen at *b*, of a dull-yellow colour at first, but changes with moultings, and is produced from yellow-coloured cylindrical eggs laid on hawthorn shoots, and rendered water-proof by a coating of strong varnish. It devours the leaves, and, while employed in this, lives in society under the protection of a silken web. Other lepidopterous insects disfigure our hedges by defoliation, such as the figure-of-8, or black-thorn moth, *Episema cærulocephala*, whose caterpillar is 2 inches long, of a yellowish-green colour, with 3 pale stripes. A still greater pest to our hedges is the brown-tailed moth, *Porthesia auriflua*, whose caterpillars, of a dusky colour, with 2 red lines on the back, and a white streak on the sides interrupted at intervals, occur sometimes in such numbers that in 1782 their webs were gathered in one day by the people of a single parish in the neighbourhood of London, that of Clapham, to the amount of 80 bushels.† The *Selandria æthiops*, an insect

* See Quarterly Journal of Agriculture, vol. ix. p. 565–72.

† Ibid. vol. xiii. p. 58, 84, and 155.

nearly allied to the saw-flies, annoys our thorn-hedges. Its whole body is shining and black, the wings obscure hyaline, with the nervures and stigma black. Length about 2½ lines. The caterpillar is always covered with a viscid slime, which it has the power of secreting in abundance, apparently from the joints of the body. It has the singular property of inflating its body. The secretion is rather glutinous, and has a strong scent, resembling that of red ink. It attacks the upper surface of the leaves, consuming the pulpy substance, but leaving the nervures, so that the leaves are completely skeletonized, appearing at a distance as if they had been scorched by lightning. Its chief season of activity seems to be during the night, as it is usually seen in the day-time resting on a leaf.* The currant or magpie moth, *Abraxus grossulariata*, also does injury to thorn-hedges, and being a beautiful moth, I have given a picture of it. In fig. 493,

Fig. 493.

THE CURRANT OR MAGPIE MOTH—ABRAXUS GROSSULARIATA.

a is the moth in flight; *b* shews the motion of its geometer caterpillar; and *c* the manner in which it suspends itself from a leaf by a thread.†

(2773.) [The *double mould-board plough* is an implement essentially requisite in the cultivation of the turnip and potato crop. When duly constructed, it is highly efficient in the formation of the drills or ridgelets for either of these crops, setting up at each turn the half of a ridgelet on each side, while the common plough, or the one-horse plough so much employed for this purpose,

* See Quarterly Journal of Agriculture, vol. xii. p. 253.

† Journal of Agriculture for January 1844, p. 242.

sets up only a half ridgelet at each turn, doing, therefore, but half the work. In a variety of forms also, it is much employed in the earthing up of the potato crop; for this purpose it is frequently made of wood, but in all cases the iron plough is to be recommended.

(2774.) Fig. 494 is a representation of a common double mould-board iron plough equipped for the purpose of earthing up. The frame-work of it is pretty

Fig. 494.

THE DOUBLE MOULD-BOARD PLOUGH.

much in form of the common plough, except that the beam *a* lies right in the central line of the whole plough. The bridle *b* is variously formed according to the taste of the maker, but always possessing the properties of varying the point of draught upward and downward, as well as right and left. The breast *d* is a shield forming part of the cast-iron body-frame afterwards described. The share *e* is plain on both sides, spear-pointed, and set upon the head of the cast-iron body-frame. The right and left mould-boards *f, f,* are hinged to the edge of the shield *d* with drawing hinge-pins, aud they are supported behind by a jointed iron-strap affixed to the back of each mould-board, and which slide through a socket in the body-frame, where the tails of both straps are secured by means of a pinching-screw, setting the mould-boards at any required width behind. The handles *g* are bolted on each side of the beam, as seen at *f f*, and are supported near the helves by the usual stretcher and bow. The dimensions of this plough are,—from the breast *d*, to the point of the beam, 3 feet 6 inches, from *d* to extremity of mould-board *f*, 2 feet 6 inches, and from *d* to end of helve 6 feet 6 inches. The height of the mould-board, where it joins the shield, is 12 inches, and at the point *f* 10 inches; length of share 16 inches. The mould-boards of such ploughs are liable to great variation in their form; some of them have little or no twist, and others variously contorted. Those of the present figure have been selected as possessing all the requisite qualifications for an earthing-up plough. At the fore-edge, where they join the shield, the surface is nearly in a straight line, and along the upper edge they are slightly convex; from these two lines they twist gradually, rounding away below towards the tail, so as to leave the furrow of a round-bottomed trough shape. Where the double mould-board is employed for forming the ridgelets, the mould-board is made to fit the shield *d*, as in the figure; it then stretches away to a length of 2 feet 6 inches along the upper edge, the point *f* being at a height varying from 11 to 14 inches above the sole-line. At this point the depth of the mould-board is only 6 inches, so that the lower edge runs off at a considerable elevation, and the surface having not more than 3 inches of twist, it is the lower edge only of the board that effects the purpose of laying up the earth to form the ridgelet. In working the plough, for the purpose of forming drills, there is frequently a marking-bar jointed to the beam immediately before the breast *d;* the bar folds to either side, and having an adjustable double-edged scraper fitted to it, a rut is drawn on the surface at the proper distance for the centre of the next furrow.

(2775.) The plough just described and represented, is convertible into a scuffling or cleaning plough, or horse-hoe. To effect this, the hinge-pins of the mould-boards are withdrawn, and the mould-boards removed when we have an implement represented by fig. 495, which exhibits the body-parts of the

Fig. 495.

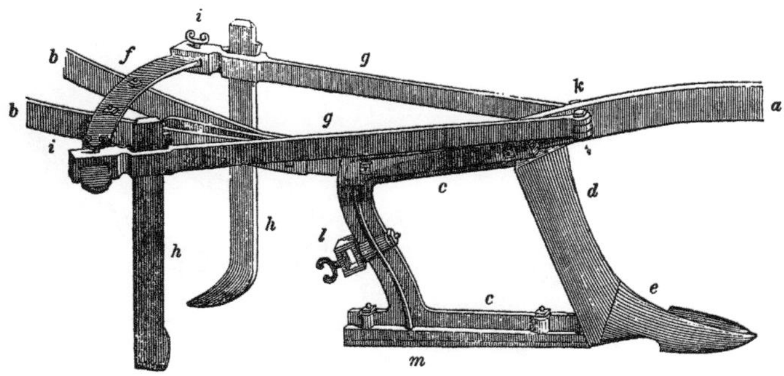

THE BODY OF THE DOUBLE MOULD-BOARD ALTERED TO A SCUFFLING PLOUGH.

same plough upon an enlarged scale, and from which, portions of the beam and handles are cut off. Here *a* is the remaining portion of the beam, and *b b* those of the handles exhibiting also their junction with the tail of the beam. The body frame *c c*, is of an irregular rhomboidal form, whereof the front bar forms the shield; its breadth from right to left behind is 3 inches, running off forward to a sharp edge, and is hollowed out hehind. The top bar *c*, by which it is bolted to the beam, is $2\frac{1}{2}$ inches broad, and $\frac{1}{2}$ inch thick; the sole bar *c* is formed flat below, to receive the plain sole-shoe *m*, and the hind bar is formed to receive the malleable iron socket *l*, through which the tails of the mould-board straps are passed, and secured by the pinching screw *l*, when these are in use. To complete the implement for the purpose of scuffling, the two wing bars *g g* are jointed to a stud that projects from the beam on each side at *k*. A quadrant bar *f*, 2 feet long, and $1\frac{1}{2}$ by $\frac{1}{2}$ inch is attached by bolts to the two stilts at *f*, and the ends of the wing-bars having a mortise formed to receive the quadrant, are moved upon this to any required width, and secured by the screws *i i*. A second mortise is punched in each wing-bar to receive the scuffling coulters *h h;* these are 2 inches broad by $\frac{1}{2}$ inch thick, thinned off to a knife-edge in front, and bent inward below till the points stand 6 inches to the right and left of the shanks. A double feathered share *e* is now fitted to the head of the body-frame, which completes this simple horse-hoe, and the change from the one state to the other is effected in a few minutes, for, in returning it to the double mould-board state, it is only necessary to remove the scufflers and the feathered share. The dimensions of the body are, height at the breast from the sole to the top of the beam 14 inches, length of sole 2 feet 6 inches, including the feathered share. The effect of this horse-hoe in the soil is to loosen the earth between the rows of drills, or if foul, to under-cut all the weeds that exist in that space, or to such breadth as the two scufflers may be set; the upright part of these coulters performing a species of paring along the sides of the two contiguous rows. If the land is in good order, and tolerably clean, stirring it with this scuffler will be sufficient; but if overrun with weeds, one or other of the drill harrows or grubber will be found necessary to prevent a re-vegetation of the weeds, and the following will be found to answer this purpose well.

(2776.) The common *drill-grubber*, fig. 496, is a light and convenient implement drawn by one horse. It consists of a central beam *a b c*, the neck part

Fig. 496.

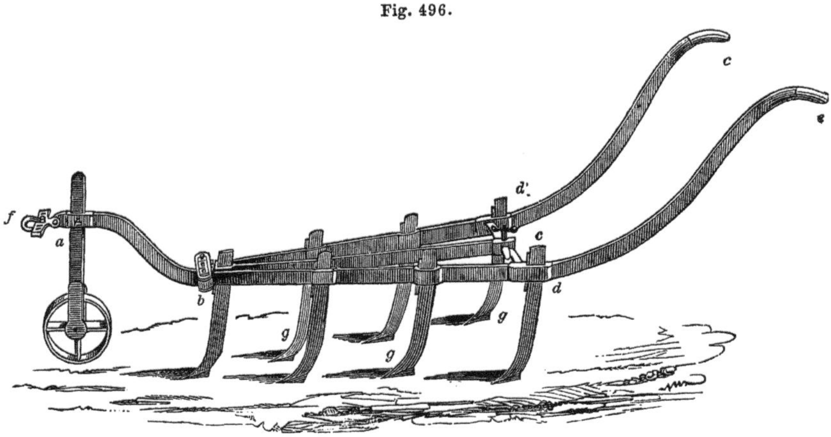

THE COMMON DRILL-GRUBBER.

of which *a b* is 18 inches long, the body part 3 feet 6 inches; and of the two wings *b d*, which are extended to *c c*, forming the handles, the length from *d* to *c* is 3 feet 4 inches. The neck part of the beam is 1½ inch square, and peened, or rounded, and this strength is carried past the first tine; the remainder of this bar, as well as of the wings, is 1½ inch deep by ½ inch thick, the handles becoming lighter backward. The beam is punched at the front for the passage of the stem of the wheel, and at *b* for the fixing of the two joint-plates for the wings, as well as for the front tine; and it is also perforated horizontally at the end *c*, for the quadrants of the wings. The wing-bars require to be very neatly forged in forming the swells, in which the tine holes are to be punched, and also for the joint at *b*, where they are hinged to the beam, between the two joint-plates, which, being riveted dead upon the beam, leave a chamber on each side for the reception of the ends of the wing-bars, and through these their joint-bolts are passed. The wing-bars are each furnished with a quadrant-bar riveted into the wings at *d d*; the tail of the quadrants passing through the mortise at *c* are secured by a pinching-screw fixing the wings at any required width. To the point of the beam is affixed a simple bridle *f* with a cross-web and shackle, giving a small range of yoke right and left; the rise of this point is 10 inches above the line of the body of the beam. The front wheel, whose office is to regulate the depth of the grubbing, is usually 8 or 9 inches diameter, set in the sheers of the stem, which may be 20 inches long, and is 1½ inch broad by ½ inch thick. The tines *g* are 15 inches long, the body being 1½ inch broad by ½ inch thick, forged with duck-feet not exceeding 2½ inches broad, and pointing slightly forward. In many localities this implement is used for all the purposes of horse-hoeing, except the process of paring or of earthing up, and having cheapness as well as utility as a recommendation, it is very generally approved of. It is, however, subject to variety in the different districts where it is employed; in some it is shortened to five tines, in others lengthened out to nine, and in many cases the tines are plain-pointed, or not exceeding 1 inch broad. It is frequently also made with the tines standing in *zig-zag* position; but except in the second pair of tines, this is of little importance, as those behind the second are sufficiently far apart to prevent them getting choked with weeds. The price of this grubber is about L.2, 10s.

(2777.) In this class of implements, we find a very handsomely constructed one, known as *Wilkie's drill grubber and harrow*, which is here represented in fig. 497. The implement is, however, of older date, and seems to have been invented by a Banffshire farmer,* the original having been constructed with wooden framing until Mr Wilkie adopted the iron instead of wood. This implement is constructed with a beam *a b*, and a pair of handles *c c* attached to the

Fig. 497.

WILKIE DRILL-GRUBBER AND HARROW.

tail of the beam, one on each side, in the way formerly shewn in fig. 495. It has no proper body-frame, but is merely a skeleton, the grubbing parts of it being the three tines or coulters *d, e, f*. The foremost of them *d* is set in a coulter-box in the beam, the two others, *e* and *f*, are continuations of the two wings, which are hinge-jointed to the side of the handles as at *g*, and where they are kneed downward at *h*, they are perforated for a quadrant bar, on which they are moved outward or inward, and are secured by the pinching screws at *h*. The front tine terminates in a double spreading feather or duck's-foot point of about 11 inches in length, and 8 inches in breadth at the heel; the two back tines are flat on the outward sides and feathered inward; their effect on the soil is therefore somewhat similar to that of the scuffler, fig. 495, paring and undercutting; but the implement is furnished with an appendage in the attached harrow *i*, which completes the operation at one turn. The harrow consists of two bars jointed to the middle of the quadrant bar which sustains the back grubber tines, and are suspended by a small quadrant attached to each of the bars *i*, from the first stretcher of the handles at *k*, and each of the harrow-bars carries three common harrow tines, but somewhat longer than usual. The harrow is, besides, capable of adjustment to depth by means of its suspenders, and to breadth by means of its two small quadrant-bars. The regulation of depth is aided by the wheel

* Agricultural Report for Banffshire, p. 135, 1811.

l, which may be from 8 to 10 inches in diameter, hung in the sheers *n*, which is jointed to the beam at *a*, and is capable of being shifted up or down upon the cross-head *o* of the beam, and fixed by a bolt at *n*, passing through the perforations of the cross-head, to which also a shackle and hook is attached for the draught.

(2778.) It requires no demonstration to shew, that the principle of this instrument is good—grubbing or scarifying to undercut all weeds, which are immediately brought to the surface by the action of the harrow, to wither and die, —but it is believed, that defects of construction exist in it that in some measure mar its utility. In stiff soils, the broad-feather shares will with difficulty be kept in the ground; and, from their great length and breadth, will have the effect of consolidating that part of it which they pass over, into a hard crust. The harrow is an important part of the implement, but adds considerably to the draught; and the implement, upon the whole, is too heavy for one horse being able to produce efficient work with it; by lightening the entire structure, and altering the form of the tine, it might be rendered a very useful horse-hoe. Its general dimensions, as constructed by Mr Wilkie, may be shortly stated. Length of beam from cross head to coulter-box 2 feet 10 inches; thence to quadrant of the back tines 1 foot 10 inches, and thence to end of handles 4 feet 6 inches; height of beam at coulter-box 16 inches, at quadrant 14 inches, and at the point 18 inches. The depth of the beam at the coulter-box is 2½ inches, and its breadth 1 inch, from which point it tapers off forward and backward; the handles are 1¾ inch in depth, by ⅜ inch in breadth. The length of the harrow-bars is 2 feet 9 inches, and the length of the tines 10 inches; the weight is 145 lb.; and the price of the implement is L.4, 15s.

(2779.) The foregoing constitute a series of green-crop horse-hoeing implements, that possess all the principal points requisite for this operation; but there are numerous varieties of all the types here exhibited, though all are referable to one or other of them. It is or ought to be an essential point in all implements of this kind, that they possess a principle of expansion and contraction, to suit the different widths of drills; and, with few exceptions, this is the case. In many of those in use, however, the mode of expansion is attended with an inconvenience arising from the shortness of the expanding wings, which throws the coulters or tines out of parallelism, and thus proves detrimental to their perfect working in the soil. This inconvenience is more felt in those implements which have their tines formed like those of fig. 497, and least of all when the scuffling tines are employed as in fig. 495. Mr Wilkie, with his usual ingenuity, has invented, some years ago, an implement of this class, in which the expansions and contractions are effected by means of a parallel motion applied to the tines, as exhibited in the following figure.

(2780.) Fig. 498 exhibits Mr Wilkie's *horse-hoe* with parallel motion; it is very similar in figure to the last, but the two back tines have their tails jointed at *a*, *b*, *c*, *d*, to two transverse parallel bars, which traverse, to a small extent, upon pivots placed in the middle of their length, attached to the tail of the beam. By moving these bars upon their pivots, from the position of a right angle with the central line of the beam, the one tine is pushed forward and the other backward, which must cause the points *a*, *b*, *c*, *d*, to approach the central line, and along with these, the points of the tines, preserving a perfect parallelism, and capable of being secured at any required width by pinching screws. This is the most perfect mode of adjustment for the tines of a hoe of this construction—three-tined—but it does not apply to those with more than three, and is, withal, perhaps too refined for a field implement. The self-cleansing form of the tine which is exhibited here,

Fig. 498.

WILKIE'S HORSE-HOE WITH PARALLEL MOTION.

and in fig. 497, in the two back tines of each, has been often dwelt upon as of much importance; but the truth of the matter seems to be, that though that particular form is beneficial in the great field-grubber, it seldom occurs, and ought never to be the case, that a drilled green crop is so overrun with weeds as to require a self-cleansing tine.—J. S.]

---

69. OF HAY-MAKING.

' —— as they rake the green appearing ground,
And drive the dusky wave along the mead,
The russet hay-cock rises thick behind,
In order gay."

THOMSON.

(2781.) It has often been alleged, that Scottish farmers display very little skill in the making of hay. Ready as I am to vindicate the general excellence of Scottish husbandry, I must own that in this respect the allegation is too well founded. As hay-making is usually performed by Scottish farmers, one would be induced to conclude that the period of conducting it had arrived before it was expected, and the time and attention spent upon it operated as a hindrance to more important work. The usual practice is, to delay cutting down the grass until it has passed its most succulent period—to allow it to lie on the ground when cut till it is either bleached by the rain, scorched by the sun, or the new growth of the mown grass has penetrated through the prostrate swathes—to put the thus weather-beaten swathes together as fast as possible, into as large ricks as can be made without the hay being heated in them—and to allow those ricks to stand on the ground till the grass under them is destroyed. The grass thus treated is then said to be good hay!

(2782.) Now, reasons explanatory, though not vindicatory, of this mode of making hay in Scotland can be given, and they are these:—There is a prevailing desire in farmers to save as much rye-grass seed as will sow the quantity of land that had been fallowed the preceding year, and which extends to the fourth, fifth, or sixth part of the extent of the farm, according to the rotation of crops pursued in it. To obtain this object, it is necessary to allow the rye-grass to attain a certain degree of maturity before it is cut down, and an easy test applied to ascertain that degree of maturity is, to sweep the hat smartly along the heads of the plants, and see whether it has caught any seeds. When rye-grass has attained this degree of maturity, the clover has passed its most succulent state, so that the crop altogether is too old before it is cut down, except when the object is to save rye-grass seed. To give time to the rye-grass seed to won, the swathes are allowed to lie spread out for some days before being gathered into ricks or colls. Another reason is, that hay-making arriving at the same time as singling of turnips, the latter receives more attention than the former, on account of its superior importance.

(2783.) These reasons, though sufficient to explain the object in view, and to describe the circumstances in which it must be attained, are no justification for pursuing so slovenly a mode of making hay, because the object aimed at can be attained by a different and better procedure; which is this:—If it is desired to raise as much rye-grass seed as annually to supply the wants of the farm, a part of the grass intended for hay should be allotted for the purpose. The quantity of rye-grass seed usually required is 1 bushel per imperial acre, and as a crop of mixed clover and rye-grass varies in yield from 25 to 40 bushels of rye-grass seed per acre, the quantity of ground to be thus apportioned may be previously ascertained; and surely this is a much more rational procedure than inflicting injury on the whole crop of hay for the sake of preserving a few bushels of seed. Nay, more, after the crop of hay has thus been injured, much of the rye-grass seed is, after all, allowed to be shed on the ground by the dilatory mode of making hay usually pursued. As to the interference of hay-making with the time of turnip-singling, it is obvious that if the grass were cut for hay earlier than it is—and it is in perfection for cutting before the rye-grass seed is formed—the interference would in a great measure be avoided. The hay, in fact, might be placed past all danger before the turnips were ready for singling, as it might be conducted between the sowing and singling of the turnips, a period generally extending from a fortnight to three weeks.

(2784.) Hay is made both of sown, and of natural meadow-grasses.

The sown grasses are employed for hay in Scotland, and of these the hay consists of red-clover, *Trifolium pratense*, and rye-grass, *Lolium perenne;* for although the white-clover, *Trifolium repens*, is sown along with the seeds of the other two, it scarcely forms a part of the first year's grass, and constitutes no part of the hay, which is always taken from the grass of the first year. As hay is thus taken from the first year's grass, it matters not whether the rye-grass made into hay is annual or perennial. The annual yields the heavier crop, but the perennial the finer quality of hay. The natural grasses constitute the hay of England and Ireland. These two sorts of hay are certainly very different in appearance, the sown grasses shewing the strong and stiff stems of the red-clover and rye-grass, and especially when the rye-grass is annual, while the hay from the natural grasses is soft and woolly to the feel, and more odorous to the scent, because the sweet-scented vernal-grass, *Anthoxanthum odoratum*, always forms a component part. In so far as their nutritive properties are concerned, if both are equally well made, there will probably be no material difference; but this fact has been established in Scotland, that the sown grasses are more nutritive for young stock, both sheep and cattle, than natural grasses, and for that reason we may hold it as true, that their hay will also be more nutritious for young stock; and in like manner, the hay of natural grass should be more nourishing to old stock than from that which is sown; and hence natural hay is best for cows.

(2785.) I have heard farmers express the opinion that sown grasses require a different treatment on being made into hay than natural grasses. If the object is to obtain rye-grass seed while hay is being made, then, of course, the two processes should be different; but if the object is to make the best hay from both the substances, then I cannot see why the processes should be different. On the contrary, the nature of all the plants employed being the same, the same treatment should produce in all the same results; and as the art of hay-making is merely to expel the water which the plants contain without injury to their texture, the only danger to be apprehended is excessive fermentation, which is easily excited in warm weather, and will proceed to a destructive extent, if not subjected to control. Still hay-making varies according to the means used for conducting it; for if manual labour alone is employed, one process should be adopted, but when mechanical assistance is received, the process should be modified accordingly.

(2786.) First, then, as to *hay-making with manual labour alone.* The implements required for the purpose are few and simple. The grass is cut with the common scythe, fig. 452; and the cutting is either let to labourers by the piece, or the ploughmen of the farm do it, should there be spare time from horse-labour between the sowing of the turnips and the hay-harvest. The grass will be better and more expeditiously cut down if let by the piece, as the contractors will exert themselves more, and work more hours, than ploughmen who have charge of horses can be expected to do. The usual cost of cutting grass for hay is 2s. 6d. or 3s. per imperial acre. I once let a small patch of 6 acres of clover, to cut for hay, to a stout fellow, who undertook to do it for 2s. 6d. the acre, but after the first half-day's work he relinquished the agreement as being too low for him. And so it really proved; for on examination the clover was found so luxuriant that it was *kneed* down, that is, its lower part was lying upon the ground, and its upper part only indicated itself as the growing crop. He agreed to take 2s. 6d. a day, and after toiling hard for his money, the cutting cost me 5s. the acre. Two or three men form a gang or head at cutting grass, or 2 heads of 2 men each, when the extent of ground is considerable. On a 500 acre farm, under the 5-course rotation, perhaps 20 acres of the 100 acres of new grass will be made into hay. On commencing to cut a field, the direction towards which the clover leans, or should it be thin and upright, the quarter of the wind, which always influences the direction of thin grass, should be attended to; and in both cases the grass should *lean away from the mower*. It always makes the best work for the grass to be mowed *across* the ridges. It is fair work for 1 man to mow 1 acre every day; and I may here remark, it is no good sign of the weight of the crop if the mowers go over more ground every day. The other implements used in manual hay-making are forks and rakes. Forks are shewn in fig. 279, and rakes are figured and described below; and of the two kinds I prefer the right-hand figure, as being the neatest and lightest.

(2787.) Allowing that the mowers have started early every morning, and there is nothing to prevent them doing so except heavy rain, the grass which had been cut down in the morning should be turned over and shaken up with the forks, or, as it is termed, *tedded*, and exposed to the sun and wind, previous to being put into small cocks before the evening. The process for putting it into cocks after the tedding is, for one field-worker to cast the swathe with a fork from the right open furrow to the crown of the ridge, going in such a direction as to have the ridges upon which the cocks are to be formed on her left hand, and which permits the working of both forks and rakes with the right hand. When the half ridge is thus cleared of the grass with the fork, another field-worker follows with a rake, and clears the same half-ridge of every stem of grass. A third worker follows the rake with a fork, and tosses the swathe from the crown to the left furrow of the ridge; and is, in like manner, followed by a fourth worker with a rake, who clears the same half-ridge. On the second or adjoining ridge, a fifth field-worker throws the accumulating swathe from the furrow to the crown, where her labour is much greater than the workers who wielded the fork on the first ridge, as she has to move the entire grass from the first ridge along with what is found upon the half of the second ridge; and she is followed by a sixth worker with a rake, who clears the ground from the furrow to the crown. In like manner, a seventh and eighth worker put the grass from the crown over the furrow of the second ridge upon the third ridge with the fork, and clear the ground with the rake; and in doing this, the worker who uses the fork is hardest worked of all; but all the rakers have the same degree and extent of labour. Thus, 8 workers are required to clear 2 ridges of grass, 4 wielding forks, and 4 rakes alternately, and the space cleared along the ridges, in this manner, is just the breadth of 4 swathes of grass, which is more or less, according to the weight of the crop, but mowers usually cut a breadth of 6 feet at each stroke of the scythe, and each breadth constitutes a swathe. In this way the band of 8 workers proceed clearing 2 ridges at a time, till they reach the end of the ridges, the grass on the headridge of which is cleared off and mixed with the other. Then they wheel round at the end of the ridges to take in other 2 ridges beyond the third ridge, and proceed in precisely the same manner, with this difference in the arrangement of the workers, that the forker who took the lead in coming up the ridges, and had the lightest share of the work, now becomes the last forker, and takes the heaviest part of the work, and the one who had the heaviest now takes the lightest work. To prevent confusion, the same raker always follows the same forker, so that the band consists of 4 pairs, each consisting of a forker and raker. Women are quite able to perform all this work, but where their number is insufficient, men take the forks and women the rakes. After the second 2 ridges have been thus cleared, the third ridge being in the middle, contains the grass of 5 ridges, which is called a *windrow*. When only a band of 4 workers can be spared to making hay, consisting of 2 forkers and 2 rakers, they must go up one ridge and down the next, to clear the 2 ridges on each side of the windrow, but fewer than 4 workers makes hay-making a very dilatory process, and more expensive in proportion to the number of hands.

(2788.) The *cocks* are raised on this middle or third ridge, and those first made are small, and called *grass-cocks*. They are put together either by the fork or the arms, with narrow bottoms, and high in proportion to their breadth, and not exceeding, perhaps, 2 feet in height. There will not be room on the ridge, at this time, to put such small cocks in a row, so they may be put up anywhere, as not to crowd upon each other, but afford room for the rakes to clear the ground around them; as it is considered very slovenly work in a hay-field, to neglect to clear ground by the rake which had been freed of its grass by the fork. The raking at this time will not occupy above 1 or 2 workers, so the rest of the rakers can be employed in assisting the forkers to put together the cocks. The field is left for the night in this state. In Ireland the grass-cocks are dexterously formed with the hands and foot, and are termed *lap-cocks*. They appear like bundles of clothes turned upon their mouths, and will defend much rain.

(2789.) Next morning the grass-cocks are shaken loosely out on the ridge, for exposure to sun and wind; and after this operation is finished, the grass which was cut in the afternoon of yesterday is tedded. In the afternoon the thrown down grass-cocks should be shaken up, after which the grass which had been tedded in the forenoon is windrowed and put into grass-cocks, in the manner just described. Before the evening, the thrown down grass-cocks are put into larger cocks, called *hand-cocks*, and which are best put together with forks by men. Hand-cocks should have small bottoms, built tapered to a fine top about 6 feet in height, and placed in a row along the crown of the ridge. The difficult part in making a hand cock is the fineness of its top, which, at the same time, should be so heavy as not easily to be blown off by the wind. The top is made in this manner:—Small forkfuls of loose hay are put above one another on the ground, and after as much is put together as can be conveniently lifted, it is firmly transfixed by the fork by one side of the heap, which, on being lifted up, is dexterously turned by the fork in a reversed position above the head, when it is planted firmly upon the top of the cock, and left there by drawing away the fork from under it; a gentle tap and dress with the fork on the outside makes it neat and firm. The ridge should be raked clean.

(2790.) The next morning, the third, the grass cut yesterday afternoon is tedded, and as much more tedded in the afternoon of what was cut early in the morning of the same day, as can be got together into grass-cocks before the evening. This is an easy day's work, and reserves strength for the greater labour of the next day, to which all hands of field-workers and ploughmen should be collected.

(2791.) Next morning, the fourth, should it prove a rainy day, let the whole field remain as it was, though the mowers may be able to continue at their work. If fine, toss over first the grass-cocks to the sun and air, then tedd the small quantity of grass that was mown after the tedding of the previous afternoon, and, last of all, throw down and scatter the hand-cocks, which, by this time, will have subsided considerably. All this will occupy, if not the whole, the greater part of the forenoon, but no more of these respective processes should be undertaken than there is force in the field to put all the hay in cocks before the evening; and of all the processes the tedding of the swathes is the most dispensable. The first thing to be done is to put the hay together that had been scattered, by putting 2 or 3 of the hand-cocks into 1 in a row along the crown of the ridge. The hay will be felt to have become much lighter in the hand; for it is surprising how soon hay wins after it has arrived at this stage, *if exposed to sun and air*. The usual practice, however, is to keep hay a long time in the hand-cock, from a belief it is safe; whereas, it is in that clammy state, which, if rainy weather prevent it being stirred about, will soon cause it to mould and contract a disagreeable smell; and the reason assigned for allowing it to remain in the cocks is, the want of time to work hay so much as it requires; but the truth is, a little persevering labour bestowed at this time will afterwards render less labour necessary, for it requires long exposure to bright suns and drying winds, to dissipate clamminess in hay. Let the hand-cocks, therefore, be first exposed and then put together at this period, 3 or 4 into 1, according to the state of the hay and the weather, and the hay will be placed beyond all danger of fermentation and mouldiness. After this operation, the scattered grass-cocks should be put into hand-cocks, however small the hand-cocks may be made; for, to allow *scattered hay to remain on the ground all night* to receive rain, would run the risk of rotting it, and even, if it escaped that catastrophe, of rendering it tasteless. Then what was tedded should be windrowed and made into grass-cocks before the close of the day's labour. All this will constitute a heavy but a good and important day's work. On such a day I have led and assisted 16 field-workers, all women, to handle upwards of 2000 stones of 22 lb. each of hay. The cocks now assume the name of *ricks* or *colls*, the latter being derived from the French, *colline*, a small hill. The colls should be gently tapered to the top, without a projecting shoulder to catch the rain, and its top fastened down with a hay-rope, twisted on the spot with the corner of a rake, or with a rope-twister or thraw-crook, taken on purpose to the field, and put across the top of the coll in the direction of the strongest wind to which the locality is subject. It is unnecessary to give directions for every day until the whole field of hay is rendered safe in colls, as they would only be repetitions of what has been said already.

(2792.) Let us next consider the making of hay with the *aid of horse-labour and suitable implements*, the employment of which makes a considerable difference in the process. The tedding-machine, represented in Plate XXXI., fig. 462, is used to ted hay, and which it best does by passing across the swathes, taking up and teasing and scattering them on the ground in the most regular manner. It is alleged that this machine is only suited to ted natural grass, but why so is not obvious, for its structure is capable of laying hold of any kind of grass. It would indeed shake ryegrass too much that is intended for seed; but I have already said it is impossible to obtain good hay and good seed from the same crop. If the object be merely to ted grass, this machine will doubtless ted sown grasses as well as natural, when they are cut at the proper age. After the grass has thus been tedded, it is allowed to dry in the sun and wind all the forenoon. In the afternoon the hay-rake, whether the common horse-rake, or American hay-rake, both of which are represented and described below by Mr Slight, is employed to rake the tedded grass into a windrow across the 4 ridges which intervene between every fifth ridge which contains a row of cocks. Where the crop of grass is very thin, the horse-rake might carry the grass into a windrow over more than 4 ridges upon the fifth ridge; but with an ordinary crop it could not perhaps accomplish this, and much less with a heavy crop. After the grass, therefore, has been windrowed across the 4 ridges, manual labour is employed to put it into grass-cocks, as in the case with manual labour. It will be observed that few people, and especially women, are required to conduct hay-making in this way, the heavy part of the duty consisting of making the cocks as often as requisite, which is best done by men.

(2793.) With regard to the *stacking* of hay, if the entire produce of the field is to be stacked at one time, as is usually done in Scotland, the colls should be put into a state to stand the weather for a considerable time; 2 or 3 being put into 1, and the large ricks thus formed are named *tramped pikes*, because they are built and tramped, a man building, and his assistant, a field-worker, carrying the hay from the fork of the carter and tramping the rick at the same time. Tramped pikes contain from 100 to 150 stones of hay each, and are commonly placed in a row at the end of the field most convenient for conveying it in carts to the stack in the stack-yard, if the hay is retained for the use of the farm, or, if disposed of, to a purchaser who stacks it for himself. The reason that hay should be piked if stacked all in one day is, that unless hay is in a state to keep, that is, not to ferment in the stack, so much cannot be put together without risk of heating. But when hay is stacked under cover, whether of temporary or permanent construction, it might be daily stacked from the hand-cocks, and the slowness of the stacking would greatly tend to render the entire hay of the stack uniform, and it would also be uniformly good, as it would be saved without long exposure to the weather, in passing through hands at this period of its condition, when it would be much more benefited by exposure than confined in a large rick. These circumstances indicate the best period for ensuring the good quality of hay; and if this condition can be secured by sheds which would place it in safety beyond the reach of weather, it should be worth every farmer's while to procure them. So slow a mode of stacking, and the use of hay implements, would enable the ploughmen to carry in the hay-harvest while the field-workers would be employed at singling and hoeing turnips, thus securing the seasonable treatment of both crops, instead of having the importance of the one pitted against that of the other. The farmers in the south of England employ *rick-cloths*, which afford but temporary shelter, but they are quite sufficient to secure the safety of the hay. Such a rick-cloth is figured and described below. As long as the stack is building, perhaps by a single cart-load a day, till the stack is thatched and placed beyond all danger, the rick-cloth is used, and the hay is secured from the bleaching effects of rain, as well as from the scorching effects of heat, which latter encrusts hay with a dry skin without winning its heart, and which state is more deceptive in its results as to fermentation than even dampness from rain. I have never known such an expedient as this employed in Scotland, though I have seen rick-cloths thrown over a hay-stack when it was building, and when the stack was so large as to occupy more than one day in building; but this is a different use from what is represented below. One instance only am I aware of a hay-stack having been built on the English principle in Scotland, and that was by Mr John Little, Carlesgill, near Langholm, in Dumfriesshire, who, instead of employing the temporary rick-cloth, erected a permanent wooden shed, which has the additional advantage of saving the thatching of the hay-stack. Mr Little has made public the particulars of the expenses of such a shed, which I give for the guidance of those who may be disposed to follow his example :—

The shed is 45 feet long, 16 feet wide, and 12 feet high, or contains, inside-measure, 7800 cubic feet, or about 290 cubic yards, which at 7 imperial stones per yard of hay, is capable of containing rather more than 2000 stones. It consists of these materials and their cost :—

14 posts 5 feet each, = 70 feet, at 1s. 3d. per foot, of larch trees 9 inches in diameter,	L.4	7	6
90 feet of wall-plate, 9 inches wide, and 2¼ inches thick, = 67½ cubic feet, at 3¼d. per foot,	0	18	3¼
363½ feet of 2¼ inch planking for roof at 3¼d. per foot,	4	18	5¾
1312 feet of ¾ inch boarding for covering the planking, at 1½d. per foot,	8	4	0
Nails,	1	0	0
Carpenters' labour for putting up the shed,	2	10	0
Levelling, sinking, and charring posts,	0	16	8
28 gallons coal-tar, at 4d. per gallon,	0	9	4
Putting ditto in roof 2 days, at 2s. per day,	0	4	0
This is exclusive of carriage from the wood to the saw-mill, and thence to the steading.			
	L.23	8	3

If covered with "double slate," instead of boards, the cost would be,	L.25	10	10½
If with "second" Lancashire slate, the cost would be,	31	15	10

Exclusive of sinking the posts, the value of the coal-tar, and the cost of putting it on.

Cost of thatching a 9 fathom rick with sprits and coarse grass is :—

Mowing thatch—drawing it in bundles—carting—and laying it on,	L.0	4	11
Ropes making and putting on,	0	1	10
Material for thatch and ropes,	0	5	0
	L.0	11	9*

The building of the rick is the same in every case, and the last item of cost of materials should, I think, be left out, as the manure afforded by it will repay its cost. To save warping, the boarding is fastened with T headed nails driven in the joinings, the cross-head of the nails overlapping the boards ½ inch, and allowing their expansion and contraction. Parallel to the joinings, and ½ inch from it, a groove of ¼ inch deep is cut along the upper surface of each plank to form a channel for rain.

(2794.) A large oblong hay-stack should be *built* in this way :—In the first place, a dry stance should be chosen, for a damp one will cause the destruction of several stones of hay at the bottom of the stack. The *stance* should be raised 1 foot above the ground with large stones inscribing the circumference, and the interior filled up with stone shivers or gravel beat firmly down. Upon this space the stack should be built by 2 men, who are supplied with armfuls of hay by a number of field-workers, whose duty is not merely to carry the hay but to tramp it under foot in a regular manner from one end of the stack to the other. The 2 men, each occupying a side of the stack, shake and build up what is called a

* See Prize Essays of the Highland and Agricultural Society, vol. xiv. p. 758-60.

dace of the hay before them as high as their breast, from one end of the stack to the other; and after half its length is built up in this manner, the women go upon it and trample it, and if they hold by one another's hand in a row, their walking will prove the more effective. The breadth of the stack is a little increased to the eaves. The hay is forked from the ground by 2 or 3 men, and when the stack has attained an inconvenient height for this purpose, there are 2 or 3 modes by which hay may be carried to greater height; one is by placing short ladders against the stack, and a man on each, some way above the ground, with his back to the ladder, where he receives the forkfuls of hay from the forker on the ground, and raises the load above his head upon the stack. Another mode is for men to carry back-loads of hay up long ladders, and empty them on the stack. A third is to erect a scaffolding of planks upon a couple of tresses of 6 feet in height, and to fork the hay off the scaffolding to the stack as it is forked upon it from the ground. Of the 3 modes, the last of the scaffold affords the most secure footing to the men at an elevation from the ground, and in the end is the most expeditious; and in all the modes 4 men will be amply employed in forking up the hay to keep 2 builders in work. The hay is forked off the ground instead of the cart, as the latter mode would hinder the horses too long to make them stand till the cart is cleared of its hay by forking. The hay is, therefore, thrown down upon the ground from the cart, and if the cart is constructed to tilt up, the deposition of hay is easily effected; and even from a whole-bodied cart hay is easily thrown off by the forkers sticking their forks under the load along one side of the cart, and pushing upwards and from them to the other side, one person holding by the wheel nearest the men to prevent the cart upsetting. The position in which the load is thrown upon the ground requires to be considerately chosen. The load should be thrown away *from* the spot upon which the men stand to fork, when the hay will easily come away with the fork, because each stratum of hay, as it was forked on the cart, then lies towards the men; whereas, when the load is thrown *towards* the forkers, the inclination of the hay abuts against them, and every forkful must then be pulled away by main force. A height of 12 feet is enough for the body of the stack, and a breadth of 15 feet is convenient for a hay-stack, and with these fixed dimensions, the length may be made more or less, according to the quantity of hay to be stacked. With these dimensions, a *new built* stack of 40 feet in length will contain about 2000 imperial stones. After the body of the stack has attained 12 feet in height, the heading is commenced by gradually taking in the breadth on each side to the ridging, which is elevated half the breadth of the stack above the eaves, and the ends are built perpendicular. One man and one woman will only find room at the finishing of the top of the stack. A few straw ropes are thrown over the stack to prevent the wind blowing off its new made top. The stack is left for several days to subside, and unless it has been slowly built and firmly trampled, it may subside in the body to the extent of 2 feet. Very probably heat may be indicated in some part of the stack a few days after it is built, by a leaning towards that part, because heating causes consolidation of the hay. A prop of wood placed against the place will prevent the stack subsiding much farther, and the handle of a rake pushed in here and there into the stack, will indicate whether the heating is proceeding upwards or to a dangerous extent. A gentle heating will do no harm, but rather good, by rendering the quality of the hay uniform, and horses do not dislike its effect. Salt has been recommended to be used in hay, and when hay is in a damp state in consequence of the weather, it is an excellent remedial measure against mouldiness, and it may be sown by hand upon every *dace* of hay laid down by the 2 builders. The proper quantity of salt to be used, according to the state of the hay, has never been correctly ascertained, and must, therefore, be left to your own judgment according to circumstances. Salted hay is very much relished by all kinds of stock, and especially by cattle.

(2795.) When the hay has fairly subsided, and the heat, if any, is no longer felt, the stack should be *thatched;* and as a preparatory operation, the sides and ends are neatly *pulled* straight from angle to angle of the stack, with a small increase of breadth to the eaves. This operation simply consists of pulling out the straggling ends of the hay, which give a rough appearance to the sides and ends, in order to render them smooth; and its use is to save the hay pulled out which would otherwise be bleached useless by exposure to rain, and to prevent rain hanging upon them about the stack. The heading or thatching consists of straw drawn straight in bundles, held on by means of straw-ropes. When a hay-stack is to be thatched, the drawn bundles of straw, and the straw-ropes, should, of course, be prepared in time; and yet it is a matter not of unfrequent occurrence for farmers to allow the hay-stack to stand unthatched until the corn harvest, for want of straw, or even to allow the hay to be left in ricks on the field till just on the eve of harvest. Straw, in some instances, may indeed be scarce, but in that case rushes and other tall grown wild plants form an excellent substitute, both for thatch and ropes. Ferns and heath are good materials for thatch. The thatching should be carried on both sides of the stack simultaneously by 2 men, and begun at the same end. The men being mounted on the head of the stack, the bundles of straw are handed up to them on a fork one by one as they are needed, and each bundle is retained in its place on the roof, beside the thatcher, by leaning against a graip stuck into the hay. The straw is first placed over the eaves, handful after handful from the eave to the top of the stack, each length of the straw being overlapped by the one immediately above it. When the thatcher feels a hollow or soft part with his feet in the head of the stack, he makes up the part by some of the hay that was pulled out of the stack, to save the wasting of thatch straw in filling up such hollows. The straw is thus laid from the eaves to the ridge of the stack to a breadth as far as the thatcher can reach at a time with his arms. When the men on both sides meet at the ridge, straw is laid along the stack upon the ridge, to cover the terminal ends of the thatch on the sloping roof, and to support the ropes which keep down the thatch. When this breadth, of perhaps 3 feet, or a little more, of the thatch is laid down, its surface is switched down smooth by the thatcher with a supple willow rod, and then a rope is thrown across the stack at its very end, and another parallel to it at 18 inches apart, and made fast at both ends, in the mean time, to the sides of the stack. Other ropes, at right angles to the first, are fastened 18 inches apart to the hay at the end of the stack, and supposing the side of the roof to be 11 feet along the slope, 6 ropes running horizontally will be required to cover the depth of the slope, leaving a space of 9 inches from the ridge for the place of the uppermost rope, and the rope at each eave is put on afterwards. Each of these horizontal ropes are twisted once round every perpendicular rope they meet, so that the roping when completed has the appearance of a net with square meshes. As every subsequent breadth of thatch is put on, the roping is finished upon it the advantage of which is, that the thatching is finished as it proceeds, and placed beyond danger from wind or rain, or disturbance from after work. If the stack stands N. and S., the E. side should have a thicker thatching than the W., as being most liable to damp, and the thatching of both sides should be thicker towards that end of the stack which is farthest from the steading, as it will stand longest, and the process of thatching should terminate at the end which will be first broken upon, that is, nearest the steading, because the thatch will come away more freely when removed in the opposite direction from which it was put on. The horizontal ropes at their termination are fastened into the hay at the end of the stack. The eave is finished by laying a stout rope horizontally along the line where the roof was begun to be taken in, and twisting it round each perpendicular rope as it occurs; when each perpendicular rope is broke off at a proper length and fastened firmly to the hay immediately under the eave, and after the eave-ropes have thus been fastened down, the projecting ends of the thatch over the eave are cut straight along the stack, and give to the heading a pretty finish. Another mode of roping the thatch is to place the ropes in a diagonal direction across the stack, and when one set of ropes cross the other diagonally, the effect is lozenge-shaped, which looks well; but roping in this fashion requires the thatcher to place all the straw upon the roof before he guides the ropes over the ridge of the stack, to do which he must stand upon the ridge and step backwards upon it—a plan which allows the wind to have liberty to blow off the thatch before it is roped at all, and also obliges him to trample down the ridge straw to a certain degree.

(2796.) On the other hand, when the stack is built under cover of a *rick-cloth* or *shed*, the hay may be led in by a cart-load at a time, employing only 3 hands, for the builder forks the hand-cocks in the field to the cart, and the carter forks the hay to the builder off the cart, as shewn in the cut below, while the same field-worker who rakes the bottoms of the hand-cocks in the field carries the hay to the builder on the stack, each forkful of hay, in this case, being thinly scattered over the stack, is easily trampled down, and has time to subside before another load is put over it on the following day. Where the rick-cloth is used the stack should be thatched on its removal; but where a shed is erected, no thatching is required.

(2797.) Hay is sometimes built in *round stacks*, which are kept of a cylindrical form for 7 or 8 feet from the ground, and then terminated in a tapering conical top, and thatched. Such stacks contain from 300 to 500 stones of hay. This form of stack is convenient enough when a whole one can be brought at once into the hay-house, but should the stack be of such a size as to be necessary to bisect it perpendicularly, the remaining half is apt to be blown over; or should its upper half be brought into the hay-house, the under part must be protected by a quantity of straw kept down by some weighty articles, and in such a case it is seldom that these are put on with sufficient care to keep out rain and resist wind. Upon the whole, the oblong form of stack admits of being most conveniently cut for use, and left at all times in safety; because a section of any breadth can be cut from top to bottom to fill the hay-house.

(2798.) The rule for ascertaining the number of stones of hay in oblong stacks is simple enough, but not so for conical stacks. To find the weight of hay in an oblong stack:—To the height from the ground to the eaves add one half of the height of the top above the eaves for the mean height, then multiply the mean height by the breadth, and then multiply the product of both by the length. Divide the gross product by 27, and the dividend will give the number of cubic yards in the stack, and that number of yards multiplied by the number of stones of hay in a cubic yard, will give the weight of the stacks in stones imperial. It is not easy to state the exact number of stones of hay in a cubic yard, as that must vary according to the compressed state of the hay, the weight actually varying from 5 stones to 9 stones per cubic yard, according to the age and size of the stack, and the part of the stack from which the hay is taken; but perhaps 6 stones may be near enough the mark of the weight in a new stack, 7 stones in one that has stood for some months, and 8 stones in one that has stood over years. The contents of a round stack with a conical top may be ascertained in this way :—Take the height of the round part from the ground to the eaves, and add to it one-third of the perpendicular height of the conical top above the eaves for the mean height of the stack. Take then the mean girth, which, if the stack is wider at the eaves than at the ground, is ascertained by taking the girth at the eaves, and also at the ground, and dividing their sum by 2. Square the mean girth, and multiply the product by the decimal .0795, which will give the area of the base of the stack. Then multiply the area by the mean height, which will give the contents of the stack in cubic feet, divide the contents by 27, which will reduce them to cubic yards, and then multiply the yards by the supposed number of stones of hay in the yard, and the capacity of the stack will be found in stones. To know the contents of a conical stack or coll, take the girth at the ground in feet, find the area of the circle in the ordinary way, and multiply the area by one-third of the height. The contents thus found in feet reduce to yards, and then multiply by the number of stones in a cubic yard. But the simplest plan in all such cases is to use any of the tables which are published for the purpose of saving tedious calculations, such as those of Anslie or Strachan, the latter of which, however, are not extended far enough to comprehend stacks of the largest dimensions.

(2799.) A crop of hay varies from 150 to 300 stones per acre, according to the season and the nature of the soil. On light gravelly soils the crop is never heavy, but its quality is generally fine, and on good clay it is usually heavy, and the plants large and strong, the clover predominating. For quantity and quality combined, a deep mellow clay loam may be regarded as the best texture of soil. On thin clay, and on thin light soil resting on retentive clay, the clover is frequently thrown out by frost in spring, and the hay then consists chiefly of rye-grass, and on the same soil the same effect is produced by severe drought in May. Good hay should consist of equal quantities of clover and rye-grass, feel pleasant to the hand, and smell fragrantly, and, when well prepared, possess a light brownish green colour ; but as it is commonly prepared, the colour is usually light brownish-yellow. Hay of natural grass, when well prepared, is darkish green in colour, feels soft, is generally of fine quality, and highly fragrant. Grass usually loses two-thirds of its weight on being made into hay.

(2800.) Of 3060 grains of white clover, as much as 2430 grains were water, 100 grains nutritive matter, and the remaining 470 grains were insoluble matter. The nutritive matter in the 100 grains consisted of 77 of starch, 2 of sugar, 7 of gluten, and 14 of bitter extract and saline matter.* "When green grass or clover, approaching to maturity," says Professor Johnston, "is first cut down, it contains a considerable proportion of starch, sugar, and gum still unchanged into woody fibre, as it would mostly be were the plant allowed to become fully ripe." Here you see the propriety of cutting down grass for

* Sinclair's Hortus Gramineus Woburnensis, p. 241.

hay before any of its seed approaches to maturity, because latterly it contains woody fibre instead of the nutritive ingredients just mentioned. But even when succulent grass is "left to dry in the open air, the circulation proceeds to a certain extent, and, under the influence of light, woody fibre continues to be formed in the upper part of each stem, until it becomes completely dry." And "it may even be a matter of doubt whether the process of change does not often proceed after the hay has been carried off the field and stacked."* All which considerations tend more and more to prove that the longer grass is allowed to stand after the plant has attained its full stature, the less digestible or nutritious the hay will become; and more than this, the longer the process of making the grass into hay is delayed, the more woody, and, of course, the less nutritious the hay will be. Every quick process of converting grass into hay is, therefore, better than any slow one. There is a very quick mode practised in Saxony, which is this:—The grass that has been cut down during the day is put into large cocks late in the afternoon. A very strong fermentation soon ensues, which continues all night until the morning, when the work-people return to the field, by which time the cocks have contracted much in bulk, and the steam rises briskly from them. They are then thrown down and scattered on the ground, and their contents allowed to remain all day exposed to the sun and air, and by the afternoon the hay is so dry and won as to be fit to be stacked, and accordingly it is gathered from the ground and carried to the stack. The new-mown grass of the day is put into large cocks in the evening, to be treated the next day in the same manner. This mode of hay-making might be followed in this country, provided we could trust our climate; but should the next morning prove a rainy or even a damp one, the contents of the cocks would inevitably be rotted. The modes of hay-making, both manually and mechanically, which I have described above, are both expeditious; as are also the modes described by Mr Little, Carlesgill, and Mr Miller of Forest, both in Dumfriesshire.†

(2801.) But as grass is known to be more nourishing than hay, acre for acre, it would be desirable could grass be prepared to retain its properties to a season when it cannot be obtained from the field. A trial of preserving grass has been lately made in Germany, in East Prussia, and it is this:—Pits are dug in the earth 10 or 12 feet square, and as many deep. They are puddled with clay, and lined with wood or brick. Into these pits 4 or 5 cwt. of grass, as it is cut, are put in a layer at a time, sprinkled with salt, at the rate of 1 lb. to 1 cwt., and if the grass is dry, that is, free of rain or dew, 2 or 3 quarts of water are sprinkled over the layer. Each layer is trodden down by 5 or 6 men, and rammed firm, especially round the edges, with wooden rammers, the object of which is the exclusion of air. A little straw is then scattered over the layer, to mark its dimensions afterwards. Layer is placed above layer until the pit is filled to the top, when the topmost layer is well salted, and the pit covered with boards or a well-fitted lid, upon which is put a covering of earth of 1½ foot in thickness. Such a pit will contain 5 layers of grass, and should be filled in 2 days. The grass soon ferments, and in about 6 days subsides to half its original bulk. The lid is examined every day, and every crack in the earth filled up, to exclude the air, which, if allowed to enter, would promote the putrefactive fermentation in the grass. When the first fermentation has ceased, the lid is taken off, and fresh grass put in, trodden down, and salted as before. The pit will now contain about

* Transactions of the Highland and Agricultural Society for October 1843, p. 59.

† See Prize Essays of the Highland and Agricultural Society, vol. xiv. p. 750 and 760.

10 tons of grass, equal to 2 or 3 tons of hay. The pits should remain shut for 6 weeks before being used, and then are used in succession. The grass thus treated has the appearance of having been boiled, and its sharp acid taste is very agreeable to cattle; and 20 lb. a-day with chopped straw will keep a cow in good condition all winter, and 28 lb. will cause a cow to give a rich and well-tasted milk.*

(2802.) As hay is usually made in a thriftless manner, and as grass is more nourishing to stock than hay, and as good food can be cooked for horses in winter without hay, I have often thought it a loss to farmers to make hay at all. The grass would pay better on being grazed, and the land would be retained in better heart. In the vicinity of large towns, it may be expedient to make hay, and yet when the crop proves heavy the price is low—the average may be stated from the tramp-rick at 8d. per stone of 22 lb. I have seen the price as low as 4d. and as high as 1s. 4d. per stone; but when the price is high, the crop is deficient, and the quality of the hay bad. Taking the heaviest crop of 300 stones at 8d., it will yield L.10 an acre, but 220 stones is nearer the mark; and yet grass lets for cutting L.12 or L.14 per acre in the neighbourhood of Edinburgh, without incurring any trouble to the farmer. I am sure if half the labour spent in making hay were bestowed in winter in cooking food for the horses, farmers would derive a profit from the exchange.

(2803.) When ryegrass is intended for seed, a part of the grass that shews the greatest proportion of ryegrass should be reserved to be cut when the seed is approaching to ripeness—not to wait till it is ripe—for ryegrass seed is easily shaken from the stalk. As it is mown, it should be tied in bundles or sheaves with thumb made straw-ropes, and set in stooks in the field to win. After a short time the sheaves should be built in hand-cocks, and, when the field should be cleared for stock, the cocks should be taken to the stack-yard, and built in one stack or in stacks according to its fitness for keeping. When thus stacked, the crop may be thrashed in spring by the thrashing-machine; but as much of the seed is apt to be blown away by the strength of the wind of the machine, and as it is troublesome to clean the machine when used in this way, and may impose this trouble at a season when every one is busily engaged, I think the better plan is to thrash the ryegrass crop with the flail. This may be done in the stack-yard, and a favourable day as to weather chosen for the operation. An outside door answers very well for a thrashing-floor; but if a quantity of this seed, and perhaps of some others, such as turnip-seed, are raised every year, it might be worth while to have a stout board which could be annually used as a thrashing-floor. The thrashing should be performed out of doors, unless it be upon an earthen barn-floor; for the stroke of the flail is apt to split the boards of an ordinary wooden floor, such as that in an upper-barn. The board is set upon 2 cushions of hay or straw, which afford a more elastic stroke to the flail than the ground. A field-gate, such as fig. 445, is placed lengthways from one end of the floor, and the large barn-sheet is spread under the gate. The sheaves are either forked from the stack on the spot, or wheeled in the corn-barrow, fig. 303, from the stack at a distance. A field-worker loosens the sheaves, and pitches them upon the floor, when wanted, with the seed end towards herself.

* I give the name of the work containing the above directions; "Verhandlung des Baltischen Vereins für Förderung des Landwirthschaft. Greisswald. 1842, p. 38." as quoted by Professor Johnston in the Transactions of the Highland and Agricultural Society for October 1843, p. 61.

Two men, one on each side of the floor, use the flail. Another field-worker stands at the junction of the floor and field-gate, and pulls the thrashed hay with a long fork towards herself upon the field-gate, over which she shakes it briskly, assisted by a third field-worker with a short fork, who tosses it from the gate on the ground to be removed. The hay, when thus treated, is not of much value, but will, nevertheless, be eaten by the feeding beasts in the hammels, or young beasts in the courts, especially after it has been damped. The thrashers occasionally clear the board of seed with their flails upon the gate, through the spars of which it accumulates upon the barn-sheet below. When the spars of the gate are filled up with seed, the seed is shovelled to one side, and riddled upon a sheet, preparatory to its being put into sacks to be carried to the corn-barn and winnowed. The seed will more quickly part with its impurities in the winnowing, after it has lain to dry and win on the barn-floor for some days. When thus thrashed in spring the quantity required for each field of the farm should be measured and sacked, to be ready for use. Instead of raising ryegrass in this manner, a not uncommon practice is to thrash the hay as it is put into tramp-ricks. The same apparatus is used as has just been described, and it is set down at the part of the field where the tramp-ricks or colls are to be built. A horse is employed to draw the hand-cocks to the thrashing-floor, which is done by hooking one end of a double cart-rope to one side of the haims, passing it round near the bottom of the coll, and hooking its other end to the other side of the haims, and, on applying his strength, the horse causes the coll to slide along the ground to the scene of action by means of the rope. A field-worker supplies the floor with small quantities of hay at a time from the transported colls with a fork; the thrashers use the flail lightly—indeed a light form of flail is best for this purpose; 2 field-workers stand one on each side of the gate with a fork, and draw away the hay upon the gate, where it is well shaken, and then thrown into a heap upon the ground, from which a man forks it to the man who builds the tramp-rick, the builder having the hay carried and tramped by a field-worker. Thus, if one part of this busy band of workers supplies that which follows with sufficient materials, the work goes on pleasantly and without collision. A field-worker is engaged in riddling the seed as it is thrashed, which, on being sacked up, is carried to the granary to be dried; and portions of hay in it, such as broken leaves of clover and stems, serve to keep the mass open and the seed more accessible to the air. When sufficiently dry, it should be winnowed in the granary, heaped-measured, and laid thicker together; and in spring, it should again be winnowed, and freed from the many fresh impurities which will have found their way into it during the winter, such as cats' and vermin's dung, cobwebs from the roof, and dust. Though freed of all these impurities, the heaped measure of the summer will tell out in an equal number of bushels of striked measure in spring. Whatever proportion of the ryegrass seed is not required for the use of the farm may be disposed of to a seed-merchant or farmer. A fair crop of ryegrass should yield about 26 bushels of seed to the imperial acre, when the ryegrass is cultivated on purpose for the seed.

(2804.) [Of the *hay-making* implements, the scythe (2615.), and fig. 452, and the hay-fork, fig. 279, having been already alluded to, I have now to notice a machine which is extensively employed by the English farmers in the preparation of *meadow*-hay, or hay prepared from the mixed natural grasses, in contradistinction from the artificial, or ryegrass and clover hay. The modes of saving or winning these two kinds of hay are different even in England, where hay-making is best understood; while in Scotland it may be held in general as decidedly bad, especially in reference to the ryegrass hay, which is the chief produce of the hay kind in Scotland. But it appears highly probable that ryegrass and clover hay might be advantageously prepared under a system differing very little from the English practice, and greatly to the advantage of the produce, as regards nutritive qualities. Though the machine I am about to describe, therefore, enters but little into that practice of the agriculture which we profess to follow; yet the machine is of that importance, and is, besides, slowly making its way amongst us, that to leave it out of this system would be a departure from duty. From amongst the few cases of its adoption in Scotland, I have been informed of its affording striking advantages, such as, in a weeping season, the hay on one property was effectually saved by the use of the tedding-machine, while on that immediately adjoining, the crop was all but lost, from delays that naturally arose from the slow hand-process of hay-making. And although it may be still necessary to adopt a distinct mode of treatment for artificial hay, the tedding-machine, from the celerity of its effects, would in all cases prove an important auxiliary in the hay-field.

(2805.) The English hay-tedding machine, which is represented in perspective in Plate XXXI, fig. 462, consists of a skeleton carriage, having a series of revolving rakes occupying the place of the body. The carriage is composed of the transverse bar *a*, 6 feet in length, into which the horse-shafts *b b* are tenoned. An iron stay-bar *c c* on each side connects and supports the shafts, and the stays are continued backward, and attached to the centre of the box that carries the axle of the carriage-wheel on each side. The length of the bars *c* from *a* to the centre of the axle-box is 3 feet 10 inches, and the bars are $2\frac{1}{2}$ by $\frac{1}{2}$ inch. The carriage-wheels *d d* are 3 feet 10 inches diameter, and turn upon arms cast on a circular box, into which the nave of the wheel, armed with a ratchet *e*, is received. The ratchet-wheel *e*, thus attached to the nave of the carriage-wheel, takes hold of the spur-wheel *f* by means of a pall, and carries it round when the machine advances, but slips hold on backing or turning. The spur-wheel *f* works into the pinion *g*, which is mounted on the end of the hollow shaft *h*, extending from side to side of the machine; and though in the figure, for the sake of distinctness, the spur-wheel and pinion are exposed to view, they are in the machine closely boxed up in a cast-iron casing, which, for perfect and safe working, is necessary to prevent entanglement from the hay falling between the wheel and pinion. A bar of $1\frac{1}{2}$ inch round iron passes through the hollow shaft *h*, and has its end fitted tightly into the outward side of the case that contains the pinion *g*, and there fixed firmly with a screw-nut on the outside of the case; and the hollow shaft and pinions being firmly connected by thin flanges (which are left out of the figure), they revolve round the central rod or shaft as one body, the rod having turned bearings where the pinion embraces it. The two rake-wheels *i i* are 2 feet 8 inches diameter, and of very light construction; they have eyes sufficiently large to pass over the end flanges, of about 6 inches diameter, of the hollow shaft, to which the pinions *g* are attached by means of their flanges; and, to fill up the large eyes of the rake-wheels, the shaft *h* is swelled out at the points of bearing. The rake-wheels, fixed dead upon the shaft *h*, are now armed with the eight rakes *k k*; these are wooden bars 5 feet 6 inches long, and $2\frac{1}{4}$ inches square, each carrying 10 light, iron teeth, about 7 inches in length. The rakes are attached to the wheels by a tumbling-joint *m m*, &c., and are held to the work by the springs *l* only; by which arrangement, when any undue resistance is opposed to a rake, such as a stone or other obstruction, the rake falls back till the obstruction has been passed, when the springs immediately return it to its working position.

(2806.) It will be observed that there is no thorough axle in the machine, as the revolution of the rake occupies the place where that member should exist; hence the axle-arms or heads are simply studs projecting from the box which contains the machinery *e, f, g*, and hence also the necessity for the connecting-bar which passes through the hollow shaft; that, together with the bar *a* and the longitudinal bars *c c*, being the only parts which constitute the frame-work of the machine. Besides the capability of backing, without turning the rakes, there is provision for disengaging them when the machine is advancing. To effect this, the pall which is attached to the spur-wheel *f*, for the purpose of deriving its motion from the ratchet-wheel of the nave, is held in action by means of a spring pressing on the tail of the pall, and the disengagement of the whole machinery from the carriage-wheel or first mover, is effected by a small tumbling lever affixed also to the spur-wheel, and fitted to bear upon and throw the pall out of geer with the ratchet of the carriage-wheel nave. The machine is also furnished with the means of elevating and depressing the centre of the revolving-rake, and, of consequence, bringing the rake-teeth nearer to, or farther from, the ground, and this is effected by turning round the circular boxes that contain the geering to the extent required, which is then fixed by means of a quadrant bolted to the bars *c, c;* a small portion of this quadrant, which is a part or flange of the geering-box, is seen with its bolt-holes at *e* on the left, and at *i* on the right of the figure. When in operation the machine is drawn by one horse, or sometimes two horses, and the result of the combination of the geering is, that the revolving rake makes 4½ revolutions for one of the carriage-wheel. The latter being 3 feet 10 inches diameter, will pass over 12 feet or thereby in one revolution, and the rakes being 4 feet 6 inches diameter over the extreme points of the teeth, will describe a circle of about 14 feet in circumference, and this revolving 4½ times for one of the other, the points of the teeth will pass through 63 feet while the carriage has moved over 12 feet, and as there are 8 rake-heads, there will be $8 \times 4.5 = 36$ contacts with the substance which is to be lifted, in a space of 12 feet, or one at every 4 inches. From this calculation it will be seen that the hay under the operation of this machine will undergo a process of teazling or tedding of the most perfect description; it will be separated and tossed about until no two stems of the plants will be left in contact, and by this exposure the drying process is effected in a period greatly shorter and more effectually than could be done by any number of hands. Thus, if we suppose the horse to walk 2½ miles per hour, and the machine to cover 6 feet in breadth, we have a surface of 1½ acres nearly covered in an hour.

(2807.) There are some variations in the mode of constructing the hay-tedder, but not differing essentially from the one here figured, which has been drawn from those manufactured by James Slight and Co. of Edinburgh, where the price is L.14.

(2808.) The following machines, though not confined to the hay-field alone, but are also employed in the grain harvest, are very frequently employed in hay-making, for the sole purpose of collecting the hay into heaps after the tedding process has been gone through.

(2809.) The *hand hay-rake* is an implement of great simplicity, but though almost elementary in its construction, it has been subjected to numerous variations, chiefly in one point, with a view to improve its construction. It consists of a head *c d*, fig. 500, or *g h*, fig. 499, of from 24 to 27 inches in length, made of hard wood $1\frac{1}{2}$ inch broad in the middle, tapering a little to each end, and from 1 to $1\frac{1}{4}$ inches thick. The head is armed with twelve or thirteen wooden teeth *a* or *e*, made of oak or ash $\frac{5}{8}$ inch in diameter, and, when first made, about $3\frac{1}{2}$

Fig. 499.

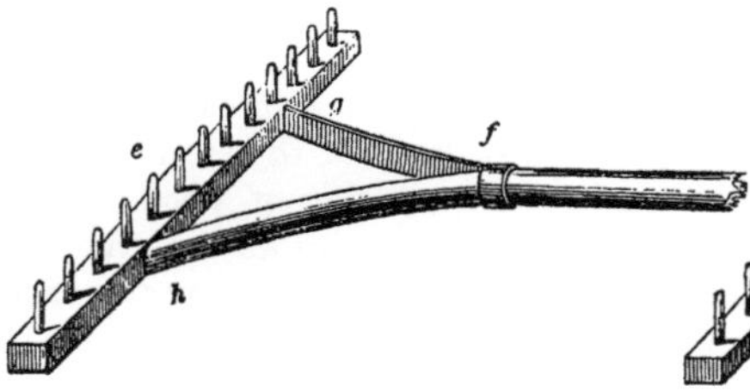

Fig. 500.

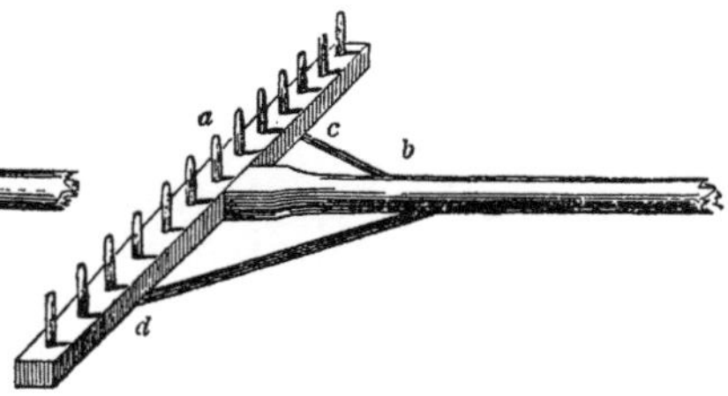

THE HAND-RAKES.

inches long. They ought to be formed with a slight swelling in the middle, or rather towards one end, the shorter end being adapted to fit into holes previously bored at 2 inches apart centres; and when the teeth are driven into their holes the swell on the tooth fills up the hole tightly, giving the tooth much greater strength than if it were quite cylindrical. The teeth are then properly secured by wedging, and the wedges and ends dressed of; the points of the teeth are likewise dressed off to a uniform length, $2\frac{1}{4}$ inches or thereby, and sharpened off from the back. The shaft or helve, which is $5\frac{1}{2}$ feet long, is usually made of ash, but as lightness is an object, its thickness ought not to exceed $1\frac{1}{4}$ inch, dressed neatly smooth and round except where it enters the head; here, and for a length of 12 inches, it is usually kept square. This part of it is either let into the middle of the head by a tenon, as in fig. 500, or it is split as in fig. 499, and enters it by two tenons. It is in this point of the construction that the variations have occurred, and from the slightest consideration it will be evident that the single tenon alone, as in fig. 500, is defective and weak; hence the grounds for the variation, to afford some additional support to the simple tenon, and fig. 499 exhibits one mode of accomplishing the object; the helve is split with a saw from the end backward to *f*, at which point it is prevented from splitting further by the application of a rivet put through the termination of the split part, or, what is much better, an iron ferule, as in the figure, is drawn tight upon it; the fork is then opened, the ends adapted to the respective mortises and secured into the head of the rake. This is apparently a very simple mode of accomplishing the object, but it is not by any means a perfect one. Were it possible to make the two parts *f g* and *f h* free of curvature, the object would be attained; but as this cannot under the circumstances be done, the head and the helve will have a very unstable attachment to each other, and a consequent weakness of parts; neither is it less expensive than other and better methods, though the rake of this form is always light in hand, which is its only commendation. A semicircular bow of bent ash wood has been often applied, passing through a hole at *b*, fig. 500, and entering the head at *c* and *d*, but this, by reason of its curved figure, is also especially defective, besides weakening the helve. Equally so is a light iron bow of the samo curvature, but, instead of passing *through* the helve, it is simply attached to it, and to the head with a nail at each of these points. This will be better in degree than the last from the greater rigidity of the iron, but is not better in principle. The true and only mode of applying the bow is that exhibited in fig. 500, where the parts *b c* and *b d* are perfectly straight, and formed of very light iron rod; it need not exceed ¼ inch diameter, flattened at the two extremities and at the point *b* into a small flat palm, and fixed to the helve and the head by a screw-nail at each point. This part of the rake is called the stay or brace, and in the last described form is as perfect as the case will admit of.

(2810.) To have said so much on an implement so simple and of such small intrinsic value may appear trifling, but the hay-rake is not so much the object here as the development of the principle, on which depends the strengthening and supporting the parts of any form of frame-work, whether of wood or iron, by the aid of diagonal stays or braces. This single member in all constructions is of such importance that no opportunity should be lost in impressing the principle on the mind of all, whose business or interest lies in those departments of mechanics where its application is required.

(2811.) The *American hay-rake*, is so called from its having been an importation from America. This implement is represented in fig. 501, in perspective,

Fig. 501.

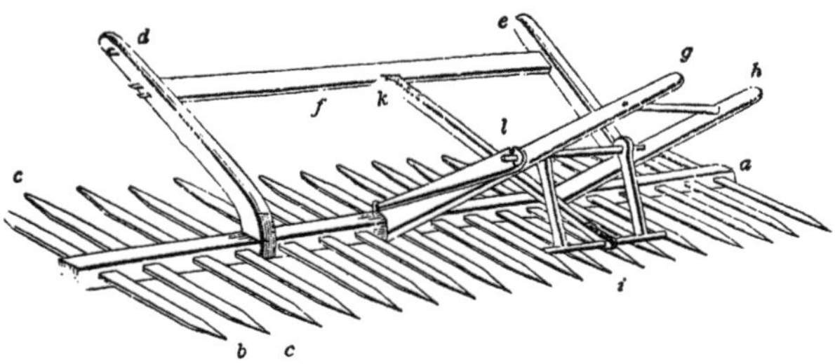

THE AMERICAN HAY-RAKE.

and lying in the working position. It consists of a beam or head *a a*, 9½ feet long and about 4 inches square; it is perforated with 18 square mortises, into which the transverse bars or teeth *b b*, *c c*, &c., are firmly fixed. These teeth are about 3 feet 10 inches long, or 21 inches on each side of the head *a a*, and are about 1¼ inch square on the body, slightly tapering to each end, where they are rounded off to a blunt point, but chiefly upon that side which is to lie next the ground, and this constitutes the body of the rake. It is drawn by a horse yoked to the draught-frame *d e*, of which *d* and *e* are two naturally bent bars of wood about 3½ feet long and 4 inches by 2½ inches at the butt-end; at the butt they are worked out in the end to a semicircle half embracing the head *a a*, which is here dressed into a cylindrical journal, and the bars *d e* are secured to the journals by a strap of iron passing round each and nailed or bolted upon the bars, leaving them freedom to traverse upon the journals. The stretcher-bar *f* is mortised into *d* and *e* at a distance of not less than 2 feet from the head *a a*. The handle frame by which the implement is guided has the two bars *g* and *h* 4½ feet long and is 2½ feet apart; the bars being attached to the head in the same manner as described for *d e*. The two bars *g*, *h*, are also connected by 2 stretchers, and upon the lower one is appended the light pendant and movable frame *i*, the depth of which from the stretcher, to which it is appended, to the bar *i* is 10 inches, and is jointed to swing freely on the stretcher. The bar *i* of this last frame is put in connection with the draught-frame, by means of the connecting-rod *i k*, which is jointed movable at both ends on round journals, and strapped as before. The stretcher on which the frame *i* is appended, is prolonged at each end to receive the catch-bars *l*, on the outside of the frame *g h*, one of which bars is seen at *l*, jointed on the prolonged stretcher, the position of which is 2 feet 3 inches from the journal of the head *a*. The catch-bars are $3\frac{1}{2}$ inches by $1\frac{1}{2}$ inch at *l*, but diminish forward to $1\frac{1}{2}$ inch square at the point, where they abutt upon an iron stud, which is presented at two of the opposite corners of the rake-head, at each side of the handle-frame, serving an essential purpose in the management of the implement.

(2812.) In the working of this rake, it lies nearly flat on the ground, and when the draught-frame is at its proper height, the connecting-rod *i k* keeps the hanging frame just within the extremity of the teeth that are then behind, and nearly bearing upon them. In this position also, the point of the catch-bars *l* is quite free of the studs of the head or beam, and by pressing down the handle-frame, the pendant *i* will come down upon and depress the teeth that are looking backward, raising at the same time those in front, such as for the purpose of passing over any obstruction. When, on the contrary, it is wished to depress the front teeth, the handle-frame is raised till the points of the catch-bars press against their studs, which will depress the front teeth; and by continuing to elevate the handle-frame, the connecting-rod, from change of position in the bars *i k*, *i l*, and *l f*, will push the pendant *i* beyond the extremity of the teeth behind, when the front teeth taking the ground, and nothing to resist the rising of those behind, the rake will immediately tilt over, the fore and hind teeth changing places; but, in other respects, every thing will be the same as before. The effects of the motion and tilting, it is evident, will be, that in the progressive state the rake collects the hay or straw upon it chiefly in the front part; and when the attendant sees that the rake is filled, he raises the handles and tilts the rake as above described, leaving the collected mass at the spot where the tilt occurs.

(2813.) A slight consideration of this implement will shew the effectual and convenient manner in which its work is performed; but it will also probably occur to the observer, that, for progressive motion, it is by no means so well adapted as for collecting and depositing the products. The heavy transverse bar, or head of the rake, is drawn forward in the worst possible position, or what is called broad-side on; and it appears very obvious, that this defect could be removed by simply applying a pair of low light wheels to the ends of the head; their diameter need not exceed 8 inches, and they should be very light. Perhaps it may be owing to this defect that the American hay-rake has of late years not been in such high repute as formerly.

(2814.) The *hay* and *stubble horse-rake* is an implement which acts on principles somewhat akin to the last, and is also frequently employed for the same purpose during the hay harvest; at present, indeed, it is much more extensively used than the American rake. The varieties of it are also numerous, but all possessing the properties of collecting to a certain amount, and then by a simple operation of the hand, dropping in a collected mass the quantity of hay or straw thus accumulated in the machine. One of the most common forms of the *horse-rake* is represented in perspective, by fig. 502, and somewhat more in detail by fig. 503, which is a longitudinal section of the body and adjoining parts, the same letters marking corresponding parts in the two figures. The body of the machine consists of a main beam *a a*, 9 feet long and 3 inches square, and another a swing-bar *b* of equal length, but only $2\frac{1}{4}$ inches square, and these are bolted together upon two side-bars *c c*, also $2\frac{1}{4}$ inches square, but only 2 feet

Fig. 502.

THE HAY AND STUBBLE HORSE-RAKE.

9 inches long, and the two first are further supported by two intermediate bars *d d*, which, in the vertical direction, are bent as seen in section fig. 503; these at their junction, with the main beam, are tenoned instead of being bolted. A

Fig. 503.

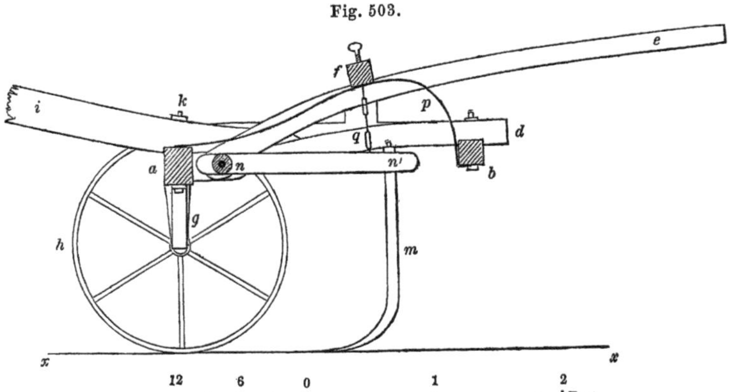

SECTION OF THE HAY AND STUBBLE HORSE-RAKE.

pair of handles *e e*, are jointed upon the thorough bolt *n*, fig. 503, and are bolted to the lifting-bar *f*, which *rests* at each end on wooden noggs tenoned into the bars *c c*, and the ends of the lifting bars are each furnished with a thumb-screw *f*, for adjusting the height of the bar. The axles *g* are kneed to stand 5 inches below the main beam, and are bolted to it directly at the tail of the axle, and also through the stud-bracket at the neck. The wheels *h* are 20 inches diameter, of light make in cast-iron, and the horse-shafts *i i* worked to such a curvature, that they shall be of the proper height for the horses' shoulders when the body of the implement stands level, are bolted to the main beam and intermediate bars at *k*; and are likewise supported by the iron stays *l*. It is also to be observed, that in both figures the horse-shafts are partially or altogether broken off for want of space. The machine is armed with 20 tines or teeth *m m*, &c. which are 18 inches high from the ground-line *x x*, fig. 503, to the upper edge of the rake-head *n n'*, each tooth being fixed into a separate head with a screwed tail and nut, while at the point they are bent forward and sharpened, but are adjusted so as not to run into the ground. The rake-heads are threaded upon the bolt *n'*, which extends from side to side of the machine, and between each 2 rake-heads a wooden stretcher-roller is applied, preserving the heads at the proper distance, and through these, as well as the handles *e e*, the bolt *n'* also passes. Seven stripping-rods *p*, bent as in fig. 503, are attached to the main beam and the swing-bar by bolts, and each of the rake-heads are attached to the lifting-bar *f* by means of a short chain *q*, whereby, when the handles *e* are lifted, the bar *f* and all the rake-tines are lifted from the ground by one operation, and when let down again till the lifting-bar or its adjusting screws rest on their studs, each tine is then at liberty to rise and fall a small space, as the inequality of the ground or any obstruction may require. In the act of working, the machine advances till the tines have collected as much hay or straw as can be contained in the bosom of the tines and heads, and when thus filled, the attendant, by lifting the handles, causes the tines to be drawn upward through amongst the stripping-rods, which discharge the contents of the rake upon the ground; the handles are immediately let down to the working position, and the work proceeds as before. As a stubble-rake, this machine is in very general use, and is highly prized from its efficiency in gleaning stubbles. The usual price of it is from L.3, 10s. to L.3, 15s.

(2815.) Amongst the many varieties of this implement may be mentioned one that has been patented by Mr J. C. Grant of Stamford. All the essential working parts of it are taken precisely from the earlier machines, similar to what is here described, and the novelty of the patent lies in the arrangement of jointed levers employed for lifting the tines and discharging the collected mass. It forms an exceedingly neat and compact machine, but is too expensive for the ordinary farmer, the price being L.8, 10s.

(2816.) The *Rick-cloth* is an important article in the English farmer's stock in trade, not so much for its value, as for its usefulness; and the extension of its use amongst farmers in general, is a matter deserving of consideration. It is of essential service to the hay-maker on the English system in the building of his large hay-stacks or ricks, since by its use, when spread over the stack, he is independent, in some degree, of changes of weather, as it forms a sure defence against rain, should that overtake him in the midst of his operations. Fig. 504 is a representation of a hay-stack in progress of erection, with the *rick-cloth* pitched over it, exhibiting the mode in which this cover is applied, and which may be described as follows.

(2817.) Two light poles or spars of a length that will rise 10 or 12 feet above the intended height of the stack, are placed one at each end, and about 2 or 3 feet distant from the ends of the stack; these are supported by 3 guy-

Fig. 504.

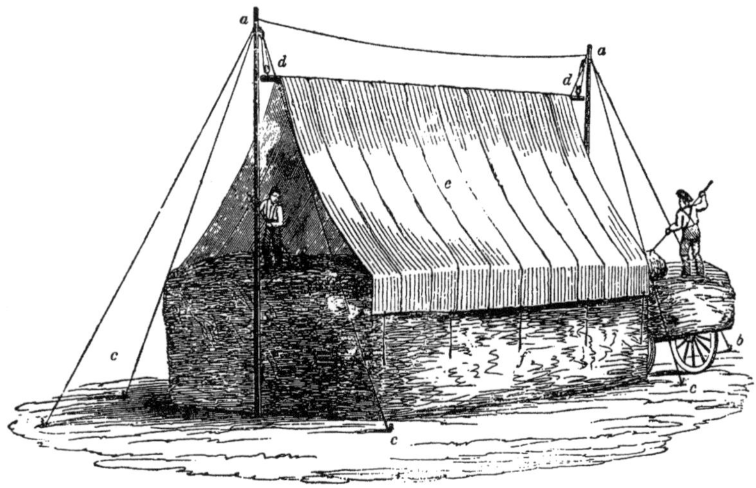

THE MODE OF STRETCHING A RICK-CLOTH OVER A HAY-STACK WHEN BEING BUILT.

ropes, two of which *c*, *c* stand opposite to each other, and the third *b*, in the direction of the length of the stack, the ends being all secured to wooden stakes, and the fourth guy *a a* is common to both poles, extending from the head of the one to that of the other. A third spar, equal in length to the distance between the poles, is laid down either on the ground, if the stack is not yet begun, or in that portion of it that has been laid down, and the two tackles *d d* being previously appended to the head of the respective poles, are hooked below to the third pole, which is to form the ridge of the cover. The canvass covering, or rick-cloth *e*, is now laid over the ridge-pole, with its middle bearing on the ridge-pole, and, if necessary, lashed thereto. The pole and cover are then hoisted up to any desired height, forming a temporary but safe and water-tight roof to the stack while it is being completed. The lower edges of the cloth are secured by bracing lines *f* either to the sides of the stack, or they may be stretched out beyond these by means of the lines, and attached to stakes or otherwise, after which, as the stack rises, another pull is made upon the tackles, hoisting up the cover to afford sufficient space under it for the builders. The figure represents one man in the act of forking hay from the cart upon the stack, and another adjusting the hay with a fork over its surface. Rick cloths of all descriptions and sizes are manufactured by Mr Benjamin Edgington, 2 Duke Street, Southwark, London Bridge, London.

(2818.) The *Flail*, fig. 505, is an implement of considerable antiquity, and of very extended application. Till about 80 years ago, it was the sole implement employed in Britain, and to this day is nearly so over a great part of Europe, for thrashing out the grain from the straw. Though falling rapidly into disuse in this country, it is nevertheless an important implement in many others, and therefore claims our attention in common with those of higher pretension.

The flail consists of two parts, the *hand-staff* or helve *a b*, and the *supple* or beater *b c*; the first is a light rod of ash about 5 feet in length, slightly in-

Fig. 505.

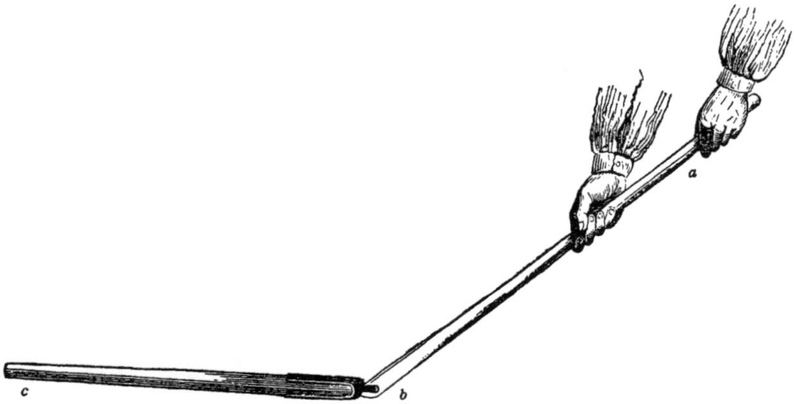

THE FLAIL.

creased in breadth at the lower extremity, where it is perforated for the passage of the thongs that bind the beater to it. The beater is a rod of from 30 to 36 inches in length, frequently also made of ash, though a more compact wood, such as thorn, is better adapted for it. If not properly applied, the ash beater will very soon separate into thin plates, which are portions of the concentric layers of the wood, and their separation arises from the beater falling upon the flat or convex side of these annular layers—or the reed of the wood, as vulgarly called. To prevent this disintegration of the wood, the beater should be constructed to fall upon the *edge* of the segmental portions of the *reed*, which is easily accomplished in its formation. The usual form of the beater is cylindrical, but frequently thickened a little towards the extreme end, the diameter being from $1\frac{1}{4}$ to $1\frac{1}{2}$ inch. For the most part, it is attached to the hand-staff by a strap of leather, or more frequently of hide untanned; when mounted in this manner, the beater is formed with two projecting ears, standing at right angles to the side on which it is intended to fall, and about $1\frac{1}{2}$ inch from the end by which it is attached, serving the purpose of retaining the end of the beater within the strap. The strap is about 8 inches long and $1\frac{1}{2}$ inch broad; it is bent over the end of the beater, and the tails brought to embrace the sides of it beyond the ears. The strap being previously perforated with four holes in each tail, it is bound by a thong of leather laced through the holes and round the neck of the beater; the upper turn of the lacing thong catching the ears, prevents the strap from sliping off. The strap, thus applied, forms a loop standing about 1 inch beyond the end of the beater; and through that, and the perforation in the end of the hand-staff, another and stronger thong is passed several turns and secured, forming thus a kind of loose swing-joint that allows free action to the beater in its gyration round the head of the thrasher, and its descents upon the thrashing-floor. Another mode of mounting the beater is by applying a strap of iron in place of leather, which is fixed to the wood by rivetting, leaving a loop as before, which must be nicely rounded and smooth, to prevent the too rapid chafing of the thong by which it is bound to the hand-staff, in the same manner as described above. The figure here described exhibits the iron strap. In constructing a flail, a very general practice prevails, which is, to have the beater club-shaped, or thickest at the furthest extremity *c*, intended, no doubt, to give the better effect to the blows; but when we consider the effects arising from the manner of wielding the instrument, any additional weight at the extremity seems misapplied. The greatest amount of useful effect will be produced by the beater when every point in its length strikes the floor with an equal amount of momentum or force; but there will be a constant tendency to a larger amount of momentum at the extremity *c* than at any other point, and a club-shaped beater will always augment this tendency, for the greater velocity of the extreme end, during the gyration of the instrument, multiplied by its greater weight, must give an undue preponderance of effect to that part of the beater, thereby lessening the general effect upon the work under performance. The opposite mode, which is also practised, to make the beater thinner towards the extremity, as exhibited in the figure, is more consonant to the laws of dynamics, and there can be no doubt that its practical effects will be equally favourable as compared with those of the club-shaped beater.—J. S.]

---

69. OF SUMMER-FALLOWING AND OF LIMING THE SOIL.

> "The bare fallow brings to teeming foyson."
>
> MEASURE FOR MEASURE.

(2819.) Although summer-fallow occupies the same division of the farm as green crops—turnips, potatoes, tares—yet it may most characteristically be regarded as the first preparation for the crop of the following year; it is a transference of a portion of the land, with the labour bestowed upon it, from one year to another; it forms the connecting link between one crop and another. But although the preparation of the soil for a part of the crop of two consecutive years are conducted simultaneously by means of summer-fallow, yet the crops which occupy the soil thus simultaneously prepared, are committed to it at very different periods, the green fallow-crops being sown early in summer, while the sowing of the fallow-crop on the summer-fallow is delayed to autumn; so that before the latter makes its appearance above ground, the former have almost advanced to maturity. Since the crop on summer-fallow is delayed to autumn—till the eve of commencing another agricultural year—the practical effect of the delay is to dispense with a crop for a whole year on that part of the fallow-break which is summer-fallowed, and, on this account, such a fallowing is commonly called a *bare-fallow*. As an entire crop is dispensed with, bare-fallowing should impart such advantages to the land as to compensate for the rest and indulgence which it receives; and so, in fact, are its advantages felt on some sorts of soils. But the truth is, after all, that bare-fallowing is a necessitous operation in soils that will not carry green or summer crops; and if such soils will not carry them, they must be operated on so as to be made suitable for a crop which they will not only bear to perfection, but will pay the expense of the operation. The sort of soils alluded to are heavy clays; and why will *they* not bear green summer-crops? A satisfactory reason cannot be given; but the fact, as developed by experience, is, that their nature is unkindly to the growth of plants commonly used in a green state; and their heavy, wet, and obdurate nature, prevents them, at any rate, from being prepared in time for sowing such plants. Could clays be altered in their nature by any means, they might be employed in raising summer-crops as well as the naturally more kindly soils; and such a change has been effected on many clay soils which were formerly incapable of rearing them; and the change has been effected by ameliorating their texture by thorough-draining, skilful tillage, and liberal manuring and liming. In this way the bounds of bare-fallow have been much circumscribed, and those of green crops as much extended. Still the heavier class of clays—the deep alluvial clays—have not yet been ameliorated to the degree of bearing green-crops profitably, so they must continue to be bare-fallowed; but more than this, part of even the ameliorated soil of almost every farm is necessitated to be bare-fallowed, for want of an adequate supply of farm-yard manure. Farms in the vicinity of large towns may be amply supplied with extraneous manure, to make up for the deficiencies of the farm-yard; but as most farms are beyond the reach of such assistance, it may be alleged that bare-fallowing, to some extent, is practised every year upon every farm; though the limits of compulsory fallowing have been much circumscribed of late years by the purchase of extraneous manure from distant sources, which are easily conveyed, and sold at prices that afford a profit. Those manures, and I only allude to them here, are bone-dust and guano. These, superadded to draining and deep-ploughing, have afforded the power to cultivate green-crops upon soils which were naturally unfit for them; and, without such auxiliaries, soils suitable for their growth would be obliged to be bare-fallowed, to allow time to collect the requisite quantity of manure to support their fertility. Until manure is, therefore, obtained in sufficient quantity, bare-fallow must exist; but whenever that desideratum shall be accomplished, many farms will dispense with bare-fallow altogether. But there is a natural obstacle to the increase of manure in farms themselves; for it so happens, that the greatest quantity of straw, which is the great source of fertilizing manure, is afforded by land the least fitted for green crops; and, on the contrary, land best fitted for green crops affords the least quantity of straw. Turnip-soils cannot supply as much straw as to manure, to the degree and in the state it should be applied to green crops, little more than ½ of the fallow-break; whereas clay soils afford as much straw for manure, in the state in which it may be applied to them in bare-fallow, as sufficiently to manure the fallow-break.

(2820.) The land subjected to bare-fallowing should have the strongest texture, be foulest of weeds, if any there be, and be situate farthest from the steading, that the carriage of turnips may be rendered as short as practicable. The winter treatment of the fallow land is the same as that for the summer crops, and this has already been described in preparing the soil for potatoes (2406.) and turnips (2499.). If one furrow —that of two-out-and-two-in, or of four-out-and-four-in—has been given to the fallow-break after cross-ploughing, it will be as much as time can afford from working the potato and turnip land; and when it is found that the fallow-break will not likely be worked for some time to come, it is better to let it lie in the rough state left by the plough, than to render it smooth with the harrow; because, should dry weather ensue, the air will more easily affect rough than smooth land, or should it prove wet, the rain will less likely render rough land tough, than land in a compact and smoothened state, and in strong soil, rain most advantageously moulders down rough clods.

(2821.) When leisure is again afforded to pay attention to the fallow break from the advanced working of the turnip-land, the *state* of the fallow-soil should be particularly examined. Should the weeds in the soil consist principally of fibrous and fusiform-rooted plants, they will be easily shaken out by the harrows in dry weather; but should the running roots of weeds be found to have threaded themselves through hard round clods, these will not be so easily detached, and it requires considerable skill in the farmer to conduct his operations so as to detach them with the least application of labour. It is inattention to *these states* of the weeds which causes so much unnecessary work in the subsequent part of summer in cleansing fallow land. If clods containing portions of the running roots of plants are knocked ever so much about in dry weather, they may be broken into smaller fragments; but the roots in them will be subdivided into as many pieces as there are broken clods, and the land will be as far from being cleared of weeds as ever, nay, the greater number of parts in which the roots are divided, so many more chances have the weeds of being disseminated over the land.

(2822.) In such a case, which is of frequent occurrence on strong land, the best plan is to allow the roots to grow for a time, and the force of vegetation will have sufficient power to break the clods, or will render them easily so by the roller, or to reduce the clods by rolling after such a shower of rain as shall have nearly penetrated them. A precaution in the use of the roller should, however, be here observed. When most of the soil is in a mouldy state, rolling the hard clods found on it, will only bury, not break them. Rolling, with a view to breaking large clods, should therefore be performed when the soil is in a firm state, against which the clods will be reduced to powder. After such a rolling, the land should be harrowed a double tine, first one way, and then across another way. The weeds and weed-roots will then be seen upon the surface. It is not expedient to gather weeds immediately on their being collected by the harrows, as a good deal of fresh soil adheres to them. A day or two of drought should intervene, and the weeds will then be easily shaken free of soil by the hand. It has been recommended by writers to gather the weeds of fallow-land by a raking implement, such as the American hay-rake, fig. 501; but every instrument of the kind will rake together clods as well as the weeds adhering to them, and if these are carried away with the weeds, the land will be impoverished by the loss of its finest soil; whereas the hand which throws the weeds into convenient heaps, can at the same time shake them free of soil. In collecting weeds, the field-workers should be ranged in a row as when weeding corn, every two throwing the weeds into the same heap; and the rows of heaps should be placed as far asunder as to allow a cart to pass between them, and take away 2 rows at a time. Many writers recommend the weeds to be burnt on the ground. No doubt, weeds will burn readily enough when dry, and the ashes of weeds constitute good manure, but, for my part, I never saw heaps of weeds thoroughly burned, and have seen their remains scattered, again to render the land foul. I agree with Lord Kames, that it is better to make a vegetable compost with weeds, than to destroy them by incineration, and with him readily demand, "What better policy than to convert a foe into a friend?" It is impossible to determine beforehand how many times fallow-land should be ploughed, harrowed, grubbed, and rolled, to render it clean; but it should be borne in mind, to incur the least expenditure of labour in accomplishing the object fully. It was once the practice to work fallow-land until it was reduced to the state of meal; but experience has long established it as a fact, to be better for the ensuing crop of wheat to preserve a good-sized clod upon the surface of the ground in winter, however much the ground may be otherwise pulverized. The land must have been very foul, the weather very unpropitious, or much time wasted, if the fallow-land is not ready for the manure by the beginning of August, before the chance of harvest interfering with the process of manuring.

(2823.) The usual mode of *laying dung on fallow-land* is to feer the ridges (829.), cart on the manure in heaps, spread them, and plough the manure in. So far as I have observed, the manure is spread over a large portion of the surface some time before it is ploughed in, when, of course, much of its moisture will be evaporated. I very much prefer another mode of ploughing in the dung, which is, to *angle-drill* the land in preparation for the dung; that is, to set up the land in *single* drills (2172.), and fig. 379, from the flat without any feering, beginning at one corner of the field, and terminating at the opposite one: the plough making the drill in one bout, the cart depositing the manure for 3 drills, 4 women spreading it immediately after the cart, and the plough following and covering up the dung in one drill in another bout (exactly as represented in potato-planting, fig. 411), is a process which so quickly and completely covers in the dung, while in possession of its moisture, that it should be universally adopted. The land remains in the drilled state until prepared for the wheat seed in autumn.

(2824.) Fallow-land is not dunged so heavily as that for green crops, not so much from fallowed soils not bearing heavy manuring, as from want of manure. From 12 to 15 tons the imperial acre is an ordinary manuring for fallow. The manure need not be so well fermented as for green crops, as there is usually sufficient time for its fermentation in the ground before the wheat is sown. If there is not sufficient time, it should be fermented before its application, as it is not expedient to sow wheat to stand the winter in soil rendered hollow by rough dung. In strong carse clay, the manure is often applied in a state little removed from wet straw, and yet it seems to answer well in such a soil.

(2825.) While treating of fallow, it is necessary to notice the *liming* of land, as lime is commonly applied at that period of the rotation of crops, though by no means applied every time the land is fallowed. It has always been a favourite practice with farmers to apply lime in as caustic a state as practicable, because, perhaps, it is then in the state of finest powder, and easily commixes with every species of soil. When obtained from the kiln or from shipboard, it is in lumps, light in weight, and are technically called *lime-shells*. There is a difference in the practice of disposing of lime-shells while in preparation for the soil. Some lay down the shells in small heaps upon the feered ridges, while others lay them in large heaps along a head-ridge. It is clear that shells cannot be laid at once upon the land, unless the land had previously been sufficiently fallowed; and as land occupies a considerable time in being fallowed in a proper manner, it is also clear, that no considerable quantity of lime can be driven, after the fallow is ready, unless the kilns are situate very near; and, at any rate, it is unnecessary to lay the lime upon the land, but at a short period before the wheat is sown. Besides, when shells are placed in heaps on the ridges, they must remain a considerable time there till reduced to powder by the air, when the lime will have lost a considerable portion of its causticity by union with the carbonic acid of the air, unless a good deal of rain shall have fallen to hasten its slaking. To preserve the shells intact, till needed, they should be put in large heaps, the outer surface of which may become neutralized by the action of the air, but the interior of which will not be so affected. Meantime the land is worked as opportunity offers, while the heaps occupy a head-ridge.

(2826.) A week or so before the lime is applied, water is poured on the large heaps of shells, in order to reduce them to a state of impalpable powder. The water will all be absorbed by the lime, which will, nevertheless, continue quite dry, thereby indicating that the water has disappeared by reason of its chemical union with the lime. A great quantity of heat is evolved during the time the lime takes to fall to powder; and when that last has been accomplished, the heaps will have swelled to more than 3 times their former size, when the lime is said to be *slaked*, and is then in its most caustic state. While the slaking is proceeding, the land that was manured in drills is cross-harrowed a double tine, to make it flat; after which the ridges are feered; and the lime is then spread along the feered ridges. The lime is spread in this way:—The frying-pan shovels, fig. 176, are the best implements for filling carts with, and spreading lime on land. A *calm* day should be chosen for the purpose, but should there be an air of wind, the single-horse carts should be so placed at the heaps as that the lime-powder which floats in air should be blown away from the horses and men. Powdered lime is heavy, but all that can lie upon a shovel is so light, that one ploughman takes a heap, and with one of his horses in a cart, for a yoking at a time, fills his own cart, and spreads the lime from it upon the land, with the shovel. The liming should be conducted against the wind; and when a number of men take from different heaps, they should so arrange themselves along the feered ridges as that the cart farthest down the wind takes the lead in spreading. In spreading lime, the man walks along the middle of the feered ridge, and casts the shovelfuls right and left from the middle towards the feering furrows, which will, of course, become, by ploughing, the crowns of the future ridges. The man who can cast the shovelfuls with either hand will spread lime better than one who is right or left handed only. The lime should be spread *evenly over the surface*, but it may be spread thicker on one part of the field than another, according to the nature of the soil. On light knolls it may be sprea thinner than in hollows, where the soil is either deeper or stronger. Progressively as the lime is spread, ridge after ridge, it is harrowed in and mixed with the soil; and immediately on the entire field being limed, the ridges are ploughed with a *light* furrow, to bury the lime as little as possible, and which constitutes the seed-furrow of the future crop. It will be observed, from what has been said on the manuring and liming of fallow-land, that the lime is spread above the dung, and sometime after its application. This relation between the two substances is held, because it is conceived, that, as dung has a natural tendency to rise to the surface, and lime to descend, this is the proper relation they should bear in the soil. Whenever rain falls, the liming should be discontinued.

(2827.) It is proper to put a cloth over the horse's back and the harness, and the men may cover their face with crape, to save its orifices being cauterised by the quick-lime. The horses, when loosened from work, should be thoroughly wisped down and brushed, to free them of every particle of lime that may have found its way among the hair; and, should the men feel a smarting in their eyes or nose, a little sweet thick cream will be felt as an agreeable emollient; and the same application will prove useful to the horse's eyes and nose.

(2828.) The *quantity* of lime that should be applied depends on the nature of the soil, the lighter soils requiring the less, and the stronger soils the greater quantity. On light turnip-soils, some think 120 bushels to the imperial acre sufficient, whilst I have used 150 bushels with benefit. I have seen as much as 510 bushels applied to the imperial acre, wheat-land, with manifest advantage. But perhaps from 150 to 240 bushels may be considered fair average quantities, according to the nature of the soil. On weak moory soils, 75 bushels are, perhaps, enough to commence its improvement with. The sort of lime should determine the quantity applied, the stronger being used in less quantity than the weak. The English lime is much more caustic than the Scotch. It is not customary to apply lime often to land, a farmer not thinking it expedient to apply it oftener than once in a lease of 19 years, on account of its expense. Its common price is 3s. per boll of 6 bushels, consequently its entire cost, at those quantities, will be from L.3, 15s. to L.6 per acre for the best sea-borne English lime, exclusive of carriage; the Scotch sells for 10s. per cart-load of 4 bolls, including carriage for 10 miles.

(2829.) Lime is applied at different periods of the year, according to the state of the land. On summer-fallow it is applied immediately before the wheat is sown in autumn. It is also used immediately after taking up the potato crop in autumn. It is applied to the land cleared of turnips by sheep, just before the sowing of the barley-seed in spring. It is also applied before the turnip-seed is sown in the beginning of summer; and it may be applied to lea immediately before being ploughed for oats in early spring. I do not say that it is immaterial to the proper use of lime to choose the season in which it is applied, convenience often determining that point as much as propriety; but experience has taught that it is used to manifest advantage when spread on land in summer-fallow, and for barley-seed immediately after the ground has been cleared of turnips by sheep.

(2830.) Lime is usually procured in summer and autumn, as the kilns are only kept in activity in those seasons, so when it is intended to apply it in spring, it is necessary to procure it in autumn, and keep it all winter; and to preserve it in the desirable state in winter, the heaps of shells should be covered with a thick coating of earth, and every crevice that appears in it should be immediately filled up. I am quite aware of the opinion of some farmers that lime is equally efficacious in the soil in the effete as in the caustic state, and Lord Kames was of that opinion; and, therefore, precautions to preserve it in a caustic state in winter may, by them, be deemed unnecessary; but as the general opinion runs in favour of quick-lime, and which opinion I support, I have treated the subject accordingly, until experience shall instruct us better. There is this advantage, however, in using quick-lime, that it is much more easily spread upon, ploughed into, and mixed with the soil than effete lime.

(2831.) It is supposed that light and heat, together with cleansing and working, have a beneficial effect upon soil. That these agencies promote fertility in some way, perhaps by affording facility to the union of oxygen with the soil, appears certain, for a smaller quantity of manure will raise as large a crop with bare-fallow as a greater quantity without it; and yet this particular result is only obtained from a peculiar class of soils—namely, the strong clays, for all turnip soils actually become more fertile by the overshadowing of a luxuriant crop of leaves than by bare-fallowing.

(2832.) The action of lime upon land seems generally well understood by farmers. They conceive that lime operates in two ways upon soil, namely, mechanically and chemically. *Mechanically*, it subdivides the adhesive portions of obdurate clay; hence it was customary in the Carse of Gowrie, when the high price of grain remunerated the outlay, to apply lime to the land every time it was bare-fallowed, that is, every 6 or 8 years according to the rotation of crops pursued. I suspect that liming is not now so frequent there as it was wont to be. *Chemically*, lime unites with vegetable matter, and assists in its decomposition, hence it has been found to act *very* beneficially on all *deaf* soils

It renders loose soils more firm. In strictly chemical language, quicklime acts aş an alkali, and this property has led some writers to assert, that green vegetables are not decomposed, but rather preserved, by caustic lime—a result which cannot be experienced bý the practical farmer in the soil; because, as caustic lime very soon becomes a compound of a hydrate and a carbonate in the soil, the property of this latter state is to accelerate rather than retard the decomposition of vegetable matter. As the consideration of this subject would lead us into a long discussion, I must content myself with quoting only a part of the account of the theory of its action by Professor Johnston, in order to shew that the opinions of farmers is not far astray on this subject, which, in some respects, is yet but obscurely understood. "Lime acts in two ways on the soil," says the Professor. "It produces a *mechanical* alteration, which is simple, and easily understood, and is the cause of a series of *chemical* changes, which are really obscure, and are as yet susceptible of only partial explanation. In the finely divided state of quick-lime, of slaked lime, or of soft and crumbling chalk, it stiffens very loose soils, and opens the stiffer clays; while in the form of limestone gravel, or of shell-sand, it may be employed either for opening a clay soil, or for giving body and firmness to boggy land. These effects and their explanation are so obvious to you, that it is unnecessary to dwell upon them. The purposes served by lime, as a chemical constituent of the soil, are at least of four distinct kinds. 1. It supplies a kind of inorganic food, which appears to be necessary to the healthy growth of all our cultivated plants. 2. It neutralizes acid substances which are naturally formed in the soil, and decomposes, or renders harmless, other noxious compounds which are not unfrequently within reach of the roots of plants. 3. It changes the inert vegetable matter in the soil, so as gradually to render it useful to vegetation. 4. It causes, facilitates, or enables other useful compounds, both organic and inorganic, to be produced in the soil, or so promotes the decomposition of existing compounds as to prepare them more speedily for entering into the circulation of plants."* In conclusion on this subject, when we consider that 1 ton of limestone is reduced in weight to a little more than 11 cwt. of shells by burning, we cannot but admire the simple art which renders so valuable a material available to the purposes of the farmer, even when it is situate at a considerable distance from his farm.

(2833.) As fallowed land is usually manured along the feered ridges by depositing the loads into heaps, I might here give a table shewing the number of heaps each cart should afford in manuring an acre with a given number of cart-loads; but as *heaps* of manure are so indefinite a standard of measure, snch a table would practically prove of little service. A much more accurate plan is to number the ridges on an acre in each field, and at every part of a field where the ridges are changed in length, as I have recommended before (478.), and try and lay down the manure on the first ridge in the proportion it is proposed to manure the acre, and by the time the second ridge is gained, the man who hawks out the dung, fig. 410, will have found out how close the hawkfuls should be laid down, and how large the heaps, if any, should be made.

(2834.) There is a species of fallowing sometimes practised, which, although I do not approve of, I must nevertheless notice—it is what is technically called

* See the whole of the xviith Lecture of Johnston's Lectures on Agricultural Chemistry, from p. 528 to p. 608, for a careful examination of the properties of lime in reference to its effects on the soil. This, in my opinion, is the most valuable monograph on lime, as a material used in agriculture, that has ever appeared.

rag-fallowing. It consists of pulverising lea ground in summer to its preparation for wheat in autumn. The lea is broken up in August, or as long before harvest as to allow time for the land to be worked ere the commencement of harvest; and as the object is to reduce the turf as much as possible, the first ploughing should be given with a shallow furrow. After the soil becomes dry, it should be harrowed repeatedly in double tines along and across the ridges in order to tear the furrow-slices to pieces, and to shake the earth from the grass, The ground then should lie some days to dry, when it may be harrowed again, if it is thought that any more earth can be shaken from the turf. After the turfs are greatly withered, the land should be cross-ploughed with a deeper furrow to bring up a body of mould, which, when dry, should again be harrowed, that the now light turfs may be brought to the surface still farther to wither and die. The land should then be ploughed in ridges for the seed-furrow, to remain till seed-time after harvest, by which time it will have consolidated and be in readiness for the seed.

(2835.) Wheat appears to grow very well in England after lea, and as it does so on one furrow, there is no need there of rag-fallowing; but the case is different in Scotland, where wheat never thrives upon one furrow after lea, and the expedient of rag-fallowing is adopted to form a consolidated mould for the plant; but in attempting to attain this end, the opposite error is frequently committed, that of rendering the soil too loose for wheat; at all events, the crop never looks promising after such a mode of fallowing, and, in truth, the expedient is never resorted to but by farmers who take advantage of their land. And yet there is no apparent reason why wheat should not succeed after lea in Scotland as well as in England. Manure, to be sure, cannot be applied upon the surface of the lea, as it would be in the way of working the rag-fallow, and the seed-furrow on lea cannot be obtained so deep as to bury any manure but in the shortest state.

(2836.) It has been ascertained, however, that by using the presser-roller, fig. 381, in compressing the first furrow on rich, mellow, tender lea, wheat may be successfully raised in Scotland; it is, therefore, surprising why this implement has not found its way into general use, though it is not difficult to discover an obstacle to its extended use as a substitute for rag-fallow, namely, the sacrifice of pasture which lea-wheat necessarily occasions, and to which may be added a circumstance that makes the sacrifice greater, that the same land which bears the best wheat yields the best pasture.

(2837.) This seems a favourable opportunity to notice a species of field-labour, namely, *spade-husbandry*, which has of late years been recommended as a substitute for summer-fallowing; not on a small scale, as suitable to the case of cottiers and small farmers only, but, on a large one, fitted for a farm of the largest class. The only farmer I have heard of, who has practised spade-husbandry on a scale commensurate with ordinary farming in Scotland, is Mr Archibald Scott, Southfield, East Lothian. "In 1831," says Mr Scott, "I determined to ascertain the difference of the expense and produce between trenching land with the spade, and summer-fallowing with the plough in the usual way. I therefore trenched 13 acres of my summer-fallow break in the months of June and July. I found the soil about 14 inches deep; and I turned it completely over, whereby putting up a clean fresh soil in the room of the foul and exhausted mould, which I was careful to put at the bottom of the trench; and this operation, I found, cost about L.3, 12s. per imperial acre, paying my labourers with 1s. 6d. per day." Here it is necessary for me to state, that Mr Scott refers to, and makes his calculations according to the Scotch acre, the usual measure of land still adhered to by the farmers of East Lothian; but the particulars about to be presented to you from Mr Scott's statement I have reduced to the imperial measure, which ought to be the imperative standard of the kingdom. "The rest of the field, consisting of about 11 acres," continues Mr Scott, "I wrought with the plough in the usual way, giving it 6 furrows, with the suitable harrowing: I manured the field in August, the trenched got 7 cart-loads per acre, the ploughed land 14. The field was sown in the middle of September, and the whole turned out a bulky crop as to straw, particularly the trenched portion, which was very much lodged. On thrashing out both, I found them to stand as under:—

By *trenched* wheat, 42 bushels per acre, at 6s. 9d.,				L.14	3	5
To 2 years' rent, at 40s. per annum,	L.4	0	0			
... expense of trenching,	3	12	0			
... seed, 2½ bushels, at 6s. 9d. per bushel,	0	16	10			
... 7 cart-loads of manure, at 3s. 9d. per load,	1	6	3			
... expense of harvesting, thrashing, and marketing,	1	4	0			
Profit,	3	4	4			
	L.14	3	5	L.14	3	5

By *ploughed* wheat, 34 bushels per acre, at 6s. 9d. per bushel,				L.11	9	6
To 2 years' rent, at 40s. per annum,	L.4	0	0			
... 6 ploughings and harrowings, at 8s.,	2	8	0			
... seed, 2½ bushels, at 6s. 9d. per bushel,	0	16	10			
... 14 cart-loads of manure, at 3s. 9d. per load,	2	12	6			
... expense of harvesting, thrashing, and marketing,	1	4	0			
Profit,	0	8	2			
	L.11	9	6	L.11	9	6"

This was but an experimental trial, and the result was certainly an encouragement to perseverance; but still it only proved, that trenching land with the spade might be a *substitute* for bare-fallowing, and nothing more; it did not prove that summer-fallowing might be *dispensed with*, so another experiment was worth the trial, to ascertain this important point. Accordingly, Mr Scott "now saw, that though it might be profitable to trench over my fallow-break during the summer months, it was by no means making the most of the system, as the operation was not only more expensive, owing to the land being hard and dry in summer, but that it was a useless waste of time to take a whole year to perform an operation that could as well be done in a few weeks, provided labourers could be had; and as, in all agricultural operations, losing time is losing money; as the rent must be paid whether the land is carrying a crop or not, so that in taking one year to fallow the land, and another to grow the crop, two years' rent must be charged against the crop, or at least there must be a rent charged against the rotation of crops for the year the land was fallowed; as I felt satisfied, that, by trenching with the spade, the land would *derive all the advantage of a summer-fallowing, and avoid all the disadvantages attending it*, I determined on trenching about 40 acres of my fallow break, *immediately on the crop being removed from the ground*, and had it sown with wheat by the middle of November 1832, and I did *not apply any manure*, as I thought the former crop was injured by being too bulky. As the crop is now thrashed and disposed of, it stands per acre as follows:—

By average of the 40 acres, 36 bushels, at 7s. per bushel,		L.12 12 0
To rent of land per acre, ...	L.2 0 0	
... expense of trenching, ...	3 4 0	
... seed, ...	0 16 0	
... harvesting, thrashing, and marketing, . .	1 4 0	
Profit, . .	5 8 0	
	L.12 12 0	L.12 12 0"

This trial was also satisfactory, because it not only proved that trenching with the spade might be substituted for summer-fallowing, but that summer-fallowing might be profitably dispensed with altogether; but it must be owned to be rather sharp practice to make the same available soil begin to bear one crop *immediately* after having borne another, and that *without manure*, and at a period of the rotation when manure is usually administered to land. Mr Scott seems quite satisfied with the system, and so he may, since it is so decidedly profitable; but it may be reasonably asked, even before experience has ascertained it, How long will land be able to bear this system of trench-fallowing with impunity? "The advantages of trenching over summer-fallow are, in my opinion," says Mr Scott, "very decided, as it is not only cheaper, but, as far as I can yet judge, much more effectual. I am so satisfied of this, not only from the experiments above noticed, but from the apparent condition of the land after it has carried the crop, that I have this autumn cultivated about 120 acres with the spade, and the crops are at present, 1834, very promising. When I first commenced, I was laughed at by my neighbours, but now, when they see me persevering in what they considered a very chimerical project, they are suspending their judgment; and several of them have made considerable experiments this year. I should think there are at least 300 acres under crop-cultivation in this way this season in East Lothian, while in 1831, when I commenced, there was not a single acre. I have, therefore, the satisfaction of knowing, that I have been the means of causing L.1000 to be spent this year amongst the labouring classes in my immediate neighbourhood; and I feel confident, that should the season turn out favourable for the wheat crop, and fair prices obtained, their employers will be handsomely remunerated for their outlay. I do not mean to say that this system will succeed on any description of soil, as it must necessarily be of some depth to admit of the operation, but there are few districts where such soil will not be found in sufficient abundance to give ample employment to the population of the neighbourhood."* I believe the adoption of this mode of fallowing land was made a question between landlord and tenant, and since the question was decided against the tenant, I have not heard of any instance of the process being persevered in. I should like to see it fairly established, that trenching might be *generally substituted* for summer-fallowing, and I should like to see the effect of trenching on land intended for green crops; but in neither case, it is obvious, should the land be taken advantage of to be made to bear a single crop without manure at the ordinary period of the rotation, as it would be better policy to apply a special manure to check the complained of exuberant tendency of growth in the straw from common farm-yard dung.*

* Mr Scott's Letter to the Rev. C. Gardner, 8th March 1834.

(2838.) [Of the pulverising implements used at all seasons, there have been already described the grubber, Plates XXIX and XXX, brake (2241.), and the common land-roller, fig. 383 ; but there remains a class of implements of the same nature, which, though not numerous, are yet of considerable importance ; they come also under the denomination of rollers, but are peculiar in their construction, and are particularly adapted for being used in summer-fallowing. The object of these machines being to break down the more indurated clods in clay soils, they are fitted to act either by abrasion, or by disintegration, or both combined. One of the simplest of them is the common plain roller, armed with strong and broad iron spikes, the latter splitting the clods, while the weight of the roller produces their further abrasion ; this implement, however, stands low in the scale of efficiency when compared with others which I have to notice.

(2839.) Crosskill's *clod-crushing roller* is one of the most efficient implements of this class, and is here represented in fig. 506, which is a view of the

Fig. 506.

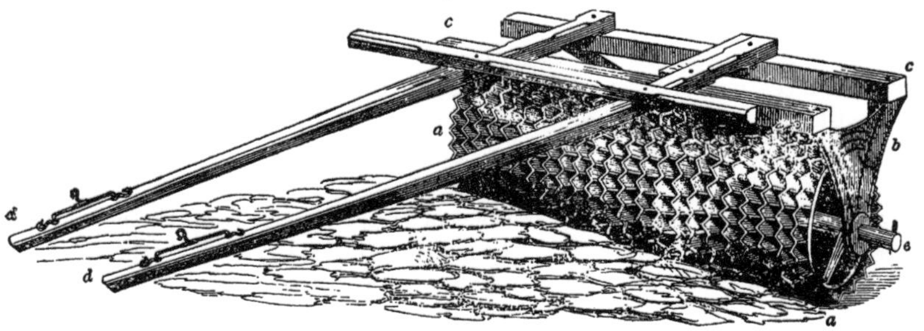

CROSSKILL'S CLOD-CRUSHING ROLLER.

machine in the working state; but to convey any idea of its construction, we must exhibit it somewhat in detail. With this view, fig. 507 is a side elevation of one of the individual wheels or plates that go to form the body of the roller, and consists of a ring or web of 30 inches diameter over the extreme points, the web is 2½ inches broad and ⅝ inch thick, formed into angular-pointed teeth. The ring is supported on the four feathered arms, and an eye formed in the centre 3 inches in depth and 2½ inches diameter, fitted to move easily on the axle of the roller. Corresponding to each tooth of the wheel, studs are cast on each side of the web, which project 1 inch from it, as seen in fig. 508, an edge view of the wheel, where *a a a a* are the projecting studs, and *b* the eye, shewing also the feathers of the arms. The wheels thus formed are threaded, to the number of 23, upon a round axle, 2½ inches diameter, upon which they are at liberty to turn separately, making up the body of the roller *a a*, fig. 506, to a length of 6 feet. A cast-iron end-frame *b*, is then placed on each end of the axle, and these are bolted to the wooden transverse bars *c c*, and to

Fig. 507.

SIDE VIEW OF ONE WHEEL OF THE CLOD CRUSHER.

Fig. 508.

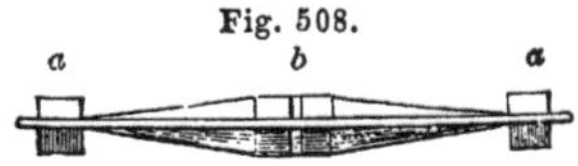

EDGE VIEW OF THE WHEEL.

these last also the horse-shafts *d d* are bolted. The axle is prolonged at each end *e* to an extent of 4 or 5 inches, forming the arm on which the carriage wheels are placed for the removal of the roller from one field to another. The carriage wheels are also of cast-iron, 3 feet diameter, plain of course on the sole; and when these are to be placed or removed, a hole is dug in the soil under each wheel, until the wheel turns freely round and can be moved on or off the axle, the roller then resting on the ground; but when the carriage-wheels are shipt on the axle and brought on level ground, the whole weight is borne by them, while the body is 3 inches clear of the surface. In this state it is travelled from field to field, and, when about to be worked, the carriage wheels are removed by the process just described.

(2840.) The effect of such a roller upon rough clay land may be easily conceived, and that where such a great number of *points* are brought into contact with the indurated clods, the result must be their reduction to a state approaching to the granular, especially if the operation is repeated. The effect is entirely different from that of the plain roller, for with it, if a clod does not crumble at once with its pressure, it is forced downward into the soil in a still solid state; whereas, with the one now described, the numerous points, acting like so many wedges, will almost infallibly split such a clod into numerous fragments, and repetitions of the process will produce a well pulverised surface.

(2841.) This implement has been but very partially used by Scottish farmers, though it is in extensive use in England; and perhaps this is to be accounted for from the much greater extent of clay soil there, and from the nature of the clays also, they are more subject to induration. Where the machine has been tried in Scotland, the results have been favourable, but the localities are not numerous where such a machine is called for. Perhaps, also, the price has been a bar to its introduction here, for that, even of the smallest size, is L.18. The weight of the implement is from 16 to 20 cwt., it forms a good draught for two horses, and frequently for three.

(2842.) A second, though perhaps less known, implement of this class is the double conical pulverizing roller, invented by J. Stewart Hepburn, Esq. of Colquhalzie, near Crieff, Perthshire.* In the construction of this machine Mr Hepburn adopts three varieties, but the leading features of the improvement is to give the roller a conical form, and add to it a series of transverse parallel flutings carried round the conical surface. Two of these conical frustæ are placed in one frame, base to base, but having their axes so inclined that the forepart of the peripheries, and also those parts which lie on the ground, shall be in a straight line. Fig. 509 exhibits this roller in the most approved form. The frame-work of it is the same as that applied to the common land-roller, *a a* being the bed-frame, surmounted by three horse-shafts, *b, b, b*, for yoking two horses abreast. The three pendant bars *c c c* carry the axles of the two cones *d d*, inclined as above described. A light frame *e* is attached to carry the scrapers, which are required to clear the flutings of the earth that may adhere to them. The effects of such a *form* of roller upon the surface of the soil are peculiar and important. While the cylindrical smooth roller acts merely by its pressure on the rough soil intended to be pulverized, one of a conical form, arranged as here described, will, besides acting by its direct pressure, produce a strictly pulverizing effect, by reason of its form, for the cones having a constant tendency to move

* Prize Essays of the Highland and Agricultural Society, vol. xi. p. 471.

outward in a circle, but being restrained by the bearings in which they revolve,

Fig. 509.

HEPBURN'S DOUBLE CONICAL ROLLER.

their surface will produce a crushing and abrading action well adapted to the pulverization of the soil.—J. S.]

---

70. OF BUILDING STONE-DYKES.

"It is by artificial calms that fields
Are warmed; and WALLS but slightly check
The sweeping blast."

GRAHAM.

(2843.) After the general principles upon which enclosures should be made have been explained in section 23, p. 355 of the 1st volume, and the treatment of hedges as a fence fully detailed (601.), it is unnecessary to dwell on the subject of field-fences when treating of *stone-dykes*, but describe at once the best mode of constructing them. It may be premised, that many dry-stone-dykes in this country are constructed on erroneous principles, the stones being laid in an irregular manner, and more with a view to give a smooth face than a substantial hearting to the wall. The coping, too, is often disproportionately large for the body of the wall, which is not unfrequently too narrow for its height. I suspect that many dry stone-dykes are built by ordinary masons, who, being accustomed to the use of lime-mortar, become regardless of bedding the loose stones of a dry dyke as firmly as they should be, and, of course, are unfitted to build one. It is true that a proper form of stones is a great assistance to the builder of stone-dykes, flat thin stones being the best: but flatness and thinness are not the only requisites; they should also have a rough surface by which they may adhere to one another in the wall; and no material, on this account, is so well adapted for the purpose as those derived from sandstone-boulders of gravel deposites, which split with the pick into flat stones of requisite thickness when first taken from their matrix, and, on being exposed to the air, become dry and hard. A builder of dry stone-dykes should, therefore, be brought up to the profession, and when he has acquired dexterity, he will build a substantial wall, at a moderate cost, which will stand upright for many years.

(2844.) Dry stone-dykes are measured by *quarters*, that is, quarters of a yard of 9 inches each. A 5-quarter dyke is the usual height of a field-fence, that is 45 inches, or 3 feet 9 inches to the under side of the cover upon which the cope-stones stand—the cover and cope-stones usually measuring 12 inches, so that the dyke stands altogether 4 feet 9 inches in height. The dyke, when finished, is measured by the rood of 36 square yards upon its face under the cover, so that every 30 yards of a 5-quarter dyke will be 1 rood in length. The usual thickness of such a dyke is 2 feet at the base, and 15 inches under the cover. But the best way to contract for the erection of stone-dykes is by the rood of 36 cubic yards, when every temptation on the part of the builder to contract the breadth, and make the heart of the dyke hollow, will be removed. A dyke that has 2 plain faces is called a *double-faced* dyke, and one with 1 face, as when built against a sunk-fence, is called a *single faced* dyke. A double-faced 5-quarter dyke requires 1 ton of stones for every square yard of its face, so that 36 tons of stones are required for every rood of 30 yards long. The expense of quarrying that quantity of stones is about 10s. the rood; the carriage of them at a reasonable distance beyond 1 mile is also 10s. the rood; and the building is commonly undertaken, when the stones are good, at 10s. a rood also; so that such a dyke costs 30s. the 30 yards, or 1s. for every yard in length. The tools of a dry-stone dyker are few and inexpensive, consisting only of a mason's hammer, a frame as a gauge for the size of the dyke, and cords as guides for the straightness and thickness of the dyke. A dyker cannot continue to work in wet or in very cold weather, as handling stones in a state of wetness is injurious to the bare hand; on which accounts, dry stone-dykes are commonly built in summer.

(2845.) The line of fence being determined on, it is marked off with a row of stakes driven firmly into the ground. The upper soil, to the depth it has been ploughed, is removed from the line to form the foundation of the dyke, and it may be driven away, or formed into a compost with lime near the spot for top-dressing grass. When driven away, it should be so immediately, and not lie to annoy the builder. When the surface consists of old firm sward, especially of moory turf, the dyke may be founded upon it; but in forming foundations, it should always be borne in mind, that dykes are apt to sink into *soft* earth of every kind to the injury of the dyke, not merely in curtailing its height as a fence, but in twisting its structure and causing it ultimately to fall; so, when the soil consists of vegetable *mould*, it should be removed altogether, and its intrinsic value in a compost will amply repay the trouble of removing it. After the foundation has been formed by the removal of the earth, the stones should be laid down on both sides *as near the line of foundation as practicable*, for it is of considerable importance to the builder that the stones be near at hand. Indeed, when stones are laid even as far as 2 yards, from the foundation, the builder incurs loss of time in throwing them nearer; but, on the other hand, no stones should be thrown into the foundation, as they will have to be removed by the builder before he commences operations. Where large boulder-stones exist, they form excellent material for the foundation of stone-dykes, and should be laid close to the foundation before the building stones are brought. The simplest mode of conveying large boulders is upon a sledge, shod with iron, and it is better for putting on and taking out than a common cart, the bottom and sides of which are apt to be injured by such boulders. Indeed, when many stones of ordinary kinds are intended to be driven for buildings, the carts should receive an extra temporary bottoming and lining with deals of common Scots fir, or deals of willow, which are better, as being softer and less liable to split. A pair of horses, yoked as in a plough, will draw a very heavy boulder upon such a sledge.

(2846.) Every preparation being thus made, the builder proceeds to his work, or rather 2 builders together, as they make the best work, and

Fig. 510.

THE FRAME AND COMMENCEMENT OF BUILDING A STONE-DYKE.

assist each other with stones which one would be unable to manage. They begin by setting up the frame *a*, fig. 510, in the foundation of the proposed line of dyke. The frame is made of the breadth and height of the proposed dyke under the cover; and it is set in a perpendicular position by the plummet *c*, attached to it. A corresponding frame should be placed beyond the point at which the dyke commences, or 2 stakes, such as *d* and *e*, driven into the ground, having the same inclination as the sides of the frame, answer the temporary purpose of an auxiliary frame. In uneven ground, a space of ½ a rood, or 15 yards, between the frames, is a sufficient stretch of building at a time; but on even ground, a rood may safely be taken in. The cords *f g* and *h i* are then stretched along the space between the frames, and fastened to each frame respectively, to guide, as lines, the side of the dyke straight, and to gauge its breadth. The frame is held upright and steady by a stiff rail *k*, having the nail projecting through one of its ends *l*, being hooked on to the top-bar of the frame, and a stone *m* laid upon its other end.

(2847.) When the dyke has a scuncheon for its end, a large boulder, such as *n*, should be chosen as the foundation-stone; and if no boulders exist, a large stone should be selected for the purpose, for no better protection can be afforded to the end of a dyke than such a foundation, especially if the scuncheon forms at the same time one side of a gateway to a field. Another boulder, or large stone, should be placed at a little distance from the first, as at *o*, and the smaller stones are used to fill up the space between them, until the space is raised to the height of the boulders. There is a great art in laying small stones, and it is, in fact, this part of dyke-building which detects the difference between a good and bad builder. In good dry building, the stones are laid with an inclination downwards, from the middle of the dyke, towards each face, as seen at *a* to *a*, and *b* to *b*, fig. 511. This contrivance causes the rain which may have found its way down through the top of the dyke to be thrown off by both sides; and, to sustain the inclination of the stones, small stones *must* be packed firmly under their ends in the very heart of the dyke; whereas stones, when laid flat, require no hearting to place them so, and may receive none, to the risk of the dyke bulging out in both faces. It tends much to the stability of a dyke to have what is called a *thorough-band* stone *c d*, placed across it at such a height from the ground as represented in the figure. In like manner the cover *e f* acts as a thorough-band at the top of the dyke; but in laying the cover, the

Fig. 511.

SECTION OF STONE-DYKES, SHEWING HOW THE STONES SHOULD BE LAID.

levelling of the dyke to form its bed should not be made of very small and very thin stones, as is too often the case, as these have little stability, being easily shifted from their position, easily broken, and, of course, constantly endanger the safety of both cover and cope. Thorough-band stones are frequently left projecting from one or both sides of the dyke by some builders, merely to indicate that they are thorough-bands; but the practice is objectionable, inasmuch as projections serve as stepping-stones for trespassers to climb over the dyke. Fig. 512 shews how a scuncheon should be formed of in-band *a a a*, and of out-band stones *b b b*, hammer-dressed, and firmly bedded upon one another. The covers *c* should project 1 or 2 inches beyond the face of the dyke, to protect the top. They should be 2 inches in thickness, and without a flaw throughout their length, which should be 2 feet at least, that their weight may keep them firm, and their size cover a large space of building. In forming the cope, a large stone should be placed at the end, as *d*, in order to keep down the cover, and act as an abutment against which the smaller cope-stones may be wedged. Other large stones, such as *e*, should be placed at short distances from each other, and upon the joining of 2 covers, to keep them both secure. Thinner stones should then be placed between these on edge, and wedged firmly, with small stones driven between them with a hammer; but the *wedging* should be delayed until a considerable length of coping is finished, which will be the better able to resist its force. Fig. 512 shews how the stones should be laid in the body of the dyke, those placed uppermost covering the joinings of those beneath them, and small thin stones are introduced here and there in the finishing to act as wedges between the large ones. The cope-stones should be nearly all of the same height.

Fig. 512.

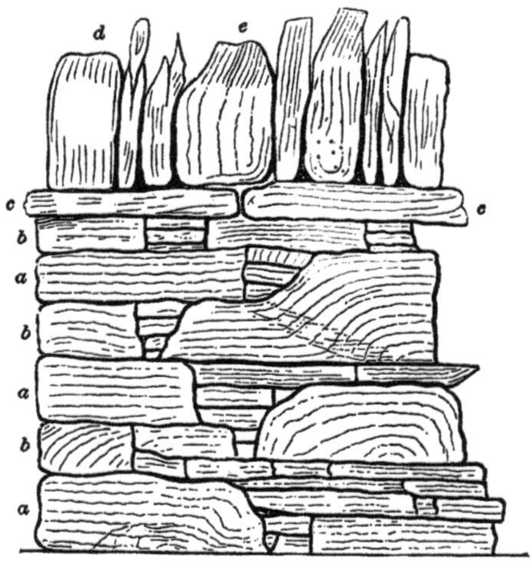

SCUNCHEON, COVERS, AND COPE OF A STONE-DYKE.

(2848.) In building a stretch of dyke, such as the rood above referred to, it is customary to carry up the building at both ends, as well as at the middle, of the stretch to the levelling of the top, before the intermediate spaces are built up, because those parts being built almost independently, act as pillars in the dyke to support the intermediate building plumb; and they are convenient for pinning the cords into while the intermediate spaces are being built.

(2849.) When a few stretches of a dyke have thus been finished, the surplus stones, if any, should be removed, and laid where they are wanted; but should there be a deficiency, stones should be immediately brought, to allow the builder to finish one stretch before he proceeds to another. The debris of stones caused by the hammer should be removed either to drains or roads.

(2850.) These are all the particulars to be attended to in building dykes for ordinary purposes; but there are a few modifications which

Fig. 513.

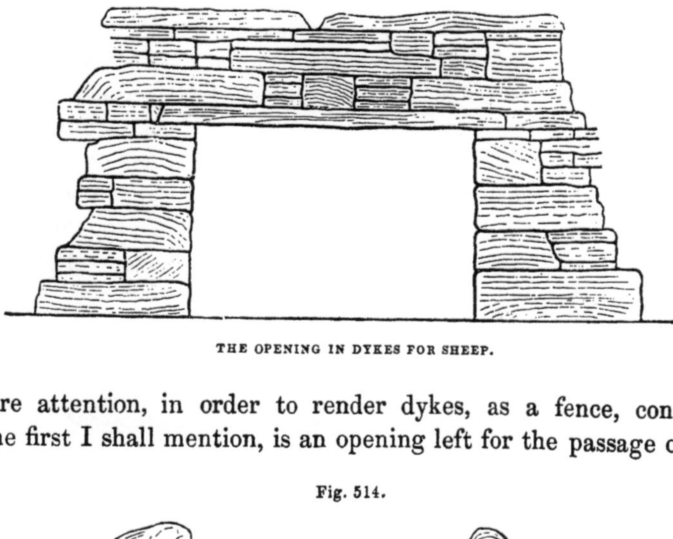

THE OPENING IN DYKES FOR SHEEP.

require attention, in order to render dykes, as a fence, convenient. 1. The first I shall mention, is an opening left for the passage of sheep

Fig. 514.

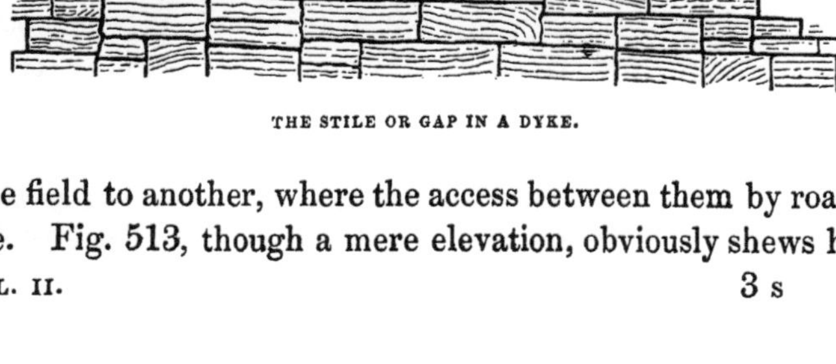

THE STILE OR GAP IN A DYKE.

from one field to another, where the access between them by road is at a distance. Fig. 513, though a mere elevation, obviously shews how such an opening is made. A bunch of thorns or whins, or a board, closes the opening when no longer needed. Such an opening gives sheep access to pasture banks, or to a grass-field from their turnip-breaks in wet weather in winter, and it allows ewes to go to a turnip-break for a few hours every day from the grass-field they are grazing in. For such purposes, an opening of from 3 feet to 3½ feet will suffice. 2. Another convenience is seen in fig. 514, consisting of a gap near the top of the dyke, which may be useful as a stile in the line of a foot-path, or a gap at the side of a cover, for hounds and huntsmen to enter with ease; and here the whipper-in may stand on the out-look for a burst. When not constantly in use, such a gap is easily fenced with a bunch of thorns or whins.

(2851.) Dykes such as I have been describing, namely, of 5 quarters in height, will fence horses, and cattle, and Leicester sheep, but will not confine Black-faced sheep. For these, higher walls must be built, or expedients used to make ordinary dykes confine them. Some of these expedients are shewn in fig. 515, where part of an ordinary dyke with its cope is seen; and the expedients consist, 1. Of cope-stones *a*, *b*, *c*, *d*, and *e*, set on edge to a considerable height. say 9 inches or 1 foot, above the ordinary cope-stones. In one case, such as that of the stones *a*, *b*, *c*, fillets of wood are laid along notches formed on their top, and wedged into them. In the case of the stones *c*, *d*, *e*, a strong rope of straw, laid

Fig. 515.

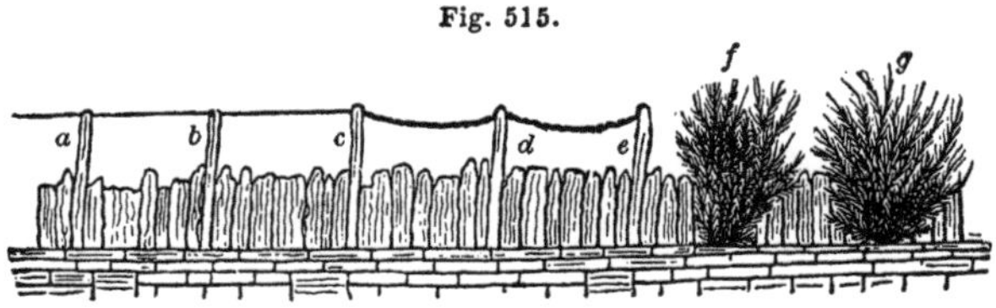

SOME EXPEDIENTS FOR INCREASING THE HEIGHT OF STONE-DYKES.

somewhat loosely over the notches, and dangling occasionally with the wind, forms a sufficient scare to sheep. 2. Another expedient, where the dyke is built against rising ground, consisting of plantation or of cultivated land, is to sow a few seeds of whin or broom in the soil behind the dyke, and cause their shoots to push forth between the cope-stones, and grow into bushes, *f* and *g*, in front of them. 3. Where good stones for covers are scarce, and where turf is tough and heathery, thick turfs, cut of the size of the top of the dyke, and laid firmly and neatly on, make very good covers, and will last a long time. Cope-stones are placed upon the turfs, which afford them a firm bed; and as heath and other wild plants, including the grasses, continue to grow in the turf, they serve to raise the height of the dyke, and enhance its appearance as a fence.

(2852.) When dykes run at right angles into one another, and are erected simultaneously, they should be built in connexion; but where a new dyke comes against an old one, the old one should not be touched, and the new built firmly beside it. Where 2 dykes cross, and the place is naturally wet, or water may be easily brought to it, a watering-pool to serve 4 fields may be easily formed; and there are two ways of making such a pond:—When the ground is firm, and the water shallow, the 2 dykes may cross, as in fig. 516, and allow the water to pass through them, and form a watering-pool in each field, such as *a*, *b*, *c*, and *d* out of a single pond. Where a pond *e*, fig. 517, already exists, and its water is too deep for dykes to traverse, the dykes must terminate at its edge,

Fig. 516.

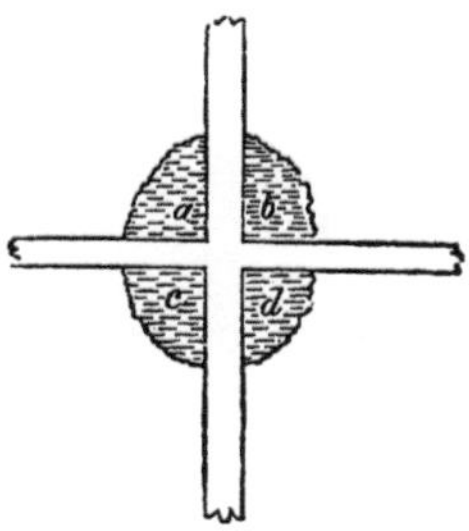

HOW TO FORM A WATERING-POOL TO FOUR FIELDS.

Fig. 517.

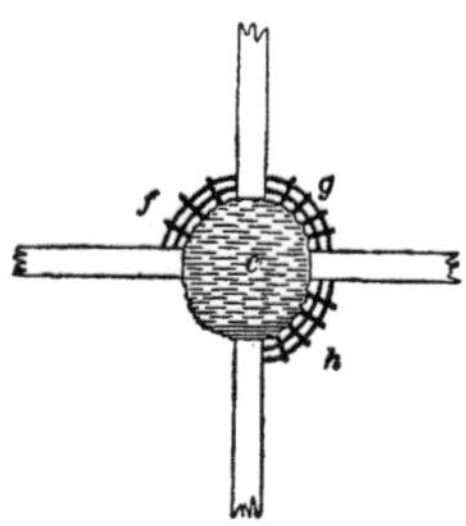

HOW TO FORM A WATERING-POOL COMMON TO FOUR FIELDS.

Fig. 518.

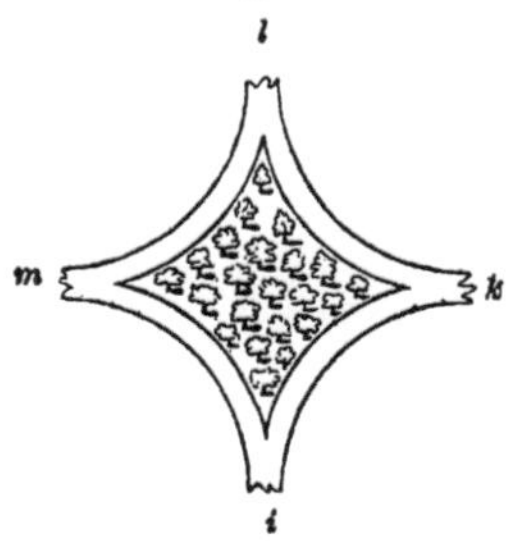

A CLUMP OF TREES AT THE MEETING OF DYKES.

and convert the pond into a watering-pool common to 4 fields. When the pond is used by only 1 field at a time, it should be fenced from the other 3 fields by means of hurdles, as at *f*, *g*, and *h*; but when it is used by more than 1 field at a time, a fence should be run across the pond, beside the hurdles in the fields not occupied by stock. Where the ground is firm, and there is no prospect of obtaining a site for a watering-*pool*, the dykes should be made to cross, and a well sunk in a corner of one of the fields, with a pump in it of such height as to supply all the fields with water in tanks by means of a spout. This expedient I used successfully on one occasion. Where the ground is firm, and no water wanted at that spot, the dyke should be built curved, as from *i* to *k*, from *k* to *l*, from *l* to *m*, and from *m* to *i*, fig. 518, and the space included between them planted with trees for ornament and shelter. There will be here little waste of land, even should it be of the finest quality, as the corners of 4 adjoining fields always contain ground that cannot be reached by the plough, while the plough can pass along such curves as near as to a straight fence. In building curvatures in dykes, builders charge ½ more per rood than for plain work.

(2853.) A stone dyke is in the highest perfection as a fence immediately from the hands of the builder; but every day thereafter the effect of the atmosphere upon the stones, at all seasons, and the accidents to which they are liable by trespasses of individuals, and the violence of stock, render it necessary to uphold their repairs frequently; and this consideration should cause the best suited materials to be selected for their original erection.

---

71. OF EMBANKMENTS AGAINST RIVULETS.

——— "that smooth the current's force,
May harmlessly, with easy flow, glide by."

GRAHAM.

(2854.) It is not my intention to enter into the subject of embankments in general, the warding off the waters of large rivers being the province of the engineer more than of the farmer to describe; but as haugh ground is subject to be overflowed by sudden floodings of the tiniest streams, when surcharged with melted snows, or dammed behind piled sheets of broken ice, at the up-breaking of a storm in winter, it may be of service to you to be made acquainted with the means, in such circumstances, of preventing small streams, when flooded, from reaching ploughed land. Where haughs are kept in permanent pasture, comparatively little injury is sustained from floodings, but rather benefit from deposition of alluvial matter; but to be prevented ploughing land, in any circumstances, from the chance of the soil being carried off by water, is a thraldom which no farmer can easily bear. In one season, 2 acres of wheat in a fine haugh belonging to myself were completely carried off, soil and crop, by the sudden eruption of the small river Vinny, in Forfarshire, caused by damming of the ice at a turn of the rivulet in the breaking up of a severe storm. The devastation seemed irreparable; yet in a short time afterwards, in spring, there being abundant depth of alluvial soil in the haugh, the large holes made, and the banks thrown up by the water, were filled up and levelled, and the soil ploughed, manured again, and sown with turnips, as if no such accident had occurred. To prevent the recurrence of similar catastrophes, I made a small embankment along the whole line of the stream, at every place where it was possible for it to overflow, and which completely defended the soil from similar harm in future. To describe the formation of such an embankment is the object I have in now entertaining the subject. Erections such as were necessary to be made along both banks of the Isla, after the Lammas floods of 1829, must be constructed under the superintendence of an experienced engineer.

(2855.) In order to determine the dimensions of an embankment adapted to the peculiarities of a locality, you should beforehand ascertain, from the best evidence you can obtain, the *highest point* to which the water of the rivulet had ever reached; and if your embankment is made 1 foot higher than that point, your land may be considered as being placed in safety. The next consideration is the distance of the site of the embankment from the brink of the water. In every place where the bed of the stream is narrow, and where, of course, the flooded current will attain the greatest height, the embankment should not only be higher, but farther removed from the edge of the stream. Where the bed of the stream, on the other hand, is broad, and there is ample space for a slow, though deep current, the embankment may be safely placed nearer the water's edge. But the safest plan in all cases is to afford sufficient room for the water, as much loss has been occasioned by overflowings of rivers, even when embanked, by contracting the channel between the embankments on opposite sides of rivers too much, from a mistaken desire to reclaim a piece of land from the river bank; forgetting that, in proportion as the river increases in depth, by confinement between the embankments on both sides, its power to do mischief is greatly increased; and it may even overcome the strength of the embankment, when the injury committed will be more serious than if there had been no embankment at all. The best policy, therefore, is to give the river sufficient room to flow, and also to remove all sudden turns in its course against which the water may strike with force, or be reflected with violence against the opposite bank. Let all curves of embankments, whether concave or convex to the river, present surfaces along which the water will flow in unbroken sweeps, not in all places conformably, perhaps, to the *natural* form of its channel; but the form of the channel should be made with easy sweeps, by taking away portions of the bank on one side, and filling up deep bends on the other.

(2856.) Fig. 519 gives a vertical section of such an embankment as may suffice to keep off the swollen water of a small rivulet. It is formed of turf derived from the river's bank, and faced with a stone-dyke to prevent its being injured by stock on the side of the field, where the dyke answers the purpose of an efficient fence. The line of the embankment should be marked off with pins, and the turf raised along the breadth of ground to be occupied by the embankment. In raising the turf, that intended to cover the *face* of the embankment next the stream should be at least 1 foot square, unbroken, and tough; and if the river bank does not afford turf of this description, it must be obtained elsewhere, and brought to the spot. The turf to build the face wall may be of any description possessing tenacity at all. The turf for the sloping-bank should be cut with bevelled edges, so that each turf may overlap 2 lower turfs with 2 of its edges—the one edge, the lowest, overlapping in the direction of the slope of the bank, the other overlapping in the direction of the flow of the water of the river. These circumstances settling the proper and relative position of the turfs, the embankment should begin to be constructed at the lowest point down the stream, and carried upwards; and it should also be begun at the water's edge, and carried upward to the top of the slope.

(2857.) I shall suppose that the turf-wall *a b*, fig. 519, shall be 4 feet in height, then a breadth of 5 feet from *b* to *c* being the base of the

Fig. 519.

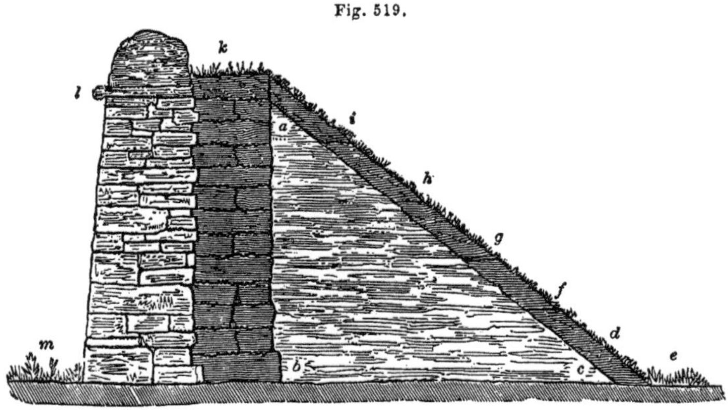

THE EMBANKMENT AND FACE-DYKE AGAINST A RIVULET.

slope of the embankment, may give sufficient stability to the structure, and slope to the face. The line *b c*, however, will vary according to the nature of the ground on the river-bank. In a steep part it may be less than 5 feet, in a gentle slope it will retain its proper length, and in a sudden and narrow hollow it may be necessary to fill up the hollow altogether, in order to make the bank uniformly even, in which case the slope may have to be built up from the very edge of the water. The first operation in the actual *construction* of the embankment is building the turf-wall *a b*, the sods of which are laid with the grassy face downwards, on the same principle as overlapping joinings in masonry. As the wall proceeds, earth is taken from the field in spadefuls to pack behind it, and to fill up the entire contents of the embankment included within *a b c.* This earth should be free of stones, and if disposed to rise in lumps, should be chopped small with the spade, and beaten firmly with a wooden beater. After a sufficient quantity of earth has been placed behind the turf-wall, the turfs of the slope *c a* are then begun to be laid at the lowest point *c*, where the first turf *d*, with the grass-side upmost, is made to grip under and abut against the sward *e* of the river bank by a notch cut out of the latter with the spade, the object of the notch being to plant the edge of the turf *through* the sward to prevent the water getting hold of it and carrying it away. Another turf *f* is made to overlap with its lower edge the upper edge of the turf *d* just laid, and the earth is brought behind it with a trowel, fig. 182, or with the hand, to the shape of the slope *c a*. In like manner, the turfs *g*, *h*, and *i* are laid one after the other, till the top of the turf-wall *b a*, and the top of the slope *c a*, are reached at the same time, when a thick turf *k*, with the grass upmost, covers the top of the wall, and finishes the slope. When the turfs are cut square, and all of the same size, which they should scrupulously be, they are quickly and evenly laid. The whole of the turfs are then beaten firmly down with the back of the spade. It will be concluded from this description, that the building of the turf-wall should proceed in advance of the laying of the turf upon the slope. In conjunction with the turf-work, the building of the stone-dyke *l m* may proceed and finish the whole embankment at once. Such a dyke as is here required is called the *single-faced* dyke, having only one finished face towards the field, with suitable covers and a strong cope.

(2858.) The cost of making an embankment 4 feet high in the wall, 5 feet broad in the base, and casting the turf for it, the materials being all at hand, is 1s. 4d. per rood of 6 yards. If the turf has to be brought from a distance, the trouble and cost of its carriage, of course, devolves upon yourself. The cost of building the face-dyke 5 quarters high, will be 8s. per rood of 30 lineal yards, and the quarrying and carriage of the stones will each be as much, or 9½d. per lineal yard.

(2859.) Such an embankment should be constructed at as early a period of the season as possible, to give the turf time to grow together before the occurrence of the earliest flood. In a very dry summer the turf may become brown, when water should occasionally be thrown upon it with a scoop from the rivulet; and in any kind of season it is possible that a turf will die here and there, when it should be removed, and a fresh one substituted. Until the turfing becomes converted into a thick and tough sward, it should be frequently inspected, and every gap in it plugged up, whether occasioned by accident, such as the feet of cattle trespassing from the opposite side, or the burrowing of animals, such as rabbits or water-rats. In the succeeding season the grass will grow luxuriantly upon the embankment, when it may be mown early in summer, to give it time to grow to a thick sward before winter. After this period the earth will become quite firm, and the embankment require nothing more than a general supervision every year.*

---

72. OF FORMING WATER-MEADOWS.

> " A free and porous soil
> Upon a gravelly bed, at all times drinks,
> Yet ne'er is quenched. Who owns a soil like this,
> If through his fields a little mountain stream,
> Not sunk in channel deep, but murmuring down
> 'Tween gently-sloping banks, a mine of wealth
> Possesses in that stream."
>
> GRAHAM.

(2860.) In a former notice of water-meadows (2051.), I endeavoured to describe the mode of conducting the irrigating water over them; and now that the proper season has arrived, I shall describe the mode of *forming* them. Several considerations should be attended to ere the formation of a water-meadow is determined on; the first and principal of which is, whether there be a sufficient supply of water in a dry season to irrigate the meadow thoroughly; and if there be not, the desire for possessing a water-meadow should be abandoned, or its extent of meadow confined to suit the water at command. Another important consideration is, whether the water can be spared for irrigation, without wasting it for other important purposes, such as the thrashing of grain, and the watering of live stock in grass-fields? If the water can be used in irri-

* Should the embankment cut off the watering of the field from the stock, a pool should be made at the lowest point down the river in each field, with its bottom below the level of the bed of the stream, and if the substratum is gravelly, the water will fill the pool, but should it prove clayey, then a float-sluice, such as that described in vol. xii. p. 137, and figured in Plate V., figs. 9 and 10, of Prize Essays of the Highland and Agricultural Society, should be used at every pool. For practical observations and illustrations on embankments of a general description, you may profitably consult Stephens's Practical Irrigator and Drainer, from p. 120 to p. 148; and also Johnstone's Theory and Practice of Draining and Embankments, from p. 157 to p. 213, quarto edition, 1834.

gation, before it is wanted for, or after it has been used by, the thrashing-machine, on the supposition that water is employed as the thrashing power, then it may profitably be employed for irrigation; but otherwise, the advantages of irrigation would be purchased at a cost beyond their intrinsic worth. A third consideration, of an important nature too, is, whether you have a right to take as much of the water of a rivulet which may form the boundary of the estate in which your farm is situate, as your water-meadows require? You can use the water of a brook which wholly passes through your farm as you please, so that it be not poisoned by noxious substances, and be permitted to find its way to the property lower down by its natural channel; but you cannot appropriate to your particular use more than half the water of a march-burn. If half the water afforded by it is not sufficient for the purpose of irrigation, you had better abandon the idea of forming a water-meadow, for a *dry water-meadow* is a contradiction in fact, as much as in terms.

(2861.) Allowing the quantity of water to be ample for your extent of irrigation, it is better to take it direct from the brook, than to erect a dam across it, to collect the water, even though you should possess the power to do so; because the nearer the bottom of the brook the water is obtained from, the better it is for the purposes of irrigation, on account of the sedimentary matter which it contains, and the nearer this matter approaches to the nature of clay and vegetable matter, the more richly it will manure irrigated plants. It may cost more to make a channel for the water obtained direct from a brook, than to construct a dam across the same brook, though that is not even probable, for unless a dam is very substantially made, so as to resist the force of the brook under every state of flood, it will cost much for repairs, besides exciting the constant apprehension of blowing from below, or bursting in the sides.

(2862.) *Sluices* should be formed to prevent the water reaching the meadows when not wanted, and also to allow one portion of the meadows to be watered at a time, while the other parts are kept dry. All sluices should be substantially and largely built with stone and lime, and their foundations sunk to a depth below what the water has any chance of reaching. The masonry in direct contact with the operating sluice-boards should be formed of droved ashlar. No doubt, sluices of this construction cost much; but unless the entire appointments in connection with water are made substantial at the commencement, and on correct principles, their repair will be incessant, and use unsatisfactory.

(2863.) The land to be converted into water-meadow should be thorough-drained, unless the subsoil consists naturally of gravel, which is rarely the case; because, if irrigating water finds its way through the soil to a retentive subsoil, it will remain there in a stagnant state, where no drains are at hand to carry it off; and the consequence will be, the sward of the meadow will, in a short time, be composed of coarse sub-aquatic plants, instead of fine meadow-grasses. The best sort of materials for filling the drains of water-meadows are tiles and soles, for these afford the quickest passage to water; and in case the meadow afterward be converted into arable husbandry, the drains should be made of a depth suited to the cultivation of the soil, namely, from 2 to 2½ feet. The drains should be placed asunder at distances corresponding to the breadth proposed to be given to the bed-work of the meadow, that every bed may have the same advantage in regard to drainage. In practice, it will be found that but a very small proportion of the water will find its way into the drains; nevertheless, it is necessary that a sufficient number of drains be present to carry off all the water that may find its way into them from any quarter; and, to insure this result, a drain should be accessible from every bed. I need scarcely refer you to the section on draining, p. 482 of the 1st volume, for the mode of making the drains here recommended.

(2864.) These preliminaries being determined, the next business is giving proper form to the water-meadow. Taking, in the first instance, the simplest case of water-meadow, that having a very gentle slope from one side to the other of the field, and also from one end of it to the other, the

Fig. 520.

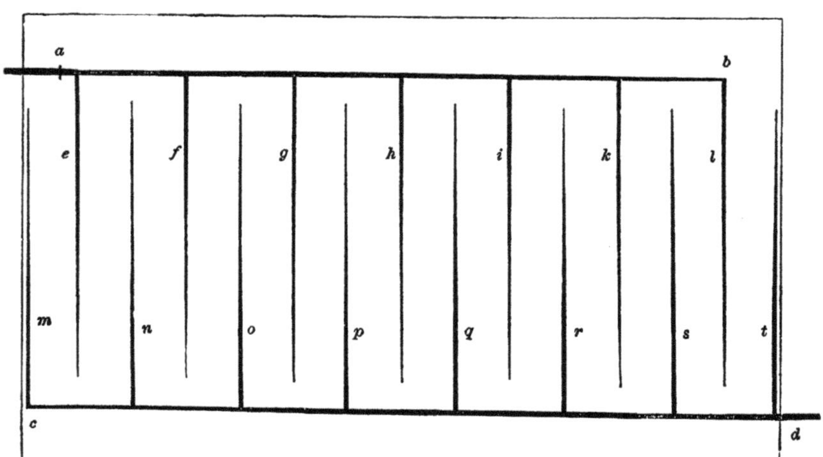

THE BED-WORK WATER-MEADOW.

first consideration is, to make the ditch, which is to conduct the water from the brook, to the *highest* corner of the field. Where this water-course or lead enters the field, a sluice *a*, fig. 520, should be put across it, to prevent the water flowing when it is not wanted. The first operation within the field, is to form the main-conductor of the water *a b* along the upper side, not on a dead level, but with a very gentle descent. The figure represents a field of uniform surface, and the bed-work in it is of the same uniform character. The *main-conductor a b*, is made sufficiently capacious to pass as much water as will cover, at one time, the entire surface of the field with running water; it should be made narrower towards *b*, because the water passing out of it into the lateral feeders, one after the other, less water will find its way to *b*, and the narrowness of the channel will keep the less quantity always at the same height. The bottom and edges of the conductor should be made all along with a uniform smoothness and inclination. The matter which comes out of the conductor, is wheeled away to other parts of the field to fill up hollows. The next channel made is the *main-drain c d*, whose province is to carry off the water out of the field, after it has served its purpose in irrigation, and, on that account, its dimensions should be equal to that of the main-conductor, and its position along the lowest part of the field. It should also have the same uniform inclination and smoothness; but as the greatest quantity of water will be in it at its lowest point, it should have the largest capacity at *d*, being at the very opposite and also lowest point from the largest capacity of the main-conductor at *a*. Whilst these two principal channels are forming with the greatest care and exactness, the intermediate ground of the field should be preparing to be occupied with another species of channels, and the preparation for them is made according to the state of the ground. If the field has been in common culture, the ground should be ploughed and harrowed, and the weeds hand-picked, as in summer-fallow, and the plough employed to gather it into ridges. The crowns of these ridges are marked by the lines *e*, *f*, *g*, *h*, *i*, *k*, *l*, and they may be at a distance of 30 feet from each other, the breadth of a couple of ordinary ridges of 15 feet. One gathering will probably not suffice, as the crowns should be 1 foot higher than the open furrows. This is all the assistance the plough can give in the making of water-meadows, and the rest should be done with the spade and wheel-barrow; and by their means the lines *e*, *f*, *g*, *h*, *i*, *k*, and *l*, are made with a uniform inclination from the main-conductor to their termination. These lines consist of channels, called *feeders*, along the crowns of the ridges, now denominated *beds*, having a width of 20 inches at their junction with, and at right angles from, the main-conductor *a*, *b*, should they extend as far as 200 yards in length, and a width of 12 inches at their termination. In like manner, the lines *m, n, o, p, q, r, s, t*, are channels, called *drains*, formed in the hollows of the open furrows; and as their province is to carry off the whole water supplied by the feeders, they should be made exactly of the same dimensions, but in the opposite direction, that is, their widest end should be at their junction with the main-drain *c d*. Both these kinds of channels being exactly parallel to each other, and of uniform inclination from end to end, the drains *m, n, o, p, q, r, s, t*, are at 1 foot below the level of the feeders *e, f, g, h, i, k, l*. The soil between them is worked smooth and even with the spade, hollows being filled up, and heights removed to a uniform surface, and with a uniform inclination, from the feeders to the drains. The ground is now ready to be sown with the seeds of natural grasses, which should always be sown without a corn crop to secure a good and early sward. The following will be found a good proportion of such seeds, per imperial acre, for water-meadows, in the different conditions of light, medium, and heavy soils:—

Botanical and English Names.	On Light Soils.	On Medium Soils.	On Heavy Soils.
	Without a Corn Crop.		
	lb.	lb.	lb.
Agrostis stolonifera, Marsh creeping bent-grass or fiorin,	2¼	2½	2¾
Alopecurus pratensis, Meadow fox-tail grass, ...	1¼	1½	1¾
Festuca loliacea, Darnel-spiked fescue-grass, ...	1	2	3
Festuca pratensis, Meadow fescue-grass, ...	2½	2½	2½
Festuca elatior, Tall fescue-grass, ...	1½	2	2
Glyceria fluitans, Floating sweet-grass, manna-grass, or float-grass, ...	2¼	2½	2¾
Lolium Italicum, Italian ryegrass, ...	6	6	6
Lolium perenne, Perennial ryegrass, ...	7	7	7
Phalaris arundinacea, Reed canary-grass, ...	1	1¼	1½
Phleum pratense, Timothy, cat's-tail, or herd-grass, .	2	3	3½
Poa trivialis, Rough-stalked meadow-grass, ...	2¾	3	3¼
Lotus major, Greater birdsfoot trefoil, ...	2	2	2
	31½	35¼	38

To protect the young plants, 1 bushel per acre of rye may be sown along with these seeds. The entire cost of these seeds is 27s, 9d. on light, 32s. 3d. on medium, and 36s. on heavy soils. "When desirable," says Mr Lawson, "the original expense of the above mixture may be decreased from 4s. to 5s. per acre, by excluding the *Alopecurus pratensis*, which is only recommended in consideration of its earliness, and half of the *Lotus major;* under most circumstances, however, it will be advisable to retain the full quantity of the latter, not only from its being the best adapted of the clover tribe for withstanding excess of moisture, but also from its attaining to full maturity at a late period of the season, when the growth of the grasses generally becomes less vigorous." *

(2865.) When the field to be converted into water-meadow has been in permanent pasture, the turf should all be taken off and laid aside for use; and the bared surface should be ploughed and wrought with the spade in a manner similar to that described above, and when the ground is quite ready, instead of being sown, the turf is replaced and beaten smooth with the back of the spade. This mode makes by far the best finish for a water-meadow, and is, in the end, most economical, inasmuch as the expense of the grass-seeds is saved, and the meadow is ready for taking on water at once, and will yield a good crop of meadow-grass the ensuing year; whereas a sown meadow cannot be watered with impunity for 2 or 3 years; and even longer, if the grass-seeds are sown down with a corn-crop. When the turfing of the meadow is completed, the water should be let into the main-conductor, and thence into each of the feeders *e, f, g, h, i, k,* and *l;* and the water overflowing these finds its way down each side of a bed into the drains *m, n, o, p, q, r, s, t,* along which it collects gradually, and is discharged from them all into the main-drain *c d,* which carries the entire water out of the field.

(2866.) This is the simplest as well as the most perfect form of water-meadow; but as examples of ground of such uniformity of surface as this is of rare occurrence, modifications should be made in the position of the feeders and drains to suit the form of ground. For example, if the ground falls more suddenly from *a* to *b*, fig. 520, than from *a* to *c*, the feeders *e, f, g, h, i, k,* and *l,* instead of being made on the middle of the beds, should be placed a little towards *a*, the higher part of the ground, making the lower side of the beds broader than the higher, because the gravity of the water will easily carry it down the broader sides into the drains *m, n, o, p, q, r, s,* and *t.* Still, in such cases, the *crowns* of the beds, that is, their highest point, should be formed where the feeders are made, wherever that may be. Should the ground fall suddenly from *a* to *c*, the water will run too fast down the feeders, and, to avoid this inconvenience, the feeders *e, f, g, h, i, k,* and *l,* instead of striking off from the main-conductor *a b* at right angles, should go off at such an acute angle, determined by levelling, as that the water shall flow in them as slowly and uniformly

* Lawson's Treatise on the Cultivated Grasses, p. 41.—1843.

as in the more level case of ground. For this purpose, the beds should be ridged up at the same angle, and the drains *m*, *n*, *o*, *p*, *q*, *r*, *s*, *t*, made to correspond with them and the feeders as in the other case, namely, to occupy the intermediate spaces with the feeders at a lower level. This form of water-meadow is more common than that in fig. 520.

(2867.) Another form of water-meadow is what is technically termed *catch-work*, from the circumstance of a lower set of feeders catching the water in its rapid descent from a higher, thus causing the same channels to act the part of feeders and drains at one and the same time. This is necessarily an imperfect mode of irrigation, and should never be resorted to but from necessity arising from irregular form of the ground; "but to give exact directions for the formation of catch-work," as well remarked by the late Mr Stephens, "is beyond the ingenuity of man; for no two pieces of land are precisely alike, which renders it impossible for the irrigator to follow the same plan in one field that he has done in another. Each meadow, therefore, requires a different design, and the construction to be varied according to the nature of the ground and the quality and quantity of water." Impressed with the difficulty of conveying useful information on this part of the subject of water-meadows, I shall refer to fig. 521 for illustrations of irregularities that may occur on some grounds; but the expediency of attempting the formation of water-meadows where the ground seems so unsuited for them, is doubtful. I conceive that the original trouble and expense of making them, and the consequent risk of injuring the ground by injudicious distribution of the water, would more than counterbalance all the benefits derived from so imperfect a structure. If the opinion of Mr Stephens, that "the benefit of irrigation depends so much upon the good management and patient persevèrance of those who have the superintendence of a water-meadow, that I do not wonder it has *so often proved unsuccessful*," is applicable to bed-work, how much more so to catch-work irrigation!

(2868.) A main-conductor, *a b*, fig. 521, is here also necessary to convey the water to the different parts of the field; and as it should have as gradual a fall in its course as that in the former case, it may have to be wound into numerous curves. On the water reaching its extremity at *b*, it will flow, on the one hand, along the feeder *b c*, and, on the other, along the feeder *b d*, both being true feeders. In its overflow the water will find its way to *e f*, which serves as the drain to *c d* in collecting its waters; but at the same time it becomes a feeder, and disposes of its water down the descent to *g h*, and so on to *i* and *k*, one after another, until the water lands in the main-drain *l m*. Part of the water finds its way to the drain *n*, which conveys it into the main-drain at *m*. In like manner, the water issues out of the main-conductor *a b* into the sub-conductor *o p*, from which it flows to the right down the feeders *q*, *r*, *s*, *t*; *f u* serving as a drain to carry their water to the main-drain *l m*; so the drain *l* serves to carry away the water from the feeders *v*, *w*, *x*, *y*, and *z*. It is obvious that the water in *c d* will impart most of its sedimentary constituents to the ground between it and *e f*, and by the time it has reached *k*, very little foreign matter will be left in it, so that the grass in the upper part of the meadow will be better nourished than that in the lower; but the sub-conductor *o p* carrying the water

Fig. 521.

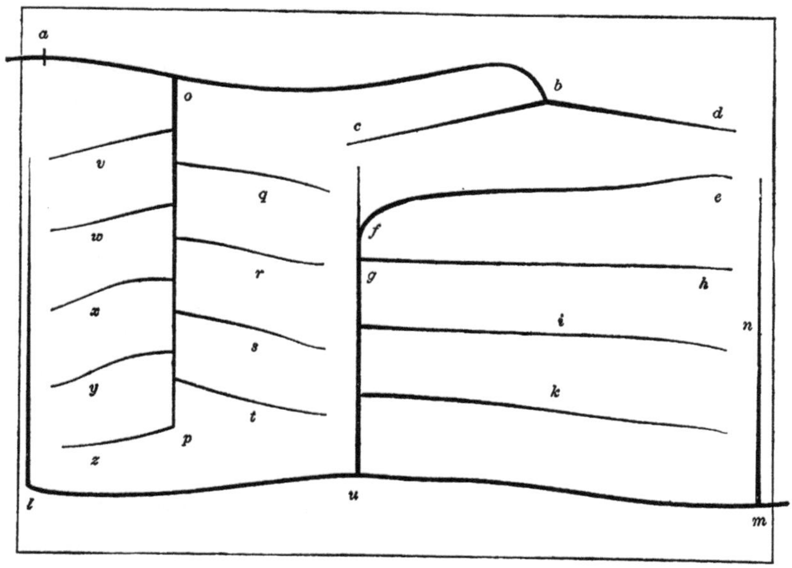

A CATCH-WORK WATER-MEADOW.

from the main-conductor direct to all the feeders in connection with the ground included from *l* to *u*, its water will bestow benefit to that portion of the meadow. In catch-work, as in bed-work, each feeder may supply water for 30 feet of ground in breadth, if its descent is gradual, but, if more sudden, the breadth may be increased to 40 feet.

(2869.) Where water flows unequally, whether in conductors or feeders, there is a plan of equalizing the flow by means of *stops*, which are obstructions placed in the channels, to retard the velocity of the water. These stops may be formed in various ways, with pieces of the natural soil left hard—with stakes driven into the middle of the channels—with sods pinned to the ground—with stones piled in heaps—and with short boards thrust across into the edges of the channels. In all cases of regular bed-work, as in fig. 520, the surface of the meadow should be so uniform in its descent, that no stops should be necessary, nor should they be required in cases even of catch-work, where the water flows as from *b* to *c*, and *b* to *d*; but where fresh water is supplied to different parts of a meadow at the same time, as down the steep sub-conductor *op*, stops are requisite to guide the water equally into the entrance of each of the feeders *q*, *r*, *s*, *t*, and *v*, *w*, *x*, *y*, *z*. But the species of stops just enumerated are all objectionable; because, stakes collect straws and sticks brought by the water; stones and turfs cause holes to be made by the water falling over them; and notch-boards injure the edges of feeders, besides causing deep holes to be scooped beyond them by the fall of water. The best form of stops consists of a piece of wood as if two wedges were united together by their bases; because, when placed in the bottom of a conductor or feeder, the water will rise over as well as slip down its inclined planes in an unbroken mass. A number of such stops of unequal breadth would fit any size of channel.

(2870.) Where the natural fall of the ground admits of the arrangement, it is quite possible to convey the water in a lead from one water-meadow to the main-conductor of another at a lower level; but, as the water would then be almost deprived of its manuring properties, where there is a large supply of water, it would be better to convey it at once to the lower meadow; and where there is no surplus water, liquid manure or other stercoraceous matter could be put into the lead, and the water, as it left the one meadow, could carry the manure into the main-conductor of the other meadow. In my opinion, liquid manure would be much more profitably applied in this way than by direct sprinkling on the soil, as the extraordinary effects produced by the foul-water irrigation in the neighbourhood of Edinburgh fully demonstrate.

(2871.) The expense of converting land into water-meadow is often very great. Where the ground is nearly level, and the surface covered with turf, the turf may be taken up, the ground properly shaped, and the turf replaced for L.3 per acre, as was instanced in one case belonging to Sir Charles Stuart Menteath of Closeburn, in 1826; whereas, in a case of Mr Lawson of Cairnmuir, in Peeblesshire, the cost was L.12 per acre. From L.7 to L.9 per acre may be taken as a fair average. Unless the advantage to be derived were considerable, such an expense would not be justifiable; but in all cases, where meadows have been well managed, the yield has been from 250 stones to 400 stones of hay per acre, of 22 lb. to the stone, with an aftermath for pasture worth 20s. the acre, from land not worth of more rent in aforetimes than from 5s. to

15s. an acre. The expense of formation is not only great, but unless attention is bestowed on the management of meadows, their produce will inevitably diminish. From the nature of the work connected with their formation, it cannot be otherwise than expensive; for, as Mr Stephens truly remarks,—"However simple the construction of a water-meadow may appear in a superficial view, those who enter minutely into the detail will find it much more difficult than is commonly imagined. It is not an easy task to give an irregular surface the equal slope requisite for the overflowing of water. It is very necessary for the irrigator to have just ideas of levels; a knowledge of superficial forms will not be sufficient. Few people unacquainted with the art of irrigation, and the regularity of form which the adjustment of water requires, have any idea of the expense of modelling the surface of a field."

(2872.) So much for the formation of water-meadows; and as to their management in autumn, preparatory to their full working, as described before (2051.), I shall avail myself of Mr Stephens' directions: —"At the beginning of the month of October," he says, "each feeder and drain should be cleansed, and the banks of the feeders repaired where they have received damage by the treading of cattle. . . . Let the beds of a water-meadow be ever so well formed, yet by some places sinking more than others, or by the ice raising the surface of the ground, although the water along the banks of the feeders have been ever so nicely adjusted, it often happens that there may be some places between the feeders and drains with too little water, where it might be advisable for the manager to make several rounds, redressing inequalities of the surface, so as to give every spot 1 inch of water."

(2873.) Great as are the benefits derivable from water-meadows in the low country, such meadows would prove of incalculably more advantage to our highland districts, where hay is the most valuable food for stock in winter that can be raised at such altitudes. That the formation of water-meadows is quite practicable in all our highland glens, is apparent from these sentiments of Mr Stephens, with which I cordially acquiesce, and earnestly press upon the consideration of such of you as may betake yourselves to hill-farming. "Fallaws meadow, on Sir George Montgomery's large sheep-farm, contains 15 acres, was enclosed from moorland in 1816, and, by collecting the water from the surrounding sheep-drains, 5 acres are partially irrigated, and the remaining 10 are top-dressed with the manure made from part of the produce, which is consumed in winter by the sheep of the farm in a wooden shed near the meadow. By this simple method of improvement, 15 acres of common sheep-pasture land give the proprietor from 3500 to 4000 stones of hay per annum, averaging 6d. per stone. In that year of drought, 1826, the hay of this meadow was sold from 1s. to 1s. 3d. per stone. What an immense advantage to a sheep-farm! By this simple process of enclosing and cutting a few small feeders and drains, the owner is enabled to provide food for his flock, when his less fortunate neighbours' sheep must either starve or be supplied from the farm-yard; but I am afraid there are very few sheep-farmers who are so fortunate as to have any hay over and above what is requisite for stock at home. Sir George fed the same number of sheep on the farm as he did before the meadow was cut off and enclosed; and I am fully persuaded that the same improvement might be made on almost every sheep-farm in Tweeddale; for, in almost all of them, there are situations where 5, 10, or 15 acres might be enclosed and partially irrigated, as in every pastoral district there are numerous rills which might be easily collected, and used to the greatest advantage, at a very trifling expense; so that, instead of being obliged, in snow-storms, to send 50,000 sheep to a milder climate of the southern parts of Dumfriesshire, where the owners are obliged to be at the mercy of their southern neighbours, not to mention the very serious injury the flocks receive by so long and fatiguing a journey, by adopting the above system of improvement, a considerable portion of the losses generally sustained would be prevented."* The testimony of a practical man, who has tried what he recommends, is always of great value. Mr Thomas M'Lean, Braidwood, near Penicuik, has found the advantages great obtained from water-meadows in an upland district, and advises others to do the same. "I am hopeful," he says, "that these results, in irrigation, will afford demonstrative evidence of the beneficial results which might follow its more extended practice in the *pastoral* districts of Scotland. In many situations in those remote glens, both the climate and the soil are more propitious than those of the one I have described; and, assuredly, water equally fertilizing is to be found issuing copiously from the sides of their hills and mountains, requiring only the industry of man to apply it judiciously, in order to procure a large amount of winter provender for live-stock, as well as an occasional early and late pasturage, which, in extraordinary seasons, would prove useful to an incalculable extent."†

(2874.) As it is impracticable to irrigate meadows in winter in highland districts, that process should be delayed till every chance of frost has subsided—until May, after which there will still be sufficient time for a crop of natural hay to grow, be cut down, and won, before the departure of summer. Such a meadow is useful in a backward spring for the support of ewes and lambs; and the sheep belonging to Sir George Montgomery would have inevitably perished in the cold and backward spring of 1826, had it not been for the grass afforded by a water-meadow from the middle of April to the beginning of May; after which latter period the meadow was irrigated, and produced nearly 300 stones per acre. The attention of hill-farmers cannot, therefore, be too strongly drawn to the subject of water-meadows. Any attempt to irrigate meadows in such situations in winter, and to pasture them in early spring, would but injure the meadow by means of frost, and, at the same time, rot the sheep; but sheep may be pastured, if necessary, in perfect safety on dry meadow-land in spring, and the meadow, on being afterwards irrigated, might yield a good crop of hay.

* Stephens's Practical Irrigator and Drainer, p. 6-82.

† Prize Essays of the Highland and Agricultural Society, vol. xiv. p. 693.

73. OF BREAKING-IN YOUNG SADDLE-HORSES.

"Which with a snaffle you may pace easy."

ANTHONY AND CLEOPATRA.

(2875.) As I have mentioned how your saddle-horse should be groomed (1622.), I would wish to say a few words on *breaking* him *in*. And, in the first place, I may remark, that, judging by the conduct of roadsters one meets with every day on the public roads, we may fairly conclude that the profession of horse-breaking is not well understood in this country; and the conclusion need create no surprise, when we observe those who become horse-breakers to be generally cast-off grooms unable to procure a permanent situation, just as we occasionally see a discarded ploughman become a spadesman, to save himself from starving; but it ought to excite even wonder to see farmers, who ought to be somewhat acquainted with the nature of horses, employ such persons to break-in their saddle-horses, merely because they demand a small fee, and undertake to finish a horse in all his paces in the course of 2 or 3 weeks. In fact, the farmer grudges the time, beyond a few days, any horse requires to be broke-in, and acts as if he conceived the animal should know the art of carrying him by instinct, when he himself may have been practising horsemanship all his days, and never perhaps become a horseman after all. Notwithstanding the folly of employing inexperienced men, I am convinced, were men of experience, address, and character, to undertake horse-breaking in a perfect manner, they would receive encouragement from the farmer. No miracle is required to break-in a horse; he is naturally docile, and may be *taught* to do any thing, as those who have witnessed the evolutions of the late Ducrow's stud with wonder and delight can believe; but the horse is naturally fearful, and, endowed with an undying memory, he never forgets any circumstance, however trivial, which may have aroused his fears. If the breaker, therefore, proceeds on the principle of subduing what he calls the horse's *temper*, by a constant endeavour to curb him, he may tame him for a time, but will never break him in; and even this subjection will only last as long as the horse finds he cannot obtain the mastery over his rider, who, on some occasion, may be a timid one; whereas, were it shewn to the horse that no irritation awaits him in the stable, or on being ridden, he will place confidence in his rider, will regard him with attachment, and will take him everywhere, even through danger, by the gentlest touch of the rein, or exhortation of the voice. Why should the Red Indian of North America never put a bridle on his horse's head, and only have a piece of cord round the under jaw, by which to pull him up in a gallop, and guide him, with the greatest nicety, in the hottest pursuit after buffaloes, by the palm of the hand against the side of the head; as much so as to enable him to point the arrow with unerring aim into the heart of his prey? Why should the Arab's horse come into his tent like one of his family, lie down and rest, and never think of running away from his master when his services are required? Does the horse of the accomplished European display the least degree of such confidence in his rider? Is it possible that a savage knows better how to break in a horse than an European? The difference between the cases can be explained in a few words; the savage makes his horse his companion and friend, the civilized man treats his as his slave. And can any doubt, were the same gentleness, kindness, attention to his wants, which secure to the savage the willing assistance of his horse, bestowed by the civilized man on his horse when young, would also inspire him with confidence? Every one has felt the satisfaction of riding a horse one knows thoroughly. It should also be borne in mind as an incentive to kind treatment, that were horses generally well broke-in, they could not be spoiled by even bad riders, for their paces would be so pleasant, the rider would have no inducement to try and mend them.

(2876.) The age of 3 years seems an excellent one for breaking-in a saddle-horse. The colt should be sent to grass in the end of May, and taken in to break by August at latest, by which time the grass will have operated beneficially upon him as medicine, and there will be sufficient time to teach him his paces and put him in working condition before the fall of the year, when horses are apt to become soft, and catch cold; but were he kept longer at grass, his condition might become so fat as to endanger his constitution, were it suddenly reduced to working order.

(2877.) The first thing, in bringing a horse into the stable, to which he should have been accustomed from his foalhood, is to give a gentle dose of medicine to clear the bowels of a load of grass. A second dose may be repeated in a week. A little new-made hay with oats is the best food as a transition from grass to hard food. The first treatment with the cavesson and bridle being the same as that described from (2455.) to (2457.) for breaking-in the draught-horse, I need not repeat it here. I may mention, however, that much lunging in a circle is not advisable for a young riding-horse, though horse-breakers are very fond of giving him this sort of exercise, because it saves themselves a good deal of travelling; the motion round the circle being apt to cause the colt contract a long step and a short. The circle is of most use in training to *canter*, when a leading foot is requisite to be used in that sort of action. The first tuition should be a straight-forward pace, as on a lea-field, and the *only* pace a *walk*, which should be taught to be both free with an easy head, and short with a tight rein. During the period of the walking-tuition, a great many useful lessons should be taught the colt, such as turning off you and to you—backing, whether quickly or slowly—leading, whether by the side of the head with the hand on the bridle-bit, or in front with a slack rein—standing still, whether for a short or long time—suffering to be tied to any object, such as a gate or tree—passing objects of terror, or of uncertainty, causing the animal to become acquainted with every thing it does not seem to recognise—becoming accustomed with the crack of the long, and the touch of the short whip—lifting the fore and hind legs when desired—and suffering the groom to go about him and arrange the breaking-harness. With all these matters the young colt will become much sooner *familiarized*, by the breaker going constantly about with him on foot as a companion on the road and the field, than when mounted on his back. In the stable, too, the same system of tuition should be followed out, such as suffering a person to go up on either side, and in any way—suffering to be groomed, and rather liking it than opposing it, as is too often the case—drinking out of a pail in the stable, and at the pump, or out of a trough or a brook—taking up with a dog in the stable or on the road—bearing, without a startle, the fall of the pail-handle, the broom, or any thing else—lifting the feet at the pail to be washed—being led by the forelock to the door, or the pump, or anywhere. These, and many other things, the colt can be taught to know in and out of the stable before he is mounted at all. Thus familiarized, he will allow itself to be mounted without any trouble; and all the assistance of boys with whips, and of men to hold down the opposite stirrup, recommended by Mr Youatt, dispensed with.* Tom Middlemiss, the horse-breaker whose name I mentioned before (2456.), never required any assistance to mount a young horse, nor did any person ever see him mount one for the first time. The truth is, no fuss should be made about the colt; but when a number of persons are about him when any thing is done, there cannot be but fuss, and he cannot fail to become apprehensive. He will soon confide in one person, the breaker who is constantly about him, but he will not confide in a number of persons at the same time, nor will he confide even in his breaker, when others are engaged along with him; and hence no considerate horse-breaker will permit any one to be near him, to distract the attention of

* Youatt on the Horse, p. 321-4, edition of 1843.

the colt, when he is subjecting him to tuition of any kind. When mounted, the colt should bear his rider in standing for some time before he is urged to walk, as it will habituate him to stand at all times when mounted until his rider is ready to move. Every one must have felt the annoyance of mounting a horse that will not stand. His first pace should again be a walk, which having accomplished well with a rider, the trot should be taught. It is said that trotting is not a *natural* pace for a horse, he either walks or starts off at a canter. However this may be, trotting is an indispensable pace on our roads. On teaching trotting, horse-breakers are very apt to degenerate the pace into a jog, the most dangerous of all paces for a young horse in causing him to trip, and the most difficult to break a horse from, when contracted. A short hitching walk, ready to break into the jog alluded to, is as bad as the jog itself, and is a favourite pace with horse-breakers in chewing off their wards as fast walkers; but in such a pace a young horse is almost sure to dig a toe into the ground, and if a stumble is *not* the sonsequence, it is not the *man's* fault. Let the walk be a sound walk, and a trot a fair trot, and let no bastard pace be permitted to spoil both. It is not easy to teach a young horse to canter from a trot in a straight line, as he is more apt to start off to the gallop; but a few lessons in the circle will give him an idea of a canter, as he will there learn to point the leading foot. There is some risk at first in making a young horse reduce a canter into a trot, the actions being so very different, he seems at a loss what to do, and would rather halt. The tightening of the rein by degrees is the only way of reducing the pace in safety, as it likewise is from a fast to a slow trot. A sudden halt might throw the colt upon his haunches, and irrecoverably bring him over upon his back, and such an accident as this the colt will never forget, and, in fear of it, may become restive when pulled up suddenly at any time afterwards. Every manœuvre that may occasion any sort of accident to the colt should be avoided by the rider with care, and counteracted with firmness. Thus day by day the young horse acquires experience in the management of himself on the road, or in the field, but this series of experiences is a work of much time to both man and horse—of much patience and perseverance to the man—of much endurance and irksomeness to the horse; and more than all this, much of the benefit derived from the horse-breaker will be in a manner lost, if the future rider of the horse does not guide him in a similar manner, and with equal care for some time to come. If considerations, such as these, do not induce the owners of horses to employ only men of skill and character in breaking them in, I don't know what stronger motive can be placed before them to do it.

AUTUMN.

> "Ye balmy breezes! wave the verdant field;
> AUTUMN! all your bounties, all your lustre yield;
> That fruits and herbage may our Farms adorn,
> And furrow'd ridges teem with loaded corn."
>
> FERGUSSON.

IN contemplating the nature of the different seasons, we have seen Winter the season of *dormancy*, in which all nature desires to be in a state of repose,—Spring, the season of *revival*, in which the returning power of nature inspires every created being with new vigour,—Summer, the season of *progress*, in which nature puts forth all her energies, to increase and multiply her various productions,—and, now, we see Autumn, the season of *completion and of consequent decay*, in which nature, in bringing the individual to perfection, makes provision for the future preservation of the kind. While, therefore, the natural action of spring and summer is single, that of autumn is of a compound character. "Thus, if we follow out the study of the autumn in a proper manner, it leads us to all the revolutions that have taken place in the surface of our planet; and in this way, a plant of which we can, in a few months, see the beginning, the perfection, and the decay, becomes to us an epitome of the system of growing nature in its widest extent, and through its most prolonged duration. This is the grand advantage while studying the productions of nature in their connexion, and the events and occurrences of nature in their succession, has over the mere observations of the individual substance and the passing moment; and it is this which gives to the law of the seasons so high a value above all the beauties of the seasons taken in their individual character."

Autumn brings fruition, in which the toilsome labours of the husbandman, for the preceding twelve months, find their reward It is the season in which hope is lost in the possession of the thing hoped for, and because of a harvest of plenty, it is the season of gratitude. "It is this which makes the principles of seasonal action thicken upon us as the year advances, and the autumn to become the harvest of knowledge, as well as the fruits of the earth. Nor can one help admiring that bountiful and beautiful wisdom which has laid the elements of instruction most abundantly in the grand season of plenty and gratitude." But grateful as the husbandman must always feel for the bounties of Providence, so much labour is bestowed, so much anxiety is felt by him, as regards the effects of the vicissitudes of the seasons, before " he gathers his wheat into the garner," that the reflections which the consummation of harvest is calculated to give rise to are, I fear, narrow, and even selfish. " For as the annual harvest which we obtain from the earth, is received by us as resulting from that in which we have a right of property, a merit in labour, or both united, we are apt to forget the part which Nature has in the productiveness of the year, and look upon the whole produce as the return of our own capital and our own skill, just as we do in any mechanical work, or mercantile speculation. That this is the true state of the case, is proved by the habitually proverbial fact, that the cultivators of the ground, for what purpose soever they may cultivate, are always complaining of the weather, as the grand enemy by which all their labours are frustrated, and all their products diminished. They are nowise at fault themselves, but the ' weary weather' never will be obedient to their dictates. What with rain, what with drought, what with heat, what with cold, each thrusting itself forward at the time when its opposite would have been by far the more beneficial, the crop they get is always ' below a fair average,' and what they do get, is gotten in spite of the weather, and not by means of its co-operation. It is in vain that the fable of the farmer—into whose hands Jupiter gave the management of the weather, and who, by having rain, and drought, and sunshine, and snow, when and where he wished, brought his land into a state of such utter sterility, that he was fain to plead more earnestly than ever, that so dangerous a power might be taken out of his hands—has stood on the record against them from remote antiquity; for the majority contend stubbornly, that all the merit is their own, and that all the blame falls upon the weather, which, notwithstanding all the examples which have been set before it, and all the experience it must have had, ' will not understand and obey the rules of good husbandry.' "*

The colours displayed by the autumnal setting sun are exceedingly rich; one form of the phenomenon not uncommon at that period is at-

* Mudie's Autumn, p. 25-7.

tempted to be pencilled in Plate XIV; but a far finer picture is to be seen in these words of a revered bard, who can feel intensely as he can describe beautifully.

> "A cloud lay cradled near the setting sun:
> A gleam of crimson tinged its braided snow.
> Long had I watched the glory moving on
> O'er the soft radiance of the lake below.
> Tranquil its spirit seemed, and floated slow:
> E'en in its very motion there was rest;
> While every breath of eve, that chanced to blow,
> Wafted the traveller to the beauteous west.
> Emblem, methought, of the departed soul,
> To whose white robe the gleam of light is given;
> And, by the breath of mercy, made to roll
> Right onward to the golden gates of heaven,
> Where to the eye of faith it peaceful lies,
> And tells to man his glorious destinies."
>
> WILSON.

Objects in the horizon—trees, houses, and ruins—are projected in bold relief against the clear deep sky of a calm autumnal evening at sunset. Such a scene as this—if gemmed, moreover, with the radiant and lustrous evening star—affects the mind to thoughtful meditation, not untinged with melancholy.

The temperature of autumn is high,—August, in Scotland, affording the highest average of the year, on account of warmth in the night as well as the day, though the sun is not more hours above the horizon than in March,—but Autumn follows the radiance of Summer, while Spring just escapes from the frigidity of Winter. Such is the heat, that it is no uncommon occurrence for reapers to be seriously affected by it in the harvest-field.

The labours of the field partake of the compound character of the season itself. Just as one crop is reaped from the ground, part of the succeeding one is committed to the earth; the autumnal wheat of two successive years being sown and reaped about the same time. The toil endured in harvest is almost incredible. Only conceive the entire bread-corn sufficient to support the population of such a kingdom as this to be cut down and carried, in minute portions, in the course of a single month! The usual season of reproduction among the animals of the farm is spring; but the most useful animal of all, the sheep, forms an exception to the rule, Autumn being the season in which the ewes are drafted, and the tup is allowed to go with them. There seems in Autumn a tendency in the animal frame to disease; sheep are liable to hepatitis; calves to quarter-ill; the horse to colic, and even inflammation in the bowels; and stallions and geldings become dull in spirit. Perhaps the feeding nature of aftermath, on which all animals live in Autumn, may cause a tendency, in the animal system, to predominant secretion of one of the fluids, and thereby predispose the system to particular complaints. If there is probability of truth in this surmise, preventive measures should be sought for and obtained; and oil-cake seems to possess this property. One preventive remedy against annoyance from parasitic insects and from cold to sheep, is bathing or smearing.

The sports of the field all commence in Autumn. The long contemplated gatherings in the hills, on the noted 12th of August, in quest of Grouse—game, *par excellence*, of which our country should be proud as its only indigene—cause every sheiling to afford shelter to many who, at other seasons, indulge in the far different enjoyments of urban luxuries. Partridge-shooting comes in September, sometimes even before the corn is cut down, and is followed by Hare-hunting in October; and after all the fields are cleared of their valuable produce, the inspiring ' music" of the pack is heard to resound through hill and dale.

The great event of Autumn—the harvest—naturally claims a preponderating share of the husbandman's solicitude; and until this important issue of all his toil is secured beyond danger, he cannot rest in quiet. He looks around him night and day, regarding the " face of the sky," and acts with circumspection. He sees his whole year's bread at stake, and feels that its safety depends on his own skill; and should he fail to exercise this aright, he would never cease to blame himself. None is more anxious to follow this advice than he :—

" The wind, the rain, the sun,
Their genial task have done,
Wouldst thou be fed,
Man, to thy labour bow,
Thrust in thy sickle now,
Reap where thou once didst plough,
God sends thee bread."

MONTGOMERY.

When every straw is safe in the stack-yard, and the stack-yard gate closed for the season, then, and not till then, is he satisfied of his task being finished, and enjoys undisturbed repose.

Now that we have surveyed all the seasons as they present themselves in this country, we must own our climate to be any thing but genial. The frequent changes to which it is daily susceptible render the culture of the soil always a difficult, and not unfrequently an irksome, occupation. Those vicissitudes, no doubt, sharpen the intellect of the farmer, and perhaps have been the chief means of eliciting the high skill which is so universally acknowledged to be exercised in the agriculture of this kingdom. Such skill will always have a field for exercise, for our insular position will subject our atmosphere to perpetual changes dependent on different conditions of heat and moisture produced by the state of the surrounding ocean. Notwithstanding the farmer is held to his task by a frowning climate, he would rather wish to have a smiling one, and sometimes envies the bright skies which he hears illumines the Continent. There is much truth in the desire expressed in the following observations, lightly as they are put together, for no one but enjoys fine weather, which is, indeed, always the source of gratulation when it occurs :—" It may be very well for a lover to declare in the presence of his his mistress that

" All seasons and their change
All please alike ;"

but to common mortals, occupied with the ordinary affairs of life, there is no truth in it. Who in his senses ever affirmed that the fogs of November were as delightful to him as the balmy breath of May ? If any one has, as I have, a horror of icicles, and who would never have the mild temperature of the air interrupted by the presence of a hoar-frost, let him migrate with the climate. Let him spend the month of January in Portugal; February in the Madeiras; March in Spain; April in Sicily; May in Lapland ; June in Italy ; July in Switzerland ; August in France ; September in England ; October among the forests of America ; November in Crete ; and December in the islands of the Cape de Verd. By this rotatory motion he may enjoy a delicious temperature, and revel in honeysuckles and roses all the year round."*

---

74. OF PULLING FLAX AND HEMP, AND OF THE HOP.

" Now pluck up thy flax, for thy maidens to spin,
First see it dried, and timely got in."
TUSSER.

" Lo, on auxiliary poles, the hops,
Ascending spiral, ranged in meet array !"
PHILLIPS.

(2878.) I have already described the method of sowing flax, and noticed the soil in which it best thrives (2293.) ; all that remains to be said of this crop is its *treatment in summer*, and the *mode of harvesting it*, for as to its treatment afterwards as an article of manufacture, that

* Note Book of an Oxonian, *John Bull*, for 5th August 1843.

is beyond my province. The only care required by the growing crops of flax in summer is *weeding*, and in its early stage of growth it will be much injured if weeds obtain the mastery. To obviate this inconvenience, and, indeed, to save altogether the trouble of weeding, it has been recommended to sow the land at the time of sowing the flax-seed with grass-seeds, or to sow the flax-seed in drills; but neither expedient is so suitable for flax itself as land kept clean by weeding a broadcast crop; for as equality of fibre is of the utmost importance to the value of flax, sowing it in drills admits the air unequally to the crop, and the fibre of the plants on the outside of the drills would thereby be much coarser than that of those in the interior. And as to sowing grass-seeds, even the low-growing white clover amongst flax, it should make no difference to the flax-plant whether it was choked by a valuable or a worthless plant, since both would equally be *weeds* in reference to it.

(2879.) Besides the common weeds which infest the soil, according to its nature, there are others specially found amongst flax; of these, one is the common gold of pleasure, *Camelina sativa*, the seed of which is imported among flax-seed, and the plant may be known by its attaining from 2 to 3 feet in height, having small yellow flowers, and very large pouches on long stalks. But a more troublesome weed than this is the flax-dodder, *Cuscuta Europæa*, inasmuch as it adheres parasitically to the flax plant, and, of course, injures its fibre; while the gold of pleasure may be pulled out before the flax is ready. The habits of the flax-dodder are these:—“ It is a plant which germinates in the ground, and sends up a slender threadlike stem, which, twisting itself about, soon touches one of the stems of the flax amongst which it is growing. As soon as this takes place, the dodder twists itself round the flax, and throws out from the side next to its victim several small processes, which penetrate the outer coat or cuticle of the flax, and act as suckers, by which *the parasitical dodder appropriates to its own use the sap which has been prepared in the flax, upon which the growth of the flax depends*. The dodder then separates itself from the ground, and relies solely upon the flax for its nourishment, producing long slender leafless stems, which attach themselves to each stem of flax that comes in their way. Thus large masses of the crop are matted together, and so much weakened, as to become almost useless. This plant produces great quantities of seed, which is usually thrashed with the flax-seed, and sown again with it in the succeeding year. Several years since I took considerable trouble to ascertain if all foreign flax-seed was mixed with that of the dodder, and was led to the conclusion, that the American flax-seed is nearly free from this pest, and that that from Russia, and especially Odessa, is peculiarly infested with it."* The weeds, when very young, are picked out by hand from the flax by field-workers, and in doing this, the kneeling down upon the flax does it no harm. If weeding be once effectually and timeously done, the weeds will not again much trouble the crop; and though it should cost several shillings the acre, the increased value of the crop will repay it all. Before leaving the subject of weeding, I may remark, that though the American flax-seed be free from dodder-seed, it is far inferior in giving a crop to that imported from Riga, the Riga-kind, which also bears the name of Belgian seed, from the Belgians sowing it. And I may also remark, that sowing flax on clean land will save much of the cost of weeding, that is, after a green crop, as turnips and potatoes, the cleansing of which will have rendered the soil comparatively clean for flax. If flax be thus cultivated in lieu of a corn-crop, its culture may be practised without much deterioration to the land; but if it is determined to regard flax as a green crop, and cause a corn-crop to follow it, the land will in time assuredly feel the scourging effects of such a system, and oblige its cultivators to abandon it altogether.

(2880.) The pulling, steeping, and drying of flax are simple enough, and are processes generally well understood; but Mr Henderson's account of managing the crop, whose sample of Irish flax obtained a gold medal from the Agricultural Improvement Society of Ireland, at their meeting at Belfast in August 1843, being the most practical, and, at the same time, succinct I have met with, I shall transcribe it. 1. And, first, as to test of *ripeness*, Mr Henderson says, "I have found the test recommended by Mr Boss to ascertain the degree of ripeness that gives the best produce, with the finest fibre, perfect. It is this: Try the flax every day when approaching ripeness, by cutting the ripest capsule on an average stalk across (horizontally), and when the seeds have changed from the white, milky substance which they first shew to a greenish colour, pretty firm, then is the time to pull. The old prejudice in favour of *much ripening is most injurious*, even as regards quantity; and the usual test of the stalk stripping at the root and turning yellow, should not be depended on. Where there is one man that pulls too green, five hundred over-ripen." 2. When properly ripened, flax should be *pulled* in this way:—"I use the Dutch method, say, catching the flax close below the boles; this allows the shortest of the flax to escape. With the next handful the puller draws the short flax, and so keeps the short and the long each by itself, to be steeped in separate ponds. It is most essen-

* Gardeners' Chronicle and Agricultural Gazette for 10th February 1844, p. 189.

tial to keep the flax even at the root end, and this cannot be done without time and care, but it can be done, and *should always be done.* The *beets* should always be small, evenly sized, straight and even, and should never be put up in stooks or windrows, but taken to the pond the day they are pulled, or the day after at longest, especially in bright weather, for *the discoloration produced by the sun on green flax will never be removed till it goes to the bleacher, and will give him some trouble also.*" 3. Next comes the *steeping*, which is a most important process, and is the one least understood by growers of flax in this country. You perhaps require to be informed of the object of steeping the flax-plant. The stem of flax consists of two parts possessing very different properties; the one, the outer, is fibrous, and affords the substance of flax; the other, the interior, is pithy, and is got rid of by fermentation in steeping, loosening its hold of the fibre. There is also much mucilage to be got quit of; and the sooner flax is put into steep after being pulled, the more mucilage will be dissolved from it. If steeping is so long continued as to affect the texture of the fibrous coating, the flax will be injured; and should it not be as long applied as the pithy matter may be easily loosened, much labour will be afterwards incurred in getting quit of it. Proper steeping, then, is an essential and delicate process, and on this account Mr Henderson's instructions are valuable :—"Flax is subject to injury from neglect in every process, but in this especially. The water brought to the pond should be pure from all mineral substances, clean and clear. The water of large rivers is generally to be preferred; but spring-water, which has run some hundred yards becomes soft, and will have deposited any mineral impurities it contained; immediately from the spring it seldom does well. If the water be good and soft, it is injurious to allow it to stagnate in the pond before steeping. I put in two layers, each somewhat sloped, with the root-end of each downwards: one layer at a time is said to be safer, and perhaps is so, though I have tried both and seen no difference. The flax should be placed rather loose than crowded in the pond, and laid carefully straight and regular. Having an abundant supply of water, I do not let any into the pond till the first layer is in. I cover with moss-sods (from the turf-banks) laid perfectly close, the shear of each fitted to the other. Thus covered, it never sinks to the bottom, nor is it affected by air or light. It is generally watered in 11 or 13 days. A good stream should, if possible, always pass over the pond; it carries off impurities, and does not at all impede due fermentation—flood and all impure water should be carefully kept off. The Dutch test of being sufficiently watered is certain and perfect, at least, I never found it otherwise. It is this :—Try some stalks of average fineness, by breaking the woody part in two places, about 3 inches apart, at the middle of the length; catch the wood at the lower end, and if it will pull (downward) for those 3 inches freely, without breaking or tearing the fibre it is ready to take out. This trial should be made every day after fermentation subsides, for sometimes the change is rapid. Flax is more frequently injured by too little than too much of the water. Great care and neatness are necessary in taking it out. Broken or crumpled flax will never reach the market. Spread the day it is taken out, unless it is heavy rain—light rain does little harm; but, in any case, spread the next day, for it will heat in the pile, and that heating is destructive. The most particular cause of injury in steeping, is exudation of water from the sides or bottoms of the pond. Stripe and discoloration are mostly imputed to the quality of the water brought to the pond; whilst in 9 cases out of every 10, the water oozing from the sides and bottom of the pond itself is the cause. Even if such water were pure, which it seldom is, it is injurious; but when impregnated with iron or other materials, it does immense harm. If such ponds must continue to be used, the injury may be partially amended by draining around the sides and ends, at 6 or 8 feet distance, and 18 inches deeper than the bottom of the pond, and filling the drains with stones. No other thing I know of does such extensive injury as this springing of water within the pond." 4. Flax "should be *spread* even, straight at its length, not too thick, and well shaken, so that there shall be no clots; indeed, if possible, no 2 stalks should adhere. I have ever found it injurious to keep it long on the grass; it is in the steep the wood is decomposed; on the grass the fibre is softened and the wood little, if at all, affected. I rarely let it lie more than 5 days, sometimes only 3; this year it had only 3 days, and I never had better flax. It should never, if possible, be spread on the ground flax grows on, it claps down, and the clay and weeds discolour it; clean lea, or lately cut meadow, is the best." 5. "*Lifting*, like all other operations, requires care and neatness to keep it straight to its length, and even at the roots. This operation is too frequently hurried and coarsely done." 6. If the steeping and grassing have been perfect, flax should require no fire; and to make it ready for breaking and scutching, exposure to the sun should be sufficient; but if the weather be damp, the flax tough, and must be wrought off, then it must be fire-dried. Such drying is always more or less injurious; but if it be put on the kiln in a damp state, it is ruinous; it is absolutely burnt before it is dry. All who can afford it should keep such flax over to the ensuing spring or summer, putting it dry into stacks, then it will work freely without fire-heat." 7. In the concluding remarks of Mr Henderson, there is much good sense.—" The proper culture and preparation of flax require more care, exertion, and expense, than the old slovenly method, and those who will not give those requisites would do wisely to abstain from growing flax altogether. Any other crop will abide more negligence. So much has been said and written of late of the advantage of flax-culture, that it is to be feared some may be led to carry it to an undue extent, and sow it on land not fitted for it; indeed this is already often done, and I know of nothing more injurious to the farmer. Flax is proverbially either the very best or the very worst crop a farmer can grow." *

(2881.) The crop of flax, after it is dried, is bulky for its weight; and yields from 3 to 10 cwt. per imperial acre of dried plants. From 30 to 40 stones, of 14 lb. each, the acre, of *dressed* flax, is considered a fair crop, and if *fine* quality, will fetch perhaps L.90 a ton; that is, from 4 to 5 acres are required to furnish 1 ton of flax, and a return obtained of from L.18 to L.22 per acre, exclusive of the expense of preparing it by beetling, scutching, and heckling, and may still leave from L.10 to L.15 an acre of profit, which is a large one; but should the flax prove coarse by improper management, or be injured in drying, much waste will be occasioned in dressing it, and the profit reduced to perhaps $\frac{1}{3}$ of these amounts. So the observation of Mr Henderson, of flax being either the best or worst crop for the farmer, thus receives corroboration; and it should also be borne in mind, that flax, like the potato crop, leaves no straw for manure to the land.

(2882.) I have also described the sowing of *hemp* (2299.), a crop scarcely cultivated in Scotland, and its culture in England is confined to the southern counties, being a plant indigenous to the south of Europe and India. Being tall of growth in comparison to flax, it receives no injury from weeds, but, on the contrary, smothers by overtopping them. This plant is best cultivated in drills. The crop is pulled and watered, and dried like flax, the weight of produce dressed being little more than flax, from 40 to 45 stones the imperial acre, and the profit derived from it, after deducting expenses, seems to be from L.5 to L.6 per acre, though Lord Somerville, in sanguine expectation of extending the culture of this plant in England, estimated it at L.8. The dried refuse of the stems of hemp, after the fibre has been separated, is used as fuel, and may be converted into charcoal fit for gunpowder. The seed yields about 3 quarters per acre, and from it is expressed an oil "employed with great advantage in the lamp, and in coarse painting. They

* Dublin Farmers' Gazette.

give a paste made of it to hogs and horses to fatten them; it enters into the composition of black-soap, the use of which is very common in the manufacture of stuffs and felts; and it is also used for tanning nets." The common hemp is raised in India, not for the sake of its fibre, but for the intoxicating quality of its seeds when eaten green and fresh; the hemp of commerce of that country is derived from a different plant.*

(2883.) The *hop* is not cultivated in Scotland but as an ornamental plant in the shrubbery, because the climate is not sufficiently warm to develop its cones, constituting the ripe fruit, which is its useful part. The culture of the hop is confined to the south of England, where, in 1835, the extent of ground occupied by it was 53,816 acres; and its culture is very different from that of usual field-crops. When a new hop-garden is formed, the ground is trenched to the depth of 2 feet; and as the plants occupy the ground several years, their roots strike to a considerable depth, where the more of better soil they find near the bottom of the trench, the better they grow, for as to the surface-soil, it can be manured at any time, and in many ways. The plants, previously raised from seed, as being more hardy in constitution than raised by cuttings, are placed in quincunx order, at 5½ feet distant each way, which give 1440 plants to the imperial acre. The hop-plant being diœcious in its nature, that is, having different sexes on different plants, 1 male is planted among every 10 female plants. The female plants only bear fruit, which has the form of solitary cones or strobiles, ovate and pendulous, composed of membranous scales of a pale-green colour, each containing one round flattish seed of a bay-brown colour, surrounded with a sharp rim, and compressed at the top. Being climbers, as the plants shoot up, they are tied to and trained along poles pushed endwise into the ground beside them; and in the first year's growth of the bine the poles used may be short; but afterwards, when shoots spring from the old stock, 3 bines are preserved from each shoot and trained up upon 3 poles placed around each plant-hill, with their upper ends divergent, that the air and sun may find their way into the centre of each cluster of plants; and the poles are besides so set as to allow the forenoon sun to reach the plants, and also to receive as little injury as possible from the prevailing wind of the district. The poles are from 16 to 18 feet the longest, and 14 the shortest; and as 3 are required for every hill, 4320 are wanted for every acre. The best poles are of yew, next of chestnut, then larch, ash, willow, oak cut in winter, Scots fir, birch, alder, beech, in the order enumerated. They last from 3 to 5 years, according to the wood, and cost

* See Wisset's Treatise on Hemp, which contains all that can be said on the subject. Quarto edition, 1808.

1s. per foot per 100 poles; that is, poles 18 feet long will cost 18s. the 100, or nearly L.39 per acre; and as about 500 poles are wanted every year to keep up the stock, their wear and tear costs about L.4, 10s. the acre. In order to lessen this great annual expense, it has lately been suggested to stretch a stout wire along each alley, to which the poles should be fastened. The advantages of this plan are said to be, that poles of much less value than those usually employed may be used, injury from gales of wind avoided, and the alleys being open to the influence of the sun and air, the mould, that fatal disease of the hop, prevented; and it is said that a garden of 9 acres at Halling, near Rochester, is cultivated in this way. Poles are carefully laid aside at the end of every hop-season. The bines, as they shoot up, are tied to the poles by women, who use dried rushes for the purpose. The ground receives culture to keep it clean and open till the season of picking or gathering the flowers arrive, which is commonly the first week of September. The hop, properly so called, is picked from the bine by the hand; and, to facilitate that process, the bines are cut over at 3 feet from the ground, and the poles raised and laid on their side in a convenient position and place for the pickers. Whole families of labourers are employed at picking, receiving 1½d. per bushel, at which rate a family of 5 will earn from 7s. to 10s. a-day. There being 1440 hills in the acre, and allowing 1 bushel of hops to each hill, and 1½ lb. to each bushel, the acre will yield 19 cwt. 32 lb.; but the crop is sometimes not ¼ of that quantity. Indeed, so precarious is the crop of hops, that, at Binstead, in Hampshire, a farmer grew 4½ cwt. on 10 acres in the year 1825, and in the following year, 1826, he realised 9 tons from the same land! This diversity of crop is greatly owing to the effects of insects, of which a considerable variety and in great numbers affect the hop-plant, and also of blight or mould, occasioned by damp or confined air. When ripe for picking, the hop is of a lightish green colour, and gummy to the feel when in the highest perfection. In a few days longer it becomes brown and strong tasted. At Farnham, in Kent, the hops are always picked in the perfect state, and every injured flower is put into a separate basket; and, on this account, the Farnham hops always command the highest price in the market. This hop is eagerly sought after by pale ale-brewers; while the brown stronger-tasted hop is better liked by porter-brewers. After being picked, hops are immediately subjected to artificial heat in a kiln to be dried, so that they may keep; and to shew the value of a delicate-coloured article in the market, the brown samples are strongly fumigated with sulphur, to give them a fairer and more equal appearance. It is surprising that purchasers who are judges of hops, and aware of such a practice, submit to it, unless they themselves employ it as a means of deceiving customers who never saw hops growing, and know not how they are treated. What would be thought of a corn-farmer, were he to fumigate the barley he had to dispose of with sulphur, in order to make the bright-coloured, overripened, and stained samples look all alike ? If light-coloured hops are indispensable to the brewer, let *him* manufacture his commodities as he pleases, but let the farmer deal only in the genuine produce of the soil. If he must have fair-coloured hops, let him pick them in due season, and exercise his skill in a legitimate way, and eschew every species of deception. The drying is effected with coke and some charcoal, the drying heat being 112° Fahrenheit; and costs, including every expense, 14s. the cwt. After being dried, hops are laid in a heap, to sweat and grow tough, and there they lie longer than merely to cool; for they must feel moist and clammy, and be squeezable in the hand, before they are bagged, when 5 lb. of fresh hops will weigh only 1 lb. when taken from the kiln. A *bag* of hops weighs 2½ cwt.; and is fixed by statute 4 feet wide, 7½ feet long, and to contain 5½ yards of cloth, weighing 5½ lb., which usually costs 6d. per yard. A Kent *pocket* is 3 feet wide, 7½ feet long, weighs 4 lb., and contains usually 1¼ cwt., but 2 cwt. of Farnham hops. Such a pocket will occupy a man from 3 to 4 hours to tread the hops into firmly; and to tread 4 of them in a day, at 9d. per cwt., is a very good day's work; and in doing which he becomes covered with yellow dust, to which powder Dr Ives ascribed the whole virtue of the plant. Hops cannot be tread too firmly into the bags, for the more the air is excluded the better; and, for this end, the use of the Bramah hydraulic press is recommended. It is not an easy matter to keep hops, when they shrink in the bags, and the air finds admission to them, and they then lose from 5 lb. to 10 lb. per cwt.; but the greater quantity of sulphur and saltpetre employed in the drying, the worse will hops keep. Damp is ruinous to them. In most cases, the price of old hops drops down to the half of that of new. Hops containing the most seed will retain their weight the longest.

(2884.) The duty on hops is 2d. per lb., and it amounted to L.409,055 in 1835, on the produce of 53,816 acres, being 49,086,600 lb., or 8½ cwt. per acre. Small as this impost appears, it made that year a direct tax upon the land on which the hop grew, of L.7 : 19 : 6½ per acre! I never could understand why hops should be subjected to direct taxation, when all other agricultural productions are exempt from it. I can see the plea upon which a duty on *malt* may be urged, of its truly being a *manufactured* article; but the tax on hops is a *direct* impost on the produce of the soil. Whatever benefit is derived from this impost to the revenue, from which nearly L.4000 a-year must be deducted for its collection, is more than counterbalanced by the spirit of gambling engendered in speculators, who make purchases solely in accordance with the probable amount of duty to be exacted. Offers made to farmers, resting on this contingency, lure them also into that vice, and frequently make them suffer; and they cannot avoid the temptation, for the casualties affecting hops are so uncertain, that the prospects of a crop may be blighted or secured in the course of a few days. Thus, in 1834, the hop was so much affected by the *aphis*, that the whole amount of duty was struck at about L.100,000 over all the districts; but on thunder showers falling and destroying the insects *just in time*, and the weather afterward proving very favourable to the growth of the plant, actually raised the duty, which was paid, to L.329,936, thus indicating an increase in the value of the apprehended crop of more than 3 times. Were there no duty, the farmer would dispose of his hops, when he realised them, as he does any other crop. The expense of forming a new hop-garden is L.15, 13s. per acre, including a half-year's rent. The yearly expense of maintaining an acre of hops is, up to picking time, including rent, &c., L.17, 8s.; picking, drying, and including duty on 6 cwt., L.15, 2s.; together, L.32, 10s. It is not an uncommon practice to let the working of the ground, the poling of the hills, and the tying of the bines to the time of picking, to labourers at L.3, 10s. per acre.*

(2885.) *Flax, Linum usitatissimum*, from the Celtic *Llin*, a thread, in the class and order *Pentandria Pentagynia*, of Linneus, and in the natural order of *Lineæ*, is a native of many parts of Europe, as well as of Nepaul, and North America, in corn-fields, and is said to be originally from Egypt. It has been cultivated for an unknown length of time in Britain, of which it is now considered a naturalized inhabitant; and it is cultivated both for its fibre and oil, and the husk of the seed, after the oil has been extracted from it, is employed in the fattening of live stock.

(2886.) "Mr James Thomson and Mr Bauer," relates Dr Thomson, "have shewn, that the *fibres* of flax are transparent cylindrical tubes, articulated, and pointed like a cane; while the filaments of cotton are transparent glassy tubes, flattened, and twisted round their own axis. A section of a filament resembles, in some degree, the figure 8, the tube originally cylindrical, having colapsed most in the middle, forming semitubes on each side, which give to the fibre, when viewed in a certain light, the appearance of a flat ribbon, with a hem or border on each edge. The uniform transparency of the filament is impaired by small irregular fissures, probably wrinkles arising from the desiccation of the tube. In consequence of this difference between the structure of linen and cotton fibres, Mr Thomson and Mr Bauer were enabled to ascertain, that the cloth

* See Lance's Hop-Farmer, edition of 1838, for a great deal of information regarding the culture of this interesting plant.

in which the Egyptian mummies are wrapt is always linen, and never cotton. It is clear from this, that the opinion entertained by some, that what is called in our translation of the Old Testament *fine linen* of Egypt, ought to be the *cotton cloth* of Egypt, is erroneous. We have no evidence from the cloth wrapt about ancient mummies, that the Egyptians in those early times were acquainted with cotton." *

(2887.) Large quantities of flax-seed, commonly called linseed, are annually imported into this country. It is imported for the purposes of growing the flax-crop, and for crushing into oil. Vitality not being certain in the flax-seed raised in this country, it is necessary to receive a supply from abroad, and of all foreign kinds, that from Holland is the preferable for seed, while our own answers for crushing into oil. "Crushing seed is principally imported from Russia, but considerable quantities are also brought from Italy and Egypt. Of 2,759,103 bushels of linseed imported in 1831, 2,210,702 were brought from Russia, 172,099 from Prussia, 106,294 from the United States, 105,448 from Italy, 98,847 from Egypt, 53,738 from the Netherlands," † &c. By the New Tariff, the duty on foreign linseed, and that from British possessions, is only nominal, being 1d. per quarter. The prices by the quotations in February 1844, are from 50s. to 60s. per quarter for English sowing, and from 25s. to 37s. for Baltic crushing.

(2888.) Linseed contains a great proportion of mucilage, and, when converted into jelly, constitutes an excellent nutriment for stock. The process of making the jelly is this:—"The proportion of water to seed is about 7 to 1. Having been steeped in water 48 hours previous to boiling, the remainder is added cold, and the whole boiled gently for 2 hours, keeping it in motion during the operation, to prevent its burning to the boiler, thus reducing the whole to a jelly-like, or rather a gluey or ropy consistence. After being cooled in tubs, it is given with a mixture of barley-meal, bran, and cut chaff; a bullock being allowed about 2 quarts of the jelly per day, or somewhat more than 1 quart of seed in 4 days; that is, about $\frac{1}{16}$ of the medium allowed of oil-cake." ‡

(2889.) The mucilage of linseed does not belong to any of the genera of gums; but its nature has not yet been ascertained. Its general composition and elementary constituents were ascertained to be these, by Guerin-Varry:—

General Composition.		Elementary Constituents.	
Soluble gum,	52.70	Carbon,	34.30
Insoluble,	29.89	Azote,	7.27
Ashes,	7.11	Hydrogen,	5.65
Water,	10.30	Oxygen,	52.78
	100.00		100.00

The ashes contained carbonates of potash and lime, phosphate of lime, chloride of potassium, sulphate of potash, oxide of iron, alumina, and silica. §

(2890.) The oil afforded by linseed is in the proportion of 22 per cent., that is, about 15 gallons of oil from 1 quarter of seed, at a weight of 7½ lb. to the

* Thomson's Organic Chemistry of Vegetables, p. 849.

† Macculloch's Dictionary of Commerce, art. *Flax*.

‡ Don's General Dictionary of Botany and Gardening, art. *Linum*, vol. i. p. 456.

§ Thomson's Organic Chemistry of Vegetables, p. 674.

gallon; the remainder is *oil-cake.* The best oil is that which is cold-drawn; and I suppose that the best oil-cake is obtained from this process, as having most oil in it. The warm-drawn oil is obtained by heating the seed by steam to a temperature of 200° Fahrenheit; and as the heat liquefies the oil, no doubt more is expressed from the seed, and the oil-cake cannot be so rich. Both the cold and heated seed are put into woollen bags, and pressed by means of the hydraulic-engine or a wedge, and the cake, on being taken out of the bag, is already quite firm; its weight at most is 8 lb., and sells for L.9, 10s. to L.10 per 1000, or 1⅓ farthing per lb., and the foreign from L.5 to L.6, 10s. per ton, or 2¾ farthings per lb. From this difference in the price, it would appear that the foreign is the more valuable oil-cake, owing, perhaps, to its being manufactured by imperfect machinery, which cannot extract so much of the oil out of it as our own; but be that as it may, I am not aware that any experiments have ever been made on their comparative fattening powers. A substitute for oil-cake, or what is called Hutchinson's patent oil-cake, is compounded, I understand, of 3 cwt. of linseed-oil and 1 ton of barley-meal, and is sold at L.12 per ton. I have given the proportions of Mr Warnes's compound, which is similar, in (1484.).

(2891.) Oil-cake, independently as an article of food, is an excellent medicine for live stock, preventing constipation in the bowels, and giving to the hide a *sweetness of coat* unattainable by other means. Mr Wilson of Edington Mains, in Berwickshire, tells me, that ever since he has given oil-cake to his calves after being weaned, they have not been affected with that fatal complaint, the quarter-ill, already described in the latter part of (2606.), and he has experienced this beneficial effect for 6 years without any external application of setons. By administering oil-cake to my cows after calving, I certainly prevented them being affected with red-water (2134.).

(2892.) It has been proposed of late, with a considerable degree of earnestness, to encourage the growth of flax in Britain. The attempt was made some years ago and failed; but in the present instance it is recommended with the view of raising flax-seed for feeding cattle in sufficient quantity to render us independent of foreign oil-cake, of which, no doubt, large quantities are annually imported, but to what extent I have not been able to ascertain. The *object* of the suggestion is laudable, but the *end*, I fear, unattainable; for if good *seed* is raised to make good oil-cake, or compounds with oil, the *flax* will be coarse, and flax of inferior quality will never pay so well as corn: and it should never be lost sight of, in considering this question, that to raise flax must bring it into competition with white crops, and not green crops, because to raise it as a green crop would be to deteriorate its quality by bringing it into immediate contact with manure; and, on the other hand, if it is raised without manure as a fallow-crop, it must deteriorate the soil materially—no species of crop being *more* scourging to the soil than flax, not even a crop of turnip-seed. There is, therefore, this dilemma in the matter, the quality of the flax or of the seed must be sacrificed. The seed separately will not pay the expense of culture. Seed is produced from 6 to 12 bushels per acre, taking the highest at 12 bushels, that is, 1½ quarter, and taking it also for granted that it will all be fit for *sowing*, and worth the highest current price of 60s. per quarter, the gross return would only be L.4, 10s. per acre. The flax-crop varies in weight of rough dried fibre, according to season and soil, from 3 to 10 cwt. per acre; and taking the high produce, 5 cwt. per acre of dressed flax, at the highest price of L.6 per ton, the yield will be L.31, from which have to be deducted the expenses of beetling, scutching, and heckling, and waste, and loss of straw for manure, and the profit will not exceed L.8 per acre; but though *such* a profit would certainly repay the expenses of cultivation, yet it presents the *most* favourable view that can be taken, even with the sacrifice of the entire loss of seed—the loss, in fact, of the greatest inducement for renewing the culture of the plant. In Ireland the case, I believe, will be the same, though much of the soil of that country, being mossy, is more favourable to the growth of flax than that of England or Scotland, yet even there it will be found impracticable to raise good flax and good seed from the same piece of ground at the same time; and if the seed is not good, the oil-cake will be bad.

(2893.) *Hemp, Cannabis sativa*, is in the class and order *Diœcia Pentandria* of the Linnæan system, and of the natural order *Urticaceæ*, or nettle tribe. Being a diœcious plant, that is, having the sexes in different plants, it is necessary to have a mixture of both sexes in the same plot of ground, if seed is desired to be raised and collected, though I presume for the purpose of raising the fibre merely that condition is not requisite; but as apparently similar seed produces plants of different sexes, a mixture of the sexes cannot well be avoided in practice. The male plant is more slender and delicate in appearance than the female, which bears the seed; and though this is the usual distinction of sexes in plants, yet, strange to say, that, in speaking of the hemp-plant, most writers denominate the plant which *bears the seed* the *male*, owing, perhaps, to its more robust appearance, in comparison of the *true* male, which is, of course, barren of seed. On this misapplication of the sexes, it has been well remarked: "We are the more surprised that botanical writers should fall into the error, or rather copy this blunder from one work into another, for so many ages, without correcting a mistake that inverts the order of Nature." *

(2894.) The principal use to which hemp is applied is the making of cordage of all kinds, the fibre being both strong and durable. A first-rate man-of-war is said to require 80 tons of rough hemp to supply her with her necessary tackle. Taking 40 stones the imperial acre as a good crop, 4 acres are required, at that rate, to raise 1 ton; so that a man-of-war consumes at least one year's produce of 320 acres of hemp for an outfit of cordage! "By this cordage," says Coles, quaintly, in his Paradise of Plants, "ships are guided, bells are rung, beds are corded, and rogues kept in awe." Old cordage is converted into paper, and should, therefore, never be destroyed.

(2895.) The *Hop, Humulus lupulus*, like hemp, stands in the class and order of Linnæus, *Diœcia Pentandria*, and natural order *Urticaceæ*. It is not a native of Britain, nor was its use known in this country till the reign of Henry VIII., in 1524,† when it was used in the composition of malt liquor, thus franking the truth of an old English distich, that—

"Hops, Reformation, Bays, and Beer,
Came into England all in one year."

It had not become a favourite with the people for many years after that period; for Walter Blith records, in 1653, this remarkable popular error, only 200 years ago, and such errors are rife in all ages of the world, that "It is not many years since the famous city of London petitioned the Parliament of England against

* Phillips' History of Cultivated Vegetables, vol. i. p. 230.

† Beckmann's History of Inventions, vol. iv. p. 339.

two nuisancies, and these were Newcastle coals, in regard of their stench, &c., and hops, in regard they would spoil the taste of drink, and endanger the people."* It would appear that this public expression of dislike against hops had prevailed in high quarters, for their use was forbidden by act of Parliament in the reign of James I., though the mandate was little attended to; and by the time of Anne, public opinion had so entirely changed, that hops were considered so superior to all other bitter principle, that brewers were prohibited from using any other, under a penalty of L.20. Hops had not thriven for some time after their introduction into England; and even so lately as 1695, enough was not grown to serve the country, for 510 cwt. were that year imported from Flanders and Holland. For many years past this country has been able not only to supply itself, but to export hops, and few foreign hops are imported. In 1832 there were 703,153 lb. exported, and only 11,167 lb. foreign imported; but even this small importation was not required for home consumption, for in the same year the much greater quantity of 50,113 lb. of foreign were exported. The English hop is accounted the best in the world. Hops may be used medicinally; a hop-pillow will ensure sleep to a patient in fever, when all other expedients will prove ineffective. The tender shoots of the hop in spring may be used as a pleasant bitter salad.

(2896.) The yellow powder of hops was named *lupulin* by Dr Ives, which Messrs Payen and Chavallier found, by analysis, to contain the following substances,—volatile oil 2, bitter principle 12.5, resin 52.5, silica 4, and of diacetate of ammonia, gum, bimalate of lime, traces of fatty matter, and some salts, unascertained portions of each. In another analysis, they found it to contain 13 per cent. of the weight of the cones, but deducting 4 per cent. of foreign matter, there was left 9 per cent. of lupulin. When distilled with water, it gives about 2 per cent. of its weight of colourless volatile oil, to which hops owe their peculiar smell. This oil dissolves in considerable quantity in water, and it seems to contain sulphur.†

---

75. OF REAPING RYE, WHEAT, BARLEY, OATS, BEANS, AND PEASE.

"Here stretched in ranks the swell'd swarths are found,
Sheaves heap'd on sheaves here thicken up the ground.
With sweeping stroke the mowers strew the lands;
The gatherers follow and collect in bands;
The rustic monarch of the field descries,
With silent glee, the heaps around him rise."

POPE'S HOMER.

(2897.) We are now arrived at the most important of all field-operations—that for which every other that has hitherto been described has been merely preparatory—the grand result, to attain which the farmer

* Blith's Improver Improved, p. 240.

† Thomson's Organic Chemistry of Vegetables, p. 919.

feels the greatest anxiety, and which, when attained, yields him the greatest happiness—because it bestows upon him the fruit of all his labour; but the fruit will be great or small, in proportion, all other things being alike, to the skill and industry he has displayed during the course of the preparatory year; for no adage conveys a stronger truth than this, when applied to the farmer, "As he has sown, so shall he reap." If he have dealt with the land in a penurious spirit, labouring with inadequate means, stinting manure, and grudging seed, he will reap a scanty produce, a deceitful crop, and grievous disappointment; but should he have done all that skill, industry, and liberality could effect, he will assuredly reap a bountiful reward, for the earth is ever grateful of kindness. Not having yet experienced it, you can hardly conceive the difference between these two positions of the farmer; in the latter case, the very rustling of exuberant straw, while the crop is handling, is delightful to his ear, and his heart is made glad when he sees every one bestowing with heartiness the severe labour required in cutting down and carrying in a heavy crop. In the former case, on the contrary, the boneless straw is easily squeezed in the hand of the reaper, the light scanty-grained heads are almost disregarded by him, and a single cart-load clears a large space of the field. Such a contrast is indeed lamentable, and unfortunately may be made every harvest; but it is true, nevertheless, that the contrast is becoming less striking yearly; for it is within my own recollection, that 9 bolls, or 54 bushels barley, are now as easily raised as 7 bolls, or 42 bushels, were 20 years ago; and should the present spirit of inquiry continue, and lead to still more amended practice, it is certain that the range of comparison between former and present crops will be still more narrowed, and to the furtherance of this there is, besides, this great stimulus to improvement of poor soils, that the effect of amended practice is far more apparent on them than on better soils.

(2898.) As harvest-work requires a greater number of labourers than usually live on a farm, it is requisite you should hire beforehand a band of reapers on whom you can rely on remaining with you all harvest, and not trust to the chance of a casual supply. No doubt farms in the immediate vicinity of large towns can obtain a great number of reapers daily, who go to their own lodgings in town at night; and the convenience of obtaining a day's work at good wages within a few minutes' walk of their own homes, will tempt most of the inhabitants of towns to prefer farms near them; and this being the case, farms so situate may not require a hired band, but depend on the chance of finding reapers from day to day. When harvest-work goes on in a regular manner through the country, this is an easy and simple mode of conducting harvest-work; but should a great proportion of the crop become sooner ripened than was expected, or the weather endanger the safety of the standing crop every where, the general demand for hands renders the farmers near towns no better off than those at a distance, for town reapers will then go any where for higher wages. The farmers near towns may, no doubt, then give higher wages as well as others; but the excitement of rising wages renders reapers unsettled, and their chief anxiety then is to inquire where the highest wages are to be obtained, not who the possessor is of the nearest farm. In such circumstances, the farmers near towns must put up with the services of the infirm, the young, and those who are burdened with household cares. The *safest* plan, therefore, for *every* farmer to follow, is to hire a band of reapers, proportioned to the extent of work to be performed, to remain with him all harvest, and to obtain auxiliary hands, as he wants them, on any hiring-day in the neighbouring town.

(2899.) As the conditions upon which reapers are hired depend on the mode of reaping adopted, I shall describe the various modes of reaping grain before mentioning those conditions. Every species of grain is cut down with two small instruments, the *scythe* or the *sickle*. The scythe can only be used by men, the sickle by both women and men. Reapers with the scythe must not only be strong men, capable of undergoing great fatigue, but they must use the instrument dexterously, otherwise they will make rough work and create confusion in the harvest-field, where every operation ought to be carried on with precision and least loss of time. The scytheman requires a person to follow him and carefully gather the corn he has mown into sheaves in bands, previously laid down for the purpose, and no person is better fitted for this office than a woman. Another person follows the woman, the bandster, whose duty, as his name implies, is to bind the sheaves made by the woman, with the bands he finds lying under them. Another person follows all these, and clears the ground of every loose head of corn with a large rake, and this person may either be a man or a woman; but as 1 scythesman cannot give sufficient employment to 1 raker, the economical arrangement is for 3 scythesmen to work together, with their followers, in what is called a *head*, and 1 raker will then clear the ground passed over by 3 scythesmen and their assistants.

(2900.) There are two ways of reaping with the *sickle*; in the one, the reapers receive day's wages, and, in the other, they reap by the piece. When receiving day's wages reapers consist of a band of 7 persons, called a *band-won*, who occupy 2 ordinary ridges, 3 persons on each ridge, and 1, called the *bandster*, to bind the sheaves on each ridge and set them in stooks on 1 of the 2 ridges. The bandster should always be a man, as a woman is not able for it; whereas, with a sickle, a woman is as efficient a worker as a man; indeed, what is called a *maiden*-ridge, of 3 young women, will beat a *bull*-ridge, of 3 men, at reaping any sort of corn, on any given day. The 6 reapers may be all of one sex or the other; but the usual arrangement is 1 man and 2 women on each ridge. There is reason for arranging the reapers thus in band-wons; because 1 man can bind the corn cut down by 6 reapers, and the entire band can cut down and stook 2 imperial acres of corn every day.

(2901.) When reapers cut down corn by the piece, each person is paid for what he cuts every day, and, to enable the overseer to ascertain that quantity correctly, each reaper, whether man or woman, is put on a ridge by himself. When 2 or more persons agree among themselves, or a whole family chooses, 1 ridge is appropriated to them, and the cost of cutting is paid to one of the party. This is a convenient mode of reaping for a family, whose members are too young or too weak to take part in a band-won, and it is also a good plan for initiating young reapers into the art of reaping. A bandster is commonly employed by the farmer on day's wages, to bind the sheaves and set the stooks of the reapers; but as every reaper occupies an entire ridge, he is not able to walk across and bind the sheaves of 6 ridges; and, in this mode, the bandster does not work so much to the farmer as when he forms a part of a band-won. In order to separate every reaper's work by itself, the corn as cut down is stooked on every ridge; and after the ridge is finished, the overseer counts the number of stooks on it, and marks it in a book. This mode of reaping is said to be done by the *threave*, which comprehends 2 stooks. In some cases where bandsters are scarce, the reapers who cut by the threave also bind the sheaves and set the stooks, and are, of course, paid for the additional work. Sometimes corn is let by the acre to be cut, but whoever undertakes the task must perform it in one or other of these ways.

(2902.) These being the arrangements of the different modes of reaping corn, you will now be able to understand the conditions upon which the reapers are respectively engaged. The *head* of mowers, and the *band-won* of reapers, are engaged for the entire harvest, however long or short time that may occupy, according to the state of the weather; but in 3 weeks, usually, all the corn is cut down, not that the engaged band may have reaped it all by themselves in the time, for much assistance may have been given, and is commonly calculated on from strangers. who are hired by the week in the market-town on Monday morning.

The engaged band receive food, bedding, and bed-clothes, in an apartment, and cash; and to those who go to their own homes at night, an allowance is made in lieu of supper. Mowers get 3s., bandsters 2s. 6d., and gatherers (women), 2s., and reapers, men and women, 2s. a-day. Their food consists of oatmeal porridge and milk to breakfast and supper, and bread and beer to dinner—wheaten bread, ¼ of a loaf, or 1 lb. in weight; and a drink, besides, of ½ quart of beer in the afternoon. This is the food used in the southern borders of Scotland. In the more northern parts, the breakfast and dinner consist of 1 loaf of bread and 1 quart of beer, and the supper of porridge and milk, or bread and beer again. The bread is made of oatmeal, baked with yeast, 10 loaves from 1 peck or ½ stone of oatmeal, making each loaf, when baked, 14 oz. weight. Some people cause 12 loaves to be baked from 1 peck of oat-meal. The meal allowed in lieu of supper consists of 2½ stones of 14 lb. for the harvest, which is equivalent to 2 oat-loaves each night, without beer. The bedding consists of 2 pairs of blankets, with 1 pair of sheets, chaff-bed and bolster to every 2 persons hired for the harvest, but 1 pair of blankets, with plenty of straw, in the straw-barn or elsewhere, is the usual accommodation of the ordinary reaper hired for the work. All engaged reapers have their food and bed on Sunday; and on many farms, broth is made as a change in the Sunday's dinner, along with bread and potatoes, instead of beer. Those who reap by the threave, receive 3d. per threave of 2 stooks for oats and barley, and 4d. for wheat; and when they also bind and stook, the allowance is 3½d. for oats and barley, and 4½d. for wheat per threave. These receive no food at all. When they cannot go to their homes every night, they are allowed bedding as above, according as they are hired for the harvest or the week; but many go home on Saturday night. In hiring, instead of fixing wages per day, it is not unusual to give a slump sum for the harvest, irrespective of the number of days it may last; though, of course, the 3 weeks are held as the criterion of calculation. When the harvest is over before that period, it is conceived the harvesters have gained an advantage, but so does the farmer, in the shortness of the harvest; and, should it exceed that time, the reapers conceive they have lost; and so does the farmer in a double sense—in the protracted harvest, which is never a favourable one—and in the extra food given to the reapers. As a slump sum for the harvest, the reaper may receive L.1, 16s., the bandster L.2, 5s., and the mowers L.2, 14s., besides food. I understand that, in England, harvesters get from L.3, 10s. to L.5, besides 4 or 5 English pints of beer a-day, for the harvest, without food; and with food, of course, much less.

(2903.) These different ways of paying reapers have their advantages and disadvantages. The simplest way is that by the threave, and its advantages are, that the reapers are paid in money for what they cut down, and there is no trouble in providing food; but the disadvantages are great, inasmuch as it is the direct interest of the reaper to make very small sheaves, and, of course, thus to augment the cost of reaping; and, when the reapers also bind and set the corn, much temptation is put in their power to deceive their employer in the size of the sheaf; such reapers being known to bind stubble in the heart of many of the sheaves. One check over such a practice is the appointment of a bandster to bind the sheaves and set the stooks, and when he proves a sterling man, he will not bind a sheaf that is less than the proper size; but it is rare to find a bandster that will act in this manner, his sympathy being easily excited to favour the poor "widow," "the lone woman," or the helpless "orphans." The only effectual check is in the steward's hand, who carries a gauge to measure the sheaf before it is set, and, by practice, he can ascertain by the eye whether sheaves are of the proper size, which is 1 foot of diameter across the band; but the steward cannot be present every where over a wide extent of field at the same time, and cannot, therefore, inspect every sheaf before it is set in a stook. The stook should contain 12 sheaves of oats and 14 of wheat, a threave of each containing 2 stooks. There is an inconvenience, too, in placing a great number of reapers on single ridges, inasmuch as different reapers having different powers of reaping, one cuts through his ridge far ahead of his nearest neighbour, whilst others delay cutting out their ridges for several days. There is, besides, the no inconsiderable trouble imposed on the steward of counting the threaves of every ridge, and marking them down in a book, and of calculating every day's reaping of those who are hired by the day. Threavers, however, have a strong inducement to cut the straw near the ground, as it the more readily increases the sheaves to their requisite size.

(2904.) There is great efficiency in the band-won system of reaping, a large number of reapers being entered for work over a comparatively small breadth of the field, each band-won soon finds its way to the end of the ridge, and as the leading one should keep foremost, the rest follow in eschellon in excellent order; but there is great desire frequently evinced for striving, or what is commonly called *kemping*, that is, the last entered band-wons desire to push on and finish their ridges before those who have entered before them, and the certain consequence is bad work, and not unfrequently bad feeling is engendered between band-wons. There is also very considerable trouble imposed on the inmates of the farm-house, in providing and cooking food every day for so many people.

The making of porridge for perhaps from 50 to 70 persons twice every day, at morning and evening, and the dealing out to each band-won its quota of beer and bread at dinner, are attended with much trouble, and no small degree of anxiety in what is done giving satisfaction to the people. Complaints may be made of the porridge being too thin, or of not being boiled enough, or of the milk being sour, or of the bread not being well baked, and the beer not well brewed; and, perhaps, of the whole being dealt out in scanty measure. The farmer has little control over the baker and brewer; but this rule should be followed in regard to these two purveyors, namely, a fresh supply of bread should be received every day, and distributed to the reapers stale, that is, at least 24 hours old, and a stock of beer laid in in barrel before harvest; so that, when a fresh supply is brought every day, or every other day, by the brewer, it will have time to ferment and settle before it is wanted for use; but commonly the demands on the brewer at this time are so numerous, that he can seldom supply his customers, but from hand to mouth, and commonly with unfermented liquor. It is, however, entirely in the farmer's own power to see the porridge properly cooked, and the milk in a proper state for use.

(2905.) Reaping with the scythe is as efficient a mode as by band-won; but the same trouble in regard to food attends it as the latter mode; and in both cases, when reapers have to be accommodated at night with bedding, the trouble is considerably increased. As a check on those who are hired for the week on taking away their bed-clothes, when their period of service expires—a circumstance which sometimes happens—is to cause them to deliver up the clothes before their wages are paid. There is another circumstance which is entirely in the power of the farmer himself to regulate, which is, giving the reapers their meals at stated times every day, regularity in this respect having a material effect in keeping the work of the field in good order; for whenever a meal arrives at the appointed hour, there will be no flagging on the part of the reapers; but should it not arrive when expected, the work is carelessly done, and the people have no heart to go on. The hours that seem to be well adapted for the appetite, and which divide the day into pretty equal parts, are, 8 o'clock in the morning for breakfast, to start to work again at 9; and to dine at 1 in the afternoon, to begin work again at 2, and have a resting-time for the drink of beer at 4, for about $\frac{1}{4}$ of an hour, and then to work till dusk, having begun at dusk in the morning in a late harvest, and at 5 in the morning in an early one. From 5 in the morning to 7 in the evening gives 12 hours of work, and 2 hours for meals, which is as long a day's work as reapers can endure for a harvest of 3 weeks, especially in warm weather. In order to keep proper discipline among the reapers, not a person ought to be permitted to leave a ridge without consent asked and obtained from the person intrusted with the superintendence; and when the food arrives, there should be no cessation of work till the word of command is given by the superintendent, who should be guided by the watch, not disbanding the reapers till the hour of repast actually arrives, and not allowing one minute to pass beyond the hour of recommencing; both which rules should be peremptorily adhered to. When reapers find no advantage taken of their own time, they will be less tempted to take advantage of the time of their master; at any rate, they are then deprived of any excuse or plea for doing so.

(2906.) *Sickles* are of two kinds, the scythe and the toothed. The *scythe sickle* is so called, because of its being provided with a cutting edge, which is kept sharp by the application of sandstone. Its mode of cutting is with a draw of the hand, or with a stroke upon the straw near the root end, in considerable quantity at a time, and the sheaf is gathered with it by rolling the cut corn against the standing with the left hand, occasionally assisting by a push of the sickle. This sickle is capable of cutting expeditiously, and, in the hands of an expert reaper, much corn may be cut down with it; but it is a rather dangerous weapon for the operator, should its blade be reflected from a clod or a stone; and as it cuts many straws which elude the grasp of the hand, it is apt to leave a foul stubble. The true form of this sickle is given in the figure below, and it costs the following prices, according to size:—

	No. 1.	No. 2.	No. 3.	No. 4.	
Common, .	9d.	10d.	11d.	1s.	each.
Patent, .	1s. 2d.	1s. 3d.	1s. 4d.	1s. 5d.	

The *toothed sickle*, as its name implies, has its edge cut into small teeth, which, when applied to straw that is held firm against it, cuts it through like the rasp of a file. From this description, it is obvious that its action is very different from the scythe-hook, inasmuch as it cannot cut a straw until the straw is held firm, either directly by the hand, or against a handful of cut corn. Its proper use, therefore, is to cut the corn in small portions. An expert hand can also cut a great quantity of corn with it, and as it requires no sharpening, it occasions no cessation of work. The figure below shews its proper form, and it costs 6d. or 7d. each, according to size. Many sickles are made so straight at the point, as to be unable to gather the corn they cut.

(2907.) *Reaping scythes* being of different forms, and having to be mounted in a peculiar manner, are figured and described below, and the mode of using them I shall soon describe.

(2908.) Before proceeding to employ reapers in the field, it is requisite you should know when corn is fit to be cut; for if cut down too soon, or allowed to stand too long, loss of produce will be incurred. It may be laid down, as a general rule, that corn in a healthy state comes to maturity first in the ear, and then in the straw; and when it does not, that is, when the straw becomes matured first at the root, then, of course, the grain suffers premature decay. Whenever a crop is observed to be in the latter state, it need not be allowed to stand longer on the ground, as it can derive no more benefit from the soil, and its grain will won as readily in the stook as on foot; and so, in like manner, whenever the ear is observed to be sufficiently ripe, the crop should be cut down, as the straw will won more rapidly in the stook than standing on the ground. The only matter of doubt, then, in the case is, when the ear is sufficiently ripe. The most ready way of judging of this, in wheat and oats, is the state of the chaff, and of 2 or 3 inches of the top of the straw under the ear; if all these are of a uniform straw-yellow colour, and feel somewhat hard in the ear, in the oat, and absolutely prickly to the hand, in the wheat, when grasped, they are ripe; or the grain itself may be examined, and should it feel firm under pressure between the finger and thumb, it is ready for reaping; or should the neck of the straw yield no juice when twisted round by the finger and thumb. Barley should be of uniform yellow colour in the grain and awns, and the rachis somewhat rigid; for as long as it moves freely by a shake of the hand, the grain is not sufficiently ripe, nor will it be of uniform colour. It is not equally prudent to reap all sorts of grain in the same degree of maturity. When wheat is reaped before it is sufficiently ripe, it is apt to shrink, and have a bluish tint in the sample; and when too ripe, the chaff opens from the grain, which is apt to be thrown out by the least agitation of the air; and some sorts of white wheat are very subject to be thrown out by the wind, even before reaching the point of maturity. When very ripe, wheat bends down its ear, opening the chaff, and becomes stiff in the neck of the straw, indicating that nature intends the grain to be shaken out. Red wheat is less liable to be shaken than white; but any kind will shake out when too ripe, provided the plant is in good health, and the grain of good quality; for as to immature grain, it is difficult to make it leave the chaff even when hardened, and the spelt wheat has so tenacious hold of its capsule, that it is difficult to be disengaged from it even by the blows of the flail, fig. 505. It may be supposed, that whenever the ear and the entire straw are of uniform yellow colour, the plant is no more than ripe, and so it is; but by the time the straw has fully ripened to the root, the ear will be rigidily bent, and ready to cast its seeds with the slightest violence. The same rule may be applied to barley as to wheat, that is, whenever the neck of the straw is ripe, it is time to cut, for when too ripe, the ear bends itself down diverging the outward row of awns nearly at right angles with the rachis, and is apt to be snapped off altogether by the wind. In regard to oats, the same rule also applies; but there is much less risk of cutting oats unripe, in comparison to allowing them to stand till perfectly ripe, as they are easily shaken out by the wind,—the chaff standing apart from the grain. When bean-straw turns black, it is fit to cut, and so is pease-straw, when the pease become firm in the pod. In every case it is much safer for the crop to be cut before it is ripe, than after it has become too ripe, as I shall shew below.

(2909.) With all these preliminaries arranged, and with your mind satisfied that the crops are in a fit state for reaping, proceed with the reapers to the harvest-field, and conduct the reaping in the best manner for each respective sort of grain. And, in regard to this, I should remark, that the harvest-field will not be properly conducted, unless a person is appointed to superintend the reapers; for to none of the reapers can such a task be deputed, as his own occupation is sufficient to occupy all his attention. The steward is the person who should undertake this duty, unless you undertake it yourself. It is his duty to mark the time to commence work, and to leave it, and of the hours of meals and of rest; it is his duty to restrain the impetuous, to urge the slow worker, and to keep every one in the best spirit for work; and it is his duty to see the ground neatly cleared of the crop, and the crop itself judiciously handled. The man who fulfills all these duties as they should be during the entire harvest, accomplishes no easy task, either of body or mind. He should not be mounted on horseback, but be on foot, ready to keep every thing right; for it should be remembered that a single minute's loss of work of a large band of reapers, causes a great loss in the gross amount of work. The farmer himself, when not superintending, may move about on horseback; but a horse is a troublesome companion to a man who has to move about close to work-people in a harvest-field.

(2910.) Let us first take the management of reapers in *band-wons.* It should be made a point to select a band-won of picked reapers and bandster to take the lead. This is necessary, both as a pattern that cannot be excelled by any other band-won, and for quantity and quality of work. The strongest band-won, taking into consideration a proportionate quantity of wheat to be cut, consists of 1 man on each ridge, supported by 2 women, 1 on each side. The man makes the bands from the crown of the ridge; and as the heaviest and longest straw generally grows there, he is the fittest person for taking the heaviest part of the work, while the women can stoop most easily to clear both the furrows. The corn-band is made by cutting a handful of corn, and dividing it into 2 parts; and, by plaiting the corn-ends of the straw together, and twisting them together, so as the ears of corn shall lie above the twist, when the band is placed extended on the ground, as is endeavoured to be shewn in fig. 521, where *a* are the ears of corn, *b* the twist, and *c c* the stubble-ends of the straw. It is the man's duty to have a band ready to lay down as soon as the one previously laid down is filled with corn sufficient to make a sheaf. In this mode of reaping, no regard is paid to the size of sheaf; and this being the case, the reapers are apt to make the sheaves much larger than they should be, indeed sometimes so full as to render it almost impossible for the bandster to make the ends of the band meet. Too full bands should not be allowed, as the sheaves are apt to burst in being handled afterwards, and the sheaves should be made no larger than easily to be lifted about. The women having nothing to do with bands, devote their whole time to reaping, and filling the bands with corn. It is the duty of the woman who reaps the right side of the ridge, that is, whose hook-handle is next the open-furrow, to clear the corn of the open-furrow as far as the furrow-brow of the ridge on her right, not only because hers is the leading ridge, but because the woman employed on the ridge on her right cannot gather the corn so well or so easily with the point of the hook, as she can with the handle-end of the hook. Clearing the open-furrow being the most troublesome part of the work, the 2 women change places at every landing, and take the reaping of it alternately. There is such an understanding among the members of a band-won, that should any difficulty occur in the reaping of one of the ridges, such as a spot of corn much laid, the reaper who encounters the difficulty is immediately assisted by her companions. Thus the band-won reap till the 2 ridges are cut down, when they return to the end of the field they began at, and commence upon new ridges. Band-won after band-won do the same till the entire number of band-wons are again placed, or *stented*, into their ridges, in the same manner as they were by the steward at commencing to reap the field.

Fig. 521.

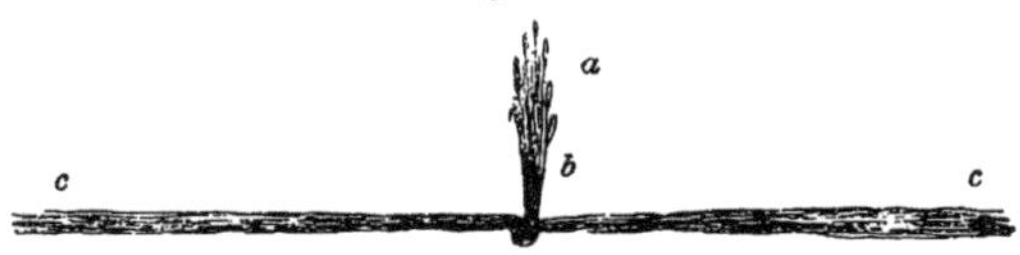

THE CORN-BAND READY TO RECEIVE THE CUT CORN.

(2911.) It is not easy to *describe* the best mode of cutting corn with the sickle. In using the scythe-hook the body is brought low, by resting chiefly on the right leg doubled under the body, and the left one stretched out to act as a stay to steady the entire frame. The right arm is stretched amongst the corn, and in drawing it towards you, near to, and parallel with, the ground, the standing corn is cut, and is received and held up with the left hand on one side, and by the standing corn behind it on the other. A creeping advance is then made of the body towards the left, and another similar cut made with the sickle, and the additional corn still gathered and supported by the left hand and the standing corn. The reaping thus proceeds *across the ridge*, in breadths measured by the stretch of the arm, until as much is cut as can well be kept together by the left hand, when the whole is lifted by the hook and left hand, and placed into the band to assist in making the sheaf. Any loose straws on the ground are then swept by the hook amongst the standing corn. The great object, in good reaping, is to make short stubble, because more straw is thereby gained to the sheaf, and less left on the field; and it is impossible to cut the stubble short, unless the body is brought so near to the ground as to allow the arm to sweep the sickle parallel to it. Reapers who only bow the body down from the haunches, and keep the legs upright, draw the sickle *up* towards their knees, causing the stubble to be cut like a series of notches, low in front, and high at the hind part. In using the toothed hook, the corn is cut in small hookfuls, retained firmly in the left hand, and collected in it as long as it can contain no more, and is then put into the band. In reaping with this instrument, the body bent forward answers the purpose, as small hookfuls can be cut near the ground; indeed, the nearer the ground the easier is the straw cut, but the straw, in this mode of reaping, is always too firmly squeezed in the hand.

(2912.) The *bandster*, as soon as one band is filled with corn, begins his operations, and he should bind the sheaves in this way:—Going to the stubble-end of the sheaf, with his face to the corn-end, he gathers the spread corn into the middle of the band with both hands, and taking an end of the band in each hand, he moves the sheaf round as much as to place the corn-end parallel to his left arm; then drawing the ends of the band together as forcibly as he can, especially with the right hand, and as close to the sheaf as possible, keeping the purchase, thus obtained, good with the side of the left hand, he twists the ends of the band round each other with the right hand, and doubles the twist under the tightened part of the band, pushing it through as far as to keep a firm hold round the sheaf. After this operation, the corn-end of the band, *a*, fig. 521, is held firm by the pressure of the sheaf against the ears of corn and the twisted part of the band at *b*. The sheaves bound on the left-hand ridge are taken to the ridge on the right on which they are all stooked ; and after a sufficient number of sheaves have been bound, the stook is set, and it is set in this way :—2 sheaves, *a* and its opposite, fig. 522, are taken by the bandster, one in each hand, and set a little apart on the ground, with the corn-ends close together, such as *b* and *c* are, in such a position that the length of the stook shall stand N. and S. ; and these are intended to form the centre of the stook. Other 2 sheaves, *d* and its opposite, are set on one side of *a*, in a similar manner, independently, and not leaning against them; other 2, *e* and its opposite, are placed in the same manner on the other side of *a*, and so on, till 10 sheaves in a double row of 5 are thus set, and which complete the body of the stook. A sheaf *f* is then split from the band towards the corn-end, and laid astride upon the top of half the number of standing sheaves, in nearly a horizontal position, and another sheaf *g* is placed in a similar manner upon the tops of the other half number of sheaves, having their butt-ends stuck into each other. These last are called *hood-sheaves*, and are intended as a protection to the others against rain, which their drooping position prevents remaining upon. There is method in setting the sheaves in the body of a stook: were the corn-knots of the bands, as *h*, set outwards in the stook, the rain might injure the corn in them, and as they bear a sensible proportion to the corn of the whole stook, the sample might be materially injured ; but in turning the corn-knots inwards, and the root-ends outwards, as seen at *h*, such an injury to the sample is prevented: but the corn-knot is placed uppermost in the hood-sheaves, and exposed to the rain at *i*; because, were the other *side* of the sheaf exposed *upwards*, where the straw is gathered into the form of a groove down the length of the sheaf, the rain would penetrate by the groove into the body of the sheaf in its horizontal position, and would thus prove more injurious than the spoiling of a corn-knot. Barley and oat stooks are almost always hooded, though, if the weather could be depended on, the precaution would be unnecessary ; but as both those kinds of grain remain a considerable time in

Fig. 522.

THE BARLEY AND OAT STOOK HOODED.

the field before being fit for the stack, it is the safest plan to put hoods upon the stooks. But wheat-stooks, one of which is represented by fig. 523, are very seldom hooded, because they require to stand only a short time in the field. It will be observed, that the wheat-stook com-

Fig. 523.

THE WHEAT-STOOK WITHOUT HOODS.

prises 14 sheaves, whereas those of barley and oats have only 12; the reason for the greater number being, I conceive, that the wheat-stook was once hooded, and the greater length of its straw required 6 sheaves to afford room for the length of 2 hood-sheaves; and, without such a reason, one cannot see why a stook of one sort of grain should contain more sheaves than that of another.

(2913.) Little more is required to be said on the subject of *threaving*, the threavers being placed 1 on each ridge. They are bound to follow the same rule in regard to clearing the right-hand furrow, and whenever the threaver finishes the reaping of his ridge, he enters upon a new one. Threavers always make low, clean stubble; and as it is their interest so to do, in order to make up the sheaf as soon as possible, they seldom require to be corrected for rough and bad work. The steward gauges the size of the sheaf with the sheaf-gauge, fig. 524, which encompasses 1 foot square between the limbs *a b c d*. The bandster sets the stooks on every ridge; and to facilitate their being carried off by the cart, they should be put on both sides of every alternate open-furrow: 1 bandster being required to every 4 threavers. After the stooks are set on each ridge as it is reaped, the steward counts their number, and marks them in a book, in which the names of all the threavers are entered, leaving out odd sheaves in the counting, but counting single stooks as half-threaves, the odd sheaves on several ridges being stooked. Threavers

Fig. 524.

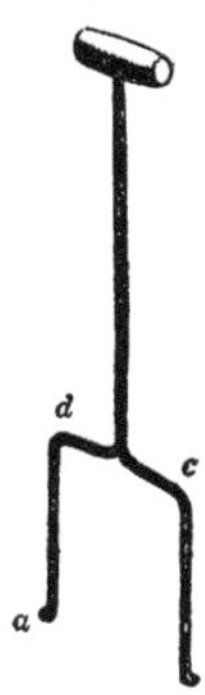

THE SHEAF-GAUGE.

themselves count their threaves, by drawing a straw from every stook in returning down the ridge. I have said *down* the ridge; because it is easier for reapers to reap up hill than down hill, and they are generally entered to their ridges at the bottom of the inclined ground. Threavers who work by the day have their accounts cast up by the steward at the end of the day's work, and paid on the spot; but they should be made to finish the reaping of every ridge entered before they are paid, otherwise confusion will ensue in the threave-book, by having the names of different individuals on the same ridge. If any individual threaver cannot conveniently finish his ridge, he must find some friend to do it for him, or remove the odd number of threaves from his newly entered ridge, to the ridge of his friend, or in any other way that will avoid confusion on the ridges. The individual powers of reaping are best exhibited in threaving. I have seen more than one young woman cut 24 threaves of oats a day, and thus make 6s.; and I have seen many cut 20 threaves, all good sizeable sheaves, for it is the old and infirm who practise tricks in filling their sheaves. When I mention, that from 8 to 12 threaves are considered a very good day's work, these exertions will be better understood. In all great feats of threaving, I have always found women superior to men, and even more enduring for a length of time; for, with the exception of one tall blacksmith, who wielded with uncommon strength an extra-sized scythe-hook in the left hand, I never saw a man who could cut the largest quantity mentioned above.

(2914.) Reaping with the *scythe* is a nice operation, and requires considerable skill. The scythes should be mounted as figured below, and made fit for work some time before being wanted in the harvest-field. There should be a number of small articles always ready in the field in case of accident, the procuring of which wastes much time, when not at hand. These are, a small hammer for fastening the wedges of scythe-ferules and of rake-handles; bits of old sole-leather for bedding the tines of the scythes upon; pieces of cord for tying anything; small large-headed nails for fixing the stays to the sned of the scythes; a large coarse file for rubbing down the turned-up point of a scythe, when it happens to come against a stone; a sharp knife for cutting bits of leather, and for removing any raggedness upon the rakes or cradles.

(2915.) The various forms of scythes are the cradle-scythe, the straight-snedded scythe, and that with the bent sned, fig. 452, already described; and the greatest favourite amongst mowers is the cradle-scythe, because it is easiest to wield by the arms, and does not twist the lumber-region of the body so much as the 2 common scythes; and, I may remark, that it is this last effect which forms the great objection against the scythes in ordinary use. And yet it is not easy to see why the use of the cradle-scythe, which is borne by the arms alone, in front of the body, and which does not admit of being balanced in one hand like the other scythes, should be less fatiguing to work with; yet there is no doubt of the fact, and, on that account more work is done with it.

(2916.) In commencing to cut a field of corn with the scythe, that side should be chosen from which the corn happens to lie, if it be laid, and if not, then the side from which the wind blows. The scythe makes the lowest and evenest stubble *across* the ridges, and then also most easily passes over the open-furrows. Other things being favourable, it is best to begin at that side of a field which is on the left hand of the mowers. If all these conveniences cannot be conjoined, as many as can should be taken advantage of. The ground should have been rolled, and all large stones removed in spring, otherwise the scythes will run the risk of being injured in the face by stones, and even by clods.

(2917.) I have already said, that reaping with the scythe is best executed by the mowers being in what is called *heads*, namely, a head of 3 scythesmen, 3 gatherers, 3 bandsters, and 1 man-raker, or of 2 scythesmen, 2 gatherers, 2 bandsters, and 1 woman-raker. On a large farm the heads may consist of the former, and on a small one the latter number. The best opening that can be made of a field for scythe-work, is to mow *along the ridge* by the side of the fence, which is kept on the left hand, from the top to the bottom of the field: and while one head is doing this, let another mow along the bottom head-ridge, the whole length of the field, and thus open up 2 of its sides. After this, the first head commences mowing at the lowest *corner* of the standing corn, across 6 ridges, or 30 yards, which is as far as a scythe will cut corn with one sharping. Suppose all these preliminaries settled, the scythesman who is to take the lead first sharpens his scythe. In sharping a scythe for cutting corn, the scythe-stone has to be put frequently in requisition, for unless the edge is kept keen, the mowing will not only be not easy, but bad; and unless a scythesman *can* keep a keen edge on his scythe, he will never be a good mower, and will always feel the work fatiguing to him. The sharping should always be finished with the *straik* or strickle. The stone need not be used at every landing, the strickle answering that purpose; but whenever the scythe feels like a drag on the arms, the stone should be used. In mowing, it is the duty of the mower to lay the cut corn or swath at right angles to his own line of motion, and the straws parallel to each other, as at *a*, *a*, *a*, fig. 525; and to maintain this essential requisite in corn-mowing, he should not swing his arms too far to the right in entering the sweep of his cut, for he will not be able to turn far enough round towards the left, and will necessarily lay the swath short of the right angle; nor should he bring his arms too far

Fig. 525.

THE MOWING OF CORN WITH THE SCYTHE.

round to the left, as he will lay the swath beyond the right angle; and, in either case, the straws will lie in the swath partly above each other, and with *uneven ends*, to put which even in the sheaf is waste of time. He should proceed straight forward, with a steady motion of arms and limbs, bearing the greatest part of the weight of the body on the right leg, which is kept slightly in advance, as seen at *b*, *b*, *b*. The sweep of the scythe will measure about 7 feet in length, and 14 or 15 inches in breadth. The woman-gatherers *c*, *c*, *c*, follow by making a band, such as fig. 521, from the swath, and laying as much of the swath in it as will make a suitable sheaf, such as *d*, *d*. The gatherer is required to be an active person, as she will have as much to do as she can overtake. The bandster *e* follows her, and binds the sheaves in the manner already described, and any 2 of the 3 bandsters, *f*, *f*, set the stooks *g* together, so that a stook is easily made up amongst them; and in setting them, while crossing the ridges, they should be placed on the same ridge, to give the people who remove them with the cart the least trouble. Last of all comes the raker *h*, who clears the ground between the stooks with his large rake *i*, described below, of all loose straws, and brings them to a bandster, who binds them together by themselves, and sets them in bundles beside the stooks. This is better than putting the rakings into the heart of a sheaf, where they will not thrash clean with the rest of the corn; and, moreover, as they may contain earth and small stones, and also inferior grain, from straws which may have fallen down before the mowing, it is better to thrash bundles of rakings by themselves. The figure exhibits the 3 kinds of scythes in operation.

(2918.) A scythesman will cut fully more than 1 imperial acre of wheat in a day. Many farmers affect to believe that the scythe is an unsuitable instrument for cutting wheat; but I can assure them, from experience, that it is as suitable as the sickle, and that mown sheaves may be made to look as well as reaped. No doubt mowing wheat is severe work, but so is reaping it. Of oats, 1 scythesman will mow fully 2 acres with ease. The oat-crop is remarkably pleasant to handle in every way; its crisp straw is easily cut by the scythe, and being hard and free, and generally not too long, is easily bound in sheaf and set in stook. Nearly 2 acres may be mown of barley; but the gummy matter in the straw, which gives it a malty smell, causes the stone to be frequently used in mowing barley, and the straw being always free, the bands are apt to break when rashly handled in binding the sheaves.

(2919.) One mode of setting up corn to dry quickly is in *gaits*, that is, the band of the sheaf is tied loosely round the straw, just under the corn, as at *a*, fig. 526, and the loose sheaf is made to stand upon the lower end of its straw being spread out in a circular form, as *b* to *c*, and they are set upon every ridge. The wind whistles through the open sheaf, and even the rain passes through, and does not hang upon it. The expedient of gaiting, however, is only practised in wet weather, and even then only should the crop, if allowed to stand, be endangered by a shaking wind. It is confined also to a particular kind of crop, namely, oats, wheat and barley never being gaited; because when wheat gets dry, after being cut in a wet state, it is apt to shake out in binding the gaits; and when barley is subjected to the rough usage of binding, after being won, the heads are apt to snap off altogether, and, besides, exposure in gaits, would injure its colour, and render it unfit for the maltster. Oats are protected by a thick husk, and the grain is not very apt to shake out in handling, excepting potato-oats, which are sel-

Fig. 526.

A GAIT OF OATS.

dom gaited, the common kinds only being so treated. But, for my part, I would not hesitate to gait any sort of oats when wet with dew in the morning, or even when wetted with rain, rather than lose a few hours work of reaping every morning, or at nightfall. Gaits, it is true, are very apt to be upset by a high wind ; but after having got a set, it is surprising what a breeze they will withstand. After being blown down, however, they are not easily made to stand again, and then 3 at least are required to be set against each other ; but whatever trouble the resetting them should create, they should not be allowed to lie on the ground, and it will be found that a windy day dries them very quickly, and secures their winning.

(2920.) *Rye* may be reaped or mown in the same manner as the other cereal crops. Its straw being very tough, may be made into neat slim bands. It usually ripens a good deal earlier than the other grains ; and its straw being clean and hard, does not require long exposure in the field, and on that account the stooks need not be hooded.

(2921.) *Beans* in drills are reaped only with the sickle, by moving backward, taking the stalks under the left arm, and cutting every stalk through with the point of the sickle. When beans are sown by themselves straw-ropes are required for bands ; but when mixed with pease, the pea-straw answers the purpose. Beans are always the latest crop in being reaped, sometimes not for weeks after the others have been reaped and carried. In stooking, bean-sheaves are set up in pairs against one another, and the stook may consist of 4 or more sheaves, as is thought most expedient in the circumstances for the winning of the crop, the desire being to have them won as soon as possible, to get the land sown with wheat. *Pease* are also reaped with the point of the hook, gathered with the left hand, while moving backwards, and laid in bundles, not bound in sheaves, until ready to be carried, and are never stooked at all.

(2922.) Farmers negligent of weeding their corn, give reapers much uneasiness, and waste much time, while getting rid of large weeds, some of which injure the hands of the reapers and of the gatherers often seriously ; and of these the corn dead-nettle, *Galeopsis tetrahit*, is dangerous, causing swellings, heat, and pain in the hands ; as is also the biennial spear-thistle, *Cnicus lanceolatus*, the spines of which breaking in the flesh, give acute pain when touched, and are very troublesome to pick out. The only remedy against these annoyances is wearing gloves of sheep-skin made on purpose, called *shearers' gloves*, which cost 1s. the pair.

(2923.) The *expenses of reaping and mowing* may be ascertained in this manner, on the supposition that the harvesters receive bread and beer at all their meals.

10 loaves out of 1 peck of oatmeal, say at 15d. per peck, and 2d. per peck for baking them with yeast, = 1¼d. per loaf.

1 loaf to each harvester at breakfast, ...	L 0	0	1¼
1 ... ... dinner, ...	0	0	1¼
2 loaves ... ... supper, ...	0	0	2½
2½ quarts of beer to each harvester at breakfast and dinner, at 1d. per quart, ...	0	0	2½
Daily expense of the food of each harvester, . . .	L.0	0	7½

At *reaping*, food of 6 reapers and 1 bandster, at 7½d. each per day, ...	L.0	4	4½
Wages of 6 reapers, at 2s. per day, . .	0	12	0
... 1 bandster, at 2s. 6d.	0	2	6
Cost of reaping 2 acres, ...	L.0	18	10½

Taking 2 acres of wheat, barley, and oats overhead, the expense at 18s. 10½d. will be 9s. 5¼d. per acre.

With loaf-bread, at 1¾d. the loaf, the daily expense would be raised to 9½d. per day, and the reaping to 10s. 6¼d. per acre.

With porridge and milk at breakfast and supper, the milk being ½ quart each meal to every person, would reduce the price below that of bread and beer of any kind; but how much I do not know, as I have not ascertained how much meal is consumed by each harvester in porridge.

At *mowing*, food of 3 mowers, 3 gatherers, 3 bandsters, and 1 raker = 10 persons, at 7½d. each per day, ...	L.0	6	3
Wages, 3 mowers, at 3s. a-day, ...	0	9	0
... 3 gatherers, 2s.	0	6	0
... 3 bandsters, 2s. 6d.	0	7	6
... 1 raker, 2s. 6d.	0	2	6
	L.1	11	3

6 acres of oats, at 31s. 3d.	L.0	5	2½	per acre.
5 ... barley, ...	0	6	3	
4 ... wheat, ...	0	7	9¾	

It should be observed, that the prices of the articles of food enumerated above are taken at the market value, which is, of course, above what they cost the farmer who raises them; and this being the case, the cost of harvest-work, as represented above, is fully above the mark.

I have repeatedly ascertained that *threaving* never costs less than 12s. per acre, including the wages and food of bandsters.

I have been informed that mowing in England costs from 8s. to 24s. per acre, according to the state of the corn and the weather; but when it is wholly done by extra hands, the cost is from 10s. to 20s. per acre, without food. Reaping is said to cost 25s. per acre, if done as low as 2 inches of stubble, and 12s. per acre, if the stubble is left 12 inches long.

(2924.) It will be observed, from the foregoing statements, that mowing is by much the cheapest mode of cutting down corn, and on that account should be universally adopted; for the item of harvest expenses is a heavy one in the farmer's books, and every legitimate means should be used to lessen it. Any plan that would deprive the inhabitants of a farm of work, I would hesitate to recommend; but when the farmer, as at present situated, is very dependent on the public market of labour to secure his whole year's produce, he is justified in the endeavour to make himself independent in this respect. It is scarcely practicable to render himself entirely independent of the market, without throwing his horses idle for a long time; and, as there is no economy in such a course, the only other left him, is to economize on the usual cost of harvest-labour. Now, the scythe offers this advantage, and so efficient is that implement, that whoever has tried it fairly, and for a whole harvest, has never relinquished it. I know a prejudice exists against it, on the ground that a mown stook does not look so well as a reaped one, and that it is difficult to build a neat stack with mown sheaves. I own there is a roughness in the appearance of mown sheaves, but I can assure every one who has not discovered what I am going to state by his own experience, that that is no detriment to the corn, and the objection disappears as work-people become *accustomed* to the implement. Such a roughness would not exist were mowers careful to lay the swathes at *right angles to their line of motion*, and were gatherers to lift the swathes even, and lay them down in armfuls even in the bands; and though it is scarcely possible for work-people to bestow utmost attention at all times, yet practice makes great proficiency in this matter, which compensate, in a great measure, for want of constant attention. I found this effect produced after employing the same hands at the same work for successive years, and who at length became as neat and clever at their respective works as I could wish them. Besides the actual economy in employing mowers, only conceive the advantage of being able to carry in the corn after it has been exposed in the air to win only half the usual time. Reaped oats require to stand in the stook a fortnight before they will keep in the stack; mown oats, in the same weather, may be carried in a week. Barley, when reaped, is scarcely fit for leading in less than 3 weeks; when mown, it may be put into the stack in safety in 10 days. And as for mown wheat, about 3 days will suffice to win it. It is an error to suppose, as is too commonly imagined, that a mown stook takes in rain;

on the contrary, I have frequently ascertained that it takes in rain less than a reaped one. In one instance I remember of a field of potato-oats being finished cutting, that heavy rain fell the next day, and continued, without intermission, for 3 days, the last of which was very windy, and the wind had changed from E. to W., when it faired. About ⅓ of this field had been reaped with the sickle, and the reason that implement was used in it at all, was to give a little harvest-work at threaving to a few elderly men and women, cottars, and hinds' wives who, having to attend to young children, could not undertake the regular work of a harvest-field. Impressed with the common belief, that a mown sheaf must take in rain, I went to the field after the rain had ceased, to ascertain the state the stooks were in, never doubting they would be soaked, while the reaped ones would be comparatively dry; but judge of my surprise when the fact was the very opposite, the mown sheaves not being wetted to the heart, while the east side of the reaped sheaves were soaking to the bands. On consideration, I accounted for the difference of the phenomena in this way:—In reaped sheaves, and especially when cut by threave, the straws are straight and hard pressed, between which the rain finds its way into the heart of the sheaves; while the straws in the mown stooks, being somewhat bent and broken and interlaced on the surface, this texture prevents the rain penetrating, and is rather serviceable in throwing it off. Besides this property, mown sheaves are evidently more pervious to air than reaped.

(2925.) An advantage of another kind obtained in mowing corn should not be overlooked, which is the very short stubble left in the field, and the larger quantity of straw carried to the stack-yard. The following statement may be depended on, as being the result of experiment:—

	Cwt.	Qr.	Lb.
Weight of straw per acre, when cut to 2 inches of the ground,	26	1	0
... ... 8	23	1	6
... ... 12	21	0	2

So that ½ cwt. of straw is left per acre on every field by every 1 inch of stubble. When we know that the value of straw is commonly about 28s. per ton, or 15s. per load of 36 trusses, we cannot but feel surprise at the barbarous mode of using the wheat-straw in some places of England, in first reaping the ears of corn, and then mowing the straw.

(2926.) It is not easy to state, with sufficient accuracy, the proportion which the straw and grain bear to each other, nor, of course, the ratio realised by the farmer, which must depend on the mode of cutting adopted by him. I have reeeived the following statement of their relative weights in the neighbourhood of Edinburgh, from Mr Andrew Gibson of the Dean Farm, whose good farming is well known :—From a crop of wheat, of 40 bushels to the acre, or of 2600 lb., at 65 lb. per bushel, the straw will weigh 9 kemples of 440 lb. each, or 3960 lb., affording just ½ more weight of straw than of grain. From a crop of barley of 60 bushels, weighing 56 lb. per bushel, or 3360 lb. per acre, the weight of straw is 7 kemples, or 3080 lb., being $\frac{1}{10}$ more weight of grain than of straw. From a crop of 60 bushels of oats, at 45 lb. per bushel, or 2700 lb. per acre, the weight of straw is 8 kemples, or 3520 lb., being ⅓ more weight of straw than of grain. These are all average quantities. In ordinary crops at a distance from towns, the proportion between the grain and the straw is *supposed* to be, for, I believe, it has not been satisfactorily ascertained by experiment, the grain ⅓, and the straw ⅔, of the entire weight of the crop.

(2927.) There is another curious inquiry connected with this subject, namely, the proportionate weight of *roots and stubble left in the field after the crop is reaped.* In reference to the roots of natural grasses left in the soil acting as manure, Professor Johnston observes, that " the same is the case, to a greater or less extent, with all the artificial corn, grass, and leguminous crops we grow. They all leave their roots in the soil; and if the quantity of organic matter which these roots contain be greater than that which the crop we carry off has derived from the soil, then, instead of exhausting, the growth of this crop will actually enrich the soil, in so far as the presence of organic matter is concerned. No crops, perhaps, the whole produce of which is carried off the field, leave a sufficient mass of roots behind them to effect this end; but many plants, when in whole or in part eaten upon the field, leave enough in the soil materially to improve the condition of the land, which, in all cases, those are considered as the least exhausting to which are naturally attached the largest weight of roots. Hence, the main reason why poor lands are so much benefited by being laid down to grass, and why an intermediate crop of clover is often as beneficial to the after-crop of corn, as if the land had lain in naked fallow." The remarks here given are not all connected with our present subject, but they shew the partial compensation which white crops return to the soil by means of their roots. The Professor then gives the results of some experiments of Hlubek, conducted on a small scale, with a view to ascertain the quantity of roots left in the soil by plants after their parts above ground had been removed; and though the experiments were made with a few of the natural grasses, after they had been made into hay, I shall transcribe them, as the results give the proportion between the roots and stems of gramineous plants.

English and Scientific Names of Grasses.	Produce in		Produce in Roots.		Weight of dry Roots to 100 lb. of Hay.
	Grass.	Hay.	Fresh.	Dry.	
	lb.	lb.	lb.	lb.	lb.
Tall fescue—*Festuca elatior*,	124	36	56	22	61
Sheep's fescue—*Festuca ovina*,	90	30	...	80	266
Timothy—*Phleum pratense*,	90	25	56	17	60
Rough cock's-foot—*Dactylus glomerata*,	202	67	...	22½	33
Perennial ryegrass—*Lolium perenne*,	50	17	...	50	300
Meadow fox-tail—*Alopecurus pratensis*,	106	35	...	24	70
Couch—*Triticum repens*	120	60	...	70	116
Annual meadow-grass—*Poa annua*,	...	...	...	...	111
Soft and smooth brome-grass—*Bromus mollis* and *racemosus*,	...	...	...	...	105
Sweet-scented vernal-grass—*Anthoxanthum odoratum*,	...	...	...	...	93

"The fourth column of the above table," observes Professor Johnston, "shews how large a quantity of vegetable matter some of the grasses impart to the soil, and yet how unlike the different grasses are in this respect. The sheep's fescue and perennial ryegrass—besides the dead roots which detach themselves from time to time—leave at the end of the fourth year a weight of *living* roots in the soil, which is equal to 3 times the produce of that year in hay. If we take the mean of all the above grasses as an average of what we may fairly expect in a grass-field, then the amount of living roots left in the soil when a 4-year old grass-field is ploughed up, will be equal to ⅙ more than the weight of that year's crop." The mean of the table gives a weight of 121 lb. for every 100 lb. of crop; and if we take this as a rule for estimating the weight of roots left in the ground by the cereal crops mentioned above in reference to their weight of straw, and if we take their respective weights of grain and straw as given above, the following will be the entire weights grown of those crops:—

				Tons.	cwt.	qr.	lb.
Of Wheat,	Grain,	2600 lb.					
	Straw,	3960 ...					
	Roots,	4779 ...					
			11,339 lb. =	5	1	0	17
Of Barley,	Grain,	3360 ...					
	Straw,	3080 ...					
	Roots,	3726 ...					
			10,166 lb. =	4	10	3	2
Of Oats,	Grain,	2700 ...					
	Straw,	3520 ...					
	Roots,	4259 ...					
			10,479 lb. =	4	13	2	7

(2928.) Although it is perfectly true what Professor Johnston says in his concluding remarks on this head, that "this burying of recent vegetable matter in the soil, in the form of living and dead roots of plants, is one of those important ameliorating operations of nature which is always, to some extent, going on wherever vegetation proceeds:—it is one by which the practical man is often benefited unawares, and of which—too often without understanding the source from whence the advantage comes—he systematically avails himself in some of the most skilful steps he takes with a view to the improvement of his land;"* it is, nevertheless, of greater benefit to the farmer to cut his crops as near the surface of the ground as possible, thereby putting it in his own power to return the straw to the soil in a state of manure best suited to the wants of the crop, than to cut the stubble high, merely because the soil will derive benefit from it, as an organic substance, when it happens to become decomposed in the soil.

(2929.) That one period of their age is better than another for reaping grain crops has been proved by very careful experiments made by Mr John Hannam, North Deighton, Yorkshire. Without entering into the details of these experiments, it will be sufficient to give only their results. Of wheat reaped at various times, the following were the advantages and disadvantages derived:—

No. 1, reaped *quite green* on 12th August, and stacked 26th August, gave a return of L.11, 17s. per acre.

No. 2, reaped *green* on 19th August, and stacked 31st August, gave a return of L.13, 6s. per acre.

No. 3, reaped *raw* on 26th August, and stacked 5th September, gave a return of L.14, 18s. per acre.

No. 4, reaped *not quite so raw* on 30th August, and stacked 9th September, gave a return of L.14 : 17 : 4 per acre.

No. 5, reaped *ripe* on 9th September, and stacked 16th September, gave a return of L.13 : 11 : 8 per acre.

Hence a	*loss* of	L.1	14	8	per acre upon	No. 1	as compared with	No. 5.
...	...	0	5	8	...	No. 2	...	No. 5.
...	*gain*	1	6	4	...	No. 3	...	No. 5.
...	...	1	5	8	...	No. 4	...	No. 5.
...	...	3	1	0	...	No. 3	...	No. 1.

Hence, also, wheat reaped a *fortnight* before it is ripe gives an advantage on every point, namely:—

In weight of gross produce of . . .	$13\frac{1}{5}$	per cent.
... ... equal measures, nearly . .	$\frac{1}{2}$	...
... ... equal number of grains, nearly .	$2\frac{1}{2}$	...
... quality and value, above . . .	$3\frac{1}{4}$	...
... weight of straw, above	5	...

Besides these, other advantages are, straw of better quality, a better chance of securing the crop, and a saving in securing it. On the other hand, wheat,

* Johnston's Lectures on Agricultural Chemistry and Geology, p. 618-20.

reaped a *month* before it is ripe, gives an advantage of 22 per cent. in weight of straw compared with the ripe, but suffers disadvantage in every other point, namely :—

In weight of gross produce, . . .	$11\frac{5}{13}$	per cent.
... ... equal measures, above . .	$\frac{1}{3}$	...
... ... equal number of grains, above .	$13\frac{1}{6}$	...
... quality and value, above . . .	$\frac{4}{5}$	...

(2930.) These may seem trivial advantages and disadvantages, when confined to the area of a single acre ; but apply the particulars to the extent of ground under wheat-culture in the kingdom, and the truth of the Scottish adage, " mony a little maks a muckle," will be clearly seen ; for, as Mr Hannam observes, " When we consider that there are in England and Scotland about 4,000,000 acres of wheat grown annually, producing 12,000,000 quarters of grain, of which $\frac{3}{4}$ are allowed to become ripe ; when we consider that by cutting this sooner we should produce an increase of $15\frac{1}{7}$ per cent. of flour, and realise an increased value of 7s. $6\frac{1}{4}$d. upon every quarter produced ; and that we should produce food for 1,362,857 persons over and above what we now produce, and an extra annual income of L.512,491 ; and when we consider that this increase would be so much added to the wealth of the country, that it is equal to the proceeds, at 3 per cent., of an estate worth L.17,083,033 ; and that the increase of our population demands an increased supply of food, I would ask, what is our duty in this case ?"*

(2931.) Upon one occasion I cut down a few stooks of potato-oats when quite green, though full in the ear, to allow carts to pass to a place destined for the site of a hay-stack, and after standing till the rest of the field was brought in, they were thrashed with the flail by themselves, and the sample they produced was the most beautifully silvery grain I ever saw ; but not having made the experiment with any view to the crop, I pursued the investigation no farther, and cannot say what effect so very early reaping had produced upon the quantity and quality of meal.

(2932.) Reaping is executed in various ways in England, and one of them is technically called *bagging*, which is performed in this manner in Buckinghamshire :—" First make a band and lay it down, then, standing in the furrow with the left hand to the standing corn, cut a handful, put the stubble-ends of this handful all even, then grasp it in the left hand 8 or 10 inches from the end, and with this assistant lay a little of the standing corn back, or from you, and with the *scythe*-hook chop off, cutting inwards close to the ground, the corn so laid off ; move the left hand forward, lay back the corn as before, and make another cut, and so proceed—moving left hand, one foot, and the scythe-hook simultaneously—across the " land" or " ridge," or half way across it, if there are two persons on it ; 4 to 5 yards is the usual breadth taken. Having reached the breadth intended to be taken, drop the corn which till now has been held in the left hand, among the cut corn which now leans against the standing corn, and commence collecting what has been cut ; for this purpose, walk backwards over the same ground, or rather a little nearer the standing corn—use the left

* See Quarterly Journal of Agriculture, vol. xii. p. 22-37, and vol. xiii. p. 170-87, for the detailed particulars of Mr Hannam's interesting experiments.

hand, the hook, and the right foot—roll over the cut corn with the hook, and, at the same time, cut some more with the point of it, and keep walking backwards and collecting all together till you reach the furrow from whence you started, when you will find you have got an armful. Lay this armful into the band, cut another left-handful as at first, and again go on cutting inwards; returning with the armful, lay it in the same band, which is then enough for a sheaf. Make another band for another sheaf, and proceed as before, cutting forward, and cutting and collecting backwards, clearing about 1 yard wide. Price —oats about 9s., wheat about 12s. per acre, to cut, bind, and shock."* In Worcestershire, bagging is executed by a tool called a bean-hook, and the straw is cut by a stroke instead of a cut of the sickle, holding or collecting the straw in the left hand. The best baggers use a wooden hook in the left hand to collect and bring together the cut wheat in a bundle-like shape to the ground.

(2933.) This bagging reminds me of the Hainault or Flemish scythe, which is held in the right hand by a handle, the forefinger supporting its weight by being introduced into a leather loop, and the blade of the instrument is kept steady by a flat and projecting part of the handle resting against the back of the arm. The corn is cut by the reaper stepping backwards as he cuts the corn with a series of strokes against the standing corn; gathering at the same time the cut corn towards him by means of a hook, which is held in the left hand, and with this and the assistance of the foot and scythe the corn is put into the band. The hook may be suspended to the wrist with a string in a slit passed through the handle. This process, you will observe, is very similar to bagging just described. I accompanied the Flemish reapers, John B. Dupre and Louis Catteau, through Forfarshire in 1825, and drew up a report of their proceedings in that county for the Highland and Agricultural Society. The impression on the farmers present was, that a saving of about $\frac{1}{4}$ might be effected by this scythe in comparison with the common sickle; but it has not made its way into this country; nor is it equal to our cradle-scythe.†

(2934.) In using the common scythe, I observe, in some parts of the country, that the swath is cut and laid up against the standing corn, and I believe the practice is nearly general in England. The corn is then gathered while in that position. Why this mode of laying the swath should be *preferred* to laying it flat on the ground, it would be difficult to conjecture, for I cannot see a single advantage the method possesses; while the irregular state in which the corn is taken away by the gatherer, compared with lifting it up from the swath, appears to me to be a sufficient objection against it.

(2935.) There are various ways of stooking corn besides those represented above in figs. 522 and 523. In Ireland, a safe plan against wind and rain is practised in clustering the standing sheaves with their tops close together; and after placing 2 hood-sheaves almost in a perpendicular position, with the stubble-end uppermost, they are lashed together by a wisp from one hood being passed into the band of the other. Stooks are also set, with the standing sheaves, in the form of a cross, across the open furrow, or sheugh, and covered with 4 hoods meeting with their butt-ends in the middle. A very effectual way of keeping sheaves dry, and, at the same time, of exposing them to the air, is adopted in Sweden, by thrusting the end of a small pole, about 6 or 7 feet long,

* Mark-Lane Express for August 1841.

† See Prize Essays of the Highland and Agricultural Society, vol. vii. p. 244.

a, fig. 527, into the ground, and after impaling 1 sheaf *b* upon the stake, with its butt-end standing on the ground, others *c c* are spitted upon the stake through the bands, parallel to, and above each other, till the stake is filled, the sheaves being inclined with their heads downwards for the rain to drop off. This plan has been tried in this country by Mr Boswell of Kingcausie, in Kincardineshire, with success; and I should conceive, in fields surrounded with woods, and where larch weedings are abundant, the plan an excellent one for winning the corn well and fast. But to notice an opposite extreme, the barley and oats in the south of England are seldom or never stooked at all, but mowed down in swathes, left on the ground to win, and carted home to a large barn like hay—a more slovenly and objectionable mode cannot be imagined, of handling so delicate a grain as barley, or one so easy to germinate as oats.

Fig. 527.

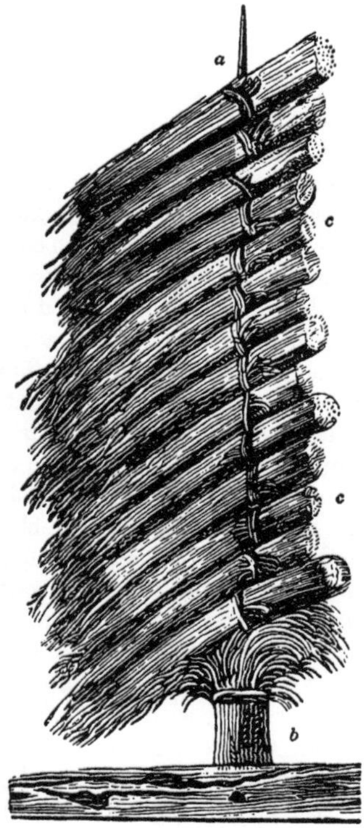

THE SWEDISH STOOK.

(2936.) [The first and most important operation of the harvest season being the reaping of the crop, I have now to proceed with the description of the machines and implements used in that interesting operation. It is more than probable, that, in the dawn of agriculture, the tiny crop of grain was plucked by the roots from the earth without the aid of any instrument other than that most comprehensive one, the human hand; but the inconveniences attending this mode must have been soon superseded by the adoption of the reaping-hook, or sickle, which is coeval with the most remote history. It is believed, that instruments or machines of a more complicated form than the sickle have been resorted to by some of the nations of antiquity; but of these we have no certain historical information; and when it is considered, that, even in modern times, with all our advantages of mechanical knowledge, and the numerous appliances of mechanical experience and skill, we have, as yet, failed in producing a really effective reaping-machine, we are led to the conclusion, that nothing of the sort in a perfect form had ever been achieved by the ancients. During the last sixty years, many attempts have been made to produce a perfect reaping-machine, but, with few exceptions, they have proved abortive, and the problem, in its fullest extent, remains yet unsolved; but, nevertheless, several of these attempts have been attended with at least partial success—a success which has extended to the reaping of grain when it stands erect, or even slightly inclined, if all in one direction; but have all failed in making satisfactory work when the crop has been lodged by rains, or contorted by winds. The step which has been made in achieving even this much, is, however, of importance, and holds out a prospect of the probability that we shall yet see a reaping-machine adapted to the reaping of at least a very large proportion of our fields, in the situation in which they are found in general, when ready for the sickle. But as none of these machines can be considered perfect in their present state, I deem it unnecessary to enter into any minute details of their construction here, and will only give a summary view of their principles, and the *leading* features of their construction, with a reference to writers, whose more detailed information may be obtained by those who desire to prosecute the subject.

(2937.) The first reaping-machine that came before the public with any claim to efficiency, was that of Mr Smith of Deanston, about the year 1814–15, when it was tried before a joint-committee of the Highland and Agricultural Society, and of the Dalkeith Farming Society, the latter having previously offered a premium of L.500 for a perfect reaping-machine.* On that trial, the machine operated with very considerable effect; but though much approved of, in a general view, was not considered so complete as to warrant the committee to award the premium. Mr Smith's machine continued for a number of years to be the only one at all effective; but as it did not possess all the requisite qualities for perfect reaping, it did not come into actual use beyond a very limited extent. The cutting principle in Smith's original machine is beautifully simple, being a plain smooth edged annular knife of about 6 feet diameter, and 6 inches broad, attached by arms to an axle, with which it revolves with a velocity that makes one revolution for every 2 that the machine travels over the ground. The cutter was surmounted by a light sheet-iron drum, or rather frustum, of an inverted cone, about 3 feet in height, which serves as the collector or gatherer of the machine; and this was attached to, and revolved along with, the cutter and its axle. The motion of the cutter and gatherer was derived from a pair of carriage-wheels fixed upon an axle, which carries a toothed bevel-wheel, and from this last by a pinion and horizontal shaft, and second pair of bevel-wheels, the last of which was fixed upon, and gave motion to, the cutter-shaft. The carriage-wheels, with their axle, bore all the frame-work and gearing of the machine, with a slight preponderance forward; and this preponderance was borne by a small roller, placed under the fore-part of the frame-work, near the centre of the cutter, serving to preserve the uniform height of the cutter from the ground; a pole projected backward from the body of the machine, to which two horses were yoked, one on each side, by a splinter bar, pushing the machine before them; and as the pole projected a few feet behind the horses, it served as a rudder to the machine in the hands of the driver. The circular cutter here adopted is, perhaps, the most perfect method of performing the act of cutting the grain, which is done by it in a very successful manner, the machine being provided with apparatus for cutting higher or lower, as the crop may require. The gathering process is less perfect; the cut grain is carried round by the motion of the drum to the left side, where, as soon as it is clear of the standing corn in front it falls outward, forming a continuous swath at one side of the track over which the machine has passed.

(2938.) At a subsequent period, Mr Smith effected various improvements on his reaping-machine; and in 1837 it was exhibited in operation at Ayr, before a committee of the Highland and Agricultural Society, where it performed to the satisfaction of all present, both in cutting and gathering; but, at the same time, it showed marks of unwieldiness, especially from its great length, which is about 20 feet, that still rendered its general application doubtful. The chief point of its improvement, at this trial, lay in the gathering-drum being now

* Prize Essays of the Highland and Agricultural Society, vol. x. p. xi., Preliminary Notice.

adapted to revolve separate from the cutter, with a velocity at its periphery equal to the progressive velocity of the machine; and in the drum also being armed with rakes, or wooden teeth inserted all over its circumference, which supported and carried the cut grain round to the proper point of discharge.

(2939.) The next machine of importance, in the order of time, is that of the Reverend Patrick Bell, now minister of the parish of Carmylie, in the county of Forfar. It appeared in operation in 1827–8, and received the sanction of the Highland and Agricultural Society in 1829, by a premium of L.50.* For several years thereafter, this machine retained considerable repute, and a considerable number of them were made, and successfully applied in the counties of Fife and Angus; but after all this apparent success, it has also fallen much into disuse, like its predecessor. Mr Bell's machine acts upon the clipping principle, the cutter being a series of scissors, the upper blades of which are immovable, and of the form of isoceles triangles, whose heights are 10 inches, and bases about 4 inches, adapted to cut on both edges. The lower blades are of a similar form, each jointed upon a pin, in a position corresponding to the space between the upper blades, and their shank or handles are all jointed to a traversing bar, put in motion by a crank; there are ten pairs of such scissors, and by the above arrangement, the whole are made to cut simultaneously right or left. The motion of the cutter is communicated from the carriage-wheels, as in Smith's machine, there being two principal wheels for that purpose, and two minor wheels to bear up the cutters and fore-part of the machine. The gathering process is accomplished by an endless web, which is placed above and behind the cutter, revolving either to right or left; it receives the grain as it falls from the cutter, and is regularly carried to one side, and dropt in a continuous swath, as before. To ensure the falling of the cut grain upon the web, there is a light four-leaved vane placed in front of the machine, and is made to revolve by means of a band; the leaves of this vane press gently upon the yet uncut grain, gathering it towards the web, until it is severed below by the scissors, when the web carries it off. For several years this machine bid fair to become a favourite; the work was well performed wherever the corn stood upright, and that it would work amongst laid corn was not expected of it; yet it is now seldom seen in the field, and the only cause to which its failure can be attributed, is the complicated structure of the cutter. A cutter consisting of such a number of parts, each of which requires the most perfect adjustment to render it effective, being besides, liable to unequal wear, renders it yet more liable to derangement; and further, as the derangement of a single blade of the cutter is sufficient to destroy the usefulness of the machine; and the rectification of the defective blade requiring a workman of superior talent, it must have become both troublesome and expensive to effect such repairs; and this, added to the risks of delay in an important season, seem to have contributed to the gradual disuse of the machine.

(2940.) Soon after the introduction of Mr Bell's reaping-machine, there appeared another competitor, Mr Joseph Mann, Raby, near Wigton, Cumberland.† Mr Mann's machine appears to have been constructed, after many years' labour, so early as 1826, but it was not exhibited in Scotland till 1832, at the Highland Society's Show in Kelso, where it was tried before a committee with a success very similar to those that had gone before it. Though not affording en-

* Quarterly Journal of Agriculture, vol. i. p. 217.

† Ibid. vol. iv. p. 250.

tire satisfaction, it nevertheless possessed some points of considerable importance. In construction, it was more compact than some of the previous machines; and, from the simplicity of its movements, would be less expensive. It was drawn by one horse, walking before the machine, and by the side of the standing corn, cutting a breadth of from 3½ to 4 feet, and would, therefore, cut nearly 10 acres in 10 hours. The cutter was on the revolving principle, but instead of being circular, it formed a polygon of 12 sides, each side of the polygon being a separate blade, easily removed and changed. The gatherer was the revolving cylindrical drum with rakes, afterwards adopted by Mr Smith of Deanston, before alluded to; but in Mr Mann's machine the drum revolves at a considerably higher velocity, making 28 revolutions in a minute, while the cutter makes about 200. In this machine, therefore, the velocity of the rake-teeth is 400 feet per minute, or nearly double that in Mr Smith's improved form. This high velocity of the rake carries away the cut grain in a thinner layer upon the rakes, but it requires the application of a comb to strip the corn from the rakes, and thus secures its being always dropt at one point in the machine. The motion of the cutter and rake were obtained from one of the carriage-wheels as in the others; but here they were communicated by pitch-chains, and the front part of the machine was supported by a castor-wheel, to the stem of which the horse-shafts were attached, the castor-wheel running by the side of the standing corn. I am not aware of this machine having ever made farther progress; though I do consider that it possesses some advantages over any that have yet been tried, and, if fully matured, might have superseded its competitors.

(3941.) In viewing the subject of the reaping-machine, three things have to be specially considered. The expense of the machine, the expedition it may effect in reaping the crop, and the saving in expense that may be effected by it. The last described machine could certainly be procured at a price not exceeding L.30. The same machine was, perhaps, not capable of reaping an acre per hour consecutively; its effect may, therefore, be restricted to 8 acres per day, being about 12½ days to every 100 acres. It will follow, therefore, that on a large farm two machines might be requisite, making an outlay of L.60. To qualify this outlay, the expense of reaping (as ascertained, by Bell's machine), has been estimated, including the expense of binding up the sheaves, at 3s. per acre; while the minimum expense of reaping by the hand is taken at 7s. per acre, and the maximum may run to 12s.,* but taking the mean at 9s., gives three times the expense of machine reaping. For 100 acres, therefore, cut by hand, the expense would be L.45, while that by the machine is only L.15, giving a saving of the entire price of the machine in one year. But suppose the price of a machine could be saved even in two years, the advantages would be very great, under the supposition that an effective machine could be procured for L.30. It is deserving of consideration, also, that although no reaping-machine is to be expected capable of cutting down the crops in every possible state, yet we may infer that, even on a farm of ordinary extent, such a breadth of crop may be found, in almost any season, capable of being cut by a machine, as would save half the expense of its purchase. To obtain an effective and durable reaping-machine, is, therefore, yet an important object to the farming interest.

(2942.) In the present times, it may be safely averred, that the only means of reaping is by the *sickle* and the *scythe*. The first is a very simple, but, at

* Quarterly Journal of Agriculture, vol. iv. p. 89.

the same time, so far as it goes, a very efficient instrument. It is employed in various states, not differing much in the general form, though exhibiting marked differences in the detail; but these varieties are confined under two very distinct forms, the *toothed* and the *smooth-edged* sickles. Fig. 528, represents the *toothed* sickle, an instrument so well known, that it requires little description here. The blade, in the common toothed sickle, is principally made of iron, but with an edging of steel; the teeth are formed by striking with a chisel and hammer, in the manner of file-cutting, the cutting being only on the lower side; but when the blade has been bent to the proper form, tempered, and ground on the smooth side, the serratures are brought prominently out on the edge of the blade; and as the striking of the teeth is performed in a position oblique to the edge of the blade, at an angle of about 70°, the serratures on the edge acquire what is called a *hook* towards the helve, thus causing the instrument to cut keenly in that direction, when drawn through the standing corn. When the blade has been thus finished, a wooden helve of the simplest form is fitted upon the pointed tine formed at its root for that purpose. The toothed sickle is made with various degrees of curvature and of weight, but chiefly as represented in the figure, and it has been the subject of several patents, chiefly depending on the formation of the blade. One of these is only of two or three years' standing, and promises to be an important one. Messrs Sorby and Son, of Sheffield, are the patentees; and the principle upon which their patent is based, is a blade of rolled cast-steel swedged into a form that gives a sufficient degree of stiffness to the blade, without the increase of weight that accompanies the thick-backed or the other patent ribbed-back sickles. In the new patent, the advantage of a small quantity of the very best material—cast-steel—is combined with extreme lightness and a due degree of strength and stiffness, the latter arising from the swedged or moulded back.

Fig. 528.

THE TOOTHED SICKLE.

(2943.) The *smooth-edged* sickle, or *scythe-hook*, as sometimes called, differs from the former in being broader in the blade, and longer withal, but in curvature it resembles the former; and its chief difference lies in being ground on both sides, to form a fine and thin sharp edge. Like the toothed sickle, the blade has undergone various improvements; and Mr Sorby's cast-steel swedged-blade is also extended to the smooth-edged sickle.

(2944.) In the formation of the sickle, the curvature of the blade is a point of more importance than to a careless observer may appear; and though the ordinary reaper is seldom qualified to judge in this matter, he may feel pleased to be informed, that there is a certain curvature that will give to the muscles of his right arm the least possible cause for exertion, while there are other curves that, if given to the blade of the sickle, would cause him to expend a great amount of unnecessary exertion in the arm, and a consequent unnecessary fatigue would follow. Fig. 529, which represents the smooth-edged sickle, has a curvature approaching very near to that which, in this instrument, may be termed the *curve of least exertion;* and throughout that portion of the sickle that performs the cutting process, it possesses this peculiar property, from the following circumstance, that lines diverging from the centre of the handle of the sickle at a, and intersecting the curve of the cutting-edge, all the diverging lines will form equal angles with the tangents to the curve at the points of intersection. This property gives to the cutting-edge a uniform tendency to cut at every point in its length, without any other exertion than a direct *pull* upon the helve; were the curvature less at any point, a pressure of the hand would be required to keep the edge to the work, and were the curvature greater at any point, or on the whole, the exertion to make the cut would be greater, as it would then become more direct, instead of the oblique drawing or *sawing* cut, which, in all cases, is the most effective, and productive of least resistance.

Fig. 529.

THE SMOOTH-EDGED SICKLE.

(2945.) A mode of using the smooth-edged sickle has of late years come into some repute, known in Scotland by the provincial term *dinging-in* (striking-in). In this process the sickle is not drawn through the straw, but is struck against it, somewhat in the manner of using a scythe; indeed, the practice originated in the attempt that was made some years ago to introduce the Hainault scythe in the harvest operations of this country, but without success. In the (dinging-in) practice, the left hand is employed, with its back towards the right, in slightly bending down the grain, and holding it to the blow of the sickle. A man practised in this mode of working will do one-half more work than is usually done in the common way; but the stubble is left less regular, and, except by very expert hands, there is a want of tidiness in the process. It is obvious that this is the same mode of cutting corn as bagging, which has been described above.

(2946.) The *scythe* is also extensively used as a reaping instrument, but more especially in England, and in some of the northern counties of Scotland, chiefly for the barley and oat crops, though in some localities also for wheat.

Fig. 530.

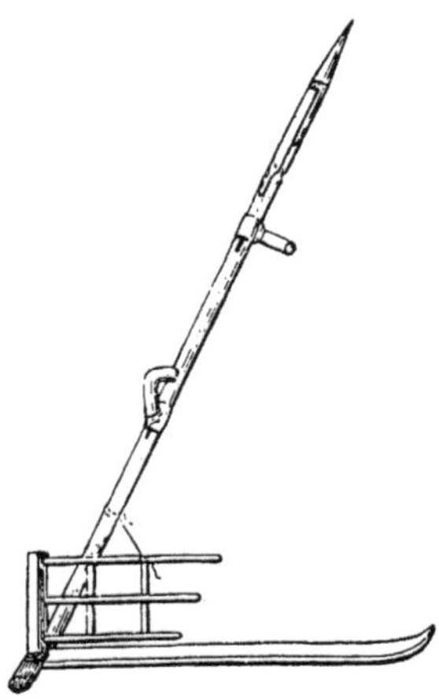

THE COMMON REAPING-SCYTHE.

Fig. 531.

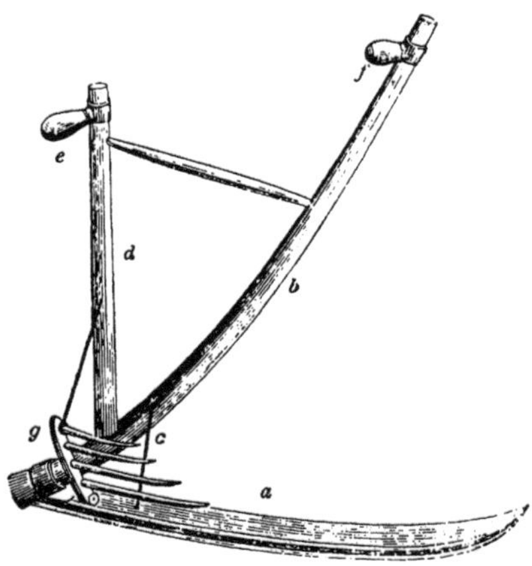

THE CRADLE-SCYTHE FOR REAPING.

The common hay-scythe is used for this purpose, in most cases, with the addition only of a very light rail or cradle, sometimes attached to the handle, and in other cases to the heel of the blade, as in the annexed cut, fig. 530, which is the common scythe fitted for reaping. In the scythe mounted in this manner, the use of the rail or cradle will be obvious, being the collecting of the swath into a compact body, and depositing the same in as regular a manner as possible, preparatory to the binding of it into a sheaf.

(2947.) For a considerable period, a scythe mounted in a different form has been in use for the purpose of reaping, and it is believed that its introduction can be traced to Banffshire or Aberdeenshire, where the scythe is rather extensively used for reaping. About ten years ago this form of scythe came more prominently before the public,* and, for a time, gained considerable repute, under the name of the *Cradle-scythe.* Of this form of mounting a reaping-scythe, there are many varieties; but they all agree in one point, that of having two short helves, the one branching out of the other, instead of the common long helve or sned. Fig. 531 is a view of the cradle-scythe in one of its most approved forms, wherein *a* is the scythe-blade, *b* the principle helve, to which the blade is attached in the usual way, the hook of the tine being sunk into the wood, and an iron ferule brought down over the tine, binding it firmly to the wood; but the blade is further supported by the addition of the light stay *c*, termed by mowers the *grass-nail.* The minor helve *d* is tenoned into the former, and the two handles *e f* are adjusted by wedges in the usual way, to the height and mode of working of the mower. The *cradle* or rake consists of a little wooden standard *g*, about 8 inches high, jointed to the heel of the blade, so as to fold a little up or down across the blade; into this is inserted three slender teeth, following the direction of the blade, and may be from 6 to 15 inches long; the head of the standard is supported by a slender rod of iron, which stretches about 18 inches up the handle, where it is secured by a small screw-nut, capable of being shifted up or down to alter the position of the standard and its teeth to suit the lay of the corn. The standard or rake head was at one time recommended to be made in the segment of a circle,† for which there seems no good reason, either practical or philosophical; but the idea was seized upon, and the cradle-scythe mounted in that form was widely distributed; but instead of this supposed improvement tending to increase the favourable opinion of scythe-reaping, the practice seems rather on the decline; and there is good reason to believe, that this malformation of the rake may have had no small share in producing a distaste for scythe-reaping as a practice; whereas, under proper management, and a judicious choice of implements, there can be no doubt of considerable advantages being attainable from scythe-reaping, as compared with the sickle.

(2948.) It is believed by very competent judges, that the figure here given of the cradle-scythe possesses all the advantages that are derivable from using this instrument; and, in addition to the details already given, I may add, that the length of the blade is from 3 feet 4 inches to 3 feet 6 inches; the left or principal helve is about 4 feet, the other 3 feet, measuring direct from the heel in both cases; the distance between the helves, where the handles are applied, is 24 inches; and in setting the blade the following rule is to be observed. When the framed helves are laid flat on a level surface, the point of the blade

* Quarterly Journal of Agriculture, vol. V. p. 106.

† Ibid., vol. vi. p. 30.

should be from 18 to 20 inches above that surface, and measuring from a point on the left helve, 3 feet distant from the heel of the blade, in a straight line, as at *b*, the extremity of the blade should be also 3 feet distant from the point *b*. Iron has, in many cases, been substituted for wood in the construction of the helves; but it does not, by any means, appear to be so well adapted to the purpose as the wooden helves; when constructed of iron, if they are made sufficiently light, there is too much elasticity in the fabric, which is fatiguing to the workman, by reason of the tremor produced at every stroke of the scythe.

(2949.) As essential appendages of the scythe, there must be kept in view the *scythe-stone*, with which it is kept in proper cutting order; a soft fine-grained sandstone is the proper material from which it should be selected; and, as for applying it, it were difficult in words to describe the proper mode of doing so, but the following may suffice to give the uninitiated some notion of the process. The scythe is placed on the left side of the mower, with the point of the blade resting on a small stone on the ground, the heel of the blade and helve being supported by the left hand; the scythe-stone, which is about 10 inches long, is grasped by one end in the right hand, and is thus applied, crossing the blade. The effect of the sharping is produced by making sliding strokes with the stone along the blade on each side alternately, upward and downward; but the downward strokes only, or those *towards* the point of the blade, should be the effective ones, the returning stroke of the stone touching the blade but lightly. The object of this mode of treatment is based on the principle, that the scythe cuts after the manner of a saw; though the teeth are not very prominent, still the fine serrature is there; and to make it cut with the best effect, it must be *hooked* towards the point of the blade, or in the direction of its motion, when cutting; and the application of the stone, as above described, produces this direction of the minute teeth in the cutting-edge. The strokes of the stone cannot conveniently be carried over the whole length of the blade at one throw of the hand, hence the sharping begins at the heel, and proceeds downwards, but from the position described, there still remains a few inches at the point untouched, and to reach this part with the stone, the mower lifts up the blade, by seizing it about the middle with the left hand, and bringing it to a horizontal position, the helve still resting partly on the ground, and partly against his body, he is enabled to complete the process of sharping. Throughout this operation it is to be specially observed, that in applying the stone, it must always be held flat upon both sides of the blade; if this is not attended to, either the edge will not be improved, or it will be rounded off, and the consequence of this last effect will be, that the scythe will speedily be unfit for cutting until it be re-ground.

(2950.) The *strike* or *strickle*, which is also used in the sharping of the scythe, principally for giving it a finer edge, though this is very frequently a mistaken notion, is formed of a piece of hard wood, about 15 inches long, one end of which is shaped into a handle; the body of it is sometimes $1\frac{1}{2}$ inch square, but in most cases it is about 2 inches broad, by $\frac{3}{4}$ inch thick. It is coated over with granulated emery, embedded in a cement such as glue; and to produce the best effects, the emery should be of a medium degree of fineness. The strickle being a light instrument, it is always attached to the helve of the scythe, and is, therefore, always at hand, ready to be applied in the event of any accidental injury to the edge of the scythe being sustained, while the stone, from its greater weight, can only be resorted to at the landings, where it lies in readiness.

(2951.) *The hand stubble-rake.* The gleanings of the stubble is an object of considerable value, and to secure it for the benefit of the farmer, different implements are employed. The principal, and the most effective of them, is the horse-rake, figured and described in (2814.), but in absence of that machine, the hand-rake is found to serve a very good purpose. Fig. 532 is a representation of this rake, and it is of very simple construction, the form being precisely that of the hay-rake, but of enlarged dimensions. The head, *a b*, is 5 feet long, and should be made of good tough ash, 2½ by 2 inches; the helve *c d* may be 6 feet in height, of the same material, and furnished with a handle *e* that can be fixed in any desired position, by means of a ferule and wedge. The helve is tenoned into the head, and supported by the iron brace *f*, *c*, *g*. The teeth are of iron, 7 inches in length, and set at 4 inches apart, but formed in the lower part, so that the bend rests on the ground, preventing the points of the teeth penetrating and mixing the earth with the gleanings. The best method of fixing the teeth is by a screw-nut, as in the horse-rake, as they are thereby easily removed in the case of being broken, without risk of injuring the head. It is also advisable to have the ends of the head hooped, to prevent splitting.—J. S.]

Fig. 532.

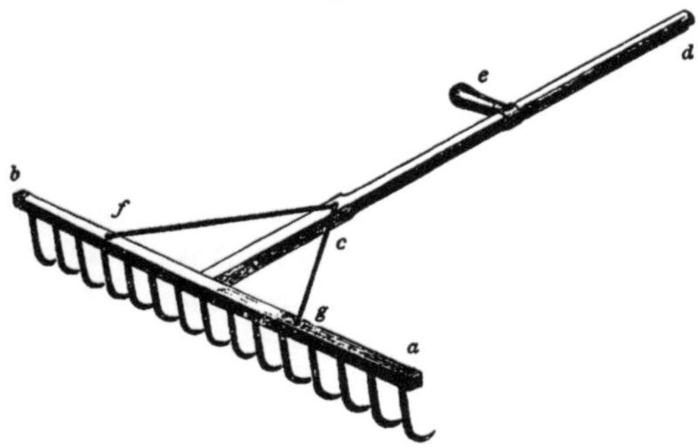

THE HAND STUBBLE-RAKE.

---

76. OF CARRYING IN, AND OF STACKING WHEAT, BARLEY, OATS, BEANS, AND PEASE; AND OF THE CONSTRUCTION OF CARTS.

> " Inwardly smiling, the proud farmer views
> The rising pyramids that grace his yard,
> And counts his large increase; his barns are stor'd;
> And groaning staddles bend beneath their load."
>
> SOMERVILLE.

(2952.) It is necessary that reaped corn remain for some time in stook in the field, before it will keep in large quantities in the stack or barn. The length of time will, of course, depend entirely on the state of the weather, for if the air is dry, sharp, and windy, the corn will be ready in the shortest time, while in close misty damp, it will require the longest time; but, on an average, for wheat 1 week, and for barley and oats 2 weeks will suffice. In this respect the superiority of the mown, over reaped, corn, manifests itself; mown wheat being ready for the stack in 4 or 5 days, and barley and oats in 8 or 10, the chief cause of the difference being the loose and open state in which mowing leaves the straw in the sheaf.

(2953.) Mere dryness to the feel does not constitute all the qualities requisite for making new cut corn keep in the stack. The natural sap of the plant must not only be evaporated from its outside, but also from its interior; and the outside may feel quite dry, whilst the interior may be far from it; and the knowledge of the latter property constitutes the whole difficulty of judging whether or not corn will keep in the stack. There is one criterion by which, whether or not a sheaf is fit to keep, may be ascertained with almost certain success, which is, that if the sheaf feel quite dry, the straws be loose and easily yield to the fingers, and the entire sheaf feel light when lifted off the ground, by the hand thrust through the middle of *the band*, the sheaf is ready; for if it present all these qualities at the band, where it is most compressed, the rest will sure to be won. While winning, it is as probable that the sap of the cereal grains is converted into woody fibre, as that of clover, on being converted into hay (2800.).

(2954.) The winning of corn is comparatively an easy matter, when the weather is dry; but in windy and showery weather, the stooks are apt to be blown down and become wet, and incur the trouble of setting up again at the first moment of calm. When the air is calm, dull, damp, and warm, every species of grain is apt to sprout in the stook before being ready for the stack. When much rain falls, accompanied with cold, the grain becomes sooner ready than the straw for the stack, and, to win the straw, the bands are not unfrequently obliged to be loosened, and the sheaf spread out to dry in the wind and sun; and in like manner the sheaf should be spread out even in dry weather, when a large proportion of young grass is mixed amongst barley-straw. When barley is mown, it is quite practicable to get rid of much of the grass amongst it, by the gatherers shaking the grass out of the swathes by the corn end, before laying the corn into the bands. This extra process of shaking will, no doubt, lose some time, but it is probably better to lose a little time in the mowing, than subject the corn to perhaps a week longer exposure in the field. It is scarcely practicable to shake out grass in reaping, as the hook would require to be laid aside while both the hands were engaged in shaking every handful; and to cut the stubble as high as the grass, would be preposterous. An extra hand, therefore, would be required for this purpose among reapers on every 2 ridges. Corn wins in no way so quickly as in gaits, fig. 526.

(2955) While the first reaped corn is winning in the field, the stack-yard should be put in order to receive the new crop, by removing every thing that should not be in it, such as old decayed straw, weeds, implements, &c.; but no such use as the latter should ever be allowed of a stackyard, as implements should be put into their proper places, when their respective season for use expires. Where stathels are used, it should be seen that they are in repair; dry straw should be at hand to form the bottoms of new stacks, where stathels are not used; and even where they are in partial use, straw is required for the stacks built on the ground, as stathels can only be placed round the utmost limits of a stackyard, unless an inordinate space of ground is appropriated to it, with a cart-road between every 2 rows of stathels, as already commented on at 7 (40.). Straw-ropes should also be provided for covering the barley-stacks, in case of threatened rain.

(2956.) The description of a *stathel* or *staddle*, is given in (70.), and a perspective view in fig. 24, which is a framing of wood put together, to remain in a permanent form upon stone props, made up of 3 parts, a sole *b*, a pillar *c*, and bonnet *d*. It has been brought to my notice, that a better form of stathel would be to have the frame in 2 pieces, to remove into a shed, and be protected from the weather, when not in use. Such a movable frame would answer the purpose as well as a permanent one; but when removed, the props should require to be removed also, otherwise they will be driven over, and perhaps broken, or at all events disjointed, and their replacement would create a good deal of trouble when wanted; and this is a species of trouble, perhaps, more serious than the decay of the wood of a frame 1 year or 2 sooner than it would otherwise be. True, the props could be made of one piece of cast-iron, which they sometimes are, and where stones are scarce, this material seems indispensable; but where stones are plenty, they make cheaper props than cast-iron; and, although cast-iron ones would be easily removed, the soles on which they rest would still be liable to be broken, and the setting them again on a level would impose considerable trouble and expense. So that, on the whole, such a stathel as is shewn in fig. 24, will, in my opinion, answer well for ordinary farms. The framing is made heavy, the scantlings being 9 inches in the side, a size which the Scots fir easily attains, because it is to be permanent in its position, and to last a long time, remaining serviceable for a lease of 20 years; but, of course, where larch is abundant, smaller scantlings will be handier, equally strong, and last perhaps longer. Stathels require no straw to render them fit for building stacks upon.

(2957.) A *stool of straw for a stack* is made in this way:—Stick a fork *a*, fig. 533, upright in the ground in the centre of the intended site of the stack. Put a quantity of dry straw around the fork, and shake it up as you would do the litter of a horse, spreading it out to about the size and form the stack should occupy on the ground. Take a long fork *b*, with the radius of the stack marked upon its shaft, embrace the

Fig. 533.

THE FORMING OF A STOOL FOR A STACK; THE LOADING OF A CART WITH SHEAVES; AND THE BUILDING OF A STACK.

upright fork *a* between its prongs, and, holding its shaft at the specified distance of the radius, push in or pull out the straw with your feet, as seen from *c* to *d*, into the shape of the circle described by yourself in walking round the circumference of the stool.

(2958.) Having made these preparations, on the supposition that the corn is ready to be carried in and stacked, the first thing is to provide an efficient person to fork the corn in the field to the carts, a man being the best for the work, as he is able not only to wield the sheaves with ease, but possesses dexterity to place the sheaves in the position most convenient for the ploughman to build them on the cart; for throwing sheaves in any manner upon the cart makes the forking no easier for the forker, while it gives the ploughman much additional trouble to turn the sheaves upon the load, where his footing is not very secure. A loss of 2 or 3 minutes incurred in this way in loading each cart makes a considerable loss upon the day's work. The steward should build the stacks unless he is specially engaged with the reapers in the field, when another man should be hired to stack; but on a large farm more than 1 stacker may frequently be required at a time. When more than 1 stacker is employed, each should have the same head of carts leading to him, even when both heads are leading from the same field, that the corn may be regularly supplied to him.

6

(2959.) The *cart* is the vehicle employed in carrying in corn. The common cart, such as is represented in Plate XVIII., mounted with a framing called *tops*, is used in some parts of the country; while, in others, the *corn*-cart, different forms of which are represented and described below by Mr Slight, is employed, the body of which is put on the axle and wheels of the common cart, and provided with ropes to hold on the load. Of the 2 sorts of carts for this purpose, I prefer the corn-cart, as it carries the load lower to the line of draught; and it is provided with a plank, called the *lead-tree*, for the ploughman to stand upon when forking off the lower part of the load.

(2960.) The horses are usually yoked *double* in leading time, as there is only 1 man to each pair of horses; and there may be inequalities in the fields and roads which require the force of 2 horses to surmount. This method of yoking is represented in Plate XVIII., with all the necessary harness; and, as I have had occasion to mention the price of the harness required for the plough, I may here enumerate those for the cart. In addition to the collar and bridle, which were noticed before, there are,—

Saddle and girth, . .	22 lb.	weight,	worth	L.1	2	0
Breeching, ...	10	...	...	1	0	0
2 Back-bands and girth,	8	...	...	0	16	0
Cart-chains, ...	14	at 7d. per lb.		0	8	2
Stretcher, ...	...	...	...	0	3	0
	54 lb.			L.3	9	2

In Forfarshire the trace-horse is harnessed in a different manner. A broad strap is hooked to the back of the upper part of the collar, and terminates at the other end in a crupper, and a haunch-strap passes down in each side to support the trace-chain. This plan supports the stretcher well, and prevents it falling so low as the hocks of the horse in the turnings, but it confines the action of the horse, by buckling the shoulder and tail together, and when the bearing-rein is passed over the harness, the horse's head is very much confined. Harness leather is best, in the long-run, though dearest at first. Double cart-ropes measure 24 yards, single 12 yards, and cost 4½d. per lb. Hemp and Manilla reins cost 5½d. per lb. There should always be double reins used with the cart, whether the horses be yoked single or double. The double reins are connected at the end with a tie and loop to hang on any hook, or slip below any part of the harness, as shewn below the haunch strap of the breeching of the shaft-horse in Plate XVIII. With the collar and bridle, the shaft-horse carries 58½ lb. of harness, and the trace-horse 34½ lb.

(2961.) In carrying the crop off the ground, the object is to do as little injury as possible to the land with the cart-wheels, especially to young grass; and for this reason, as well as forming an excellent guide, the horses should walk in the open-furrow between the ridges in every alternate furrow, while the wheels pass along the furrow-brow of the ridge on each side. When corn is cut with band-won reapers, the stooks of 2 ridges being placed on 1, the cart clears the produce of 2 ridges, and the same may be the case with mown corn, provided the bandsters are instructed to set the stooks in the same manner upon the ridges while working across them. In threaving, the stooks being set on every ridge, the forker is obliged to go from one ridge to another to clear 2 ridges, which occasions much loss of time.

(2962.) A corn-cart is loaded with sheaves in this way :—The body *e*, fig. 533, is first filled with the sheaves lying with their butt-ends towards the shaft-horse's rump at one end, and the back-end of the cart at the other. When the corn is on a level with the frame or *shilments* of the cart, the sheaves are then laid across the body of the cart in a row along both sides of the frame, with the butt-ends projecting as far beyond the frame as the band, as at *f* and *g*; the sheaf on each *corner of the frame* being held in its place by being transfixed upon a spike attached to it. Another row of sheaves is placed above the first, and the corner ones kept in their places by a wisp of corn, held fast by the band being placed *under* the adjoining sheaf. Sheaves are then placed along the cart with their butt-ends to both its ends, in order to hold in the first laid sheaves, and to fill up the hollow in the middle of the load. Thus row after row is placed, and the middle of the load filled up till as much is built on as the horses can conveniently draw, 12 large stooks being a good load. It should be ascertained that the load is neither back laden nor too heavy upon the horse's back; and, indeed, if the cart has been evenly laden according to its form, there is no risk of either inconvenience being felt by the shaft-horse. A load thus built will have the butt ends of all the sheaves on the outside, and the corn-ends in the inside.

(2963.) The *ropes* are then thrown over the load to keep it from jolting off the cart upon the road, and in necessarily crossing gaw-cuts in the fields. There is usually a ring in the back-bar of the cart to which the rope is fastened by the middle, from whence it is wound round each back-stay; but fastening the rope round the back-stays will answer without a ring. The ropes are then thrown across the load diagonally to the opposite angles in the front of the cart, and one end is made fast to each shaft, with as much force as the forker can exert on the ground, with the assistance of the ploughman on the top of the load. The crossing of the ropes prevents the load splitting asunder over the sides of the cart, while it is jolted along the road. Some ploughmen profess to shew their dexterity in building loads of corn, and of bringing them to the stackyard without the assistance of ropes; but there is no use of running the risk of losing time by breaking the load and strewing the road with sheaves. Such a fate attended the first load I tried to build even with the assistance of ropes. When the corn is mown, a woman is usually employed to rake the ground on which the stooks stood, because they were set upon the ground at the time of mowing before it was raked.

(2964.) When corn is fit for stacking, the carrying is continued from break of day to twilight, provided there be no heavy dew at morning or night. From a little after sunrise to a little after sunset, the corn may be taken in with great safety. It is customary, in some parts of the country, to keep the horses in the yoke when employed at leading, and feed them with corn from nose-bags while the carter is taking his dinner, and to give them green food, such as tares, while the cart is unloading at the stack. In other parts, the horses are taken out of the yoke, watered, and put into the stable, where they receive their corn while the men are at dinner. This is the best way for the horses, though it usually puts off 1 hour of the best part of the day before the horses again take the road, whereas ½ an hour in the other case is quite sufficient for the men to dine, and the horses to eat their feed of corn. Some horses are apt to take fright, owing, no doubt, to the mode in which they had been broke-in, when the bridles are temporarily slipped off their head for the purpose of taking the bit out of their mouth, to allow them to eat the tares with freedom; and, to avoid such an accident with a horse known to be easily frightened, its bit should be made to fasten with a small strap and buckle to the near side of the bridle.

(2965.) A load of tares is brought to the steading fresh in the morning for the horses. They are not fit for the use of horses until the pods are pretty well filled with grain, prior to which state they are apt to purge and weaken them, when subjected to much work in the cart, as in the time of carrying in corn. For cows it matters not how succulent tares may be, as they yield much milk on them; and, indeed, no other green food is so productive of milk.

(2966.) In setting a loaded cart to the stool or stathel of a stack, it should be studied to let the ploughman have the advantage of any wind going in forking the sheaves from the cart. The stack should be built in this way:—Set up a couple of sheaves against each other in the centre of the stathel, and another couple against them in the other direction. Pile others against these in rows round the centre, with a slope downwards towards the circumference of the stathel, each row being placed ½ the length of the sheaf beyond the inner one, till the circumference is completed, when it should be examined, and where any sheaf presses too hard upon another, it should be relieved, and where there is slackness, another sheaf should be introduced. Keeping the circumference of the stack on the left hand, the stacker lays the sheaves upon the outside row round the stack, putting each sheaf with his hands as close to the last set one as he can get it, and pressing each sheaf with both his knees, as represented at *h*, fig. 533. When the outside row is thus laid, an inside one is made, with the butt-end of the sheaves resting on the bands of the outside row, thereby securing the outside sheaves in their places, and at the same time filling up the body of the stack firmly with sheaves, as seen at *i*. A few more sheaves may be required as an inmost row still, to fill up and make the heart of the stack its highest part. It is of immense benefit to a stack to have its centre well filled with sheaves, as it is the heart-sheaves which retain the outside ones in their places with an inclination downwards from the centre to the circumference; and this position of the outside sheaves is necessary to prevent the rain finding a passage along the straw into the very heart of the stack, where, of course, it would find its way easily, were the sheaves to lie with an inclination downwards to the centre of the stack, and where it would soon spoil the corn. The sheaves are then said to take in *pen-wet*. The number of rows required to fill the body of a stack depends on the length of the straw and the diameter of the stack. In crops of ordinary length of straw, such as from 4½ to 5 feet, a stack of 15 feet diameter is well adapted; and 1 inside row, along the bands of the outside one, with a few sheaves laid across one another in the centre, will make the stack completely hearted. Where much wheat is raised, and which often reaches from 5 to 6 feet in length, the stack should be 18 feet in diameter, to give room for hearting. In a stack of 15 feet, the breadth is rather too much for the ploughman to fork heavy sheaves across to the stacker, when the stack attains a certain height, and when the load in the cart becomes low; and the stacker should always receive the sheaves within his reach, as he cannot rise from his knees to take them without much loss of time, and risk of making bad work. To expedite the building, a field-worker *k* should stand on the stack, to pitch the sheaves with a short fork to the stacker in the position they are wanted by him, to save him the trouble of turning them, as the sheaf *l* is shewn in the cut. By a little management, the field-worker may receive every sheaf as the ploughman pitches it from his fork upon her fork, and prevent it falling upon the stack until she throws it in its proper position beside the right hand of the stacker; and in doing this, she should not give him more sheaves than 1 by 1 as he builds them; nor should she receive them faster from the ploughman than the stacker can build them, for no time is gained, but lost, in piling more sheaves upon the stack than the stacker can dispose of. The proper relative positions of the ploughman *m* and the field-worker *l*, for thus giving and receiving the sheaves, are endeavoured to be represented by those figures in the cut.

(2967.) As every cart is unloaded, the stacker descends to the ground, by means of a ladder, and trims the stack, by pushing in with a fork the end of any sheaf that projects further than the rest, and by pulling out any that may have been placed too far in. It seems to be considered necessary to make the stack swell out as it proceeds in height, if we may judge from common practice, as is shewn in the cut below, in which the ladder is seen to lean against such a form of stack; but it is *not* necessary to swell it out in the building, in order to throw off the drops of rain from the eaves, for the spreading of the eaves by the stack itself, in settling down after it has been built for a few days, is quite sufficient to throw off the drops. The leg of the stack may, therefore, be carried up nearly perpendicular.

(2968.) The *ladder*, figured and described below, at fig. 539, is an indispensable implement for the builder of stacks; not only to allow him to get upon and from the stack when he chooses, with ease to himself, and safety to the stack, but to enable him to measure the height of its eaves, so that all the stacks of the same diameter may be built of a uniform height. A couple of 10-feet, a couple of 15-feet, and a 24-feet, ladders, will suffice for all the purposes of the stack-yard, and repairs of the roofs of the steading and houses.

(2969.) As the stack rises in height with cart-load after cart-load, the trimming cannot be conveniently done with a fork; a thin flat board about 20 inches in length, and 10 inches broad, nailed firmly to a long shaft, is an appropriate instrument for beating in the projecting ends of the sheaves, and giving the body of the stack a uniform roundness. A stack of 15 feet in diameter should ultimately stand 12 feet high in the leg, and an allowance of 1 foot, or 1½ foot, for subsidence, after the top is finished, according to the firmness of its building, is generally made. The height is measured with the ladder, and allowing 2 feet for the height of the stathel, a 15-feet ladder will just give the required measure of the height of the leg before the top is set on. The eaves of the stack are formed according to the mode in which it is to be thatched. If the ropes are to be placed lozenge-shaped, the row of sheaves which forms the eaves is placed a little within the topmost outside row, and after the top is fully finished, its slope will be the same as that of a roof, namely, 1 foot less of perpendicular height than half the diameter. In finishing the top, every successive row of sheaves is taken as much farther in as to give this requisite slope, and the bevelled bottoms of sheaves, as they stand in the stook, answer this purpose well; the hearting being particularly attended to in every row, till the space in the centre of the stack is limited to an area, upon which 4 sheaves can stand with their tops uniting, and their butt-ends spreading out to give a conical form to the top; and these sheaves are kept firm in their place against gusts of wind with a straw-rope wound round them and fastened to the sheaves below If the thatching is made with heavy ropes, running from the top of the stack to its eaves, the eaves-row of sheaves is made to project 2 or 3 inches beyond the upper row of the body of the stack, and the slope of the top is formed from the point of projection of the eaves, and the top finished as in the other mode. The field-worker remains on the top as long as she has footing to hand the sheaves to the stacker. Thus, as one stack is finished, another is founded on the adjoining stathel, and as one is finished on a stool of straw, another stool should be ready to have another founded upon it.

(2970.) There is seldom leisure to thatch stacks as long as there is corn to carry in, and the finer the weather the less leisure will be found. A damp day, however, which prevents leading, answers very well for thatching, as the straw is not the worse of being a little damp; but in heavy rain it would be improper to thatch and cover up so many wet ends of sheaves as the top of a stack contains. But before thatching can be carried on, preparations should be made for it some time before, that is, straw should be drawn in bundles, and ropes twisted ready to be used; and a rainy day in harvest cannot be better appropriated than to such purposes. The straw is drawn in handfuls out of one another, till the individual straws become parallel when the handfuls are carefully laid together till as much as a thick bundle is drawn, and the bundle is then tied in the middle with a piece of straw-rope, which is the quickest mode, or with a thumb-rope.

(2971.) Straw is twisted into rope with different instruments, and in different styles. The simplest instrument is the old-fashioned *throw-crook*, fig. 537 below. It is used in this way:—The left hand holds by the ring at the end of the shank; and round the point of the head is received a wisp of straw from the person who is to let it out to be spun. The right hand holds the middle of the shank loosely, and causes the head to revolve round an axis, formed by the imaginary line between the head and ring, and the twister walks backward while operating with the instrument. The person who lets out the straw sits still on a stool, or on bundles of straw, and using the left hand nearly closed, restrains the straw in it till sufficiently twisted, and then lets it out gradually, while the right hand supplies the straw in equal and sufficient quantities to make the rope equal throughout, the twister taking away the rope to the requisite length as fast as the spinner lets it out. The spinner then winds the rope firmly on his left hand in an ovoidal ball, the twister advancing towards him, as fast as the spinner coils the rope, with a hold of the end which secures the ball firm. In the Borders the spinner lets out the straw with both hands, while stooping with his head down and his back turned to the twister; but the rope made in this manner is always thick and rough, compared to the other mode. A woman is usually employed as the twister, and a man as the spinner. It is great ease to the left hand of the twister to have the crook fastened round her middle, with a piece of stout straw-rope. The best sort of straw for rope is that of the common or Angus oat, being soft and pliable, and it makes a firm, smooth, small rope. An ordinary length of a straw-rope may be taken at 30 feet. Counting every interruption, a straw-rope of this length may take 5 minutes in the making, that is, 120 ropes in 10 hours: a man's wages 20d., and a woman's 10d., making together 2s. 6d., will make the cost of twisting a single rope just 1 farthing. Straw-ropes are twisted in quite a different way, with a machine similar to the one used by rope-makers to twist their cords; which is figured and described below. In using it the twister sits still, while the spinners carry the straw under their arm, and move backwards as they let out the straw. The spinners then wind the rope upon the left hand, and advance, during the winding, towards the machine, where they are ready to begin to spin again. Usually 3 spinners let out to 1 twister, and as they can spin as fast with this machine as with the crook, the cost of making each rope will be less than that given above; but an inconvenience attends the use of this twister—when one of the spinners breaks his rope, he is thrown out of work till the others begin a new rope; and all the spinners must let out with the same velocity, otherwise one will make a longer or a harder twisted rope than the other.

(2972.) It is the duty of field-workers to fetch bundles of straw and straw-ropes to the thatchers as they are wanted.

(2973.) Having the materials ready—drawn straw and straw-ropes, the covering or thatching of a stack is done in this manner:—On the thatcher ascending to the top of the stack by means of a ladder, which is immediately taken away, a bundle or two of straw is forked up to him by his assistant, a field-worker and which he keeps beside him behind a graip, as noticed in covering the hay-stack (2795.). The straw is first laid upon the eaves, beyond which it projects a few inches, and then in an overlapping manner upwards to the top. Where a butt-end of a sheaf projects, it should be beaten in, and where a hollow occurs it should be filled up with a little additional straw. In this manner the straw is evenly laid all round the top of the stack to the spot where the thatcher began. Suppose he has laid the covering on the top of the stack, fig. 534, all round to the line from *a* to *b*, before closing up which he makes the top *a*, consisting of a small bundle of well-drawn long straw, tied firmly at one end with a piece of cord; the tied end is cut square with a knife, as shewn at *a*, and the loose end is spread upon the covering, and forms the finishing to it. To secure the top in its place, a straw-rope is thrown down by the thatcher from *a* to *d*, the end of which his assistant on the ground fastens to the side of the stack. After passing the other end of the same rope round the top, he throws it down in the same direction, where it is also fastened to the stack. In like manner, he throws down both the ends of a rope from *a* to *c*, where they are also fastened by the assistant. These 2 ropes are seen at *e* and *f*. Having thus secured the *top*, the thatcher closes in the covering from *a* to *b*, when the ladder is placed to receive him. Taking the ladder to *c*, he inclines its top over the covering of the stack, and secures its lower end from slipping outwards by a graip thrust against it into the ground. He then mounts and stands upon the ladder at the requisite height above the eaves at *c*, and there receives a number of ropes from his assistant, which he keeps beside him, between the ladder and the stack. Holding on by the end of a coil of rope, he throws the coil from where he stands at *c* down in the direction of *d*, to his assistant, who, taking it in

Fig. 534.

THE LOZENGE MODE OF ROPING THE COVERING OF A STACK.

hand, allows the thatcher to coil it up again upon his hand, without ruffling the covering of the stack, till of sufficient length to be fastened to the side of the stack. The thatcher then throws the other newly coiled end in the same direction of *d*, where, on his assistant taking hold of it, the thatcher retains the rope in his hands by the double, and places it in its position at *g*, a little way below *e*, and keeps it in its place till the assistant pulls it tightly down, and makes it fast to the stack like the other end. Thus the thatcher puts on every rope below *g*, till the last one on that side has reached *h*. He then takes the ladder to *d*, where he puts on every rope below *f*, till they reach the last one, *i*. Ropes thus placed from opposite sides of a stack cross each other in the diamond or lozenge-shape represented in the figure. It will be seen that a windy day will not answer for laying on the covering of stacks. To give the thatch-straw a smoothness, it should be stroked down with a long supple rod of willow; and to give the ropes a firm hold, they should receive a tap here and there with the fork, while the assistant is pulling the last end tight. The thatcher is obliged to throw down the rope at first *coiled*, and to coil up again the second end before it is thrown down, because the loose ends of straw-ropes would not descend within reach of the assistant. The ends of the ropes are fastened to the stack by pulling a handful of straw from a sheaf a little out of the stack, and winding the rope round it, and the knot, thus formed, is pushed between the rope and stack and keeps the rope tight. On such a stack, which is 15 feet diameter at the base, 17 feet diameter at the eaves, 12 feet high in the leg, and 6½ feet high in the top, 10 ropes on each side are quite sufficient to rope it.

(2974.) It is seldom that the covering of a stack is finished when the straw and ropes are first put on; the stack being placed beyond danger, others are covered to the same state, and the finishing is left till more leisure is found, and, indeed, all the stacks to be early thrashed, seldom receive the finishing work at all. Many farmers only finish the outside rows of stacks. However, as I am speaking of the subject just now, I shall describe the finishing process at once. An eaves-rope, *k l*, fig. 534, is spun long and strong enough to go round the stack. Wherever 2 ropes from opposite directions cross at the eaves-rope, they are passed round it, and being cut short with a knife, are fastened to the stack, immediately below the projecting part of the thatch over the eaves. Thus the 2 ends of all the 20 ropes are fastened to the stack, and the thatch is cut with a knife round the eaves, in the form shewn from *d* by *b* to *c*.

(2975.) Another mode of roping the covering of a stack is shewn in fig. 535, where the covering being put as described above, ropes are put over the crown of the stack all round, from *a* to *b*, *a* to *c*, *a* to *d*, &c.

These ropes, at their crossing over the top, are fastened together by a rope, which is tied above them, and cut off in the form of a rosette, as at *a*. The ropes which cross these are either put on spirally from the top *a*, till they terminate at *d*, or put on separately in bands, parallel to the eaves-rope *e f*. In either case they are twisted round each crown-rope, from the top to the eaves, as *h* is seen twisted round, *a d*, *a c*, and *a b*, by *g*, till it reaches the point *h* again. The crown ropes, such as *a b*, *a c*, and *a d*, are made to pass round the eaves-rope *e f*, and fastened to the stack under the projecting covering, and the covering is cut short round the eaves, from *e* to *f*, in the manner described above. This mode of roping, though requiring fully more ropes than the last, perfectly secures the thatch against the strongest winds; though the crown-ropes may be fewer in number than shewn in the figure.

Fig. 535.

THE NET-WORK MODE OF COVERING AND ROPING A STACK.

(2976.) There is still another mode of roping the covering of a stack, and it is applicable to the heads of all stacks whose eaves are formed of a row of sheaves projecting beyond the body of the stack. It is shewn in fig. 536, and is in common use in Berwickshire. The first thing done, is to put a strong eaves-rope round the stack, below the projecting row of sheaves from *a* to *b*. The straw is then put on much the same way as described above, but rather thicker, and it projects farther beyond the eaves. The tops of the finishing sheaves of the stack are pressed down, and a rather large, hard bundle of short straw is placed upon them, to serve as a cushion for the ropes to rest upon, and which is put on after most of the covering has been laid on. The thatcher then perches himself upon the top of the hard bundle, where he receives the ropes as they are wanted, and on being thrown to him, he catches them readily on the point of a long fork. Some dexterity is required to throw a coil of straw-rope to the top of a stack. The best position to do it is to stand as far from the

Fig. 536.

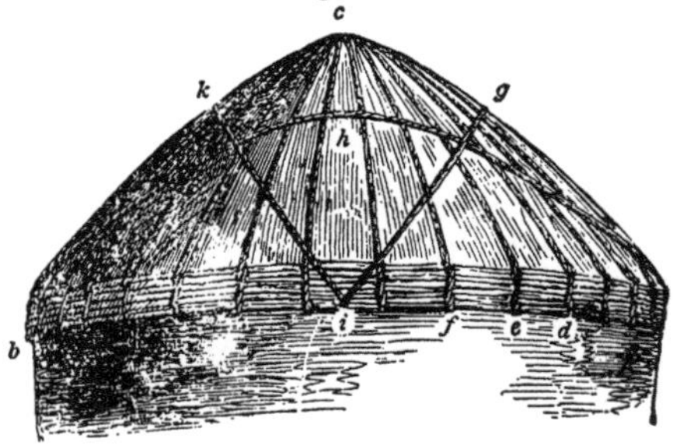

THE BORDER METHOD OF COVERING AND ROPING A CORN-STACK.

stack as to see the thatcher fully in the face, and clear of the head of the stack; then, to take the coil by the small end, pitch it upwards with a full swing of the arm towards the thatcher's feet, and he will catch it firmly on the prongs of the fork; if aimed to a higher level the fork will be almost certain of missing it, the critical position of the thatcher not allowing him to turn his body, but only to move his arms. Uncoiling the half of a rope, the thatcher throws it past the eaves to his assistant, who fastens that end, and while this is doing he throws the other half of the coil down in the opposite direction, across the top of the stack, to another assistant, who fastens it to the stack at the opposite side; 1 assistant may suffice, by tying first one end of the rope, and then the other, but with 2 assistants this mode of roping is conducted with great celerity. Thus, rope after rope is thrown, to the number of 30, before the top of stack is sufficiently roped, as seen in the figure, from *c* to *d*, *e*, *f*, &c. The ropes, where they cross at the top, are tied together with a piece of straw-rope, to prevent their slipping off. The difficult part of this mode of roping is in finishing the eaves, which, if well done, looks remarkably neat; but, if otherwise, is apt to have a slovenly appearance. In finishing, the ends of the ropes are loosened from the stack, and passed between the eaves-rope and stack, and on being brought upwards, are passed behind the ropes themselves, about 8 or 9 inches above the eaves-rope. The end of the first rope, suppose *c d* thus fastened, with the part of the covering which projects beyond the eaves, are brought along the face of the eaves, and the second rope *c e* is placed over them, before it is passed below the eaves-rope and turned upwards, and passed behind itself. The end of the first rope, that of the second, and the projecting covering, are then all brought along the face of the eaves, and the third rope *c f* is placed over them, and treated like the 2 preceding, and thus all the way round the stack with both ends of the 30 ropes. The last finish is made by bringing the ends of the 2 or 3 last ropes along the face of the eaves, behind the 2 or 3 first ropes. It will be observed, that while the end of one rope fails to go beyond its length, the end of another rope is gained, so that the band of ropes along the face of the eaves remains about the same breadth round the stack. The ropes between the eaves and top are held down by 4 or 8 ropes, *a h b*, *g i*, *k i*, which are so thrown across as to quarter the top of the stack, and their ends are fastened to the eaves-rope.

(2977.) There are still other modes of thatching stacks, such as sticking in handfuls of straw into the butt-end of the sheaves, and keeping them down with stobs of willow, or with tarred twine, both in imitation of the thatching of cottages; and they are modes in which the use of straw-ropes are dispensed with altogether. Whichever may be the plan adopted, the chief object should be to secure the corn in the stack in a dry state, with the least trouble and expense. Of all the modes in practice, I see none more efficient and better-looking than the lozenge-shaped roping of fig. 534.

(2978.) Where rough grass is found on a farm, such as on a bog which is partially dry in summer, I would suggest its being mown and sheafed, for covering stacks. A day or two spent in mowing such grass, after the harvest is over, are well spent, even at the rate of wages and food of ordinary harvest-work. Not only does this stuff save the drawing of good straw where it is scarce, but of itself forms good covering for stacks which are soon to be thrashed; and by the time it has served the purpose of thatch, it will be dry enough to litter courts, and thereby add to the manure-heap. Reeds might be used in the same way, where they do not find a profitable market as thatch for cottages.

(2979.) These are the usual modes of stacking and covering stacks of wheat, barley, and oats, in good weather; but in wet weather many expedients in stacking are tried to preserve the corn from heating, and it is necessary you should be made acquainted with them, to put them in practice when required. The most common expedient is to erect a pyramid of 3 small trees or weedings of larch or Scots fir, tied together at the top, and fastened together by the sides with thin fillets of wood. Around this pyramid is the stack built, and its use is to form a hollow space in the centre of the stack, into which the air may have access. These structures are commonly called *bosses*. When placed on a stathel, the air enters them directly from below but when on the ground, it is requisite to form an opening from the outside of the stack to the boss, by means of a low tressle; and if a tressle is placed on each side of a stack, and the position of both chosen with regard to the direction from which the wind most prevails, a complete circulation of air may be maintained through the stack. The greatest inconvenience felt in the use of this form of boss is, that, on the stack subsiding, its sharp point is apt to pass to one side of the centre of the top, and thereby to give a high shoulder to the stack; and it is well known that rain is almost certain of entering a stack by such a shoulder.

(2980.) There is another form of boss which I much prefer to this. It consists of 3 cuttings of trees—of weedings, too—7 feet long, joined together by fillets of wood, in the form of a prism, whose side is 3 feet in breath. The prism is set on end, and on a stathel only requires to be nailed to it at the bottom; but as a farther precaution for its stability, a spur from each tree should be nailed to the stathel within the body of the prism. On the ground, it requires a tressle as well as the other sort of boss; but it has the advantage over the other kind, of supporting the top of the stack evenly, when it subsides upon the upper end of the prism, relieving the body of the stack of the weight of its top.

(2981.) Another expedient is to set the upright sheaves, which form the foundation of the stack, round a long cylindrical bundle of straw, made firm by straw-rope being wound about it; and as the stack rises in height, to draw the bundle through its centre to the top when it is removed, the effect of the hole left in the stack being to create a ventilation of air through it, and to allow the heated air to escape, while the cool air enters from below.

(2982.) In wet weather corn is built in small stacks even in the stack-yard; and should the weather prove settled wet, a dry moment should be seized to put 2 or 3 stooks into what are called *hand-huts* in the field, that is, small stacks built by hand, by a person standing on the ground. Sometimes corn is built on a head-ridge of the field, instead of being carried to the stack-yard, as the same strength of men and horses will stack more corn in this way in a single fine day, than carry it to the stack-yard; and the stacks derive more benefit from the air in the field than in the yard. These stacks are also thatched in the field, and carried to the thrashing-machine some time during the winter. It is not an uncommon practice of some farmers to build a portion of their crop in the field every year; but the practice is not commendable, for, besides the trouble and dirt created in carrying straw for thatch to the field, as much confusion and dirt are created in carrying the corn to the steading in winter, when some of it cannot fail to be shaken out of the sheaves, and when the stacks wanted cannot, perhaps, be brought in for a track of bad weather or through deep snow. A scheme may be justifiable under peculiar circumstances, which would be wrong in ordinary practice; and this is one of them.

(2983.) Of the 3 species of cereal grains, barley is most liable to heat in the stack, partly owing to the soft and moist quality of its straw, and partly because clover is always mixed with it; and, on these accounts, it is advisable, in most seasons, to make barley-stacks smaller than others, both in diameter and height, and always to support them with prismatic bosses. Much care should be bestowed, not only by the use of these expedients, but on building barley-stacks safely with proper hearting, to prevent heating; for the least touch of heat spoils it for malting purposes, and malting barley always carries the highest price in the market. Besides injuring the grain, heating compresses barley-straw very firmly together, and, in fact, soon rots it. When a single stack only is seen to heat, it may be instantly carried into the barn and thrashed, the air attending which will cool both grain and straw; but when a number shew symptoms of leaning to one side, about 24 hours after being built, or exhibit a depression in the top, a little above the eaves, you may suspect heating not only to have commenced, but to have proceeded to a serious degree. An incipient symptom of heating is when moisture is observed on any part of the top of a stack early in the morning, delicately indicated by cobwebs, before the sun has evaporated it; and, when heated air is felt, or steam seen to rise from the top of a stack, the symptom is unequivocal. Heated barley lubricates the parts of the thrashing-machine with a gummy matter. Oats are less apt to heat than barley, though their heat is stronger. If there is the least sap remaining in the joints of the straw, oats will be sure to heat in the stalk. Heating gives to oat straw and grain a reddish tinge, and renders the straw quite unfit for fodder, and the grain bitter and unpalatable, both to horses, in the shape of corn, and to people in that of meal. Wheat seldom heats, but, when it does, the heat is most violent. I never saw stacks of wheat heated but once, when it was foolishly led into the stackyard the day after it was reaped. Partial heating will be induced in stacks by leaning over after being finished in building, when the compressed part may be expected to heat; and to avoid the tendency of a stack leaning to one side or another, a safe practice is to set props loosely round a stack, to guide it in its subsidence, especially if it has been rapidly built; but it should be borne in mind that if one prop is pushed harder in than the others, it will cause the stack to swerve from it. Some stacks begin to sway the moment the top is put on, and such should be supported with props to keep them right.

(2984.) In filling a stackyard, respect should be had to the convenience of taking in the stacks to be thrashed as they are wanted. As barley is the sort of grain first disposed of in the market, the barley-stacks should be built nearest the barn, except those intended for seed. Wheat, except in very fine seasons, is seldom firm enough to be presented to market till the spring, and on that account may be built on the outside of the stackyard, and wheat-stacks grace a stackyard well. Oats are wanted at all times, for horses and fodder, and should, therefore, be always at hand, as the stackyard becomes cleared. Consideration in these particulars saves a great deal of trouble and confusion in bringing in stacks to be thrashed.

(2985.) With regard to the carrying of beans and pease, they are usually the last of being brought into the stackyard, and particularly beans. The bundles of pease are turned in the field till they are won, when they are rolled up in smaller bundles, and tied by a wisp of their own straw. Pease-straw is very apt to compress in the stack, and, of course, to heat, and should therefore be built with bosses, either in round stacks or oblong ones, like a hay-stack. The largest stack of corn I ever saw was one of pease, which was 150 yards in length; but there was an opening in it, in which any person could have walked upright through the entire length of the stack. When pease become very dry in the field before they are led, the pods are apt to open and spill the corn, particularly in sunny weather; and to avoid this loss, the crop is usually brought quickly into the stackyard, and built in ventilated stacks. Beans are a long time of winning in the field in calm weather. As it is desirable to have the land they grow on ploughed up for wheat, they are not unfrequently carried to a lea-field and stooked upon it, till ready to be stacked. Being hard and open in the straw, they keep pretty well in small stacks, though not quite won, and the risk of keeping is worth running in dry weather after much rain, when the pods are very apt to burst and spill the corn on the ground. In building pease and beans, the sheaves are not laid with nicety, nor do the stacks receive much trimming, the pease none at all, the beans with the back of a shovel. Thatching pease and bean stacks is conducted in the same manner as described above; but less pains are bestowed in finishing them off. As, however, a good deal of corn is exposed on the outside of those stacks, the thatching is not unfrequently brought down their legs, and kept on by ropes.

(2486.) In connection with this latter subject, I may mention, that when corn is mown in a slovenly manner, a good deal of heads of grain may be observed exposed to the air on the outside of the stacks; and to save these, the stacks are shaved down with the blade of a scythe, fastened lengthwise on a shaft, upon the barn-sheet laid upon the ground to receive the shaved-off corn. This process not only saves the grain, but gives a smoothness to the appearance of the stack, and prevents the mixing of sprouted grain among good samples from the body of the stack, and which mixture would inevitably occur, were the grain on the outside allowed to be exposed to the weather. The shaving, it is obvious, should be done immediately after the stack has subsided, and even before, if there is appearance of rain; and it is easily done, and does not occupy much time. The shaved appearace of a stack may be seen at *c*, fig. 539, page 1104.

(2987.) These are all the particulars which occur to me to say on bringing in the crop. In conclusion, I may advise you not to imitate those farmers, who, because they are gratified on having their crop safe in the stackyard, seem regardless of the state of the stackyard itself, and leave it for a long time littered with the refuse of the thatching straw, which, when it becomes wetted with rain, is not only useless as litter elsewhere, but soon heats, and causes an unpleasant odour around the stacks. The spare straw should be removed immediately after it is of no use in the stackyard, either to the straw-barn, if it is drawn and bundled, or, if loose, into the sheds of the hammels, where it will be ready for littering the cattle that may soon be expected to occupy the hammels for their winter quarters. The ground should be raked clean, the air will then become sweet, the stacks have free circulation of it amongst them, and the poultry will have the opportunity to pick up every particle of grain that had fallen upon the ground. This act of cleanliness being done, the stackyard gates may be shut, and the labours of the farm, in so far as they concern the crop, may be said to have been brought to a termination.

(2988.) [The *throw-crook*, fig. 537, is an instrument that has been long in

Fig. 537.

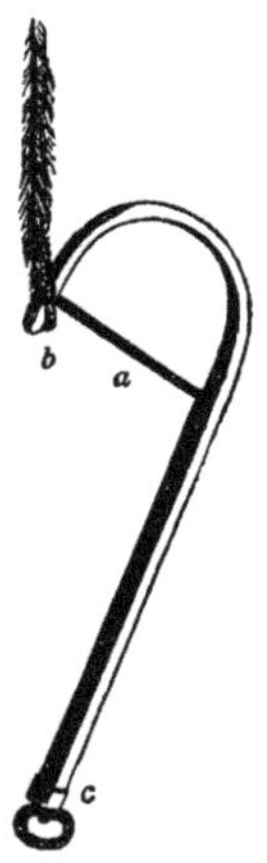

THE THROW-CROOK.

Fig. 538.

THE STRAW-ROPE SPINNER.

use for the purpose of spinning or twisting straw-ropes, and is one of those primitive inventions that required only the cutting of the first crooked sapling that might come to hand; for though our figure represents an artificially formed implement, any piece of bent material will answer equally well, all that is required being such a form as will give it the character of what we now denominate a *crank*, and to have a swivel-joint at the end, to allow it to turn freely and independent of that appendage by which it is attached to the body, or to the left hand of the person who turns it. The implement represented in fig. 537, is made of a piece of tough ash, about 3½ feet long; the bent part is thinned off until it is capable of being bent to the curve, and is there retained by the iron stay *a*, the part *b* being left projecting beyond the stay, for the attachment of the first end of the rope that is to be made. The end *c* is furnished with ferule and swivel-ring, by which it is attached to the person, by a cord passed round the waist. In using the implement, the rope-maker is stationary, usually sitting beside the straw, and the spinner moves backward as the rope extends.

(2989.) The *straw-rope spinner*, fig. 538, is a machine of recent introduction to the operations of the harvest season, and is of considerable importance in facilitating the process of straw-rope making. Comparing it with the old instrument, the *throw-crook* just described, the advantage is considerable; for with that two people must be occupied in the making of one rope, whereas with the spinner four people only are required to make three ropes, being a saving of one-third of the time occupied by the old practice. The spinner has been constructed in various forms, though exhibiting but two distinct varieties of the machine, the first distinguished by the spinner being stationary, and the ropemaker moving away from it; the second, by the ropemaker being stationary, and the spinner moving away from him. The first kind is found to be the best in practice, and I have therefore chosen an example of it for illustration. Fig. 538 is a view of the machine, consisting of a sole frame, *a*, *a*, measuring about 2 feet each way, with an upright post *b* tenoned into the sole, and carrying the cross-head *c d*. The cross-head is a hollow box or case, adapted to contain the machinery of the spinner, consisting of 5 light spur-wheels, about 6 inches diameter, placed as seen in the case *c d*. Of these, the central and the two extreme wheels are mounted upon axles, which terminate in the hooks *e*, *e*, *e*, the other two wheels being merely placed intermediate, to produce revolution in the three principals in one direction. A winch-handle *f*, is fixed upon the axle of the central wheel, on the side opposite to the hooks, and to prevent the machine from moving with the strain of the ropes, a few stones, or other weighty substances, are laid upon the sole-frame. The machine is then put in operation by the driver turning the handle, and the three ropemakers, each with a quantity of straw under his arm, commences his rope by binding a few straws round the hook appropriated to himself. He then proceeds backward, letting out the straw as he advances, and the rope takes the twist, until the length required is completed, when each man coils up his rope into an oval ball.

(2990.) Some machines of this form are mounted on wheels, thereby coming under the character of the second kind, when the rope-maker is stationary; but great inconvenience must attend any attempt at working in this manner. Another form of the machine adapted to work, as one of the second kind, is that which is strapped to the body of the driver, he moving away from the stationary ropemakers. This method also is attended with inconvenience, especially to the driver, who, having the machine strapped in front of his body, the handle being at the end, and the machinery consisting of bevel-gear, having the external form of the cross-head alone of fig. 538, the handle is brought so near to his body that much of the muscular force of the arm is lost by this misapplication. By using a well-constructed machine for straw-rope spinning, not only is there a saving of expense effected, but the ropes are much better twisted, and, of course, stronger than those made by the old implement. The price is from L.2, 5s. to L.2, 10s.—J. S.]

(2991.) *Ladders* are best formed of tapering Norway pine spars, sawn up the middle. A useful form of ladder for farm purposes, is shewn in fig. 539, where the rounded parts of the divided spar *a a*, are placed outmost, and are connected together by steps *b*, of clean ash driven through auger-holes, and firmly wedged into the spars. The steps are 9 inches apart, and are 16 inches long at the bottom, and 13 inches at the top, between the spars in a ladder of 15 feet in length which is the most useful size for the use of a stack-yard. To prevent the ladder from falling to pieces, in consequence of the shrinking and loosening of the round steps, a small rod of iron is placed under the upper, middle, and lower steps, where one of its ends is passed through each spar, and it holds the spars firmly together by means of a shoulder and nut and screw on each of it ends. When properly finished and painted, such a ladder will last many years.

Fig. 539.

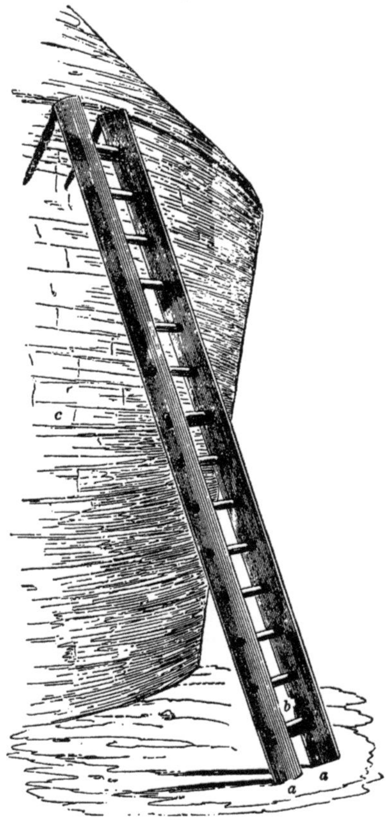

THE LADDER.

(2992.) Some dexterity is required to set a long ladder on end, as also to carry it from one place to another. To place it in a perpendicular position, its lower or heavy end should be placed against any object capable of resisting its slipping upon the ground; and on its light end being elevated arms-length above the head, the position is kept good by a step being placed between the prongs of a fork, by means of which the end of the ladder is still more elevated, while it is still increased, by pushing the arms against one step after another, till the perpendicular position is gained. A long ladder is carried from one place to another in this way, provided the distance to be carried is short. Set the perpendicular edge of the ladder against the right shoulder, and then take hold of a step with the right hand, and raise the ladder steadily by it from the ground, while to retain the perpendicular position, grasp a step above the head firmly with the left hand, and then walk steadily forward. Some can carry it steadily by grasping a step with both hands, with the edge against the shoulder; and some even are so powerful in the arms, as to carry a ladder by the steps at arms-length, with one arm above, and another below, the head. A ladder may be moved on the ground a short distance, while standing in a perpendicular position, by holding a spar in each hand at arms-length, and then moving first one foot of the ladder in advance, and then the other, till the spot is gained. This sort of motion, when applied to moving a large stone, is technically named by masons *cutting*. A long ladder is brought down from the perpendicular to the horizontal position, by placing it against a stack, or any other object which will resist its foot slipping on the ground, and allowing it to rest in an inclined position upon the arms, with the hands stretched above the head, the ladder approaches the horizontal position the farther you recede from its lower end, the upper end being supported by another person with a long fork. When not in use, a ladder should be laid on the ground along the side of a stack, or the stackyard wall. I have said more, perhaps, on the subject of handling ladders, than seems necessary; but the truth is, ladders are more frequently destroyed by being brought to the ground in a careless manner, and by being blown down by the wind, while resting against a stack, than by fair use.

(2993.) *Cart-ropes* last according to the care bestowed on them. When used with the corn-cart, they should never be allowed to touch the ground, as earthy matter, of whatever kind, soon causes them to rot. On being loosened when the load of corn is to be delivered to the stacker, they should be coiled up before the load is disposed of, and the coil suspended from the back part of the cart quite clear of the ground. A soft rope holds much more firmly, and is less apt to cut than a hard one.

---

77. OF DRAFTING EWES AND GIMMERS, OF TUPPING EWES, AND OF BATHING AND SMEARING SHEEP.

"But rather these, the feeble of thy flock,
Banish before th' autumnal months: ev'n age
Forbear too much to favour; oft renew,
And through thy fold let joyous youth appear."

DYER.

"The ewes, being rank,
In th' end of autumn turned to the rams."

MERCHANT OF VENICE.

(2994.) When last speaking of sheep, the lambs were weaned and buisted. One of the processes amongst sheep in early autumn, in the beginning of August, is *drafting ewes and gimmers*, that is, separating those to be disposed of from those to be kept. Drafting, however, applies only to a standing flock of ewes. By a *standing* flock is meant a fixed number of ewes, which are made to rear their lamb year after year. Instead of having a standing flock, some farmers are in the custom of buying every year a flock of ewes big with lamb, receiving the lambs from them, and disposing of both ewe and lamb at such a season as best suits the market for those respective sorts of stock. The ewes are sold in autumn to be fed on turnips, and the lambs are disposed of, after being weaned, to rearers of sheep who breed none. Such a one is called a *flying-flock*. Of course, flying-stock require no drafting; where all are disposed of, none obtain the preference of being kept.

(2995.) There are various marks of deterioration which determine the drafting of *ewes*. *Bareness of hair on the crown of the head*, which makes them obnoxious to the attacks of fly in summer—*deficiency in eye-sight*, which prevents them keeping with the flock, and choosing out the best parts of pasture, and best points of shelter—*ill-shaped teeth and jaws*, which disable them from masticating their food so well as they should—*want of teeth* from old age, when, of course, they cannot rop sufficient food to support their lambs—*hollow neck*, which indicates breeding too near akin—*hollow back*, which implies weakness in the vertebral column, thereby rendering them unfit to bear lambs to advantage—*flat ribs*, which confine the space for the fœtus within the abdominal region—*a drooping tail-head*, which affects the length of the hind-quarter, a space occupied by superior flesh—*bad feet*, which prevent travelling with ease along with their companions—*round and coarse bone*, which indicates coarseness of flesh—*thin or short coat of wool*, which lessens the clip and the profit of wool—*diseased teat or udder*, which diminishes the supply of milk for the future progeny—*scarcity of milk*, by which lambs, not obtaining sufficient nourishment in the early period of their existence, are stinted in growth, and weakened in constitution—*carelessness of disposition*, which induces neglect of the lamb, particularly one of twins, which is in consequence ill nursed—*producing worthless lamb*, by which profits are much diminished—*missing being in lamb*, a failing which is apt to recur in any future year—*casting lamb*, a propensity likely to recur every year—*rottenness*, which is, of course, objectionable in every animal that produces young—*shortness of breath*, which prevents them seeking their food, and eating so much of it, as they should—*tendency to scouring*, or *the opposite*, the former imposing weakness, the latter inducing inflammation—*delicacy of constitution*, which disables them from withstanding the ordinary changes of the weather—*diminutive stature*, or *inordinate size*, which destroys the uniformity of the flock. This is a long list of faults incidental to ewes, and yet every one may be observed, and which every breeder of sheep is desirous to get rid of.

(2996.) It is not at all probable that any flock of ewes present *all* these objectionable qualities in one season; but, notwithstanding this favourable circumstance, it is not in the power of the breeder to draft *every* ewe having an objectionable property every year, because, the farm supporting a stated number of ewes, the extent of their draft depends on the number of *good* substitutes which may be obtained from the gimmers; for it is obvious no good object is attained by drafting a bad ewe, and taking in its stead a bad gimmer. The number of gimmers fit to be transferred to the ewe flock, should therefore be, in the first instance, ascertained, and a corresponding number of the worst ewes drafted.

(2997.) In *drafting gimmers*, many of the above faults may be observed in them also, though every fault arising from lamb-bearing cannot possibly affect young sheep. The faults incidental to gimmers are, bareness of hair on the crown of the head, ill-shaped teeth and jaws, hollow neck, hollow back, flat rib, low tail-head, bad feet, round coarse bone, thin and short fleece, rottenness, shortness of breath, tendency to scouring or otherwise, delicacy of constitution, and diminutiveness of stature and inordinate size. These faults are numerous enough, but not likely to be all observed in the same year, and less likely in the same individual. Most of them may be got rid of by rejecting females which have more than one of them, and by employing tups free of them all. When the external form is improved, the constitution is also strengthened. Gimmers, when they become ewes, are moreover likely to be deficient in milk, careless of their young, and produce small lamb; but these faults disappear in the succeeding year; and should this not be the case, the ewe, though young, healthy, and fresh, should be drafted. Thus ample drafting can alone ensure a sound, healthy, well-formed, young and strong-constitutioned flock of ewes.

(2998.) After being drafted, the ewes and gimmers are buisted with tar (2635.), those to be retained, on the near rib, and those drafted, on the far.

(2999.) The *flock-ewes* and *flock-gimmers* may be put together, or kept separate, as convenience may determine. If tupped together, they may go together all the season; but if not, which is always the case on large farms where more than 1 tup is employed, they may be grazed separately, to save the trouble of separating them afterwards. Both should be kept on moderate pasture, to prevent attaining too great fatness before the tupping season, which commences the second week of October; but good fresh aftermath of grass, or, what is still better, rape, for about 2 or 3 weeks before the tups are put amongst them, will not only make them fresh, but will cause the season to come on them more strongly and equally than without such assistance.

(3000.) The *draft-ewes* and *gimmers* should be put on the best grass the farm affords, immediately after being drafted and buisted, to fatten as soon as possible, as they are usually sold before the time arrives for putting lean sheep on turnips. Indeed, the ewes which have missed lamb, called *tup-eild ewes*, having had no lamb to rear, will be fat in the early part of summer; and when kept to the end of autumn become very fat, not less than 30 lb. per quarter, if of the Leicester breed. The draft-gimmers, and ewes which have borne lamb may reach, by the latter period, 20 lb. per quarter. The gimmers should fetch the highest rate of price, the tup-eild ewes should also yield good mutton, but old ewes will be dry in the flesh. Drafts are ready for sale in September. The great fair in Northumberland, dedicated to St Ninian, held on the 28th September, for the sale of draft-ewes, I have seen contain 70,000 ewes, partly fat and lean.

(3001.) *Tups* require but little preparation on being put amongst ewes. If their skin is red in the flanks when the sheep are turned up, they are ready for the ewes, for the natural desire is then upon them. Their breast, between the fore legs, is rubbed with *keil* or ruddle, on being placed amongst ewes, that they may mark their rump on serving them. It is the duty of the shepherd to mark every ewe so served, that he may observe whether or not the season returns upon her, and to be prepared for the day of her lambing when it arrives. The period of gestation of the ewe is 5 months, and as the tup is usually put amongst the ewes from the 8th to the 11th October, the first lambs may be expected to appear on the 8th or 11th of March following. A young active tup, a shearling, will serve 60 ewes, an old one 40 ewes. Tup-hoggs are not used, not having attained maturity in any particular, though one is sometimes put amongst ewes to make an old tup more active. When tups are too fat, they are apt to become lazy, and will only willingly serve a very few ewes; and when this is the case, it is better to put him in a field by himself with a few ewes, selected to suit his particular qualities, than to urge him to over-exertion by means of a tup-hogg, because several of the ewes served by him in such case may miss being in lamb. A tup-hogg so empoyed, is seldom allowed by the tup to serve a single ewe, being driven off wherever he is seen to go with one; but as a sure prevention of his serving, a piece of cloth is sewed to the wool on the under part of his belly. Tups, when too heavy, are apt to contract spavin in the hock joints, in consequence of the great weight of the carcass upon the hind legs, in the act of serving ewes. Most of the ewes will be tupped during the second week the tup is amongst them; and, in the third, they will all be served. It is likely that some of the first served ewes will return in season, and these should be specially noticed by the shepherd, as it is not improbable they will become tup-eild. The ruddle requires to be renewed almost daily, and even oftener, when the tupping is active.

(3002.) Tups are not selected for ewes by mere chance, but according as their qualities may improve those of the ewes. When ewes are nearly perfect, they may be selected for breeding tups. A good ewe flock should exhibit these characteristics:—*strong-bone*, supporting a roomy frame, affords space for a large development of flesh,—*abundance of wool of good quality*, keeping the ewes warm in inclement weather, and ensuring profit to the breeder,—a *disposition to fatten early*, which enables the breeder to get quit of his draft-sheep readily,—and *being prolific*, which increases the flock rapidly, and is also a source of profit. Every one of these properties is advantageous in itself, but when all are combined in the same individuals of a flock, that flock is in a high state of perfection. Now, in selecting tups, it should be observed whether or not they possess one or more of those qualities, in which the ewes may be deficient, in which case their union with the ewes will produce in the progeny a higher degree of perfection than is to be found in the ewes themselves, and such a result will improve the state of the future ewe flock; but, on the contrary, if the ewes are superior in all points to the tups, then of course, the use of such tups will only serve to deteriorate the future ewe flock.

(3003.) After 3 weeks have elapsed from putting the tup amongst the ewes, he should be withdrawn; because lambs begotten so long after the rest, will never coincide with the flock. Tups should, after serving, be put on good pasture, as they will have lost a good deal of condition, being indisposed to settle during the tupping season. The ewes and gimmers may now be classed together, and get such ordinary pasture as the farm affords. During the autumnal months, they will find plenty of food in the pasture fields, but in winter, rather than be fed on turnips, a rough pasture-field should be reserved for them. When they have no such pasture, they will require to be put on a break of turnips for 2 or 3 hours every day; but it should be borne in mind, that a *fat* ewe always bears a small lamb, and is very subject to inflammatory fever after lambing, from the recovery of which, if she ever recover, the probable result will be scantiness of milk. Swedish turnips will produce this effect on ewes more readily than other kinds, therefore a few white turnips should be reserved for them as long as practicable, should they receive turnips at all.*

(3004.) Immediately after the arrangement for tupping the ewes are made, part of the sheep-stock undergo a preparation for being put on turnip, and the preparation consists of bathing them with a certain sort of liquid. I have said that sheep are affected by a troublesome insect, the keb or ked, or sheep-tick, fig. 456, which increase so much in numbers, as the wool grows towards autumn, as to become troublesome to sheep; and were means not taken to remove them, the annoyance they occasion the sheep would cause them to rub themselves upon every object they can find, and in thus breaking their fleece, deteriorate its value to a considerable extent. Another reason for bathing sheep is, that on

* See paper by me on Drafting Sheep in the Quarterly Journal of Agriculture, vol. p. 1005—11.

experiencing so great a change of food, as from grass to turnips, cutaneous eruptions are apt to appear on the skin, even to the extent of the scab, which would deteriorate the fleece even more than the rubbing occasioned by the ked. The liquid, then, which would be of service, should combine the properties of killing the ked with certainty, and of preventing eruptions on the skin, without injury to the staple of the wool; and these effects are attained by the use of tobacco-liquor and spirit of tar, the former instantly destroying the ked, and the latter acting as a preservative to the skin. The former precaution is necessary to be exercised on all classes of sheep, but the latter is the more necessary on sheep bought to fatten on turnips, as travelled sheep are almost always affected with cutaneous eruptions, and especially Black-faced sheep direct from the hills, after they have been on turnips for some time. As a matter of safety, then, for your own flock, however clean it may be, every sheep that you buy from another flock, whether intended for feeding on turnip or for augmenting your own flock, should be bathed immediately on its arrival on your farm, and before it can possibly have had time to contaminate your own sheep.

(3005.) The materials used in the *bath* are tobacco, spirit of tar, soft soap, and sulphur vivum. The tobacco is best used in the state of leaf, but I understand it is illegal for tobacconists to sell tobacco in the leaf. Being used in the proportion of 1 lb. of tobacco to every 20 sheep, it is put into a boiler with 1 quart of water for every 1 lb. of tobacco, and boiled gently for several hours. The tobacco is then wrung out, and returned again into an empty boiler with ½ quart of fresh water for every 1 lb., and boiled as long as any colouring matter can be obtained from it, when it is wrung out and thrown away. The entire water will have boiled in to 1 quart to 1 lb of tobacco. This forms a decoction of tobacco which is much better than an infusion. The soft soap is used in the proportion of 1 lb. to every 20 sheep, and is dissolved thoroughly in a sufficient quantity of warm water. The sulphur vivum is pounded fine and mixed with the soap in the proportion of ¼ lb. to 20 sheep, and it combines with the soap in some degree, but when both are mixed with the tobacco-liquor, the sulphur is apt to be thrown down in the original state of powder. I don't know of what particular use the sulphur vivum is, for the greater proportion of it is certainly precipitated to the bottom before it is used, unless its sulphureous property may serve to prevent the ked breeding again for a time; but if this is its only use, the flour of sulphur should be more efficacious. The tobacco-liquor is put into a tub, and the solution of soft soap and of sulphur vivum are intimately mixed with it. The bath thus made is administered by means of a tin-flask capable of easily holding 1 quart, and provided with a long spout, by which to pour it along the shedded wool of the sheep. The spirit of tar is poured into the flask of liquor when about to be used, in the proportion of ½ a wine-glass to 1 quart, and stirred. Some people use stale urine over and above these materials for the professed purpose of making the bath stronger, but I cannot see how it can strengthen any of these ingredients; for, as to any caustic effect upon the skin, a small addition of the spirit of tar will have a much greater effect than the urine. If the urine is employed with a view to its ammoniacal vapour destroying the insects, that may be a good plea for its employment in summer to prevent the extension of insect vermin; but in autumn, when the keds are entirely destroyed by the spirit of tar, and will not again appear till spring in a new progeny, the urine seems useless in winter.

(3006.) This is an effectual bath, and is not expensive, the tobacco being 3s. 6d. per lb., a bottle of spirit of tar 1s. 6d., soft soap 6d., and sulphur vivum 1s. per lb., give a cost of 5s. 6d. for 20, or 3¼d. per sheep. But it should be mentioned, in regard to this bath, that though in very common use it tinges the colour of the wool. and is more expensive than some other baths to be adverted to hereafter.

(3007.) An useful implement in bathing sheep is the *bathing-stool*, fig. 540, which is made of the best ash. It consists of a seat *a*, for the shepherd to sit on while bathing the sheep, 1 foot square; the sparred part is 3 feet long, and 30 inches wide in front from *b* to *c*, its greatest width being across at *d*. The legs *e*, *e*, &c., are 18 inches high, attached by means of iron rods passed through their upper part and the frame of the stool, and secured with nut and screw.

Fig. 540.

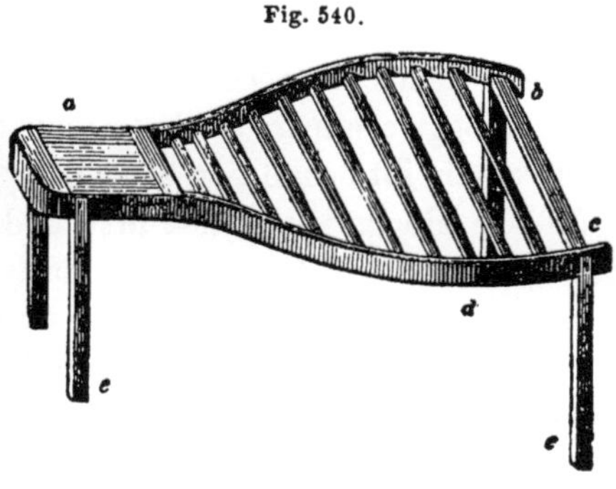

THE BATHING-STOOL FOR SHEEP.

(3008.) The *bathing* is conducted in this way. A sheep is caught and placed on the stool on its belly, with its 4 legs through the spars, and its head towards the seat *a*, on which the shepherd sits astride. The wool is shed by the shepherd, with the thumbs of both hands, from one end of the sheep to the other, and when he has reached the farthest end of the shed, an assistant, a field-worker, pours the liquor from the flask equally along the shed, while kept open by the shepherd with both hands. The sheds made are 1 along each side of the back bone, 1 along the ribs on each side, 1 along each side of the belly, 1 along the nape of the neck, 1 along each side of the neck, and 1 along the counter. From these sheds the bath will cover the whole body. The sheep is turned on the sides and back to obtain easy access to all these several parts. Additional liquor is put on the tail, head, scrotum, inside of the thighs, brisket, root of the neck, and top of the shoulder, because these are parts most likely to be affected by scab, and are chiefly the seats of the nidi of insects. The shepherd and his assistant will bathe 40 sheep in a day. Dry weather should be chosen for bathing, else the rain will wash away the newly applied bath; and coarse clothes should be worn by those who administer the bath, as it is a very dirty process. When the sheep are lying on their back on the stool, their legs are not tied, so the assistant should be aware of receiving a kick from the hind feet on the face, or on the flask.

(3009.) Shortly after bathing, the keds may be seen adhering to the points of the wool dead; and when sheep are much infested by this vermin, the fleece may be seen speckled thick with their bodies. Sheep are differently affected with keds; and those which recover from a lean or stinted state, to one in better condition, on a change of food, are most liable to be overrun with them, as cattle are with lice when improving in condition on turnips. This being the case, the ked may be expected to increase rapidly on sheep that have been some time on turnips, and hence the necessity of bathing sheep before being put on turnips. Hoggs are most liable to their attack, because, perhaps, they get most rapidly into condition after being some time weaned.

(3010.) Hoggs are bathed first, and then the ewes, though not at the same period; for if hoggs are put early on turnips, say the middle of October, the ewes will not have been tupped by that time, and they should not be bathed till after being tupped, as the smell of the bath might interfere with the effluvium of the season, and deceive the tup; and its effects upon the skin may also interfere with the coming of the season upon the ewe. Be the effects of bath what they may, the safe practice is not to bathe flock-ewes till after having been tupped; and on this very account, the bathing of ewes ought to be conducted with very great care, as a twist or rack given to the body in catching, and lifting them on and off the stool, may cause them to cast lamb, and, for fear of such accidents, the sooner they are bathed after being tupped the better, the body being then not much under the influence of the fœtus. I don't know that any case of ewes casting their lamb can be traced to this particular cause, but it is a fact, that injury to the body of any animal in the period of gestation, is liable to cause abortion; and there is no reason why injuries sustained at bathing should not produce that effect, as well as other injuries, though I believe this particular source of injury has been entirely overlooked by farmers. The tups should be bathed immediately before or after the ewes.

(3011.) A syringe has been recommended for use in the application of bath to sheep, because it is imagined it might be introduced amongst the wool without breaking the consistency of the fleece, which shedding always does. No doubt, fleeces which have been shedded, that is, divided into portions indicated by the locks, are more apt to be blown asunder by the wind, but this is only for a short time, after which they recover their coherence; and the uncertainty attending the knowledge of whether the entire body is covered by the bath when applied by a syringe, more than counterbalances any advantage it may afford by keeping the fleece entire. From the viscid nature of the bath, too, it is not improbable that a syringe will not part with the liquid at all times with the same facility. Such an instrument, besides, in the hands of a rude operator, may tear off piles of wool unhidden to view, and even abrade the skin. I am much disposed to believe that the manual operation of bathing will not easily be superseded by mechanical means.

(3012.) Instead of bathing sheep in this manner, which is the old-fashioned one, it has of late years been recommended to dip them in tubs containing liquids, that answer the same purpose as the bathing materials I have spoken of. It is obvious, that a liquid, to be applied with certainty to the entire body of the sheep through its wool, must be as limpid as water; and, accordingly, all the dipping compositions are dissolved in large quantities of water. A solution of corrosive sublimate would easily kill keds and harden the skin, and a solution of it may be made as limpid as water; but although I do not know the composition of dipping solutions, many varieties of which are now sold, I am told that not a few of them contain arsenic; which being the case, I would not employ them myself, or advise you to do it, as I have often said that the use of any substance which contains arsenic ought to be scrupulously avoided on every farm on which live-stock are reared. I observe that Bigg's sheep-dipping composition, which is used by some farmers of my acquaintance, is sold at 9d. per lb., or in casks of 100 lb., sufficient to dip 500 sheep, for L.3, 10s., or rather more than 1½d. per sheep, which is certainly a cheap application; but I do not know what it is compounded of. Mr Wilson of Coldstream has also contrived a dipping mixture, which answers well for killing vermin in sheep, as also for improving the wool. Farmers of my acquaintance, who have used it for 15 or 16 years, speak highly of it as a comfort to sheep, and an improver of wool. The mixture is sold in packets. Each packet is dissolved in 2 imperial gallons of boiling water for 10 minutes, and poured into a large tub, containing 40 gallons of cold water. Add 4 lb. of soft-soap to every packet, and the quantity will bathe 50 hoggs. It will destroy all kinds of vermin in sheep in ½ an hour, and, according to Mr Wilson's statement, costs not more than ½ of any other bath now in use, but the exact amount of its cost I do not know. The dipping apparatus consists of a box for holding the bath, to one side of which is attached a drainer, upon which the bathed sheep are placed, and the liquid squeezed out of their wool. From the drainer the sheep are slid down a short inclined plane into a pen in which the bath drips from them, and runs away by a channel, and is collected in a vessel for use again. The apparatus is 11½ feet in length, 2 feet 9 inches in height, and 4½ feet in width. The bath-box contains 125 gallons, and the pen 10 or 12 hoggs at a time. The sheep are dipped in this way :—Every sheep is held by 2 men, 1 holds the head with the right hand, and the 2 fore feet with the left; the other man holds the 2 hind legs. They dip the entire body of the sheep, with the back undermost, in the bath, with the exception of the head. They then place the sheep upon the drainer, squeeze some of the bath out of the wool with the hand, and this part of the bath returns immediately into the box. The sheep is then slid down on its side on the inclined plane into the pen, where it remains for a time to drip its wool of superfluous bath. The pen is provided with a door, by which the bathed sheep walk out. In this way 3 men and 1 boy can dip 28 scores of hoggs in 7 hours. This is quick work, and is certainly a great saving of time and labour, and also of bath. Should experience speak in future as favourably of this process as it has hitherto done, it will, no doubt, entirely supersede the old bath of tobacco-liquor and spirit of tar described above.

(3013.) On *carse* land, and on *farms in the neighbourhood of towns*, no ewe-flock is kept, a few wethers being perhaps fed on turnips on the latter. Ewes are not kept on *all pastoral farms*, some being better adapted for feeding than rearing sheep; and even where kept on land subject to rot only in wet seasons, a flying stock is preferred to a standing one. Hard short pasture is best for a standing flock of ewes on pastoral farms, because, though the lambs may be small on such ground, and would continue so were they retained on their native pasture, they are sound and hardy, and never fail to grow and thrive when removed to a more genial climate and luxuriant pasturage.

(3014.) Ewes of a standing flock are, of course, drafted on a hill as on a low farm, and the same principles of drafting apply to both. The season of tupping is a month later, near the middle of November being chosen for that matter, as the middle of April is early enough for lambs to be dropped on the hills.

(3015.) Hill flocks are also prepared for winter by an application on their skin, not merely, however, for the purpose of killing vermin, but of protecting them against the severe effects of inordinate cold and wet. For this latter purpose a thin liquid would not suffice; it must have such a consistence as to withstand melting by the natural temperature of the body, and of being washed away by the most drenching rains. The substances supposed to possess those properties in a high degree were tar and butter. The tar itself would have effected both purposes, as its pitchy quality would prevent its melting at a low temperature, and rain has little effect upon it; but butter was added with a view to neutralizing the caustic effect of tar upon the skin, and at the same time to encourage the growth of wool. When the skin of sheep is covered with such substances, they are said to be *smeared*.

(3016.) Now this practice, contrary to most practices in farming, is unsupported by theory, and indeed by common sense; for why should it be imagined that plastering over the skin with an adhesive substance, thereby artificially stopping every pore in the skin of an animal susceptible of perspiration, and rendering it uncomfortable in feeling, should protect its body from cold, while its natural clothing of wool, the most non-conducting of all substances, is incapable of affording it that protection? Does not common sense suggest the natural consequence of plugging up the pores of an animal's skin, to be injurious to its health? Why should any external application to the skin promote the growth of wool, when that substance derives its entire support from the body of the animal? No doubt, owners of *very lean* sheep may have their flocks overcome with the united effects of rain and cold, but they might also have observed that sheep in such condition are not sufficiently covered with wool to protect their lean bodies from the elements. Common sense suggests that a thicker covering of flesh and fat on the bodies will withstand cold, and throw off rain as well, at least, as tar, while the natural functions of a clean skin would be preserved; that a thick covering of flesh and fat will also promote the growth of wool as well, at least, as butter; and that a thick covering of wool, and flesh, and fat, will ward off cold and rain as well, at least, as any substitute that man can apply. Food and shelter, then, are alone wanting to free hill-flocks from the filthy process of smearing; and it is cheering to observe hill-farmers arriving at this conviction, by their daily experience of the inefficacy of smearing, as an equivalent for food and shelter. Of course mountain sheep, like all others, should be relieved from the annoyance of vermin.

(3017.) But another circumstance, extrinsic of the farm, tends to force the same conviction on hill-farmers, namely, the desire of woollen manufacturers to obtain wool free of extraneous matter, not merely because it suits their purpose better, but because there is less *waste* in manufacturing it than the ordinary wool, as supplied from hill-farms. The waste in scouring and dyeing wool that has been smeared at all is $\frac{1}{2}$, and when it has been smeared with tar and butter, it amounts to $\frac{5}{8}$ of its bulk. This being the result of the wool, as managed by the hill-farmer, he need not be surprised at the low price offered for it by dealers. It might be impolitic to relinquish smearing entirely, until other means are substituted to render it useless; because the constitution of hill-sheep may have been so affected by the influence of smearing, that to abandon it suddenly might endanger their health; still the day should be, with anxious hope, looked for, when smearing shall be regarded as the relic of a barbarous system; and were the hill-farmer, in the mean time, to try experiments on a portion of his flock, whether health and vigour could not be attained without smearing, by means of stores of food in winter, to be used in cases of emergency, and of increased warmth in lairs, by means of shelter—could these points be proved by farmers on actual experiment, and then by actual experience, landlords would then have a strong inducement to assist their tenants to raise a sufficiency of winter food either of turnips, or of irrigated meadow-hay in localities best suited to each, and to plant extensively in such situations as to afford the greatest extent of shelter.*

(3018.) There are many varieties of salves in use, and trials of new ones every year evince the dissatisfaction of hill-farmers in them all; and after all trials have failed, and I have no hesitation in affirming my belief that they will all fail, as substitutes for flesh and fat, to raise wool worth the manufacturer's consideration, then we may expect to witness the entire abandonment of the smearing system; though it should be remembered, that, even after smearing shall have been abandoned, some application will be requisite to destroy vermin on sheep; and from some remarks of Mr Boyd, of Innerleithen, on the efficacy of a new salve approved of by several hill-farmers, it would appear that its chief quality consists in destroying vermin.† That vermin ought to be destroyed at all hazards, is obvious, from the circumstance, when sheep of any sort, and especially those on the hills, are annoyed by them, they will not settle to their food, they will rub against any object they can find, a stone, tree stump, earthy bank, or roots of heather, and they thus tear their wool off, and expose themselves more and more to the effects of the weather. They also run about impatiently, and become heated, and then become chilled by the first blast they afterwards encounter. Rain makes vermin more active, and the constant annoyance created by their increased activity, is almost fit to render sheep frantic.

(3019.) Salves are so rife, that I am not sure I have even heard all their names; but these I have noticed:—Common tar and butter, in the proportion of 30 lb. of butter and 20 quarts of tar; common tar 20 quarts, and 30 lb. of palm oil; white tar 20 quarts, and butter 30 lb.; white tar 20 quarts, and hogs' lard 30 lb.; melted butter applied as oil; hogs' lard applied as salve; whale-oil; Taylor's salve; Wilson's dipping mixture; tar and grease; turpentine and grease; arsenic, black-soap, butter, and fish-oil; arsenic, butter, black-soap, and turpentine; oil and tallow. Of all these, every one raises his voice against the use of common tar and butter, or common tar and grease, and white tar and butter partake of the same condemnation. Hogs' lard, oil, and butter, applied by themselves, though liable to be washed away with heavy rains, are held in considerable esteem. The 3 following salves are well recommended, and are considered improvements, by the farmers of Peebles and Selkirk shires, on those in common use.

¾ lb. crude white arsenic, at 7d. per lb.	L.0	0	5¼
28 lb. butter at 5¼d. per lb., or in wholesale 9s. 6d. per stone, of 22 lb., ...	0	12	0
5 lb. black-soap, at 4d. per lb., ...	0	1	8
10 lb. rough turpentine, at 1¾d. per lb., ...	0	1	5½
	L.0	15	6¾

* See some rational observations and convincing experiments of Mr M'Turk of Hastings Hall, in Dumfriesshire, on the impossibility of smearing to promote the growth of wool, in vol. xiv. p. 652-62 of the Prize Essays of the Highland and Agricultural Society.

† Prize Essays of the Highland and Agricultural Society, vol. xiv. p. 62-3.

This compound, with 60 quarts of water, forms a salve for 100 sheep, and costs something more than 1¾d. per sheep. The water being heated, serves to keep the salve in a liquid state during the time it is applied to the sheep; and too much attention cannot be bestowed on stirring the mixture, as the arsenic is apt to fall to the bottom, and equal attention is required to the spreading of the salve evenly on the skin. This salve was proposed by Mr Ballantyne, Holylee, in Peebleshire, and differs from that adopted in 1833, by the addition of the turpentine, which, having adhesive properties, it is said to keep the wool closer over the sheep. Another salve consists of the following ingredients:—

1 lb. arsenic, at 7d. per lb.	L.0	0	7
12 lb. butter, at 5¼d. do.,	0	5	3
3 lb. black-soap, at 4d. per do.	0	1	0
2 bottles of best fish-oil, at 3s. 4d. per gallon,	0	0	6¾
	L.0	7	4¾

These, mixed with 60 quarts of water, will bathe 100 sheep, at a cost of under 1d. per sheep. The recipe for making this salve is this:—"To 12 quarts of water, add 3 lb. of black-soap; and when it comes to the boil add 1 lb. arsenic, and let the whole boil together for 10 minutes. Then add 12 lb. of butter, and the 2 bottles of fish-oil, and boil the whole 5 minutes longer, stirring the mixture all the time. Then pour in as much water as to make the 60 quarts. When used, a little is heated at a time in a pot, as it is too thick for use when cold. It should be poured on the sheep out of a tin tea-pot, and a long-handled tin-ladle should be used for stirring up the mixture in the pot, and pouring it into the tea-pot. The last quantity of the mixture should have a little water added to it, as it will become too strong by reason of the deposition of the arsenic to the bottom." The deposition of the arsenic, like that of the sulphur vivum, alluded to above, should render the efficacy to be derived from its use equally problematical.

(3020.) This salve, when liquid, is put on the sheep in this manner:—"The sheep should be laid on the smearing-stool, fig. 540, in such a position, as when the first shed is drawn from the ear to a little below the loin-bone, the mixture may run towards the middle of the back. This shed will be along the upper part of the ribs, and about 5 or 6 inches from the back, and into it should be poured a double quantity of the liquid. This will serve to saturate all the wool left unopened towards the back. Other 3 long sheds on the side, 2 on the edge of the belly, 2 on the shoulder and neck, with a little on the breast, between the thighs, and the fore part of the hip, will suffice for one side. Then turn up the other side of the sheep, and lay it exactly in the same position as at first, that is, with the back lowermost, to facilitate the liquid mixture running towards it, which it will readily do, Then pour on this side as on the other, always taking care not to carry the sheds towards the tail farther than in a line drawn from the top of the loin to the middle of the thigh. The descent from that line being towards the tail, the liquid will find its way to that part. The principal object to be attended to in the above process is to keep the exposed parts of the animal as free of sheds in the wool as possible, which will thus be done, and yet the whole of the wool be sufficiently saturated with the mixture."* The hinder

* Prize Essays of the Highland and Agricultural Society, vol. xiv p. 67-8.

part of a sheep bears the coarsest wool, and as it always exposes that part to the storm, and as the top of the back is most exposed to falls of rain, these parts being specially kept *free of sheds*, the wool will, with greater difficulty, be thrown open by the wind. When using this liquid, the shepherd should protect his legs with a leather apron, and 1 man and 1 boy will pour 70 sheep in a day; but when every thing is conveniently placed, and no delay incurred, they will pour 80 in that time. It should be remarked, however, that all *animal* oils give a brown tinge, and moss-water, when employed to wash sheep, gives a blue tinge to wool, both of which have to be got rid of in manufacture, though neither may injure the texture of wool.

(3021.) The third salve alluded to, was proposed by Mr Joseph Stewart, of Leslie, in Fife, a practical shepherd. A lengthened experience in the *smearing* of sheep has convinced him that the greater number of the substances usually employed for that purpose, are more or less injurious, both to the sheep themselves, and, of course, to their wool. Such he considers to be the case with tar, turpentine, tobacco-juice, and arsenic; and as to arsenic he regards it as so deleterious as to be wholly inadmissible in any shape whatever into sheep salve, an opinion in which I am quite ready to coincide with him. He has known sheep, after being smeared with a salve, of which arsenic formed a portion, remain in a dull and unthriving state all winter; and when the use of such a salve was persevered in for 3 consecutive years, the sheep so treated almost invariably lost their teeth. To avoid such evils, he had recourse to simpler substances, and found that a mixture of oil and tallow, in equal proportions, answers the purpose exceedingly well. These may be used alone, or in admixture with a small quantity of tar; but he conceives that the oil and tallow of themselves form the best salve for sheep that has hitherto been tried, and is convinced, that if the sheep-farmers of Scotland would use that salve alone, they would find their advantage in obtaining about ⅓ more money for the wool on account of its superior quality, as a natural consequence of the improved condition of their sheep.

(3022.) In applying tar-salve to sheep, when thick, it may be taken up on the fore and middle fingers of the right hand, and spread along the shed, and worked amongst the wool; and when thin, the entire hand in a hollow shape is used for lifting and pouring it on, and working it in. The stool is employed for laying the sheep on, though this convenient implement is not used at all places.

(3023.) Another method of protecting sheep in the hills from the effects of weather, is *bratting*, that is, covering them with a cloth; and as a protection from cold, and a means of keeping the fleece dry, it must be infinitely superior to all the salves that have been invented, at the same time, it will not dispense with a bath for killing vermin; but as a protection to wool, it is as effective as any salve, as has been satisfactorily proved by Mr M'Turk, who bratted one side of 8 sheep, and salved the other side, and completely bratted 2 sheep, without any salve or oil at all, and found the wool derived as much advantage from bratting alone, as from the salve; proving that protection from weather is alone requisite to preserve the wool on sheep in a desirable state. It must be owned, however, that bratting, as formerly practised, imposed a great deal of trouble on shepherds to keep the cloths on, and which, if not done, would expose the sheep to still greater risk from weather, than if they had not been protected at all.*

* Prize Essays of the Highland and Agricultural Society, vol. xiv. p. 660.

"The practice was to sew the brat to the wool of the animal, which, in hands little accustomed to the use of the needle, was both awkwardly performed, and attended with great trouble and loss of time. It was also very inefficient; for the wool to which the threads were attached was liable to be pulled out, and when one part of the brat got loose, it tended to detach the other parts, while, if the sheep happened to be attacked with scab or vermin, it seldom failed to displace its covering, by rubbing against almost every object that came in its way. Even a more frequent cause of displacement arose from the shedding of the wool* in spring, especially if the sheep happened to be in a reduced condition."

(3024.) With regard to the kind of cloth fit for brats, that is very well suited for the purpose which is "made from the refuse wool of carpet manufactories, equally thick and warm as a blanket, and can be got for 6d. per yard. If cloth, such as sacks are made of, be employed, it may be had for 4d. per yard. When intended for bratting hoggs, it should be ¾ wide, and 2 feet will be sufficient to cover 1 hogg. When intended for old sheep of the best description, the brats may be made larger by applying the cloth the long way, and we have then 27 inches of width to cover the back, instead of 24 inches, and it can be cut off as long as the largest sheep will require. The brat should always come as far down the sides as to cover the widest part of the ribs, and all the back, from the tail to the back of the neck. The most expeditious plan of selecting brats is to take an average-sized sheep, and measure the cloth for it."

(3025.) "When the cloth has been applied to the individual animal intended to wear it, the parts should then be marked to which the different straps and strings are to be sewed, to hold the brat in its proper place. A strap must then be fixed to one of the front corners, in a diagonal direction, so as to pass beneath the throat, and sewed to the other corner after it is put on, in the same way as the other straps which are intended to pass through beneath the legs; these must be sewed only at the one end, till the covering be put upon the sheep, and then the other end can be sewed so as to make the brat fit. The straps should be of some soft material, that they may not chafe and injure the skin when the sheep is in motion. If, when made, the brats are dipped in coal tar, it will enable them the better to resist the wet, and to prevent them rotting. If taken care of, they will answer the end for 5 seasons. They ought to be made early in summer, in order that the tar may be dried before the time of using them arrives in November. They ought not to remain on the sheep longer than the beginning or middle of April, according to the state of the weather, and the condition of the flock at the time. We recommend that considerable care should be bestowed in attaching the strings to the proper places in the one first made, so as it may fit well, and serve as a pattern to make the others by. A person accustomed to the needle will make 1 brat in 5 minutes, and it may be put on in less than other five minutes."

(3026.) A brat of woollen cloth, with the strings, will not cost more than 5d., and a flaxen one about 3½d.; but the former will last longer, and answer the end better. The cost for brats will thus be 1d. a-year per sheep. "In

* This phrase, which means the natural falling off of wool in spring, which it will do when sheep have been affected in their health in winter, should not be confounded with the term shedding of wool, in the process of bathing or smearing, which means a temporary separation of the locks of wool, in order to give the liquid access to the skin.

order to prevent them being stolen, and to enable the shepherd to distinguish the flock under his charge, they should all be marked with the buisting-iron, fig. 546, *a*, dipped in white paint."

(3027.) "Some days before the brats are put on for the winter, the sheep must be poured with one or other of the bathing mixtures, to destroy vermin. When the wool has regained its usual appearance after this process, the brat should be put on; and the mode in which we have recommended this to be done, completely obviates the inconveniences arising from the method which has hitherto been practised, of sewing it to the wool; for it will be observed, that the brat is wholly attached to the body, and not at all to the wool, and it, therefore, possesses this decided advantage over the old method, that it cannot be rubbed off, while, at the same time, it preserves the wool from being so, in case of the pile getting weak from disease, or from low condition."

(3028.) "We have found, from our own experience, and we have never heard the fact doubted by any one conversant with the management of sheep, that no salve which has hitherto been tried will afford a degree of protection *equal* to bratting, when it can be sufficiently secured on the animal. We are, therefore, entitled to conclude, that by this treatment the flock will be in higher condition, and if so, the clip will be greater, and the loss by death considerably diminished. When the brat is taken off in April, the wool will be found to have retained the *yolk*, and will appear quite yellow. When examined, it will be perceived to be sappy and sound, and quite free from the defect that wool-staplers call *husky* and *pinny*, that is, dry and brittle, which occasion much loss in the manufacture. When washed, its natural whiteness is quite unimpaired, we would even say increased, from the soap which had been used in pouring, and from the yolk which is retained." *

(3029.) *Letting of tups.* Breeders of tups, besides disposing of them at stated prices, by private bargain, usually appoint a day in autumn for letting, by auction, to the highest bidder, their spare tups for the season, on the condition of being returned in good health and condition at the end of the tupping season. Days I have seen, when L.50 have been obtained for the use of a Leicester tup for the season, when 60 ewes were the most he could serve; now L.5 is nearer the mark, not so much because tups were then so much dearer, but because fine tups are now more generally diffused through the country. There can be no objection urged to this mode of disposing of extra stock, provided the sheep are presented to the notice of breeders in their natural state. Let them be washed and cared for in all possible ways, as to food and shelter, but as to *trimming* sheep, for the special purpose of super-imposing qualities on them which they really do not possess, it is a practice which cannot be too much deprecated; and I am happy to say that most of our Scottish tup-breeders possess greater integrity than to have been lured into the practice. The English alone have fallen into the snare; and in England it is carried even to an absurd pitch, so glaringly absurd, indeed, that not a novice even, of the points of a sheep, but could detect a trimmed sheep. It is necessary to explain to you, that trimming, or dressing, consists of clipping away with the shears the points of the locks of wool on every part of the body where they are considered to injure the appearance, or to affect the symmetry of the sheep; and this is carried to so great a length, that, on close-woolled sheep, such as the Southdown, the trimming is practised over the entire

* Transactions of the Highland and Agricultural Society for July 1843, p. 84-6.

body, so as to produce acknowledged fine points of symmetry, in parts of the sheep in which they are naturally deficient. To call things by their right names, this is nothing less than fraud, intentional fraud, to take in the ignorant and unwary; for no *judge* can be deceived by it, and no one aware of the practice but must detect it. The *eye* may be deceived at first, but the hand will dispel the illusion instantly. For the sake of fair dealing, it is hoped that this fraudulent practice will be abandoned.

(3030.) The *diseases* incident to sheep in autumn are yellows and rot, both arising from a disordered state of the same organ—the liver, and pining. The *yellows* is jaundice, exhibiting yellowness of the eye, the mucous membranes, and the urine. Bleeding, and purging with aloes and calomel, are the appropriate remedies. Fat draft-ewes which have fed some time upon aftermath are most liable to the disease. Inflammation of the liver is the cause, in which the pain of the affected part is very obscure, and the natural language of the sufferer not very expressive, nor is the symptomatic fever marked. Here a striking analogy is noted between the lower animals and man, inasmuch as there is generally a sympathetic pain in the right shoulder, so strongly marked, as often to be mistaken for the principal disorder, and treated accordingly. Whenever you observe, therefore, a lameness of the right leg of any fat sheep on foggage, you may suspect the existence of yellows, and examine the inside of the eyelid, and observe whether any yellowness exists there.

(3031.) The *rot* is a far more serious disease, causing the death of numbers of a flock in a very short period. Deficient food in summer, and a flush of rank wet grass in autumn, occasion ruin to the health and constitution of sheep. In the wet and cold season of 1817, when sheep could not obtain a mouthful of good food in summer, and when the autumn arrived, accompanied with a flush of wet herbage, I knew a farmer in Cheviot who lost 300 Cheviot ewes in the course of 2 or 3 weeks. The early symptoms of rot are very obscure, a circumstance much to be lamented, as in the first stage alone does it often admit of cure. "The animal is dull," observes Mr Youatt; "lagging behind his companions, he does not feed so well as usual. If suspicion has been a little excited by this, the truth of the matter may easily be put to the test; for if the wool is parted, and especially about the brisket, the skin will have a pale yellow hue. The eye of the sheep beginning to sicken with the rot, can never be mistaken; it is injected, but pale, the small veins at the corner of the eye are turgid, but they are filled with yellow serous fluid, and not with blood. The caruncle, or small glandular body at the corner of the eye, is also yellow. Farmers, very properly, pay great attention to this in their examination or purchase of sheep. If the caruncle is *red*, they have a proof, which never fails them, that the animal is healthy. If that body is *white*, they have no great objection or fear—it is generally so at grass; but if it is of a *yellow* colour, they immediately reject the sheep, although he may otherwise appear to be in the very best possible condition; for it is a proof that the liver is diseased, and the bile beginning to mingle with the blood. There is no loss of condition, but quite the contrary; for the sheep, in the early stage of rot, has a great propensity to fatten. Mr Bakewell was aware of this, for he used to overflow certain of his pastures, and when the water was run off, turn those of his sheep upon them which he wanted to prepare for the market. They speedily became rotted, and in the early stage of the rot they accumulated flesh and fat with wonderful rapidity. By this manœuvre he used to gain 5 or 6 weeks on his neighbours."

(3032.) It is alleged, that when sheep have access to salt, they are never known to be affected with rot; and I have as little doubt, that were oil-cake put freely within their power in such a season as I have alluded as having occurred in 1817, that they would escape the malady. Change of pasture from a wet to a dry situation may be the means of even curing the rot, when the change has been effected at an early stage of the disease. All land that has been irrigated in summer, and produces a rank growth of grass in autumn, should be avoided by sheep, as being the very pest-house of rot. Some land, in its natural state, will affect sheep with rot when grazed upon it, such as soft spungy soil, having a clayey tendency, and never free of moisture. In so far as this last cause of rot is concerned, draining has had the effect of rendering land sound which was known to have rotted sheep in a short time. In the prospect of such an evil, when inevitable, sheep cannot be long kept on the same farm, but must be sold in the course of a few months; in short, tenants in their fear, are obliged to keep a flying-stock, especially of ewes, for these are the most easily affected portion of the flock. The application of lime has rendered land sound which was subject to rot, even after it had been drained. Sudden frost and thaw, alternately, in spring, produce rot, according to the old proverb—

"Mony a frost, and mony a thow,
Betaken mony a rotten yow."

(3033.) The liver of rotten sheep always contains the well-known animal the *fluke*, so named from its striking resemblance to a flounder. Its nature has not yet been satisfactorily examined. It was named *Fasciola* by Linnæus, and *Distoma hepaticum* by Rodolphi. Its intestinal ducts contain great numbers of grains of a pale red colour like sand, which are supposed to be its eggs; and as no difference of sex has been observed, it is believed to be a hermaphrodite. It is supposed that its eggs find their way to the grass, from which sheep receive them into their stomach, and thus are supposed to find their way into the liver. The eggs are found in the biliary ducts, in the intestinal canals, and even in the dung of healthy sheep, and they swarm in the dung of rotten ones. Much more could be said on the subject of rot, but I must refer you to authorities.*

(3034.) "*Pining*," says the Ettrick Shepherd, "is a very descriptive Scotticism, from the verb; for no creature can have a more languishing and miserable look than a sheep affected by this malignant distemper. Well may I describe it, for in the last 9 years I have lost upwards of 900 sheep by its ravages. It is quite a new disease on the Border; for I was 20 years a shepherd and never saw an instance that I can remember of with certainty, nor did I ever hear its name save from Galloway, where it was called the *vinquish*, and where it has been prevalent for ages. It was likewise known long ago in some of the districts of the middle Highlands. It is a strange distemper. On the genuine pining farms, the sheep do not take it by ones or twos, but a whole flock at once. It is easily distinguished by a practised eye, the first symptoms being lassitude of motion, and a heaviness about the pupil of the eye, indicating a species of fever. I wish I had science to describe it in a pathological manner, which I have not, and therefore shall not attempt it; but at the very first the blood is

Youatt on Sheep, p. 445-62; Quarterly Journal of Agriculture, vol. v. p. 503, and vol. vi. p. 331-4; Parkinson on Live Stock, vol. i. p. 419.

thick and dark of colour, and cannot by any exertion be made to spring; and when the animal dies of this distemper, there is apparently scarcely one drop of blood in the carcass. It lives till there does not seem to be a drop remaining; and even the ventricles of the heart become as dry and pale as the skin. This simple fact may, I hope, enable men of science to comprehend its nature. It is most fatal in a season of drought; and June and September are the most deadly months. If ever a farmer perceive a flock on such a farm having a more than ordinary flushed appearance by rapid thriving, he is gone. By that day 8 days, when he goes out to look at them again, he will find them all lying, hanging their ears, running at the eyes, and looking at him like so many condemned criminals. As the disease proceeds, the hair on the animal's face becomes dry, the wool assumes a bluish cast, and if the shepherd have not the means of changing the pasture, all those affected will fall in the course of a month."

(3035.) The rationale and cure of this fatal disease, is thus attempted to be given by a writer. "The disease called *pining*, seems to arise from an enervated and costive habit of the animal, which may be produced by a want of proper exercise, in conjunction with feeding in pastures of an astringent nature. The principal districts in which this disease is found to prevail, are the green pastures of the Cheviot mountains, the chain of hills running through the S.W. parts of Roxburghshire, the pastoral districts of Selkirk and Peebles shires, and some other districts of Scotland, as Galloway. Under the old practice of keeping the sheep in flocks, as they are termed *hirsels*, of weaning the lambs in the months of July or August, and afterwards of milking the ewes for 8 or 10 weeks, the pining was unknown in most of these districts; but under this mode of treatment, the sheep were frequently subject to diarrhœa,—a disease diametrically opposed to pining. The farmers of those pastoral districts have generally improved upon the older methods of keeping their sheep. They find it to be more profitable to allow the whole flocks to pasture together indiscriminately and undisturbed. The lambs remain unweaned, until they wean themselves, which generally does not take place till the month of December. By this mode of management, the ewes and lambs are found to be of a higher comparative value than all the sum realized by the sale of cheese made from the milk of the ewes; besides, the ewes are not subject to various accidents arising from so frequently collecting them together for the purpose of milking. But under this undisturbed state of management, in all cases where dry astringent pastures are produced, such as on the syenitic porphyry of the Cheviot range, the pining made its appearance. That such pasture promotes this disease, is further strengthened by the fact, that it is more common in dry than in wet seasons; and most so at that season of the year when, by the influence of the sun, the plants are less juicy; or early in autumn, when the grasses which have pushed to seed become less succulent. This disease is not known on the whole of the clay-slate range of the Lammermoor Hills, where heath prevails, occasionally interspersed with green pastures, and where the *hirseling* practice is pursued. Nor is it known to exist in general on green succulent pastures, or even heaths, growing on calcareous or sandstone grounds, where the nature of the food, and the exertion of the animal in procuring it, appears to counteract the progress of the disease, arising from the inactivity of the digestive function. If we suppose these to be the predisposing causes of this disease, the prevention or remedy will suggest itself, either under the head of food or treatment. Should it be deemed inexpedient to adopt the mode of keeping the flocks in hirsels, a change of place, and, consequently, of food, is necessary to accomplish this purpose. The salutary effects of a variety of food on the animal system are well known. When sheep affected with this disease are put upon a heath, it has frequently the desired effect; but when the animal is much overcome with the disease, its state of languidness may prevent it from taking such a quantity of food as will produce a reaction upon the animal functions. The most effectual cure, therefore, in all cases, is a change to a more rich and succulent pasture."*

(3036.) I may here take the opportunity of mentioning what I should have done in (2606.), when speaking of the diseases of *cattle* in summer, that " inflammation of the larynx frequently takes place, the disease, at the same time, spreading from the delicate lining membrane to the nearest parts. In this way lymph is effused, and the play of the parts impeded. Sometimes the smaller cartilages themselves are altered, being thickened and contorted, and small tumours are apt to be produced, both within the tube and without it. The marked symptoms are local pain, difficulty in breathing and swallowing, and general fever: the treatment required is venesection, and the other parts of the antiphlogistic regimen. Tumours occurring in this locality in cattle constitute the disease called *clyers*, which, though it may not for a time interfere with fattening, yet speedily injures health.†"

---

78. OF LIFTING AND PITTING POTATOES.

" If planted in fair rows,
They marshalled grew, the plough will best perform
The reaping task: amid the tumbling soil,
The vegetable mine, exposed to view,
The gatherer's basket fills."

GRAHAM.

(3037.) The harvest-work of a farm cannot be said to be completed until the potato-crop is taken out of the ground and secured against the winter's frost. By October the potatoes may be expected to be ready for lifting. The fitness of potatoes for lifting is indicated by the decay of the haulms; for as long as these appear at all green, you may conclude the tubers have not yet arrived at maturity. In an early season potatoes will ripen before October; and, though the weather should continue fine, the best plan is to let them remain in the ground till the corn-crops are entirely harvested. Immediately after the fields are cleared of corn, the potatoes should be taken up and secured, to allow the land to be ploughed up for wheat; indeed it will happen, under all

* Quarterly Journal of Agriculture, vol. ii. p. 697, and 707.

† Dick's Manual of Veterinary Science, p. 76-7.

circumstances of weather, that the corn will be ready to be cut down and carried in before the potatoes are fit to be taken up.

(3038.) There are 2 modes of lifting potatoes, namely, with the plough and with the potato-graip. The plough is the most expeditious, though I believe the ground is best cleared of tubers with the graip. With either instrument a large number of people to gather the potatoes are required, each of whom should be provided with a small semi-spheroidal-shaped basket, with a bow-handle, to gather the potatoes into, and then to put them into sacks or close-bodied carts. When a farmer lifts potatoes on his own account, they are usually put into a cart and carried direct to the pits. When he lifts them on account of a purchaser, or a number of purchasers, they are measured on the spot from the basket, and put into sacks, in which they are easily delivered. When lifted for shipment to the London market, they are first riddled into sizes, then measured or weighed on the spot, and put into carts, and taken to the ship's side. The potato-riddle is made of wire, with meshes from $1\frac{3}{4}$ inch to $1\frac{1}{2}$ inch square, and, if rimmed with oak, costs 2s. 6d. each. The riddlings, or small potatoes, are used on the farm. Potatoes are usually sold by weight, and a given weight represents the measure of a boll; which boll, again, differs in weight in different parts of the country, thereby rendering the imperial weights and measures act, in this instance, nugatory. A boll is 20 stones of 14 lb. to the stone, in some parts of Scotland, and it is 40 stones in others; and, to contradistinguish them, the small boll is called a *single*, and the large a *double* boll. The produce of a crop of potatoes varies amazingly,—from 30 to 100 single bolls per imperial acre. It is singular that the price does not vary nearly so much; from 4s. to 6s. the single boll, being the limits between dear and cheap prices. Taking 60 bolls as the medium produce, and 5s. as a medium price, the gross return from an imperial acre of potatoes will be L.15. If the expense of lifting the crop is taken at 30s. per acre, the return will be L.13, 10s.; a large sum certainly, but then it should be remembered that potatoes leave no straw for manure, and require, on the contrary, a large quantity of manure to raise even a tolerable crop. They incur considerable trouble in their delivery; and, being a perishable commodity, cannot be kept beyond a given time.

(3039.) In employing the *plough* to take up potatoes, the common one, with 2 horses, answers well; but as the potatoes run the hazard of being split by the coulter when it comes in contact with them, it should be taken out, the sock being sufficient to enter the plough below the drill, and the mould-board to turn them out of it. The plough in going up splits one drill, and in returning splits the next, but no faster than a band of gatherers, of field-workers, if numerous enough, but if not, assisted by hired labourers, can clear the ground of them into the baskets. In free soil potatoes are easily seen and picked up ; heavier soil is apt to adhere to them, in which case it is a good plan to make a stout field-worker shake those portions of the earth turned up by the plough, which still adhere in lumps, with a potato graip, and expose the tubers. Potato-gathering should not be continued so late in the evening as the tubers cannot be easily seen ; nor should it be prosecuted in wet weather, which causes the earth to adhere to them, and renders them undistinguishable from the earth itself. Of course every one, the smallest tuber, should be taken off the ground, not only on the score of economy, to realise the whole crop, but to remove them as a weed from among the succeeding crop. After the field has been gone over in this manner, the harrows are passed a double tine to bring concealed tubers to the surface, when they are gathered by the people, and to shake the haulms free of soil. These after-gathered potatoes are usually reserved for pigs and poultry. Whenever the field is cleared of the crop, the haulms are gathered by the field-workers and carried to the compost stance, to be converted into manure, and these are the only return which the potato crop makes to the soil.

(3040.) A simple instrument, fig. 541, which may be substituted in the plough for the mould-board, for turning potatoes out of the drill, was contrived by Mr John Lawson of Elgin. It consists of 6 malleable iron bars, the outer ones ⅝ of an inch square, the inner ½ an inch in diameter, joined together in the form of a brander, 26 inches long from *a* to *b* ; 5 inches in breadth from *b* to *c*, at the fore part, where is a plate of iron ; 27 inches in length from *c* to *d* ; and 18 inches in breadth from *d* to *a*. This brander is attached to the right side of the head and stilt of a plough, in lieu of the mould-board, by the screws *e*, *e*, the fore-end *b c*, being placed close behind the sock, as seen at *a*, fig. 542, which shews the plough mounted with the brander, having its upper angle at *a*, fig. 541, 8 inches, and the plane of its face so bent down as to have the lower angle *d* only 4 inches above the sole of the plough. The openings between the rods will thus be rather more than 2½ inches at the widest end of the brander, between *a* and *d*. The mode of operation of the brander is, that while the earth partly passes through it, and is partly placed aside by it, the potatoes are wholly laid aside, so there are few of them but are left exposed on the surface of the ground.

Fig. 541.

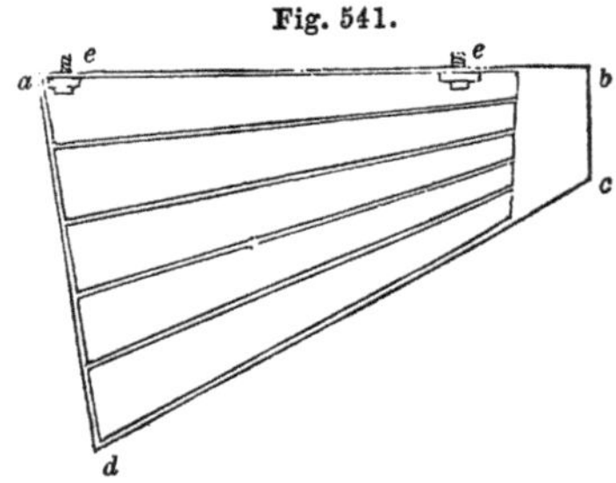

THE POTATO-RAISER, OR BRANDER.

(3041.) "This plough," says Mr Lawson, "is drawn by 1 pair of horses in the same manner as the common plough. In working it, the plough-

Fig. 542.

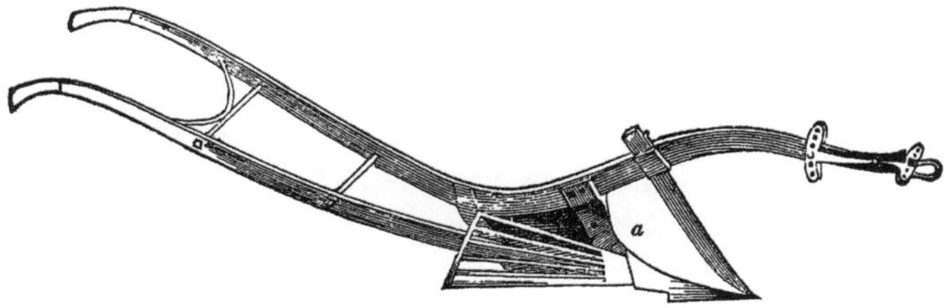

THE POTATO-RAISER ATTACHED TO THE PLOUGH.

man inserts it into the potato drill so as to have the whole of the potatoes on his *right-hand side.* He then proceeds along the drill, splitting it up in the common way. The earth is thus thrown to the right-hand side, and the potatoes lie scattered on the surface of the ground behind the plough. Women follow, provided with baskets, into which they gather the potatoes, and throw the stems upon the drill which lies to the right hand of the one from which they are gathering the potatoes. The reason why the potato-stems are thus removed, is, that as soon as the potatoes which lie on the surface are gathered, the plough returns, and again proceeds as before, through the part of the drill in which the potatoes lay, still turning the earth to the right-hand side. This second operation raises to the surface any stray potatoes which the first may not have turned up, and the potatoes thus raised are gathered by the women who attend for that purpose. The second operation may be delayed till about 12 drills are turned over by the first operation, and the potatoes gathered. The plough may then be put through these 12 drills the second time. By this there will be a saving of labour, as a smaller number of women will gather the potatoes by the second operation, while the greater number of them may remain with another plough in its first movement. If the stems of the potatoes be very *strong and luxuriant,* a few of the women might be directed to go along the drill and pull them out of the ground, at the same time plucking off any potatoes that may adhere to and come along with the stems. If this is done, the plough will turn up the greater part of the potatoes by going only *once* through the drill; but, in going twice, it will do it in the most satisfactory manner. A man with 1 pair of horses will thus pass over the ground as quickly as with the common plough. In light soils this plough performs its work in a very efficient manner. It pulverises the soil in an extraordinary degree, and scarcely leaves a single potato in the soil. I have never before been able to clear my fields of potatoes so effectually as by this implement, or at nearly so small an expense."*

(3042.) When potatoes are taken up by manual work, it is done with the *potato-graip*, fig. 543, the prongs of which are flattened. Being rather severe work to use this graip, men are employed for the purpose, 1 man taking 1 drill, close beside that of his fellow-workmen, while 2 gatherers to every man are ready to pick up the potatoes he turns out into the baskets. In using the graip, it is inserted into the side of the drill, and below the potatoes, with a push of the foot, and the graipful of earth thus obtained is turned on its back into the hollow of the drill, exposing the potatoes to view on the top of the inverted earth, from whence they are gathered. The men then pass the prongs of the graip here and there through the inverted graipful and the soil on the drill, to detect and expose to view every tuber lurking beneath the soil. In this manner, 1 man and 2 women will take up, of a good crop of 80 bolls of 20 stones per acre, 20 bolls a day, which will cost 3d. per boll at the following wages:—1 man 2s., and 2 women at 1s. 6d. each per day, without food, = 5s. per day, or 20s. per acre. Such graips cost 2s. 3d., and when handled 3s. each.

Fig. 543.

THE POTATO-GRAIP.

(3043.) The apprehensions excited by the failure in the potato crop, have caused farmers to bestow particular pains in procuring proper seed for their crop. Those who believe the failure to be a consequence of the degeneracy of the constitution of the potato try a frequent change of seed, and inquire for new varieties of approved properties, and those who believe the attainment of great ripeness to be a means of weakening the constitution of the plant, will endeavour to lift the part of his crop intended for seed at an earlier period than the rest, before, in short, it shall have attained maturity. This can easily be done, and the potatoes intended for seed stored in such a way as to preserve them from decay. I am in the belief, that the *small* potatoes of a crop of large tubers would answer the same purpose, because, being the latest growth of the crop, they are formed by the latest efforts of the tuber-bearing fibres, and are, therefore, latest at arriving at maturity. I particularly allude to small potatoes among large ones, because a crop of uniformly small potatoes may be as fully ripened, and be tainted with failure as readily as large ones. Instances are not wanting to prove the value of small potatoes producing large crops.

* Quarterly Journal of Agriculture, vol. viii. p. 551-2.

(3044.) Where farm-servants have potato ground granted them as part of their wages, their crop is taken up at the same time as the rest of the field, and the cost of taking it up, of course, falls upon their master. Where servants are supplied with an annual stipulated quantity of potatoes, instead of a certain space of ground, it is delivered to them as taken up from the field. The quantity usually stipulated for is 7 or 8 bolls, of 40 stones each. The quantity should, of course, be measured or weighed, but the body of the cart is made usually the instrument of measurement; and, I believe, in all such cases, servants prefer taking their chance, well knowing they are generally dealt with in no illiberal spirit. Of the two modes of paying servants, in seasons prolific of the potato, the servants who have ground planted are well supplied with this wholesome vegetable, but in bad seasons they suffer considerable privation, and they bear it patiently, because they know the deficiency to have arisen from no circumstance over which their master has any control; and those, on the other hand, who receive a stated allowance of potatoes every year, also suffer in a bad year, by deterioration in their quality. The former class of servants have a direct interest in the quantity, but both have a direct interest in the quality of the crop. While alluding to allowances to servants, I cannot help hinting, that a few potatoes, no more than will fill the body of a cart, will prove an acceptable gift, at this season, to the poor lone cottar who works in your fields, or whose daughter fulfils the useful office of field-worker.

(3045.) In regard to the *storing* of potatoes, there is no difficulty in the early part of winter, when a low temperature prevails, and vegetation is lulled into a state of repose. Potatoes may therefore be kept in almost any situation in the early part of winter; but then, if damp is allowed to surround them for a time, it will inevitably rot them, and if air finds easy access to them at all times, the germ of vegetation will be awakened in them at the first call of spring. To place potatoes beyond the influence of those elements as long as convenience suits, they should be stored in a dry situation, and be covered up from the air; and no mode of storing affords most ready means for both those requisites than the ordinary forms of pits in dry soil.

(3046.) Fig. 544 shews the 2 different forms of ordinary potato-pits, the one being conical, the other prismatic in shape. The conical form is usually employed for pitting small quantities of potatoes, and is well suited for small farmers and cottars; the prismatic is the form commonly adapted for storing large quantities. For both sorts, a situation sheltered from the north wind should be selected, and the ground should either be so dry of itself as to absorb the rain as it falls, or so inclined as to allow surface water to pass away quickly from the site of the pits. The site should be conveniently situated for opening the pits and admitting carts to them, and so near the corner or side of a field as not to interfere with its being wrought in winter.

(3047.) A *conical* pit of potatoes is formed in this manner :—If the soil is of ordinary texture, and not very dry, let a small spot of its surface be smoothened with the spade. Upon this spot let the potatoes, as they are taken out of the cart, be built by hand in a conical heap, not exceeding 2 feet in height; and the breadth which a cone of that height will occupy, so as not to impose much trouble in piling up the potatoes, will not be less than 4 feet, and is more likely to be 5 feet. The potatoes are then covered with a thick thatching of dry clean straw. Earth is then dug with a spade from the ground in the form of a trench around the pit, the inner edge *a*, fig. 544, of the trench being as far from

Fig. 544.

THE CONICAL AND PRISMATIC FORMS OF POTATO-PITS.

the pile of potatoes as the thickness of the covering of earth to be put upon it, which is considered sufficient at 1 foot. The first spadeful is laid around the potatoes on the ground, and the earth chopped fine and beaten down with the spade, in order to render the earth as impervious to cold as possible, and the drier the earth is, the less effect will frost have upon it, and the less distance will it penetrate through it. Thus spadeful after spadeful of the earth is taken from the trench and heaped on the straw above the potatoes, until the entire cone *a b c* is formed, which is then beaten smooth and round with the back of the spade. The top of the cone at *b* will then be about 3 feet 3 inches in height, and the width of the cone from *c* to *a* about 7½ feet. The trench round by *a c* should be cleared of earth, that no surface water may lie near the pit, and an open cut should be formed from the lowest side of the trench to allow the water to go away most freely.

(3048.) When the soil is naturally thoroughly dry, the site of the pit may be dug out of the solid ground a spade depth, for storing the potatoes into what would then really be a pit, and then the apparent height of the pit above the surface of the soil will be small; but unless the soil is as thoroughly dry as sand or gravel can make it, the potatoes should be piled upon the natural surface of the ground.

(3049.) The *prismatic*, or *long* pit, *d e f g*, fig. 544, is formed exactly in the same manner, with the exception, that the potatoes in it are piled in a straight line along its sides *d h*, instead of round, as in the case of the conical pit *a b c*. The height of the pile of potatoes should not exceed 2½ feet, and its breadth will spread out to about 7 feet, and allowing 15 inches for thickness of straw and earth, the height of the finished pit will be 3 feet 9 inches, and breadth 9 feet 6 inches. The direction of a long pit should always be N. and S., in order to place both its *sides* within reach of the sun's rays.

(3050.) It is considered that when fresh potatoes are heaped together in large quantities, that a certain degree of fermentation ensues, which increases the temperature of the mass so much as to awaken vegetation in the tubers, and the existence of long sprouts so frequently found covering the top of the heap of potatoes when a pit is opened in spring, is adduced as a proof of the effects of such an increase of temperature; and the thick covering of dry straw usually placed above potatoes in pits is also considered a great means of retaining within the bounds of the pit the heat evolved from the potatoes, and hence a plan has been suggested, of having openings left along the upper parts of pits through which such heated air may escape. The openings are left by drawing small bundles of long straw, tied at one end with a piece of cord, and cut square like the rosette on the top of a corn-stack *a*, fig. 534, and which are placed upright, and project upwards at short intervals along the top of the pit, before the earth is thrown upon the straw near its ridging, and the earth is put round, and beaten down, and finished smoothly, beside the projecting part of each bundle. These vents may be seen at *i* and *k*, in fig. 544.

(3051.) If these apprehensions in regard to the heated state of potatoes in pits have any foundation, the only way to remove them is to place potatoes either in small independent quantities, such as in conical pits, or in a low heap in a long pit; and there are many substances beside straw by which potatoes may be covered in a pit. The use of straw is not solely to keep the earth which covers them from the potatoes, but to act as a non-conductor to frost, when it should happen to become severe, and when it would penetrate through even 1 foot of earth; and if straw is useful in this latter capacity, it will, of course, act as strongly as a non-conductor to internal heat as against external cold. But, perhaps, such apprehensions in regard to the effects of frost in a pit of potatoes are groundless; for, to preserve potatoes cool during winter, it has been recommended to mix earth amongst them in the pit, and to have no covering above them but earth, in imitation of the state potatoes happen to be left in the soil all winter, and found in spring quite fresh and good, and in which state they make the best seed of any. The rain, the frost, and snow, it is alleged, would thus bear the same relation in winter to potatoes in so prepared a pit as they do to those left by chance in the field. Potatoes, I know, have been pitted amongst earth, and they have kept well enough all winter, but the experiments were conducted on a small scale. How a whole crop of potatoes could be conveniently stored in that manner on a farm admits of doubt; besides, the analogy of this mode of pitting, and of preservation in the field all winter, is far from strict, when the different conditions of the potatoes are taken into account. In the pit, one layer of potatoes lies above another, with intervening layers of earth, a relation in which, that if one potato rotted it could not fail to affect those lying immediately above and below it; whereas in the field, every potato lies singly and in one layer, surrounded by earth, independently of others, and whether it becomes rotten or continues sound, it can not possibly affect any other potato in the field. Nor is the state of potatoes improved when removed to cellars or outhouses; for the probable dampness in the cellar will cause many potatoes to rot, and constant exposure to the air, if not directly conducive to vegetation, will cause evaporation of the water in the tuber, and its consequent shrinking, Until, therefore, experience proves the superiority of a better plan, the old one must be followed; and if certain bad effects have of late years been experienced by implicitly following the old plan, such as heating, in consequence of the heaps having been made too large, modifications should be adopted in the construction of pits so as to avoid the evils complained of. I think that if potatoes are not placed together in larger heaps than I have mentioned above, there will be little risk of their heating in the pits.

(3052.) And yet those who doubt the tendency of the potato plant to degenerate, have never been able to explain the different effects which the same treatment produces now that it did 20 years ago,—namely, that let the pits then be formed as large as you please, and the cut-sets of potatoes then heaped on the barn-floor as high as you please, and for as long and as short a time as you please, before they were planted, not a word was then heard of the disease called the failure; but if there is any truth in the conjecture formed by many persons, that the failure is to be ascribed to the common mode of pitting the potato, the conclusion is irresistible, that some change must have come over the potato, since there is none in the pitting. Nevertheless, I do not deny that potatoes heat in the pit, and in heaps of cut-sets in the barn; but what I want to know is, *why* potatoes heat now in the pit, and in heaps in the barn, when no such effect was known to exist 20 years ago? Let the treatment of keeping the potato in winter, and of the cut-sets, by all means, be changed to suit the altered state of the tubers; but the adoption of a safe change in practice should not induce us to shut our eyes to the necessity which caused the adoption of so great a change, or to content ourselves in the belief of secondary causes, because the investigation, to be satisfactory, happens to be difficult.

(3053.) A foreign cultivator "found, that 5 weeks after he had harvested a crop from a field planted with diseased potatoes, they began to undergo a dry corruption, and that, even if, externally, they had a sound appearance, they had *internally* a number of the *blue spots*, called stagnation spots, which, when the potatoes were *boiled*, *remained hard*, were rejected by cattle, and which could not be used for the manufacture of brandy, as besides being unsuited for the purpose, the potatoes would not go through the crushing-mill."* It will be remarked, that these are precisely the effects of the disease called the failure as recognised in our own potatoes, and they may be regarded as infallible means of detecting whether tubers are diseased or otherwise.

(3054.) It is truly remarked by Sinclair, that "the varieties of the potato are very numerous, and the confusion of their names inextricable." Without attempting to particularise any one variety, I would say generally, that as an article of food, the potato is now universally cultivated, and this, no doubt, chiefly from the facility with which it may be raised, the pleasantness of its taste, the simplicity of its cookery, and the nourishment which it affords. According to a statement of Sinclair's, it appears that in several varieties of potato fit for human food, the nutritive matter varies from 200 to 260 grains in 64 drachms; that those quantities of nutritive matter contained from 169 to 204 grains of starch, and from 31 to 61 grains of albumen, mucilage, and sugar.†

(3055.) As compared with grain, "Cobbett's assertion, that wheat produces more nutritious matter per acre than potatoes, is now completely disproved. Later experiments have shewn, that it is very near the truth to estimate the proportion of the nutritive power of wheat to that of potatoes as about 7 to 2; or, in other words, 2 lb. of wheat affords as much sustenance as 7 lb. of potatoes, though it may be doubtful whether it affords so much nourishment; for, by cal-

* Translation in the Journal of Agriculture for March 1844, p. 395, from the *Livlundische Jahrbucher*, an agricultural periodical, published at Dorpat and Moscow.

† Sinclair's Hortus Gramineus Woburnensis, p. 409.

culating the produce of the two crops, it has been determined that 1 acre of wheat will produce sustenance for 3 persons, as long as 1 acre of potatoes affords it to $6\frac{5}{8}$ persons. This is on the supposition that the power of nutrition of a plant is only in proportion to the quantity of farinaceous and glutinous matter contained in it. But this is by no means certain. We have not been able to discover what it is that renders one substance more proper for food than another . . . . In estimating the amount of aliment afforded by potatoes and grain, we should rather calculate according to the mass of vegetable matter capable of satisfying a full grown-person. . . . . I may add, that potatoes, from some peculiarity in the mixture or degree of condensation in their nutritive matter, seem to possess an advantage over all grains as constant food. It is well known that potatoes and water alone, with common salt, can nourish men completely; we have hundreds of instances, in many parts of Ireland, where the people have lived constantly on this diet from necessity, and yet have been robust, healthy, and long-lived as persons fed plentifully on animal food. Now, I know of no meal derived from grain used as the general food of any nation, without being mixed with other kinds of alimentary matter, as oil, fruits, whey, or milk."*

(3056.) The potato has been subjected to careful analysis by different chemists, and these are a few of the results :—

KINDS.	Starchy Fibrin.	Starch.	Vegetable Albumen.	Gum.	Acids and Salts.	Water.	Analyst.
Red Potato, . . .	7.00	15.00	1.40	4.1	5 1	75.0	Einhoff.
... ... germinated,	6.80	15.20	1.30	3.7	...	73.0	...
... ... sprouts, .	2.80	0.40	0.40	3.3	...	93.0	...
Kidney	8.80	9.10	0.80	...	...	81.3	...
Large red	6.00	12.90	0.70	...	...	78.0	...
Sweet	8.20	15.10	0.80	...	...	74.3	...
Peruvian	5.25	15.00	1.88	1.87		76.0	Lampadius.
English	6.83	12.91	1.04	1.70		77.5	...
Onion	8.38	18.75	0.90	1.66		70.3	...
Voigtland	7.13	15.41	1.25	1.95		74.3	...
Paris	6.79	13.30	0.92	3.3	1.4	73.12	Henry junr.

Of these constituents of the potato, it will be observed how large a proportion the water bears; and the remainder is chiefly composed of fibrin and starch. Hence, potato differs essentially from wheat and barley, by containing no gluten, and approaches in some measure to the nature of rye.† Of the dry constituents, starch is frequently used in the manufacture of wheaten bread, the potato-flour giving to the loaf, when stale, a tendency to excessive crumminess. Potato-

* Quarterly Journal of Agriculture, vol. v. p. 341-2, a paper enumerating the various uses to which the potato may be applied.

† Thomson's Organic Chemistry of Vegetables, p. 840-2.

flour does not injure bread as an article of food; but still it is an adulteration, and its use constitutes a fraud on the public, when the same price is demanded for the loaf, partly made of this cheap material, as for one of wheat alone.

(3057.) Potato-starch may be converted into *tapioca*. The tapioca of commerce is derived from cassava, a preparation made in the West Indies, the tropical regions of America, and on the African coast, from the root of the manioc, *Jatropha manihot*, an euphorbaceous plant. "The Cassava cakes sent to Europe, which I have eaten with pleasure," says Dr Ure, "are composed almost entirely of starch, along with a few fibres of the ligneous matter. It may be purified by diffusion through warm water, passing the milky mixture through linen cloth, evaporating the strained liquid over the fire with constant agitation. The starch dissolved by the heat thickens as the water evaporates; but, on being stirred, it becomes granulated, and must be finally dried in a proper stove. Its specific gravity is 1.530—that of the other species of starch. The product obtained by this treatment is known in commerce under the name of *tapioca;* and being starch very nearly pure, is often prescribed by physicians as an aliment of easy digestion. A tolerably good imitation of it is made by beating, stirring, and drying potato-starch in a similar way."* An acquaintance of mine, a farmer in Forfarshire, Mr James Scott, Beauchamp, instead of disposing of his potatoes, of which he used to raise large quantities every year, converted them into tapioca; thereby saving a great deal of carriage which the delivery of potatoes occasions, and retaining the refuse of the manufactured potato, upon which he fed a large number of pigs, and which, moreover, afforded a good mash to the horses.

(3058.) "When potatoes are boiled, they lose from 1 to 1½ per cent. of their weight. The juice which may be separated from them is sweet-tasted. The meal is insoluble even in boiling water, though potato-starch forms a transparent solution with hot-water. Thus, it appears, that by boiling, the albumen, fibrous matter, and starch, combine together, and form an insoluble compound." Simple as the matter appears, it is not every cook who *can* boil a potato *well.*

(3059.) "When potatoes are exposed to the action of frost, it is well known that they become soft, and acquire a sweet taste. This taste is succeeded by a sour taste, owing to the rapid evolution of acetic acid, and the root soon passes to putrefaction. From the experiments of Einhoff, we learn that the sugar is formed at the expense of the mucilage; for the other ingredients were found in potatoes sweetened by frost, in the usual proportion. He considers this sweetening process as connected with the vegetative powers of the root." "Dr Peschiez of Geneva has described the presence of mucous sugar and of gum in the potato. This explains why it is capable of undergoing the vinous fermentation." The acids contained in potatoes in the natural state were ascertained by Einhoff to be a mixture of the tartaric and phospheric acids. He also obtained from 1820 parts of dried potatoes, 96 parts of a greyish white ash. Of these, 64 parts were *soluble* in water, and 35 *insoluble*.† A minute analysis of the ash of potatoes is given by Professor Johnston, including that of the haulms. In 10,000 lb., there were found of—

* Ure's Dictionary of the Arts and Manufactures, art. *Cassava*.

† Thomson's Organic Chemistry of Vegetables, p. 840.

	In the Roots.	In the Tops.
Potash,	40.28 lb.	81.9 lb.
Soda,	23.34	0.9
Lime,	3.31	129.7
Magnesia,	3.24	17.0
Alumia,	0.50	0.4
Oxide of Iron,	0.32	0.2
Silica,	0.84	89.4
Sulphuric Acid,	5.40	4.2
Phosphoric Acid,	4.01	19.7
Chlorine,	1.60	5.0
	82.83	308.4

It will be observed what a large proportion of potash, lime, and silica, the tops of the potato plant yield. "These roots," as observed by Professor Johnston, and the observation applies to turnips, carrots, and parsnips, as well as to the potato, "contain very much water, so that, in a dry state, the *proportion* of inorganic matter present in them, is very much greater than is represented by the above numbers. I have, however, given the quantities contained in the crop as it is carried from the field, as alone likely to be of practical utility. The crops of these various sorts vary very much in different localities, being in some places twice, and even thrice, as much as in others,—every 10 tons, however, which are carried off the ground, contain about 9 times the weight of saline and earthy matters, indicated by the numbers in the table."*

(3060.) *Solanina* was discovered by M. Desfosses in the fruit of the common potato, and M. Otto discovered it also in the potato itself, after it had been allowed to germinate, and this substance is an acrid narcotic poison. "Its existence in the potato after germination," says Dr Thomson, "is an interesting fact, and should lead to the cautious use of that important vegetable after germination has commenced."

(3061.) "It is well known that a spirit can be extracted from potatoes. From this spirit Messrs Bertillon and Guietand extracted a volatile oil, being a colourless limpid liquid, having a strong smell, hot acrid taste, and being very soluble in alcohol."†

(3062.) The particles of potato starch are irregular ellipsoids, varying in size from $\frac{1}{300}$ to $\frac{1}{3000}$ of an inch; those of wheat being separate spheres of $\frac{1}{1000}$ of an inch; so that it would be an easy matter to detect potato-flour amongst wheaten flour with the microscope.

* Johnston's Lectures on Agricultural Chemistry and Geology, p. 322-3.

† Thomson's Organic Chemistry of Vegetables, p. 264-5, and 481.

79. OF SOWING AUTUMN WHEAT, AND OF THE CONSTRUCTION AND PRINCIPLES OF AGRICULTURAL WHEEL CARRIAGES.

> "How ceaseless is the round
> Of rural labour! Soon as on the field
> The withered haulms and suckers crackling blaze,
> And with their far-extending volumes, load
> The wings of Autumn's latest lingering breeze,
> The WHEATEN SEED-TIME all your care demands."
>
> GRAHAM.

(3063.) How ceaseless, indeed, is the round of rural labour! No sooner does the farmer secure his crop, the progress of which towards maturity has excited his most lively solicitude during the course of a whole year, than he begins to sow the succeeding one, and strives to prepare as much land for it before winter sets in as he possibly can secure. The crop usually sown in autumn is *wheat*, that plant being able to bear the vicissitudes of winter in our latitude, though barley and tares are also sown at the same time in some parts of England, where they stand the winter well, but neither can withstand a Scottish winter.

(3064.) You have just seen (3059.), that the constituent parts of the haulm of potatoes consist of large proportions of potash, lime, and silica—ingredients useful to many of the crops which follow in rotation after a summer-fallow; and though incineration is the means by which the chemist discovered those substances in potato-haulms, yet I would advise you to make a compost of them, as I did before (2040.), rather than convert them into a "crackling blaze," as described in the motto, inasmuch as their inorganic constituents can be secured equally well by fermentation as by incineration, while their organic constituents are not dissipated, but thereby preserved for use.

(3065.) You will recollect that we left off the working of summer-fallow after the land was dunged, (2823.), and when the land was to receive no lime. It is now our business to finish the summer-fallow, by the sowing of *autumn-wheat*, the crop for which the land was specially prepared by fallow. The first process is the levelling of the drills which cover the dung, by harrowing them across a double-tine; and, unless the land is of very strong clay, 1 double-tine will be sufficient for the purpose. After the land has been harrowed level, any root-weeds that have been brought to the surface should be removed, but surface-weeds will soon wither in the sun and air. The land should now be feered, to be gathered up (824.), fig. 133, into ridges, and if it has been thorough-drained, or is otherwise sufficiently dry, 1 gathering-up will make a good seed-bed; but should it be too wet for once gathered-up ridges to lie safe all winter, the ridges should be twice gathered-up (839.), fig. 139. The second gathering-up, however, should not be done immediately after the first, but after such an interval of time has elapsed between them as to allow the land to subside, and the subsidence will be much hastened by a fall of rain. Should the fallow-land have been dunged in the usual way, that is, the dung spread upon the surface, and ploughed in, with feered ridges, the feerings should have been so made as to leave a ½ ridge at the side of the field, that now, when the land is about to be ploughed for the seed-furrow, the ½ ridge is converted into a whole one, and the field thereby made fit to be continued in that form.

(3066.) But a practice has come into use since the introduction of the grubber, Plates XXIX. and XXX., and from (2428.) to (2448.) inclusive, which possesses some advantages, on strong land, in certain circumstances, which is to cover in the sown wheat with the grubber, instead of first ploughing the land for a seed-furrow, and to finish the work with a slight harrowing. When the grubber is contemplated to be thus used, the land, in summer-fallowing, should be feered in ridges, as if for the seed-furrow, when covering in the dung, as the subsequent grubbing will not alter the form of the ridges further than to reduce their crowns, and fill the open-furrows a little. When a tough waxy clod is expected to arise on ploughing strong land, rather wet below, for a seed-furrow, or when there is appearance of unsettled weather, the grubber will be found useful in keeping dry the meliorated soil on the surface, and also in getting quickly through with the wheat-seed.

(3067.) Land that has grown *beans* is usually gathered-up for the seed-furrow at once, and sown immediately, as the season is far advanced by the time the bean-crop is carried in and stacked; but the seed-bed would certainly be in a better state for wheat, were the land allowed to subside for a few days before the seed is sown. Where the land is strong, and the ridges are sufficiently rounded, the grubber may be used on bean-stubble, as on fallow-land, for covering in the wheat sown broadcast on it, but for this purpose a 4-horse grubber is required, the surface being finished with a slight harrowing. The grubber will succeed in this very well, in as far as the wheat is concerned, and it has a great advantage, in a late autumn, in getting through the work expeditiously; but on strong land, not thorough-drained, and in a comparatively flat state, grubbing will not succeed on bean-stubble; at least, the wheat seed would incur considerable risk of being rotted in that state of the soil all winter, and the land itself of becoming soured; and the land would not escape the effects of such an injury, even though it it should be fallowed the succeeding year.

(3068.) The land which has grown *potatoes* should be ploughed for a seed-furrow, because, having been laid flat for the culture of this summer crop, the grubber cannot be employed with advantage in the case, as that implement leaves the land as flat as it found it. The land only receives 1 furrow after potatoes; and it should have time to subside a little, though the usual practice is to sow the wheat upon it as soon as ploughed. The reason why I have frequently recommended the subsidence of the land before being sown with seed is, that wheat thrives better in soil having a little firmness in it than in the loose state the plough leaves it.

(3069.) When land is naturally strong enough to grow wheat, but is somewhat soft, and so wet below as to make it apt to throw out the wheat plant in spring, the best plan is to make a seed-bed by *ribbing* with the small plough, fig. 376 and (2185.). The wheat is sown broadcast over the ribs, and harrowed in with a double-tine along (2208.). The ribbing gives the wheat a deeper bed in the soil than mere harrowing, and a deeper hold of the soil in spring, and it has also the advantage of stirring only the dry surface soil for the seed-bed. It can only be practised, however, on land that has been ridged up for a seed-bed for a considerable time, as the small plough does not make good work on new ploughed land, it, small as it is, going too deep, and making the drills too wide; and it is never employed on land that has not been ridged, being unfit to turn up land in a hardened state.

(3070.) Another mode of preventing the throwing out of the wheat plant in spring on soft land, otherwise well adapted for wheat, is, first, to feer the land into ridges, sow the wheat broadcast between the feerings, and cover it in with a light seed-furrow, leaving the land unharrowed and rough all winter.

(3071.) Whether harrowed or not, it is of great importance to leave wheat land *rough* all winter, that is, with a round large clod upon the surface. These clods afford shelter from wind and frost to the young plants, and when gradually mouldered by frost, serve to increase the depth of the loose soil, and protect the roots of the plants from frost. Indeed, whenever the land is harrowed as fine as meal with autumn wheat, the rain never fails to batter its surface into a crust, and the frost heaves it up in spring like fermented dough; and this action raising the plants along with the soil, they are exposed on the surface after the frozen earth has subsided, on being thawed by sunshine (2211.). Such an effect, however, can only occur where a considerable quantity of moisture exists under the surface of the soil, ready to be acted on by frost. Draining, then, is the only safeguard against the young wheat plant being thrown out of the earth in spring. As the ground is desired to be left in a rough state in autumn, no use is, of course, then made of the roller.

(3072.) Autumn wheat is almost always sown broadcast in Scotland, except, perhaps, in the neighbourhood of large towns, where it is sown with the drill, Plate XXVII., figs. 381 and 382; and the reason why the drill is used in that particular locality is the facility afforded by the drilled rows to hoe the land free of surface weeds, which invariably make their appearance where street manure is used. In England wheat is very generally drilled, for the reason just given, and surface weeds seem more generally numerous and troublesome throughout England than Scotland. Sowing broadcast is represented by fig. 378, and is fully described from (2193.) to (2203.). The machines for sowing grain in drills will be found fully described by Mr Slight from (2357.) to (2366.).

(3073.) Wheat is *pickled* at this season, as well as in spring, before it is sown, and the reasons for treating it thus, as far as they are known, as well as the best method of doing so, will be found described in (2191.) and (2192.), and (2204.).

(3074.) Of the varieties of wheat suitable for sowing in autumn, many exist—some comparatively new, others which have been known for many years. Of the oldest esteemed varieties in Scotland, *Hunter's*, of a yellow colour, stands perhaps at the head. It yields both good grain and straw, is prolific, hardy, and held in estimation by bakers. It seems, in the sample, to consist of mixed varieties, and it is perhaps on this account that it is so esteemed by bakers. It will grow on any kind of soil. For deep mellow clay loams, the white varieties of England, such as are sown in the counties of Kent and Middlesex, are admirably suited. They are short, plump, thin-skinned grain, very beautiful in sample, and very fine in the straw. They are very liable, however, to many casualties, such as smut, rust, and to shake out, and on that account ought only to be sown on the best wheat soils, and in the most favoured situations; and whoever attempts to raise them in other circumstances, will assuredly suffer loss. Of this character is the *Pearl* wheat, attracting particular attention at present, and especially the *Golden Drop*. The Chiddham white wheat seems a favourite at present, and has certainly yielded prolifically in many instances. Hickling's white Prolific was a favourite a few years ago. Of the red varieties, the old *Creeping* is still a sound, prolific, hardy wheat for inferior soils and situations. It is an oblong grain, thin in the skin, and yields a rich coloured strong flour. The *Blood-red* is becoming a favourite wheat, and seems hardy. The *Belgian* red, a professed new variety, possesses all the characteristics of the old *Lammas-red*, a good wheat, and a great favourite 30 years ago. The bakers are very fond of high mixed wheat from Dantzig, usually of fine quality and small grain; but being a mixture of so many varieties, it would not answer for seed in this climate; there being a chance of the varieties not ripening at the same time.

(3075.) Almost every year some new variety of wheat, of great reputation, is offered to public notice, and no doubt a good variety may thus occasionally be obtained. But the safe plan for you to follow is to adhere to that variety, or those varieties, which have succeeded in the district in which you farm; because one variety may succeed well in that locality, and not another. It is right to try experiments in growing new varieties which are well recommended, and you should give such experiments every justice; but until the fact is proved by yourself or your neighbours, that the recommended wheat is a good one for your locality, it should not be extensively cultivated. The good properties of wheat are, heavy weight, fine sample, prolific per acre, and plenty of clean straw. A good weight is from 62 lb. to 66 lb. per bushel; it is now, however, as easy to obtain 65 lb. as it was a few years ago to get 63 lb. a bushel. Years ago 24 bushels per acre were considered a fair crop on ordinary wheat land; it is now as easy on the same soil to raise 32 bushels. The greatest difference in these results has been realized in the course of years, from the inferior, not the superior classes of soil.

(3076.) There are other modes than those I have mentioned of sowing wheat on fallow ground. One of these is *dibbling*, and there are various ways of dibbling wheat. One is to make a hole not exceeding 2½ inches deep, with a dibble not so thick as that used for planting potatoes in gardens, to drop 1 seed or 2 into it, and to cover them with earth with the foot; the holes being made 4 inches apart, and 7 inches wide between the rows. But this is a very slow process. A more expeditious plan is to use an implement made of a cylinder of wood 6 feet long, 4 inches in diameter, and divided lengthways by the middle, to make into 2 dibbles. Pins of wood of a conical form, 3 inches long, are driven perpendicularly at 4 inches apart into the apex of the curved side of the split cylinder. This implement forms a number of dibbles, by being laid along the ground with the pins downwards, which are pushed into the ground with the pressure of the foot, to make as many holes as there are pins. The implement being removed by means of a handle attached to its flat side, boys or girls drop 2 seeds into each hole, and cover them with earth. Another and more certain plan of dibbling is this:—a flat thin board of wood is provided with holes 4 inches apart in the row, and the rows 7 inches asunder. This is laid flat on the ground, when small dibbles are pushed through the holes to the requisite depth of 2½ inches into the soil, the depth being determined by a shoulder on the dibble; 2 seeds are then dropped into the hole as each dibble is withdrawn; and when the board is lifted up from the space it occupies to another space in advance, the earth is brought over the holes and seed by the foot. It is asserted by those who have sown wheat by dibbling, that about 1 bushel is sufficient seed for 1 acre, and that the produce will be 5½ quarters per acre, that is, the produce bears a proportion of 44 to 1 of the seed sown.*

* See Mark-Lane Express for February 1843.

(3077.) Another mode of sowing wheat has a similar effect in the appearance of the growing crop as ribbing with the small plough, and this is accomplished by using the *common plough with a single horse*, and depositing the seed, and along with it, if necessary, any species of manure dust, such as rape-dust, in the furrow. The seed is dropped out of a hopper placed in the bosom of the plough, and the quantity is regulated by a grooved axle, made to revolve by a small wheel, which receives its motion by being carried along the ground with the plough. The immediate effect of this operation is to cover the seed-wheat with the plough-furrow, to prevent its being thrown out by the frost in spring in soft and spongy land, and to cause the crop to grow in rows 9 inches asunder. The pulverized manure is sown at the same time and in the same manner as the seed, out of the same hopper, in which a division is formed to separate the seed from the manure, and both fall through spouts, one placed behind the other. Such a drill has already been described by Mr Slight in (2309.) and (2310.). The advantages resulting from this mode of sowing wheat on spongy clay soil are, that the horse does not tread on the seed, and the seed requires no covering in by the harrow. It is necessary, however, to caution you in the use of rape-dust and guano in contact with seed, as both are apt to affect the vitality of seeds, without the intervention of a little soil, or the previous mixture of a little earth.*

(3078.) Another mode recommended for cultivating wheat is *transplanting*. It is proposed to sow a small portion of ground with seed early in the season, and to take up the plants as they grow, divide them into single plants, and transplant them. By thus dividing the plants as they tiller into single plants at 4 periods of the season, a very small quantity of seed will supply as many plants as would cover a large extent of ground. Though wheat no doubt bears transplanting very well, yet as the scheme implies the use of much manual labour, it is questionable if it will repay the expense. The proposed method has been tested by experiment, and the question of comparative expense stands thus:—Suppose 440 grains of wheat are sown on the 1st of July, by the beginning of August each seed will afford 4 plants, or in all, . . 1,760 plants.

At the end of August these will produce . . .	5,280	...
In September these again	14,080	...
And in November these last will produce . . .	21,120	...

The *time* occupied in sowing the 440 grains, and dividing and transplanting the produce of them, stands thus:—

					Hours.	Minutes.
" July, sowing		. . . . . .	440	grains,	0	20
August (beginning),	taking up	. .	440	plants,	0	20
... ...	dividing into	. .	1,760	...	1	10
... ...	planting	. . .	1,760	...	3	30
August (end),	taking up	. .	1,760	...	1	28
... ...	dividing into	. .	5,280	...	3	30
... ...	planting	. . .	5,280	...	10	33
September, ...	taking up	. .	5,280	...	4	24
... ...	dividing into	. .	14,080	...	9	23
... ...	planting	. . .	14,080	...	28	9
November, ...	taking up	. .	14,080	...	11	44
... ...	dividing into	. .	21,120	...	14	4
... ...	planting	. . .	21,120	...	42	14
					130	49"

Nearly 13 days.

* See Mark-Lane Express for November 24. 1842.

Of these 13½ days, 4½ days may be reckoned for women and boys occupied in taking up and dividing the plants, which, at 10d. per day, will cost 3s. 9d. The remaining 9 days are for men transplanting, at 10s. per week, or 1s. 8d. a day, which will cost 15s. more; in all, 18s. 9d. per acre. As the saving of seed was estimated at 18s. 6d. per acre, the expense of transplanting would thus be greater than the saving effected. It appears, besides, from Mr Palmer's small but well-conducted experiments, that 10 plants out of 48 died, which is a large proportion, leaving more than ⅕ of the ground blank, whilst the return of the produce does not warrant the supposition that the remainder would make up the deficiency; the 2½ lb. of grain from the space of ground experimented on is not more than 20 bushels per acre, and still, after adding ¼ to this for the destruction of seeds occasioned by birds, the whole amount, only 25 bushels, would not be a full crop. Taking 2d. an hour as the fair cost of man's labour, the cost of transplanting would be L.1 : 12 : 9, to set against the 18s. 6d. of seed saved; and supposing the plants to be set as far apart as 1½ foot, the cost would still be L.1 : 4 : 7 per acre. "In my opinion," says the experimenter, "the only way of executing this plan is to dibble in the seed, 2 grains in a hole, about 4 inches from each other, the plants to be taken up when they are in a proper state, and divided into 5, which would be as many on an average at that time as could skilfully be made, and then planted out at once, where they are to remain, thus getting rid of all the intermediate dividings. The number of grains of wheat required for 1 rood would be 42,240, which would not exceed 1 pint at farthest, as I have ascertained by actually counting that quantity; and, consequently, 1 pint of wheat will plant 21,120 holes. If each hole gives 5 plants on an average, which may be reasonably expected, there would be at the disposal of the farmer 105,600 plants, a quantity sufficient to plant 5 acres, at 17½ inches apart. If farmers were to adopt a system of this kind, there is little doubt but it would soon gain ground in the country. Seed for 5 acres, sown in the usual way, would cost about L.4 : 12 : 6, whereas, 1 pint of wheat would only cost about 1 penny. The plants may be taken up, separated, and planted again by 50 people in one day; 25 taking up, dividing, and supplying 25 planters, allowing each to plant 4,224 per day, and this might be wholly done by boys and girls at 6d. per day; consequently, the farmer's saving, exclusive of the original cost of dibbling 1 rood, would be L.3 : 7 : 5 per every 5 acres, as thus:—

" Cost of seed-wheat for 5 acres				L.4	12	6
Cost of 1 pint of seed-wheat,	L.0	0	1			
Dibbling it in ¼ of an acre,	0	0	0			
Taking up and planting,	1	5	0			
				1	5	1
				L.3	7	5"*

(3079.) The object of these various modes of sowing wheat, is the *saving of seed*—a great object, certainly, when it is borne in mind, that $\frac{1}{14}$ of the whole grain grown in the country is consigned to the earth in seed. From the statement just given in regard to the transplanting of wheat, it appears that 4 English pints of wheat are capable of supplying a sufficient number of plants for 1 acre. By ordinary drilling 3 pecks are enough seed for 1 acre, according to

* The Gardener's Chronicle for October 1843.

a statement of Mr Hewitt Davis, whose wheat crops average 4½ quarters per acre on poor soil, which is manured when the green crops are sown, and by sheep eating green crops off the land. His profits from farming seem to arise solely from the economical system of management he pursues. "The soils I cultivate," he says, "are naturally very poor. Two of my farms are principally gravel, and most of one is a moorish boggy gravel, extremely spongy in winter, and very apt to burn in summer, and formerly in the state of, and similar to, the heaths at Bagshot and Blackheath. The other is a very hilly farm, with but few inches of soil, on chalk. These farms have been greatly improved by the use of the subsoil and trenching ploughs; but they are *only kept in profitable tillage by the general economy of the system.*" *

(3080.) But practice has always tended against the use of a small quantity of seed; and the practice is sanctioned by the fact, that though large quantities of seed are usually sown, in many seasons the young plants come up rather scanty. There is a circumstance attending the germination of seed, which may account for much of the seed being destroyed by the insect tribe, especially if the weather be such as to cause its germination to be tardy. The circumstance I allude to is the nutritive matter of the seed, which constitutes its largest proportion, speedily becoming fluid after being sown, by absorption of moisture, then milky in its aspect and sweetish in its taste. In fact, a good deal of saccharine matter is always evolved in the process of germination. Now, as all creatures are known to be fond of sugar, the germinating seeds become a ready prey to hosts of insects constantly existing in the soil; and should the progress of germination be retarded by any cause external to the seed, we can easily conceive that injury derived from insects may soon render the seed useless for the purposes of vegetation; for if the nutritive matter be extracted, there is nothing to nourish the rudimentary plant, as it is not then sufficiently developed to draw nourishment from the soil.

(3081.) Fig. 545 represents a grain of wheat magnified, and so dissected as to shew its component parts. It consists of 2 skins, an outer and an inner, *a a* the outer, and *b* the inner, skin; *b* is where the nutritive matter, called the *albumen*, is situate, and it constitutes the whole seed, and is mostly hid beneath the skins; *c* is the little scale or cotyledon through which the nutritive matter passes in the sweetish state, when the grain is germinating, and by which it is rendered most fit for the nourishment of the little plant; *d* is the rudimentary plant, at the base of which 3 tubers may be seen, from which as many roots or stems, or both, will afterwards proceed; and *e* is the point where all the 3, the nutritive matter, the little scale, and the rudimentary plant are united. All these parts are essential to the growth of the seed, for, when any one of them is absent by accident or design, the seed fails to spring.

Fig. 545.

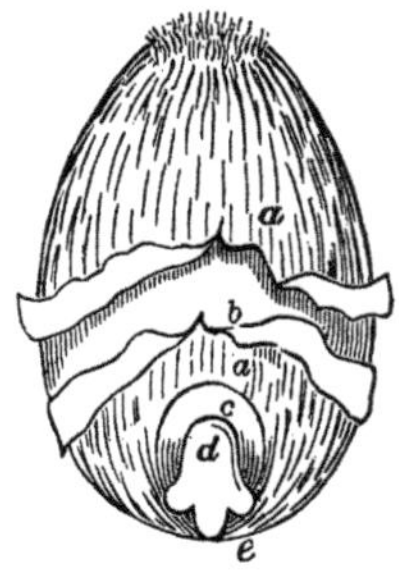

THE COMPONENT PARTS OF A GRAIN OF WHEAT.

(3082.) The seeds of almost every species of plant naturally possess such a structure that only 1 stem can proceed from them; but in the grasses, and more particularly in the cereal grasses, that is, those which yield corn, there is observed a remarkable departure from this form of parts. In them the em-

* Bell's Weekly Messenger for September 1843; and see also Mr Davis' pamphlet on the same subject.

bryo plant is usually thickened towards its base, and so organized, that, instead of 1 stem, 3 or 4 usually spring from 1 grain: and in other cases, a number of stems so great has issued as almost to exceed belief. The peculiarity mentioned may be observed at *d*, fig. 545, which is the rudimentary plant, having 3 projections in the lower part, while in another seed there would have only been 1; and from each of these a rootlet or a stem, or both, will proceed when the grain is placed in the soil. Fig. 546 represents the grain in a state of germination, 1 shoot *a* having left the sheath, another *b* is just evolved, and a third *c* remains unevolved; and *d d* are the rootlets. It is this peculiarity of structure which compensates, in some degree, for the loss arising from the destruction of seed. The germination of autumn wheat takes place in about 17 or 19 days. The force of the vegetation of a single seed is so great, as to be able to raise a weight of 200 lb., as has been proved by the process being made to split hollow balls, in the same manner as the Florentine academicians measured the expansive force of freezing water.

Fig. 546.

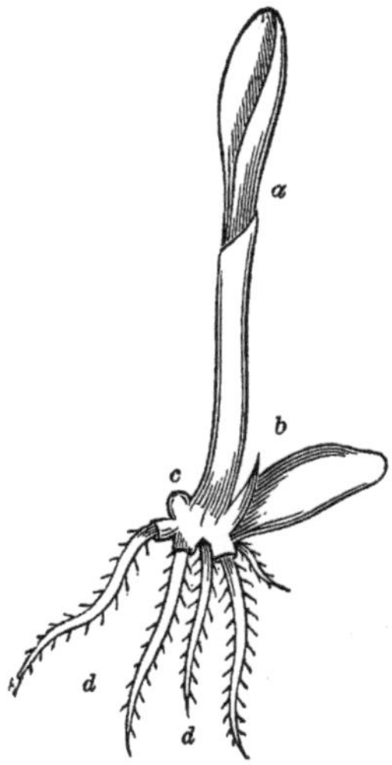

A PLANT OF WHEAT IN THE STATE OF GERMINATION.

(3083.) There is a valuable property in the growth of wheat which I must not omit to mention, which is, that the wheat plant pushes out 2 sets of roots, the one from the seed, and the other from the crown of the plant, from which the leaves originate. The *seminal* roots as *d*, fig. 546, push out from the seed wherever it may have been sown, provided the depth does not exceed 7 inches; but the *coronal* roots always push out as at *a*, fig. 546, very near the surface of the soil, not much exceeding 1 inch below it. The existence of those 2 sorts of roots clearly indicates the nature of the plant to bear sowing both in winter and spring; and the relative position of the roots indicates the differences of culture which should be given to the plant in winter and spring. In winter wheat-seed should be pretty well covered with soil, say 4 inches, in order to be placed beyond the influence of frost, but in spring it should be sown nearer the surface. "The first or *seminal* root," says a writer in the Georgical Essays, quoted by Colonel Le Couteur, "is pushed out at the same time with the germ, which, together with the farina, nourishes the plant, until it has formed its crown." "In the spring, when the crown has become sufficiently large, it detaches a number of strong fibres, which push themselves obliquely downward. These are the *coronal* roots. A small pipe preserves the communication between them and the seminal roots; and it makes an essential part of the plant, and is observed to be longer or shorter, according to the depth that the seed has been buried. It is remarkable, however, that the crown is always formed just within the surface; its place is the same, whether the grain has been sown deep or superficial. I believe I do not err when I call this *vegetable instinct*. As the increase and fructification of the plant depend upon the vigorous absorption of the coronal roots, it is no wonder that they should fix themselves so near the surface, where the soil is always the richest. From an attention to this circumstance, we are led to explain the operation of *top-dressings*." . . . "Wheat sown superficially," continues the same writer, "must be exposed to the severity of the frost, from the shortness of the pipe of communication. The plant in that situation has no benefit from its double root. On the contrary, when the grain has been perfectly covered, the seminal and coronal roots are kept at a reasonable distance. The crown being well nourished during the winter, sends up numerous stalks in the spring. On the *tillering* of the corn, the goodness of the crop *principally depends*." Spring-sown wheat having no time to tiller, may therefore be sown nearer the surface, and also thicker than autumn-sown. Viewed in this light, drilled wheat in spring should grow more equally than seed covered in with the harrow; for, as this writer justly observes, "a field of wheat, dibbled, or sown in equidistant rows by the drill-plough, always makes a better appearance than one sown with the harrow. In the one, the pipe of communication is *regularly of the same length*, but in the other *it is irregular, being either too long or too short*." I would say, with Colonel Le Couteur, that the exact depth at which wheat should be sown is a question of local experience, bearing relation to the nature of the soil and climate.

(3084.) As to the natural tillering of wheat, some varieties evince extraordinary powers in that respect. Colonel Le Couteur mentions that "one plant from a single grain of a downy variety, in 1833, threw out 32 tillers; all produced ears, with an average of 50 grains to each, or 1600 grains from 1—an enormous produce," as the Colonel regards it, but judiciously adds, "which no field cultivation could be fairly expected to attain, as it is not the extraordinary quantities which art *may* produce, either by *extreme care, subdivision, and transplantation*, that should be brought under the consideration of farmers; but the *fair and legitimate mode of husbandry*, attainable to all who will devote to its pursuit that industry and inquiry without which their art is a mere mechanical operation, throwing in a little seed into the ground, and leaving nature to do the rest." *

(3085.) Early sown wheat, when growing luxuriantly in autumn, is sometimes affected by slugs in damp weather. Mr Wentworth, of Harlow, gives this recipe for destroying them. "Provide, fresh from the kiln, as much lime as will allow 5 bushels to the acre. Slake it 2 days before sowing; choose a calm and mild morning; commence sowing early enough to finish before daylight; and 1 man can sow 1 acre per hour, sowing 2 yards at a cast." Turnip-leaves have been recommended to be laid upon wheat, and as slugs readily take shelter under them, they can thus be collected and destroyed. Of the 2 modes of destruction I should conceive the quicklime to be the more efficacious, especially in moist weather.

(3086.) [The philosophy, as well as the practice of the construction and application of wheel-carriages, has been so frequently discussed by writers in this branch of mechanics, that I had, at one time, deemed any attempt to enter upon such a subject altogether uncalled for, more especially when it is considered that many of those writers have been, and are, men of the highest acquirements in the walks of mechanical philosophy. On a closer examination, however, it has been found that, in some cases, philosophical knowledge has been applied to this subject under a theoretical view, and the powers of mathematical reasoning

* Le Couteur on Wheat, p. 29-35; and it is but justice to Mr Lawson of Elgin to state, that he detected this peculiarity in the germination of wheat when too deep sown, so long ago as 1827, as may be seen by referring to No. vi. p. 15 of the Old Series of the British Farmers' Magazine.

brought to bear upon it before a sufficient stock of experimental data had been obtained, upon which a theory could be satisfactorily built up. In others, obsolete practices, and those which even common sense condemns, have been dragged into the arena as examples of errors that existed only in the imagination of the author. In others, again, we find misconceptions brought forward in support of particular views ; these could not be supposed to have sprung from ignorance of principles, but rather from inattention in the application of those principles to a hasty view of a practical subject. In some of our most popular works even, there are to be found some examples of an erroneous application of thoroughly established principles of mechanical philosophy to this particular branch of mechanics. For these reasons, I am induced to venture upon some practical considerations of the construction and application of wheel-carriages, not tying myself to any particular theory, or to the demonstrations of any previous writer, but keeping as much as possible to facts and experiments, or to such geometrical demonstrations as may seem requisite to elucidate particular points of the subject.

(3087.) It is, perhaps, unnecessary to enter into the oft-repeated imaginary history of wheel-carriages, beginning with the simple roller without an axle, and rising through the gradations from that to the most elaborately constructed modern wheel, not omitting war-chariots, &c. There is something, however, in the first mentioned—the simple roller, that deserves notice, from a peculiar property which it possesses when employed as a vehicle of transportation ; and so little is this property known to the bulk of practical people, and even those of higher grade, that I have seen it, within these few years past, proposed as the basis of a patent machine for achieving most important results, and held out as an entirely new discovery,—a new *mechanical power*.

(3088.) The *roller* is the simplest and most accessible auxiliary in the hands of the common labourer or workman, to aid his energies when he has occasion to move heavy masses, as beams of wood or iron, or blocks of stone, over short distances ; and, how few of those, who are every day employing this mechanical agent, are likely to consider that, by means of it, they are moving the substance under transportation, over *twice* the *space* that the roller itself has passed through. When the substance moved is borne *upon* rollers having an axle passing through them, as in the case of a wheel-carriage, the transported body either rests upon the axle, or on a carriage attached to it ; hence the velocity of its rectilineal motion must be the same as that of the axis of the roller, which, as is well known, passes over a space equal to the circumference of the roller or wheel in every revolution of the same. When the substance simply rests on the periphery of the roller or wheel, it will be easily seen, that for every revolution that the roller makes it will move over a space on the surface of the ground equal to its own circumference, and that while doing so, the substance which it bears along, will, in the same time, have applied to the roller a portion of its length equal to that circumference, and which will have passed *over* the roller ; but the substance being entirely borne by the roller, must also partake of its progressive motion through space, and that at the same rate as the roller itself ; hence the substance must have passed over double the space that the roller has done in a given time. This result will, perhaps, be more clearly understood by considering increments of the motions.

(3089.) Let *a b*, fig. 547, be the surface of the ground on which a roller, represented by the circle *c d e*, is to roll, bearing the beam *f g*. If the beam be pushed in the direction of *g f*, every point in the arc *c d* of the roller will successively come in contact with corresponding points in the surface of the beam lying between *c* and *d′*; during this process the circle will have moved

Fig. 547.

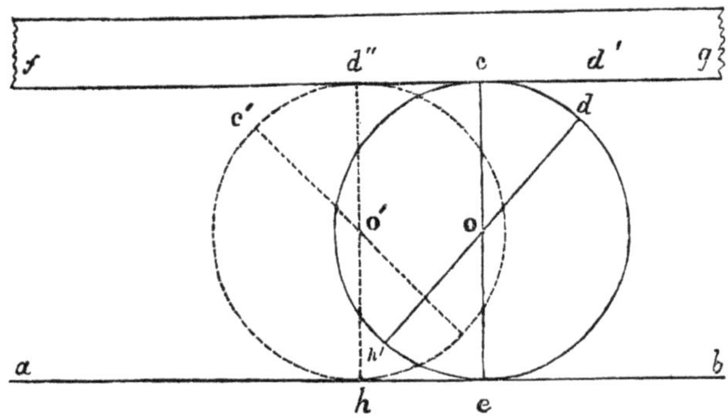

THE EFFECTS OF THE SIMPLE ROLLER.

in the direction *b a*, over a space *e h*, on the ground equal in length to the arc *e h′* or *c d*, and the centre *o* of the circle will have moved over a like space *o o′*, but the diameter *e c* will now have assumed the position *e′ c′*, and *h′ d*, the perpendicular position *h d″*, the point *d′* having moved to *d″* or through twice the space *o o′*, or twice the arc *c d*; and the beam *g f* has in like manner moved through twice the length of the arc *c d*, equal to the space *d′ c d″*, which is double of *o′ o* or *e h*. It results from this conclusion, that the body must very soon pass beyond the roller, hence a constant succession of rollers must be applied in front of the moving body, and hence, also, the application of this species of carriage must, in general, be limited to short distances. But from the extreme simplicity of the apparatus, and if the bed on which the rollers move is of a uniform texture, an almost entire absence of friction is obtained, whereby very heavy masses may be moved with a comparatively small force, and we accordingly see this simplest form of carriage in almost daily application. It is further well known that this principle has been applied in modern times to the transport of immense masses, larger, perhaps, than ever were moved by the engineers of antiquity, however colossal their works may have been. The mass of rock that forms the pedestal to the statue of Peter the Great at St Petersburgh, stands a noble monument of modern engineering, especially as regards its transport over a distance of four miles, achieved by the application of this principle; the rollers, however, in this case, were spheres of metal about 6 inches diameter, and not cylinders, but this does not alter the principle of the motion. The subject is one of much interest, but it would be out of place to follow it up here, our object being more of an agricultural than an engineering character.

(3090.) Though it is probable that the simple roller had succeeded the sledge, as a medium of transport for heavy bodies, in the early stages of society, it appears evident, from historical records, that the wheel and axle had been resorted to at a very remote period; for both in sacred and profane history we can trace their use through a period of not less than 3000 years. It appears very probable, also, that in the early application of the axle, it was adopted at first to combine two short rollers, which would have an advantage over the long roller, by avoiding, in their progress, much of the obstruction that the continuous roller must have met with. The transported body would then lie upon the axle, and to prevent its travelling over it, which it would do, though at a much slower rate, than under the arrangement shewn by fig. 547, the transition is easy to round the axle, and confine it to turn in a box or bush attached to the body under transportation, which now, in fact, becomes the carriage. There is every reason to believe that this has been the original form of wheel-carriages; that is to say, the axle and the wheels moving together, the former being confined to turn within bearings attached to the body of the carriage, and the wheels short but solid cylinders of wood. This form of wheel and axle is not confined to antiquity; it has existed in Europe, in Britain, within the last century, and is yet to be met with in some remote regions. It is worthy of remark, that the wheel and axle of almost all railway carriages, whether of early or late construction, are upon this plan; and the most perfect of all wheel carriages—the railway locomotive—is the same.

(3091.) It might be urged from this last circumstance, that this being held as the best form of wheel and axle for railways, and especially the locomotive, that it ought also to be the best for all other wheel carriages; but there are circumstances which render it applicable in the one case, and inadmissible in the other. In the railway carriage, it has been found by experience that the wheels and axles are less liable to derangement when turning *together* within a box or bush, than with the *wheel* turning upon the axle, though a carriage mounted in that way is not so well adapted to turn in a curve of short radius; but to obviate this, railways are now laid out with curves of very long radius. On the other hand, wheel carriages for common roads and all domestic purposes, would be extremely inconvenient if their wheels moved together, as they have very frequent occasion to turn in very narrow limits, which could only be done under great disadvantages were the axle and wheel to move together; but with wheels turning upon the axle, the greatest possible freedom of motion is afforded for turning in a small space; and, but for this difficulty, it appears yet undetermined whether or not the ordinary cart would be better adapted for its work, and the wheels more durable, were they fixed upon the axle. Be this as it may, the step from the fixed wheel and axle to the fixed axle and revolving wheel, has been easy and obvious.

(3092.) Though in some barbarous cases, the rude form of wheel and axle is to be found in later times, there can be no doubt that in those countries where the arts flourished at an early period, the wheel and axle were employed under the same principle as we now see them in the present day; and this is another instance of the really small progress of human invention during a period of from two to three thousand years; but it must be admitted, that though we have not brought any new principle into operation, we have greatly improved in the details of the construction of wheel carriages.

(3093.) In considering the action of carriage-wheels in general, we have seen that, when applied as simple rollers, if the surfaces on which they roll were level, perfectly smooth, and hard, and the roller also possessing the two latter qualities, there would be little or no resistance from friction, and the carriage or body would move with the least possible force. These circumstances not being attainable, and, from considerations of convenience, the wheels being placed upon an axle, it comes to be viewed under a change of circumstances; thus, though we may still conceive the wheels to move upon a surface that might present no obstruction to its forward motion, we have an element of resistance now introduced,—the friction caused by the wheel turning upon its axle; which, along with certain peculiarities connected with the subject, is first to be considered.

(3094.) If we take the wheel as rolling upon a level plain, and its circumference being supposed a perfect circle, it presents several peculiarities. Let the circle *a*, fig. 548, represent a wheel rolling from *b* towards *c* on the level

Fig. 548.

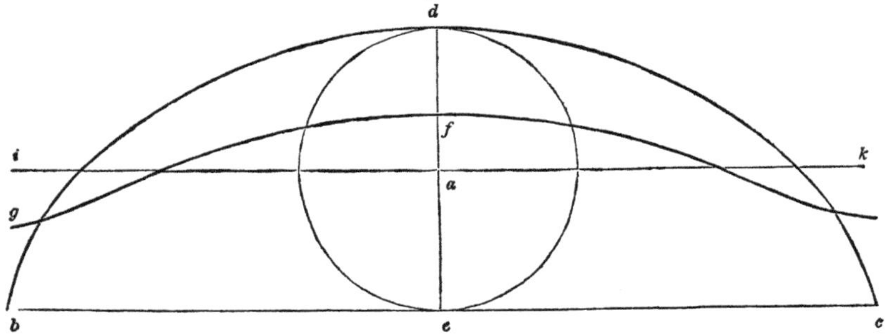

THE CURVES GENERATED BY A POINT RESPECTIVELY IN THE CIRCUMFERENCE OF A WHEEL AND ITS NAVE DURING ONE REVOLUTION OF THE WHEEL.

line *b c*; when it has made one revolution, its centre *a* will have moved through a space, and will have described a straight line *i k* in that space, in a vertical plane, equal to the circumference of the wheel; that point of the circumference, also, from which it started on the level plane will have described the beautiful and highly important curve the cycloid *b d c*, having its base *b c* equal to the circumference of the wheel, and its height *d e*, or axis of the cycloid, equal to the diameter, while the length of the curve *b d c* has been found to be equal to *four times* the diameter of the wheel. When the describing point is taken anywhere within the circumference of the circle or wheel, the curve described, though a modification of the cycloid, assumes a somewhat different character. Suppose a point in the circumference of the *nave* of the wheel, as at *f*, the curve generated by this point will be inflected, as *g f h*, crossing the straight line described by the central point of the circle twice in the course of one revolution of the wheel; but the base of any curve that can be so generated will be always equal to the circumference of the primary circle. It is foreign to the objects of this work to trace the remarkable properties of the cycloid; suffice it to say, that they are highly important in the theoretical consideration of dynamics, and have been investigated and determined by the most eminent geometers; and we can thus see, that even the mean and homely object—a cart-wheel, may have afforded data upon which some of the finest intellectual discoveries of the human mind have been based; for the investigation of this curve has occupied the attention of such men as Des Cartes, Pascal, and Huygens, to the latter of whom we owe the discovery of the remarkable property, that in a cycloid (in a position inverted to that of our fig. 548) bodies let fall from *any* point in the curve will reach the lowest point of it at the same time.

(3095.) As regards the theory of the wheel, applied practically in agricultural carriages, we have, first, to notice the manner in which it is affected by the draught, the resistance of the road, and the friction at the axle. We have seen that wheels in moving along a level road have their centres always preserved at the same height; and, from the manner in which the carriage is attached to the axle, and the draught to the carriage, we know that the ultimate effect of the latter results in its direction being nearly in a straight line drawn from that point of the shafts where they are suspended from the back of the horse on to the centre of the axle. This line may, in some cases, be horizontal, as in the case of a horse of low stature; but in most cases it forms an angle with the horizon of from 5° to 12°, and its true position is such as to bring that point of the shafts to which the draught-chains are attached, into a straight line between the centre of the axle and the hook of the horse's collar, and in the present consideration, it will be sufficient to assume that the line is horizontal. If we now suppose a carriage loaded to the ordinary extent, and bearing upon the axle, which again rests upon the lower side of the box or bush of the wheel, we have the following results.

(3096.) Let fig. 549 exhibit the above arrangement, *a b c* is the wheel, *d* the axle, *a e* the horizontal plane, on which the wheel moves in the direction of the line of traction *d f*. If the plane *a e* is truly level and smooth, the wheel will be put in motion with a very small force; and, if we suppose all friction at the axle destroyed, it goes on without stopping, as there would be no resistance; but such a state is not attainable in practice, and that from external causes,—the resistance of the air, the impossibility of procuring a surface devoid of inequalities, and, consequently, any attainable surface will be productive of friction or resistance. A wheel, with a cylindrical edge, rolling on a plane, whether horizontal or inclined, as it does not *slide* upon that plane, but makes simply a succession of contacts, cannot be said to meet any resistance from friction; the chief resistance it encounters, then, is from inequalities of surface, which may induce a constant succession of assents of the wheel and its load over these inequalities, with intermitting descents of the same, or, as in the case of a soft and yielding surface, it will amount to a constant assent, but still without *friction*, a term which can only be applied to that species of resistance arising from the motion of one body *sliding* upon, or through another. The chief source of friction, then, in the wheel, is that which is generated between the axle and its bush, the amount of this friction will depend upon the relative diameter of the wheel and its axle, and upon the nature of the materials of the axle and the bush, and also the state of their surfaces. As regards the first, if the radius of the wheel *d a*, fig. 549, is taken at 28 inches, and that of the axle *d i* at $1\frac{1}{4}$ inch; and a force applied to the axle in the direction *d b*, while the whole pressure of the load rests on the point *i*, where the friction is produced. If the axle is thus drawn forward by the force *d b*, the wheel is resisted by its contact with the ground compelling it to roll along, and thus there is produced the sliding motion at *i* upon the axle; but the radius *d a* acts in this case as a lever of advantage, in favour of the force *d b*, in the proportion of *d a* to *d i*, or the circumference of the wheel to the circumference of the axle. For when the entire circumference of the wheel has been applied to the surface of the ground,

Fig. 549.

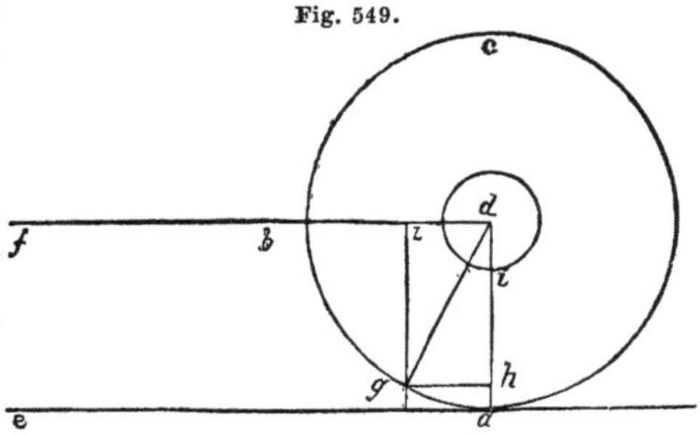

THE FRICTION OF CARRIAGE WHEELS.

that of the axle only, has been applied to the surface of the bush; and as the diameter of the axle may be taken at $\frac{1}{22}$ of the wheel, the amount of friction acting in resistance to *d b* the draught, will be only $\frac{1}{22}$ of the entire resistance, whatever that may be.

(3097.) Much uncertainty prevails as to the value or amount of friction in relation to the load, while some experimenters have found it not to exceed $\frac{1}{40}$, with a surface of iron running upon brass, and well oiled, others have made it as high as $\frac{1}{9}$ or $\frac{1}{10}$ of the load; and as cart-grease is not usually the most unctuous of substances, we may with safety take it at $\frac{1}{16}$ for cart axles. Let the load then, including the cart, be 25 cwt., $\frac{1}{16}$ of this is 175 lb. for the absolute resistance from friction at the axle; but I have shewn, that the leverage of the wheel and axle reduced it to $\frac{1}{22}$ or $\frac{175}{22} = 8$ lb. nearly, as the actual value of the friction of the axle, resisting the force in the direction *d b*, fig. 549, or, in other words, the draught; but here it must be understood, that the axle and bushes must be in the best condition, the former steeled on the lower side and turned, or turned and case-hardened, and the bush smooth and hard. We see from this reasoning, that the friction of the axle produces but a small portion of the resistance; for taking the draught at an average of 160 lb., which is known, both from theory and experiment, to be about the amount of a horse power, moving at $2\frac{1}{2}$ miles per hour, the resistance of the axle forms only $\frac{1}{20}$ of this draught.

(3098.) It is from the surface of the road then, that we are to find the remaining $\frac{19}{20}$ of the resistance; and this will appear sufficiently obvious from a consideration of fig. 550. The resistance to the motion of the wheel, it has been stated, arises from inequality of the surface over which it rolls, and this may be shewn by taking the point *b* as an obstacle coming under the wheel, and over which the wheel and load must be raised. Most of those who have treated this part of the subject, have usually demonstrated only the relation of one wheel to another, in terms of their diameter or their radius, and the resistance respectively encountered in passing over any obstacle, viewing the wheel simply as a roller, within which the whole load is concentrated. Thus, if the circle *c a d* represents a wheel rolling on the horizontal line *c h*, and opposed by an obstacle at *b*, while it is drawn forward in the direction *o a*, by a certain force, this force will act with a leverage equal to *o e* in overcoming this obstacle, while the weight of the load on the wheel opposes it with a leverage that is represented by the line *eg*. By construction, *b e* is equal to $\sqrt{dc} \times \sqrt{ec}$;* and supposing the obstacle to be the same for all heights of wheel, *b f* = *e c* will be constant, while *d e* varies as the diameter, but

Fig. 550.

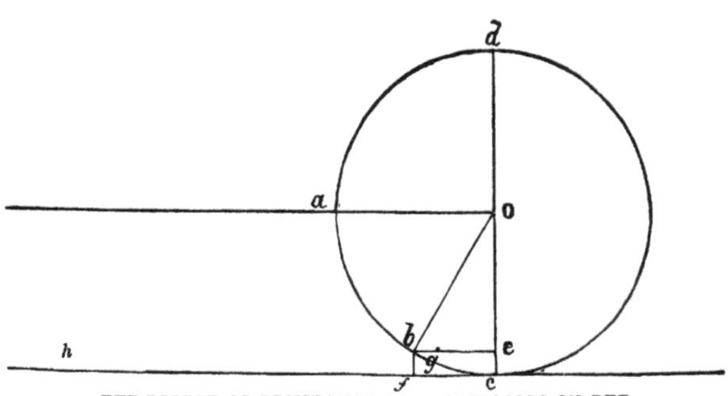

THE EFFECT OF RESISTANCE FROM OBSTACLES ON THE SURFACE OF THE ROAD.

* Euclid, III. 35.

b e will increase only as the square root of *d e*, which, in all ordinary cases, may be held as equal to the diameter, and therefore the force required to move the wheel over the obstacle *b*, will be inversely as the square root of the diameter of the wheel, nearly; for example, taking the common cart-wheel of 56 inches diameter, its square root is 7.5 nearly, which may be called unity; and take another wheel of 68 inches, or one foot higher, the square root of this is 8.2, and compared with the first, is 1.09 to 1. If a force of traction, therefore, of 300 lb. were required to overcome this obstacle with wheels of 56 inches, a force of 275 lb. would overcome it with wheels one foot more in diameter, for 1.09 : 1 : : 300 : 275; this is supposing, that in both cases the force is applied horizontally from the centre of the axle.

(3099.) The foregoing example is still only applicable to the case of a roller or wheel having the load concentrated within it, or otherwise placed right over its axle; but as this does not apply with exactness to a two-wheeled cart having the load disposed in such a manner as to throw its centre of gravity about 5 inches before the axle, I shall now take a more practical case, and consider it under this last circumstance.

(3100.) Referring still to fig. 550, suppose that the wheels are moving over a surface that yields to the pressure so as to sink one-half inch into the road, which may hence be considered as presenting a constant obstacle of one-half inch in height. This will give us *d e* equal to $55\frac{1}{2}$ inches, and *e c* one-half inch, the $\sqrt{55.5} \times \sqrt{.5} = 5.2$ inches $= b\,e$, and if the force of traction is applied at *o* in a horizontal direction, it gives the lever $o\,e = \frac{d\,c}{2} - c\,e = \frac{56.}{2} - .5 = 27.5$ inches, while the resisting arm is 5.2 inches, making the power to the resistance as 27.5 to 5.2. The entire load being supposed 25 cwt., of which, from its disposition on the length of the lever formed by the body and shafts of the cart, from $\frac{1}{18}$ to $\frac{1}{20}$ is borne by the horse, the remainder by the axle; the load on the horse may therefore be taken at $\frac{25}{19} = 1.3$ cwt. nearly, leaving on the axle 23.7 cwt.; and this increased by the weight of the wheels and axle, will make the amount of resistance about 28 cwt.; but as the power of the lever *o e* is to that of *e b* as 27.5 to 5.2, or 5.3 to 1 nearly, then $\frac{28}{5.3} = 5.283$ cwt., or rather above $5\frac{1}{4}$ cwt. for the force of traction during the *first instant* of raising the wheels with this load over an obstacle $\frac{1}{2}$ an inch high. This is supposing that both wheels meet the obstacle, and that it is perfectly incapable of being depressed. If only one wheel meet it, then the resistance is reduced one-half, and the traction one-half, and if the obstacle arises from a uniform sinking of the surface by the pressure of the wheels, then both meet the obstacle; but as it suffers depression while the load is in the act of being raised, the maximum of resistance will again not exceed one-half of the result quoted above, or 2.625, and will vary downward according to the state of the road, till it reach the minimum force of traction, which cannot be estimated at less than 120 lb.

(3101.) From the foregoing remarks we have seen that the friction at the axle is inversely as the diameters of the wheel and the axle, but, at the same time, that friction is so small in amount, that any practicable addition to the diameter would produce such a mere fraction of reduction upon the total resistance, as to be of no importance in practice. On the other hand, though the resistance arising from impediments on the surface of the road affect wheels to an amount greatly beyond the friction at the axle, to not less than 20 times the amount of the other, and though the reduction of resistance increases in a much lower ratio with the diameter, being inversely as the square roots nearly, yet even this, as we have seen, would only increase the advantage by $\frac{1}{12}$ of the whole resistance, or as 1 : 1.09 as before, with wheels of one foot greater diameter, an advantage not to be compared with the many inconveniences that would attend such an increase of height in the wheels.

(3102.) *The construction of wheels.*—The early construction of wheels has already been adverted to generally, but there remains to point out something of the principle which directs the modern wheel-maker. With but few exceptions carriage-wheels of all sorts are constructed principally of wood, but bound and supported with iron ; such a wheel for one-horse carts consists of the *nave,* a turned block of wood, usually elm, 12 inches long, and from 10 to 12 inches diameter, bound at each end with iron hoops; it is pierced with twelve mortises equally divided around it, into which are firmly inserted a like number of spokes, always made of oak. A ring of wood in six segments or felloes is mortised upon the extremities of the spokes, and this is surrounded by the hoop or tire of iron either in one entire ring, or in segments. Of wheels so formed there are several varieties ; that without the *dished* character may be held as a nonentity, except in carriages of one wheel, or of three wheels, where the third or front wheel may be found without *dish,* and in some cases where a double inverted dishing is adopted. The dished wheel, or that which has its spokes so placed in the nave, that if laid with the end of the nave upon an extended flat surface, the point of each spoke would be farther distant from that surface than the root, where the spoke enters the nave, and the spokes of the wheel thus arranged, when put in revolution would describe a cup or dish-form figure, concave above, forming the front of the wheel, and convex below, forming the back. The class of dished wheels are separated again into two great divisions, the conical dished wheel and the cylindrical; the first is known chiefly in England, and is that in which the tire, properly so called, is applied in segments, and sometimes, as in the wheels of very large waggons, the tire is composed of two, three, or four breadths of iron. In wheels of this kind the tire is sometimes applied in one ring or hoop, but its object as a binding-hoop is in a great measure frustrated by the conical surface to which it is applied, and hence the necessity of such solid tires being secured by strong rivet-bolts passing through it and the felloes, to prevent its dropping off the wheel. The cylindrical wheel, like the last, may have any degree of dishing given to the spokes; but the sole of the felloes or periphery is worked off at right angles to the plane of the felloes, and consequently cylindrical wheels of this kind are always hooped with an entire ring of iron, applied when red hot ; its contraction, therefore, when cooled, brings the felloes into close contact with themselves and the iron, and affords the greatest possible strength to the wheels. The sole or tire of conical wheels varies in breadth from 4½ inches to 16 inches, the latter being found only in the largest class of carrier's waggons, while the former, and intermediate sizes, are to be seen on two-wheel carts and all intermediate carrying vehicles. The sole of the cylindrical wheel is seldom above 6 inches, generally 5 inches, but the great bulk of them is 2½ to 3 inches broad. Wheels of this form, when above 3 inches broad in the sole, are run upon straight axles, the plane of the felloes being vertical, hence they move without any tendency to deviate from a straight line. Those with a sole of 3 inches and downward usually run upon bent axles, as afterwards more particularly described, and hence partake a little of the tendency of the conical wheel, the plane of the felloes standing oblique to the vertical.

(3103.) A great deal of irrelevant matter has, from time to time, been put forth, on the subject of dishing and conical wheels, in some cases deviating so far from the subject as to mistake the one quality for the other, and exhibiting such a want of information on the English practice of using conical wheels, as to represent a highly dished and conical wheel running in a waggon, with the plane of its felloes standing vertical, evidently intended to exhibit their destructive effects on the roads, by cutting them up, in a way that never occurred except in the mistaken imagination of the writers. Similar incongruities are to be found respecting the dishing and form of the wheel, where the conical character is confounded with that of dishing, probably arising from the absence of a true conception of the terms as applied in practice. It is pleasant, however, to see that men of talent are still found, who will devote their energies to the elucidation even of this every-day subject, and that with a degree of simplicity and perspicuity not hitherto to be met with on this subject. Thus, in the volume of "The Library of Useful Knowledge" chiefly devoted to the "Horse," there is a treatise on Draught, in which the principles and the defects of *conical* wheels are so clearly illustrated, that nothing remains to be said in condemnation of the practice. Instead, therefore, of spending time in attempting to elucidate their construction and defects, it seems better to allow them to take their place among the things that *have been*.

(3104.) To proceed, then, in the consideration of the *cylindrical wheel*, and, in connection with it, to consider also the formation of the axle, it must be observed, that there are two varieties, the inclined and the upright, though both varieties are *dished* wheels; and of these the first to be considered, as being the oldest, is the inclined cylindrical wheel, as it is seen in section in fig. 551. In this figure *a e* is the nave, with the axle *b* passing through it. The spokes *c c* are represented as they are inserted in the nave, and as they pass through the felloes *d d*, and these are encircled by the iron-hoop or tire, seen in section at *g* and *g*, but when complete, serves to bind and compress the entire ring of felloes into one mass. This common cart-wheel is

Fig. 551.

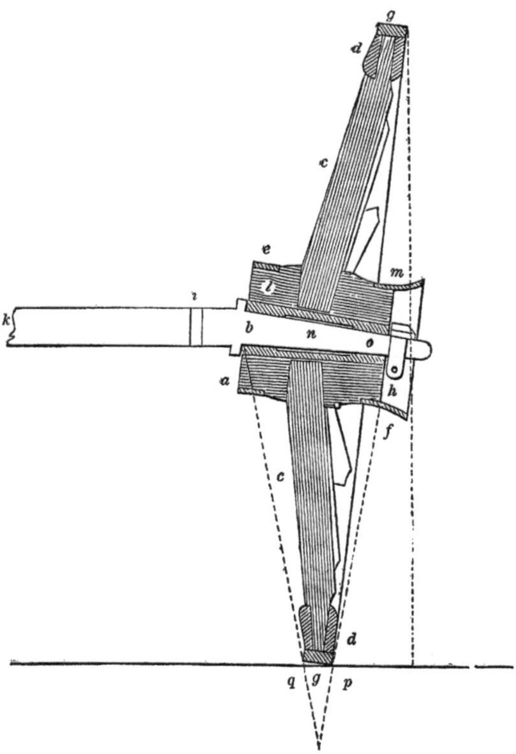

SECTION OF THE INCLINED CYLINDER CART-WHEEL.

seldom made below 54 inches in diameter, nor exceeding 58 inches, measuring over the iron hoop. The ordinary length of nave is 12 inches, and its diameter varies from 10 to 12 inches. In an average diameter of 11 inches in the middle, the inner end of the nave is 10 inches, and the outer end 9 inches; in the former, it is bound by an iron hoop *a e* of 2 inches in breadth by $\frac{3}{8}$ inch in thickness, and on the outer end by a hoop *f m* of 4 inches in breadth, about one half of which projects beyond the end of the nave, forming a shield to defend the end of the axle from accident and from mud. The iron for this last hoop or band is manufactured for the special purpose, the edge of the bar intended for the outward part being $\frac{3}{8}$ inch in thickness, while the inward edge is not more than $\frac{3}{16}$ inch. When a block has been selected for the nave, an auger hole is bored through its axis, into this an iron bar or mandril is driven, and upon it the nave is turned to the proper shape and dimensions, and lines drawn round it limiting the length of the intended mortises; its circumference is then divided into twelve equal divisions, and upon these divisions, as a centre, an auger hole is bored upon each in the direction of the radii of the circle—or if machinery is applied in the manufacture, the nave is suspended between proper centres, furnished with a dividing index, and the vertical drilling machine brought to bear upon the nave, dividing and boring it at one operation. The enlargement of the hole is then performed either with the mortising-chisel and hand mallet, or by the mortising-machine, until the mortise is of the proper size and direction. In forming the mortise, its *sides* ought to be parallel, the width from 1 inch to $1\frac{1}{4}$ inch, the front end must be cut to the slope of the intended *dish* of the wheel, while the back of the mortise approaches nearly to a right angle with the axis of the nave. A practice has been long prevalent to make the mortises wider at the front than at the back, and in support of the practice it has been alleged that it tends to keep the spoke firm in its mortise, and at the proper inclination. Little consideration is necessary to shew that the practice can neither accomplish the one nor the other, but it involves a condition much worse than this negative quality, which is, that it leaves less substance of wood, in the bridge or partition between the mortises, at the front than at the back, thus inducing an unnecessary weakness in that part of the nave more than at the back, and though this may not exceed $\frac{1}{8}$ or $\frac{1}{4}$ of an inch in each partition at the surface, it forms a more sensible proportion of it towards the bottom; and as the nave is the basis on which the whole wheel depends, every means ought to be adopted to secure its uniform strength.

Fig. 552.

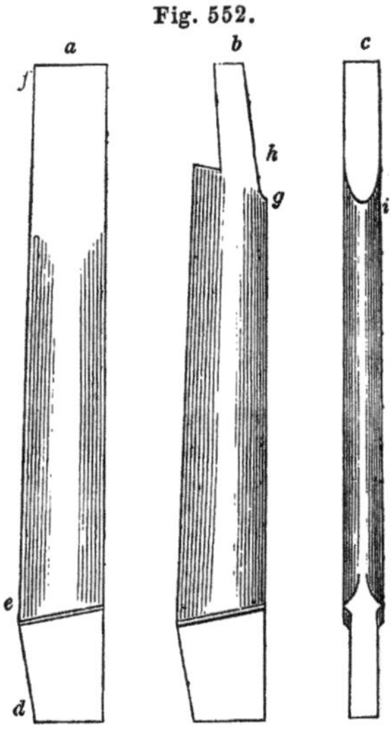

THE SPOKES OF A WHEEL.

(3105.) The *spokes* of the wheel ought to be prepared of the best oak-root, thoroughly seasoned, for which purpose they require at least three years preparation. The length of the rough spoke is 27 inches, and when dressed to enter the nave, the breadth at the root is from $3\frac{1}{4}$ to $3\frac{3}{4}$ inches, the thickness of the body ranging from $1\frac{1}{2}$ to $1\frac{3}{4}$ inches. Fig. 552, *a* is a side view of a spoke when prepared to be driven into the nave; *b* the same when it has been finished in the wheel; and *c* a front view of the same in the finished state. In figure *a*, the form of the root tenon is exhibited, wherein the front edge is uniformly straight, but on the back the tenon part is wedge-formed, the slope *d e* deviating from the line *e f* a quantity nearly equal to the dish which the spoke is intended to have in the wheel, whereby the back of the mortise approaches to a right angle to the axis of the nave. In wheels where the dishing is carried to an excess, this rule would give the tenon so much of the wedge-shape as to render it very liable to be dislodged from the nave; and hence, as one evil begets another, arises a necessity for giving the back of the mortise an inclination, or what is termed being *under-cut*, which induces the further necessity of weakening the partition behind by sloping the back of the mortise, or the alternative of giving it less width behind than in front; but every one of these conditions should be avoided as much as possible. The shoulders of the tenon, instead of being cut in at right angles to the faces of the tenon, are usually cut at an angle of not less than 45°, as seen at the figure *c*; these, when the spoke is driven home, fill up firmly the parts which are cut away for this purpose in the two edges of the mortise, giving to the spoke a more stable bearing on the shoulders, and, at the same time, increasing the strength of the tenon at this essential point. Figure *b* is exhibited finished at the head, though this operation is performed only after it has been driven into the nave. The face-edge *b h* of the head tenon has its plane dependent upon the amount of dish, for the part from *b* to *h* is always cut off at right angles to the axis of the wheel, being also parallel to the intended face of the felloes. From the point *h* towards the body of the spoke, the face is usually curved to *g*, as in the figure, the principal use of which is to facilitate the formation of the curve given in the finish of the face of the spoke, as seen at *i* in figure *c*. The back part of the head tenon is quite arbitrary, but it is proper to give it greater breadth at the root than at the point by about $\frac{1}{4}$ inch. The fitting of the mortises of the nave to the spokes is an operation requiring considerable care; the front of the spoke must be set at the proper angle of dishing, which should not be under $82\frac{1}{2}$° with the axis of the wheel, nor should it exceed 84°, or nearly $1\frac{1}{4}$ inch, and 1 inch on each foot of the length of the spoke. It is difficult, and unnecessary, to determine the degree of force requisite to drive the spoke into the nave, but it may be affirmed, that there may be as much injustice done to the fabric by having the spokes too firmly driven into their mortises, as by having them too easy; the danger of the former arises from overstraining and crippling the partitions of the nave, and of the latter, from the spokes becoming loose in the mortises; a due mean between these, which can only be judged of from experience, will always produce the most lasting results.

(3106.) *The double shoulder of the spoke.*—It is deserving of remark, that for some reason now lost in the obscurity of time, the tenon at the point of the spoke is almost invariably formed with a shoulder or bearing on one side only, as in fig. 552, *b*. From the manner in which the pressure comes upon the felloe and the spoke, being that of compression on the end of the latter, it would certainly appear more natural that it should be *shouldered on both sides*, but more especially in the wheel with broad felloes. Here, I should say, it is indispensably necessary to the due stability and support of the wheel; nevertheless, it is seldom, if ever adopted, except in the practise of a few wheel-makers.

(3107.) The *felloes*, in a wheel of ordinary size, are always 6 in number, and are usually either of ash or elm; they ought also to be well seasoned, and of sound quality; the breadth, from front to back, is from 3 to $3\frac{1}{2}$ inches, and their depth $3\frac{1}{2}$ to 4 inches. Fig. 553 is a view of a felloe square-dressed and mortised, being in the state in which they are first prepared previous to being put on the spokes; *a a* are the mortises, and *b* is the dowel-hole, as bored in each end of the felloes. The dowels are usually of oak, ¾ inch diameter, barrel-shaped, and their chief use is to keep the joinings of the felloes fair and even while finishing, and until the tire is put on. The felloes, previous to their being put on the spokes, are partly rounded off to the form seen in section, fig. 553, *a'*, in which the dotted lines represent the mortise, *f* the face, *e* the back, and *d* the sole, of the felloe. After being put on the spokes, dowelled and wedged up, the felloes are dressed thoroughly off either by turning, or by the plane, and other hand-tools, the periphery, or sole of the wheel, being dressed off *square* to the face of the felloes without being at all conical. In fitting the end joinings of the felloes, which is done by running a saw several times through the meetings, and in finishing the joints, it is proper to leave them open outwardly, or towards the circumference, because from the felloes being generally cut out of straight wood, the parts *c c*, of two contiguous felloes, when presented to each other, have the grain of the wood abutting obliquely, and hence liable to compression when brought to their bearings; this occurs when the tire or hoop is applied, and to admit of the full effect of the hoop as a binder, the end joinings of the felloes are all finished slightly open.

Fig. 553.

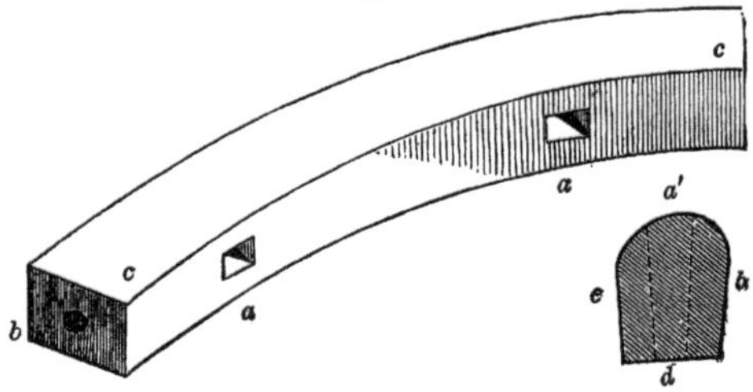

THE FELLOE OF A WHEEL.

(3108.) The application of the *tire* or *hoop* is the next process in the construction of the wheel. For narrow wheels, which are now under description, the iron of the hoop is from 2½ to 3 inches broad, seldom the latter, and the thickness from ¾ inch to 1 inch. The bars are first bent in the cold state to the curve, either by passing them through a three-roll machine, or what is simpler, they are bent over a curved block of somewhat smaller diameter than the intended wheel; when brought to the circle the ends are welded, but previous to doing so, the length of the circumference of the wheel, and the length of the inside of the hoop, are compared; this operation is performed by measuring the length of the circumference by means of a simple odometer, a small iron wheel turning upon a pivot, which is held in the hand; with this the length of the external surface of the felloes is measured, and also the internal surface of the hoop, and when welded up, the latter should be 1 to 1½ inch shorter than the other, proportioned to the allowance that has been given for the closing of the felloe joints. The hoop having been set tolerably correct to the circle, and also out of twist, it is heated either in the forge fire, or in a pile; but what forms the most perfect and economical mode of heating, is the hoop-furnace, in which, as well as in the pile, two, four, or more hoops, may be heated at once. In this operation the heat required is only such as will produce that degree of expansion in the iron, as to make it pass freely over the wheel, and for this purpose a dull red heat is sufficient. The wheel is laid on a hooping bed of stone, or of cast-iron, with its front downward; the heated hoop is brought over it, if sufficiently expanded and true to the circle, it passes on to the wheel with little trouble, and at most, with a few twitches of the hooked lever, or a blow of the hammer, it takes its place. Cold water, in abundance, is immediately applied to the iron, serving at the same time to prevent the burning of the felloes, and to expedite the contraction and tightening of the hoop ; so soon as it begins to bite, the wheel is turned up, and the wood and the iron are adjusted to each other. The iron is now cooled down to prevent all deterioration of the wood, the wheel being thus strongly compressed on all points, the shoulders of the spokes against the felloes, and the joinings of the latter brought also into close contact. This operation, by compressing the joinings of the felloes, while they are resisted by the spokes from contraction of the wheel's diameter directly, has always the effect of increasing the dishing of the wheel, by the spokes yielding forward in obedience to the contraction of the iron, and the compression of the felloes. It is necessary, therefore, in the first construction of the wheel, to make allowance for this ; and to this end 2½ inches of dish will be sufficient before the iron is applied, as it will then be increased to nearly 3 inches. In wheels hooped in this manner, the only fixing given to the hoop beyond that of contact, is a spike driven through it into each felloe, and the only use of the spikes is to prevent the hoop falling off, in the case of felloes shrinking in a long course of dry weather.

(3109.) The *bushing* of the wheel is the concluding process. The bushes are of various sizes, suited to all purposes ; for wheels, such as we are here describing, they are 2½ and 1½ inch diameter, or 2¾ and 1¾ inch ; the cart-wheel bushes in common use being always tapered in these proportions. They are cast with a *chilled* interior surface, that is, cast upon a turned iron core or mandril, which gives them a tolerably smooth surface, with considerable hardness. In setting the bushes in the wheel, they must be placed so to have their axis in the centre, and at right angles to the face of the wheel ; and to insure the attainment of these points, the wheel, with the bush fitted in, should be suspended on the axle, and the bush adjusted with wedges, so as to make the wheel revolve truly.

(3110.) The foregoing description has reference to narrow wheels ; but in the case of the common Scotch broad wheel, the only point of difference lies in the greater breadth of the felloes and the hoop; the dishing is usually the same as for the narrow, though, according to the ordinary views of the case, the dishing ought to be less, or to vanish entirely ; this, however, would weaken the wheel much more than any ordinary amount of dishing could produce ; and this for reasons that will immediately be shewn. In the broad wheel the felloes have the same depth, or nearly so, as before ; but the breadth varies from 4 to 6 inches. The spokes are usually finished in the same manner as before ; but the mortises of the felloes are placed farther from the face, so as to throw the bearing on the spoke, nearly in the middle of the breadth. The hoops are, of course, suited to the breadth of felloes, but the thickness is always less than in those of narrow wheels,—seldom exceeding ¾ inch, and descending to ½ inch.

(3111.) *Of the axle.*—Cart-axles are always prepared in the form of blocks under the forge-hammer, and are afterwards finished to the true form on the anvil, with the hand and sledge-hammer. Fig. 554, is a view of an ordinary

Fig. 554.

THE CART-AXLE.

cart-axle in the finished state, and adapted to narrow wheels. The body part, from *a* to *b*, is 1½ inch thick, and 2½ to 2¾ inches in depth; it is thickened at *c* and *d* to a breadth equal to the diameter of the arm, for the purpose of admitting the opening of the caddy-bolt holes *c* and *d*; the arms of this example being 2¾ inches diameter at the shoulders *e e*, and diminishing to 1¾ inch at the linch-pin holes *f f*. In preparing the axle on the anvil, an incision is made along the lower side of the arm in the parts which will bear upon the bushes, and a strip of steel is laid thereon and welded into and upon the arm. When finished and hardened, this presents a smooth and durable surface on those parts of the axle where the greatest pressure occurs, thereby at once lessening the friction, and increasing the durability of the axle.

(3112.) In axles of the common construction, the arms are simply rounded and dressed a little with the file till they enter freely into the bushes. In connection with this rude finish, an opinion prevails among workmen, that a transverse section of the arms should not be a circle, but an oval, or approaching to the figure of a longitudinal section of an egg, the lower side on which the pressure exists corresponding to the end of the egg. This form, it is alleged, by allowing the bush to touch the axle on a line passing along its lower side, and again on the fore-part towards the upper side, produces less friction than if the surface of the arm were a perfect cylinder, or rather cone; and having also the advantage of a space before and behind the line of bearing in which the grease has a lodgement, and ready at all times to lubricate the bearing part of the surface. It must be admitted that there is some show of truth in the latter part of the allegation, but the former is assuredly one of the vulgar errors; for if it is true, as shewn by experimenters, that with the same load, friction is not increased by an extension of the rubbing surfaces, then there can be no extinguishing of friction from a decrease of the bearing surface on the axle; and even to admit, according to some experimenters, that there is a small increase of friction by an extension of surface, it would form such a small amount, the whole being but $\frac{1}{20}$ of the draught (3097.), that any reduction of that friction cannot possibly be otherwise than of very small importance.

(3113.) The more refined mode of finishing the axle-arms by turning is unquestionably the best adapted to destroy friction, especially when the surface of the bush has been also bored, or otherwise rendered accurately round and smooth. The retention of the grease is also partially accomplished by the chamber formed in the central part of the bush, seen at *n*, fig. 551. But the best proof of the advantages of turned and well-fitted axles is derived from the performance of the Royal-mail and stage coaches, whose axles fit their bushes so exactly as to be almost air-tight. In these, it is true, the constant lubrication is effected in a very perfect manner by the bush being converted into an air-tight chamber, which retains the grease without risk of its being squirted out. The want of some more secure means of retaining the grease in the cart-axle is a serious inconvenience and loss; for if the grease is so *thin* as to be completely useful, it is squirted out of the open bush in a very short time, causing a loss of the material; to obviate which, it is compounded in such a manner as to give it a consistency that will prevent its too free discharge; but by the same means it is rendered less serviceable as a lubricator, and a consequent small increase of friction ensues from that circumstance.

(3114.) It is satisfactory to observe, that of late years the utility of the patent axle has so far caught the attention of cart-proprietors, that we occasionally see carts mounted in this manner; and it has been ascertained that though this mode of mounting the axle is attended with an additional expense of L.2 to L.3 on each axle, the saving of grease alone in a short time repays the outlay, while the uniformity of the lubrication reduces friction, and gives such a free action, compared with the common mode, that the same horses are able to take an increase of load, which has been calculated at 10 per cent. above the ordinary load. A considerable saving of time and trouble is also effected; for it is not necessary to apply fresh grease oftener than once in 8 days, and the wheels may not require to be taken off more than once in 3 months.

(3115.) There are two kinds of such axles and bushes especially in use; that known by the name of the Mail-coach patent axle, employed chiefly for the Royal mail-coaches; it retains the principles of the tapering and bent axle, and is exceedingly simple and effective, though not quite so elegant as that known as Collinge's patent, used for light carriages, but the former is best adapted for such vehicles as carts. A section of the *mail-coach axle and bush* is here exhibited in fig. 555, wherein *a* is the arm of the axle and *b* the bush, *c c* being the nave in which it is inclosed. The bush is cast on a turned mandril and thereby hardened, a chamber is formed in the mouth of it, to which the axle is accurately fitted, the collar *d* filling exactly the chamber at the mouth. The outer end of the bush is solid, and it is rendered grease-tight at the mouth by means of a leathern packing or washer applied to the end of the nave and embracing the axle; this is again covered by the iron plate or collar *e*, which, besides retaining the solid collar of the axle in the chamber of the bush, is brought home to the end of the nave and bush by three bolts, which, passing through the plate and the nave, are tightened at the outer end of the latter by screw-nuts, until the collar of the arm has just freedom to turn in its chamber. The reservoir for the grease is the open space between the point of the axle-arm and the bottom of the bush, and the charge of grease is introduced to this chamber through a small opening at *b*, which is closed by a screwed plug *f*. Axles of this construction require a peculiar mixture of grease, which shall have such a consistency as to prevent its escaping too freely by the side of the collar *d*; hence any pure oil is not suitable for the purpose, because of its too great fluidity causing it to be speedily exhausted. A composition that is well adapted to these axles consists of equal parts by weight of whale-oil and lard, with about $\frac{1}{5}$ part of vegetable tar. The mixture, when cold, is of a consistency similar to honey, requiring to be heated to render it so fluid as to pour into the orifice of the bush, where it is again congealed, until by the motion of the wheel a sufficient temperature is produced to keep the grease in a state capable of lubricating the surfaces without becoming so fluid as to escape by the collar.

Fig. 555.

THE MAIL-COACH AXLE.

(3116.) Collinge's patent axle is formed with cylindrical arms, but is also bent; and there are others upon similar principles, all of which are, perhaps, more elegant in their arrangement and finishing, but are too elaborate to be recommended for the purposes we have here in view; I therefore deem it unnecessary to go into their details.

(3117.) *Of the inclination of the axle-arms.*—On this part of the axle much has been said and written, and demonstrations innumerable produced, to shew that axles should have certain forms that are, in fact, impracticable, and others that never can, or never have existed, except in the diagram of the mathematician. Certain points in the subject have apparently been taken up originally under theoretical, and even mistaken views, and the conclusions drawn from these have been, of course, participations in the original error; but what is more unfortunate, they have been received and handed from one to another with all their faults, the subject, as it would appear, being considered by succeeding writers not worthy of the trouble of being again taken up at the fountain-head of experiment, or of resorting to the induction that may be derived from every-day experience. Without entering, therefore, into, or following any of those learned discussions, I shall take a simple practical view of the case, referring, when necessary, to previous conceptions of the subject.

(3118.) It has been held by some writers who have treated this subject, that the position of the axle-arm should be such as to have its lower or bearing side at right angles to the position of the spokes that are at the moment bearing the load of the axle, the bearing-spokes being, by hypothesis, at the time vertical; it being supposed, under this view, that the dishing of the wheel is adapted to bring out this result, and that any deviation from this rule will produce an undue pressure of the end of the nave and bush against the collar of the axle, if the arm droop at the point, or against the linch-pin, if the arm is turned up from the horizontal line at the point. It will be observed that this proposition implies that the lower or bearing side of the axle should always be horizontal, and this without reference to the degree of taper given to the arm. Now, such a proposition might possibly have applied at a time when wooden axles were almost exclusively employed, the taper of these having been much greater than is required for those of iron. But were it essential that the lower spokes should stand perpendicular, and the principle carried out to the iron axles now in use, whose taper amounts to 1 inch on 12, or $\frac{1}{2}$ inch on each side, it would follow that the dishing of the spokes would amount to only $\frac{1}{2}$ inch on a foot, or about $1\frac{1}{4}$ inch of entire dish; and from what has been said (3108.), this is too little to afford sufficient support and strength to the wheel. In practice, we find that all the above conditions may be deviated from with impunity; for with dished wheels, and the arms of the axles bent downward, the most perfect freedom of action is found to exist, and not only in this case, but with broad-dished wheels running upon straight axles with arms tapering in the same degree as the former, a like freedom of action is obtained. It appears, therefore, that the points hitherto considered by writers on this subject as of essential importance in the formation of the axle, and the relative position of the wheel, may be departed from without inconvenience, and that the requisite freedom of action of the wheel upon the axle-arm is to be accounted for from other causes than have hitherto been recognised.

(3119.) Having pointed out what appears to me as fallacies existing in the hitherto received opinions for the form of the axle, I have now to add, on this part of the subject, a few remarks on the common practice mode of making the cart-axle in Scotland, chiefly as regards the *setting* or form of the axle. The practice has long prevailed of bringing the wheels of vehicles, intended for load, to one standard of width, where they touch the surface of the ground, and this standard, or gauge, has, by long established use, been fixed at 4 feet 4 inches between the tread of the wheels, where they rest on the ground, and as the breadth of the hoop, in narrow wheels, is 2½ inches, the gauge over all is 4 feet 9 inches. By a corresponding usage, the length of the axle-bed, or that part of it lying between the collars, *e e*, fig. 544, has been fixed at 3 feet 6 inches, thus affording two fixed conditions in the problem. Were all wheels made with one uniform degree of dish, we should then have one simple standard for the *set* of the axle-arms, which would be this—Let the concavity, or true dish, in the front of the wheel, and measured at the surface of the nave, be 2½ inches, the entire deflection from the base of the nave *a e*, fig. 551, to the front of the felloes, will then be 10 inches, and were the axle straight, the wheels would stand at the distance of 3 feet 6 inches, + 10 + 10 inches, or 5 feet 2 inches over all, but as they must be brought to the gauge of 4 feet 9 inches below, each wheel will be brought to an inclination of 5 inches from the perpendicular upon its entire height, or it is 1.12 inches nearly on each foot of its height. It is evident, that since the axis of the axle-arm must stand at right angles to the face of the wheel, whatever degree of inclination the wheel makes with the vertical line, the axis of the axle-arm must make an equal degree of inclination with the horizontal line, or its deflection will be also 1.12 inches upon a foot of its length. The axis, or central line, of the arm will therefore *hang*, or be bent downward at the point 1.12 inches below the straight line, passing along the central line of the body of the axle; but as it is easier in practice to operate upon the lower side of the axle-arm than upon its axis, deduct half the difference of the diameter of the two ends of the arm, which, in ordinary cases, is ½ inch, and this deducted from the deflection due to the axis or central line, we have .62 inch, or ⅝ nearly, remaining for the deflection of the lower side of the arm. This quantity is, however, subject to variation from two causes, the dish of the wheel, and the diameters of the axle-arm, but the rule will apply to all cases; and though it does not produce a perfectly correct form of the axle, being defective in principle, yet it approaches so near the truth, that we are surprised to find such a close approximation produced simply from experience, for it does not appear that the *principle* has ever been detected on which the practice has probably, by accident, been based. In most of the attempts made to establish a principle on which the form of the axle ought to depend, the demonstrations have failed by a misconception of the direction in which gravity acts through the medium of carriage wheels.

(3120.) The rule above referred to may be stated more concisely thus—*To the length of the axle-bed add twice the deflection of the wheel, measured from the base of the nave to the face of the felloes; from this sum subtract the intended overall width of the wheels at the ground, and the remainder will be the deviation of the face of the wheel from the perpendicular. Divide the last acquired number by the diameter of the wheel for the deviation per foot, which will also apply to the central deflection of the axle-arm, and subtracting half the difference of the inner and outer diameters of the arm from the rate of deviation thus found, will leave the amount of deflection for the lower side of the arm.*

(3121.) *Example.*—Let the axle-bed be 42 inches long, the wheels 4 feet 6 inches diameter, the deflection, or total dish, measured from the base of the nave to the plane of the felloes, 11 inches, width of gauge 57 inches over all, and the diameter of the axle arms 2¾ and 1¾ inches, then

$$(42+11+11)-57=7 \text{ inches},$$

which, divided by the whole diameter, is $\frac{7}{4.5} = 1.55$ inches, the deviation for the face of the wheel from the perpendicular, or of the axis of the axle from the line of its body. Then half the difference of the diameters of the arm $\frac{2\frac{3}{4}-1\frac{1}{3}}{2} = \frac{1}{2}$ or .5, and $1.55 - 5 = 1.05$ inches, is the amount of deflection in 1 foot of the axle-arm.

(3122.) The axle of broad dished wheels, with a cylindrical sole, can only be formed straight, or without bend in any direction; any deviation from this is a positive disadvantage, and though the adoption of cylindrical arms for broad wheels has been sometimes recommended, it can only be applicable when the wheels not only stand vertical, but they must have no *dish;* and it is an error to suppose that broad dished wheels could work with advantage upon straight axles with cylindrical arms.

(3123.) Of the practical utility of dishing, but especially of giving the wheels an inclination outward above, there can be but one opinion—that it is a necessary evil; producing advantages which more than counterbalance any disadvantages that may arise by thus deviating from just theoretical principles. An upright wheel without dish, and running upon a cylindrical axle, is unquestionably perfect in theory; but the various considerations that practice brings to bear on the economy of their application, decides the question in favour of a deviation from strict theory. It has, therefore, been found better to sacrifice a little to utility, and by *dishing* to gain additional strength to the fabric. By inclining the wheel, also, a greater width is obtained in the cart; and the inclination affords another advantage—though certainly not of vital importance—that of throwing off mud from the carriage and the load. The bending and tapering of the axle-arm, consequent upon the dishing, will perhaps appear the most objectionable point in the combination; but this, also, has its compensating quality, for though we have here the friction arising from a conical surface, it is acting against a similar surface revolving round the axle, and the result is only a slight increase of velocity in the revolving surface towards the base of the cone; but as friction is little, if at all, affected by velocity, there will, consequently, be no increase of the former at the larger end of the axle, as compared with the smaller. In such a case, therefore, the effect is totally different from that of a broad conical wheel rolling upon the extended surface of a road, and constrained to move forward in a straight line. In this case, were the wheel allowed to travel in the circle which the conical surface of the wheel would generate, it would produce nothing beyond the natural resistance of the rolling surface; but being constrained to move right forward, if we suppose the middle part of the tire to have the mean velocity and to roll without sliding, then those parts of the tire lying towards the base of the cone would be accelerated, and would have a tendency to tear up the surface of the road backward, while the parts lying towards the apex of the cone, would have the opposite effect, thus producing great and unnecessary friction, all of which is avoided in a conical bush turning round a conical axle.

(3124.) A doubtful practice has long existed in the formation of cart axles, that of giving the wheels a *gather*, as it is called, in the fore parts. This is done by bending the axle arms slightly forward, so as to make the wheels run a little nearer to each other in front than behind, the difference varying from one to three inches. There seems no good reason whatever for this peculiarity, though enough of questionable ones are adduced by those who uphold the practice. If any advantage is to be gained from it, the common bent axle, with the cart in yoke, possesses it in a sufficient degree, from the manner in which the horse is yoked to the cart; for, as we shall see, the shafts of the cart, taken at the point where the back-band is attached, is always higher than the axle, and as the axle is attached sufficiently firm to make it move with the cart, and if, when the shafts are horizontal, the wheels are found to be equally distant, before and behind, their relative position will be changed by any movement up or down in the shafts; thus, if the point of the shafts is raised, which is their working position, the wheels will be brought nearer to each other before than they are behind. It is highly probable that this natural result, by being observed in a well working vehicle, has given rise to the opinion.

(3125.) In regard to the defective state of the question as to the principles that govern the formation of the axle, and adaptation of the wheels to bring out upon sound data the most perfect action of the wheels on their axle with a minimum resistance, I have, after some investigation of the relation of the parts to each other, and from careful observation of numerous cases, arrived at what I conceive to be a principle, which is sufficient to account for the phenomena under consideration. The views that have usually been taken of carriage-wheels, when the object was to investigate the effects of the load upon them, would lead us to imagine that the wheel is composed, and not only composed but continues to act, and to be acted upon, as if it were a number of disjointed parts; but such an assumption is not fact,—a wheel built as has been here described, forms one strongly compacted piece of iron-bound carpentry, wherein every individual part assists in supporting all the others, and the entire frame becomes a rigid structure. Instead, therefore, of supposing that any one spoke that may happen to be in the lowest position at any one instant, and that in a vertical position, supports the whole load, it is more correct to view the wheel as a solid of revolution, generated by the quadrilateral figure *b o p q*, fig. 551. In this figure, formed about the extreme points of the lower or bearing half of the section of a wheel, we have the side *b o* forming the lower side of the axle-arm, and *p q* the sole of the wheel, the whole load on that wheel will be distributed over the line *b o*, and transferred to *p q*, or that part of the wheel which rests upon the ground. If we now conceive the axle *k b o* completed, and the opposite arm supporting its wheel, affording a similar solid, whose section will exactly resemble *b o p q*, upon these two sections the indefinite load on the axle will be equally divided, while the sole of each wheel, or the side *p q* of each, will be pressed to the ground with a force equal to, or a little more than, half the load, and the friction caused by this pressure will prevent the sole from all lateral motion on the surface of the ground. If we now consider the effect of gravity upon the sole *p q*, and its supposed opposite wheel, its natural tendency is in the vertical direction, but its *effects* may be diverted from this into any other direction.

(3126.) *Example.*—In fig. 556, let the straight line *a b* represent a rigid bar supported on two struts *c d* and *e f*, the surface of the ground being represented by *g h;* let a weight *w* be appended to the bar *a b*, gravity will then act in the direction *w w*, but the beam and load being supported by the struts *c d* and *e f*, the pressure is borne equally by them, each strut sustaining a pressure of nearly half the weight, together with the beam, and this transferred pressure will be in the direction of the two struts, and not in the vertical line. The construction here repre sented will come under the denomination of a tottering equilibrium, which it is easy to obviate, but is not of importance under the present consideration; and it

Fig. 556.

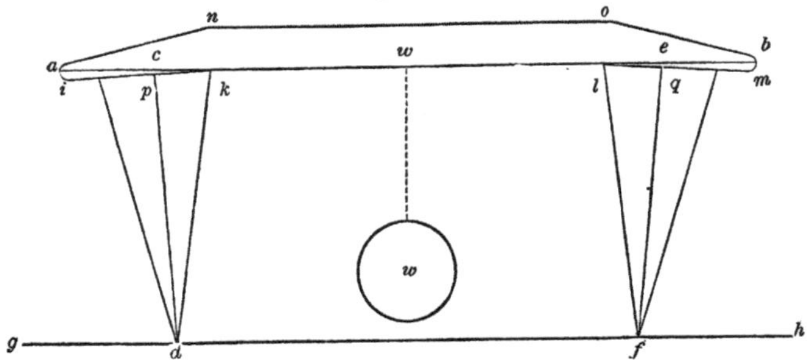

THE PRINCIPLE OF THE BENT AXLE.

will be evident that, besides, a tottering equilibrium, the head of each strut will have a tendency to *slide* outward. To prevent this last result, let the lines *i k* and *l m* be drawn at right angles to the struts, and draw also the lines *a n*, *n o*, *o b*, the space included between these lines and *i k*, *k l*, *l m*, represents a longitudinal section of the cart-axle, with bent arms; the lines *i k* and *l m* being the lower side of the arms, which rest upon the bushes of the nave represented here by the same lines *i k* and *l m*. The struts which are now supposed to be shortened, and instead of giving their support at *c* and *e*, they support the axle at the points *p* and *q*, and being now at right angles to that part of the bar or the axle, at which they give their support, the struts have no tendency to slide either outward or inward, so long as the bar itself is prevented by any means from being affected by motion in the direction of its length. If the triangles *d*, *i*, *k*, and *f*, *l*, *m*, are now completed, we have figures exactly corresponding to the prism of revolution, when its sides have been extended to *r*, fig. 551; and it is to be observed, that the triangles *d*, *i*, *k*, and *f*, *l*, *m*, fig. 556, being isosceles, the sides *d i*, *d k*, being equal; and if it be supposed to represent the section of a solid, there is no tendency, though supported on a mere point, to move in one direction more than in the other; hence, if wheels and their axles were constructed to fulfil these conditions, they would possess the greatest possible freedom of action, and free from all undue pressure either upon the collar of the arm, or upon the linch-pins. I consider myself warranted, therefore, in offering this as the true principle on which we are to depend for producing the minimum of friction and resistance from the action between wheels and their axles.

(3127.) It is evident, that all wheels having the same quantity of dish, and their axles having the same degree of taper, will come within the condition now specified; and it is to be remarked, that the position of the axle-arms makes no change in the conditions; for whether it be straight, as in those of broad-dished wheels, or inclined in any degree, as in narrow ones, the rule still holds good. But to bring out the advantages of this principle of formation in its utmost perfection, we have to attend to the following points:—That the line of the true or virtual dish of the wheel have the same inclination to the face or plane of the wheel, which is at right angles to the axis, as the inclination of the taper on the axle-arm or the bush makes with the axis of the arm or the bush; and what I here call the virtual dishing, is measured by a line, supposed to be drawn from the middle of the axis of the nave, through the middle of the sole of the felloe, as from n to the middle point of the sole $p\,q$, fig. 551, or to the apex r. In the common bushes this inclination is ½ inch on a foot, and as the radius of a wheel is 2¼ feet, the amount of virtual dishing would be 1⅛ inch, which is too little for the stability of the wheel, and should be at least 1 inch on a foot; but this increase of dishing involves the necessity of a like increase in the taper of the axle-arm, which would now be required to have its two ends 1¼ and 3¼ inches diameter, and this would be attended with various inconveniences. In practice, however, it is not only desirable, as a matter of opinion, that the wheel should have a tendency, while turning upon the axle, towards the shoulder or collar of the axle, but it is demanded as a matter of safety; for if a linch-pin were to drop out, and the tendency of the wheel to be outward, the wheel would infallibly drop of; but when the bias is towards the shoulder, the wheel may keep its place even in absence of linch-pins. It follows, therefore, that though the method here promulgated is mathematically true in theory, we may, by a slight deviation from the rigid rule, retain the advantages of the combination, and, at the same time, render greater safety to the vehicle. This is accomplished by retaining the bushes commonly in use, and instead of giving the wheel the low dish due to my calculation, let them have a dishing of 2½ to 3 inches, or about 1¼ inch on the foot. It has been ascertained, from a long series of observations, that with these last proportions, complete freedom of action is obtained as well as safety; and the existing practice is, in general, happily sufficiently near to the true theory, to give every advantage that can be desired; but, nevertheless, I would strongly impress on wheel and axle makers, not to allow the dishing to exceed 3 inches on the wheel when hooped, and to bear in mind, that when the dishing and the taper of the axle are adjusted to each other in the proportion stated above, it matters not, so far as the free working of the wheel is concerned, what degree of bending is given to the axle-arm. It may be straight, or even bent upward, yet the free action of the wheel is not impeded, neither is there any undue strain upon the spokes; for, in the section, fig. 551, in whatever position they stand, the strain or pressure upon them is always one of direct compression, whilst they are mutually aided in sustaining the load by the contiguous parts of the wheel. It must, however, be admitted, that such undue spreading of the lower part of the wheel as would arise from bending the axle upward, would produce a tendency in the wheels to displace the materials of the road, by the thrust of the wheels outwards; but in respect of the action of the wheel upon the axle, all that is requisite is the equality of the angles b and o of the triangle $k\,l\,n$, fig. 551, or, in practice, a near approach to it.

(3128.) The practice now becoming general of employing *half-round* iron for wheel tires in fast carriages, and partially so for cart-wheels, removes much of the inconvenience arising from flat iron tires as applied to inclined wheels. Flat iron tires, as will be seen in fig. 551, will at first, if running on perfectly level roads, have the greatest wear on the outward edge, but the tire, in all cases, ultimately acquires the half-round form; and as this form brings the strain more equally upon the sole, it seems desirable that this should be adopted from the first.

(3129.) Besides the wheels here described, which are those generally employed on the farm, there has of late years been introduced a variety of forms of construction in malleable iron wheels, some of which, by a misnomer, have been called wheels on the suspension principle, under which name it was alleged, that, instead of the load being supported on the lower spokes, it was constantly transferred to those above the axle, and the upper portion of the hoop was supposed to bear the load, on the principle of suspension from the hoop as from an arch. In the strict sense of suspension, nothing could be more fallacious than this idea; but in a more limited sense, taking the wheel as a rigid and immutable structure (3125.), it has some semblance of truth, for all the parts do bear a portion of strain at all times when loaded, but the parts below the axle must always bear incomparably the largest share.

(3130.) Numerous patents have been obtained for the construction of iron wheels. In those most approved of, the spokes and rim or hoop are of malleable iron, and the nave of cast-iron. The hoop is usually of the class denominated broad, being 5 or 6 inches, and is formed with a strong rib standing inward. The spokes are solid round rods, their ends welded into the rim, and spreading at their base into two distinct rows, giving an outline to the sides of the wheel of a lenticular form. The spokes having been duly welded to the rim, and adjusted to proper form at their base, the nave is then cast upon them, and it is afterwards bushed with brass. Wheels of this construction are now common in London, but their price is high, being about double, and from that to treble, the price of wooden wheels, which circumstance of price is adverse to their introduction for agricultural purposes. From the nature of their construction also, especially in the welding of the spokes into the rim or tire, there will be great difficulty in effecting a solid and secure junction of the parts; and unless this is performed in the most perfect manner, the jolts and shocks to which wheels are subject, will render them liable to frequent disruption; and the repairing of such structures will be attended with great difficulty. A material source of the difficulties attending the perfect construction and preservation of iron wheels, made on this principle, must arise from the disparity of the amount of substance contained in a spoke and the tire to which it is welded, as the most careful manipulation will not prevent defects at or near to these weldings, which may cause the wheel to fail in those points where repair becomes the most difficult. In locomotive and railway carriage-wheels constructed of malleable iron, some of the difficulties above referred to are avoided, by the nearer approach to uniformity in the quantity of substance contained in the parts; but such uniformity, if extended to cart or waggon wheels, would render them greatly too heavy for their purpose.

(3131.) The introduction of the cast-iron nave for wheels, about 40 years ago, was hailed as an important discovery; and, from the extreme durability and unchangeable nature of that material in a dry state, much was to be expected from it. Experience—the only sure test of all such innovations—has, however, decided against it; and the cast-iron nave is again nearly discarded from practice, except in wheels for light purposes. The cause of this disappointment seems to arise chiefly from the unyielding state of the iron, and more or less from the comparative hardness of the oaken spokes, combining to present a difficulty to the wheel-maker in fitting the one to the other, so as to have a perfectly uniform bearing in the mortise of the nave. If this is not accomplished in the first instance, the tenon of the spoke bearing, as it were, on a few points, and leaving interstices more or less open, the spoke is exposed to two sources of deterioration,—from bearing on a few points, it is subject to abrasion on these points, and so to become loose, and apparently chafed or cut by the iron; and, wherever such interstices occur, water will insinuate itself, and hasten the destruction of the spoke within the mortise. A suggestion has been thrown out by a writer on this subject, to remove the bad effects of moisture entering the mortises. It is, to "bore a small hole in the socket, to allow the moisture to escape." How these holes are to be made we are not informed; but it will be a difficult matter to place them so as not to let water in, as well as let it out. I am inclined to think, that it were better and safer to adopt a measure to *prevent* all access of moisture; and perhaps nothing could better effect this purpose, than that substance lately brought before the public—the *marine glue*. If this substance does really possess the qualities held out for it, a due application of it to spokes on their insertion into cast-iron naves, would render these naves far more efficient than they have hitherto been, and might altogether supersede those of wood, with a prospect of durability much exceeding any wheels with wooden naves, and would greatly facilitate the application of the upright lenticular wheel with wooden spokes and felloes.

(3132.) *Of the carriage.* Agricultural carriages are either four-wheeled waggons or two-wheeled carts; and as the Scotch practice, which we profess chiefly to follow, admits, with very few exceptions, the two-wheeled cart only, the following observations are chiefly confined to that implement.

(3133.) Though the cart, in general, is a vehicle very much diversified in structure to suit the numerous purposes to which, in a commercial country, it is applied; yet for the purposes of the farm its varieties lie within narrow limits, and may be classed under two principal kinds, the *tilt* or *coup* close-bodied cart, and the close-bodied dormant cart; but these, again, vary as to size, forming *single* and *double-horse* carts, which are merely varieties of the first. A third and less important kind, is the corn or hay cart, used chiefly in the seasons of corn and hay harvest; and there are others not required on every farm, but are important to some, such as the cage-cart, for carrying lambs and other live-stock to market, and the water and liquid-manure carts already described, fig. 366, and (2045.), and (2046.).

(3134.) *The tilt-cart*, fig. 557, is the most important vehicle of transport on the farm, and is employed for nine-tenths of all the purposes of carriage required in the multifarious operations throughout the year. It is employed to convey manure of all kinds; to convey stone and other materials for draining and other operations; leading home turnips and potatoes, and for carrying produce of all kinds to market. For some of these operations the tilt-cart is pre-eminently adapted, such as carrying and distributing of manures, or other matters that can be safely discharged by tilting. The dormant cart, on the other hand, is sufficiently commodious when substances have to be carried that require to be discharged from the cart by lifting, such as grain in bags, and many other articles requiring to be conveyed to and from the farm.

(3135.) Fig. 557 is a view in perspective of the common one-horse tilt or coup cart, of a simple and much approved construction, and consists of the following parts. The wheels *a*, *a*, which are of the usual height, 4 feet 6 inches, are of the dished construction, with cylindrical tread or sole, and are inclined from the vertical to bring them to the standard gauge below. The axle, which is of the bent order, with 2½ inch arms, is only seen as it protrudes through the nave. The body of the cart *b b*, with its bolsters, one of which is seen at *c*, by which it rests upon the axle, and to which the shafts are jointed by means of a joint-rod that passes through the bolsters and the ends of the shafts. The shafts *d* are secured to the body by means of the lock seen in the figure in front; and they are here represented resting upon a tressle to keep the cart upon a level; and, lastly, the top-sides *e*, *e*, which are fitted to ship and unship as occasion may require. But in order to convey a thorough conception of this important agricultural vehicle, the following detailed description, with figures, is

Fig. 557.

THE SINGLE HORSE TILT-CART.

given in geometrical plan and elevation, and which, except when otherwise stated, are all to one scale, $\frac{5}{8}$ inch to a foot.

(3136.) *Construction of tilt-cart.* The body-frame of this cart is represented

Fig. 558.

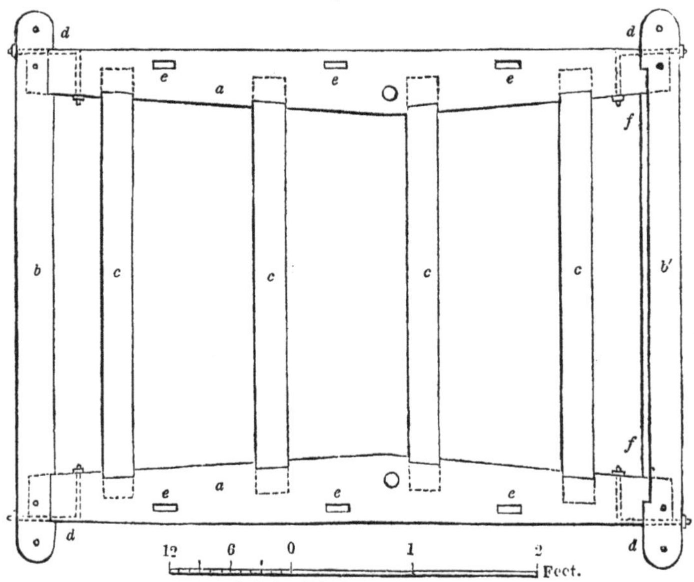

Scale for all the figures of the Cart-body.

THE BODY-FRAME OF THE TILT-CART.

in fig. 558, where *a*, *a*, are the two main bearers, *b* the fore cross-head, *b* the back cross-head, and *c*, *c*, *c*, *c*, the slots or floor-bearers. The main bearers and cross-heads are made of ash or elm, the former 2½ inches in depth, 6 inches in breadth at the point where they are pierced for the caddy-bolts, and 4½ inches at the ends. The fore cross-head is also 2½ inches in depth, by 4 to 4½ inches in breadth, and the other 3½ inches in depth by 4½ inches in breadth, and the slots are from 3½ to 4½ inches by 1½ inch, and should always be made of oak. In order to prevent too great a loss of strength in the frame-work, the mortises for the slots are taken to a depth not exceeding 2½ inches; the tenon of the slots is formed on one side, or single-check tenons, whereby the upper surface becomes flush with the surface of the bearers, and the check is rabbeted into the edge of the bearer, as seen in the annexed fig. 559, where *a* is a cross-section of the bearer, and *b* a slot with its tenon rabbeted into the bearer. The tenons at the ends of the main bearers pass, in like manner, only half through the cross-heads, as shewn by the dotted lines, in fig. 558, and these joinings are screwed and drawn hard up by means of the screw-nuts of the hook-bolts *d*, *d*, *d*, *d*, which are seen on a scale of double the size in fig. 560. The main bearers are mortised at *e e e*, *e e e*,

Fig. 559.

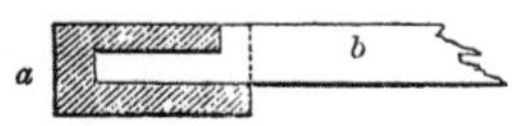

THE JOINING OF SLOTS AND BEARER.

Fig. 560.

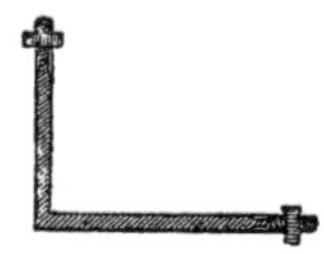

A HOOK-BOLT.

for the wooden standards that form the sides, three on each side, besides the fore and back iron-stays to be afterwards described. These mortises are so placed as to produce a width *within* the sides of 3 feet 6 inches at the floor-line. In framing the back cross-head, which is 1 inch thicker than the bearers, the difference of thickness is brought upward, and the rabbet *f f* is taken out an inch each way, so that its bottom is level with the main-bearers, and the ends of the flooring-boards are received into the rabbet, thus bringing the surface of the floor and of the cross-head to one level. To defend this cross-head from the constant wear to which it is subject, it is necessary to inlay upon its surface a bar of iron not less than 2 inches in breadth by ¼ inch thick, extending the whole length of the cross-head, or, as sometimes done, to make the cross-head 3¼ inches thick, and covering it all beyond the rabbet with an iron plate 3 or 3½ inches in breadth; but various modes are adopted for the finishing of this part of the cart, such as the application of an iron bar alone, 2½ by ¾ inches, in place of the wooden cross-head, and its iron fender, in which case the last slot must be put in close upon the iron cross-head, in order to support the ends of the flooring. This method, which is of late introduction, cannot be considered as equal in security and durability to the first described method. The body-frame of the cart, finished as above described, varies in its dimensions according to taste and circumstances, but a good medium size for a one-horse cart is 5 feet 4 inches in length, and 3 feet 10 inches in width over all.

(3137.) The standards of the sides are set with a slope outward of about 3 inches on a foot of height, and the head has usually the same degree of slope, while the door of the cart stands nearly perpendicular, or not exceeding 1 inch of slope outward, making the top dimensions about 5 feet 8 inches in length, and 4 feet 7 inches in width, over all. Fig. 561 is an elevation of the side of this cart, in which the horse-shafts are broken off at *a*; *b b* is the edge view of

Fig. 561.

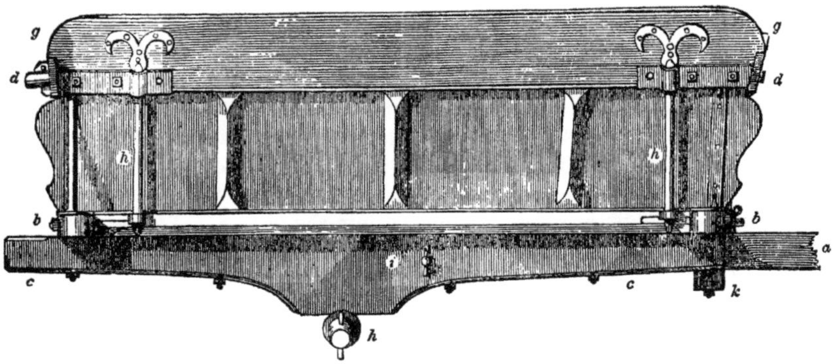

THE SIDE ELEVATION OF THE TILT-CART.

the body-frame; *c c* one of the bolster or limbers, which serve to support and bind the body, while they serve also as the medium of connection for the horse-shafts. The bolsters terminate forward at the face of the cross-head, where they are $3\frac{1}{2}$ inches deep, and $2\frac{1}{4}$ inches in breadth; at the bearing on the axle they are 7 to 8 inches in depth, and 4 to $4\frac{1}{2}$ inches in breadth; and at their termination behind, where they project 5 or 6 inches beyond the cross-head, they are $3\frac{1}{2}$ inches square. The bolsters should be made of ash or elm, but as a means of saving expense, they are frequently made of beech-wood, which, in point of durability in any exposed work, is much inferior to either; but this, and many other inferior substitutes, are, of necessity, resorted to by the agricultural mechanic to enable him to produce his manufacture at the low market price of the article. The end of the axle is seen at *h*, and *i* is the joint-bolt on which the horse-shafts are jointed to the bolsters, *k* is the end of the lock-bar framed into the shafts, and by which they are hooked to the body. The figure exhibits also the three oak standards, which are $2\frac{1}{2}$ inches in breadth at bottom, tenoned into the body-frame, and at top into the top rails, or shelvements, *d d*; these are also of ash or elm, and are $2\frac{1}{4}$ or $2\frac{1}{2}$ by 2 inches. The height of the sides measuring on the slope is 17 inches over all, or it is 14 inches inside measure; but this dimension varies more than any other, depending on the purpose to which the cart is to be applied, and many other circumstances. The top sides *g g*, are usually made of deal 1 inch thick; they are 5 to 6 inches high, and are furnished with the iron arms *h h*, which slide into the eyes *g g* of fig. 562, their lower extremity being also received into eye-bolts fixed into the edge of the body-frame *b b*, fig. 561.

(3138.) The shelvements are furnished at the extremities with iron mounting, which is seen in plan in fig. 562, on a scale of double the size; in this figure *a* is the fore-end mounting, and *b* that for the back-end or door; *c c* is the wood of the rail broken off; *d d* the bolts by which the irons are fastened to the wood; *e* is the screwed tail, by which the forehead-bar is secured to the top rail, and *f* is the pin upon which the door is passed and secured by a cutter dropt into the cutter-hole at *f*; *g* is the eye forged in the irons for receiving the arms of the top sides. The fore, as well as the back end of the top-rails, are frequently supported by iron stays, to the strap of which the ends of the side-boarding are also

Fig. 562.

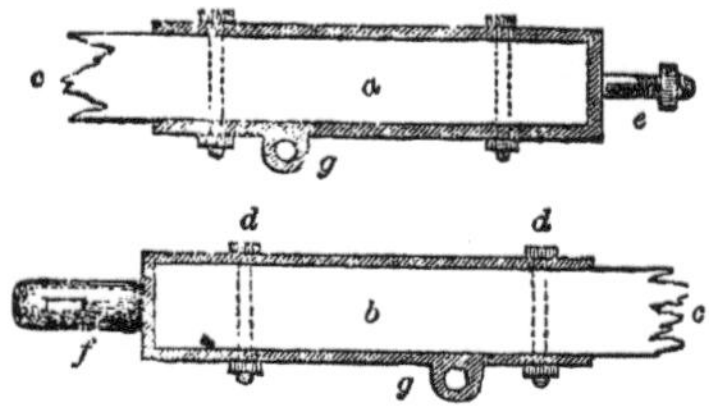

THE IRON ENDS OF THE TOP-RAILS.

Fig. 563.

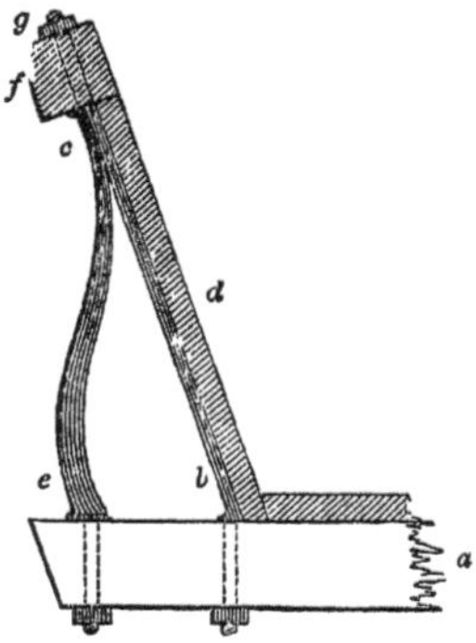

THE IRON STAY.

secured by rivets; but it is essential that the stay be applied at the back end. Fig. 563 is an end view of the stay, on a scale of double size; *a* is a part of the cross-head broken off, *b c* that part of the stay that supplies the place of a standard; it is 2 inches broad, and ¼ inch thick, and the boarding *d* is fixed to it with riveting. The stay proper *e c*, is welded to the standard at *c*, where a tail-bolt is formed that passes through the top-rail *f*, and secured by the nut *g*, the bottom of the standard and stay being both also formed into tail bolts, that pass through the cross-head, to which they are secured with screw-nuts below.

(3139.) A *bottom plan* of the tilt-cart is represented by fig. 564, in which *a*, *a*, *a*, *a* is the body-frame, *b b*, *b* the fore and back cross-heads, *c*, *c*, *c*, *c*, the slots

Fig. 564.

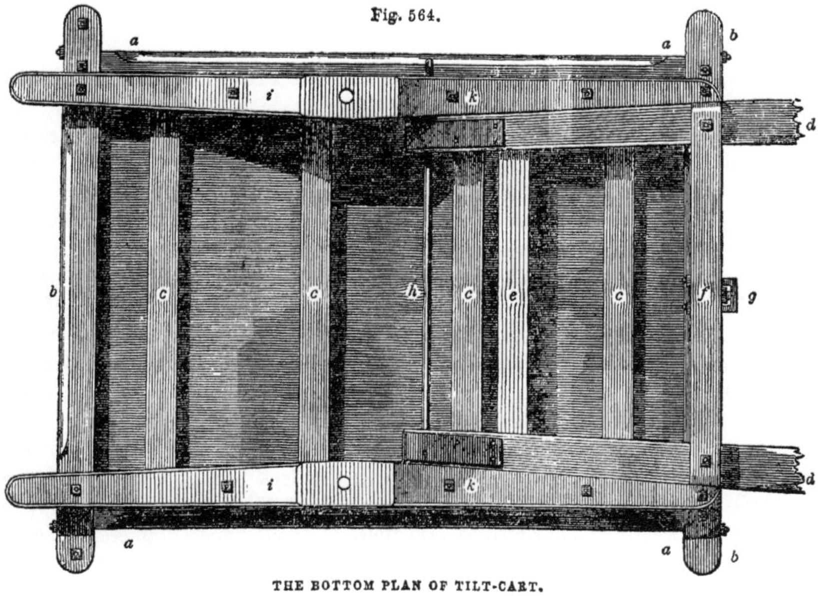

THE BOTTOM PLAN OF TILT-CART.

of the body-frame, *d d* the shafts broken off, *e* the slot, and *f* the lock-bar of the shafts, *g* a part of the lock, and *h* the joint-bolt passing through the bolsters *i i*, and the ends of the shafts, which are here armed with plates of iron to strengthen and support the wood. In forming and placing the bolsters, attention must be paid to having a proper width in front for the shafts, which should not be less than 3 feet 2 inches between the bolsters at the front, making their over-all width at that point 3 feet 7 inches; at the caddy-bolt *k k* the width over the bolsters may be 3 feet 4 to 3 feet 6 inches, and behind, the precise width is unimportant. Equally so is the width of the shafts at the joint-bolt, as that dimension is adapted to the limbers, but the distance from the bolt to the front of the cross-head is 2 feet 6 inches. The bolsters are secured to the body-frame with four or five half-inch screw-bolts with nuts.

(3140.) An elevation of the *front* of the cart is exhibited in fig. 565, with the axle attached to the bolsters. *a a* is the axle slightly bent in the arms, as

Fig. 565.

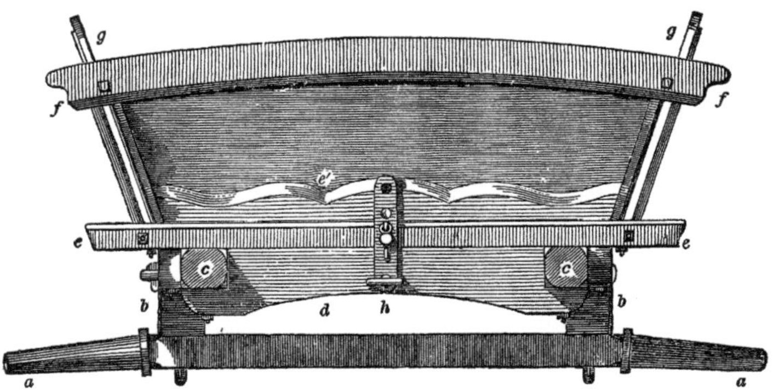

THE FRONT VIEW OF THE TILT-CART.

for narrow dished wheels, *b b* the bolsters, *c c* the shafts in section, and *d* the lock-bar. The cross-head *e e* is 4 feet 6 inches long, forming part of the body-frame, as before described. The lower head-rail *e'* is 4 inches in height, and 1½ to 1¾ inch in thickness; it is fitted in between the side-boardings, and spiked or bolted to the cross-head, upon which it lies about 1½ inch within the front edge. The upper head-rail *f f* is 5 feet 2 inches long, 4½ inches deep, and 1⅝ inch thick; it is secured by the tail-bolts of the end-mounting, of the top-rails, as described (3138.); and here it may be remarked, that in low-priced carts, this head-mounting is only a single strap and tail-bolt, in place of the double gland, fig. 562, *a*. In a front finished in this manner, with lower and upper head-rails, the upfilling is done with boards standing on end, making it strong, simple, and durable. In connection with this, and above the head-rail, are seen the ends of the top-sides *g g*. The lock *h*, in this example, is an iron bar, 2 inches broad, ⅜ inch thick, and about 10 inches long; it is hinge-jointed to a bolt near the lower end, that is fixed in the lock-bar, and the lock has a bent tail or stop, descending 2 inches below the joint, which permits it to open to a limited extent. It has also several holes of about 1 inch diameter punched in its length, as seen in the figure, and a pin-headed bolt is inserted into the front of the cross-head projecting about 2 inches from the head. The holes of the lock are adapted to pass easily upon this pin, and when so placed, a cutter or forelock is dropt into the cutter-hole of the pin. The several holes in the lock serve a number of useful purposes; thus, if the load is too heavy forward upon the horse, by shifting the pin one, two, or more holes higher in the lock, the centre of gravity of the load is thrown backward, and lightens the pressure upon the horse; again, in distributing common bulky manures upon the field in small heaps, if the front of the cart is raised and sustained by means of the lock, the manure is more easily discharged with the dung-drag.

(3141.) Fig. 566 is a section, at double size, of the cross-head and lock-bar, giving a side view of the *lock*. In this figure, *a* is the lock-bar, shewing the position of the lock and its joint-bolt, *b c* is the lock, *b* being the joint, and *b d* the bolt, *e* is the bent tail, which allows sufficient range to the motion of the lock on its joint to pass over the lock-pin *f*, before being stopt against the face of the bar. The lock-pin has its tail-bolt *f g* secured into the cross-head *h*, and the cutter *i*, attached by a chain, prevents the lock leaving the pin till it is relieved by the withdrawing of the cutter *i*; the parts *k*, *m*, *n*, are respectively the floor, and the upper and lower head-rail in section, and *o*, *p*, *q*, *r*, are in like manner, respectively the horse-shaft, side-boarding, shelvements, and top-side, all broken off.

Fig. 566.

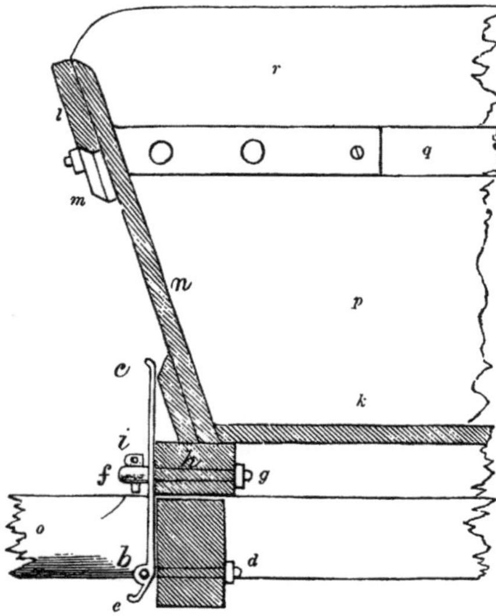

THE LOCK, AND SECTION OF CONTIGUOUS PARTS.

(3142.) The *shafts* of the tilt-cart are generally made of ash—seldom of oak, but of late years they are frequently made of the American birch and elm, either of which, when fresh and sound, are, from their tenacity and elasticity, well adapted to the purpose. The length of the shafts is about 9 feet 2 or 3 inches, of which 6½ feet go for the yoke; they are worked off and lightened towards the point, and finished according to the taste of the maker. In that part which passes into the cart-body, they are framed upon one slot, *e*, fig. 564, which is tenoned into the shafts, and by the lock-bar *f*, which is simply bolted to the shafts; it being of great importance not to weaken the shaft by any mortise at this point, where the greatest strain occurs.

(3343.) In regard to the scantling of shafts, various opinions have been broached, some advocating the principle that they should be considered in the relation of a beam that has to carry a load, implying that their largest dimension should be in the vertical direction; while much of the general *practice* places them in the other direction, or, at most, making the two dimensions equal. It is well known, that in fixed beams of equal length the strength increases *directly* as their breadth and as the *square* of their depth. Thus, if a beam of any given length, and 4 inches square, will carry 10 cwt., another beam of the same materials and length, 4 inches broad and 8 inches deep, will have an increase of strength in the proportion of 4^2 to 8^2, or 16 to

64, being 4 times stronger than the first, or capable of bearing 40 cwt.; but by laying the last beam flatways, its strength will be only double of the *first*, or capable of bearing 20 cwt. This is an important consideration in the management of beams; but I have grave doubts whether it is desirable to apply the principle to such a movable beam as the shaft of a cart. In stationary beams rigidity is almost always a chief object; but where one end of the beam is supported on the back of a living and moving animal, rigidity, as appears to me, should be avoided as much as possible, and elasticity made the desirable object. The rigidity of shafts that are formed on the principle of permanent beams, from the severe shocks that they must communicate to the back chain of the horse, and thence to the whole frame of the horse itself, must be very injurious to the animal, and which might be very much mitigated by adopting an elastic medium. Shafts, then, of tough ash, or other elastic woods, I conceive to be the most desirable, and in order to obtain the principle as far as possible, their breadth should rather exceed their depth; and general practice approaches this, as we commonly find such shafts about 3½ inches square at the lock-bar; but let them be made 4 by 3 inches, which is nearly the same sectional area as the former, and the horse would be a gainer, in point of comfort, compared with what he would be were the same shafts placed on edge. Neither would the shafts be much, if at all, more liable to fracture if placed flatways; for a simple beam that has no elasticity is more liable to fracture from sudden shocks, than one of the same absolute strength, but possessing a certain degree of elasticity; and cart-shafts are peculiarly liable to such shocks.

(3144.) The yoking-geer of the shafts is seen in fig. 567, which is a portion of the off-side shaft enlarged, and broken off at *a*. It consists of the

Fig. 567.

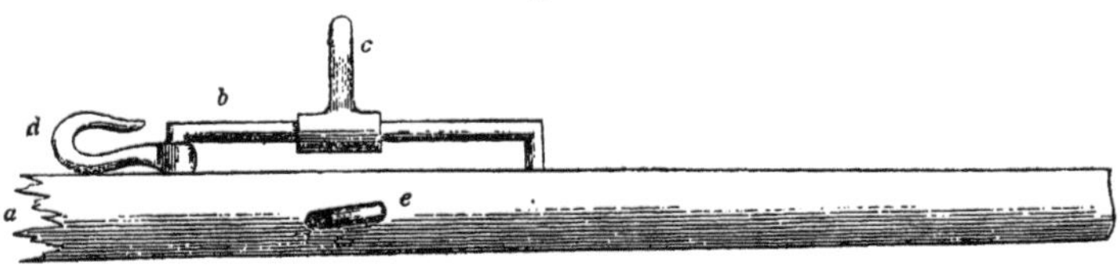

THE YOKING-GEER OF THE SHAFTS.

runner-staple, *b*, which is from 8 to 10 inches in length, its centre being placed at about 4 feet 6 inches from the lock-bar. The back-chain-hook, *c*, is fitted to slide from end to end of the staple, and the breeching-hook, *d*, takes a fixed position on the back leg of the staple; while the draught-hook or staple, *e*, is generally placed obliquely on the inner side of the shafts, its position there bringing it more directly in the line of traction than if it were attached to the runner-staple, in which position it is sometimes, though improperly, placed.

(3145.) The *back* view, and *door* of the cart, are seen in fig. 568, where *a a* is again the axle, *b b* the end view of the bolsters, *c* the lock-bar and its connections, as seen under the body, but at the further end, and *d d* is the cross-head, *e* is the bottom-rail of the door, and *f f* the top-rail, the two being connected by plain boarding standing on end, *g g* are the top sides, and the arms on which they are supported are seen at *h* and *h*. The door is furnished with the bottom-catches *i i*, which are strongly riveted and bolted to the bottom-rail of the door, their points being received into eye-bolts fixed in the cross-head, which secure the lower edge of the door. The upper part of the door is received upon the iron ends of the top-rails as seen at *h h*, where it is secured by the cutters dropt into the eyes of the end pins.

Fig. 568.

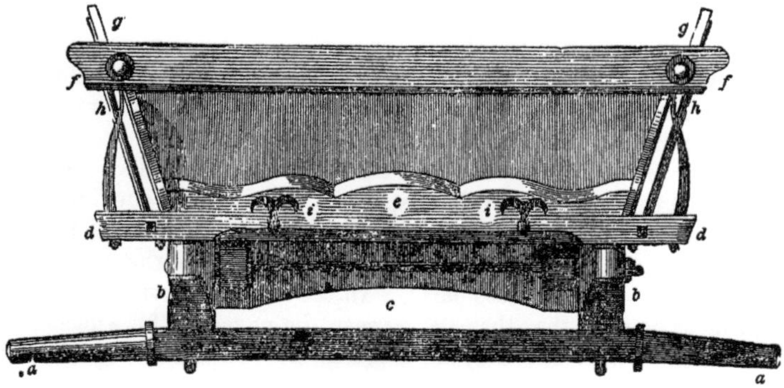

THE BACK VIEW AND DOOR OF THE TILT-CART.

(3146.) The *dormant-bodied cart*, of which fig. 569 represents a side view, has its body constructed in all respects similar to that of the tilt-cart; descrip-

Fig. 569.

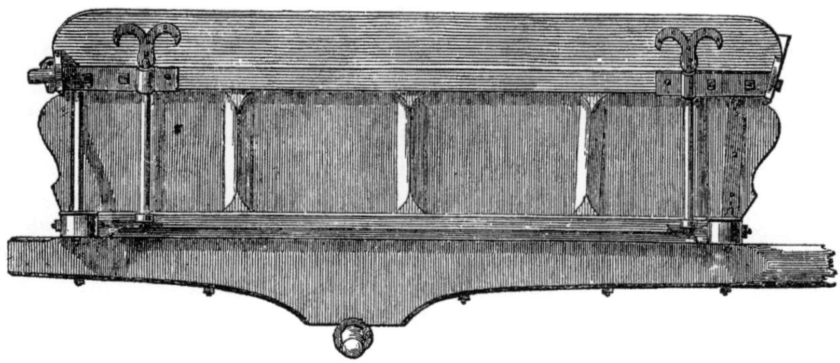

THE DORMANT-BODIED CART.

tion of it is therefore unnecessary, except as regards the shafts. This cart requires no bolsters, but instead thereof the shafts are prolonged backward, taking the place of, and serving most of the purposes, of the bolsters; and at a slight glance, when viewed on the side, the cart has every appearance of a tilt. The entire length of the shafts is about 12½ feet; those parts that fall under the body are fashioned off, as in the figure, similar to the bolsters of the tilt-cart, and of the same dimensions; of the parts before the body, the dimensions and finish are the same as in the tilt.

(3147.) A great variety of other forms of construction are followed in the formation of the dormant-bodied cart. In many examples the *body-frame* of the tilt-cart is dispensed with, the slots being then tenoned directly into that part of the shafts which fall within the length of the body, and the side-standards are tenoned into the upper surface as in the tilt. In others, while the slots are framed with the shafts, the side-standards, instead of being tenoned also into them, are attached to the *outside* of the shafts by means of double clasp-bolts. This last is a simple and durable construction, well adapted for the carriage of heavy articles, such as coal, and the like; and it gives a greater width of body inside than the former. In a fourth form, the slots, instead of being tenoned into the shafts, are simply laid upon them as bearers, and bolted. The side-standards are fixed with clasp-bolts, as in the third variety; and in all these four varieties the front of the body is formed by a bottom head-rail bolted to the shafts, the top head-rail being applied as in fig. 565.

(3148.) It will readily be observed, that in the first described variety of the dormant-bodied cart, the length, but especially the width, is equal to that of the tilt-cart, which is wider than the whole distance between the naves of the wheels, the body-frame being projected a little over these; hence the necessity of the deep bolsters and shafts of these carts. In the succeeding varieties alluded to, the width must be such that they lie between the naves of the wheels; the two varieties, however, with the clasped-standards, acquire a width inside of 3 or 4 inches greater than the second variety.

(3149.) The *two-horse* agricultural cart differs only from the one-horse tilt, fig. 557, and its details, in being of larger dimensions, but especially in depth; the length is also increased a few inches, while the width remains nearly the same, and the limbers are stronger; but all the dimensions are variable, according to the tastes and objects of the owners.

(3150.) In all carts of the descriptions here noticed, the cladding or boarding of the floor and sides is an important point,—very fine and straight-grained deal should be avoided, because of its liability to split. Of the woods best adapted for the purpose, I may name the common saugh or willow, the larch, the common Scots fir, and others of the pine tribe; and the more they abound in sound knots, so much the better are they adapted to the purpose, not only preventing the splitting of the boards, but adding to the durability of the material. The nails used for fixing the boarding should always be the common *cart-nail*, which is distinguished from other common nails, by its diminished length, increased thickness, and being chisel-pointed, qualities that adapt it for being driven into hard wood, while its thickness gives it the requisite strength to resist the hard usage that such machines are always liable to.

(3151.) Next in importance to the carts already described, come the corn and hay carts, of which there are many varieties; but in many situations, and under certain systems of management, a substitute for these is adopted, in the application of the hay-frame to the common close-bodied cart; and this, though somewhat injudicious in principle, is rather extensively adopted.

(3152.) The *hay-frame* is a light rectangular piece of frame-work, as represented in plan, by fig. 570, and in section, by fig. 571. It consists, first, of two main-bearers *a* and *b*, which are fitted to lie across the cart, the one *a* to the fore part, slightly notched upon the top-rails, and leaning against the upper head-rail, the top-sides being at the same time removed; the other bearer *b* is fitted in like manner to the back part, leaning against the door. A pair of light side-rails *c* and *c* are applied on each side, crossing the bearers, and notched upon and bolted to them with screw-bolts. These are again crossed by two rails *d* behind, and by three more *e e* in front; and as these last project over the back of the horse, they are made arch-form, to give freedom to his motions, the whole

Fig. 570.

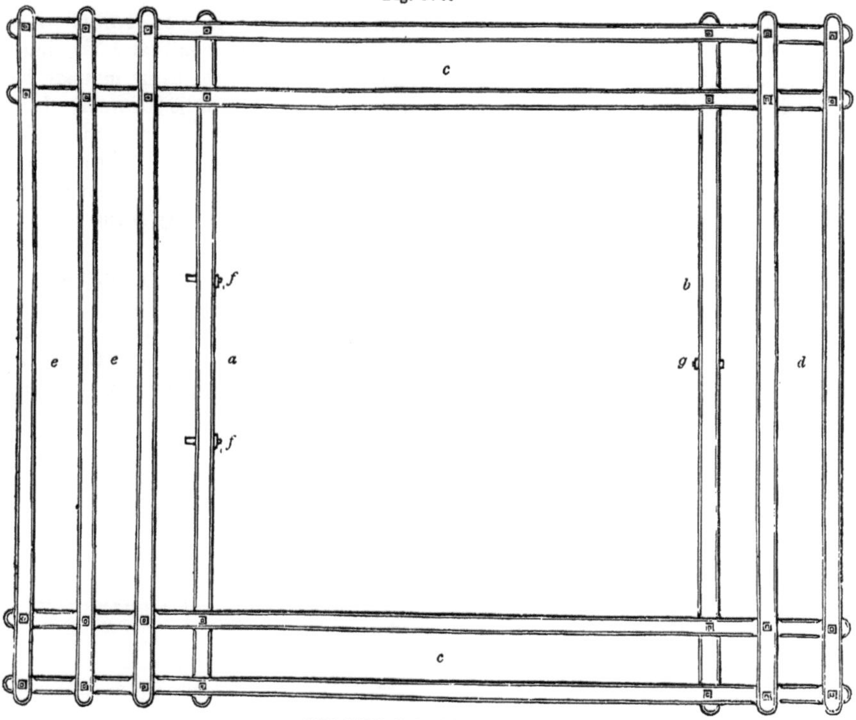

THE CORN AND HAY FRAME.

being bolted to the longitudinal rails. Fig. 571, the transverse section, exhibits the form of the bearers, and their notchings *a* and *b* upon the top-rails or

Fig. 571.

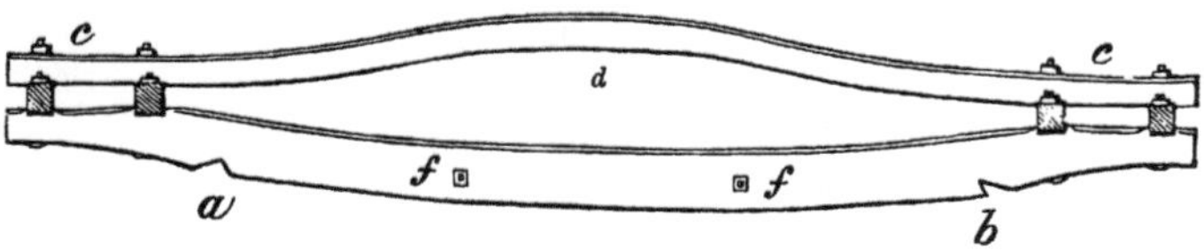

THE TRANSVERSE SECTION OF THE HAY-FRAME.

shelvements. These bearers are about 6 inches in depth by $2\frac{1}{4}$ inches in breadth; the parts projecting beyond the top-rails being tapered off from below to about $3\frac{1}{2}$ inches in depth, and the extreme length 7 feet 6 inches. The section shews also the longitudinal rails *c c*, as cut across by the section and the arched cross-rail *d*. The extreme length, from outside to outside, of the front and back cross-rails is usually about $10\frac{1}{2}$ feet, and the breadth in the same manner about $7\frac{1}{4}$ feet. Both longitudinal and cross-rails are from 2 to $2\frac{1}{4}$ inches square, and all made of hard wood, usually ash. Various methods of securing the hay-frame to the cart are adopted. These figures exhibit one simple and effective method, by means of the bolts *f f* in the front bearer, which pass through it and the upper head-rail, and by the single bolt *g* in the hind bearer, which passes through it and the door. Another method is by hook stanchions of iron, two of which are appended to eye-bolts in the crossing of the fore cross-rail and the inner longitudinal rail, and two similar stanchions appended in the same manner behind. These four stanchions have their lower ends adapted to hook into eye-bolts fixed in the horse-shafts before, and in the end of the bolsters behind. This method has the advantage of serving as supports to the extremities of the frame, as well as binding it to the cart, though the latter is not so well performed as by the simple bolts shewn in the figures. The price of the hay-frame is from 40 to 50 shillings.

(3153.) The *corn and hay cart*, being adapted to a specific purpose, is a more efficient vehicle than the common cart and hay-frame, and is of very simple construction. A side view of this cart is given in fig. 572, and a transverse section in fig. 573; in these the same letters mark the corresponding parts of the two figures. Lightness being an object in this construction, the shafts *a a* are

Fig. 572.

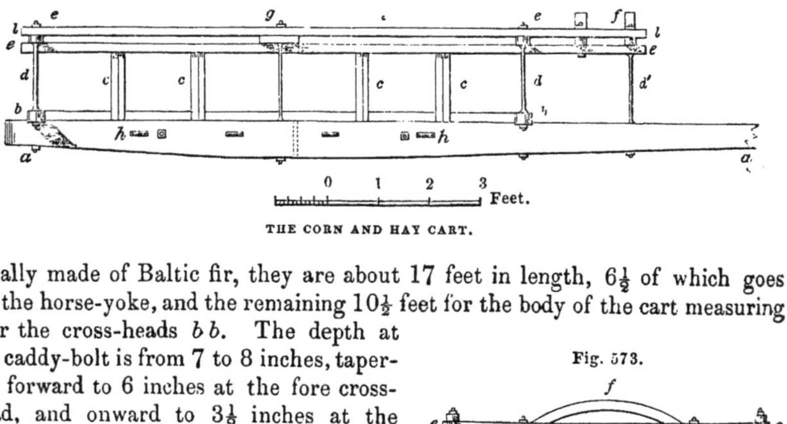

THE CORN AND HAY CART.

usually made of Baltic fir, they are about 17 feet in length, 6½ of which goes for the horse-yoke, and the remaining 10½ feet for the body of the cart measuring over the cross-heads *b b*. The depth at the caddy-bolt is from 7 to 8 inches, tapering forward to 6 inches at the fore cross-head, and onward to 3½ inches at the point; a similar diminution being carried backward, making the depth 5 inches at the back cross-head. The thickness throughout the body is about 4 inches, the yoke part being diminished to 3 inches at the point. The shafts are framed upon the slots *h*, and are further strengthened by two thorough-bolts passing from side to side of the body. The cross-heads *b b* are 7½ feet long, 4½ inches in breadth, and 2½ inches in depth, secured to the shafts by the tail of the iron standards passing through them and the shafts. Each side side is formed with four oak standards *c*, *c*, *c*, *c*, and four of iron; two of the latter *d d* are those that pass through the cross-heads, serving to bolt them to the shafts; one is placed in the centre, and the remaining one *d′* before the fore cross-head, to support the overhanging parts of the upper works. The inner top-rails are mor-

Fig. 573.

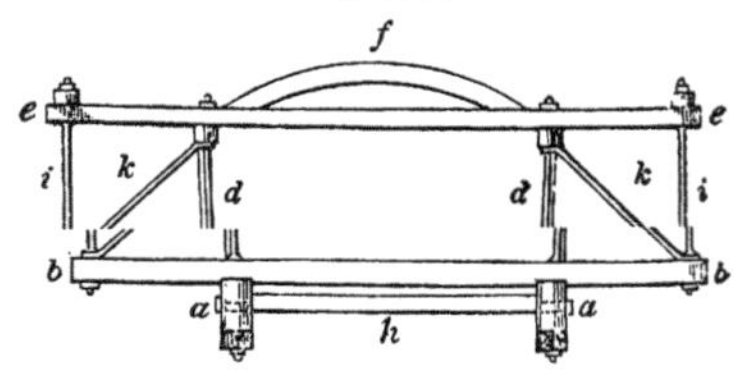

THE TRANSVERSE SECTION OF THE CORN AND HAY CART.

tised upon the oak standards, and are pierced by the upper extremities of the iron ones. The rails are 12 feet in length, usually made also of fir, and are about $2\frac{1}{2}$ inches by 2 inches. Two upper cross-heads *e e*, together with two arched rails *f*, and a middle or load-rail *g*, are next applied. The two first, and also the two arched rails, are always made of hard wood, ash or elm. The first are about 3 inches by $2\frac{1}{4}$ inches, the arched rails $2\frac{1}{4}$ inches square, and the middle rail 9 inches by 2, the whole being 7 feet in length, and are secured by the extremities of the iron stays passing through them, with screw and nut above. As seen in the section, fig. 573, the ends of the upper cross-heads are supported by an iron standard *i* planted near the end of the main or lower cross-head; and between each of these and the principals *d d*, a diagonal stay *k* is applied, to give lateral support to all the upper works. The outer top-rails *l l*, fig. 572, are of the same dimension as the former, and are supported on the extremities of the cross-rails, along with which they are secured to the outer iron standard by screw-nuts, the bearing on the middle rail being secured by a simple screw-bolt. The extreme breadth over these outer rails is about 7 feet.

(3154.) The yoke-geer of this cart, is of the same construction as that of the tilt-cart; and for the purpose of collecting any shaken-out grain, the body is usually close-floored, besides having a ledge-board of three or four inches high, running along the inside of the standard, as seen from *b* to *b*, in fig. 572. The corn and hay cart is seldom furnished with axle and wheels, exclusively attached to it, but is adapted to those of any of the common carts upon which it is placed when required. In piercing the shafts for the caddy-bolts, the same rule is observed as for those of the tilt-cart described (3159.), taking the extreme length over the cross-rail, behind the front arched-rail, and dividing this in two equal parts, setting off $\frac{1}{14}$ of the whole length backward from the middle point, gives the position of the bolts.

(3155.) A corn and hay cart, simple in construction, but possessing complete efficiency, and greater safety than the former, was contrived by a farm-servant, by name Robert Robertson, and was introduced in 1832* in the west of Fifeshire and neighbouring counties, and of which fig. 574 is a view in perspective, with its wheels and axle in full working state. The shafts and body-frame of this cart may be considered as identical with that already described, which, without the upper works, is the simple dray-cart. Upon this body-frame is placed the fore and back cross-heads *a a* and *b*, projecting beyond the body, their extreme length being $7\frac{1}{2}$ feet. The other and lighter cross-rails are applied one before and another immediately behind the wheels, and the whole bolted to the shafts. Upon these are laid a longitudinal rail $3\frac{1}{2} \times 2\frac{1}{2}$ inches on each side, and not more than 2 inches beyond the body-frame, and two similar portions of longitudinal rails are also laid on each side, extending from the fore and back cross-head to the wheel-rails; over these longitudinal rails is laid another light cross-rail behind, and the parts all secured with bolts. A light frame *d d* is raised upon the fore cross-head *a* to a height of 2 feet, with two iron stanchion rods at each, and these surmounted by an arched rail, which is supported against the pressure of the load, by two iron stays from the shafts. The outer longitudinal rails being cut by the wheels as above described, are connected again by the arched iron bars *e* and *e*, which are bolted at the ends to their respective rails, and these are connected by the broad cross-rail *f*, the arches

* Prize Essays of the Highland and Agricultural Society, vol. xi. p. 395.

rising sufficiently high to allow the wheels to have freedom to turn below the rail *f*. A side-board *g* is also raised on each side upon the body-frame, and under the arch-rail, extending a little before and behind the wheels, thus preventing

Fig. 574.

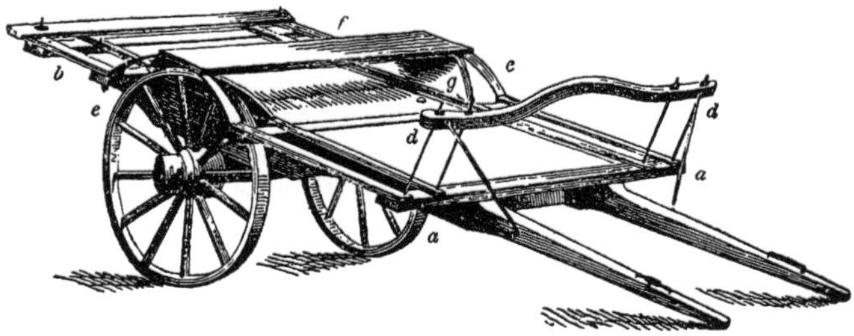

ROBERTSON'S IMPROVED CORN AND HAY CART.

the load from coming in contact with the wheels. The body-frame is floored over in the usual manner, and the space between the body and the inner longitudinal rails is filled up with hinged flap-boards, which are required about 4 inches broad, thus preventing the loss of grain that may be shaken out in the cart.

(3156.) Carts of this construction possess several advantages; from their simplicity is derived cheapness; and from the load assuming its full breadth over nearly the whole floor of the cart, at the lowest possible position; the centre of gravity of the whole load will be very considerably lower than in that of the formerly described cart, and still more so than on the hay-frame. This last quality produces greater stability, and reduces the risk of upsetting, besides affording a greater facility of loading. There is also the advantage of its easy conversion into an open dray-cart for carrying timber or the like, which is done by unbolting the cross heads and rails, and removing the upper framework in a mass. The latter quality, it may be observed, is frequently connected with the common corn and hay-cart, by having the whole of its upper works based upon two longitudinal rails corresponding to those of fig. 574, and these being attached to the two main cross-heads, the whole upper works can be lifted off the body-frame, and again secured by the insertion of four bolts.

(3157.) *The disposition of the load.* It is always of importance to husband well the energies of the horse; and in no case is it more necessary than in the cart-horse. The disposal and management of the loaded cart has frequently occupied attention; by some it has been treated in a becoming manner, though frequently too abstruse for the agricultural mechanic, or general agricultural reader.* In others, the lucubrations are neither rational nor mathematical, though professing to be the latter; but exhibit an attempt to set at nought the received statical laws, and partially to annihilate matter, or at least its effects, as evinced in the principle of gravitation, a curious example of which is to be found in a recent periodical.† I do not stop to point out the errors here alluded to, but proceed to a simple consideration of the subject.

(3158.) To facilitate the arrangement of the load in the two-wheeled cart

* Quarterly Journal of Agriculture, vol. iv. p. 683.

† Journal of the Royal Agricultural Society of England, vol. ii. p. 73.

here described, the practice has been to place the cart upon the axle, in a position that places $\frac{4}{7}$ of the body before the axle, and $\frac{3}{7}$ of it behind. Whether this has been deduced from calculation or experiment, cannot now be determined; but one thing is certain, that the above proportion seems to suit all purposes, and what is more, it yields by calculation and experiment, a result which loads the horse in the shafts with a fair degree of pressure, and such as he is quite capable of supporting through a moderate journey. The amount of this load on the back of the horse has frequently been very much overrated; and few practical people having a clear conception of its amount, it may be satisfactory to many that the true state of the matter be made to appear.

(3159.) The position of the caddy-bolts is usually determined in accordance with the above practice, or the following division affords the same result: To the length of the body-frame add half the slope of the front, or 2 inches, divide this whole length into two equal parts, and subdividing the back half into seven equal parts, the caddy-bolt holes will be in the first of these subdivisions, or at $\frac{1}{14}$ part of the whole length behind the central division. Thus, in the cart whose body-frame is 5 feet 4 inches, add 2 inches for the half slope of the front, making 5 feet 6 inches, or 66 inches in all. Half this is 33 inches, $\frac{1}{7}$ of which $= 4.71$ inches, and $33 + 4.71 = 37.71$ inches, the length of that portion of the body that lies before the caddy-bolt, while the remaining 28.29 inches lies behind, being $\frac{4}{7}$ and $\frac{3}{7}$ as before. When the cart is loaded, therefore, there will be a preponderance of $\frac{1}{7}$ of the load before the axle on which the cart-body rests.

(3160.) Taking the whole load at 21 cwt., $\frac{1}{7}$ of this is 3 cwt., or 336 lb., the preponderance of the load forward, on the supposition that it is uniformly distributed over the cart-body; but as the common practice places it more towards the front, the preponderance may, with more accuracy, be taken at 400 lb. Taking the weight of the cart-body, exclusive of the wheels and axle, at 5 cwt., and taking $\frac{1}{7}$ of this also, or 80 lb., and adding to this the effect of the shafts, with the iron mounting, which may amount to about 60 lb., and is to be considered as acting at the centre of gravity of the preponderating part of the load, we have an aggregate preponderance of $400 + 80 + 60 = 540$ lb. This quantity is to be supposed as distributed over that part of the body lying before the axle, and from the disposal of this part of the load, its centre of gravity will be at the distance of about 20 inches before the axle. Again, the shafts in this arrangement form a lever of the second order, whose fulcrum is the axle, and the load is applied at the distance of the centre of gravity, 20 inches from the fulcrum, while the power (the horse's back) is applied at $4\frac{1}{2} + 3$ feet, or thereby $= 90$ inches from the fulcrum, giving the power a mechanical advantage of 90 to 20, or 4.5 to 1, the horse therefore bears $\frac{1}{4.5}$ part of the preponderating load, which, as we have seen, is 540 lb., $\frac{1}{4.5}$ part of which is 120 lb. for the pressure on the back of the horse. This calculation agrees perfectly with experience, for by experiments on loaded carts with a dynamometer, the strain on the backc-hain of the horse has been found to range from 100 to 140 lb., when the shafts were in the due working position.

(3161.) From the want of a due consideration of the effect of the load as it presses on the horse's back, the results are frequently much exaggerated, and much uncalled-for pity and commiseration excited in behalf of the horse, by those who, not taking time to investigate, are led by their feelings into erroneous conclusions, as to the amount of suffering which, in too many cases, is inflicted most unnecessarily upon that valuable animal by his careless and unthinking driver. But the direct load on the horse's back here established, is that due to travelling on a level road, and it is liable to variation when travelling on inclined roads, always to the inconvenience and suffering of the horse. In such cases, the variation depends much on the nature of the substance composing the load. If it is of a nature to load high, the variation becomes the greater, in consequence of the centre of gravity being proportionally high; and the inclination of the cart-body and shafts following the rise or fall of the road, the position of the centre of gravity is thrown backward or forward; backward in ascending, thereby lessening the pressure on the horse's back; and forward in descending, increasing the pressure. These variations, though, no doubt, a source of inconvenience, have also been exaggerated; for I have found, from actual experiments, as well as from calculation of the effects, that in descending an inclination of road so great as 1 in 8, that the *increase* of pressure on the back of the horse is not more than $\frac{1}{3}$ of the pressure on a level road, the ultimate strain on the back-chain arising from such inclination being about 160 lb. This is, besides, a very extreme case, for few roads are now to be found with such high gradients. There is good reason, therefore, to conclude that the additional strain on the back will seldom amount to 30 lb., except with loads of corn and hay, in which the centre of gravity is necessarily high. In ascending acclivities, the pressure on the back is *reduced* in nearly the same proportion as the increase upon declivities.

(3162.) The inconveniences thus attending the variation of load in ascending and descending inclined roads, has given rise to numerous proposals for methods to equalise the pressure on the back. Hitherto, however, nothing of value has been discovered, in which the trouble and inconvenience attending the adjustment of the apparatus proposed, does not more than counterbalance any advantage that could be obtained from it. One of the most feasible plans proposed to achieve this, is to have the cart so placed on the axle that the whole cart-body, with its load, shall be made to slide a few inches backward or forward upon the axle, by means of a screw placed at the back of the cart, and may be turned by the hand. Such an apparatus may, no doubt, be rendered capable of effecting the intended purpose; but it would so encumber and break in upon the simplicity of the machine, besides requiring considerable attention, and all these to an extent that, it is to be feared, would not be compensated by advantages of equal value.

(3163.) It will always be on descending inclinations of road that the horse will be most distressed; and, from the comparatively small increase of load that is brought upon him in the descent, as has been already shewn, it seems probable that what he suffers arises more from the cart with its load pressing him *forward*, than from the direct effect of gravity upon his back. If this view is correct, and there appears much reason to believe it, it would then be of greater moment to obtain a ready and efficient means of *retarding* the forward tendency arising from gravity, by the application of a *drag*, than by any attempt at shifting the load. This has been attempted by the action of the horse in *holding back*, through the medium of the breeching-harness acting upon friction-blocks pressing against the edge of the wheel; but it seems very questionable if the method can be rendered sufficiently effective, with safety to the horse. As a convenient means of dragging, there appears no method more simple and efficient than that which we have copied from our French neighbours, and is now applied with such success to stage-coaches, namely, the lever, or the lever-and-screw drag, applied to the periphery of the wheels. This mode possesses most of the advantages required in an efficient and convenient drag, is easily managed, and not very expensive, neither would it incommode the ordinary working of the cart; and, as it can only be required in hilly districts, or where roads are not well levelled, it could be laid aside or applied, according to the nature of the journey that might be in contemplation. Attempts have also been made to apply a drag on this principle to the nave of the wheel; but this has been, and always will be, a failure, owing to the comparative smallness of diameter, and consequent low velocity of the revolving surface against which the friction-blocks are pressed, affording but a small degree of friction, unless the levers are made very strong, and to act with great mechanical advantage. But these conditions involve two other disadvantages—the additional weight of a stronger apparatus, and the risk of fracture, arising from greater pressure being requisite to give it effect.

(3164.) The *direction of the line of draught* is a point which is also deserving of consideration in its application to wheel-carriages; and in this view the structure of the horse, as an animal of draught, should be kept in mind. From the configuration of the skeleton in the horse, and his muscular system, we perceive, at a glance, that his form is adapted to produce a much greater effect in dragging a load, than in carrying upon his body. In man the case is different. From the circumstance of his skeleton, in the lower extremities and trunk, being composed of parts superimposed, and forming a column, which, though furcated below, is only thereby rendered the more stable, he is capable, though greatly inferior in weight and mass, to bear a vertical load even surpassing that of a horse, while in dragging he is greatly inferior; and it is only by inclining his body from its upright natural position, that he is able to exert beyond a very moderate force in a horizontal or slightly inclined direction. It is true, that, if his body is once so far inclined as to admit of his gravity to bear with its maximum effect, while his limbs and vertebral column are brought to act in concert with his weight to resist a force of traction applied to his shoulders, he will, in such position, produce a much greater effect, taking weight for weight, than a horse; and while a man will travel with a load on his shoulders equal to $\frac{2}{3}$ of his own weight, a horse will not do the same with more than $\frac{1}{3}$. But a horse will perform a journey with equal ease, and drag a load of twice his own weight, if placed in a two-wheeled cart. And hence we see the vast importance of the modern system of good roads and wheel-carriages, as compared with the state of things only a century ago, when roads, if existing at all, were in a wretched state, and, consequently, much of the conveyance of goods and produce was performed by pack-horses—the load laid on the back of the horse. Now, the useful effect of a horse may be estimated at six times that of the back-load practice.

(3165.) In achieving these great improvements, advantage has been taken of the physical properties and conformation of the horse, especially that of his being more adapted for drawing than for carrying. But it has also resulted, that, from his powers of draught depending, in some degree, upon his weight, as well as upon his muscular energy, it is found advantageous to give to his weight an increased effect, either by direct application of a limited load upon his back, as in the case of the two-wheeled cart, or artificially, by causing the direction of the line of draught—as in the plough—produce indirectly a force in the direction of gravity, by the oblique direction of the draught. Thus (808.) we see, that, in drawing the plough, while the horse is exerting a force of draught of 160 lb. to draw forward the plough, he is bearing, as a part of this result, a vertical pressure, at or near his shoulders, arising from the direction of the trace-chains, which here lie at an angle of about 18°. Though this angle, in working the plough with horses of ordinary stature is found convenient and effective, it does not follow that the same is requisite in all cases of draught; for, with due consideration to the comfort of the horse, it should vary with the weight and the supposed muscular powers of the different horses that may be employed; though, from the practical difficulties attending any attempt at such adjustment, it becomes necessary to adopt general principles, and to be satisfied with approximation. In no case, however, can it be otherwise than disadvantageous to the horse to make him draw horizontally; and much worse were the line of draught to incline upwards from his shoulders, as either of these would tend to deprive him of part of his natural weight.

(3166.) It appears evident, that in order to give the horse full advantage of his own weight in pulling, that the line of direction in which his force is applied should lie in a plane that would pass through the centre of gravity of his entire

Fig. 575.

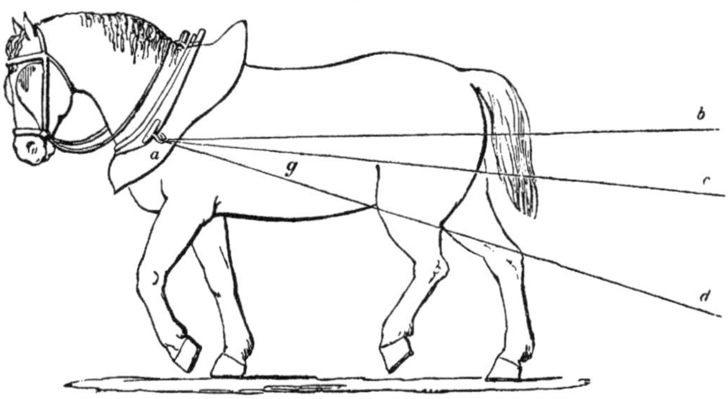

THE PRINCIPLES THAT DETERMINE THE ANGLE OF DRAUGHT.

mass. This will be more clearly evident from fig. 575, representing a horse with lines of draught extending from the hook of his collar, and which will serve to elucidate our position. The centre of gravity of the horse will lie about the position *g*, coinciding nearly with the line *a d*, forming an angle of 18° with the horizon, which is nearly the angle of draught in the plough. Hence, in the plough-yoke, the traces being at 18 , the principal condition will be very nearly fulfilled, and there will be no undue tendency either to take from or add to the effect of his natural weight. If he is made to draw at a higher angle, such as would cause the line of draught to fall still beyond the position of his hind feet, he will feel an unnatural pressure towards the earth on his fore-quarters; and if the angle is still increased till the line fall *within* the position of his hind feet, the unnatural pressure would extend to both fore and hind quarters: if the contrary, at a lower angle, as at *a b*, the line of draught lying thus above the centre of gravity, the effect will tend to lift the fore-quarters from the earth. The latter takes directly from the power of the horse; the former, though it will not reduce his power directly, will produce a necessity for greater exertion in his muscular system throughout, by having to move with what is equivalent to an increase of his weight.

(3167.) In the two-wheeled cart we can observe a very neat adaptation of the principle here brought forward; and though it appears justly adapted to suit the case, it has probably been realised more from chance than from reasoning, but most probably from a long course of observation, without reference to causes and their effects. In the cart with wheels of the height commonly employed, we have the line of draught determined by the radius of the wheel, the height of the hook in the horse-collar from the ground, and the distance from that hook to the centre of the axle. The height of the hook may be taken at 4 feet 6 inches, the radius of the wheel at 2 feet 3 inches, difference 27 inches, and the horizontal distance 7 feet 6 inches, giving an inclination to the line of draught of about 12°, corresponding to the line *a c*, fig. 575. The plane of this line lies higher than the centre of gravity of the horse, and for that reason would have the tendency of taking from his weight; but to obviate this, the necessity arising from the nature of the vehicle, that a small portion of its weight be thrown upon the horse's back, comes aptly in to restore the equilibrium. By throwing a portion of the pressure of the load upon the back of the horse, we do in effect what is laying an extraneous body upon that part of the animal, which, for the time, becoming in effect a part of his entire mass, raises his centre of gravity, and if all is properly adjusted, it will be again in the plane of the line of draught, and the fore-quarters of the horse will be nearly as free to move as before. There will, however, be an additional pressure thrown upon his hind-quarters by this extraneous weight; but this is again compensated by the average force of traction in the cart being less than that of the plough, in the proportion of 120 to 160. Although, from the nature of the subject we cannot reduce these data to any minute calculation, there appears to be sufficient grounds to establish the foregoing remarks, as being based upon the legitimate principles that govern the determination of the angle of draught. It is certainly evident that the nearer we can approach to the principle of coincidence of the centre of gravity of the horse and his load with the plane of the line of draught, so much the nearer are we to removing all unnecessary exertion of the animal.

(3168.) From these considerations, it follows that the angle of draught, so far as regards the horse, should not exceed 18°, but that it may range from that to the horizontal line, or even to ascend above it, provided that duly proportioned extraneous loads be placed upon him. We see here also a corroboration of the advantages of two-wheeled carts over four-wheeled waggons; for if the foregoing views are correct, waggons having fore-wheels of the same height as common cart-wheels, the shaft-horses in such waggons never can give out their full effect, or an effect equal to what they would do in carts; and it is curious to remark, that giving to waggons high front-wheels, which is now becoming prevalent, is rendering them less effective than under the old system of low front-wheels, so long retained for the convenience of turning. Waggons with low fore-wheels brought the line of draught to nearly the same angle as in the plough; hence the shaft-horses would give out their maximum effect, and what is more, the higher angle of draught must have served to counteract the disadvantage under which low wheels must always lie in passing over obstacles, or in soft roads producing a continuous obstacle. We see here also the disadvantage of employing two, three, or more horses in line, whether in the plough, the cart, or the waggon, for in all such cases every horse before the one in the shafts must of necessity draw upon a line nearly horizontal; and it is not too much to say, that, so yoked, a horse will not give out more than $\frac{4}{5}$ to $\frac{5}{6}$ of his full working effect. This conclusion agrees with well established practice under the Scotch system of managing horse-labour; and it is, moreover, seen to be gaining favour over the still older practice of the four-wheeled waggon; for in a report of experiments on the large scale of the reduction of horse-labour by the use of single carts, by Mr Hannam of Burcott, Oxfordshire,* a saving of expense, on the carting alone, of a farm, amounts to 19 per cent., as compared with team-labour in waggons, and no less than 47 per cent. when the pair-horse system is extended to all the operations of the farm.

(3169.) I have thus endeavoured to establish some data for the angle of the line of draught, shewing that it is not necessary to be alike in every case, and that beyond a certain limit either way it is disadvantageous to the horse. In the cart, as well as in the plough, it is confined practically within a small range of angle; thus the height from the ground to the centre of the axle gives nearly an invariable point. The hook of the collar is subject to small variation, arising from the height of the horse and the distance between the hook and the axle; and in some descriptions of carts, is liable to more considerable variation; but, on the whole, the range will be between 8° and 13°; and it is to be kept in view, that the straight line lying between the hook of the collar and the centre of the axle is invariably the true line of draught. It is not affected by any bolstering up of the cart upon the axle, nor by lowering it, though these two conditions have an influence on the load, as they tend to raise or lower its centre of gravity; and to keep that low should be an invariable rule. In yoking the horse in the cart, the back-chain should be so adjusted to length as to place the shoulder-slings or draught-chains in the *straight line towards the axle;* every deviation from this places the horse at a disadvantage. If the back-chain is too long, so as to make the draught-chains range towards a point below the axle, the horse will be unnecessarily loaded on the back, causing undue exertion both before and behind; if the contrary, he will be deprived of part of the benefit of his weight, and his fore-quarters will not have their due share of effect on the draught.

(3170.) For horses employed in light carriages going at a high speed, the draught-chains or traces are not required to be at such a high angle as for heavy loads at a low speed, the amount of draught in cases of high speed becoming very small as the speed increases, and the effect of the angle of draught has little influence on the amount of exertion, and in all such cases it is better to approach as much as possible towards the horizontal line.

(3171.) This leads me to consider the effect of springs on loaded carriages. Of the efficacy of springs in fast carriages there can be but one opinion, whether we regard safety, comfort to man and beast, and ease of draught. For all vehicles also employed for carrying the lighter description of loads, the advantage of springs seem equally well established; but for all such purposes they should possess the property of admitting undulation in the vertical direction only, with as little horizontal motion as possible. The latter, especially in the direction of the draught, produces constant inequalities in the resistance to the draught, from the swinging motion of the vehicle, whereby the momentum which at any

* Journal of the Royal Agricultural Society of England, vol. ii. p. 73.

one instant may have been acquired by the carriage, may be destroyed by a swing of the suspended body in an opposite direction. The grasshopper-spring, in some of its modifications, is the one to be adopted in all such cases. The effect of the grasshopper-spring being to produce vertical motion almost solely, has the beneficial effect of greatly lessening the jolts and shocks that vehicles are subject to on the very best of roads, and in the case of small obstacles being presented to the wheels these springs perform a very important part. Instead of the whole mass of the carriage being instantaneously elevated upon the obstacle, the wheel and the axle, or, it may be, the bed-frame of the carriage, are in the first instant raised, and are allowed to do so by the yielding of the springs before this upward motion is communicated to the suspended part of the vehicle, thus dividing the upward impetus between the two. In descending from any such obstacle, the reverse of this process occurs, the fall being in like manner broken by the elasticity of the springs; but this is not all the advantage, for in thus passing over a stone, or any obstruction that occupies but an instant of time, the yielding of the spring before elevating the load, will, in effect, reduce the height of the obstacle to an extent proportioned to the elasticity of the spring, thereby requiring less exertion of the power to take the carriage over it.

(3172.) The useful application of springs to heavy loaded carriages, has also been frequently recommended by those who may be competent judges; and it must be admitted, that they might be applied with beneficial effect, in so far as regards the draught and ease to the animals of draught; but a serious obstacle intervenes, the expense,—and especially if the expense incurred would balance any advantage that might accrue from their adoption. The cost of a set of springs for a single-horse cart, supposed to bear a load of 1½ ton, cannot be less at the present time than L.3, which adds more than 25 per cent. to the price of the cart complete without springs; and we are yet without any certain data, by which we can calculate that a horse would take a greater load upon a cart so constructed. There can be no doubt, that roads and horses would be benefited by the introduction of springs to all carriages; but it is yet an unsolved problem, whether, on the whole, it presents that amount of economy to make its adoption an object to the farmer.

(3173.) There is one other point yet in connection with the draught which is deserving of notice, and it bears also some relation to springs. I have said, that if springs are applied to carriages, they should be so constructed as to afford no elasticity in the longitudinal direction. Such elasticity destroys the impetus that may have been generated in the carriage, in the case of the wheels being checked by any obstacle; and instead of that impetus being directed to the carrying of the vehicle over the obstacle, the sudden shock allows the whole of the momentum to expend itself in causing the suspended body of the carriage to be swung forward, while the whole mass is nearly, or altogether, brought to rest. This is, without doubt, true, both in theory and practice; but the theoretical part of it has been carried further than this, for we find a clever writer, already quoted,* asserting, that not only should there be no elasticity longitudinally between the wheels and the carriage, but it should also be entirely banished from between the wheels and the horse. This is certainly not agreeable to practice; and, what is more, it is contrary to a well-known principle, that impetus or momentum will enable a body to overcome obstacles, which it could with difficulty accomplish if starting from a state of rest. A horse, there-

* Library of Useful Knowledge, the Horse, p. 445.

fore, at the first instant of his motion, having not yet acquired any momentum, but moves simply with his initial force, will not have an effect upon the load on a carriage equal to what he would acquire in the succeeding instant, when his initial force has been multiplied by his velocity. I apprehend, therefore, that there ought to be a distinction between that longitudinal elasticity that may exist in a carriage betwixt the wheels and its body if suspended on springs, and that betwixt the wheels and the animal of draught, or the power, whatever it may be. Every-day experience teaches us this fact; for who has not observed that the draught chains of a horse that have stood all the ordinary strains of dragging the same loaded cart, plough, or machine, for days or weeks, when the horse is allowed to take a step in advance before they have been brought to their bearing, snap asunder; and what can have produced this fracture but some power more intense than the chains have hitherto undergone—the increased momentum of the horse? Such being the case, there must be some advantage in having a degree of elasticity interposed between the shoulders of the horse and the ultimate resistance—the wheels of the carriage. Independent of the occasional advantages afforded to the horse for overcoming obstacles, such an arrangement will yield much ease and comfort to the animal; and, fortunately too, nature has provided the means for this compensation, both in the animal structure and in the physical condition attendant upon inert matter. Thus, in the animal frame, those parts upon which the actions under consideration depend, are more or less yielding and elastic; and again, the substances that form the media through which the force of the animal is applied, both from their inherent nature and the external laws under which they act, necessarily partake of the principle of elasticity. For example, if a chain is employed as a medium, its *gravity* has a perpetual tendency to make it assume the *catenarian* curve, and the change of form, from the curve towards the straight line, produces a beautifully elastic spring. In the plough this is particularly the case, as has been before observed (805.), and it exists more or less in every arrangement of strain or draught, from the chain-cable of our floating bulwarks, through the tow-line on the canal, the buffers, connecting links, and springs of the locomotive, down to the meanest office in which either inert matter or *animals* of draught are employed as a medium of traction. Where a less rigid material than iron is employed, such as hide, leather, hemp, and other substances, though they partake less of the catenarian elasticity from their lower specific gravity, a compensation is afforded by their inherent elasticity. We may here conclude, that where nature has so largely provided a principle which appears to be on all occasions advantageous, and conservative of both living and inert matter, it were folly in us to attempt counteracting it, and much more so when, from its abstraction, we should earn nothing but disadvantages.

(3174.) The inference to be drawn from the above remarks is, that instead of endeavouring to render the *yoke* of a horse, or all that intervenes between his shoulders and the cart-axle as rigid as possible, we should, on the contrary, give it elasticity, and by so doing we shall not only render the labour of the horse more endurable, but we prevent innumerable accidents, by breakage, to the substances that are employed as the medium of draught—chains, traces, swing-trees, and the like.—J. S.]

(3175.) This seems a favourable opportunity for describing the *yoking* of horses to the cart. The harness required for work-horses, in double and single carts, has been already enumerated and described in (2960.); and you will find a pretty good representation of a double-horse cart *in yoke* in Plate XVIII. On placing that plate before you, you will more easily understand what is about to be said. The *shaft-horse* requires the following harness to be fully equipped,—bridle, collar, haims, saddle, and breeching. The use of the bridle, collar, and haims, have been already described in (1115.) to (1117.), both inclusive; and these constitute the harness common to both plough and cart. The breeching is buckled to the back part of the wooden tree of the saddle, at such length of strap as suits the length of the horse's quarter. The saddle—as saddle and breeching together are commonly called—is placed on the horse's back immediately behind the shoulder, and strapped firmly on, in case of slipping off in the yoke, with the belly-band, which can scarcely be seen in the plate; the breeching being put over the horse's hind-quarter. Time was when a crupper was a general appendage to the breeching; the effect of which was to place an undue pressure upon the root of the horse's tail, when the saddle was pressed forward by the back-chain, on the cart descending a declination. Now that the comfort of animals is better attended to, by the removal of annoyances to the work-horse, that of the crupper is not the least. The back-chain is fastened to the back-chain-hooks of the shafts of the cart, and there gets leave to remain constantly. The shafts are held up with their points elevated; the horse is *told* to turn and back under them, which he does very obediently, and even willingly; they are then brought down on each side of the horse; the back-chain is adjusted along the groove of the saddle, to such length as that the draught-chains, when extended, shall be in a straight line to the axle; the shoulder-slings, or draught-chains, are linked to the draught-hook of the cart (3144.), at such length as to be extended in the above line; the breeching-chains are linked to the breeching-hooks (3144.), of such length as to allow the breeching to hang easily upon the hams of the horse—not to chafe the hair—in his motion forward upon level ground, but so tight as that before the back-chain-hooks, with the back-chain, slip as far back as they can upon the runner-staple, (3144.), the hams of the horse shall press against the band of the breeching sufficiently to keep the cart back, and to keep the cart back completely, before the horse's rump shall touch the front of the cart. The cart belly-band is then buckled round the near shaft under the runner-staple, just as tight as not to press against the horse's chest on level ground, and only when he goes up-hill. All these adjustments of parts are made in a short time, with even a new horse, cart, or harness, and they require no alteration afterwards.

(3176.) The harness of the *trace-horse* is simple beyond the collar, haims, and bridle, consisting only of 2 back-bands, belly-band, and trace-chains. The back-band is placed where the saddle should be, and is fastened to the trace-chains on either side with a buckle-like long-tongued hook. The trace-chains are linked to the draught-hook of the haims at one end, and fastened by a hook at the other end to a staple in the under side of the shafts; the point of which hook is always placed in the inside, to put it out of the way of taking hold of any thing passing near the shafts of the cart. The trace-chains are usually in 2 pieces, divided at the stretcher, the stretcher, and *short-ends*, as the pieces attached to the cart are usually called, being left attached to the cart when the horses are unyoked. A hook on each side of the stretcher attaches the short-ends to the other part of the trace-chains. The use of the stretcher is solely to prevent the trace-chains chafing the hind-quarters of the trace-horse. The trace-chains being distended from the haims to the shafts, the back-band is hooked on to them, so as always to lie firmly on the horse's back; and the belly-band is also hooked in like manner to the same part of the chains, to keep both ends of the back-band firm. The rump-band is hooked on to the trace-chains, so as to lie easy on the rump when these are distended; and the position of this band may vary farther or nearer to the loins or rump, as it may lie best, its use being solely to keep the chain and stretcher from falling on the horse's heels in turning. The horses are now ready to start, in as far as the harness is concerned; but no cart, single or double, should be allowed to be used without double reins, as already mentioned in (2960.). In some parts of the country, as in Forfarshire, the trace-horse is harnessed with back-strap, rump-strap, and crupper. The back-strap is hooked on to the upper part of the back of the collar at one end, and is held firm by the tail by means of a crupper at the other. The rump-strap proceeding backward from the back-strap, suspends the stretcher upon the chains immediately behind the horse's hams and under his tail; and in this case there are no short-ends, because the trace-chain extends the whole length, and, when the horse is unyoked, is unhooked from the cart, and hooked upon the chain at the end of the stretcher, to keep it from trailing on the ground. The effects of yoking the trace-horse in this manner is to confine his action to an unreasonable degree, for, when the bearing-rein is placed over the top of the haims, his mouth is drawn up and held tight by the crupper through the means of the back-strap, and the stronger he pulls, and the more the lower part of the collar is drawn back by the strain on the trace-chains, the more is the upper part of the collar drawn forward, and the strain upon the crupper increased, by which the root of the tail is unduly pressed against, while a part of the shoulder a little above the draught-hook of the haims is also unduly pressed upon by the collar. All this annoyance is inflicted upon the trace-horse for no apparent purpose but to prevent the stretcher sinking down upon the horse's hocks when the chains are slacked; and the contrivance certainly effects this advantage, but at much annoyance to the animal.

(3177.) To loosen the horses is to undo what has been done in yoking; the reins are first taken off and coiled up; the stretcher is unhooked from the chains, and it and the short-ends brought over the head of the shaft-horse and laid upon the shafts of the cart behind him, and the trace-horse is then free; the cart belly-band is then unbuckled; the draught-slings and breeching-chains are unhooked; and on the shafts being raised up, the shaft-horse is free; and on the bearing-reins being slipped over the top of the haims, the horses' heads are also free to take a drink of water, or to shake themselves. The cart should always be under cover in a cart-shed when not in use, as when not so accommodated, and being an implement composed of many parts, the weather soon has an injurious effect upon its upper works. When backed into the port of a cart-shed, the shafts are easily put out of the way of the horse by hanging the back-chain upon a hook suspended by a chain from the balks of the shed, when not floored above, but when so, the hook is suspended from a joist.

(3178.) The *grease* used for farm-carts is commonly a mixture, melted together in equal parts, of tallow or train-oil and common tar. It is kept in a deep narrow tub, and applied with a broad pointed stick. The tub should have a cover, but is usually without one, and subject to collect dust in the cart-shed. When a cart is to be greased, the linch-pin and washer are removed from the projecting point of the axle, fig. 557; the upper part of the wheel is then pulled with such a jerk away from the cart, the lower remaining in the same spot of ground it was, that the point of the axle-arm shall lean upon the inner end of the back of the nave. The grease is then spread upon the upper side of the axle-arm with the stick, the wheel pushed back to its place, and the washer and linch-pin respectively restored to their proper places in the projecting point of the axle. The groove of the saddle is also greased, to lessen the friction of the back-chain when playing over it.

(3179.) [I have in an earlier part of the work made reference to weighing-machines for in-door purposes, and, in particular, for those of the barn (1893.) and (1894.) and fig. 356.; but in close connection with the foregoing subject, stands another weighing-machine, the cart-steelyard, which, by judicious arrangement, may be applied to many other objects in the weighing department. It is not in the weighing of carts, nor of grain alone, that the farmer should be interested; there is nothing that the farm produces, nor that is received or delivered upon it, that is not deserving of being weighed and registered. Not that every particle of produce, nor every cart-load of manure should be precisely weighed, but large samples of both might be taken, upon which a fair average could be struck; but in many cases, such as the produce of a particular field that has been undergoing a certain routine of management, the whole produce may be taken with comparatively little trouble if the means of doing so were at hand, and the certain result would be a source of high satisfaction to the active and industrious farmer. In all manufacturing establishments, the raw material consumed is carefully ascertained as well as the finished goods. Farming is but another name for a manufacture of goods, the most important of all others—the staff of life, and why should the farmer not adopt the same correct system in conducting the details of his business, that is every day followed in the establishments of those who are employed probably in the re-manufacturing of his produce? It is not meant to be urged, that he could gain any thing directly by adopting the practice of weighing all his outgoings and incomings of manures, produce, &c.; but the satisfaction of knowing at all times what he is putting into this field, or drawing from that, in tons and hundredweights, is a matter of no small importance; and, what is more, it would form part of a system based upon close attention and vigilance towards the details of the business in hand, that is the mainstay of all sound management.

(3180.) It happens unfortunately for the dissemination of the practice here recommended, that there is no single machine at present in use that is adapted to all the purposes requisite for the out-door weighings of the farm. Machines there are in abundance admirably adapted to all the necessary weighings of the barn, and, in like manner, are they procurable and equally well adapted for the weighing of loaded carts; but there are numerous occasions always occurring in which the weight of bodies and substances, intermediate to a bag of grain and a cart-load, such as the weighing of stock of all kinds when in the progress of fattening, and the quantities of food spent upon such animals, for some of which purposes the present cart steelyard is not well adapted, and for others of them quite unfit. I do not here take upon me to point out the remedy for these defects, but simply to shew that they exist, and I have no doubt that were a really serviceable weighing-machine, of more general application and of reasonable price brought out, that every farmer who values his own interest would soon adopt it. In the mean time, it may be stated that the principles of the cart steelyard are such as would appear capable of being converted into such a machine as is yet wanted, and under this impression the following description and figures of it are here placed among the other machinery and implements of the farm.

(3181.) The cart-steelyard or weigh-bridge, which is here exhibited by fig. 576, a geometrical plan—fig. 577, a longitudinal section—and fig. 578, a transverse section, is a machine in which a combination of levers are employed to effect, in a commodious way, the weighing of bodies of considerable weight, and which would require the common Roman steelyard of most inconvenient dimensions, or a balance equally cumbrous, besides the inconvenience of a great mass of movable weight. The combination consists of two double-fulcrum levers of the second order, combined with a single lever of the first order. The relation of the arms of the first are 3½ to 1, and of the second 8 to 1, making the ultimate ratio 28 to 1, so that every cwt. placed upon the platform of the machine is balanced by 4 lb. on the scale-board attached to the second lever.

(3182.) In describing the construction of this compound steelyard, we have in fig. 576, the ground or geometrical plan, the bed-frame *a a a a*, which is

Fig. 576.

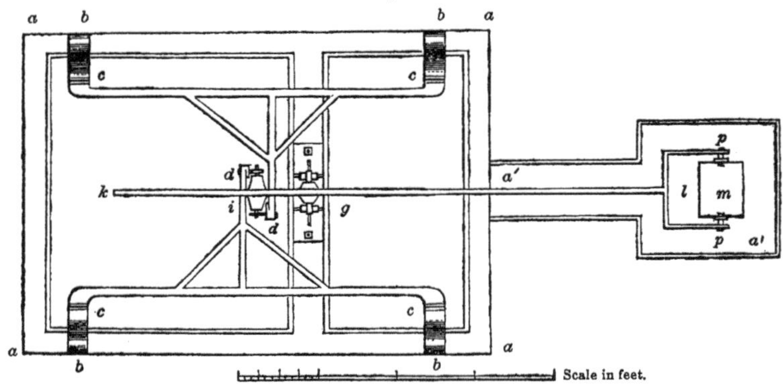

PLAN OF THE CART-STEELYARD.

6 feet in length by 4 feet in breadth, the bars being about 5 inches broad and 1 inch thick, except the middle bar, which is 6 inches, the whole surrounded by a stand-up flange 2 inches in height. This frame is laid in a pit formed of masonry, *x x*, figs. 577 and 578, adapted to the size of the frame, and having the surface of its foundation course laid level, with a footing inwards, at a depth of 1 foot 8 inches under the surface of the ground, sufficient to bear the sole-frame, and upon which it requires to be solidly bedded. Four blocks *b'*, *b'*, *b'*, *b'*,

Fig. 577.

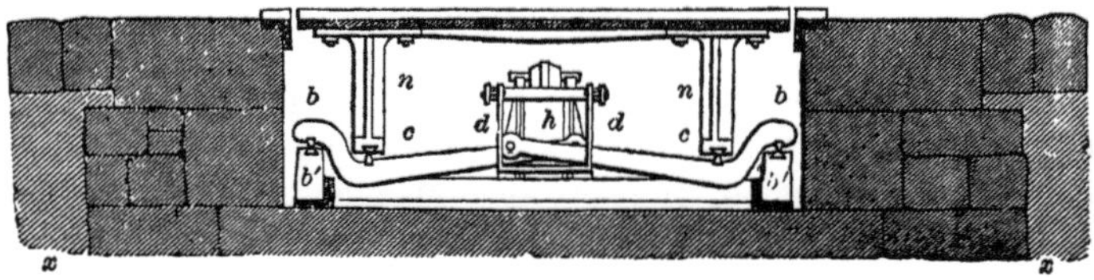

TRANSVERSE SECTION OF THE CART-STEELYARD.

are cast upon the upper surface of the frame, standing at a height of 6 inches above it, as seen at *b′ b′*, in fig. 577; they are faced on their top-surface with a cradle of steel, forming the dead fulcra of the levers. The two first levers *b c d*, *b c d*, fig. 576, seen also in profile at *b c d*, *b c d*, in fig. 577, are so formed in the horizontal direction, as to bring their points of bearing at *b*, *c*, and *d* to the requisite position; and in the vertical direction, as seen in fig. 577, to bring the *centres b c d* into one plane. The *centres* of these two levers at *b* and *c* are steel pieces, worked off to a knife-edge, and fixed into a dovetailed seat in the levers, while that at *d* is a steel pin passing through the end of the lever, the *centre* being formed on the pin at one side, as seen in fig. 576; and when the levers are duly placed, their ends *d* pass each other about 6 inches, being thus suited to the *centres* of the second lever. The second lever *k l*, seen in figs. 576 and 578, has its main centres *g* supported upon the two arms of

Fig. 578.

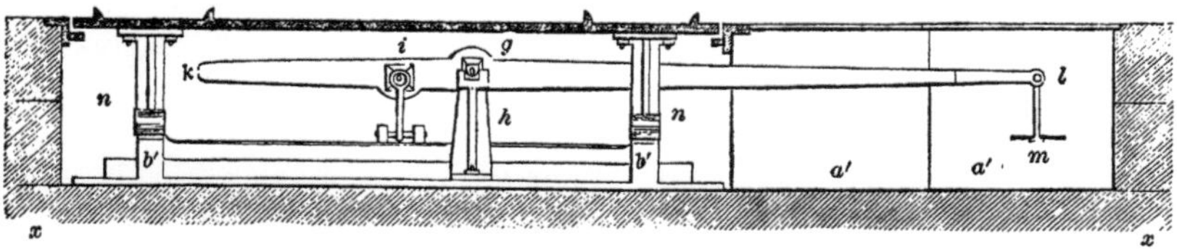

LONGITUDINAL SECTION OF THE CART-STEELYARD.

the standard *h*, raised upon the middle bar of the sole-frame, as seen in figs. 577 and 578. From the *centre g* to *i*, the distance is 8 inches, and from *g* to *l* it is 64 inches, or 8 to 1; the extremity *i* to *k* being for the purpose of adjusting the equilibrium of the machine. Upon the centres *i*, of which there are one on each side of the lever, links are appended, which, in their lower *bend*, receive the centres *d d* of the two first, and the extremity *l* is formed into the fork, *p p*, fig. 576, upon the centres of which the scale-board *m* is suspended. The chamber *a′ a′*, in which the second lever vibrates, is of the same depth as the main pit, and may be formed with masonry, or a casing of cast-iron, with covering plates to inclose it. The platform, which is left out in fig. 576, but is seen in profile in figs. 577 and 578, is also a frame of cast-iron, whereof the end bars are 15 inches in breadth, with raised ledges, as seen in the section fig. 578, to guide the cart-wheels when being placed upon the platform. Four pendent pillars *n*, figs. 577 and 578, are bolted to the lower side of the platform, and steel cradles are fixed in their lower ends, which are adjusted to bear equally on the centres *c* of the first levers. When the platform is thus placed on its centres, and the scale-board suspended on its centres at *p p*, fig. 576, if the equilibrium is not perfect, it is to be adjusted by adding to or taking from the back end of the second lever. A light cast-iron frame is also laid into the masonry round the edge of the pit, having guides to lead on the cart-wheels, and also studs projecting inward, coming under the platform at the wheel-tracks, these last serving to bear not only the weight of the platform when unloaded, but to receive the shocks of the load when coming upon it, thereby saving much of the tear and wear of the centres.

(3183.) In this form of weighing-machine, it will be observed that the platform can have no lateral motion, which would be a defect if it had a great range vertically; but this being extremely small, the want of lateral motion does not affect the accuracy of the indications. In those machines having slight lateral motion, the principles of the levers are the same as here described,—the chief difference being this, that the platform is suspended upon links, to accommodate which the pillars *n n* are lengthened downwards, and hooked under the first levers, so as to rest upon the links suspended from them.

(3184.) A more elaborate form of the machine is in frequent use where the business is extensive; it consists in the application of a third lever, which is so arranged as to stand within doors in the weigher's office. This lever is graduated, so that *one weight* adapted to slide along upon its edge gives the result of any weight that comes upon the platform down to a certain denomination, suppose stones, or $\frac{1}{8}$ of a cwt.; and by the aid of a second and smaller weight, properly adjusted, the result in lbs. may be obtained, and so on. There are still other machines used for the purpose of weighing loaded carts, but they are less perfect, and therefore not to be so much depended on. The price of cart-steelyards, as here described, range from L.15 to L.25.—J. S.]

---

80. OF EGGS.

> "Mine honest friend,
> Will you take eggs for money?"
>
> WINTER'S TALE.

(3185.) The treatment which every kind of poultry should receive in autumn is that recommended in spring (2485.), to be followed in summer. Geese, ducks, and turkeys, should only hatch in spring and the early part of summer; for late hatchings of these never produce birds worth rearing, as they cannot attain in the same season a tolerable size and degree of fatness. But common fowls may be hatched throughout the summer, and even to a late period in autumn, and the chicks be reared to a useful state. What can constitute a more delicate dish than a chicken, boiled, roasted, or broiled, at any season, but especially when the productions of the garden are in the highest state of perfection in summer and autumn?

(3186.) Hens are fond of making their own nests and bringing out broods in corn-fields, and at the roots of hedges and shrubs; and *when hens have their liberty during the day*, it is impossible to prevent their following this inclination, which is common to all, under the best regulated system; and so high a value do I set upon liberty to these creatures, on the score of health and strength of body, and flavour in flesh, that I would rather run the risk of losing a few broods in the year, by the fox and polecat making free with a self-set hen or two upon nests of their own seeking in corn-fields, than see them cribbed in summer within a court of the largest dimensions.

(3187.) One of the daily cares of the hen-wife in summer and autumn is the *gathering of eggs*. Whenever a hen is observed to shew a desire to lay, a nest should be provided for her in a quiet and convenient place, and if she is directed to it at the commencement of her laying, she will continue to frequent it ever after, if undisturbed; but a nest is not required for every laying hen, as several will lay in succession in the same nest, some hens laying earlier in the day than others; and so tenacious are they of their right to particular nests, that two hens will not unfrequently occupy the same nest at the same time.

(3188.) Every place is not equally suitable for a hen's nest. In other places than the hen-house, hens are not fond of laying their eggs on a level with the ground, though a quiet corner in a shed under some shelter is not unfrequently selected by themselves for the purpose; but they usually prefer to be elevated above the ground, such as in the mangers of stables, or in a trough of a shed or hammel, or upon the top of the wall of a stable, byre, or outhouse, immediately below the roof. When nests are made in such places as hens would themselves prefer, they are much more likely to be frequented than when an opposite determination is taken. One reason, perhaps, for their preference to the manger of the work-horse-stable is, that in daily picking up morsels of food dropped there by the horses, while the latter are at work in the field, the manger presents a convenient place on the spot when the pressure for laying overcomes them.

(3189.) Ducks are very careless layers, dropping their eggs wherever they seek for food, and these, when discovered by the pigs, are champed up as the most delicate morsels that fall in their way. Common rooks watch for such stray eggs, and carry them off even out of the courts of the steading. To secure the eggs of ducks, the birds should be examined before being let out in the morning, and those indicating hard with egg should be confined in the hen-house till they have laid, and afterwards let at large.

(3190.) It should be the hen-wife's duty to visit every nest and collect the eggs from them every day, and the period of creating the least disturbance to the poultry in this duty is the afternoon, say between 2 and 3 o'clock, before the birds begin to retire to roost. A nest-egg should be left in every nest; because it is an old established fact, that all domesticated birds, at least, prefer to lay in nests containing eggs to those which are empty. Eggs are most conveniently collected in small hand-baskets; and a short light ladder will give easy access to all nests elevated above reach from the ground.

(3191.) Whether eggs are retained for use at home, or disposed of to the egg-merchant, it is of importance to keep them in a fresh state for some time. This end may be attained by preventing the air penetrating the pores of the shell, and the yoke coming in contact with its inside. A simple and effectual way of preventing the air penetrating the shell is to rub the egg over with butter when taken warm from the nest; and as simple a way of preventing the yoke adhering to the inside of the shell, is to roll the egg from one side to the other every day. This treatment eggs should daily receive, whether kept for your own use or sold to the dealer; and it will preserve them in quite a fresh state for several weeks. But I need hardly inform you that this is not the general mode of treating eggs in farm-houses, whether intended for use at home or for sale, they being commonly kept in promiscuous heaps, and used in the state when taken from the nest; and if they are habitually used on the day they are laid, no unnecessary trouble is required to be taken with them. When trouble is desired to be avoided with eggs, they are sold to the dealers every week in spring and early summer, when most abundant; but the price is then very low, not above perhaps 4d. per dozen—a price unremunerative for the least degree of care bestowed on fowls. If it is desired to render eggs a remunerative item of farm economy, they should be preserved in a fresh state until the season of scarcity arrives, when, of course, they would realize a fairer price. It is easy to preserve eggs to that season by simply smearing them with butter immediately from the nest, and setting them on their small ends among salt in barrels in a dry room. They may be kept in this manner for 7 months, at least I have found them *quite fresh* at the end of that period, so much so, that they contained the very *milk* which constitutes the criterion of an egg being fresh when boiled; and how much longer they would have remained in that state I cannot say, as I never tried the experiment farther. I have heard of an instance of eggs packed in salt becoming so salted as to be useless; but in this case the salt might have been damp, and the eggs might not have been greased at all. Both circumstances are material, for without grease air cannot be excluded, and I know that new-laid eggs will soon become salt when floated in *brine*. Eggs will also keep fresh in a barrel of lime-water; but the dry salt, in a dry chamber, is a much more agreeable material to put the hand amongst every day. In short, they may be kept in any way that will exclude the air entering the shell; but whatever other additional expedient is adopted, that of greasing them with butter immediately from the nest should never be neglected.

(3192.) Hens begin to lay about the beginning of March, and continue to the beginning of October. They do not lay every day, that is, every 24 hours, some laying every other day, and some missing 1 day in 3. They lay about 2 dozen of eggs at one period, then cease for 2 or 3 weeks, and again lay other 2 dozen, and so on for the number of months mentioned. Of all these months, however, they lay most constantly in March and April. After each period of laying they are inclined to sit on the eggs, and when it is not desirable for them to incubate, it is difficult, in most cases, to drive them from their propensity to *clucking*, as it is commonly called.

(3193.) Many cruel expedients are practised by country people to prevent hens clucking, such as ducking them in water, dipping them in water for a few seconds, pulling feathers in a particular manner out of particular parts of their body, and such like barbarities ; all of which I believe to be ineffectual, at least I never saw a single instance of their success. The only effectual plan I know, without giving bodily pain to the animal—for if it is desired to have hens to lay eggs in preference to hatching chickens, it is requisite to remove from them the desire to sit—is to place them in darkness, and there deprive them of food and water, for 2 days and 2 nights, and, in difficult cases, for the 3d day. The simplest means of accomplishing this is to procure a number of light-made tubs, each just large enough to hold 1 hen within it when standing on her feet, and to have its top and sides pierced with holes to afford air, but so small as exclude light, and to give them such an inclination as the hen may not see through them. Such a tub is placed mouth downward over a hen, in a quiet place, such as an out-house, and not in the hen-house. In this position, the desire to sit will be removed from her in 2 days, and in obstinate cases in 3 days. It is not an uncommon practice to whelm an ordinary tub over a number of hens, with one side of its mouth raised a little from the ground, the effects of which contrivance are to allow as much light to enter the tub as to let the hens see to fight each other, and in many such cases they are removed with their scalps bared to the bone, and even pecked and trampled to death.

(3194.) Neither dogs nor children should be allowed to run after laying hens, as such treatment obliges them to lay their eggs before they are provided with the shell. Guinea fowls are incessant chasers of hens.

(3195.) Ducks are great layers, laying an egg almost every day. They commence at the beginning of April, and cease at the season the bean usually comes in bloom—in July.

(3196.) Autumn is the time to select hens for laying the eggs in the ensuing winter. They should all be young, but of different ages, that a succession of layers may be maintained during the season, as I have mentioned before in (1708.), and the food they should receive I have also mentioned in (2485.). Like the eggs collected in summer, those dropped in winter should also be smeared with butter when taken from the nest. Hens readily take to the nests made for them at this season, evincing little desire to make them in the fields for themselves. Autumn is the season of moulting with fowls. The only care requisite under this periodic visitation, is to keep the creatures warm at night, for chillness of air may be looked for at nights in October.

(3197.) *Pigeons* will produce a young brood in every month in autumn, and in this season of heat and abundance of food, both old and young become full sized, and in fine condition. They should be fed in common with the other poultry; but though this be regularly practised, they are fond of going to the fields in search of variety of food. Fields of new-sown pease, oats, barley, and wheat, are eagerly visited by them in quest of grain that has escaped burial by the harrow. Even the turnip-seed is eagerly sought for in a new sown field; and where the wheat crop is laid by the weather, there they will congregate in flocks, and pick out every kernel within their reach. For wheat and pease they will fly miles to obtain. It is, no doubt, provoking to see corn nearly ready for the sickle destroyed in the field by any kind of animal; but a farmer regards the depredations *of his own pigeons* in a pardoning mood.

(3198.) Turkeys and geese commit sad havoc in corn-fields situate near the steading, for a week or two before harvest—a sight intolerable to the farmer, not so much on account of the quantity of grain actually consumed, as for that scattered about and trampled down. When corn happens to be growing near the steading, all the poultry should be confined within court-yards for a time, and supported entirely by hand. A few sheep-nets stretched over the court-walls of hammels, afford an easy means of confining a large assortment of poultry, and of subdividing them into classes.

(3199.) The egg, being a remarkable object, and produced in such abundance, and so universally esteemed as an article of food, a few particulars obtained by men of science of its nature and constitution may prove interesting to you.

(3200.) "The *eggs* of all birds," says Dr Thomson, "so far as they have been examined, bear a striking resemblance to each other. They consist of 4 parts, *first*, the *shell*, which is white in the eggs of the common fowl, and of many other kinds; but it is often coloured or spotted of various colours, so as to give it a beautiful appearance: *second*, the *membrana putaminis*, a thin transparent pellicle, immediately within the shell; at the great end of the egg this membrane is detached from the shell, leaving a certain distance between them, which is filled with air: *third*, the *white* or *albumen*, a glairy liquid, consisting of albumen dissolved in water, and contained, like the vitreous humour of the eye, in an extremely thin membrane, divided into cells: *fourth*, the *yolk*, a thick and almost solid yellow matter, inclosed in a peculiar membrane; this membrane, by 2 ligaments, called *chalazæ*, is tied to the membrane of the albumen, and thus the yolk is kept in the centre of the egg."

(3201.) In regard to the constitution of these various parts, it appears from an analysis of Dr Prout, which was later than that of Vauquelin, that the *shell* of the common fowl consists of

Carbonate of lime, with a little of carbonate of magnesia,	97	parts.
Phosphate of lime and magnesia,	1	...
Animal matter,	2	...
	in 100	..

The *membrana*, according to Hatchett, consists of coagulated albumen. The *white*, or *albumen*, coagulates into a firm white solid, when heated to 159° Fahr.: and when evaporated to dryness, leaves about 14 per cent. of albumen. Dr Bostock has shewn that it contains also a little mucus. The constitution of the white, according to him, is

Water,	80
Albumen,	15.5
Mucus,	4.5
	100.

Dr Prout obtained, by combustion, the following fixed constituents in 1000 grains of the white of egg, from 3 different eggs:—

Sulphuric acid,	0.29	0.15	0.18	grains.
Phosphoric acid,	0.45	0.46	0.48	...
Chlorine,	0.94	0.93	0.87	...
Potash, soda, and carbonates of potash and soda,	2.92	2.93	2.72	...
Lime, magnesia, and the carbonates of lime and magnesia,	0.30	0.25	0.32	...
	4.90	4.72	4.57	...

M. Mulder has proved that the sulphur and phosphorus are in the state of sulphur and phosphorus, and not in that of acid; and this was to have been expected, from the well-known alkaline reaction of the white of an egg. Dr Prout's analysis of the *yolk* of an egg which was hard boiled in distilled water, and weighed 316.5 grains, gave these results:—

Water,	170.2	grains, or	53.78	in 100	parts.
Albumen,	55.3	...	17.47	...	...
Yellow oil,	91.0	...	28.75	...	...
	316.5	...	100.00	...	...

According to Planche, 1000 parts of yolk of egg furnish, at an average, 180 parts of oil. This oil consists of stearin 10, and of elain 90 parts; the stearin is white and solid, and does not stain paper like oil. He found this stearin and the fat of fowls to agree very nearly. The elain possesses the character of a fixed oil. Chevreul found 2 colouring matters in the yolk, the one *red*, and the other *yellow*. Lecanu, besides the stearin and elain, extracted from the yolk a crystalline matter, which melted at 293° Fahr., and which he considered as of the same nature with cholesterin from the brain. Dr Prout determined the quantity of fixed constituents in 100 grains of the yolk, by incineration, in 3 different eggs, thus:—

Sulphuric acid,	0.21	0.06	0.19	grains.
Phosphoric acid,	3.56	3.50	4.00	...
Chlorine,	0.39	0.28	0.44	...
Potash, soda, and the carbonates of potash and soda,	0.50	0.27	0.51	...
Lime, magnesia, and the carbonates of lime and magnesia,	0.68	0.61	0.67	...
	5.34	4.72	5.81	...

Whether the sulphur and phosphorus exist in the yolk in the state of acids, or as sulphur and phosphorus, is unknown. When we compare the fixed constituents of the white and yolk, we cannot avoid being struck with the difference. The *white* contains a much greater quantity of fixed alkalies than of any other fixed constituent; while in the *yolk* the most abundant constituent is phosphoric acid, which amounts to from 3.5 to 4 grains; or if we suppose it to exist as phosphorus, it varies in different yolks from 1.55 to 1.77 grains.

(3202.) The specific gravity of a new-laid egg varies from 1.080 to 1.090; an egg, therefore, is heavier than sea-water, the specific gravity of which is 1.030. When kept, eggs rapidly lose weight, and become specifically lighter than water. This is owing to the diminution of bulk in the contents of the egg; the consequence of which is, that a portion of the inside of the egg comes to be filled with air. Dr Prout kept an egg 2 years, and found that it lost weight daily, at an average rate of 0.744 grains. The original weight was 907.5 grains, and after 2 years' exposure to the atmosphere, it weighed only 363.2 grains. The total loss amounted to 544.3 grains, or considerably more than half the original weight. The loss in summer was somewhat greater than in winter, owing, no doubt, to the difference of temperature. When an egg is, therefore, employed as a test of the strength of brine, the newer it is, the stronger is the brine that floats it. The relative weights of shell and membrane, albumen, and yolk, are very different. Supposing the original weight of the egg to be 1000 grains, Dr Prout found the relative proportions, in 10 different eggs, to be as follows:—Shell and membrane 106.9, albumen 604.2, and yolk 288.9 grains. When an egg is boiled in water, it loses weight, particularly if it be removed from the water when boiling, and be permitted to cool in the open air. The water will be found to contain a portion of the saline constituents of the egg. The loss of weight from boiling is not constant, varying from 20 to 30 grains, supposing the original weight to have been 1000 grains. The quantity of saline matter obtained by evaporating the distilled water in which an egg was boiled, amounts, at an average, to 0.32 grains. It is strongly alkaline, and yields traces of all the fixed principles found to exist in the egg;

but the carbonate of lime is most abundant, and is obtained by evaporation in the form of white powder.

(3203.) The source from whence is derived the bones of the chick while in the egg, is still an object of research. At the full term of incubation important changes in the constitution of the egg are completed. "The albumen," as Dr Thomson observes, "has disappeared, or is reduced to a few dry membranes, together with earthy matter. The yolk is considerably reduced in size, and is taken into the abdomen of the chick, while the animal has attained a weight nearly equal to the original weight of the albumen, together with that lost by the yolk, minus the loss of weight sustained by the egg during incubation. The alkaline matters and chlorine have diminished in quantity, while the earthy matters have considerably increased. . . . During the last week of incubation, the yolk has lost most of its phosphorus, which is found in the animal converted into phosphoric acid, and, combined with lime, constituting its bony skeleton. This lime does not exist in the recent egg, but is derived from some unknown source during the process of incubation. Mr Hatchett made the curious remark, that, in the ova of those tribes of animals the embryos of which have bones, there is a portion of oily matter, and in those ova whose embryos consist entirely of soft parts, there is none. In what way the oily matter contributes to the formation of bone it is impossible, in the present state of our knowledge, to conjecture. Nor can any source of the lime of the bones be pointed out, except the shell; and it would be difficult to determine whether the shell loses lime during the process of incubation."*

(3204.) M. Raspail, in investigating the nature of animal albumen by the microscope, as exemplified in the white of the egg, observes, that "the albumen of the pullet's egg is composed of an insoluble and regularly-organized texture, which contains in its cells a soluble substance much more susceptible of alteration than the texture is. Chemists had previously acknowledged the existence of an albumen soluble in water, and of another which was insoluble; but they had not remarked that these 2 sorts of albumen existed together in the white of eggs, and they had considered this substance as a variety of the insoluble albumen. . . . But the insoluble substance of the white of an egg is rendered apparent only by degrees, and accordingly there is a period when it can scarcely be distinguished in this respect from the soluble substance, and this is when the egg is fresh, that is, recently laid. Hence, as I have already pointed out regarding the vegetable textures, the textures are formed by the aggregation of the particles of the soluble substance, or, in other words, the soluble substance is converted by solidification into the parietes of cells. All these circumstances establish a complete analogy between *gluten* in vegetables and *albumen* in animals."†

(3205.) Of the nature of egg and seed, the origin of animal and vegetable individuality, and of the natural analogy between them, M. Raspail thus expresses himself: "The egg and the seed are cells detached from the texture of the mother, in consequence of an influence of an opposite kind. This influence may proceed from an external body which we call the male, or from an internal cause which we altogether neglect to notice. The egg and the seed may more particularly attract our attention, in consequence of their forms and dimensions; but the slice of a polypus, which becomes an entire animal, and the fragment

* Thomson's Chemistry of Animal Bodies, p. 446-55.

† Raspail's Organic Chemistry, p. 243-4.

of a potato, which produces a complete plant, are sufficient to teach us that the generative faculty is preserved by the whole organic system, and that the whole organized being is complete in any one of its cells."*

(3206.) The duty on the importation of foreign eggs is fixed by the new tariff at 10d. per 120, and 2½d. from British possessions. The trade in eggs is carried on to a very great extent in this country. The importation from France is large, and the supply obtained in the west coast of England and Scotland from Ireland is also great. About 20 years ago and upwards, L.30,000 worth of eggs used to be sent from Berwick-on-Tweed to the London market in the course of the year, but since the interference of large importations from France and Ireland, that trade, I believe, has almost dwindled away. The price of eggs is never high in the country, but in towns it is almost always so. In the most abundant season eggs, that can be relied on as fresh, are never below 7d. per dozen in Edinburgh, and in winter, especially at Christmas, when the confectioners use large quantities, they are as high as from 14d. to 18d. per dozen.

This subject brings me to the end of the farmer's year, the subject following being the ploughing of stubble, which I described at the beginning of winter, as the first operation of the year, and the first preparation for the next year's crop. In describing each operation as it fell to be executed in its own season, and in its proper relation to the one which preceded, and the one which was to follow it, I have endeavoured to make the description as minute and in as explicit terms as to enable you to execute it in the way it ought to be in the fields; or at least, to afford you data to compare any operation you see executed in the fields with the mode described here. In the endeavour to make the descriptions as explicit as to be useful, I find I have occupied a much greater space than I intended; but my sole object being to afford instruction in agriculture to the inexperienced beginner, I conceive it is better to have erred in teaching on the side of prolixity than of epigrammatic obscurity. Nothing, to my feeling, is so provoking to a reader, desirous of obtaining full information on a particular subject, as to find the author treating it only in the most general terms. To avoid this fault, I found it necessary to adopt the style of instruction as if of young noviciates; and to give the descriptions point, I found it further necessary to refer to a specific and acknowledged good system of agriculture; and as the *mixed* system, as it is called—because it embraces the treatment of plants and of animals in intimate relation to each other—is, in my opinion, the best, I adopted it as the subject of illustration, whilst other systems, such as the pastoral, which devotes its entire attention to stock, and the arable, which is exclusively appropriated to the raising of

* Raspail's Organic Chemistry, p. 75-6.

corn, have not been overlooked. Inasmuch, therefore, as the whole yearly operations of the farm have now been described, I might finish the work here ; but there are still many other subjects, connected with every system of farming, upon which you require instruction, before you can be prepared to conduct a farm on your own account. I feel it incumbent on me to apprise you of those subjects, because they are really important, and would have gladly gone with you into their consideration in the same manner as I have treated the subjects already discussed, but that their proper elucidation would occupy a much larger space than I can with propriety extend this work further, so I must treat them in the most succinct manner. The subjects I allude to are, the rotations of crops—the fertilizing of the soil by means of manures—the points of live-stock most desirable to be obtained in breeding—the making of experiments in agriculture—the destruction of vermin—the looking out for a farm—its rent—its lease—its stocking—and the amount of capital required to carry it on—the improvement of waste land—of farm book-keeping—and of the conveniencies and comforts of the cottages of farm-servants. To each of these subjects our attention shall now be cursorily directed.

81. OF ROTATIONS OF CROPS.

"Otes, rie, or else barlie, and wheat that is gray,
Brings land out of comfort, and soon to decay;
One after another, no comfort between,
Is crop upon crop, as will quickly be seen."

TUSSER.

Experience has demonstrated, not that "crop upon crop," as expressed in so very general terms by Tusser, but that one crop after another of the same kind, greatly reduces the fertility of all soils. This obvious conclusion might also be drawn from reason as well as experience, because it is reasonable to suppose that crops of the same kind require and will appropriate to themselves, the same sort of food out of the same soil. Experience has also demonstrated, that one crop after another of a different kind, does not materially reduce the fertility of soils. This conclusion might also be drawn from reason as well as experience, because it is reasonable to suppose—at least equally so as in the other case—that crops of different kinds require and appropriate to themselves different sorts of food out of the same soil. From these two facts, derived from experience, this deduction seems fair, that the fertility of the soil is best maintained by taking different crops after one another; and, moreover, as every crop, though of different kinds, derives support from the soil, and in so far assists in exhausting it, it is a conclusion as obvious as any we have drawn, that a limit must be fixed to the number of crops that should follow one another. Accordingly, in practice, a limit is placed on the number of crops taken in succession, of whatever kind, and this number and succession is called a *rotation* of crops.

Though all crops derive support from the soil, one kind appropriates its food in a very different degree from another, and even the same crop requires food in different quantities, according to the use to be made of its product. In practice, different crops are cultivated for very different purposes. One class are cultivated entirely for their *seed*, and these are called corn-crops, such as wheat, rye, barley, and oats; and in this class may also be placed beans and pease, as these are also cultivated for their seed, though, from difference of structure in the straw and difference of habit, they require a different sort of food. Other kinds are solely cultivated for their roots and leaves, and are therefore called *green* crops, such as turnips, potatoes, carrots, mangel-würzel, parsnips, clover, tares, lucerne, and saintfoin.

It is, I believe, allowed by botanists, that a herbaceous plant, when suffered to ripen its seed, extracts from the soil a much greater quantity of nourishment than it proportionally does before the development of the flower stem; and, I believe, they also allow that narrow-leaved plants depend more upon the soil for nourishment than those having broad leaves. On these accounts, a crop of corn of any kind which ripens its seed and straw, should take from the soil much more nourishment than another crop which is carried away after its leaves only have been developed. If these statements be correct, and I believe them to be facts, there will be little difficulty in ascertaining which of the crops usually cultivated on a farm require most nourishment from the soil. Wheat, rye, barley, and oats, are suffered to ripen their seed and straw, and it is requisite they should do so, to render their produce useful; and as they all have narrow leaves, which are also deciduous before the seed becomes ripe, they may justly be regarded as those cultivated plants which take the greatest quantity of nourishment from the soil. Pease and beans are also suffered to ripen their seed, but having broad leaves, compared with those of corn plants, and which are persistent, and their straw being cut down while in a comparatively green state, their exactions from the soil are not of a severe character. On the other hand, the natural pasture grasses, although identically of the same structure and habit as corn plants, deteriorate the soil comparatively little, because being constantly cropped by live-stock, the greatest proportion of their produce is returned to the very soil which supports them, in the altered state of dung and urine. Forming, besides, a close sward over the soil, they protect it from the vicissitudes of the atmosphere, and the exciting action of heat, which tends to exhaust putrescent manures when applied to bare soil. The sown grasses, as clover and rye grass are usually called, even when combined, exhaust the soil but little when pastured; but of the two, the clover being a broad-leaved plant, has less dependence on the soil than the rye-grass, which is a true grass; but when both are allowed to develop their flower stems, and to be converted into hay, the rye-grass, like the corn plants, derives much support from the soil; and as it is too commonly allowed almost to ripen its seed before the hay is cut, and in some seasons almost constitutes the entire hay-crop, the clover having been thrown out in winter, it is clear that such a crop of rye-grass hay is nearly as severe on the land as the corn crops. The class of plants which supply bulbs for the use of live-stock in winter, such as turnips, mangel-würzel, carrots, and parsnips, though yielding large and heavy crops, do not exhaust much of the manure in the soil; because, besides having expanded and large leaves, which elaborate much subsistence from the atmosphere, they are biennial, and are consumed in their first year while the leaves and bulbs only are developed. The bulb seems to be a repository for supplying food, besides the roots, to the flower-stem and seed, in the second year; and as such plants produce a large quantity of seed, they are, perhaps, the most exhausting of all crops when allowed to perfect it. The potato, though a tuberous plant, and bearing little true seed, is, nevertheless, allowed to ripen what it does bear, as also its haulm, and thus far it is an exhausting crop; but its numerous leaves modify the demand of the tubers and roots upon the soil, by deriving a large proportion of subsistence from the atmosphere. The flax plant having very few leaves, and being suffered almost to ripen its seed, may be regarded as one which exhausts the soil next in degree to corn plants.

Having thus considered the nature of the various plants usually cultivated on the farm, inasmuch as they affect the manures which nourish their roots in the soil, it will be useful to arrange them in the order of their probable power of exhausting the fertility of the soil. 1. The *corn crops* I would place first in order, and these, perhaps, successively, thus: wheat, oats, barley, rye. 2. I would next place *flax*. 3. Then the *potato*. 4. Next the sown grasses when made into *hay* in the green state they ought to be. 5. *Pease*, and then *beans*. 6. The *root-bearing* plants next, and these in this order: carrots, parsnips, Swedish turnips, yellow turnips, mangel-würzel, white turnips. 7. Then tares and clover when cut, as also lucerne, saintfoin, and crimson clover, (*Trifolium incarnatum*). 8. Next the sown grasses, clover and rye-grass, when pastured. 9. And the least exhausting of all I would regard permanent pasture of the natural grasses.

With a choice of so many plants, and possessing so very different powers of exhaustion, there would seem no difficulty of arranging such a succession of them as should least deteriorate the soil, and of selecting a sufficient number in succession as should suit every kind of soil. This is the abstract view of a rotation; yet it cannot be put into practice, as the expediency of cultivating many plants is entirely dependent on local circumstances. For example: in the neighbourhood of large towns, potatoes, turnips, carrots, are cultivated more with a view to supply the wants of the people, than to suit the nature of the soil; and, for a similar reason, few stock are pastured there in summer, the grass being usually sold for cutting as green food, or made into hay. But though circumstances thus operate to modify the rotation in certain localities, the general principles upon which every rotation may be based can be followed everywhere. For the great object in adopting a rotation at all, being to save the land subjected to it as much from deterioration as possible, and as corn crops must be raised in every species of arable soil, not merely because they supply the chief food of man, but because they also supply provender and litter to live-stock in winter, in this cold and damp climate—the best way of preventing the land from deterioration will be by counteracting the exhausting power of those crops. The only practical mode known of counteracting this power, is to raise them between other crops which extract less nourishment from the soil than they do; and as all crops exhaust the soil less or more, it is necessary to support the soil with manure at one part of the rotation at least, and an enriching course may still further be pursued, by raising the less exhausting corn crop at the end of the rotation farthest from that at which the manure is applied. This course of practice is of general application.

Notwithstanding these explanatory remarks, you will not thoroughly understand *why* a rotation should be useful to the soil, until you are made acquainted with the theories which have been offered in explanation of it; and these I will now relate, and they consist only of two.

No theory on the subject attracted public notice until that offered by the late celebrated De Candolle, the eminent botanical physiologist of Geneva, in 1832, founded on the investigations of M. Macaire Prinsep on the radical excretions of plants, in support of the reputed discovery of Brugmans. M. Macaire found that when plants were placed in distilled water, they emitted certain substances which were detected by reagents to be of very different qualities from the pure water itself; and when his experiments produced the result, that beans languished and died in water which contained the matter previously exuded by the roots of other plants of the same kind, whilst wheat throve in the water charged with the matter from leguminous plants, De Candolle considered such distinct results sufficient to establish a theory of the rotation of crops. Accordingly, his theory was, that the roots of plants imbibe soluble matter of every kind from the soil, and thus necessarily absorb a number of substances not adapted for their support, which are subsequently expelled by the roots, and returned to the soil, as excrements. As plants cannot subsist on matter which they eject, the more of this matter the soil contains, the less it becomes fit to support plants of the same kind. These excrementitious matters, however, may be taken up by other kinds of plants, which will thus resume them from the soil, and render them again fit for supporting those of the first kind; and if those other kinds also expel substances from their roots which can be appropriated as food by the first class, both kinds will improve the soil in two ways.*

The power of plants to excrete noxious matters from their roots has since been denied by many chemists, though it is admitted that they do emit from their roots substances soluble in water; but that these are nothing else than the simple sap of the plant. That plants do excrete certain substances from their roots cannot be denied, otherwise the peculiar odour imparted by certain plants to the soil where they have grown cannot be explained. Who has not felt a sensible odour from the ground in digging out an ash-tree by the root, or in removing the roots of raspberry plants in a garden, or in exposing the roots of the common mint from black mould? Nay, in taking up a crop of potatoes or turnips in the field, who is not sensible of the peculiar odour of the earth, distinctive of those crops, when ploughed immediately after their removal? Whether the odours alluded to arise from true excrements, or only from the excretion of the ordinary sap of the plant, is immaterial, provided the substance thus excreted proves injurious to succeeding plants of the same kind. Mr Gyde of Painswick's opinion, founded on experiments made by himself, with a particular view to the solution of this very question, is, that though plants have no power of selection, but take into their texture any solution offered to their roots, they have little or no power of again excreting it; that any excretions are only of the true sap; and that plants watered with excretion receive no injury from it.†

In the first edition of his work, Professor Liebig entirely adopted the theory of De Candolle, and adduced many proofs, derived from chemistry, in support of it;‡ but if the excretory theory is untenable, as seems to be the opinion of botanists, all the ingenious hypotheses suggested by De Candolle, and supported by Liebig, fall to the ground. Liebig withdrew his support of De Candolle's theory from his last edition, and yet seems to adhere to a modified view of it, when he says, that "Transformations of existing compounds are constantly taking place during the whole life of a plant, in consequence of which, and as the results of these transformations, there are produced gaseous matters which are excreted by the leaves and blossoms, solid excrements deposited in the bark, and *fluid soluble substances which are eliminated by the roots*. Such excre-

* See De Candolle's Physiologie Vegetale for an account of Macaire's Experiments, in tome i. p. 248-51; and for De Candolle's own views on the rotation of crops, see tome iii. p. 1474–1520.

† Transactions of the Highland and Agricultural Society for October 1843, p. 80.

‡ Liebig's Organic Chemistry in its application to Agriculture and Physiology, p. 155-74 Edition of 1840.

tions are most abundant immediately before the formation and during the continuance of the blossoms; they diminish after the development of the fruit. Substances containing a large proportion of carbon *are excreted by the roots, and absorbed by the soil. Through the expulsion of these matters, unfitted for nutrition,* the soil received again with usury the carbon which it had at first yielded to the young plants as food, in the form of carbonic acid. The soluble matter thus acquired by the soil is still capable of decay and putrefaction; and, by undergoing these proprocesses, *furnishes renewed sources of nutrition to another generation of plants;* it becomes humus." *

There is another doctrine which shews why a rotation of crops should be advantageous to the soil, without adopting or rejecting that of the excretion of plants, which is, that the same kind of plant requires the same sort of food to support it. It is reasonable to believe, for example, that all the varieties of wheat will affect the same soil in which they are made to grow in a similar manner; and when it is admitted that a number of crops of wheat grown on the same piece of land, will in time exhaust in it the entire supply of the particular sort of food which wheat appropriates to itself, all that is requisite for a proper explanation of a rotation of crops seems granted. The same conclusion, by the same process of reasoning, may be arrived at in regard to every other species of crop. So that, if we wish to derive from any piece of land all the benefit it can bestow on a particular crop, we must change that crop before its food is exhausted; and if we wish the soil to supply food to that crop for a long period, we must place such another crop before and after it, as both will take different sorts of food for their support. Still the time will arrive, when the food of all will be exhausted, and must be replenished by a new supply of manure. Either theory brings us to the same conclusion,—namely, the necessity of supplying manure to the soil even in the most favourable rotation of crops.

Having now before you the nature of the different plants, as regards their power of exhausting the fertility of the soil, and the reasons why rotations of crops are advantageous to the soil, you will be easily convinced of the correctness of the following conclusions, namely,—that the highest class of exhausting plants, such as wheat, oats, barley, and rye, should not follow immediately in succession; that the less exhausting plants, such as the green crops, turnips, mangel-würzel, cabbage, carrots, or potatoes, should intervene; that the tuberous and fusiform rooted green crops, such as potatoes, carrots, parsnips, being the most

* Liebig's Chemistry in its application to Agriculture and Physiology, p. 33–4. Edition of 1843.

exhausting, should be associated directly with fresh manure; that the bulbous rooted green crops, such as turnips, mangel-würzel, should also be directly provided with fresh manure, but will thrive with less than the preceding class of green crops; that forage plants, such as tares, clover, may be regarded as substitutes for bulbous plants, but being less exhausting, may be placed farther from the manure; that the pasture grasses, being the least exhausting of all crops, may be placed farthest from the manure, and the longer they remain in pasture, the soil receives the greater amelioration; and that manure should be applied at the commencement of a succession of crops, and though this is the usual place for it, it may be applied with advantage at any period of the rotation.

On fixing on a rotation of cropping, permanent pasture should be left out of consideration, because, being an unchangeable condition of soil, it cannot be embraced in a rotation, which implies a frequent return of the soil to the plough. Where a large proportion of permanent pasture exists, it will have considerable influence in determining the length of a rotation, inasmuch as, in such a case, there is no use of having the arable land long under grass, as is the common practice in England; but it should be borne in mind, on the other hand, that where a considerable proportion of the land is under the plough, a large quantity of manure should be applied at a time, and that frequently.

Nothing farther occurs to me to say in explanation of a rotation. I shall now enumerate a few of the rotations practised on different soils, interspersed with such critical remarks in favour of or against them, as the circumstances of the case may call forth; and in indicating the proportion which each member of the rotation bears to one another in each rotation, reference will always be kept in view to a farm of 500 acres.

Rotations for clay soils.—This 2-course rotation is practised on some fine strong clays in England:—

1. Wheat, with manure, constituting	½	of the arable land	= 250 acres.
2. Beans, without manure, ...	½	... ...	= 250 ...
			500 ...

The grass is permanent meadow. The land is dunged every other year with the wheat, but being naturally fertile, does not require a heavy manuring so often. With so much wheat straw, and refuse of the bean crop and hay, the meadow receives a manuring for the hay. Unless the land is of the richest description, and has a natural stamina, it could not endure such a succession of wheat crops even with manures.

There is a 3-course shift in the following terms:—

1. Fallow.	Bare fallow, Potatoes, Turnips, Tares, Beans,	⅓ of arable land	= 166⅔ ac.	Potatoes,	10 ac.
				Tares,	10 ...
				Turnips,	40 ...
				Beans,	40 ...
				Rag fallow,	66⅔ ...
2. Wheat and Barley,		⅓	= 166⅔ ...	Winter wheat,	126⅔ ...
				Spring wheat,	10 ...
				Barley,	30 ...
3. Clover and Grass,		⅓	= 166⅔ ...	Clover,	20 ...
				Hay,	50 ...
				Pasture,	96⅔ ...
			500		500

This will answer on a poor clay-loam, and on still inferior soil oats should be taken in lieu of wheat. The beans on the fallow-break should be raised in drills, and manured along with the fallow-crops. The fallow-crops will be easily raised, and not require a heavy manuring after the clover lea. The whole rotation being an easy one for the soil, may be practised for a few years on worn-out land, until it is brought into heart, when it may be relinquished for a better one. This rotation places ⅓ of the land under fallow, ⅓ under corn, and ⅓ under grass.

Both these are uncommon rotations, but the following 4-course is a very common one in England. The land is dunged in the fallow division.

1. Fallow,	Bare fallow, Potatoes, Turnips, Tares, Beans,	¼ = 125 acres,	Potatoes,	10 acres.
			Tares,	10 ...
			Turnips,	30 ...
			Beans,	25 ...
			Bare fallow,	50 ...
2. Wheat and barley,		¼ = 125 ...	Winter wheat,	95 ...
			Spring wheat,	10 ...
			Barley,	20 ...
3. Clover,		¼ = 125 ...	Hay,	40 ...
			Clover,	15 ...
			Pasture,	70 ...
4. Wheat and feed oats,		¼ = 125 ...	Wheat,	105 ...
			Feed oats,	20 ...
		500		500

Wheat is a very common crop after clover in England, but in Scotland is rarely ever attempted, partly, no doubt, from the inferior state of the climate, and partly from the sure profit derived from a good crop of oats. The beans are here converted into, and manured as, a fallow-crop. On account of the oats, this rotation is changed in Scotland to the following, where a 4-course shift is practised on rather strong soil, and where it can only be supported in the neighbourhood of large towns, whence manure can be obtained in abundance. The dung is applied to the fallow division, and the beans treated as a fallow-crop.

1. Fallow,	Bare fallow, Potatoes, Turnips, Tares, Beans,	$\frac{1}{4}$ = 125 acres,	Potatoes,	10 acres.
			Tares,	10 ...
			Turnips,	40 ...
			Beans,	20 ...
			Bare fallow,	45 ...
2. Wheat and barley,		$\frac{1}{4}$ = 125 ...	Winter wheat,	85 ...
			Spring wheat,	10 ...
			Barley,	30 ...
3. Red clover and annual rye-grass,		$\frac{1}{4}$ = 125 ...	Hay,	30 ...
			Clover,	15 ...
			Pasture,	80 ...
4. Oats,		$\frac{1}{4}$ = 125 ...	Oats,	125 ...
		500		500

There is no way but this of modifying the 4-course shift, excepting in converting it into a severer one, by making more fallow, and having only a few acres of potatoes and turnips for domestic use; taking oats instead of barley, and having more hay. In all 4-course shifts, $\frac{1}{4}$ of the farm requires to be manured every year, and that again every 4 years, an extent of manuring which no farm could supply from its own resources. Such a shift, therefore, is only practicable for sustaining the fertility of the soil near a large town, or other sources of extraneous manure, such as an unlimited supply of sea-ware. A 4-course shift divides the land into $\frac{1}{4}$ fallow, $\frac{1}{2}$ corn, and $\frac{1}{4}$ grass.

On *clay loams*, a 5-course shift is universal.

1. Fallow,	Bare fallow, Potatoes, Turnips, Tares,	$\frac{1}{5}$ = 100 acres.	Potatoes,	10 acres.
			Tares,	6 ...
			Bare fallow,	24 ...
			Turnips,	60 ...
2. Wheat and Barley,		$\frac{1}{5}$ = 100 acres.	Winter wheat,	40 ...
			Spring wheat,	20 ...
			Barley,	40 ...
3. Red clover and annual and perennial rye-grass,		$\frac{1}{5}$ = 100 acres.	Hay,	30 ...
			Clover,	5 ...
			Pasture,	65 ...
4. White clover and perennial rye-grass,		$\frac{1}{5}$ = 100 acres.	Pasture,	100 ...
5. Oats,		$\frac{1}{5}$ = 100 acres.	Potato oats,	60 ...
			Common oats,	40 ...
		500		500

The dung is applied on the fallow division. Modifications might be made on fine land, by dispensing with the naked fallow altogether, and substituting a green crop instead, such as beans, when manure is at command. This would not have the effect of contracting the extent of winter wheat, nor would it, were the potato crop extended as a substitute for the entire bare-fallow; but were any other green crop substituted for the bare-fallow, the extent of winter wheat would be contracted, but in such a case the spring wheat might be extended to make up the deficiency. The hay might be confined to the wants of the farm, and the pasture thus increased in the first year of grass. In the last year only as much oats might be grown as to satisfy the demands of the farm, and the remainder of the lea rag-fallowed for wheat. Where this modification could be adopted, it would prove a profitable rotation. In all 5-course shifts only $\frac{1}{5}$ of the land is manured every year, and that proportion only once in every 5 years; so that good land which grows wheat freely, would be quite able to keep itself in good heart, with the occasional assistance of lime, from its own resources. A 5-course shift gives $\frac{1}{5}$ of the land in fallow, $\frac{2}{5}$ in corn, and $\frac{2}{5}$ in grass—$\frac{1}{5}$ in new and $\frac{1}{5}$ in old grass.

On good clay loams a 6-course shift is a favourite one, and it is easy in its demands on the resources of the farm; for the manuring is only once in every 6 years, and then only over $\frac{1}{6}$ of the surface, unless a sprinkling be applied in the 4th year. On fine land the cropping may be divided in this manner :—

1. Fallow. { Bare fallow, Potatoes, Tares, Mangel-würzel, Turnips, }	$\frac{1}{6} = 83\frac{1}{3}$ acres.	Potatoes, . . .	8 acres.
		Tares, ...	6 ...
		Bare fallow, . .	$19\frac{1}{3}$ ...
		Mangel-würzel, .	10 ...
		Turnips, . . .	40 ...
2. Wheat and Barley, . . .	$\frac{1}{6} = 83\frac{1}{3}$ acres.	Winter wheat, .	$43\frac{1}{3}$ ...
		Spring wheat, .	15 ...
		Barley, . . .	25 ...
3. Red clover and annual rye-grass, ...	$\frac{1}{6} = 83\frac{1}{3}$ acres.	Hay, ...	20 ...
		Clover, . . .	5 ...
		Pasture, . . .	$58\frac{1}{3}$ ...
4. Wheat and oats, . . .	$\frac{1}{6} = 83\frac{1}{3}$ acres.	Wheat, . . .	50 ...
		Oats, ...	$33\frac{1}{3}$ ...
5. Beans and pease, . . .	$\frac{1}{6} = 83\frac{1}{3}$ acres.	Beans, drilled, .	$53\frac{1}{3}$ ...
		Pease, broadcast,	30 ...
6. Wheat, ...	$\frac{1}{6} = 83\frac{1}{3}$ acres.	Wheat, . . .	$83\frac{1}{3}$ ...
	500		500

The modifications which might be introduced here to suit local circumstances, are an increase of potatoes, mangel-würzel, or of turnips by a few acres, say 10, and a diminution to the like extent of the bare fallow. On somewhat inferior soil, an increase of bare fallow would be necessary, as the supposed inferior soil would not produce sufficient manure for green crops. This last modification would, to its extent, increase the quantity of winter wheat in the following year; and the land being supposed inferior, the spring wheat would be reduced or withdrawn, and the barley increased. In the 4th year there would be oats entirely, and no wheat at all, after the clover; and in the last year there might be some oats after the beans and pease, instead of wheat entirely. In either plan of this 6-course shift, upon strong land, it would be desirable to dung the bean and pease stubble for the succeeding wheat and oats—not a full dunging, however, but perhaps ½; and in regard to second manurings in general, it should be borne in mind, that, besides doing good to the crops after the beans and pease, the fallow crops after it will be more easily raised with ordinary manuring.

It has been found in the neighbourhood of towns, that the clover plant when sown every 4 years in the rotation has almost ceased to appear, and as this is justly regarded as a calamity in localities in which a constant demand for clover exists, an alteration has been introduced in the rotation, with a view to improving the condition of the soil for that plant, and it is gratifying to know that it has in consequence re-appeared in almost its former vigour. These 2 modifications, one on a 5, and the other on a 6-course shift, have been found to answer the purpose just explained in regard to clover in the vicinity of Dundee.

1. Turnips.	Dunged.	1. Turnips.	Dunged with 10 double-loads, or 15 tons of dung.
2. {Barley. / Oats.	Half-Dunged.	2. Barley.	
3. Potatoes.	Dunged.	3. Potatoes.	Dunged.
4. Wheat.	Half-dunged.	4. Wheat.	Half-dunged.
5. Grass.	Cut for forage and hay.	5. Grass.	Cut for forage and hay.
		6. Grass.	Pastured.

Though the clover still reverts in 5 and 6 years, the frequent manuring keeps the soil always in a fresh state; and it is well known that clover thrives best on fresh land. Both these rotations are open to the objection of taking grass after potatoes, a species of fallow crop not congenial to grass, especially on soils tending to clay. In as far as concerns the grass, it would be better to make the turnips and potatoes exchange places; but then the turnips would not all be removed in time for wheat to be sown in autumn, and every situation is not suited to spring wheat. Where spring wheat may be sown with impunity, and where barley is desirable, transferring the places of turnips and potatoes would be of advantage to the clover; but all the difficulties of the case would be removed by beginning the rotation with potatoes following with wheat, and then taking turnips followed by barley, which last crops would then precede the grass. A 6-course shift on strong land places $\frac{2}{6}$ in fallow, $\frac{2}{6}$ in corn, and $\frac{2}{6}$ in grass, that is, $\frac{1}{3}$ of the land in each species of cropping.

Rotations may be extended beyond 6 crops on strong land. In the

Carse of Gowrie this was a rotation very generally followed some years ago, and, I believe is so still:—

1. Fallow. Dunged and limed.
2. Wheat.
3. Barley.
4. Clover and rye-grass.
5. Oats.
6. Beans and pease. Stubble-dunged.
7. Wheat.

It will at once be observed, that 2 white crops follow in the barley and wheat, and the reason given, in justification of this deviation from good farming, is, that fallow wheat is there too strong to sow down with clover, whereas clover is always good after barley; and that the land is naturally so fertile, that barley always proves a good crop, and shews a clear and beautiful sample after wheat. No doubt, clover succeeds better after barley than wheat, and samples of barley after another corn crop always look well; still the rotation cannot be recommended on principle, and even in practice it would be better, I think, to change it to this:—

1. Fallow. Dunged and limed,	$\frac{1}{8} = 62\frac{1}{2}$ ac.	Potatoes,	10 ac.
		Bare fallow,	$52\frac{1}{2}$ ...
2. Wheat,	$\frac{1}{8} = 62\frac{1}{2}$ ...	Winter wheat,	$62\frac{1}{2}$ ..
3. Beans, turnips and tares. Turnips dunged, beans and tares half-dunged,	$\frac{1}{8} = 62\frac{1}{2}$ ...	Beans,	$29\frac{1}{2}$ ...
		Turnips,	25 ...
		Tares,	8 ...
4. Wheat and barley,	$\frac{1}{8} = 62\frac{1}{2}$ ...	Winter wheat,	$37\frac{1}{2}$ ...
		Spring wheat,	10 ...
		Barley,	15 ...
5. Clover,	$\frac{1}{8} = 62\frac{1}{2}$ ...	Hay,	$29\frac{1}{2}$ ...
		Cutting,	8 ...
		Pasture,	25 ...
6. Oats,	$\frac{1}{8} = 62\frac{1}{2}$ ...	Oats,	$62\frac{1}{2}$ ...
7. Beans and pease. Oat-stubble half dunged,	$\frac{1}{8} = 62\frac{1}{2}$ ...	Beans, drilled,	40 ...
		Pease, broadcast,	$22\frac{1}{2}$ ...
8. Wheat,	$\frac{1}{8} = 62\frac{1}{2}$ ...	Winter wheat,	$62\frac{1}{2}$ ...
	500		500

The wheat after beans and tares half-dunged will not be too strong to sow down with clover. This rotation requires good soil to bear so much corn crop, and a large supply of straw to dung the land so often. An 8-course shift, such as this, places the land under $\frac{3}{8}$ of fallow or dunged crop, $\frac{4}{8}$ of corn, and $\frac{1}{8}$ grass. The land should not get sick of clover once in 8 years.

Here is a rotation practised by Mr Henderson, the successful competitor for Irish flax, exhibited at Belfast, in 1843, under the auspices of the Agricultural Improvement Society of Ireland. His own words, in its recommendation, are these:—"Without method there cannot be success. Different soils require difference in rotation, and suit different crops. I will speak only of that I use. My farm is a strong and pretty deep clay croft, and has proved well suited for flax; therefore I use that crop more frequently, say twice in the course, than will be generally found advisable.

1. Potatoes or turnips. Dunged and limed.
2. Winter wheat.
3. Flax.
4. Clover and rye-grass. Cut for hay, being top-dressed with soot.
5. Pasture.
6. Pasture.
7. Oats.
8. Flax.

"After the wheat 1 ploughing is sometimes sufficient, but 2 are generally safest, 1 in autumn and 1 in spring. After lea-land oats 2 ploughings are indispensable, and a 3d frequently advisable; for the land must be perfectly pulverised and cleared of all roots of every sort, or no crop. I do not plough deeper than the mould." I have not before referred to flax as a member of any rotation; but regarding it as an exhausting crop next to corn, wherever it is cultivated it should, of course, come in the rotation in place of a corn crop. Recommended after a corn crop, as in this case, and especially twice in the course of a rotation, the practice is reprehensible, in as far as the condition of the soil is concerned; sowing down clover with flax after a white crop is as bad husbandry as can be, and were it practised on land which has grown clover for a long time, clover would soon cease to grow. In Ireland, where the culture of clover is but of recent introduction, it may grow well for some time under any treatment; but the Irish farmer should beware of the nature of this plant, and rule his practice by our experience, which would warn him against putting so useful a plant to the trial here recommended. Flax will no doubt grow of finer quality after a white crop on land in good condition, or on soil naturally fertile, as on such soils it would be coarse, and apt to branch, if grown after a manuring; and if the main object of the Irish farmer is to desire to cultivate flax of the finest quality, it would be better to acquaint him at once of the deteriorating effect of flax, thus cultivated, upon the condition of the soil, than to encourage him to make the other crops he raises subservient to flax, and to inculcate in him a wrong opinion. There seems another wrong opinion to be abroad in Ireland in regard to flax, that that really valuable plant is *neglected* in its culture, and that were it not for the neglect the Irish farmer would be much wealthier than he is. He should be distinctly assured that in making money in cultivating flax, he must do it either at the sacrifice of corn or deterioration of soil, for *both* corn and soil he *cannot* continue to have in *perfection* along with *fine* flax.

Rotations for light soils. The 4-course shift cannot be practised on light soils, even within command of manure, their texture being easily injured by too frequent application of the plough, and their improvement is best effected by consolidation under grass. The 5-course shift, therefore, is what is most commonly used on light soils, such a one as is specified above for clay loams, the wheat being curtailed, or withdrawn altogether, the barley extended, and the fallow conducted almost wholly under green crops, particularly turnips. On all light soils live-stock thrive well, so that turnips and pasture are the crops which should receive most attention on them. In recommending the above 5-course shift, it may prove useful to you to particularize the proportions of the several crops most applicable to light soils, as thus:—

1. Fallow.	Bare fallow, Potatoes, Turnips, Tares	$\frac{1}{5}$ = 100 acres	Potatoes	15 ac.
			Tares	6 ...
			Turnips	69 ...
			Bare fallow	10 ...
2. Wheat and barley,		$\frac{1}{5}$ = 100 ...	Winter wheat	31 ...
			Spring wheat	10 ...
			Barley	59 ...
3. Red clover and perennial rye-grass,		$\frac{1}{5}$ = 100 ...	Hay	20 ...
			Clover	5 ...
			Pasture	75 ...
4. White clover and perennial rye-grass,		$\frac{1}{5}$ = 100 ...	Pasture	100 ...
5. Oats,		$\frac{1}{5}$ = 100 ...	Potato oats	60 ...
			Common oats	40 ...
		500		500

On light soils, this rotation will not supply manure to go over the entire fallow division. On gravelly turnip soil, I have found 300 acres farmed in this 5-course shift, unable to manure more than 40 acres, out of 60, the fallow division, even with a good crop; and as summer fallowing actually injures such land, extraneous manure to go over 20 acres was required to be purchased and brought on the farm every year. But a preferable rotation to this, on such a soil, is a 6-course one obtained by extending the grass division to 3 years, the effect of which will reduce the fallow one from 100 acres to $83\frac{1}{3}$ acres, to extend the grass from $\frac{2}{5}$ to $\frac{1}{2}$ of the farm, and to dispense with bare fallow altogether, curtailing the potatoes to 10 acres, and extending the proportion of turnips to $67\frac{1}{3}$ acres. In adopting this modification, it will be advisable to diminish, a little, the number of cattle bred on the farm, and to increase that of the sheep, as some of these are nearly independent of turnips, and the whole almost independent of straw, in winter; and this latter point is a consideration, as the number of acres of straw will now be curtailed from $\frac{2}{5}$ to $\frac{2}{6}$ of the extent of the farm, but nevertheless this ameliorating rotation may, after all, sustain the gross amount of straw, and of that of the green crop too. A 6-course shift, on light land, gives $\frac{1}{6}$ fallow, $\frac{2}{6}$ corn crop, and $\frac{3}{6}$ grass.

This rotation possesses the advantage of being extended to any length by repeating the grass; but an inordinate extension in this way interposes a barrier against itself, by increasing too much the summer, and decreasing too much the winter, food of live-stock; in short, the fallow division might be reduced to nothing, and the grass occupy the whole farm. I know farmers who extend the grass to 5 years, and where 2 farms lie contiguous, the one farmed in 5 years' grass, and the other in the ordinary 2 years, and where the rotations are made to alternate on those farms, both of them will be kept in good heart, and supply the finest pasture for live-stock that can be obtained; the grass having all the *strength* of old meadow, whilst its *youth* secures it against being spoiled with moss. On breaking up the grass after 5 years, it is not unusual to take 2 crops of oats, and extend the rotation to 9 years; but though the second crop of oats may be good, and even better than the first, it is bad farming, and perhaps not profitable after all; for the turnips succeeding it will require to be manured, whereas they might be raised after the first crop of oats without manure at all, or with very little assistance from one of the extraneous manures. I remember an old lea in Berwickshire being broken up for oats, and the farmer was recommended to take another crop of oats, to rot the turf, and it would be a better one than even the first; but he preferred rearing Swedish turnips without dung, and was wise in this determination, for they yielded him L.21 per acre when eaten off with old sheep, at only 6d. per head per week. A heavy crop of oats may be obtained on breaking up old lea, provided it be ploughed early in winter. When the old grass-parks belonging to Lord Panmure, at Panmure, in Forfarshire, were ploughed up a few years ago, to be laid down again, the first crop of oats yielded, over the entire extent of about 250 acres, 77 bushels, or rather more than 9½ quarters, per imperial acre.

While referring to oats, I may say that grass seeds never succeed so well after them as after barley or wheat, and the practice should therefore be avoided; but it must be borne in mind, that in high and exposed parts of the country, where oats alone can be grown with impunity, the practice must be pursued.

Before concluding the subject of rotations, I cannot impress too strongly upon your minds, never to abandon a regular course of cropping. So long as you adhere to it, the labour of the farm will be conducted with regularity and despatch; and whenever you forsake it, and adopt one which suits the circumstances of passing events, from that moment you will be enmeshed in difficulties, out of which you will be unable to extricate yourself. You will not then be able to foretel the proportion between your white and green crops,—you will never be sure whether you have too few or too many stock for the food on the farm, until some misfortune occurs, such as pasture-grass running to seed, or what is more likely, the stock seized with disease—the victims of low condition—and you will never be able to calculate the resources from which you are to derive your rent. In short, it would require a man of uncommon foresight to manage a farm well that is not in a regular course of cropping,—an ability far beyond what is requisite for a farmer. On the other hand, a slavish adherence to any particular rotation evinces want of judgment. The judgment should be exercised according to the circumstances of the times and season, and such modifications introduced in the rotation as will benefit both the land and its occupier. One legitimate mode of deviating from a *strict* rotation is this:—The field which grew a crop more exhausting than the rest in the course of one rotation, should be made to bear a crop of ameliorating character in the following rotation. For example, where potatoes grew in one rotation, turnips should be substituted in the next, and potatoes may be taken anywhere after turnips. In like manner, an interchange of soil should take place between the different kinds of turnips, because Swedish turnips are more severe upon the land than the white turnip, so these should alternate. The bare fallow should be made to alternate with a green crop. So also with barley and wheat, where the latter is felt to be severe upon the land. Even a severer course is at times justifiable, such as taking wheat after lea, where you have reason to suspect that oats will fail. Another root crop of a different nature, such as mangel-würzel, or carrots, or even cabbage, may be raised for a season, rather than allow the same root to be too often grown on the same soil. When any crop fails, which clover sometimes does, it should be ploughed up, and another of *a different kind* taken in its stead. Potatoes sometimes fail; they should be ploughed up and turnips substituted. Sometimes Swedish turnips are destroyed by insects; try white turnips as a substitute on the principle of a change, and because the season may still be not too far gone for them; but should it be too late for turnips, try a late cutting of tares, or bare-fallow the land for autumn wheat. In short, whenever one crop fails, another useful one should be substituted in its place; for if you do not cause the soil to be occupied with a crop, nature will soon cover it with plants which we regard as weeds. Nay, farther, where a change is thus forced in the rotation, a field may be miscropped in order to bring it the more quickly into the legitimate rotation of the farm; but in treating a field in this manner, it should be brought as soon as convenient into the fallow division, and receive *more* than its proportion of manure. Graham has truly said, that—

> " Change is the very life
> And soul of husbandry ; 'tis change of crops,
> By some *rotation* termed, that makes the ground
> Perform its task with unexhausted power."

---

82. OF FERTILIZING THE SOIL BY MEANS OF MANURES.

> " The earth's a thief,
> That feeds and breeds by a composture stolen
> From general excrement."
>
> TIMON OF ATHENS.

The great object for which farming *ought* to be pursued, whether in pastoral or arable districts, is increase to the fertility of the soil. The object, however, commonly kept in view in cultivating the soil, is constantly to derive the largest amount of produce from it. Though these two ends are diametrically opposed, as regards the condition of the soil, enlarged crops cannot be obtained but from increased fertility. Strange to say, that this truth seems only to have been discovered lately, and it is its adoption now as a rule of practice that constitutes the great difference between the agriculture of the present day, and that of former years. Not many years ago, cultivators were so irrational as to believe that they might continue to reap bulky and weighty vegetable crops from the soil, without having to return to it an equal weight of vegetable matter. Their practice implied the belief, that a virtue naturally exists in the soil, which enables it to yield crops out of its abundance ; and the belief certainly receives support from the fact of soils of natural fertility yielding largely with very inadequate culture ; and such a belief is naturally clung to with great tenacity, by farmers who cannot conceal from themselves the mortifying fact, that the ordinary resources of ordinary farms are unable to afford a return of as much matter in support of the soil, as the weight of the crop obtained from it. To make up for the deficiency, many expedients are resorted to —such as bare-fallows, changes of rotation, lime, and at length draining, which are all tried in succession and in co-operation, in order to sustain the soil in good heart; but useful as all these auxiliaries are, they are found to be no substitutes for the one indispensable source of fertility—*farm-yard manure*. It is admitted, on all hands, that without this manure, or some equivalent, if such there be, it is impossible for the soil to continue, for a series of years, to yield abundant crops; and it should also be admitted, that where the soil is not manured to the degree to call forth its *best* energies, a large amount, both of time and produce, is lost by a state of cultivation which is inefficient.

It was, and still is, a very natural desire in the farmer to be able to conduct his farm upon its own resources. True, he carries off to market a great weight of its produce every year; but it is equally true that the farm is, as it were, a field of creation,—where is raised every year what never existed elsewhere before. There is no unreasonableness in the supposition at first, that the application of all the disposable manure of the farm, together with skilful culture, might sustain, or even increase the fertility of a portion of its soil. It is easy to suppose, that, in addition to manure, skilful culture on exposing the soil to the atmosphere, by the action of the plough, the harrow, and the roller, may tend to increase its fertility by pulverisation; because observation affirms, that where the *natural* productions of the soil are most luxuriant, the soil is deep, and in a pulverised state. It is easy to conceive, when the soil is thus exposed by mechanical means, that a mutual chemical reaction takes effect between its constituents and the component parts of the air; and that the influence of rain, and heat, and light, may so alter the tone of the soil, imparted by the last crop, as to render it better for a succeeding one. It is easy to imagine, that, when superfluous water on land in winter is provided with channels, through which to flow awav easily, and not remain to consolidate, refrigerate, and acidify the soil, that the soil will become warmer, more easily pulverised, and more congenial to vegetation. With all these means of melioration, and with experienced skill, conjoined with the enriching quality of every animal and vegetable manure available, together with such a rotation of cropping as to render those means effective to the greatest degree; it is, I say, very natural in farmers to expect the soil, in such circumstances, to yield an increased produce. Yet, after all, melancholy experience has shewn the unreasonableness of the expectation, and has proved, beyond doubt, that no farm is able to *sustain* the fertility of its soil by its own resources, far less to *increase* it. The disclosure is useful, because, though disheartening, it has not dissuaded the farmer going in quest of assistance, nor has a knowledge of his wants deterred others presenting to his notice an almost innumerable host of succedanea. The difficulty with him now is, in choosing from among these what is really a useful manure.

Before directing your attention to any of the substitutes for farm-yard dung, which are now-a-days so rife, let us consider, in the first place, the extent of the resources which a farm of mixed husbandry possesses in supplying itself with manure; and to forming a just estimate of this inquiry, I regret to say, little information is to be found on which much reliance can be placed. It is a species of information, however, worthy of being ascertained by experiment on every class of soil, and in every system of husbandry.

Resources of the Farm.—The entire resources of a farm consists of the straw of the grain crops, all the green crops, whether of forage, tubers, or bulbs, all the grass and hay, all the dung of animals, whether confined in the steading, or at large in the fields, all the weeds picked off the fields, and every other refuse, such as coarse grasses, scourings of ditches, &c. Now, on looking at (1970.), it will be found from data adduced there, that a return of 1 ton of straw per imperial acre, at an average, from all the crops usually cultivated, is above the mark for Scotland. The English authorities, Arthur Young and Mr Middleton, estimated the average at from 1 ton 7 cwt. to 1 ton 5 cwt. per imperial acre. The late Dr Coventry estimated the average for Scotland at 1 ton 1 cwt. Judging from the produce in the neighbourhood of Edinburgh, I should say that 1 ton per acre was quite high enough an estimate for Scotland. Taking 1 ton as the average, the question is, What quantity of muck will this afford? and in considering this question you should remember, that, in the system of husbandry adopted for illustration, 200 acres are every year in corn, 200 acres in grass, and 100 acres in fallow. So that the whole dry straw of a farm of 500 acres would only weigh 200 tons. Dr Coventry estimates it as probable, that straw, after it has been wetted by the dung and urine of animals in courts and stables, and by the rain that may have fallen upon it, will weigh four times more than in the dry state; but that wet litter is reduced $\frac{1}{3}$ of its weight by fermentation before it is applied to the soil in the shape of manure. The other $\frac{2}{3}$, together with the pulse crops, as pease and beans, and the refuse of the corn crops, such as chaff, &c., he supposes may supply 4 tons of manure from every acre of straw, or 800 tons in all. Supposing the hay crop to weigh $1\frac{3}{4}$ ton per acre, and treated in the same manner as fodder-straw, will afford 6 tons per acre of manure, which over 20 acres of hay, will afford 120 tons of manure. The moist part of the turnip crop may be considered as computed in the additional weight acquired by the dry straw, after the turnips have been used by the live-stock in the courts and stables, still the firm portion of the crop will yield a great return, and, besides, improve the quality of the entire bulk of manure. Supposing that 24 tons is a fair crop of turnips per acre, and that ¼ of this is available for manure, 6 tons per acre will be derived from this source, as assumed by Dr Coventry, and if there are 69 acres of turnips, the dunghill will be increased in weight, if not in bulk, by this means, to the extent of 414 tons.* These are the chief resources of available manure on the farm, and they afford an aggregate of 1334 tons, which, at 15 cwt. per cart-load, gives about 1778 loads of dung. Of these, the potatoes require 20 loads per acre (2411.), 15 acres = 300 loads. The 69 acres of turnips, according to the dunging specified in (2500.), would be divided into 30 acres of Swedes, at 20 loads per acre = 600 loads; 10 acres of yellow, at 16 loads = 160 loads; and 29 acres of white, at 13 loads = 377 loads, in all for green crop 1437 loads; and as 10 acres of bare fallow and 6 acres of tares have to be dunged, which at the least will require 16 loads per acre, 256 loads will be required for this purpose (2824.). This calculation leaves 85 loads over after dunging the fallow division to an ordinary degree.

I suspect that the quantity of manure derived from the farm, as I have just stated it, and which is founded on the data furnished by Dr Coventry, gives too favourable a view of the farm, and is not in conformity with the experience of most farmers. I remember when in Berwickshire, on a farm of near 700 acres of land of good stamina for corn, no manure was ever bought for it. It was farmed on the 5-course shift, the fallow-break comprehending 130 acres, and about 90 stacks of 15 feet diameter used to be built within and without the stackyard in a good season. Though the turnips, occupying about 80 acres, were well dunged, keeping in view that a part were to be eaten off by sheep, I must own that the bare-fallow-break, consisting commonly of 40 acres. the remaining 10 being in potatoes and tares, were but lightly manured; and, no doubt, had manure been as plenty as is represented above, the bare-fallow land would have received more than it did. To sustain the stamina of the land, what was bare-fallowed in one course was made to bear turnips in the next. I may mention, in explanation of the circumstances I have stated regarding this farm, that it was situate 10 miles from the market-town, and neither bone-dust, nor any such manure, was in vogue in those days. With the facilities now existing for obtaining

* Coventry's Notes on the Culture and Cropping of Arable Land, p. 24.

manure, farmers may conduct their rotations with comparative ease, and as they please. On a 300 acre farm of turnip-land in Forfarshire under a 5-course shift, which I referred to in the rotation of cropping light lands, I could not, for the first few years, manure from its own resources more than 30 acres of the fallow-break of 60 acres; and even after 8 years of improving culture, the quantity never exceeded 40 acres, the remaining 20 acres being dunged with extraneous manure, partly with bone-dust and partly with cows'-dung, purchased at 5s. per ton, or 7s. the double-horse load. So great is the diversity of results obtained by farmers in regard to the proportion which the straw really bears to the crop, that little reliance, I fear, can be placed on Dr Coventry's estimate, as one for general application, even though we should be made acquainted with the premises from which he drew his conclusions. A limited experiment would afford no satisfactory results on this subject. By way of illustration, I may just mention the result of some experiments which were tried by Colonel Le Couteur with 4 different sorts of wheat with a view to ascertaining the quantity of straw afforded by each, and he obtained these very different results under the same circumstances; namely—

		Bu.		lb.			Straw.		
The White Downy,	yielded	48	and	4557	of straw	=	95 lb.	per	bushel.
... Jersey Danzig,	...	43½	...	4681	...	=	107	...	...
... Whittington,	...	33	...	7786	...	=	236	...	...
... Belle Vue Talavera,		52	...	5480	...	=	105½	...	...*

The quantity of straw to the bushel in the Jersey Danzig and Belle Vue Talavera is nearly the same, though the gross amount of produce, which is the source of manure, is very much in favour of the latter; while with the other two varieties of wheat, the quantities are very dissimilar and disproportioned, both of grain and straw; and on taking the gross weight both of grain and straw, the diversities and disproportions are just as great,—thus:—

			lb.		lb.				
Of White Downy,	the grain	weighs	2976,	the straw	4557=	little more	than	½	times.
... Jersey Danzig,	...	...	2740,	...	4681=	...	...	⅔	...
... Whittington,	...	...	2013,	...	7786=	...	...	3⅚	...
... Belle Vue Talavera,		...	3172,	...	5480=	...	...	1½	...

It is clear, therefore, that any results on this subject that should command general credence are yet to be derived from experiments conducted on a large scale throughout the country.

* Journal of the Royal Agricultural Society of England, vol. i. p. 123.

Farm-yard dung.—The acknowledged universal applicability of farm-yard dung to every other article of the kind, may arise from its very complex composition affording nourishment to every kind of plant raised on the farm. It is a compound of straw of various kinds, of horse-dung, cattle-dung, pigs'-dung, of the urine of those animals, of whatever dung the poultry may have dropped in their peregrinations through the different court-yards, and of rain-water, but of the sorts of dung, much the largest proportion consists of that of cattle. Analyses, I believe, have been made of farm-yard dung, in the state it is applied to the land, but as portions vary in composition, according to the proportion of the different sorts of dung and urine it contains, it will be more satisfactory to give the analysis of each component part, than of the whole together, though it is the aggregate which plays the important part in the economy of a dunghill. Of the various constituents of straw you have already been made acquainted, in (1965.) and (1966.).

The composition of cattle and horses' dung and urine is as follows:—

COWS' DUNG.

	Hailden.
Phosphate of lime,	10.9
... ... magnesia,	10.0
Perphosphate of iron,	8.5
Lime,	1.5
Gypsum,	3.1
Chloride of potassium, ... copper,	traces.
Silica,	63.7
Loss,	1.3
	100.0

COWS' URINE.

	Brande.
Chloride of potassium and sal ammonia,	15.
Sulphate of potash,	6.
Carbonate of potash,	4.
... ... lime,	3.
Urea,	4.
Water,	650.
	682.

HORSES' DUNG.

	Jackson.
Phosphate of lime,	5.00
Carbonate of lime,	18.75
Phosphate of magnesia,	36.25
Silica,	40.00
	100.00

HORSES' URINE.

	Vauquelin.
Carbonate of lime,	11.
... ... soda,	9.
Hippurate of soda,	24.
Chloride of potassium,	9.
Urea,	7.
Water,	940.
	1000.

I am not aware of any analysis of pig's-dung, but Spreugel examined pig's urine, when the animal was fed on corn offal, and found it to consist of—

Water, ...	92,600	in 100,000 parts.
Urea, with very little mucus, albumen, and colouring matter, ...	5,640	... ...
Salts, as common salt, muriate of potash, gypsum, carbonate of lime, and sulphate of soda, . .	1,760	... ...
	100,000*	

Of the origin of all these substances in the urine and dung of animals, and of the use of them as a manure to the soil, Liebig thus expresses himself in his own peculiar manner: "It has been shewn," he says, "by an examination of fæces, and of urine, that the mineral ingredients of the food—the alkalis, salts, and silica—are eliminated in these excrements. Urine contains all the soluble mineral substances of the food, while the fæces contain the ingredients insoluble in water. As the food is burned in the body just as it would be in a fire-place, the urine may be said to contain the soluble salts of the ashes, and the fæces the insoluble salts. These analyses shew, as nearly as can be expected from experiments of this kind, that all the constituents of the ashes of the food are again obtained, without alteration, in the solid and liquid excrements of the horse and cow. The action produced upon our fields by the liquid and solid excrements of animals ceases to be mysterious or enigmatical, as soon as we have attained a knowledge of their mode of origin."† Here, then, a mutual reproduction goes on between the food and the dung of animals; whatever ingredients animals consume in their food, those only they can and do void by their dung and urine, and these again constitute the best manure for raising the food upon which the animals feed. It follows that the ingredients afforded by straw, hay, turnips, and potatoes, are voided as dung and urine by the animals which feed upon them, and that the dung derived from them makes the best manure for raising the same crops. It follows also, that the farm itself is the best source of the manure that should be applied upon it. Also, that could the whole food consumed on the farm be returned again to the soil, in the shape of dung and urine, it would continue to yield without diminution; but this is impracticable, because the animals which are fed, take away, in increased size of body, and the animals wrought, in muscular energy, much of the ingredients of the food they consume, so that the soil must be supplied with manure from other sources to be

* Journal of the Royal Agricultural Society of England, vol. 1. p. 492.

† Liebig's Chemistry, in its application to Agriculture and Physiology. p. 176-7. Edition of 1843.

able to sustain its fertility, and much more so to increase it. This conclusion, which reasoning may have arrived at, is that derived from experience.

Assuming this to be the best general theory that can be given of the source of manure for a farm, we may make the same remark which Professor Johnston does, when speaking of the particular crops of a rotation: "It may be said that this explanation seems to imply that the same kind of crop may be reaped from the same soil for an indefinite number of years, by simply adding to it what the crop carries off. This is certainly implied in the principle; and *if we knew exactly what to add for each crop*, we might possibly attain this result, except in cases where the soil undergoes some gradual chemical alteration within itself, which it may require a change of treatment to counteract."* In connection with this view of the subject, practice appropriates the several sorts of dung in a determinate way. For example, horse-dung is preferred for potatoes, cow-dung for turnips, and care is taken not to apply pig's dung to potatoes, as it will inevitably impart a strong disagreeable taste to them.

Farm-yard dung is always applied in Scotland to the soil when it is under the operation of the plough, that is, it is always buried *under* a portion of the soil; and the object of this treatment is to secure all its volatile ingredients, as well as its more solid constituents. In England, however, it is extensively employed in top-dressing old meadow-land, which is made to produce hay every year; and, no doubt, if well fermented, and applied in moist weather, the soil will derive much benefit from it, and some such application is necessary, when the entire produce of the grass is carried off, as is the case with the hay crop. But it cannot admit of doubt that this practice occasions much waste of manure; very much of its volatile part must be dissipated, and much of its solid part dried by wind and heat. The practice is indicative of bad farming, for two reasons which ought to be conclusive with a good farmer. The first, as I have already stated, is the waste, to whatever extent, of valuable manure which it occasions; and the other reason is, that as old meadow-land is not included in the rotation of the rest of the farm, the manure it receives is so far a robbery of the arable farm, while it may return no manure at all, as all the hay may be sold and carried off. The rotation usually followed in England, in conjunction with old meadow-land, is, as I have already mentioned in the preceding section, the 4-course shift, a course which it is impossible to uphold on any farm without the

* Johnston's Lectures on Agricultural Chemistry and Geology, p. 719.

assistance of extraneous manure. It is evident, therefore, that top-dressing old meadow-land with farm-yard dung from another portion of the farm which is in a different course of management, is a scourging system for any arable land, and is, on that account, bad farming.

Farm-yard dung is also used in conjunction with other manures. Bones and guano are used along with it in the raising of turnips; and I am satisfied this is the best way of raising turnips, whether they are to be partly eaten off with sheep, or entirely carried away, and, at the same time, of maintaining the *stamina* of the soil, that is, its power of endurance under any system of cropping.

The durability of farm-yard dung is its great recommendation as a manure. Doubtless it is applied in large quantities, not less than from 10 to 20 tons per imperial acre, but a great proportion of this weight consists of water, even of well fermented dung; and were it practicable, or even proper to evaporate this, and thereby greatly reduce the weight, I am doubtful that the efficacy of the manure thereby would be impaired. I am persuaded that the first evaporation from a dunghill under fermentation consists entirely of water, and that not only a strong fermentation, but one conducted in an advanced part of the season, say not before April, is required before the constituents of a dunghill are begun to be dissipated. It is only after a *strong* smell is emitted, that a decomposition of parts is accomplished; for as to ammoniacal vapours, flying off ammonia has too strong affinity for water to leave the dunghill before it becomes dry enough.* For there is much virtue in the *sap* of dung, as the experience of every dry season confirms; and it is very difficult to evaporate the entire sap from a well-mixed dunghill, as the state of such dung shews even after fermentation has ceased in it.

Dung is applied at the commencement of every rotation of crops with the fallow green-crops, and with bare fallow; and when applied at any other time, it is near the termination of a long rotation. A rule for the quantity of farm-yard dung to be applied according to the length of the rotation, as given by Dr Coventry is, that 5 tons per acre are required every year to sustain the fertility of soil; and therefore land which is dunged every 4 years in a rotation of 4 courses, should receive with the fallow-crop 20 tons per acre; in a 5-course shift, 25 tons; in a 6-course shift, 30 tons, and so on.† These quantities constitute, no doubt, a sufficient manuring to ordinary crops; but it appears to me to be reversing the order of propriety, to give land under the severest shift—

* Professor Henslow's suggested experiments in Suffolk, may in time clear up this subject.

† Coventry's Notes on the Culture and Cropping of Arable Land, p. 4.

a 4-course one—the smallest modicum of manure, when it should receive the largest; for there is surely truth in the observation, that land grazed with stock becomes ameliorated in condition—actually increased in fertility. A 6-course shift, therefore, having 3 years of grazing, should require less instead of more manure even at a time than a 4-course one on land of similar quality.

Human fæces.—The food of man being of the richest and most varied description, human fæces and urine should contain valuable and numerous ingredients as manure; and if the principle be sound, which Liebig maintains, that animals fed on a certain kind of food void excrements best suited as manures for raising that food, then the food of man should best be raised from his own excrements manuring the soil. The analysis of Berzelius of human urine and fæces gives the following constituents in 1000 parts:—

HUMAN URINE.	
Urea,	30.10
Free lactic acid, lactate of ammonia, and animal matters not separable from them,	17.14
Uric acid,	1.00
Mucus of the bladder,	0.32
Sulphate of potash,	3.71
... ... soda,	3.16
Phosphate of soda,	2.94
... ... ammonia,	1.65
Chloride of sodium,	4.45
Muriate of ammonia,	1.50
Phosphate of magnesia and lime,	1.00
Silica,	0.03
Water,	933.00
	1000.00

HUMAN FÆCES.	
Phosphate of lime,	100.
... ... magnesia	
Traces of gypsum,	
Sulphate of soda,	8.
... ... potash,	
Phosphate of soda,	
Carbonate of soda,	8.
Silica,	16.
Carbonaceous residue and loss,	16.
	150.

In regard to what man returns to the soil from which he extracts his own nourishment, it is thus represented by Liebig:—"The importation of urine or of solid excrements from a foreign land is quite equivalent to the importation of corn and cattle. All these matters, in a certain time, assume the form of corn, flesh, and bones; they pass into the bodies of men, and again assume the same form they originally possessed. The only true loss that we experience, and that we cannot prevent on account of the habit of our times, is the loss of the phosphates, which man carries in his bones to the grave. The enormous quantity of food which man consumes during the 60 years of his life, and every constituent of it that was derived from our fields, may again be obtained and restored to them. It is quite certain that it is only in the bodies of our youth, and in those of growing animals, that a certain quantity of phosphate of lime is retained in the bones, and of alkaline phosphates in the blood. With the exception of this extremely small proportion, in comparison with the actual quantity existing in the food, all the salts with alkaline bases, and all the phosphates of lime and magnesia which animals daily consume in their food—in fact, therefore, all the inorganic ingredients of the food—are again obtained in the solid and liquid excrements."*

Human fæces constitutes a most efficient manure in the raising of turnips, but its tenacity renders it very difficult of application to the soil; and this is the case, whether it be commixed with a common dunghill, or with earth, chaff, or saw-dust, because none of these substances unite with it readily. It may be mixed with any of these ingredients, or applied alone, and if so, sparingly. As to the offensiveness of its odour, which many work-people stickle at, it may be overcome, by sprinkling occasionally over it, when being removed, a solution of the chloride of lime. This solution may be purchased in quart bottles at only 1s. each, and it should be diluted with 14 times its bulk of water when used. There is great waste of this valuable manure near dwelling-houses and farm-steadings; and though necessaries were erected, they would remain neglected. I think, however, their neglect greatly arises from their bad construction and position. They are usually no larger than a sentry-box, and devoid of convenience, far less of comfort; and they are generally placed so close to the dwellings of the people, or in so conspicuous a place, that they become a nuisance repulsive to the most insensitive. To make them really available for the purpose intended, there should be 2 sets of them, 1 for women and 1 for men. Each set should consist of a house sufficiently large to contain 2 seats with covers, for adults, and 1 seat for children; the floor should be elevated at least 1 step from the ground, and covered with stout pavement or boards; the walls and ceiling should be plastered and whitewashed with size to prevent the washing being rubbed off; there should be a small glazed window, partly movable to let in fresh air; and a stout door, provided with a strong latch-lock, for entering by a key from without, and a wooden bar to fasten it from within. A latch-key of different construction, should be provided for each necessary to be used respectively by the men and women. The sites of those houses should not be close to, nor far from, the dwelling-houses, but in a retired place, sufficiently screened in front with shrubs, and each set should be provided

* Liebig's Chemistry, in its application to Agriculture and Physiology, p. 178—18. Edition of 1843.

with a dung-court behind, having easy access for carts. The cattleman should sprinkle a little rich earth, set down for the purpose, over these courts every day. Were such commodious necessaries provided on every farm, with an understanding that each family was responsible for their 2 keys; were each family bound to keep the 2 houses clean by turns for a week; were the duty imposed on the steward to see that the houses are kept clean every week, and to demand a sight of the keys when he chose; were such regulations enforced for a short time, *as a strict injunction from the master*, I am satisfied they would be adopted and acted upon with as much willingness as other duties are usually fulfilled by country servants. Besides these houses, small tanks, easily removable, placed near the dwellings, would be a means of collecting an immense quantity of urine in the course of a year.

Human fæces is mixed with other ingredients, and sold under various denominations, such as pondrette, animalized carbon, desiccated compost, and the like. When such a composition is honestly formed, it cannot fail to make a powerful manure; but the farmer has no security against adulterations, and it is well known he is plundered at all hands by the imposition upon him of useless compounds. For my own part, I can say that when the animalized carbon first came to this country, about 20 years ago, it raised turnips as well and as cheaply as bone-dust; but it soon fell far short of its first exertions, though it rose in price as it fell in value. So with desiccated compost; I have tried it in comparison with farm-yard dung, pigeons' dung, and rich vegetable mould, and so far was it from being a manure at all, that even the black mould taken from the bottom of an old stone-dyke raised better turnips. Indeed, it scarcely afforded a better result than some drills which were not dunged at all, but were sown with turnips, by way of contrast, and as a standard of comparison. I am sure many farmers have been grievously deceived in the purchase of manures, and this being the case, every *compound* he wishes to try, he should mix for himself at home with the genuine ingredients of which it should consist. Such a precaution is necessary, for to be deceived in the particular of manure, is, in effect, to incur the loss of a whole year's crop, and such a loss involves not merely individual, but national interests.

Bone-dust.—The composition of this substance, which is of so much worth to the farmer, I have already given in (2529.), and, on account of its containing so great a variety of constituents, it is a true and valuable manure. It is now believed that the phosphate of lime, with which they most abound, is the most valuable ingredient in the manure of bones. Bone-dust exhibits, however, a peculiarity in its effects, as a manure, which seems inexplicable, namely, that a given quantity produces a maximum effect. Thus, I have tried 12, 16, 20, and 24 bushels per acre with white globe turnips, and found the crop to improve with 12 to 16 bushels; but what is remarkable, neither the 20 nor the 24 bushels gave a greater crop than the 16. This, no doubt, may be explained from the probability of the turnip requiring only a certain quantity of nourishment, which the 16 bushels supplied, and this may account for the amount of the turnip crop received; but it cannot account for the insensible effects upon the succeeding crops, for neither the barley, the grass, nor the oats which followed the turnips in the rotation, were in the least more increased in bulk and quantity with the 20 and 24 bushels than with the 16, though the 16 yielded better than the 12. We cannot conceive that the soil received no greater benefit, as regards condition, from 24 than from 16 bushels, yet the crops indicated no difference whatever. It is true I did not measure and weigh every bushel and ton of the produce, but I had the same means of judging them all,—namely, by minute inspection. I knew that the respective quantities of manure and seed were applied during the entire rotation on very similar soil in quality and situation, in the same field, and on the same day. Nor were these comparative experiments conducted on a very small scale, such as ⅛ of an acre; for each portion comprehended 4 long ridges of 15 feet in width, containing not less than 1½ acre. There were other results brought out by this experiment. The turnips were all carried off the ground, that is, none were eaten off with sheep, as the ½ should have been, and in so far the clearing of the field after bone-dust was an act of bad farming; but the robbery was committed from necessity, as there was a deficiency that season of dunged turnips for the cattle, whilst the turnips raised by bones were more extensive than the sheep I had could overtake. Though an act of bad farming, the experiment proved two important particulars; *first*, that bone-dust of itself benefits the whole crops of a rotation; the barley, grass, and oats, that followed the turnips, were all good; and, *second*, they were equally good, turnips included, with similar crops raised in the same field, and on the same soil, with 16 tons of well-made farm-yard dung. Indeed the grass was in quality much finer. So we may conclude, that 1 bushel of good bone-dust is equal in effect upon crops, during a 5-course rotation, to 1 ton of farm-yard dung. It does not follow, however, from this result, that that small quantity of bone-dust will sustain the enduring fertility of soil for many years like dung.

Perhaps there is no way of applying bone-dust so efficaciously—and certainly there is none in my estimation—as upon farm-yard dung. Drill the land for turnips, say with 12 cart-loads of dung, and then sow the seed with 8 or 10 bushels of bone-dust. The bone-dust secures the early progress of the plant, and the dung sustains it after the roots strike into it. Such turnips eaten off with sheep, should put and keep any land in good heart. With 8 bushels of bones, and finely riddled coal-ashes at pleasure, an excellent crop of turnips may be raised for sheep.

Guano.—This is a foreign substance, which has only recently been introduced into the country as a manure. It is just the dung of birds, and is perhaps no better manure that that of our own sea-birds would be, could it be preserved; but no sooner is it voided, than the rain and snow, and waves of the ocean, wash it away; whereas, in the tropics, whether in America or Africa, the heat desiccates and preserves it immediately on being voided. It is a compound containing many ingredients, as may be seen from the following analyses:—

	By Bartels.	By Völckel.
Muriate of ammonia,	6.500	4.2
Oxalate of ammonia,	13.351	10.6
Urate of ammonia,	3.244	9.0
Phosphate of ammonia,	6.250	6.0
Waxy substance,	0.600	...
Sulphate of potash,	4.227	5.5
... soda,	1.119	3.8
Phosphate of soda,	5.291	...
... magnesia and ammonia,	4.196	2.6
Chloride of sodium,	0.100	...
Phosphate of lime;	9.940	14.3
Oxalate of lime,	16.360	7.0
Alumina,	0.104	...
Residue insoluble in nitric acid,	5.800	4.7
Loss, consisting of water, ammonia, and organic matter, not estimated,	22.718	32.3*

In the short time since the introduction of guano, it has proved itself a true and valuable manure. When tried on turnips against farm-yard dung, at the rate of only 3 cwt. per acre, it produced 20 cwt. 6 stones, on a similar piece of ground, that 18 cubic yards of dung per acre produced 19 cwt. 2 stones. Tested against bone-dust, at the rate of 16 bushels, and coal-ashes 8 bushels, together 24 bushels per acre, which produced 19 cwt. 2 stones, guano, at the rate of 3 cwt. per acre, yielded 23 cwt. 2 stones. Against bone-dust alone, at the rate of 16 bushels per acre, which produced 24 cwt. 7 stones, guano, at the rate of 2 cwt., produced

* Liebig's Chemistry, in its application to Agriculture and Physiology, p. 181. Edition of 1843.

31 cwt. 4 stones.* Guano is very efficacious for turnips, along with a little farm-yard dung. Its fame as a manure is now established, though as a substance which would make a good manure, it was known and examined by Sir Humphry Davy more than 40 years ago. Such is the demand for it, that its price is about L.10 per ton, and it reached, in the summer of 1844, to L.14. The value of bone-dust, in consequence, has fallen to 1s. 9d. per bushel. In the use of guano precaution is requisite, as it is apt to affect the vitality of seeds sown in contact with it, so that a little earth between it and the seed is necessary.

Pigeons' dung.—This manure, I have no doubt, would be as valuable as guano, could it be obtained in sufficient quantity. I have tried to raise turnips with it, and succeeded to admiration; and one season, 1823, I raised Swedish turnips with 4 double cart loads. The quantity was applied in the drill with a shovel by guess, but having the desire to make it go as far as possible, I suspected that I had stinted the land of manure. The seed was afterwards sown upon the drill, which buried the dung, and the crop throughout the season was very superior to that from farm-yard dung or bone-dust. The bulbs proved large, and a heavy crop; but I had not leisure at the time to attend to particulars. Next season the dove-cot only yielded 1 double load of dung, but so far as it went, I was equally successful in raising Swedish turnips. Tanners, I believe, will give a high price for pigeons' dung, as I have been offered 16s. per ton for it; but I would advise you rather to use it at home for Swedish turnips. I have seen it stated somewhere, that 50 bushels of pigeons' dung; or 40 bushels of pigeons' dung with 8 bushels of rape-dust; or from 12 to 15 bushels of pigeons' dung, with 12 to 15 bushels of bone-dust, are sufficient to raise turnips equal to a good dunging of farm-yard manure. My opinion is, that pigeons' dung is as efficacious as guano, or at least as bone-dust, and that, therefore, those quantities of pigeons' dung are much too great per acre imperial. When pigeons' dung is wetted with water, it ferments rapidly, and in a few days may be riddled and mixed with equal quantities of ashes, and sown for turnips, at 32 bushels per acre. When this mixture is spread in January or February, out of carts, as a top-dressing on new grass, it is said to make it fit for cutting 14 days earlier than the ordinary time. Whether these statements are strictly correct, I cannot say from my own experience, but they are worth testing by experiment.

Pigeons' dung has been chemically examined. "The excrements of pigeons," says Sprengel, "have been chemically examined by Sir Hum-

* Transactions of the Highland and Agricultural Society, for October 1843, p. 70-2.

phry Davy and myself. Davy found in 100 parts by weight 23 parts of substances soluble in water, consisting of urea, urate of ammonia, common salt, and some others. According to my own experiments, pigeons' dung half a year old contained only 16 per cent. of bodies soluble in water, consisting of very little urea, but of a large proportion of carbonate, sulphate, and muriate of ammonia, common salt, and sulphate of potash. The other 84 parts insoluble in water consisted of coarse siliceous sand, silica, phosphates of lime and magnesia, traces of alumina, and oxides of manganese and iron. The abundance of soluble substances explains the quick effect of pigeons' dung, and also shews us once more the great value of mineral manure."* Hence the propriety of applying pigeons' dung fresh, or of strewing the floor of the dove-cot with soil abundant in humus, for the ammonia of the dung to combine with the humic acid of the earth.

Fish garbage.—In fishing villages, where fish are smoked or salted, a considerable quantity of fish refuse may be obtained, and it constitutes an efficient manure for every kind of crop. On the east coast of Scotland, 30 barrels of fish heads and guts, half of cod and half of haddock, are enough of manure for 1 acre. The barrel contains 30 gallons, and 4 make a cart-load. The refuse sells at 1s. 6d. per barrel, and so does liver and oil refuse. In preparing fish refuse for manure, it is emptied from the barrels on a headridge of the field to be manured, and mixed with a quantity of earth sufficient to cover the refuse completely. It is driven fresh to the field whenever a supply can be obtained from the fishers. In 2 or 3 months the compost is ready for use; and as a manure for turnips is superior to farm-yard dung, and equally beneficial on light and heavy soils. When used for turnips, the compost is spread with shovels out of the cart along the drills, at the rate mentioned; over which the drills are split, and the seed sown along the drills by the machine. Of course, it may be applied to bare fallow for wheat, as well as for green crops. It is sometimes laid on as a top-dressing in autumn upon lea, and ploughed in; and, as may be expected, the succeeding oats prove an excellent crop. Swedish turnips are afterwards taken with the ordinary manuring of farm-yard dung; and, in the circumstances, they never fail to yield abundantly, while the soil is put into the finest condition. From 400 to 600 barrels of this refuse are obtained by a farmer during the season; but those whose farms are nearest the villages have the best chance, unless a special agreement be made with the fishers. Fish refuse may, therefore, be regarded as a true manure.

* Journal of the Royal Agricultural Society of England, vol. i. p. 493.

See (2401.) 11. In regard to sprats, as a manure, Mr Cuthbert Johnson relates, that "the farmers of Essex and Suffolk purchase these fish by thousands of bushels at a time, and carry them in waggons 10 or 15 miles into the inland districts. The quantity applied per acre varies from 25 to 45 bushels, the poor gravelly soils requiring more than the loamy lands. They are spread by hand from seed-baskets, and on winter fallows intended for oats, on which, especially if the summer is not too dry, it produces most luxuriant crops, of a peculiar dark green colour, yielding 10 or 11 quarters per acre, and that on land of a very second-rate description. The effect of the application, however, remains only for 1 crop. They produce an equally good result, if mixed with earth, and suffered to remain and dissolve for some time in the heap, before they are carted on the land. In this way they answer exceedingly well for turnips. They are usually obtainable at the rate of 6d. or 8d. per bushel."* The refuse of pilchards and of herrings are, of course, of equal value to those mentioned, where they are obtainable.

Sea-ware.—To farmers situate on the sea-coast, this manure is a valuable acquisition, so much so, that, on the east coast of Fife, I have heard it stated that as much as 10s. per acre are offered for farms that command a large supply of sea-ware more than for others not so fortunately situate. On many of the farms in East Lothian, from 100 to 120 imperial acres are annually manured with sea-ware; and when I mention that 30 double-cart loads are spread on 1 acre, you may conceive the labour incurred in carting from 3000 to 3600 loads during a short season; for it is only in winter that the ware is cast ashore by storms, when the plants have arrived at maturity, and are more easily detached from the rock by a heavy sea. The collecting and driving are calculated in Fife to cost from 1s. to 1s. 2d. per cart-load. Sometimes when a bank of sea-ware has been driven on shore, and there is risk of its being washed away again by the waves, all hands are employed, men, women, and horses, to land as much as they can above high-water mark, as long as the danger of losing it exists. In Fife, 16 loads per acre of ware are supposed equal to 20 loads of farm-yard dung, but this seems an exaggeration. There is no doubt, however, that it makes an excellent top-dressing for the aftermath of a crop of hay. It is likewise spread on lea, and affords the means of yielding a fine crop of oats. It is also ploughed in with the oat-stubble, in preparation of the land for turnips. In all cases it is ploughed in as fresh a state as possible; and to assist the plough in burying the long leaves and tangles, a field-worker follows the plough, and rakes the ends of the ware into the furrow with the

* Johnson on Fertilizers, p. 114.

small dung-spreading graip, fig. 151. The composition of sea-ware, and a few remarks on its natural history, will be found in (2041-2-3.).

Cow's urine and dung are obtained by farmers from the cowfeeders in town, on payment of 5s. per cow for the year, and the expense of driving, when the cows are in the byre, and not in the fields in summer; or, if paid for in kind, instead of money, ½ kemple of 16 stones, of 22 lb. to the stone, per annum for each cow. Cow-dung is sold at 5s. per ton, or L.4, 15s. per cow per annum. The market gardeners in the neighbourhood of Edinburgh manure their garden-ground with cow-urine, to the extent of 40 tons per imperial acre. This quantity, though raising large crops of vegetables, is found to exhaust the soil so much as to become effete, and were it not stimulated with ordinary manure for some time, the vegetables would not arrive at perfection. On fields cow-urine may be applied with advantage in wet weather on clover aftermath that is intended to be taken up for oats, to the extent of 12 to 15 tons per imperial acre; but it has been found to injure oats after *rye-grass*. Fig. 366 shews the construction of a liquid-manure cart, out of which the cow-urine is distributed over the ground by means of a disperser furnished with holes, (2045.), as also fig. 367.

The substances I have mentioned may all be regarded as true manures, that is, as possessing a composition, the particulars of which contain the substances requisite for the maintenance of all the plants cultivated on a farm; but there is a class of substances which, until very lately, were never regarded as essential to the well-being of plants, namely, their *specific* constituents, which are *inorganic* or *mineral*. The vegetable organic structure, which forms the body of the plant, is so obvious, that its maintenance has only hitherto attracted the attention of cultivators, and the nature of its minute constituents has been overlooked by men of science. True, hints have been thrown out, that, in consequence of the want of success in cultivating plants in particular circumstances, particular substances may be required to supply the peculiarities of their composition; and several years ago Mr Grisenthwaite expressed his opinion, that it was by their special constituents that plants were alone contradistinguished from each other, the organic structure being alike in all; and therefore recommended a minute analysis of all the cultivated plants to be undertaken, in order that the peculiar constituents of each might be ascertained. His reasoning on the subject was in these terms: "Elements, as the very term implies," he observes, "are now known to be incapable of being changed into each other. They admit, when considered *per se*, of no alteration but as regards magnitude and figure; and all the variety of matter discoverable in the world is produced by combination of these elements in different proportions. From this fact we are immediately led to deduce the following important conclusion: That when out of one substance another is to be formed, as alcohol or acetic acid out of sugar, or, to confine our views to agriculture, grain out of manure, it is obvious that the elements of the first must be contained in the second; as if they be not, the conversion cannot take place. This is a truth which applies with peculiar force to the doctrine of manures, and renders it imperatively incumbent upon the agriculturist to investigate the constituents both of the crops he grows, and the manures he employs to make that growth successful. It is very reasonably to be feared, that many failures, quite inexplicable to the farmer, may be explained upon these principles. He has, very frequently perhaps, some grain upon land which has not contained the elements necessary to the production of the crop, and therefore the crop has failed; and he continues to suffer a recurrence of the same loss year after year, because he is unacquainted with the causes upon which it depends. If all crops were composed of the same elements, this reasoning, this discrimination, among manures, would not apply, nor be necessary to be regarded by the agriculturist; and it is upon such a supposition that the practices of husbandry have been uniformly conducted, and are at the present day conducted.

"To illustrate the preceding reasoning, we may select the wheat crop as an example, which, while it is doubtless the most important to mankind, is also better known in its constituents than most other grain. If we examine the straw of wheat we shall find it to be composed of common vegetable matter; or of oxygen, hydrogen, and carbon. This I call *common vegetable matter*, because the elements are common to every known vegetable substance. If we examine the grain, we shall find its constituents to be starch and gluten; and if we carry our researches still farther, we shall find that the elements of starch are precisely the same with the elements of common vegetable matter, viz., oxygen, hydrogen, and carbon; but the elements of the gluten, besides consisting of the three just named constituents, contain nitrogen also, an element not common to vegetable substance, but composing a large part of most animal matters. Now, from what has just been stated, it is clear that the same manure which is employed in the production of the straw and the starch of the wheat crop, cannot possibly produce the gluten also. For this depends upon the presence of a distinct element, an element which cannot, as far as our present knowledge extends, be formed out of other elements, either by the operations of art, or by the processes of nature, both of which are in reality the same. This is a fact which has never, I believe, been regarded by writers on the theory, or men engaged in the practice of agriculture; and yet upon it depends the successful cultivation of this most important crop."*

This extract really explains the entire motives by which the agriculturists at present desire to obtain the assistance of chemistry in raising larger crops and of better quality; and it contains the entire rationale of the doctrine of specific manures, the desire to apply which to field-culture has given the peculiar bias at present to the agriculturist's mind. It was reserved for Liebig to point out what those specific substances are which contradistinguish the plants usually cultivated in the fields; and this knowledge he has acquired by the very means pointed out by Mr Grisenthwaite, namely, by laborious analyses of the plants and of their products. His investigations in this difficult and interesting field of inquiry have enabled him to determine that ammonia is the most valuable food of plants; that supplies of it may be obtained for them by the decomposition of the various salts of ammonia; that other salts are required, if not directly, for yielding essential ingredients, at all events indirectly, for assisting in the decomposition of the ammoniacal salts; and that the ashes of plants indicate the peculiar mineral or minerals which each plant takes, in greater or smaller quantity, into its composition.

The employment of specific manures, recommended by theory in the first instance, and urged by the successful researches of chemical investigation, is now prescribed to the farmer as a practical operation; and it must be owned he has received the solicitation in a very confiding spirit, much more so than any subject I remember him to have received, which had not the previous sanction of his own experience. He has evinced a desire to try every suggestion offered, and has even gone the length of requesting a chemist of established reputation to examine the results of his experiments, and to suggest further experiments upon them, with a view to ultimately obtaining useful results. What those results may ultimately prove, time alone can determine; and as every experiment in agriculture takes one year at least for its completion, that time must yet occupy several years. A great problem is evidently at work at present on this subject in the field of agriculture; and, as its object is decidedly good, I cannot but hope, for the sake of the country as well as the farmers, that it will be successfully solved. So long as it is under solution, however, I think the best plan for me is to decline entering into the subject of specific manures, because the mere enunciation, and much more the recommendation, of results, as yet untested, is as likely to lead you into error, as to guide you towards truth; for repeated and extensive trials have

* Grisenthwaite's New Theory of Agriculture, p. 161–4. Second edition. 1830.

yet to be made ere facts can be established; and without the establishment of indisputable facts no general conclusions for your guidance can be arrived at. The best service I can afford you at present is, therefore, to point out to you the best papers that have been written by the most extensive experimentalists; and as the subject admits of improvement by every new experiment, the accounts of the most recent experiments should possess the greatest interest. To open up the entire subject, would, besides, occupy a much greater space than I have to spare; and if entered on at all in its present unfinished state, it could only consist of relating the particulars of what every experimenter had observed; and these you will appreciate far better in the experimenters own words.*

There are other substances employed as manure, very different from those just referred to, as well as from ordinary manures, and which can only be obtained in quantities in certain localities. For example, *soot* can only be obtained in large quantities from large towns, and it makes an excellent top-dressing for one season on grass. The quantity employed is about 40 bushels per acre, and the cost is from 1s. 3d. to 2s. 3d. per bushel. As this is a very disagreeable substance to sow with the hand, a machine has been in use for some years, for the purpose of distributing it equally over the surface of the grass-land, a description and figure of which, by Mr Slight, will be found below. The effect of soot is to promote the growth of the leaves of plants, and particularly of grass, and to impart to them a dark green colour. I have heard it stated that cow-feeders object to graze cows on pasture that has been top-dressed with soot, in consequence of the taste which it imparts to milk; and they

* See Transactions of the Highland and Agricultural Society for March 1844, pp. 161-204, by Mr John Hannam, North Deighton, Wetherby, in Yorkshire. For patient investigation, accurate observation, clearness of detail, and intelligent deduction, this paper, in my opinion, is a perfect model of an account of agricultural experiments. In the Number for July 1844, pp. 227-49, the experiments of Mr A. F. Gardiner, overseer to Mr Fleming of Barrochan, in Renfrewshire, are very well related. In the same Number, from pp. 250-4, the conclusions by Mr Lumsdaine of Lathallan, in Fifeshire, are correctly drawn from the experiments, and are in themselves important. In the Number for July 1843, pp. 28-36, the account of the experiments made by Mr Maclean at Braidwood, near Penicuik, in Mid-Lothian, with 28 different substances, at a considerable elevation above the level of the sea, are worth perusal; as well as some experiments by Mr Carstairs of Springfield, near Penicuik, on the effects of some special manures on moss-land, which are curious and encouraging to those who possess similar soil, in the Number for July 1844, pp. 277-9. Mr Thomas Bishop, land-steward at Methven Castle, in Perthshire, gives an account of experiments made with a few uncommon substances, such as grass-weedings, cocoa-nut dust, carbonised saw-dust, exhausted cow-dung, wet wasted straw, compared with known fertilizers, in the Number for October 1843, pp. 64-7; and in the Number for October 1844, p. 304 and onwards, will be found one paper by Mr John Finnie, Swanston, Mid-Lothian, and another by Mr Charles Stevenson, Redside, East-Lothian. The Appendix to Professor Johnston's Lectures on Agricultural Chemistry and Geology is wholly occupied with accounts of experiments with special manures, made in different parts of the country, with such remarks upon, and suggestions from, them, as the circumstances of each case called forth.

even object to purchase the hay for cows that has been saved from grass top-dressed with soot. The effects of soot are evanescent, not enduring beyond one season. It should be applied in spring when the grass is damp, and in calm weather. When applied in dry weather, it is apt to scorch the grass.

Woollen rags make an excellent manure for potatoes, when chopped small, and strewed along the drills, at the rate of from 3 to 4 cwt. per acre on light, and 12 cwt. on strong soils. It is mostly used, however, for the manuring of hop-grounds. Trifling as this article may seem, 20,000 tons are annually used in England, as high as 5 guineas per ton.*

Green-weed of delicate variety, "found alone in protected situations in the estuaries of our rivers, is used in the upper parts of the Forth, and still more especially so in the Eden. Mr Meldrum of Bloomhill, near St Andrews, besides collecting the weed on his own shores, rents that of his neighbours. He frequently applies from 300 to 400 cart-loads in a single year, and reckons 10 cart-loads good, and 15 heavy, manuring. When laid on in winter, and ploughed into the furrow-ground, it produces a fine pulverising effect. With this alone a wheat crop of 6 quarters an acre has been produced, with a heavy crop of beans the year after without additional dung."†

Shell-fish and Shells.—I have known ground mussel and oyster shells used as manure for turnips; but double the quantity did not produce the same effect upon the crop as bone-dust; perhaps it would require 40 bushels to produce the same effect as 16 bushels of bone-dust. One use made of this shell-dust is to adulterate bone-dust therewith. It has been lately stated that common shell-fish, such as whelks, cockles, and mussels, to the extent of 16 bushels per acre, have been employed with success to raise turnips, the bushel weighing 1 cwt. To those near the coast, with a rocky shore, such manure is obtainable.

Shell-marl.—In some parts of the country, such as Forfarshire, this substance is found in considerable quantities associated with peat. It occurs in beds in deep peat-bogs, lined above and below with a layer of very fine unctuous clay. It is taken out of the bogs by means of a boat mounted with a dredging apparatus. When of fine quality, and in a dry state, it is as white as lime, not crumbling down into powder like quicklime, but cutting something like cheese with the spade, and adhering in large lumps when spread. It is applied at the rate of from 40 to 50 bolls per imperial acre, the boll containing 8 cubic feet, and selling at 9d., making the cost of manuring from L.1, 10s. to L.1 : 17 : 6

* Johnson on Fertilizers, p. 124.

† Quarterly Journal of Agriculture, vol. xi. p. 308.

per acre, exclusive of the cost of carriage. When applied to land as a calcareous substance in moderation, it assists the action of ordinary manure; but it is too often applied solely as a manure, and in the above quantities, namely, from 35 to 45 cubic yards per acre, when it never fails to do mischief. It does not injure fresh land, it is true; on the contrary, it seems to stimulate it greatly, causing it to exert itself, and thereby soon becoming exhausted. When repeated frequently as a sole manuring, I have seen the land reduced to such a state of fermented dry pulverization, that with a stamp of the foot, the leg has been driven into the ground as high as the ankle, and a dust raised by the stroke. "Applied to lands followed by severe cropping," remarks Mr Headrick, "it has reduced them almost to a state of utter sterility, which they have not recovered to this day." *

Besides those substances, which attract the attention of most farmers, there are numerous others which may be used as manure, that are nearly overlooked by him, and these have been denominated *waste* manures. They comprehend all matters allowed to waste themselves on the farm; the sewerage of towns, which are allowed to run waste to an enormous extent; the waste of manufactures, such as shoddy, flax waste, sugar waste, tanners' waste, and the like; and local wastes, such as peat, weeds, ashes, &c. In regard to the importance of these substances as manures, trifling as they may seem, Mr Hannam observes, in his preface, that, "while pointing out the waste of manure which too commonly takes place throughout the country, and suggesting available means for its prevention, the author has endeavoured to call attention to the subject, as of an equal importance to the farmer individually, and the public generally; for though to make that which is useless to the farmer valuable to him, and to give him an efficacious and economical agent by which he may augment his produce, *is* one means by which he may *reduce his expenditure and increase his income*, at the same time it is one from which the public will reap an *increased supply* of food at *a decreased* cost."†

Of the important part which every manure of the most trifling nature may play in the economy of husbandry, may be learned from these observations of Liebig: "It is certainly the case, that we could dispense with the excrements of man and animals, if we were able to obtain from

* Headrick's Agricultural Survey of Forfarshire, p. 406. In enumerating these substances, I have confined my observations to those which are within the reach of many farmers. For manures from more distant farms, I refer with pleasure to Mr Hannam of North Deighton's Essay on Rape-Dust and Hand-Tillages.

† Hannam on the Economy of Waste Manures, p. vi., an excellent little treatise.

other sources the ingredients on which depends all their value for agriculture. It is a matter of no consequence whether we obtain ammonia in the form of urine, or in that of a salt from the products of the distillation of coal, or whether we obtain phosphate of lime in the form of bones, or as the mineral apatite. The principal object of agriculture is to restore to our land the substance removed from it, and which the atmosphere cannot yield, in whatever way the restoration can be most conveniently effected. If the restoration be imperfect, the fertility of our fields, or of the whole country, will be impaired; but if, on the contrary, we add more than we take away, the fertility will be increased."*

[The theory of the action of manure has been more carefully and extensively studied by chemists than any other branch of agricultural science, and to a certain extent their labours have been completely successful. We are enabled now with absolute certainty to state what the result of the application of most manures will be, but we have yet to determine the various steps of the process by which the result is brought about. Owing to the necessity of condensing my observations, I shall confine myself to the mere statement of a few general principles upon which the successful application of manure depends.

1. The great object of applying manure is for the purpose of rendering soil more fitted for the production of the largest possible crops, and this may be effected in three different ways :—

First, By giving to the soil, in the shape of manure, certain ingredients which it did not previously possess, but which are essential to vigorous vegetation.

Second, By applying certain substances which will call forth the *latent* energies of the soil itself, although these substances themselves may be useless to the crop.

Third, By applying substances capable of altering mechanically the condition of soil when this chances to be defective.

2. The increased fertility of soil by the three different methods above detailed are capable of being effected by a great variety of means, the choice of which is determined by economy; and as the great object of farming is to convert refuse matter into crops, and crops into stock, it follows that the most economical procedure that can possibly be adopted is to use every means for collecting and saving every species of refuse that can be made available as manure. It is only when the refuse of the farm is not sufficient for the culture of the soil that the farmer is warranted in purchasing manure, or, in other words, it is only those farmers who, by skilful tillage, have raised the capabilities of their soil above the standard of fertility, that should ever require to look beyond themselves for the means of keeping up the fertility of their farms.

The simplest method of describing the action of manures is by arranging the more common ones in a table according to their mode of action; and in reference to this table, it will be necessary to state that nitrate of soda, saltpetre, and sulphate of ammonia, are now considered as indirectly yielding *organic* matter to plants, by furnishing, through their decomposition, the azote required for its production.—H. R. M.]

* Liebig's Chemistry in its application to Agriculture and Physiology, p. 177. Edition of 1843.

	Manures which act by yielding *organi;* matter to plants.	Manures which act by yielding *earthy* and *saline* matter to plants.	Manures which act upon the organic matter of the soil.		Manures which act by altering the *texture* of the soil.	Manures which act specifically upon certain crops.		Manures which act as stimulants.
			By Fermentation.	By Chemical solvents.		From containing peculiar *saline* matter.	From peculiarity of *organic* matter.	
VEGETABLE.	Rape-cake. Malt-dust. Steepings of flax and hemp. Green plants. Straw. Woody fibres. Tanners' spent bark.	All, more or less.	All slightly.					
	Farm-yard dung.	Farm-yard dung.	Farm-yard dung.					
ANIMAL.	Dead animals. Fish. Blubber. Excrements. Urine. Horn. Hair. Woollen rags. Feathers. Guano,	All, in some degree.	All, more or less.			Horn.	Rape-cake. Malt-dust ? Horn.	
MINERAL.		Chalk. Gypsum. Marl. Saltpetre. Common salt. Kelp. Nitrate of soda. Sulphate of soda. Sulphate of ammonia.		Kelp. Lime (hot.)	Marl. Chalk. Lime (mild.)	Gypsum. Chalk. Kelp. Common salt. Saltpetre.		Saltpetre. Common salt.
MIXED.	Bones. Composts.	Bones. Ashes. Soot. Compost.	Bones. Composts.	Ashes. Composts.		Bones. Ashes. Soot Composts.	Bones.	Soot.

[The experience of every succeeding day tends to shew the extension of the application of granulated manures to the soil; and the powerful influence of some of them, coupled with the high price of the article, renders the necessity of a uniform and sparing distribution of them imperative. Machines being better adapted to the performance of a uniform distribution than a number of individual hands, and as they can likewise be adjusted to deposit either a larger or smaller quantity with uniformity, the demand for such machines must increase with the extending application of the manures.

It is found, however, that any one form of machine is not fitted to deposit all the varieties of such manures, and that to some extent each kind has required its own peculiar construction of the distributing parts of the machine. The late introduction of guano has increased this difficulty; for in the more early application of it, the article was purchased always in a very damp state, and care was taken to impress on the mind of the purchaser that a state of natural dampness was requisite to retain the virtues of the manure until it was put into the soil. This circumstance caused much trouble to those who attempted to distribute the manure with machines, for by reason of its dampness, aided by the action of the revolving parts of the machines, the guano was converted into a paste, and the machine ceased to perform its wonted duties. But it soon became evident that the dampness so much recommended by the vender was only the effect of a liberal administration of water applied by him to increase the weight of the article, and the amount of his own profits; and this detection once made, we now see the article publicly recommended for its dryness. The previous state had, in the mean time, called forth a degree of attention to the means of distributing that and other roughly granulated substances through the machine usually employed, and this has extended our knowledge in this department of agricultural machine-making.

The results of these inquiries and experiments have been to produce a conviction, that the simple principle of distribution so successfully employed in the broadcast, and grain drill sowing-machine, Plate XXVI. and (2357.), may, with very slight modification, adapted to the particular object, be rendered available for almost all granulated manures, more especially now that the important article of guano will generally be procurable in a dry state, which hitherto has been the most difficult to accomplish. Soot, as a manure, has presented the same difficulty in its distribution from the ordinary machines; and as an example for both of these articles, it may be stated that a machine, such as seen in Plate XXXI. and (2544.), which is adapted for bone-dust, will neither deliver guano nor soot in a regular manner. Hence, also, the following machine, constructed for the distribution of soot as a top-dressing, has been greatly improved by the application to it of the principle of the broadcast machine.

The *soot-sowing machine*, from the limited supply of the article which it operates upon, can never be ranked amongst the most important class of machines on the farm; still, owing to the powerful effect of the manure itself, its due distribution is of importance, and, from its extreme lightness, it cannot, without disadvantage, be sown by the hand. The machine here described was the production of Mr Main, factor to the Earl of Dalhousie.* Fig. 579 is a view, in perspective, the horse-shafts being broken off; and fig. 580 a transverse sec-

* Prize Essays of the Highland and Agricultural Society, vol. xii. p. 535.

tion, with the shafts also broken off, the letters of reference applying to corre-

Fig. 579.

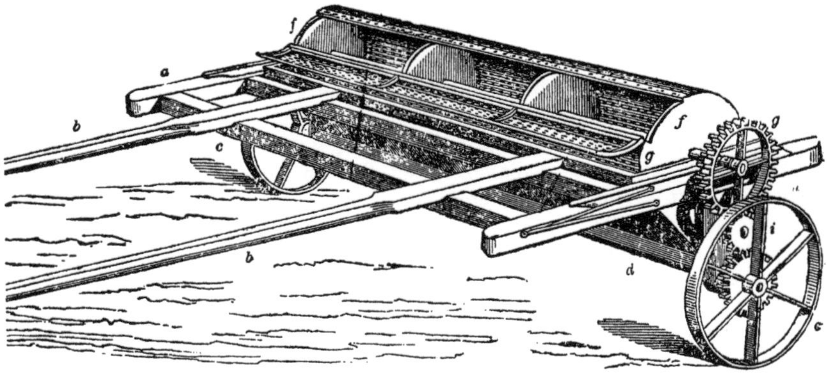

THE SOOT-SOWING MACHINE.

sponding parts in the two figures. The machine consists of a bed-frame *a a*, to which the horse-shafts *b b* are attached, and is mounted on a pair of low wheels

Fig. 580.

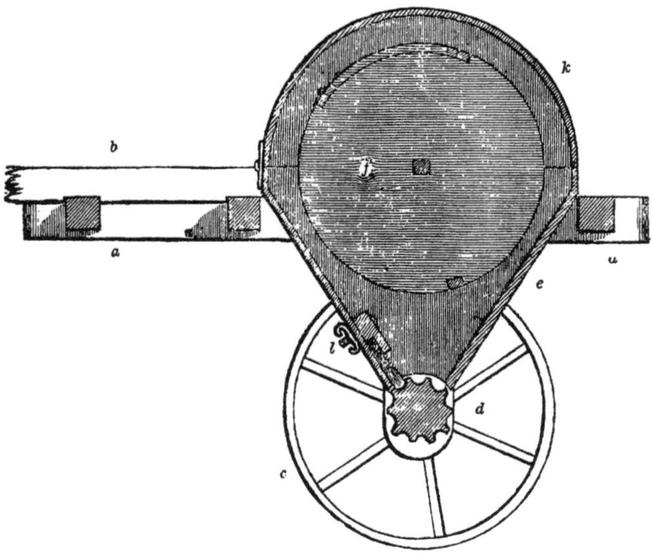

TRANSVERSE SECTION ON THE SOOT-SOWING MACHINE.

c c, of 22 inches diameter, fixed upon, and turning with, the axle, around which there is built a wooden cylinder *d*, about 6 inches diameter and 6 feet in length, fluted longitudinally. A chest *e*, 6 feet in length, is appended to the body-frame, and descends so far as to half embrace the cylinder *d*, and is surmounted by a semicylindrical cover, which is left out in fig. 579, but is seen in section *k*, fig. 580. In the interior of the chest there is placed a cylinder of sheet-iron *f*, 22 inches in diameter, perforated all over with holes of ½ inch diameter, and as much apart, giving to it the character of a riddle. The cylinder is closed at both ends, and has a trap-door on one side, as seen in fig. 579, hinged, and secured at each end with hook and eye. An axle of iron passes quite through the cylinder, having journals that rest in two jointed bars *g g*; and on one end of the axle, produced beyond the bar *g*, is mounted a wheel *h*, of 18 inches diameter. The axle of the carriage-wheels *c* carries also a wheel 9 inches diameter, and the two are connected by means of the intermediate wheel *i*, thus producing motion in the perforated cylinder, as well as in the fluted one that is carried by the axle. The purpose of the perforated cylinder into which the soot is first delivered, is to separate stones or other hard substances that may be mixed with it; that of the fluted cylinder being the distribution of it from the machine; and the hinged cover prevents it flying off during the agitation by the first cylinder.

The operations of the soot-machine are effected thus:—a charge of soot is put into the cylinder, the chest closed, and the machine put in motion. By the revolution of the upper cylinders, the soot is separated from the stones and refuse with which it is always mixed, and so passes into the lower part of the chest, from whence, by the revolution of the fluted cylinder, regulated by a brush *l* extending the whole length of the cylinder, it is distributed in an equal manner upon the ground. When the soot has been discharged from the upper cylinder, the cylinder is raised from the chest by means of the knee-jointed bars *g g*, and when so elevated, the trap-door is opened, and the stones and other refuse discharged, preparatory to the next charge of soot.

The machine constructed as above described, has been found liable to the inconvenience before pointed out, but which has been effectually rectified by the adoption of the broadcast distributing-wheels, in place of the fluted roller; the bottom of the chest is consequently closed, except the orifice for each wheel, all the other parts of the machine remaining as they were; or by a proper adjustment, the intermediate wheel *i* is left out of the construction. It is also to be observed, that the distributing orifices for the soot require to be about 1½ inch diameter.—J. S.]

---

83. OF THE POINTS POSSESSED BY THE DOMESTICATED ANIMALS MOST DESIRABLE FOR THE FARMER TO CULTIVATE.

"From fairest creatures we desire increase,
That thereby beauty's rose may never die;
But as the riper should by time decrease,
His tender heir might bear his memory."

SONNETS, BY SHAKESPEARE.

The most difficult branch and the highest aim of farming is the breeding and rearing of live-stock, so as to produce the most perfect animal that shall yield the highest profit. It is very easy to maintain a flock or a herd that will propagate its kind; as the actual condition of many flocks and herds in the country abundantly testify. It is even easier to do this than to cultivate arable land for the purpose of raising corn; for the latter demands, and must receive, more trouble, at all events, if not attention, than a flock or herd of neglected animals, who will keep themselves alive if food be within their reach, and who will propagate their kind even in due season, if the sexes are not kept asunder.

I am not aware that the principles upon which the proper breeding of animals should be based are understood by our farmers; for I hold that our stock have been brought to their present state of perfection, merely from each farmer exercising his individual taste and judgment—by gratifying his eye, and satisfying his mind—and not in pursuance of any fixed principle he is acquainted with on the subject. There can be no doubt, however, that the symmetry, disposition to grow and fatten, and the relative proportions of the different parts of animals, are entirely dependent on laws which govern animal life; and could these be understood, so as to render their application easy, the breeder of stock would no longer entrust his success to uncertainty. The botanical physiologist and the chemist have been very dilatory in proposing suggestions to the farmer, but the animal physiologist has hitherto almost entirely neglected him. The physician and surgeon, on the other hand, have not; for the veterinarian, now a man of science, administers his prescriptions and performs his operations with the confidence and ease of a proficient. Pity the physiologist should have almost neglected so extensive and fine a field as a breeding and rearing farm presents for his peculiar pursuits, and where the functions of the living and growing structure, might so easily be observed, and their effects afterwards as easily demonstrated by *post mortem* examinations. The results obtained from such investigations would afford invaluable data, with which to compare the functions of the human structure; and, fortified by the investigations of the comparative physiologist, the human physician would treat diseases with increased skill. The anatomist acknowledges the great importance of comparative anatomy to his profession; and the surgeon owns to have received many useful hints from the veterinarian, in the treatment of wounds, and the conducting of operations of magnitude and difficulty; and so, in like manner, no doubt, would the physiologist also acknowledge the importance of the observations he might make on a breeding farm, were he disposed to turn his attention to the subject.

Meantime the breeder of stock must go on, as he has hitherto done, by the light of his own genius.*

From what I have said of what is known of the principles of breeding stock, you will not expect from me so concise an explanation of them as I have given of the methods of conducting the operations of the field; and were it even possible, I could not find space here for the discussion of so extensive a subject. Suffice it to give you a few of the rules which guide the breeder in the treatment of his stock with a view to its improvement; and for your better ability to follow me in a subject which is yet confessedly in a state of confusion, I have selected a numerous group of illustrations, in whose good points you should have confidence, because they were all faithfully taken from the life, from individuals considered excellent by competent judges.

The great aim of breeders is, that their stock shall possess fine symmetry—*shape*, as it is commonly called—robust constitution—and a disposition to attain early maturity, which insures good quality of flesh, as well as of wool in the case of sheep. Let us consider what each of those properties means, and also the *points* by which the existence of each is indicated in the animal.

Symmetry or Shape.—The shape is the first object to be attained; for without fine symmetry no animal looks well, however passable it may be in other respects; but, in fact, when symmetry is awanting, so are the other good particulars. In fine symmetry the outline of the figure viewed, whether on the side, or at either end, or from above, should be *rectangular*. Fig. 273 represents the side view of a fat shorthorn ox; and it will be observed that the body very nearly fills up the rectangle *a b d c* inscribed about it. A fat ox will fill up the rectangle more fully than a lean one; but still a lean ox should have the rectangular outline. But it is not enough for an ox to present the rectangular outline in the side profile only, it should have it in other positions, such as when viewed from behind, as in fig. 274, where the same sort of outline may be observed within the rectangle *a b d c*. True, this is a different rectangle from the other, approaching near the square, while the other is oblong. I do not know whether the rectangles of the side and the ends should bear a given proportion to each other, as the outlines of the ox have

* The only physiologists I am aware of who have made observations on breeding stock, were the late Mr Knight, and his valuable suggestions will be found in the Philosophical Transactions; and Mr Walker, whose original views on physiology have been made public in several works; for as to the value of the experiments of Sir John Sebright, his scientific acquirements were not of so high an order as to inspire practical men, at least, in the accuracy of his observations, and correctness of his conclusions.

never, I believe, been subjected to proportional measurement, though these could easily be ascertained by employing an adjusting frame; but, judging from measurements taken with the view of ascertaining the weight of beef in oxen (2589.), it is probable that a fine symmetry gives, on the side profile, a rectangle of double the length of that of the end profile. The end profile, viewed from behind, gives a rectangle equal, of course, to the depth of that of the side profile. Another point of a well proportioned ox is, that the breadth across the hooks from *a* to *b* fig. 274, should be equal to the depth from the hook *a* to the hook *c*; that is, the hind profile should be inscribed in a square. Now, an ox may be rectangular in its side, and across its rump, and yet be of different shape across its shoulders, when viewed from the front of the animal; —it may, for example, be narrow at the top of the shoulder, and wide below at the brisket. In such a case the ribs will be flat, that is, fall quickly downwards, and not project square from the back-bone, as they always do in fine symmetry. Fig. 275 shews the front end of a fat round-ribbed ox filling up the rectangle; but though the top of the shoulder fills up the rectangle as well as the lower part of the body across the shoulder points *e f*, yet the rectangle may not be a square; that is, the ox may be narrower in front than behind; and this is no uncommon case in many breeds of cattle in this country, and many breeders maintain that this form constitutes perfect symmetry, at least for certain purposes, as for milk. This mode of reasoning, however, is not satisfactory; for the idea conveyed by perfect symmetry is an equilibrium of parts. Now, we cannot conceive an equilibrium of parts to exist in an ox whose fore-end is narrower than its hind; and where this disparity is found there is evidently something wrong in the symmetry. For observe fig. 276, which gives the top view of the body of an ox,—a view which is never seen unless sought for on purpose. Here, again, is the long rectangle as of the side; but when an ox is narrower before than behind, this figure would not be a rectangle, but a trapezium—a figure no way associated with symmetry: so that, for fine symmetry, it is not enough to have the front outline filling up a rectangle, for that rectangle should be equal to that of the hooks, as may be seen in comparing figs. 274 and 275; and as we have seen, in fig. 274, that the breadth of the hooks is equal to the depth of the hock, and *its* rectangle is therefore a square, so we see, by fig. 276, that the breadth and length of an ox, when seen from above, forms a rectangle equal to that of its side, as seen in fig. 273. Perfect symmetry in the ox, therefore, implies, that all the horizontal, as well as vertical lines inscribing its outlines, are respectively parallel to each other; and that these two sets of parallel lines are at right angles to one another. On dividing the entire area of the side view of an ox into 5 parallel and horizontal parts, $\frac{3}{5}$ will be found to be occupied by the body, and $\frac{2}{5}$ below the body.

This is the abstract view of symmetry, as applied to the ox; but it is not to be supposed that every ox fulfils these conditions; for, even in the figures referred to above, it will be observed, that, in the side view of fig. 273, the brisket falls below *b*, and the flank rises a little at *e*. The brisket *h* may be observed to drop below the horizontal line *c d* in the front view of fig. 275; and in the hind view of fig. 274, it frequently happens that the hocks, *c* and *d*, do not descend so far as to constitute a square figure with the hook *a b*. Still the aim of the breeder should be to attain all the points of the abstract model, and he should never rest satisfied until he attain them. Nevertheless, it must not be imagined that the configuration of any ox in any view should fill up the *angles* of the rectangles, inasmuch as the invariable roundness assumed by the animal frame forbids such a form, which would, in fact, destroy the beautiful blending of curved lines abounding in the body, and deprive us of the highest enjoyment in looking at a symmetrical animal.

Of the states of *cattle* on a farm, that of the full-grown ox approaches nearer than any other to perfect symmetry of form. If you compare the picture of the ox in Plate VI., though that animal was then only 1 year 11 months old, with that of the bull or cow, by imagining a rectangle inscribed around each figure, you will find his symmetry more correct than that of the others. The bull in Plate XXXII. has the brisket drooping even lower down than the ox, while his neck rises in a crest behind the head. In Plate XV., though the side view of the cow is pretty similar to that of the ox, the hind view is proportionally broader across the hooks, while the front view is somewhat narrower across the shoulders. If the broad hook-bones, with the narrow chest, are not entirely occasioned in cows by calf-bearing, the disproportion between them is, at all events, greatly increased by it.

So far in regard to the symmetry of the outline of the ox; and as to the filling up of the outlines, the remarks offered in (1529.) will apply to cattle in all states, though there specially referring to the points of a fat ox. There should be a straight back, round ribs, and full muscles in every state of cattle, and the objectionable deviations from these points, mentioned in (1531.), are as much to be deprecated in the bull, the cow, the heifer, and the calf, as in the ox.

Apply the rectangle to the figure of the *sheep*, and it will fit the more closely the nearer the animal approaches to perfection. In fig. 462, the rectangle *g h i k* is applied to the new clipped body of a Leicester tup, and though an old tup, and deviations from correct form may be expected in old age, still the figure fills up the rectangle very well. Apply it over the wool to the picture of the Leicester tup in Plate XXXIII., and it will fit still better; and carry it to the picture of the Leicester ewe in Plate XIV., and it will still apply; though in this particular animal the wool grew more than usually rank upon the top of the rump. Even the figures of the lambs in the same plate will bear the application of the rectangle. In regard to the end and top views of the figure of the sheep, the ribs of the Leicester breed are so very much rounded, that they form curved lines down the sides of the rectangle, inscribing the hind and fore profiles; and along the sides of the rectangle, inscribing the view of the back from above; still the flatness across the back, and the straightness across below the chest, fill up the lines of the rectangle. In other breeds, such as the Black-faced and Cheviot, as may be seen in Plate XXVIII., the rectangle may still be applied.

To the figure of the *horse* the rectangle may also be applied, as the picture of the draught-gelding in Plate VII. may be imagined to be inscribed, and of this class of stock the gelding is the best state for comparison, as the ox is that of cattle. The picture of the draught-stallion in Plate XVI., will bear the same application; though, in this, as in all cases of stallions, the neck rises to a high crest. The figure of the draught-mare, in Plate XVII., warrants the same application. But the end views, in the case of the horse, bear different proportions to those of cattle, the hind view being always broader, and broadest in the mare, approaching more to the square, or like that of the ox, than the front view, which is always narrow. The square-like hind view is very well seen in the left hand dark figure of a mare, in Plate XVII., and the narrow front view is as well observed in the right hand dark figure of a draught-horse, in Plate VII. The comparative breadths of the hind and front views approach nearer each other in the draught-horse than in any other breed; his drawing powers lying in the shoulders, the muscles there become developed to a great degree, while, in the saddle-horse, the propelling power is in the hind quarter, where the muscles become developed, and as the fore-quarter is only required to support the fore-hand of the animal in action, the muscles there are all light. For these reasons, it is obvious that the view of a horse from above will not form a rectangle but a trapezium. In perfect symmetry the void below the body is of the same depth as the body to the top of the rump; the length of the body from the shoulder point through the centre of gravity of the horse to the hams is 2 times the depth of the body;

the length of the neck to the withers is equal to the depth of the body from the withers; and the distance from the fore fetlock to the elbow joint is the same as from the latter point to the withers.

There are peculiarities in some of the points of the horse which deserve to be mentioned. A horse with ears set high on the crown of the head is of lively temperament, but is hasty of temper, and easily made afraid; while a horse with ears set wide apart is dull in action, but sagacious, and easily taught.

Following the same rule, the rectangle will be found to apply to the

Fig. 581.

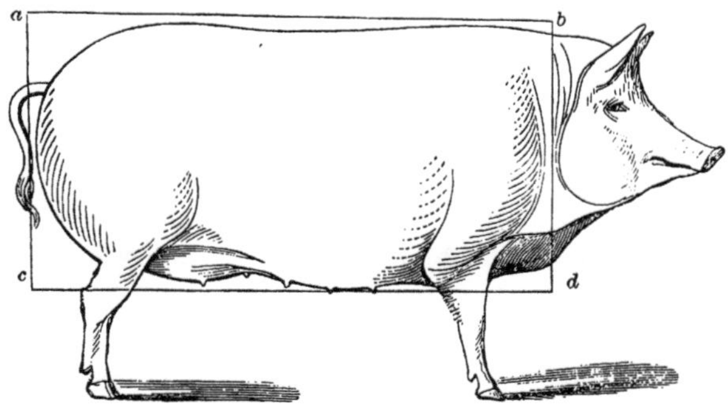

THE RECTANGLE TO ILLUSTRATE THE SYMMETRY OF THE PIG.

shape of the *pig* as to the other domesticated quadrupeds, as may be seen by *a b d c*, in fig. 581. The end views of the best formed pigs are quite rectangular, though the generality of those met with in the country have a tendency to a larger rectangle in the fore than in the hind view. The view from above is also rectangular, except in the case of pigs having thick shoulders.

Robustness of Constitution.—Having obtained a good shape, the next object of the farmer with his stock, is to impart to them a strong constitution, that is, to impart to them physical powers to do their work with ease, to eat their food with good appetite, to digest it within a reasonable time, to feel no fatigue with necessary daily exercise, and to withstand the weather, whether hot or cold;—in short, always to enjoy good health, and be in a growing condition. The following are indications of robust health which cannot be mistaken. The skin loose upon the body, feeling soft and mellow, and covered thickly with long soft hair. The eyes full and clear, and in the case of cattle, sheep, and pigs, the skin of the nose always bedewed with moisture. The bones of the leg strong, broad, and flat, with the sinews thick, strong, and distinctly developed; the whole being closely covered with the skin. This condition constitutes what is termed *clean* limbs. The horn of the feet and horns smooth and shining, and indicating healthy growth at its junction with the hair. The hair of the tail in cattle, and of the mane and tail of horses, long, shining, and with a tendency to crisp; the wool of sheep waved, greasy, interlaced, and of fine quality. Along with these physical properties, the animal spirits are alive to every passing event, the senses acute, and the instinct sagacious.

All these indications of a sound constitution are obtainable by selecting parents which possess them; and they are sustained by kind treatment, comfortable lodging, and abundance of good food. Neglect of a proper selection of parents, will produce, perhaps, the very opposite results. The skin will become thin and covered with scanty hair and open wool. With such a covering, the animal will easily be affected by the least change of weather, be overcome with the heat of summer, feel uneasy in the cold of winter, and be chilled by every shower that falls. A very small bone, however clean, is always accompanied with reduced size of carcase, though indicative of great disposition to fatten. Should the bone become thick and round, it gives a clumsy form to the limbs and head, and is invariably accompanied with dulness of spirit, and in the horse, in particular, with want of action. Those conditions of the skin and bone are not so much objectionable in themselves, as hurtful in their consequences. A thin-skinned and a thick-boned animal is ill to maintain in condition. Thin-skinned animals are easily affected by the weather; a fit of indigestion in consequence frequently overtakes them, and their condition alternates between worse and better; they are, in fact, tender in constitution. Thick-boned animals never seem to relish their food, and shew carelessness about every thing that concerns them.

Disposition to attain early maturity.—But the farmer's aim is not confined to a good shape and robust constitution; there must be a fineness given to every point—every part must be finished off, as it were, with a polish. This fineness can only be superimposed by parents derived from a race which have long possessed all the fine properties—individuals whose pedigrees can be traced—whose breeding is accounted pure. No matter whether the breeding be of the plebeian caste of the draught-horse, or of the aristocratic family of the racer,—nor whether the animals be horses, cattle, or sheep; the breeding of all should be pure of its kind.

The immediate effect in employing parents of pure breeding, is not so obvious upon the carcase as upon the extremities. The head and limbs first become finer; the fat disappears from them; the muscles expand into a thin and broad form; the tendons increase in strength; the bone becomes harder and of finer quality; the blood-vessels large and full of blood; the skin thinner, and drawn more tightly over the bone, so as to exhibit its irregularities, and the tracings of the blood-vessels distinctly; the hair is smoother and closer; the nostrils more expanded; the eyes more prominent and clear; the ears thinner, more capacious, set higher upon the head, and more quickly moved about; the action of the limbs quickened; and the animal spirits raised. All these changes may take place without materially affecting the shape of the body or the strength of the constitution; but the improvement of the extremities is rapidly followed by that of the body; the skin, though not thinner of itself, floats upon a thicker mass of cellular tissue, which imparts to it the agreeable sensation of the *touch*. The hair multiplies in numbers, becomes finer in quality, more fixed in colour, longer, and softer. The proportion of flesh to bone increases. The fat is more generally and equally diffused over the body. The texture of the muscles is finer and firmer in the fibre, and the tissue between them becomes loaded with fat. In a word, such a total change is effected upon the entire animal, that from being coarse, dull, clumsy, and sluggish, it becomes fine, spirited, handsome, and active; and as it improves in personal appearance, so it increases in its tendency to grow, and to come sooner to maturity in stature and condition.

Selection of Parents.—You must not imagine, from this uninterrupted description of the improvement of parts in animals, that all those changes are, or can possibly be, effected during the lifetime of one individual parent. The entire of those changes can only be effected in the course of generations; but as all domesticated animals produce rapidly, that is, within the course of every year, it is possible to effect great improvement in the offspring during the lifetime of a female parent, by employing a better male every year. To hasten the improvement more rapidly, better parents of both sexes may be chosen as soon as the points of the progeny shew themselves, and which are appreciable in 2 years.

Of the 2 sexes, the properties of the male are more obviously imprinted on the offspring than those of the female; and it is on this account that males are in most request among breeders for the improvement of their stocks. High prices have been given, and high premiums are annually offered, for superior males, whether stallions, bulls, or tups; and it is fortunate for the more rapid extension of the improvement of stock, that the influence of the male thus bears sway in the propagation of his kind, as by permission of polygamy 1 male may serve many females, and of course extend his influence in the exact proportion to the number of the latter; whereas a female produces usually 1 in the course of a year. A stallion serves from 60 to 80 mares, a bull 60 cows, and a tup 60 ewes in a season, not once or twice only, but as often until the female prove with young. Generally the female conceives at the first service of the male, and at all events at the second; but should her desire continue beyond that time, it is better to withdraw her from breeding altogether, than run the risk of having a late progeny, or none at all; for a late calf, lamb, or foal, loses 1 year of its progress, compared with its earlier born companions. But the part which the male and female respectively undertake, in the improvement of the progeny, differs widely, and is entirely dependent on their state of breeding—the head and extremities being improved by the highest, and the carcase and procreating powers by the lowest bred. The highest bred imparts the small head, prominent eyes, thin ears, wide nostrils, clean limbs, full blood-vessels, action, spirit, and intelligence; in short, all the parts which go to develop the mental qualities; while the lowest bred improves the shape of the carcase, by straightening the back and rounding the ribs, extending the muscles, laying on the fat, enlarging the capacity of the pulmonary organs, and decreasing the size of the abdominal. On distinguishing the parts more immediately imprinted by each parent, it is obvious that the effects of the highest bred are more striking than those of the other. Hence, a fine head, well set upon the neck, a lustrous eye, a gay look, an engaging manner, and free action, are more readily appreciated than a fine mould, sound constitution, touch, and disposition to grow.

Few farmers commit a mistake in the choice of the male for improving their stock. Breeders themselves pay particular attention to the production of the male, and farmers who purchase or hire males never grudge paying a high price for them; and hence the male-breeding market is always at a premium. But many mistakes are made by breeders in the choice of the females to breed from. In the case of horses particularly, any sort of mare is considered good enough to produce a foal, and if she happen to be unfit for work, or has met with an accident, such as having swung her back, or has a bad leg, she is the one selected for breeding the future labour-stock of the farm. The best stallion, it is true, will be put to her, most probably the prize-stallion of the local agricultural show; but as to the mare herself, it is considered a great sacrifice of physical power to breed from young fresh mares. Now, in acting thus, the farmer commits a fatal blunder; he is depriving himself of the use of superior farm-horses, which would not only work with ease to themselves, and satisfaction to him, and at the same cost of keep, but would do him credit wherever they went. Instead of pursuing this system, he should select 1 or 2 of the best mares in his possession to breed from, and if he has none possessing youth and beauty of mould, let him buy 1 or 2 at any price, if such are to be purchased. Let them not be under 5 years of age when the horse is first put to them; for they will not have acquired their full stature until that age. From that period let them bear a foal every year till they attain 12 or 13 years, when they should cease to breed; for *old* mares, any more than old females of other classes of animals, cannot produce a vigorous progeny. For those 7 years the fresh young mares, stinted to the best stallion that can be secured, will each produce, and bring up, a foal every year, and perform, at the same time, their part of the work with ease. Such foals will be strong in bone and constitution, and be always in high condition, because they are the offspring of young mares themselves, high in condition, and overflowing with milk; and having been brought up together will work better together. I speak not in this tone for the purpose merely of condemning the practice usually followed by farmers in breeding farm-horses, but having myself tried the system I recommend, I can do it with confidence. I purchased from a dealer a very excellent Clydesdale mare in St John's Market at Perth, for L.42, a large price for the time. She was 5 years old off, and was altogether such a mare, for mould and strength, as is seldom to be found in a market. Next year she was put to a prize stallion, also of the Clydesdale breed, and bore a foal every year till she reached 10 years of age, when, much to my disappointment, she ceased to breed. Her first 2 foals were fillies, and, when 4 and 5 years old, were sold at the sale, when I declined farming, for L.108, to convert into brood-mares. Her other 2 foals, colts, at 4 years old, realized L.38 and L.44. I had other 2 mares which began to bear foal at 7 years of age, 1 of which bred draught-horses, which realised from L.35 to L.40, and the other was put to a coaching-stallion, and produced excellent harness-horses; 1 of them a mare, was purchased by a friend in Ireland, and proved one of the fastest goers in harness I ever saw. I am perfectly certain the same results will be realised everywhere by breeding from excellent young mares; and though 3 mares out of 10 horses, which wrought my farm, bore foals every year, the work never fell behind its season; and they had as rough and heavy work to do for several years, whilst improvements were in hand, as could be encountered on any farm. The system pursued in regard to the breeding of cattle and sheep is generally much better than that for horses, though I think that cows are not generally so well selected as they should be; and one reason is this:—whenever a cow happens to be a good milker, she is kept to breed as long as she is worth keeping for that purpose, without regard to any other point of excellence; and her calves may be good or bad, in proportion as the points of the bull which begot them are so. Those who keep a stock of cows for breeding bulls, of course, do not follow this hazardous plan of producing valuable calves; but I believe by far the greatest number of cattle bred in this country are produced in this careless manner; and the simple recital of the system is quite sufficient to account for the great number of inferior cattle to be met with in all our public markets.

The system of breeding I have recommended is applicable to every kind of stock; but modifications are allowable in it, and in fact expedient, according to the particular object the breeder has in view. If his object is simply to breed cattle and sheep for ordinary markets of fat and lean stock, a useful lot of cows or ewes, with a good tup or bull, are all that are requisite; and once having obtained a desirable stock of either, and not wishing them of finer quality, his solicitude will be confined to sustaining their character, by preventing the least tendency to deterioration. In such a case the females are commonly bred by the farmer himself, and the male is purchased from a breed of superior class to his own. So far the practice is good; but when a good cow, or a lot of good ewes, of the same or superior character as the stock itself can be picked up now and then, the opportunity should not be lost in infusing a *fresh* strain of blood into the stock. A frequent change of the male is advisable in such a system, as the stock will soon become too near akin to him. On the other hand, those whose professed object is to breed animals to propagate their kind, must bestow great care and skill to sustain their breeding-stock in the highest state of perfection. The least defect in the female, whether arising from accident or inherent weakness, should be the instant signal for her removal from the stock, and as to the male, the very best that can be found should be purchased at any price. The best male in reference to the particular case may not be the best animal of the stock from which it is desired to purchase, because it should pre-eminently possess points which will either improve or sustain those of the females which he is destined to serve. The properties, therefore, which a male should possess are entirely relative and not absolute; for absolutely good points ought to be possessed by the females before the breeder should attempt to produce breeding-stock from them. To manage this part of his business well, is a very difficult task for the breeder; and a single mistake committed in this respect, for a single season, may cause him more trouble to rectify, and incur greater loss, than if he had been content with his own stock in the state it was. It must be owned, however, that whenever a breeder discovers a defect in any point in his stock, his desire to remedy it is natural; and whenever he finds an animal superior to his own, it is as natural he should endeavour to possess it. If, for example, he thinks his own stock somewhat slacker behind the shoulder than they should be, the remedy, he conceives, is to use a male which is free of that deficiency. The usual practice is to select a male for this purpose, which is *rather full* behind the shoulder; because, what more natural than to fill up a deficiency with a redundancy? Such an expectation, though natural enough, may lead to unexpected results; for the use of the *redundant* point may give such an impetus in the stock to rectify their deficiency, that its tendency to fill up may proceed too far, even in one generation, and cause redundancy where there was deficiency before; but this is just the result which ought to have been anticipated, if there is any truth in the breeders' maxim, that "like begets like." The employment, therefore, of a symmetrical shoulder to rectify a deficient one is the practice most consonant to reason and principle; for what *more* is wanted than to *fill up* the deficiency behind the shoulder; and how should that be *best* attained but by using a model which is exactly in the state which you wish the deficiency to assume? The want of forethought to consequences of this sort has been the cause of deterioration to many a good stock. Too many farmers attempt to produce breeding-stock, and they are tempted to proceed with the system as long as they observe the progeny improve upon their parents; and as long as they are observed to improve, purchasers will be found to pay good prices—for they like to purchase from a thriving stock—till a point is reached beyond which improvement does not succeed; and this is considered the point of perfection, though it may be very far short of it. The breeder's skill may be unable to carry him farther; and, on account of his want of skill, will be induced to present his breeding-stock to the attention of other breeders with the greater earnestness. Herein is concealed another source of error; for it is clear, that as long as a stock is itself far short of perfection its offshoots cannot afford a certain means of improving other stock. Trial may be made of it; but the result may be disappointment to the purchasers, and disgust to the self-sufficient breeder himself. One circumstance alone, I conceive, would justify a breeder in offering imperfect stock for breeding, which is, being situate in a country where crossing with other breeds might be carried on with advantage. The conclusion of the whole matter, then, is, that a stock producing offshoots fit to breed from should have arrived at perfection, at least to such a degree of perfection as the best stocks known; and the object of the owner of such a stock should not be so much to *improve* its best animals—for that seems impossible—as to improve the inferior, and *sustain* the quality of the best; and before a stock can be brought to that state of perfection, the pedigree of its best animals—for no stock contains all its animals equally good—might be traced back for many generations. Doubtless, this is a severe criterion by which to judge the character of a breeding stock; but when it is considered that its members are proferred with a view to improve, or at least sustain, the breeding of members of another stock equally well bred as itself, it is necessary to fix a high standard of comparison, and nothing short of this will afford satisfaction. There are many stocks of Short-horn cattle and Leicester sheep, whose character will bear the strictest scrutiny by such a criterion.

Breeding in-and-in.—No wonder when high-breeding produces such an improvement in stock as to render the head small, fine, and beautiful, the extremities elegant, the form handsome, and the disposition so accommodating as that the animals grow and fatten without feeling disturbed at what passes around; in short, become so prepossessing, as to make their owners mistrust the stocks of others, and employ only their own to increase its own numbers. It was this feeling which actuated Bakewell to breed only from his own stock, after he had brought the Leicester sheep and Long-horn cattle to perfection. For a time Mr Mason of Chilton pursued the same course; and there are breeders in England at the present time who maintain that it is the best system, and will follow no other. Perhaps a stock brought to the highest state of perfection, and at the same time possessed of sound constitution, may be supported free of deterioration for many years by the peculiar skill of its owner; and I can conceive a high-bred stock, full of young blood, such as Bakewell's was during his whole lifetime, to be increased and supported by its own members though bred in-and-in, that is, supported by near kindred or consanguinity. There was one valid reason for Mr Bakewell employing only his own stock, which was, that there was no other so good as his own to select from; and it would have seemed extraordinary in him, of all men, to have employed any animal of acknowledged inferiority to his own; but I suspect no such liberty may be taken, with impunity, with a stock as highly bred, but not so youthful in blood. At all events, many instances have occurred, in which many a fine stock have been ruined in character, and have entailed irreparable loss on their owners simply by being bred in-and-in. The immediate effects of employing parents nearly allied by blood to propagate their kind are remarkable. The bone becomes very small, of condensed texture, and fine quality. The skin is so thin as to receive the appellation of *papery*, and so open of texture as to be sensible to the least change of temperature; and hence animals bred in-and-in are very susceptible of catarrhal affections, and on which account they are liable to consumption and clyers (3036.). The carcase is much reduced in size, and the disposition to fatten increases to such a degree that the animal may be said to be always in a condition to be slaughtered; and it is, perhaps, this tendency to fatten, which has proved the great inducement with many breeders to tolerate the in-and-in system. The hair is short, smooth, and thin-set, and the wool short, thin-set, and watery; and both hide and fleece lose a large proportion of weight. The body assumes a change of form, the barrel being beautifully rounded, but seems stuffed, as it were, within the skin. The extremities are very fine, the head and hoofs small, the ears thin and broad, and the head of the sheep is almost bare of hair, of a blue colour, very liable to be scalded by the heat of the sun, and attacked by the fly. The neck of both cattle and sheep are thin, and droop with a downward curve from the head to the top of the shoulder. The points just enumerated shew the unprofitable state into which a stock may be brought by being bred in-and-in. Mr Mason's fine Short-horn stock latterly shewed symptoms of the bad effects of this system; and Mr Robertson's stock at Ladykirk, which contained at one time by far the finest Short-horns in Scotland, suffered after his demise from the same cause, as was apparent on the animals presented at the sale which dispersed them. Only cattle and sheep have been subjected by farmers to be bred in-and-in, for their draught mares are usually covered by stallions obtained from a distance; and of cattle and sheep, the injurious effects of the system have only been observed in Short-horn cattle and Leicester sheep. The injurious effects of the system have been felt, it is true, in the racing stud; and are evinced by the racers of the present day being unable to carry heavy weights, and run the long distances of the horses of old. At present nothing but speed is regarded.

Now, that high-bred stocks exist in every district of the kingdom, there is no excuse for pursuing the in-and-in system of breeding; and the attempt is the more inexcusable from the remarkable fact, brought to light only since the distribution of high-bred stock, that its progeny, after being distributed for a time, may be brought together to propagate their kind, and their offspring will exhibit no symptoms of in-and-in breeding. Such a result would seem to indicate that change of soil and situation renovates the animal as well as the vegetable constitution.

It may be proper to mention what relations are considered allied in blood, in breeding in-and-in. Connection of the sire with his grand-dam, dam, and sister, is improper. So it is even with his aunt, niece, and cousin-germain. One should imagine that a breeder would feel repugnance at connecting so near a consanguinity as the former; yet the union is too often permitted to take place in the latter cases.

Crossing.—The union of different breeds of the same sort of animal is a favourite scheme with many breeders, and, under certain conditions, produces good results. Those conditions are, that the male employed in the crossing shall have the superior breeding of the 2 parents, and that the situation in which the cross-progeny shall be brought up shall be suitable to it. The first condition is usually complied with, but the second is as commonly disregarded; and the consequence is, that the crosses attempted to be brought up in situations quite unsuited to their nature have proved a failure.

In the crosses commonly attempted in this country, among cattle the Short-horn bull, and among sheep the Leicester tup, have been employed to serve the ordinary breeds of cattle and sheep. The effects have proved satisfactory; for though the progeny could not but be expected to be inferior to the sire, they are superior to the dam. The effects are, an enlargement of the carcase, a finer skin, longer hair and wool, cleaner bone and finer head, and the disposition to fatten greatly accelerated—all of which are superior to the ordinary points of the breed; but, of course, the better the dam is, the more decidedly improved will those points be. In effecting such a cross, it has been found that the higher bred the male is the finer is the cross, that is, the nearer it will approach his properties; and even an *over-bred* male, that is, one shewing symptoms of having been bred in-and-in, may be used with advantage in crossing. Where a superior cross-male happens thus to be produced, a strong desire is evinced by breeders to keep him entire for service, and even to use him as a sire, instead of expending money in the purchase of a high-bred male. A more short-sighted step than this cannot be taken by a breeder, because, from such a male he has no assurance that the progeny will be superior to the dam. Indeed, he has no guarantee of what the state of the progeny will be at all, for it may be worse than either sire or dam; and if such a disappointment overtakes him, he has himself to blame, having left the whole matter to chance. The employment of a high-bred sire, on the other hand, will never lead to disappointment, as the progeny will assuredly be superior to the dam. It is this assurance which affords much satisfaction to the breeder of high-bred stock, by which he can anticipate the quality of his forthcoming young stock; and it is with high-bred stock alone that the maxim of "like producing like," is realised. Nevertheless, there are situations in which high-bred stock cannot be maintained *as a breeding stock*, and in which nothing but crossing can be practised when improvement is desired; but the desire for improvement has been carried by some breeders beyond the bounds of prudence; they have crossed the Black-faced ewe with Leicester tups, in situations where the enlarged lamb has been unable to subsist in winter; and on this account the policy of changing the Black-faced breed of sheep in high localities seems doubtful. In lower situations, the Cheviot ewe, which inhabits the middle range of green pasture, may be crossed with the Leicester tup with advantage. Every crossing, however, should be prosecuted with caution, because the result may overstep the intentions of the breeder. It is clear that if the crossed stock is retained as females, which, in their turn, are served by high-bred males, the time will arrive when the character of the original stock will be entirely changed, and become unsuited to their native climate and pasture, and will, in fact, have become the same breed as their high-bred sires. It is quite possible to originate a race of Short-horns and Leicester sheep any where suited to their nature, by constantly employing a high-bred bull and tup to serve cross-bred heifers and gimmers, generation after generation; and were this practice generally adopted, the time would arrive when the original breeds which were crossed would disappear altogether. Such a result would prove injurious to the breeder himself, inasmuch as the pasture would be unsuited for the stock he had caused to be produced; so that his best plan is to preserve the original breeds in the higher parts of the country, and take the crosses to the low country to be fed off. The temptation of larger profits has already caused the Cheviot to drive the Black-faced breed from the lower pastures to the highest, while the cross-bred Cheviot, with the Leicester, have descended, on the other hand, to the low country, and there have met the true bred Leicester. This result, upon the whole, has done good, as it has increased the quantity of mutton in the market; and the skilful pasturage which the hills have received since a regular system of breeding has been introduced, has caused them to yield a larger quantity of finer grasses. The crossing of the Black-faced sheep has undergone a change; the Leicester tup is too heavy to serve Black-faced ewes on the hills, and to bring those ewes to the tup in the low country is attended with trouble and expense. Instead, therefore, of the Leicester tup being so employed, the Cheviot tup has been substituted; and though the cross is inferior—for nothing can exceed the beauty of the lamb produced between the Black-faced ewe and Leicester tup—it is a good one, and has enlarged the Black-faced mutton. So long as crossing is conducted with the breeds in their natural state, it will go on without confusion, but the moment cross-bred tups are employed as improvers of stock, their interference will produce confusion in the crosses, and throw discredit upon crossing altogether.

I purpose giving a short description of the animals whose portraits are given in the plates illustrative of this work, and shall make such remarks on the form of the animals as to give you an idea of their leading points, whether good or bad. In order to render reference to the figures in the plates more easy, wood-cuts of their points have been provided, that they may be the more distinctly indicated.

Horses.—Among horses I shall take the draught-horse, fig. 582, and Plate VII., as the standard of comparison. This gelding is of grey colour, was bred by Mr Curry at Brandon, in the county of Northumberland, and is the property of Messrs Howey and Co., the great carriers from Edinburgh into England. He is not a thorough-bred Clydesdale, having a dash of coaching blood in him, a species of farm-horse very much in use on the Borders, and admired for their action and spirit. This gelding exhibits such a form as to constitute, in my estimation, the very perfection of what a farm-horse should be. His head *a* is small, bone clean, eyes prominent, muzzle fine, and ears set on the crown of the head. His neck rises with a fine crest along the mane from the trunk *b h* to *a*, and tapers to the head, which is beautifully set on, and seems to be borne by the neck with ease. His limbs taper gradually from the body, and are broad and flat; the knee *k* is straight, broad, and strong, and the fore-arm *i* broad and flat—all excellent points in the leg of a draught-horse, giving it strength and action. The back of the fore-leg, from the fetlock joint *l* to the body *o*, is straight, indicating no weakness in the limb—a failing here causing the knees to knuckle, and rendering the horse unsafe in going down-hill. The hind-legs *m*, as well as the fore ones *k* and *l*, stand directly under the body, forming firm supports to it. The body is beautifully symmetrical. The shoulder slopes backwards from *h* to *b*, the withers at *b* being high and thin. The sloped position of the shoulder affords a proper seat for the collar, and provides the muscles of the shoulder-blade *g* so long a lever as to cause them to throw the fore-legs forward in a walk or trot; and with such a shoulder a horse cannot stumble. The back, from *b* to *c*, is short, no longer than to give room to the saddle. The

Fig. 582.

THE DRAUGHT-HORSE.

chest, from *b* to *o* is deep, giving it capacity for the lungs to play in, and room for the muscles required in draught. The top of the quarter from *c* to *d* is rounded, the flank, from *c* to *n*, deep, and the hind-quarter, from *f* to *e*, long. On looking on the side profile of the entire animal, the body seems made up of 2 large quarters, joined together by a short thick middle, suggesting the idea of strength; and the limbs, neck, and head, are so attached to the body, as to appear light and useful. In a well-formed horse, I may remark, that the line from the fetlock joint *l* to the elbow joint, at *o*, is equal to that from the joint *o* to the top of the withers *b*. In a low-shouldered leggy-horse, the line *l o* is much longer than the line *o b*; but in the case of this horse, the body *b o* is rather deeper than the leg *l o* is long, realising the desideratum in a farm-horse of a thick middle and short legs. The line across the ribs from *g* to *f* is, like the back, short, and the ribs are round. He is 16 hands high, measures from *a* to *b* 35 inches, from *b* to *c* 33 inches, from *c* to *d* 19 inches, being in extreme length 7 feet 3 inches. Length of the face 25 inches, breadth of face across the eyes 10 inches, length of ears 6½ inches, breadth across the hook-bones 22 inches, girth behind the shoulder 80 inches, girth of fore-arm 23 inches, girth of bone below the fore-knee 9½ inches; —the breadth of this bone shews the strength of the fore-leg of every horse;—girth of neck at the onset of the head 32 inches, girth of muzzle 21 inches, width of counter 19 inches, and height of top of quarter from the ground 63 inches. In a draught-horse the collar causes the muscles to enlarge upon the shoulder, and the neck to become thin. This horse's name is Farmer, his walk is stately, and he can draw 3 tons on level ground, including the weight of the waggon. He is a well known animal in Edinburgh, and is generally admired.

Fig. 583, and Plate XVI., is the portrait of the black draught-stallion, Champion, bred by Mr James Steedman, Boghall, in the county of Mid-Lothian. He is of the true Clydesdale breed. He gained the first prize at the Highland and Agricultural Society's Show at Glasgow in October 1837, and obtained premiums elsewhere. He is a sure foal-getter. He is fully 17 hands high, and though otherwise a large animal, being 8 feet 7 inches in length, his action is high and uncommonly light. On comparing him generally with the gelding just described, though his body is longer, both hind and fore quarters

Fig. 583.

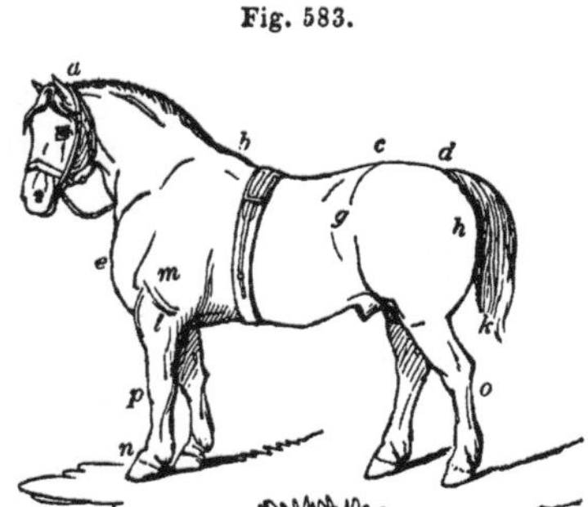

THE DRAUGHT-STALLION.

are long and deep, and exhibit a large display of muscle. His middle is somewhat small, as is almost always the case with stallions which have served many mares. Like all stallions, his neck rises beautifully from his body *b e*, in a full crest from *b* to *a*, evincing that castration has the effect of reducing the size of the muscles of the neck of all geldings. The shoulder slopes well back from *e* to *b*, giving freedom of action to the fore-legs, while the muscle at *m* being fully developed, assists in imparting power to that action. The hind-quarter, from *g* to *h*, is long and deep. The fore-leg is straight, and short from knee to fetlock, *p* to *n*, the bone under the knee strong, and the fore-arm *l* flat and broad. The hind-legs *o* are remarkably handsome. The sweep of line from the crown of the head along the back to the tail-head is truly elegant, giving a very fine top to the quarter, and the plenitude of hair in the tail *d k* indicates great strength of back. His eye is good, though somewhat small, the ratch of white down his face is against his cast of countenance; and having the 2 hind legs white is also against his general appearance. His disposition is remarkably docile, and his whole demeanour harmless. His constitution is good, and he is an excellent traveller. These are a few of his dimensions:—from *a* to *b* 51 inches, from *b* to *c* 30 inches, from *c* to *d* 22 inches; in all 8 feet 7 inches. Length of face 26½ inches, breadth of face across the eyes 11 inches, length of ears 6½ inches, breadth across the hook-bones 30 inches, girth behind the shoulder 90 inches, girth of fore-arm 28 inches, girth of bone below the fore-knee 12 inches, height of top of quarter from the ground 67 inches, girth of neck at the onset of the head 39 inches, girth of muzzle 24 inches, and width of counter 22 inches.

Fig. 584, and Plate XVII. is the portrait of a brown mare belonging to Mr George Bagrie, Monkton, near Dalkeith, Mid-Lothian. She gained the first premium at every show of stock she was ever exhibited. The white ratch down her face, and so much white on her legs, detract from her general appearance; but notwithstanding these drawbacks, she is an exceedingly handsome mare. You have only to look at the plate to observe the beautiful flowing lines of her whole contour, and also the great substance of both fore and hind quarter. The rise and crest of her neck from *b* to *a*, and from *e* to *a*, are remarkably fine. The back from *b* to *c* is some-

Fig. 584.

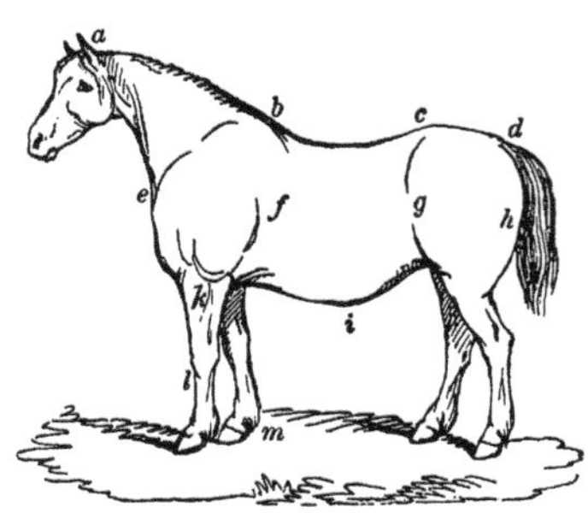

THE DRAUGHT-MARE.

what hollow, and there is a corresponding depression of the belly at *i*, both the consequences of foal-bearing, as well as a slackness of the flank in front of *g*, the usual deficiency there of brood-mares. The top of the rump from *c* to *d* is also very fine. The shoulder slopes well from *e* to *b*, indicating good action; the muscles are well developed on the fore-quarter from *e* to *f*, indicating power in draught; and the ribs are round, and long from *f* to *g*, a favourable configuration in the brood-mare for giving room for the growth of the fœtus. The hind-quarter from *g* to *h* is long. The legs are placed directly under the body, the fore-knee *l* being broad and strong, the back of the fore-leg from the fetlock *m* to the body straight, and the fore-arm *k* broad and flat. I have no measurement of the dimensions of this mare, which I regret, as losing a means of comparison with the gelding and stallion. Beside roundness and length of rib, a brood-mare should be wide across the hook-bones and the pelvis, to afford room for the growth and subsequent egress of the foal, as is better shewn in the dark small figure of a mare on the left hand side of Plate XVII.

I have chosen a *black* stallion, *brown* mare, and *grey* gelding, as illustrative of the 3 colours most commonly seen among farm-horses. A black stallion seems the generally favourite colour, and a brown mare is not uncommon, but the grey colour is less in vogue than it was 20 years ago; but why I cannot say. It is said that the feet of grey horses are more tender than of horses of other colours; and, for the same reason, whatever that may be, it is alleged that white feet are more tender than any other colour. I once corresponded on this subject with the veterinary surgeon of one of the regiments of Life Guards, the horses of which are black, and his statement was, that he had not observed any remarkable defect in those horses which happened to have white feet; but that his attention had not before been particularly drawn to the subject. It might be worth while to ascertain whether or not it is found that the horses of the Scots Greys are more liable to tender feet than those of other corps of a different colour.

Cattle.—I have already said so much on the points of cattle, that nothing more remains to be advanced on that subject; so I shall merely allude to the particular points exhibited by the animals represented in the plates. And, first, I shall take, as a standard of comparison, the Short-horn ox. fig. 585, and Plate VI. This was an ox from the herd of Mr Wilson of Cumledge, in Berwickshire, who has long been known as a successful breeder of Short-horn steers. His stock comes to maturity, and is fattened off, at 2 years of age, when they commonly attain 70 stones imperial. The ox in fig. 585 was 1 year 11 months old when his portrait was taken; and he was afterwards shewn at the Highland and Agricultural Society's Show at Berwick-upon-Tweed, in October 1841. He was of a roan colour, with a good deal of white. His head was remarkably fine, with a pleasant countenance, full eyes, and small slouching sharp-pointed horns. Besides a fine head, he had a straight back, round rib, deep flank, and full neck-vein. His principal measurements were, from *e* to *a* 27 inches, from *a* to *b* 32½, from *b* to *c* 21½ inches, in all 6 feet 9 inches; and the girth of the body behind the shoulder at *f* 7 feet 3 inches. His measurement for beef was 4 feet 6 inches in length, by 7 feet 3 inches in girth, equal to 56 stones imperial; so I should say from these figures, that this ox was rather too short for a perfectly symmetrical figure.

Fig. 585.

THE SHORT-HORN OX.

Fig. 586, and Plate XXXII. is a portrait of a red and white Short-horn bull. This animal was bred by the late Mr George Brown at Whitsome Hill, in Berwickshire. He was got by a red and white bull belonging to Mr Robertson of Ladykirk, named Valentine.* At that period, Mr Robertson's stock of Short-horns was in its glory. The dam of this bull was got by a red bull, never named, bred by Mr Thomas Smith, now at Buckden, then at Grindon, in Northumberland, and was a son of his old roan bull Duke; and at that period few farmers had so high a bred stock as Mr Smith. The grand-dam was 1 of twin quey-calves produced by a heifer, purchased in calf by Mr Brown from Mr Mason of Chilton. One of the twin calves, when a 2-year old quey, Mr Brown sold to the late Duke of Buccleuch for 50 guineas, and the other he retained for himself. I purchased this bull when 1 year old from Mr Brown for 20 guineas, and kept him at Balmadies, in Forfarshire, for 8 years, during which time he proved himself a sure and excellent calf-getter, and evinced a gentleness of disposition to every person who approached him, in a remarkable degree for a bull. He had many good points—small head, lively eye, small, fine, white horn. He was well filled up behind the shoulder, at *f*, fig. 586, a point in which many bulls are deficient. He had a long quarter from *g*, a difficult point to attain in a bull, car-

* For Valentine's pedigree, see Coates' Herd Book, vol. i. p. 141, (661.).

rying the flesh to the hocks *d*; a thick flank *e*; ribs round, forming a *straight line* from the shoulder point in front of *f*, past *g*, to the margin of the round above *d*. His fore-arm *h* was very strong; neck vein *i* full; and the crest of his neck *a* fine and not lumpy, as is too often the case in bulls; his hooks and back were remarkably straight and broad, measuring across the hook-bones at *b* 36 inches; the rump between *b* and *c* was full and round, and the tail-head *c* was remarkably level and fine, shewing no undue development of muscle, as is often the case here—a deformity too generally admired, and in so far shews a prevalence of bad taste. His neck and shoulders were strewed over with curled locks of long hair, the entire body being thickly covered with long soft hair; the face was garnished with curled hair in a line down in front of each eye; and the roots of the horns were hidden with long hair falling over the forehead. His hide was thick and mellow, and the touch fine. He had a most robust constitution, never having had a day's illness in his life of 9 years. Unfortunately I had no measurement taken of him before he was killed fat, when the butcher, Mr Johnston of Arbroath, informed me he weighed 139 stones imperial, sinking the offals. His flesh was fine, much liker ox than bull beef.

Fig. 586.

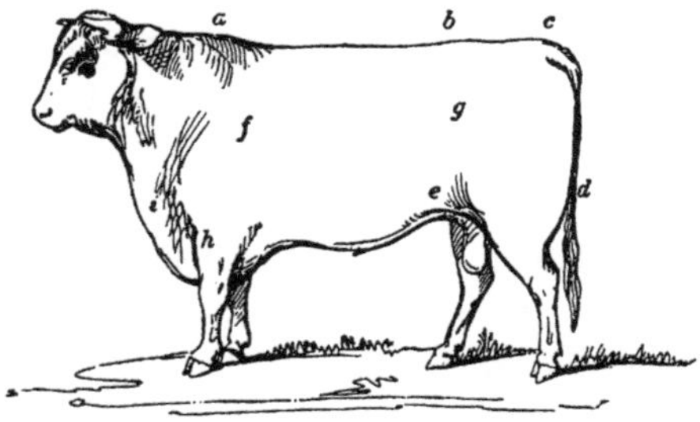

THE SHORT-HORN BULL.

Fig. 587, and Plate XV., are the portraits of 3 short-horn cows belonging to the Duke of Buccleuch, at Dalkeith Park, in Mid-Lothian. They are all of pure blood, being descended from Mr Robertson of Ladykirk's stock, when Mr John Rennie had a part of them in his possession at Phantassie in East Lothian. The cow, a *roan*, whose side view is given,

Fig. 587.

THE SHORT-HORN COWS.

was in a lean state when the portrait was taken; but the likeness is the more valuable on that account, in giving an idea of the skeleton, which in her case is very fine. She is the beautiful Kilmeny, got by Match'em. The hook-bone *b* is finely prominent, and at the same time level; the ribs *e* round; the shoulder-blade *d* sloping; the top of the shoulder *a* broad; the neck-vein at *g* fine; the muscle at *f* well developed; the tail-head *c* level with the hooks; and the udder *h* hemispherical, the teats being pendent at equal distances. The uncommon half-slouching half-projecting form of her horns, and a staring of the eyes, give her countenance, though her head is otherwise fine, a somewhat more austere aspect than cows generally have. The front view of the *red* cow in the middle is intended to shew the breadth of the fore-quarter from *p* to *r*, the roundness of the ribs at *p*, the depth of brisket at *s*, and the width of space upon which the fore-legs stand. The hind view of the *white* cow, on the right, shews the broad space between the hooks from *l* to *k*, and the same across the pelvis from *m* to *n*, 2 points essential in the cow, as a safe breeder of full-grown calves. The form of the udder behind at *o* shews the proper position of the 2 hind teats, and exhibits the loose soft skin above the udder, which is characteristic of good milkers.

The 3 colours, roan, red, and white, were purposely chosen, to shew the common colours which the Short-horns bear. Though thus divided into 3, only 2 colours are, in truth, sported by Short-horns, namely, red and white, the roan being a mixture of the other 2. The roan is a handsome colour, and is, I believe, the general favourite now; the fancy for colour having gone from the red to the white, and is now settled on the roan. Dark-red usually indicates hardiness of constitution, richness of milk, and disposition to fatten; light-red indicates great quantity of thin milk, and little disposition to fatten; but the red in either case is seldom entire, being generally relieved with white on the sides, and the belly, as in the case also of the bull described above. White was considered indicative of delicacy of constitution; and to get quit of it, and at the same time avoid the dulness of red, the roan was encouraged, and now prevails. I think I have observed that white animals shew the symptoms sooner than any other colour of being bred in-and-in.

A single *black* hair on the body, and particularly on the nose; or the slightest *dark* spot on the flesh-coloured skin upon the nose, or around the eyes; or the horns tipped with *black*, at once proclaim, that a Short-horn sporting either of these is not pure bred—all attestation to the contrary being of no avail.

To shew that careful breeding improves the *head* of every breed of cattle, specimens of the heads of 3 celebrated breeds are given, one being the *Long-horns*, a breed not confined to England, as it extends over Ireland, though it is only in England that fine specimens of the breed are to be seen. This is the breed which the famous Bakewell improved in Leicestershire, a few years before the Collings improved the Short-horns, by which they established to themselves a fame which eclipsed that of Bakewell as a breeder of cattle; not because they understood the principles of breeding cattle better than he did, but, fortunately for them, they selected a better subject to deal with—the Short-horns. Their fame, however, was not established at the cost of that of Bakewell, for he had already acquired an imperishable name, as a breeder, in the improvement of the old Leicestershire breed of sheep; but it must be owned, that, when the Collings advanced beyond Bakewell as breeders of cattle, the advancement was the more meritorious in being made in competition with so very formidable a rival. The Long-horns were originally called Leicesters; but having long horns, they were so named in contradistinction to the Durham breed, whose horns were short; and, besides, Bakewell's improved sheep were as often called Leicesters as the Dishley breed.

The Long-horn bull, a likeness of whose head is here given in fig. 588, belonged to Mr R. Horton, in Warwickshire, and was shewn, and obtained the first prize of his class at the Show of the Royal Agricultural Society of England at Oxford, in July 1839, when he was 4 years 2 months old. It will be observed that the muzzle is fine, the eye large and expressive, the horns fine, tapering, and sharp-pointed, and the entire countenance agreeable. His colour was light brown, brindled with black stripes. The skin of the nose and around the eyes dark-flesh colour. The slouching position of the horns is very common in the Long-horn breed; they are brown, with a few reddish streaks, and tipped with brownish-black.

Fig. 588.

THE HEAD OF A LONG-HORN BULL.

Another breed to which much attention has been paid, is the *Hereford*, which has long been famed for its excellent steers. Fig. 589 is the portrait of the head of a Hereford ox which belonged to Mr S. Druce of Ensham, in Oxfordshire, and was shewn at Oxford, at the Show of the Royal Agricultural Society of England, in July 1839, when

4 years 4 months old. It will at once be observed that the muzzle is fine, the eye large and full, and the horns small, tapering, and sharp-pointed. A white face is quite common in the Hereford breed, with white horns, and brownish-red points. The body is either dark or light-red and white, a common colour, or a dark rich chesnut-brown, which is becoming fashionable. The skin on the nose and around the eyes is fine flesh-colour.

Fig. 589.

THE HEAD OF A HEREFORD OX.

The *West-Highland* has long been famed in Scotland as a superior breed of cattle. They have most of the points of the Short-horns in the body, which is covered with shaggy hair, that bids defiance to the keenest blasts and the most drenching rains. Fig. 590 gives an idea of the head of an ox belonging to Mr Campbell of Jura, which was shewn with another, as a pair, at the Highland and Agricultural Society's Show at Inverness, in October 1839. It will be observed that the muzzle is fine, eye large and full, and the horns small, tapering, sharp-pointed, white, and tipped with black. The colour of the body is usually black, sometimes red, and not unfrequently dun. The black coloured, in my opinion, makes the most profitable animal to the feeder. The skin on the nose, and around the eyes, is always black.

Fig. 590.

THE HEAD OF A WEST-HIGHLAND OX.

There is a breed of cattle extensively cultivated in Scotland, the importance of which consists in affording milk for the purposes of the dairy. It is a remarkable circumstance in the history of the breeding of this race for the dairy, that the very opposite points have been cultivated by its breeders, which I have described as being essential in the opinion of the best breeders of every other race. The breed I allude to is the *Ayrshire*. The points considered good in an Ayrshire bull, by the breeders of that species of stock, are a broad short head, the horns spreading from the side of the head a little in front, and turning upwards. The top of the shoulder sharp, back narrow, ribs, of course, flat, hooks confined, hams thin, tail-head somewhat drooping, belly enlarged, and legs very short. These are all points contrary to those of a good Short-horn; and the points in which they agree are a straight back, loose mellow skin, large eye, and small horn. The cows are best liked for a very sharp shoulder and wide hooks and pelvis, in which case the ribs are flat and the belly large. The udder is desired to be hemispherical, well forward, and provided with loose soft skin behind.

I am satisfied that the points thus desiderated by Ayrshire breeders are not necessarily promotive of the principal object they have in view, namely, a large quantity of milk; for though it cannot be denied that Ayrshire cows are generally good milkers, it is attributable, in my opinion, more to the great length of time they have been devoted to that particular purpose, and which property has now become inherent in the breed, than to the form which the breeders promote; and my reason for thinking so is this:—It is well understood, in large dairies in large towns, that the Short-horns prove the most steady milkers, that is, they continue to yield milk in large quantities for a longer period than any other breed. When they are not allowed to bear calves, they will continue to yield milk until they are fat enough for the butcher; and when allowed to bear calves, they will give milk to within 5 or 6 weeks of the time of calving. Now it is averred by Ayrshire breeders, that the points cultivated in Short-horns are for the purpose of yielding flesh; and the averment is quite true; but if the points which yield flesh are also favourable to yielding milk, as the experience of dairymen in towns has proved, and their preference of Short-horns to others testify, it is clear that it is *not* in consequence of cultivating points in the Ayrshire breed of entirely an opposite character to those of the Short-horns, that the Ayrshire yields so large a quantity of milk; but that it must depend upon some other quality, otherwise the Short-horns should yield little or no milk, but entirely flesh—which is found not to be the case. I had a light red Short-horn cow, which gave 17 Scots pints of milk at the height of the grass season, and would scarcely go dry before calving; and have seen another roan cow, which gave 30 pints a-day for 3 months, and had to be milked for relief 5 times a-day.

Horns of Cattle.—There is much to be observed in the set and form of the horns of cattle. Small, short slouching horns on a 2 or 3 year old steer give a grave and majestic cast to the countenance. Horns rising outward from the side of the head, and looking up, and bending backwards, never fail to impress one with the conviction that their bearer is an ill-tempered brute, and ready to use them offensively on all occasions. Horns curving latterly and horizontally forward, give a finished appearance to the top of the head, when viewed in front, as exemplified in the figure of the middle Short-horn cow in Plate XV. When horns are long, and rise outward, forward, and elevate their points, they impart a very majestic mein to the ox, as shewn in fig. 590 of the West-Highland ox. Some horns are set looking backward behind the head, and give an idea of malformation. A horn thick at the root never looks well, neither does one that is blunted at the point; and both kinds are associated with dull feeders: nor do those which spring outwards and then turn downwards, look well, as shewn in the left hand Short-horn cow in Plate XV. A good horn is small where it emerges from the head, and tapers gradually to a fine point. A white horn is better than a dark-coloured, and a finishing of brown or black at the tip, according to the breed, is a good outset, though many Short-horns, especially of white colour, have their horns all white, and being short, do not seem obtrusive —indeed, in most Short-horns the horns serve more for ornament than purposes of defence. Oxen with spreading horns are better feeders than those which contract towards the front. Horns indicate the age of cattle. At 3 years of age the horn has attained its state of uniform growth; that is to say, it is uniformly smooth from the root to the tip. Every year after the horn is protruded from the head, with a notch on it, so that by counting the number of notches, and adding 3 to the number, the age of the animal may be ascertained. Tricks are practised by fraudulent dealers, by filing down the notches, to make the animal appear younger than it is; but a slight inspection of the horn will easily detect the fraud; and the period of the year, whether late or early, in which the animal was born, will have some effect on the notches of the horn.*

Sheep.—Of the numerous varieties of sheep cultivated in Great Britain, I have only selected 3 for illustration here, as these include all the breeds brought up in Scotland, though another has been partially introduced. The 3 varieties alluded to are the Leicester, the Cheviot, and the Black-faced, and the supplementary variety, as it may be called, is the Southdown. The 3 varieties are well suited, not only for the climate of Scotland, but also for the peculiar zones, as they may be termed, into which the pasturage of that country is subdivided. The

* See a paper on this subject in the Quarterly Journal of Agriculture, vol. iii. p. 660.

Leicesters are adapted to heavy pastures and sheltered fields in low lands. The Cheviot is equally well fitted for the middle range of green pastures to be found in the pastoral districts of the south and north of Scotland. While the Black-faced derive their chief support from the heathy pastures of the mountains; at least that range of country forms the nurseries of this hardy race.

Plate XXXIII. contains the portrait of a *Leicester tup* belonging to the Duke of Buccleuch. This animal exhibits the peculiar properties of the breed to which he belongs; the principal of which are, a white face and legs covered with hair, hornless head, and body enveloped in long wool. The individual characteristics of this tup are, rectangular carcase, round rib, as exemplified in the fore and hind views of the figures on the right and left hand in the back-ground of the plate; small bone, fine head, small muzzle, large full eye, and expressive countenance; but his ears are much shorter than usual. The head of the tup is broader across the eyes than that of the ewe or wether, and the skin becomes wrinkled upon the nose when he gets aged. The wool is thick-set, long, of good quality, and the fleece covers the entire body above and below,—a mark of sound constitution, and a great means of preserving the animal from the bad effects of the weather above, and of the dampness of the ground below. A level broad back from neck to rump, and across the ribs, is characteristic of the Leicester, and on being turned up, a broad chest and fulness on the inside of the hams. The touch should be equally mellow along the back, a hardness in any part indicating a defect. In ordinary condition, the flesh above the tail-head is nicked, which may be easily felt with the points of the fingers; but when in high condition, which they should be at tupping-time in autumn, the nicking should extend all the way from the shoulder-top to the tail. The rib should also be well covered with flesh and fat, and, indeed, a *round* rib is almost always so.

Fig. 591 and Plate XIV contain the portraits of a *Leicester ewe* and *twin lambs* belonging to Mr Brodie, Abbey Mains, in the county of Haddington. The Leicester ewe's head is generally long, narrow, and clean, with fine muzzle, prominent eyes, and long, broad, thin ears. The bone is small and fine. In this particular instance the body is well-woolled and formed. The counter *g* is full; the shoulder well filled up behind at *a*; the rib at *f* round and full; and the loin at *e* not hollow, as is sometimes the case, particularly after ewes have borne a number of lambs. In regard to the wool it is full behind the ears on the top of the neck at *k*, thus keeping these organs protected; it is also full towards the cheeks at *h*, which keeps the throat warm; the belly is well covered with wool below at *b*; and so is the flank *c*; the rump *d*

Fig. 591.

THE LEICESTER EWE AND LAMBS.

has rather a redundancy, and it affects even the shape; but gimmers, that is, ewes of their first lamb, often produce a large quantity of wool on this point, which afterwards reduces itself to an equality of the rest of the body; but it is a good property in a ewe to have plenty of wool on the rump, to protect the anal and vaginal passages below the tail, and it is also an indication that the tail-head is placed nearly in a level with the back. It is no uncommon occurrence in a Leicester ewe to bear twin-lambs, nor is it uncommon to wean twins to the extent of 50 per cent. of the whole flock, some of the ewes bearing twins every year, while others only occasionally.

The Leicester breed of sheep has been cultivated with very great care in Great Britain, since the days of Bakewell, who brought them in his own lifetime to a very high degree of perfection. Many of the flocks in the kingdom can trace their pedigree to that of Bakewell, with as much truth as the purest stud of the race-horse, or the purest herd of Short-horns, though no Flock-book has been kept to record the names of the most renowned sheep; a circumstance which excites in me much surprise The breed is deserving of the utmost attention that can be bestowed upon it, as it possesses many valuable properties. The individuals are in themselves handsome, displaying a beautiful contour, with a pleasant aspect. Their disposition is so amiable that they have no desire to overleap a fence; and this disposition, no doubt, is fostered by their aptitude to fatten. The result of their condition is a large proportion of flesh to bone, and of useful parts to the offal. Many people affect to dislike the mutton, as being too fat and flavourless. The muscle is certainly larger grained than that of the other 2 breeds, and the fibres are intermixed with fat; but though the mutton may be disliked, for these reasons, by the higher orders, the breed being generally of robust constitution the meat is always wholesome, and acceptable to work-people. Colliers, who eat a large quantity of meat, will have none other. For these reasons Leicesters are more profitable to the farmer, where they can be reared, than any other breed. The wool is of the most valuable description, not on account of the *fineness* of its quality, for many short-woolled sheep have much finer wool; but its great length, as well as its tolerably fine quality, renders it useful in the manufacture of all fabrics which require combing wool, and in which worsted is employed. This wool seems peculiar to the British climate; for in no other country have sheep breeders succeeded in raising it of the same quality; the pile becoming shorter or coarser in warm countries. So long, therefore, as peculiar fabrics are made from Leicester wool, the breeders of Leicesters need not fear the want of a steady, if not a high market for their wool; and it must always be in demand, since no other country can compete, in raising it to the same degree of perfection.

Fig. 592 represents the head of a *Cheviot tup* which gained the first prize of his class at the Highland and Agricultural Society's Show at Aberdeen in 1840, and was shewn by Messrs Craig, Bighouse, Sutherlandshire. It will be observed that its face is longer than that of the Leicester, muzzle not so fine, eye not so full, ears set not so high and handsomely upon the top of the head, and there is a rugosity of the skin across the bridge of the nose. In the white face, and want of horns, the Cheviot resembles the Leicester. The wool is short, thick-set, and of fine quality, fit for the manufacture of inferior qualities of broad-cloths. The carcase is usually unequal, the fore-quarter being lighter than the hind—narrow in front, with the fore-legs set near, and a want of depth and breadth in the counter. The flesh is fine-grained, often well intermixed with fat, and is generally esteemed for the table. The disposition of the Cheviot is somewhat suspicious, with an inclination to rove; which disposition renders the breed rather unkindly to feed, at least at an early age.

Fig, 592.

THE HEAD OF A CHEVIOT TUP.

The Cheviot, as their name implies, had their origin in the Cheviot Hills, in Northumberland. They occupy almost all the pastoral hills of the south of Scotland, especially from the centre of the country to the eastward. They are localised in some of the best parts of the Grampian mountains, and are to be found as far north as the hills of Caithness and Sutherland. They may, therefore, be regarded as a hardy race, and are well suited, on that account, for the middle green pastures of the mountainous parts of our country.

The *Southdowns* are little known in Scotland; but what is known of them is favourable to their character. Like the Cheviot they are covered with short, thick-set, fine wool, which is of a dusky brown colour, and it also affects the hair which covers the face and legs. They are a hornless breed. In symmetry of body they are much superior to the Cheviot, bringing their quarters, like the Leicester, to an equality. Their flesh is fine-grained, and, as high-flavoured mutton, is preferred to that of the Cheviot in the London market. They have also a gentler disposition, and are in consequence better feeders. The only doubt with the Southdowns, on their introduction into Scotland, was their ability to withstand the damp climate of our subalpine pastures. The experience of several years has proved that they are capable of enduring any climate with the Cheviot; and this being the case, with their other superior qualities, they bid fair to rival, and perhaps ultimately to displace, that breed. Hitherto, however, they have only been tried in a few places. Mr Hugh Watson has had them at Keillor, in Forfarshire, for more than 20 years, and they have thriven with him upon the pastures of the Sidlaw Hills. The Duke of Richmond has them in Morayshire. Other small flocks are scattered through the country.

Figs. 593 and 594 represent the heads of a *Black-faced ram* and *ewe*. The ram, fig. 593, was shewn by Mr Robert M'Turk, Hasting's-Hall, in Dumfriesshire, at the Highland and Agricultural Society's Show at Berwick-upon-Tweed in 1841, where it obtained the first prize of its class. As indicative of the long period and care with which the breed has been cultivated, we have only to look at the tapering face, small muzzle, and full eye, exhibited by the specimens in the figures. The Black-faced ram has always an arched nose, expressive of boldness and courage. The face and legs are covered with black or mottled hair, mostly the latter. The head is horned; and the horns are considered as picturesque an object as is exhibited by any animal of this country. The wool is long and coarse, which render it of small value as an article of manufacture, and being thin-set exposes the body to the inclemency of the weather. It is to assist the animal for this latter defect in the fleece that the filthy operation of smearing is resorted to. The carcase is well formed, carrying its depth forward to the brisket better than the Cheviot; but still the entire body is narrow, owing to the flatness of the ribs, which gives too much lightness to the whole carcase; or want of *substance*, as it is commonly called. The flesh is fine-grained, high-flavoured, greatly esteemed, and can be fed sufficiently fat on the turnips and pastures of the low country. The breed is very hardy, frequenting the highest parts of our heath-clad mountains, and in summer require little care from the shepherd.

Fig. 593.

THE HEAD OF A BLACK-FACED RAM.

As with cattle, the *horns of sheep* afford facilities for ascertaining the age of the animal. In fig. 593, the age of the tup is distinctly marked, the 1st year's growth being evidently the space from the point of the horn to the letter *a*; the 2d year is from *a* to *b*, the growth of the dinmont being stronger than that of the hogg; the 3d year's growth is marked from *b* to *c*; the 4th, from *c* to *d*, which great growth shews the vigorous state which the animal had attained at that age; and this is no doubt the most vigorous period of the life of a sheep; and the 5th year's growth is shewn from *d* to *e*. The respective ages of the ewe fig. 594, and of the wether, may be traced in like manner, but the horns not attaining the full development in them as in the ram, the yearly marks cannot be discerned without minute examination. As to this figure of the ewe it is evidently that of a young one, probably a maiden ewe or gimmer, or at most with the first lamb; but the wether in Plate XXVIII. is at least a 4-year old

Fig. 594.

THE HEAD OF A BLACK-FACED EWE.

Plate XXVIII. exhibits a group of *fat wethers* of the 3 native breeds we have been considering and illustrating; but all the animals composing the group are by no means alike favourable specimens of their respective breeds. The Black-faced wether is a good specimen, and many such are brought fat to the Edinburgh market every winter. They are fed in Forfar, Fife, and Perth shires, on turnips in winter, but the finest fat are those fed on turnips in East Lothian—that county affording by far the best feeding land for stock in Scotland. The Cheviot is evidently not a wether, being only a dinmont, and not pure bred; and on both these accounts is not a ripe specimen. The letter P is left on the rump to shew the tar-mark of the farm on which, or the name of the farmer by whom, it had been bred. The lightness of the fore-quarter, characteristic of this breed, is however very well shewn in the figure. It is rare to meet with Leicester wethers now-a-days, the dinmonts attaining a sufficient degree of fatness for all useful purposes. The figure in the plate is a dinmont, and not a favourable specimen, not being pure bred, and too small for the breed. The characteristic distinctions of the 3 breeds are, however, sufficiently marked in the *heads* of the figures given in the plate.

A desire seems to be spreading in this country for the naturalization of the *Alpaca* from South America. It is the opinion of those who have seen the Alpaca in their native country, that it would thrive well on our highest mountains, upon the coarsest fare that can be obtained—upon what, in fact, is refused by even Black-faced sheep—and at the same time yield a coat of fine wool, far finer and longer than any grown in this kingdom. Could this opinion be established by experiment, the cultivation of the animal would be worth attempting on a large scale; and no doubt sufficient supplies would be obtained of it in a few years. As yet, the subject is only under discussion.*

Swine.—There are many breeds of *swine* existing in the country; and whatever denomination they may pass under, if they do not possess the points exhibited in the adjoining fig. 595, and in the Plate, which I regret to observe has been neglected to be numbered, they may be pronounced defective. Here the same rules apply, not only of symmetry, as we have seen in fig. 581, but points of breeding. The head *a* is small, the face tapering to the muzzle or snout *i*, which is short and fine, the ears, set on the crown of the head, being broad, thin, long, and so

* A little work, named The Alpaca, by Mr William Walton of Liverpool, published by Blackwood in 1844, gives an interesting account of this animal, both historical and natural. The author seems to know the habits of the animal; considers that it would thrive to a profitable extent in this country; and details the particulars of a practicable plan by which a regular and sufficient supply of it, at a remunerative rate, might be obtained.

mobile as to indicate quickness of perception. The value of the head, as an article of food, is indicated by the enlargement of the muscle upon the cheek *h*. The neck from *a* to *b* is full; the back from *b* to *c*, broad; the rump from *c* to *d* full and round, and the roundness descends to the hams; the ribs *f* are round; the space behind the shoulder at *g* filled up; and so is the flank *e*; the shanks *k* are small and short, and finely tapered. A pig with these properties is always in a condition for use, from the state of a pig sucking milk, through its progress of porkling and shott till it attain the full size for bacon and hams. Such a breed never requires feeding, and as it is always in condition, it must only have time to grow to the size wanted, when a little firming of the flesh by corn is all that is required to prepare it for slaughter.

Fig. 595.

THE BROOD SOW.

The slouch-eared breed that prevailed in the country is fast being supplanted by the one I have been describing; because, wherever such a character of ear is seen, it is universally accompanied with length of leg, length of nose, narrowness of back, and dilatoriness in feeding. I believe the prick-ears, short snouts, and full cheeks, may be traced to an improvement derived from the Chinese breed, which possess those points even to a deformity. The cross with our old bony breeds has been the means of disseminating through the country a race of beautiful, profitable, delicate-fleshed pigs.

Fig. 596 represents the head of a *boar* belonging to the Duke of Buccleuch, and the *brood-sow* represented on the unnumbered Plate, is also from His Grace's stock. The boar, though evidently full-grown and large, bears the same character of pricked ear, tapering face, short nose, and full cheeks, and the flesh of the neck is seen to be fully able to support the fulness of the head. These pigs are, moreover, strong-constitutioned, and covered with plenty of white hair and valuable bristle. Their temper is generally docile, and the animals seldom wander far from the steading, or engage in mischievous pursuits.

Fig. 596.

THE HEAD OF A BOAR.

Teeth.—The teeth are far more imperfect organs to the domesticated animals than the horns. The horns are the instruments of defence and of attack; and, in the situation in which those animals are placed, are seldom put in requisition, and the more seldom the better; but the teeth are the instruments by means of which their food is broken and masticated in winter, and cropped and masticated in summer; and this being the case, the condition of the animal mainly depends upon the state of soundness in which the teeth may be preserved. There is one similar property between the horns and teeth of animals;—both furnish data by which the age of the animal may be ascertained. You have already seen how the horns are indicative of the age, we shall now advert to the manner in which the teeth may be examined for the same purpose.

Fig. 597 represents the left half of the head of an *adult horse*, viewed internally, and so figured as to shew the origin of the 5th pair of nerves, and the nervous branches which go to the teeth; and for this reason the figure may be consulted with equal advantage for a knowledge of the distribution of the nervous system and of the teeth. The milk-teeth of the horse consist of 12 incisors, 6 in each side of the head; and of molars 16 in number, 8 on either side of the head,—in all 28 teeth. The teeth of the second dentition are 40 in number, of which 28 have replaced the

Fig. 597.

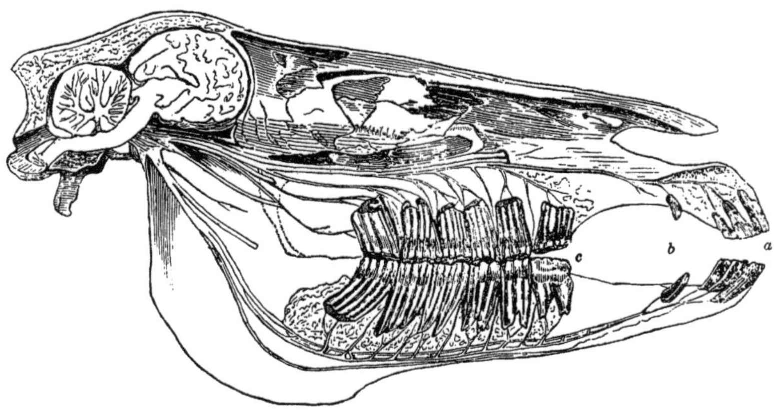

THE VERTICAL SECTION OF THE HEAD OF THE ADULT HORSE, SHEWING THE TEETH AND THE NERVOUS SYSTEM IN CONNECTION WITH THEM.

milk-teeth. Those between the incisors and molars, called the canine or tusks, do not appear along with the teeth at an early age. Others complete the arch by occupying the room made by the growth of the jaws; these are new molars which come out of both jaws. The full set consists of 12 incisors, 4 canines, and 24 molars, in all 40 in number. In fig. 597, *a* are the incisors, *b* the canine or tusks, and *c* the molars.

"This is the order of coming out of the second or permanent dentition of the horse," observes M. Rousseau. "The first permanent molar, which is situate behind the last milk-molar, presents itself before any of the milk-teeth have fallen, and makes its appearance upon the maxillary arch from the 11th to the 13th month after birth; it will be, by numerical number, the 4th persistent molar, when all the milk-molars have fallen. The 5th permanent molar, which is situate behind the preceding tooth, breaks the edge of the socket from the 14th to the 20th month. During this time the decaying teeth die from their roots, and wear down their crowns to such a degree, that the hollow which characterises the surface of the incisors at certain periods cannot be observed, so that the veterinarians call them lost-mark. The central incisor or pincer is ordinarily of the 9th or 11th month; the lesser incisor from the 11th to 13th month; and the lateral incisor, or corner-tooth, from the 14th to the 20th month. Once these teeth cease to have mark they bear upon their surface a smooth trace, brown and indelible, which diminishes the more that the teeth approach their fall. The first permanent molar replaces the 1st and 2d milk-molar from 2 years to 2½ years. The central incisor appears upon the edge of the socket after the coming out of the 5th permanent molar, from 2½ to 3 years. The 2d permanent molar replaces the 3d milk-molar a little after the same term, or 1 or 2 months of difference. The 3d permanent molar replaces the 4th decaying molar at 3 years; at this time also appears the 6th and last molar. The lesser incisor from 3½ to 4 years. The canine or tusks appear from 4 to 4½ years. At last, the second dentition is ordinarily terminated by the lateral incisor or corner-tooth. It must not be thought, however, that the coming out, as I have endeavoured to indicate as the most ordinary, is without variation; this would be to give to nature too regular a march. All teeth in general are the more developed that they belong to a large and robust subject."

Fig. 598 gives a similar representation of the dental system of the *adult ox*, and of the nervous system connected with it, that the preceding figure gives those of the horse. The milk-teeth of the ox are, 8 incisors on the lower jaw, and none on the upper, and 12 molars, 3 on each jaw. In the adult ox are 8 incisors on the lower jaw, and none on the upper; and there are 24 molars, 6 on each jaw. In the figure *a* are the incisors, and *b* the molars, and the same configuration exists in the sheep. "In the second dentition, these teeth shew themselves upon the edge of the socket in the following order," says M. Rousseau. "The 4th permanent molar comes out from the 4th to the 6th month after birth. and commences the second dentition. The 1st or central replacing incisor from the 15th to the 22d month. The 5th, or penult molar, from the 18th to the 22d month. The 2d replacing molar, as

Fig. 598.

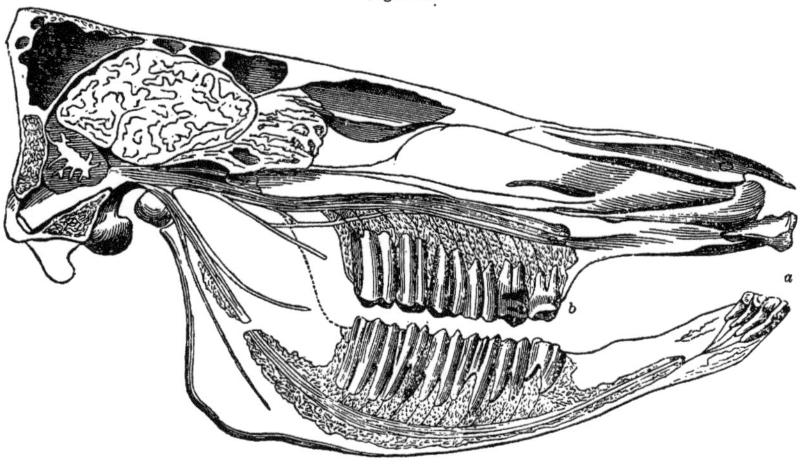

THE VERTICAL SECTION OF THE HEAD OF THE ADULT OX, SHEWING THE TEETH AND THE NERVOUS SYSTEM IN CONNECTION WITH THEM.

also the 2d incisor of this order, appear from the 28th to the 32d month. The 3d replacing molar, as also the 3d incisor, come out very near at the same time, that is, from the 38th to the 48th month. The 6th, or last molar, from the 44th to the 52d month. At length the 4th permanent incisor tooth terminates the second dentition, which is ordinarily completed when the animal has not yet attained its 5th year."

Fig. 599 gives a section of the head of a *wild boar*, in which the den-

Fig. 599.

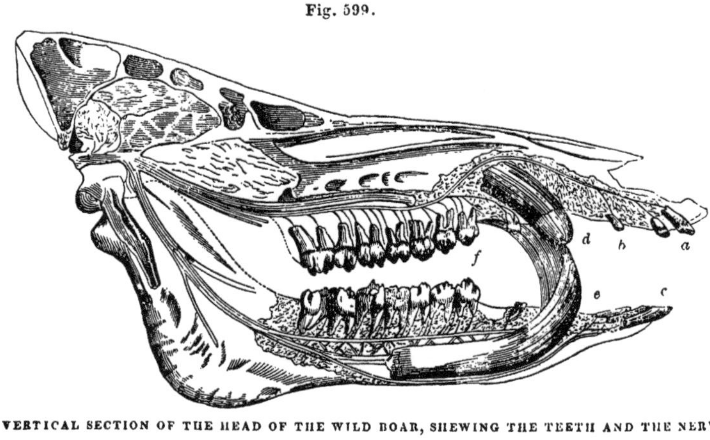

THE VERTICAL SECTION OF THE HEAD OF THE WILD BOAR, SHEWING THE TEETH AND THE NERVOUS SYSTEM IN CONNECTION WITH THEM.

tal and nervous systems are distinctly delineated; *a* are the superior incisors; *b* the superior lateral incisor; *c* the inferior incisors; *d* the upper canine or defence tusk; *e* the inferior canine or defence tusk, the origin and form of which may be easily traced; and *f* are the molars. I have selected the head of the wild boar for illustration, because the character of all the teeth, and particularly that of the tusks are more strongly developed than in the domesticated boar. The tusks of the sow are comparatively short and weak. The milk-teeth of the ordinary pig are 32 in number, namely, 12 incisors, 4 canines or tusks, and 16 molars, half of which numbers are on each side of the head. The second dentition is only completed as soon as all the milk-teeth have fallen; and these are not only replaced, but 3 other molars on each jaw rise up, one after the other, until the whole dental arch is completed, when the entire complement is 44 teeth, of which 22 are on the upper, and 22 on the lower jaws, and they are divided thus:—12 incisors, 4 tusks, and 28 molars. These teeth are composed of 2 substances, the one bony, the other enamelled. "The tusks are each enclosed in a socket, filled with a substance analogous to the marrow of the long bones; it is most remarkable, and most abundant in the inferior tusks. These teeth are only provided with enamel upon the external face of the permanent tusks." *

In regard to the indications of age by means of the teeth, in the horse the marks on the crowns of the front teeth on the lower jaw are almost always obliterated, or at least cannot be depended on after 9 years of age. After that period, the only means of judging is by the quantity of matter ground off the top of the teeth, and by the distance between the teeth; the older the horse gets, the wider the space between them becomes. In regard to the grinding down of the teeth, however, you should know that pasturage on sharp land, and support on hard meat, such as unbruised corn and beans, will wear down teeth much faster than pasturage on soft land and prepared food. The same remark applies to cows which have been pastured on sharp or soft land, and been fed on straw and turnips, or on boiled food, and it is only applicable to them, for as to steers, their age cannot be mistaken from the growth and condition of the body. The teeth of cows also stand wider as they advance in years.

A horse's mouth is easily opened for the purpose of examination, by introducing a finger by the side of the mouth into the space between the incisor and molar teeth, where, in short, the bit of the bridle lies, upon the tongue, when the horse will play with his mouth, to get quit of the finger, and shew as much of the lower teeth as to satisfy your curiosity. Some sulky horses require to have their lips held asunder; and vicious

* Rousseau's Anatomie Comparée du Systême Dentaire, p. 205-230 – a very complete and elegant work.

ones will even strike out with the fore feet when their mouth is meddled with. I had a cart-mare, the moment her mouth was attempted to be held for examination, she would wheel round quickly, and kick with the hind-feet at the person attempting it. A cow's mouth cannot be examined, without first taking hold of her nose, elevating her mouth, and drawing down her under lip. Some cows will not allow themselves to be taken by the nose, and the thing can only be done in such a case by stratagem. Some, again, have such a power in the nose by curling up the nostrils when held, that its squeeze against the holder's fingers renders them soon powerless; but a steady pinch of the thumb-nail against the septum of the nose will make any cow give way, provided the person has strength to hold her firmly at the first burst.

I am strongly tempted to give you an account of the physiological remarks of Mr Walker on the breeding and crossing of the domesticated animals, but find that even an epitome of his observations would occupy a considerable space. I must, therefore, content myself by referring to his very curious, instructive, and not less entertaining work.*

---

84. OF MAKING EXPERIMENTS ON FARMS.

> "Pleased by each trial, by no failure vex'd
> And always certain to succeed the next."
>
> CRABBE.

The farmer may be regarded as a great and constant practical experimentalist; for although all his operations are supposed to produce known results, yet the actual results obtained are often not as anticipated. When he works his land for wheat or for turnips, the operations being directed to a special end, he is certain of reaping a crop of wheat or turnips, and thus far he does not run the risk of an experimentalist; but, after all, the state of the actual crop obtained may be very different from what he expected—it may be better, or it may be worse. He

* See Walker on Intermarriage, p. 1–4, on the Classes of Organs and Functions of the Body; p. 146–242, on the Laws of Resemblance; p. 289–306, on the General Principles of Breeding; and p. 307–354, on the Application of Natural Laws to the Breeding of Horses, Cattle, and Sheep. And you may consult Prichard's Researches into the Physical History of Man, vol. i. p. 105–9, for the meaning attached to the terms, Species—Genera—Varieties—Permanent Varieties—and Races; and p 339–74, on the Colour, Skin, Hair, Form, and Temperaments of Animals. See also in p. 177, vol. xv. of the Transactions of the Royal Society of Edinburgh, an interesting Memoir on the Origin of the Domesticated Animals, by Mr Stark of Edinburgh.

may work his land in the most favourable circumstances, and after all reap but a scanty and ill-conditioned crop; or his operations may be frequently interrupted by the weather, and he may, notwithstanding, be rewarded at the end of the season with a bountiful return. "He may sow in tears, yet reap in joy." Uncertainty of result attends not one, but all crops;—it is not confined to the operations of one, but extends to those of all seasons. In this view, there is no profession which depends so much on future contingencies as farming. Unless the manufacturer wishes to make an alteration in his machinery, or in the pattern, fabric, or colour of his manufacture, he may go on producing the same results for an indefinite length of time. The farmer, on the other hand, cannot produce any determinate result;—he is more likely either to fall short of, or to overshoot, his mark Whatever he does, then, on his farm, may be regarded in the light of experiment. Yet though all his ordinary operations are attended with uncertainty in their results, they are not regarded or conducted by him *as* experiments, their uncertainty being anticipated. *His true* experiments, like those of other persons, are trials made to discover unknown effects, or to confirm uncertain results. Such experiments have constantly been undertaken by farmers, not as a class, but individually; and they have been conducted on a small scale, quietly, independently, unnoticed; and the knowledge gained by such trials has rendered his experience superior, and its effects has only been exhibited in his improved practice. The present period, with farmers, is pre-eminently one of *experiment.* It is universally known that experiments in farming are being made over the whole kingdom. They are conducted by farmers as a class, and have not originated so much with themselves, as at the suggestion of others; and the reason that *experiments* are more universally undertaken now than heretofore, is, that many substances have been recommended to farmers as valuable manures, of whose properties they are entirely unacquainted. What the ultimate result of using those substances may be, is yet uncertain, and perhaps may tend more to the advancement of science than of agriculture. Farmers thus stand to their own operations, in as far as those experiments are concerned, in an uncommon, and, I may say, anomalous, position; and it is this circumstance which suggests to my mind a doubt of the result being so favourable to them as is expected by the suggesters of the experiments. The manipulations of the experiments will be conducted *in bona fide*, the facts will be observed, though, perhaps, not at the time, and with the kind of observation they require; but a difficult part of the business is, that, owing to the experiments being suggested by others, with substances with which the farmer is unacquainted, the conclusions which should be legitimately drawn from the facts, cannot be satisfactorily arrived at by him, and any arrived at by another person, will not carry that conviction to his mind, as to induce him to continue the experiments on his own suggestions. The true way for the farmer to feel a personal interest in such matters, is to obtain a competent knowledge of chemistry; and the only way for the scientific man to know what experiments to suggest is, to acquire a competent knowledge of agriculture. The united efforts of both parties thus qualified might be expected to produce results which their efforts singly may never be able to attain. As you will, no doubt, be desirous of trying experiments, and may perhaps be placed in the position I have described, in regard to conducting experiments suggested by others, it may assist you in obtaining more satisfactory results, to be made acquainted with a few rules by which all field experiments should be conducted.

Every experiment should be made so as to be *comparative*, that is, conducted in the same circumstances as to field, soil, situation, time, and labour, as the crop raised in the ordinary way with which its results are desired to be compared. Without instituting means of making such a comparison, no conclusion arrived at can be satisfactory; because there will be no given point from which to measure the gain or loss sustained in the experiment. The experiment, moreover, should be of the same nature as the ordinary crop of the field, in which the comparison is to be made. For example. If the field is in lea, which it is intended to plough up for *oats*, then the proposed experiment should be made on oats, not on wheat, after lea; for no points of comparison could be instituted between wheat and oats. The ground also should be treated in the same manner. If the lea is to be ploughed, then it should be ploughed in the same manner, for the experimental and the ordinary portions of the crop to be compared together. This is a point of greater importance than may be imagined. I have known, in a field of lea, of strong soil, the ridges which were gathered-up, yield a better crop of oats than those cast together; and, still more, I have seen gathered-up ridges free of grub, while cast ones were affected by that complaint. No reason could be assigned for the difference; but the difference was sufficiently observable to shew that one method of ploughing land affects it in a different manner from another. I have heard it stated that the crop on the sides of ridges whose furrows lie to the west, is better than on those whose furrows lie to the east. There is no doubt that some effect is produced by the aspect of ridges, for a ridge lying east and west bears a better crop on its south than on its north side; and this effect might have been expected, for the sun's heat on the south side of ridges must be better for a crop than the aspect of the north on the other side. So, in like manner, a western aspect in a north and south ridge may be better than an eastern. Another point of importance is to include as much land in each experiment as to allow it to be properly and similarly ploughed. Land is usually ploughed in 2 or 4 ridges. If gathered-up, cast, or ploughed crown and furrow, 2 ridges will suffice; but if ploughed 2-out-and-2-in, or even 4-out-and-4-in, then 4 or 8 ridges should be appropriated to each experiment. It will not be a *comparative* experiment to use ridges whose furrows lie towards the east, against those whose furrows lie to the west. In all cases the ridges should have the furrows in the same direction. Instead, therefore, of previously fixing the *quantity* of land to be apportioned to each experiment, the *number of ridges* should be appropriated to each experiment *according to the mode in which the land had previously been ploughed;* and the *quantity* of land thus determined will, of course, depend on the length of the ridges into which the field is laid off. In this way the quantity occupied by each experiment may be $\frac{1}{4}$, $\frac{1}{2}$, $\frac{3}{4}$, or 1 acre or more of land. Minuter fractions of land, such as $\frac{1}{8}$ of an acre along part of a ridge, can give no satisfactory results in an experiment. In allocating portions of fields for experiment, as I have described, a part of the ground should be left in the ordinary state between each experiment; for the *comparison* will not be so fair nor so obvious over the entire field, if confined to one part of the ordinary state of the soil, as when extended over different portions situated at stated distances across the field. Besides, when experimental portions are thus separated from one another, the lines of demarcation of each experiment are more distinctly marked, and the risk of confounding the experimental plots lessened. For all these reasons, the experimental portions of ground should occupy ridges, and not rectangular patches across the middle of the field. Such *patches* may be differently affected by the weather—whether by rain, or sunshine, or wind—than the common ridges of the field, which stretch from one side to the other, and present different aspects at different parts; and as the entire experiments are instituted professedly for the purposes of the field, they should be conducted in as similar circumstances as practicable to ordinary operations.

If the experiment to be tried is with ordinary manures, the land should be manured with the same kind, in the same state, applied on the same day, and at the same period of the day; for I have observed different results obtained from similar manure laid on in the forenoon and afternoon of the same day, especially as regards the potato, and even turnip crop. But should specific manures be employed, then the comparative results should be obtained from ground treated in a similar manner. For example. It will not do to try different specific manures upon grass which has been laid down after potatoes, against that laid down after turnips; nor on a crop after turnips which had been eaten off with sheep, against even the same kind of crop on land from which the turnips had been carried off altogether. Comparative experiments might, no doubt, be made on these different conditions of grass, or of a crop after turnips, with the *same* specific manure; but *different* specific manures will not give *comparative* results in different circumstances. In like manner, it will not do to apply *different* specific manures to *different* sorts of wheat, barley, or oats; as each variety of grain may possess such an idiosyncrasy as to be very differently affected in similar circumstances, and, of course, the results would not *then* be comparative. Experiments, no doubt, may be made on different varieties of crop in different circumstances without reference to comparison at all, but unless the *results of experiment are compared with ordinary practice*, there is no use of making the experiment. Again, manures of whatever kind should be applied to the soil in the same manner, that is, one should not be applied by hand and another by machinery; one in a dry state, another in a state of solution, if it is desired to make the experiment comparative; but if the experiment is intended to be compared with ordinary practice, the states of the manures may be varied in themselves, though they should all be applied in the same circumstances as in ordinary practice, to render the comparison of any value. Experience has taught, that mineral manures, which are usually employed in small quantities of a few cwts., should be mixed with a little earth before being applied by hand, and the mixing may be effected in the preparation of the land. Thus, when such manures are to be applied to land that is to be drilled, the drills should first be set up before the manures are sown by hand, out of a rusky, fig. 371, and then split, or should they be sown on the flat ground, they should be harrowed in before the land is drilled up, because drilling does not *mix* soil and manure.

Every article applied to the soil in experiment should be weighed. There is no difficulty in weighing specific manures which are of little bulk, but ordinary manures cannot be weighed in a field without the assistance of a steelyard; and the want of a portable one for ascertaining the weights of bulky articles is a difficulty which meets every farmer who is desirous of ascertaining the minutiæ of what he is doing. I have often experienced this inconvenience in the field, but could not remedy it even with considerable trouble. It is scarcely practicable to erect a steelyard of the common construction upon the soft earth; and to send every cart-load of dung to the steading to be weighed is not only an intolerable hinderance, but may so affect the time allotted for arranging the *experiment*, as to render its results entirely nugatory. The produce may be weighed with greater ease; for as it has to be carried to the steading to be handled, it may be taken at once, where it can be weighed on the spot. Straw in bundles, and grain in bags, can be weighed in the barn with the barn-steelyard, fig. 356, and turnips, potatoes, hay, and forage, weighed by the large steelyard, fig. 576. The desire for accurate weighing of what is given to, and taken from, the soil, is no whim, but a necessary element of knowledge, if we wish our experiments to lead to useful results. The only *practicable* way of ascertaining the quantity of farm-yard manure to be applied, where no steelyard is at hand, is to fill the cart with an ordinary load, and let special notice be taken how far the body of the cart is filled, in the loose state, before it is sent to the steading to be weighed; and if all subsequent loads are filled to the same degree, by the same people, and with the same sort of manure, the weight of one cart with another will not differ materially. If notice of the bulk of the cart-load is *not* taken *before* it is sent off to the steading to be weighed, the compressed state of the load, occasioned by travelling on the road to and from the steading, before and after being weighed, will give a wrong idea of the weight of the future loads. Every different cart employed should have the weight of its first load ascertained, as well as its own weight. These apparently trifling matters are too apt to be overlooked, and yet, it is through such minutiæ that the convincing power of an experiment is most strongly transmitted.

In gathering the produce of experiments, every article should be carefully weighed or measured. Corn should be stooked in the same manner, allowed to remain in the field the same time; and, if stacked, should be put in the same size of stack, built in the same manner. There is more importance in this latter remark than it may seem to possess; for I knew a stacker, the corn from the stacks built by whom always came out in a raw condition till late in the season. On comparing his work with another stacker, I found that he did not heart his stacks sufficiently (2966.); and my suspicion was raised on observing that he required less corn to make his stacks apparently the same bulk as those of the other stacker. On taking up experimental turnips, the whole break should be taken up and weighed, for considerable risk of error will be incurred by taking up and weighing only a portion, and then measuring the ground occupied by the whole; because it is not unlikely for the land-surveyor, employed in such a case, for the sake of accuracy, inadvertently to measure the entire space cleared from one drill to the other which is left, instead of the space exactly occupied to the top and hollow of the drills from which the turnips had been pulled. In removing the experimental crop, of whatever kind, an equal extent of the ordinary crop should be removed at the same time, and not a fractional part merely; and all calculations should be made from the entire extent of each portion of ground subjected to experiment, and not merely from a fractional part of each portion. Attention to all these particulars involves much time and trouble, and creates no little anxiety of mind; but the time must be spent, the attention bestowed, and the anxiety endured, if we wish to arrive at the truth.

From all these considerations, it is obvious that the proper conducting of experiments is attended with much trouble and thought; and such should be the strictness of character possessed by every experiment, that unless every particular connected with it has been treated with the utmost impartiality, the results, however interesting in themselves, should be regarded with comparative indifference. The farmer, specially, has peculiar difficulties to contend with in conducting experiments. He cannot manipulate himself,—he must depend for assistance on his people; and though he must regard them as *assistants*, they are so entirely ignorant of the object of the experiment, as to require not only to be taught how to conduct it, but to comprehend its very object; and after all the enlightenment they may receive in its prosecution, they may commit such an inadvertent error, even at the very consummation of the experiment, as to render it entirely nugatory; and an error is the more likely to be committed, if any part of the operation has to be performed in a manner different from what they are accustomed to. I have no doubt that many farmers, who have been experimentalists, have experienced disappointment from such inadvertent mistakes as I have alluded to, and been obliged to suffer, by the thoughtlessness or stupidity of a moment, the loss of the knowledge that might have been derived from results which had cost them a whole year's expectation. Delay the application of a specific manure, or the sowing of a variety of grain, till the next morning, when it should have been executed this evening,—mix a single sheaf from the produce of one portion of ground to that of another, and the experiment, in so far as the portions alluded to are concerned, is rendered worthless. Incessant superintendence, therefore, on the part of the farmer himself, can alone prevent the occurrence of such mistakes.

85. OF DESTROYING AND SCARING VERMIN ON FARMS.

"How now! a rat!
Dead, for a ducat, dead."

HAMLET.

Rats are the most troublesome vermin that harbour in steadings. They not only render every place they frequent dirty, thereby creating disgust; but the mischief they do in cutting holes in boarded floors, in undermining stone pavements, gnawing harness, and destroying every edible thing, is very considerable. The old black rat of the country (*Mus rattus*) is now nearly extirpated, and the fiercer, dirtier, and more mischievous brown rat (*Mus decumanus*) has taken its place.

Many plans have been devised for the destruction of the rat. The box-trap has long been in use, but I suspect to little purpose; for though scarcely a steading is without one, it is usually out of repair, and to be most frequently found amongst the lumber. A box-trap, when newly made, may secure a few rats at first, but after the wood has acquired the disagreeable odour characteristic of the rat, scarcely another one will enter it; or, at all events, the trouble incurred in setting it for days and nights, and keeping the place quiet where it is set, and in the end only securing a single prisoner, is much greater than will be long undertaken.

As prevention is better than cure in every case of evil, rats would be much less numerous in steadings were means generally taken to prevent their obtaining harbour in buildings. Were the tops of the partition walls of the steading and outhouses to be filled up to the slates, as recommended in (73.); were ground-floors of wood to be laid as directed in (16.), and shewn in fig. 6; and were pavement floors made to rest on a foundation as described in (73.), rats could find no lodgement in places which otherwise they select to afford them protection.

Of the many other plans recommended for the destruction of rats in farm-buildings besides the use of the box-trap, such as poisoning their food with arsenic; rubbing the mouth of their holes with treacle and arsenic, which they lick off and thereby poison themselves; frying bits of cork or sponge in butter, which swell in the stomach on water being drank; mixing barytes with grease, which excites such a thirst, that they go in quest of water, and die the moment they partake of it; they have all failed more or less, and their great number indicates the uselessness of every one of them. Of late, a mixture of phosphorus and grease has been recommended by a German gentleman, M. Meyer, which is said to excite intense thirst, and, on being allayed by water, death immediately ensues; and Dr Ure refers to a prescription for preparing a poison by an English gentleman resident in Germany, which "preparation," he says, "consists essentially of phosphorus mixed with flour and sugar." It has been tried by a friend of mine in Derbyshire, who has a most extensive farm, and found to answer the purpose very well; but there is a great difficulty in preparing it, from the insolubility and even immiscibility of phosphorus in water, attended with no little danger of fire. The process I have found to succeed perfectly is as follows:—Melt hogs'-lard in a bottle plunged in water heated to about 150° Fahrenheit; introduce into it ½ oz. of phosphorus for every 1 lb. of lard; then add 1 pint of proof spirit or whisky, cork the bottle firmly after its contents have been heated to 150°, taking it at the same time out of the water-bath, and agitate smartly till the phosphorus becomes uniformly diffused, forming a milky-looking liquid. This mixture being cooled, with occasional agitation at first, will afford a white compound of phosphorus and lard, from which the spirit spontaneously separates, and may be poured off to be used again, for none of it enters into the combination, and it merely serves to comminute the phosphorus, and to diffuse it in very fine particles through the lard; this fatty compound, on being warmed very gently, may be poured into a mixture of wheaten-flour and sugar, incorporated therewith, and then flavoured with the oil of rhodium, or not, at pleasure. The flavour may be varied with oil of aniseed, &c. This dough being made into pellets, is to be laid in rat-holes: by its luminousness in the dark, it attracts their notice, and being agreeable to their palates and noses, it is readily eaten, and proves entirely fatal. They soon are seen issuing from their lurking-places to seek for water to quench their burning thirst and bowels; and they commonly die near the water; they continue to eat it as long as it is offered to them, without being deterred by the fate of their fellows, as is known to be the case with arsenical doses. My friend in Derbyshire bought a pot of M. Meyer's rat-poison, and found it to be an analogous phosphoric preparation: the present mode of preparing it is the result of my own experiments, made with the view of diffusing phosphorus through a mass of flour and sugar, without the risk of fire."* These are all means of causing death in rats, which would, of course, cause death in other animals which partook of the poison; and in so far are all dangerous expedients to be used where animals are going about at liberty. As a means of scaring them away from a place, it has been

* The Gardeners' Chronicle and Agricultural Gazette, for 20th July 1844, p. 489.

recommended to rub the mouth of their holes frequently with common tar, or coal-tar, which, smearing their coats, annoys them to such a degree as to cause them to leave the place.

Together with building up the tops of partition walls, I believe there is no way of scaring rats and mice from a steading so effectually as by *cats*. Let 1 or 2 cats be brought up in different parts of a steading, according to its size, if it is situated at a distance from the dwelling-house, and they will become vigilant guards against vermin, and if near, the house-cats will frequent it sufficiently often for the purpose; but when kept in the steading, let each receive every day, at its own place, and at a stated hour, say 11 o'clock A.M., a good mess of new milk and porridge, when it will attend to receive its food as regularly as the hour arrives; and let each have a soft, warm, comfortable bed made for it in some quiet corner. At night, and early in the morning, they will each watch and hunt in its own beat, and in the course of a short time, provided access is afforded to every apartment of the steading, the vermin will be seldom, and more seldom seen, until they disappear altogether. It is not uncommon to observe cats in the steading and stables; but they are generally neglected of food, on the erroneous idea, that if fed they will become lazy, and not hunt. So far from this being the case, that a regularly fed cat makes the best hunter, because it then hunts for sport; and not feeling hunger, it will watch in a single place for hours; and being in stout condition from its daily wholesome food, will feel itself strong enough to encounter any vermin, and will destroy any number in the course of a day; whereas a starved cat, which hunts for food, eats the first prey it catches, and then lies down to rest, in accordance with the habits of the race to which it naturally belongs. Nay, neglect the creature of food; let it depend for it entirely upon its own shifts, and hunger will prevent its watching—the great use of the animal in scaring away vermin, and oblige it to leave the steading to go to some dwelling for food—not unlikely to the hen-house for an egg, or to the hatching-house for a young chicken. The truth is, most people will not take the trouble to feed a cat daily and regularly in the steading; and the consequence is, that none will *remain* in it that can get a livelihood elsewhere.

Of all the modes I ever witnessed rats being killed, none equalled that of a Yorkshireman, of the name of John Featherston, by means of steel-traps. He had 21 small steel-traps, which he kept clean and bright. He soon traced the tracks of rats along the floor to a corner, or on the tops of walls, leading commonly by the corners of apartments to the partition wall, which they surmounted between it and the slates; the very place which I have recommended the filling up, to break off such communications. After he had discovered the different runs of the animals, he made a number of small firm bundles of straw, which he placed against the bottom of a wall where the run was on a floor, and upon its top where the run was to the roof. He used 7 traps at one place at a time, and a sufficient number of bundles of straw was used to conceal that number of traps at each place, employing the entire number of traps in 3 places, at a little distance from each other, and in different apartments. The traps were set, but not allowed to spring at first, and baited with oatmeal, scented with oil of rhodium, and placed in a row, with a little chaff over them, in the run behind the bundles of straw. The traps were baited for 2 days, the baits being replenished as soon as it was discovered, by inspection, that a bait disappeared. On the 3d day the traps were baited as before, but the restriction was removed from the spring, and then began the capture. In all the 3 days, people were prevented as much as possible from frequenting the apartments in which the traps were placed, and dogs were entirely excluded. Removing the check from the spring, from one set of traps after another, armed with a short stout stick, and furnished with a bag slung from his shoulders, Featherston put himself on the alert, and the moment he heard the click of a trap he ran to it, removed the bundle of straw, knocked the rat on the head if alive, threw it out of the trap, set it again, replaced the bundle again, put the rat into the bag, and was again on the watch from one place to another. In the course of the 3d day, from morning to the afternoon, he had collected 385 rats in the bag, and allowing all the traps to have done equal execution, each had caught more than 18 rats in the course of a single day. He bargained for 1d. a rat and his food, and in 3 days he earned his food and L.1 : 12 : 1—such was his expertness. It was not supposed that all the rats were cleared off by this capture; but they received such a thinning, as to be comparatively harmless for years after. Featherston's first business, on the day following the capture, was to clean each trap bright, before setting out on his journey; for he seemed to place greater reliance on the clean state of his traps than on any other circumstance—that the suspicion of the rats, I suppose, of the danger of the traps might thereby be allayed. The brown rat burrows in fields, and commits ravages on growing crops, whether of corn or turnips. I have seen many burrows of them in Ireland, and assisted at routing them with spade and terrier, but have never heard of their having taken to the fields in Scotland.

Weasels (*Mustela vulgaris*) frequent steadings, and do both good and harm. They do good in destroying rats and mice, by sucking their blood; and when a weasel takes up its abode in a corn-stack, where it can do no harm, not a mouse dares remain in it. They do harm in killing young poultry, by sucking their blood; a chick, a duckling, or a gosling, is in an unsafe place at the bottom of a dry stone-wall facing the south. They steal eggs too. One day I observed a weasel crossing a road at some distance from the steading, rolling an egg before it, with its fore-paws. On allowing it to proceed, I traced it to a lot of felled trees at the road-side, amongst which it had accumulated a store of 17 hens' eggs. Pole-cats (*Mustela putorius*) will visit steadings under night, and if a hole in the door, or a slit in the wall, by which poultry enter, be left open, they will creep in and commit sad havoc among the grown-up fowls, sucking their blood till they die. It is only by negligence however, that they can find access to the hen-house. If pole-cats are suspected of frequenting the hen-house, they may be caught by placing a steel trap immediately behind the door in the inside, in such a way, and with such an opening, as that the intruder cannot escape the trap. The trap should not be baited, but covered with a little chaff, and it will spring on the animal pushing itself along the ground, through the hole of the door into the hen-house. The common fox (*Vulpes vulgaris*), visits steadings at night, but can find no entrance to the poultry but by negligence; though in the quiet of a summer afternoon, when all the people are in the field at work, and just before poultry go to roost, he frequently pays a visit to the steading, and snatches up a goose or a turkey, and runs off with it to his earth. The abduction is so quietly done, that the fowl may not be missed until next morning, unless it happen to be a particular bird, such as the principal cock, oldest gander, turkey-cock, or peacock. *Such* a loss must be put up with; for, whatever precaution the farmer may sedulously use in spring for the protection of his new dropped lambs (2368.), he would lose character amongst sportsmen were he to lie in wait with a gun for a fox for the sake of a goose. Indeed, such watching would only be loss of time; for the fox is too cunning to return to the same place for a length of time to try his chance at a second capture. More than this, it will take care to keep off a rival from a distance, and it will be found when a bird is amissing in this manner, that it has been stolen by a fox from the nearest covert. Foxes frequently scent the hens or turkeys which have set themselves in the field, and carry them off their nests.

The famous question of the rook (*Corvus frugilegus*), whether it does most good or injury to the farmer, has yet to be settled—its advocates insisting that it does not touch a single grain in the field, and only devours insects; whilst the farmer maintains that it does him considerable injury. I think the question admits of an easy answer in so far as *facts*, in regard to the rook, are concerned. There is no doubt that rooks support their young entirely on animal food ; and as they hatch them at the time the oat-seed is sowing in spring, they naturally frequent those fields most which are most actively operated upon by the harrows, because that implement, by stirring the soil much more effectually than crows can with their bills and claws, brings to the surface chrysalides, larvæ, and live insects, which the crows eagerly devour, and however actively they may be employed in mischief at other times, they certainly are not so at that time. They therefore *follow* the plough and harrow in order to find the animal food for their young ; and it must be owned that they exhibit most exemplary industry in searching for that food. After the oat-seed is sown, should the grub (*Tipula oleracea*) shew any symptoms of activity amongst it, the rook is there, late and early, industriously turning over every turfy clod it finds, and scratching the furrow-brows to make more clods. So long, therefore, as the young rooks are directly dependent on their parents for support, the old ones search for animal food, and for animal food alone, upon which they entirely support their young ; and so long they cannot be said to injure the farmer in the slightest degree, but, on the contrary, may do him much good, indirectly, by removing from his fields what *may* injure his crops ; and I express myself contingently, for it does not follow that the insects in the soil must of necessity injure crops, for the soil is the habitation of their young at that season every year, and their appropriate food no doubt exists in it independently of the casualty of man sowing a crop in it ; besides it is notorious that they are not always injurious, since the crops are not injured by them every year. My belief is, that it is only when the number of insects increase, by favourable circumstances, beyond what their usual food in the soil, can support, that they subsist *mainly* upon the corn which is sown amongst them, and become injurious to man ; for that a portion of the seed sown is annually devoured by insects in the soil there cannot be a doubt, when a comparison is made between the number of seeds sown on a given area, and that of plants which actually grow upon it.

But whenever the rook gets quit of directly supporting its young, its taste undergoes a material change, as it then becomes omnivorous as to food. It still continues to eat insects of all kinds, and in all states. It will pull a worm gently out of the ground by one end, and before eating it, will strip it through its toes to squirt the earth out of its body; it will pick the meat clean off a bone, and I have seen them even pick horse-flesh as long as it was fresh off the skinned carcase of a dead horse ; it will carry off eggs from out of the court-yards, and leave the shells ; it will eat boiled potatoes and oatmeal porridge set down for the poultry; and were a bowl of barley broth set within its reach, it will soon empty it, and the thicker the barley is, the better for its taste; it will eat boiled barley and peas out of the horse mash-tub; it will dig up and carry off sets of potatoes after they begin to germinate, and when, of course, they are sweet to the taste; and it will pull up young plants of turnips just thinned in the drill. In regard to these latter subjects, it is said to take potato-sets and turnip-plants in search of insects; but as respects potato-sets, I am certain it also takes them for the sake of the sets themselves, for I have scared away crows on purpose to examine the sets they were eating, and found them quite sound and free of insects; but as to turnip-plants, I should suppose it is in search of insects that they are pulled up. It is, however, small consolation to the farmer to be assured that he would have lost his turnip-plants by insects, even if the rook had not pulled them up, for, at all events, a plant has a better chance to live as long as it is in the soil than after it is pulled up. Rooks will eat geans off the trees, and select the best; and here, at any rate, it cannot be said that they eat them in search of insects. They will alight upon laid corn of all kinds, and shake and scratch out more than they can eat; and they will also alight on stooks, and carry off heads of corn. In autumn, after the crop has been gathered, and the potatoes taken up, they will fly a great distance for the crow-berry (*Empetrum nigrum*), ripe at that time, and return in the evening to their usual haunts; and in winter, pressed by hunger, they will break the thatch of stacks to get at the corn. Mr Warburton is not very accurate in his conclusions on this subject, when he observes, that—"In winter the rook will attack the corn-stacks which have lost part of their thatch by a gale of wind. He is a slovenly farmer who does not repair the damaged roof immediately; and still we have farmers in Yorkshire of this description;"* for the fact is, that the wind breaks a stack always at the eaves, where it obtains the greatest purchase against it, and where there is no corn, whereas the rook always breaks in at the apex of the cone, where most corn is to be found; so it does not wait for the wind to render it assistance in this work. All I have said just now, in regard to the habits of the rook, I maintain are facts; for they are the results of my own observation for a number of years, specially directed with the purpose to ascertain what substances really constitute its food at different seasons; and my conviction is, that it is strictly insectivorous in spring, but omnivorous at all other seasons. As to the destruction of the rook,

* Journal of Agriculture for July 1844, p. 505.

I suppose that no class of farmers desire its extirpation ; all that they desire is, I presume, that its increase may be kept in check ; and the question, whether or not it is too numerous just now in this country, receives affirmation from the fact of its being almost in a state of starvation in winter, which could not be the case were its numbers not greater than its food is plentiful. That any one can believe that the rook does not eat grain, is beyond my comprehension; and why, let me ask, did Linnæus give it its trivial name of *frugilegus*, "picker or gatherer of corn?"

I have already mentioned a mode of scaring rooks from potato-fields in (2426.) ; and to scare them from corn-fields before harvest, I have tried a plan which I may designate as much better, at all events, than any scarecrow. It consisted of a circular board, 18 inches in diameter, and 1 inch thick. Upon its circumference were planted pieces of wood so as to form 24 embrasures at equal distances from each other ; and in each of the embrasures was planted a brass cannon made firm to the board by means of a hesp, removable at pleasure. The wood-work was surmounted by a conical tin top, exactly like the cover of a street-lamp, with an eaves projecting as much beyond the embrasures as to prevent the wind driving rain into them. The cannon were loaded with fine gunpowder, and wadded with woollen rag to prevent its ignition, and setting fire to the lying corn, when it fell amongst it. A soft thread of cotton dipped in a solution of saltpetre acted as a match, and was held upon the touch-hole of each cannon, by a pointer of copper-wire attached to the hesp. The match-thread was made as long as to suit the time fixed on between the discharge of each cannon ; and in order to lengthen the time at pleasure, the circular board behind the cannon was covered with tin, provided with separate upright partitions, which formed numerous alleys, and by laying the match along the alleys, it could not only be lengthened to any extent, but its burning end was prevented igniting any other part. The circular board and its cover were placed upon a tripod, which was fixed into the ground where the crows most frequented, and where it could best be seen. Suppose that the guns were loaded and the match lighted at 5 in the morning, and that by 8 at night it was time to cease firing, that is, in 15 hours, 37 minutes would require to elapse between a discharge of each of the 24 guns. For regularity, these discharges were much more to be depended on than the watching by any herd boy. Besides the discharges of the guns, a piece of woollen rag rubbed over with a paste of gunpowder, and dried, and set fire to, was placed in an iron cup suspended immediately within the lantern at the top of the cover, and the smoke arising from this still further intimidated the rooks; for they cannot abide the smell of gunpowder. The discharge of the guns will frighten pigeons also; but after making a few wheels in the air they will alight again on the place they were. Its position should be changed every day. It may be set up amongst potatoes, or in a plot of turnip-seed. I named this instrument the *rook-battery*.

I have already alluded to the destruction of the raven (*Corvus corax*) amongst young lambs, and of the only means of destroying it (2644.).

The wood-pigeon (*Columba palumbus*), has increased very much in numbers within the last 30 years. They do much injury on every farm they frequent, though their retired habits cause their depredations to be nearly overlooked. They eat the young plants of clover in spring, pick the leaves of young turnips in summer, frequent all sorts of corn-fields in autumn, and pick holes in turnips in winter. They are thus busy at mischief at all seasons. There is hardly any possibility of getting a shot at them but by watching in a wood at nightfall as they come to their ordinary haunts, which they, however, enter so quietly, and only few at a time, that their numbers would be but little thinned in this way. Of mischief to the farmer, from the rook and wood-pigeon, that occasioned by the latter is by far the more serious.*

OF LOOKING AT A FARM—ITS RENT—ITS LEASE—ITS STOCKING—THE CAPITAL REQUIRED FOR IT.

> "Doubt not that there
> A native pith of soil, a native warmth
> And kindliness resides; rely that there
> Grain, pulse, or root, whate'er the crop, will yield
> An early and exuberant increase."
>
> GRAHAM.

When you wish to look at a farm, and to have it valued, with a view to taking it on lease for the usual term of years, it is impossible for you to undertake such a task, until you have farmed practically, whether on your own account, or in trust for another, for a number of years. The only safe course you can pursue in such case, is to obtain the assistance of an experienced friend, who is well acquainted with the part of the country in which the farm you wish to take is situate; and should you have no such friend, one well versed in farming may answer the purpose;

* See Macgillivray's British Birds, vol. i. p. 259.

and, at all events, his experience will enable him better to make inquiries in regard to the capabilities of the farm than you yourself can.

And here let me mention, at the outset, that it is considered amongst farmers a dishonourable act to look at a farm, until you are, in the first place, assured that it is in the market. To do so, until you certainly know that the tenant in possession is to leave it, or, at any rate, until it is advertised in the public prints, or otherwise declared to be in the market, whether the possessing tenant wishes to take it again or not, is an unfeeling act, and regarded as equivalent to telling him that you wish to take the farm over his head. Such an act would be as unbecoming as to intrude yourself into a house in town, which you think would suit you, to look at its internal arrangement, before you are aware the possessing tenant is leaving it, by the usual announcement of the ticket.

Perhaps the best season to look at a farm which is well known to you, is just before harvest, when every species of crop is in the fullest luxuriance; because in this case you do not require so much to know the nature of the soil, as the condition it is in; and if it really be out of condition, the crops will exhibit unequivocal symptoms of its poverty. When the farm is unknown to you and your friends, the best season to look at it for the first time is in spring—in March—in dry weather—when the largest proportion of the soil is turned over by the plough, and when its natural state, in regard to dryness and wetness, cannot be concealed. This is the season which best shews whether or not the soil is in need of draining.

There are many points independent of the soil, connected with the situation of a farm, which require consideration, and which are too often neglected by those who value farms, and are the cause of much discontent to the tenant, after he has discovered them by dear-bought experience. The *temperature* of the locality has a considerable influence on the crops. Professor Playfair assumed that the lowest temperature at which corn will vegetate is 40° Fahrenheit, and that corn will not ripen below a temperature of 48° He proposed to date the vegetating season from 20th March to the 20th October, and considered 56° as the mean temperature of a good vegetating season.* It may, therefore, be assumed, that if the mean temperature of a place, between March and October, is below 56°, it is not likely to bear good crops. The *altitude* of a place affects its temperature materially. An altitude of 300 feet makes a difference of 1° of mean temperature; so that the effect of ele-

* Transactions of the Royal Society of Edinburgh, for 1800.

vation is the same as that of latitude. This is a point which is very liable to be overlooked in the interior of the country, where elevation is insensibly gained much beyond belief. The country may appear pleasant, and every thing indicate it to enjoy a good climate, but, on inquiry, it may be found to be 600 or 800 feet above the level of the sea; an elevation in which wheat will not ripen, and at which even barley will be a precarious crop, in many seasons. Indeed, at such an elevation it is not improbable that 1 or 2 crops may be lost in the course of a lease of 19 years (467.). In all such situations, the *daily* range of the temperature is great, descending low at night, after it has indicated a pretty high degree during the day; and it is well known that the temperature during the night has most effect on the crops; for warm nights have not only the effect of doubling the number of warm days, but the continued existence of heat saves plants from the injury they receive when checked in their growth in cold nights. In travelling at night in England in summer, there is no circumstance so striking to a Scotsman, as to find the air so warm, as compared to what it usually is in his own country. Hence, the harvests in England are always much earlier than in Scotland; and this superiority of climate will more than counterbalance the benefit derived to crops by superior skill. Besides altitude and temperature, other circumstances are worth attending to, though possessing entirely a local character; such as the distribution of *rain* in the vegetating season, that falling frequently being less favourable to vegetation, than in greater quantity at longer intervals; also, whether the locality is affected by *vapour*, thereby experiencing more cloudy than clear days. The lowness or highness of the *dew point* will have a material effect upon crops (464.). The study of the relation between local climate and the growth and productiveness of the different crops you thus see is deserving of much attention. What effect these have upon the money-rent of land it is not easy to determine; but that land so situate is of less value than that which is not so affected by local influences, cannot admit of doubt. There are still other circumstances which deserve attention, and, where they exist, affect the value of the best soil; such as long distance from markets, from coal, lime, manure, mills, and good roads. What effect, again, individually and collectively, these have upon the money-value of land, it is also difficult to determine.

All the circumstances I have yet mentioned affect the locality generally, but there are circumstances which affect the farm individually. The land may have such a steep inclination, as to require increased strength to work it. The nature of the soil, whether it is excessively strong or loose—its state, in respect to dryness or wetness; and its condition in respect to its artificial state of richness or poverty—the state of the fences, whether in themselves efficiently constructed, or only in dilapidation, occasioned by negligence—the state of the fields in respect of access to water in *summer;* I say in summer, for a farm may apparently be sufficiently watered in winter or spring, that may be very ill or not watered at all in summer, from the supply being only temporary, as from surface-water, instead of permanent, as from springs—the state of the buildings, whether capaciously constructed to contain all the stock requiring winter quarters, or commodious, even though capacious; and whether in an efficient state, or dilapidated, from negligence—the position of the farm, in respect to exposure to the south or north, and whether the slope of the fields is in one or more directions—and whether the farm is exposed to an open country all around, or sheltered on one or all sides, by natural or artificial objects. These are all circumstances which affect the money-value of land, but to what precise degree it is not easy to determine; and if all, or any of them, are overlooked in valuing a farm, the tenant may feel their effects when he is least able to bear them.

The most desirable appointments for a farm of mixed husbandry to possess are these:—Extent, from 200 to 800 acres. Soil, deep light clay-loam, capable of bearing turnips and wheat, incumbent on a *naturally* porous subsoil. A turnpike and a parish road crossing at its centre. Fields rectangular, and comprehending from 20 to 30 acres each. Fences of thorn-hedges. Ground gently sloping, or undulating to the south. Elevation not exceeding 200 feet above the sea. Water from springs or rivulets, accessible to every field. Steading and other buildings situate near the centre of the farm, capacious enough to contain all the stock in winter, and convenient enough for man and beast. Paddocks, 2 or 3, near the steading for calves, &c. Comfortable farm-house and neat garden, not far from the steading, and a public road. Shelter by high land or woods from the N., whence comes cold, and from the SW., whence blows the strongest and most shaking wind. Market-town of moderate size, not exceeding 8 miles in distance, and not more from coal, lime, and extraneous manure. Grinding-mill of wheat and oats in the vicinity.

It is barely possible for one farm to possess all these advantages, and it is not possible for every farm to possess them all; but, of course, the more that can be found conjoined the better. As, however, most farms will present a soil of mixed character, it is but right you should be made acquainted with the method of judging of soils. As to what the soil and subsoil are, and the constitution, texture, and colour of all kinds of soil, I have already said enough from (492.) to (496.).

In confining my observations to the case of a particular farm, I may notice, in the first place, that the soil will probably vary considerably over the farm. Part of it near the banks of rivulets, or on haughs, will be deep and mellow, and capable of yielding all the cultivated crops. Its position, depth—ascertained by making small pits with the spade—and uniformity of texture and appearance, stamp its superiority. Some of the soils may be thin hard clay, resting on retentive clay, with a small stone embedded in it. This is a very common soil, is not easily worked but in certain weather, when neither too dry nor too wet; though, when well wrought and manured, it yields abundantly and truly. This soil is very much improved by thorough-draining, deep-ploughing, and liming; and, after these operations, will grow turnips, especially Swedes, admirably. Part of the soil may be thin and loose, and still upon retentive subsoil. This is what is called a hungry soil, that is, requiring much and frequent manuring, and after all yielding but scanty crops, not exceeding ½ of the first mentioned, and ⅔ of the last named soil, nor is its straw so strong in the reed, or so sweet to the taste. This sort of soil will be very much benefited by thorough-draining, deep-ploughing and manuring, and much improved in texture by sheep eating turnips upon it. Part of the soil may be deep enough, in as far as the operation of the plough is concerned, but if of black peaty colour and deaf, it will be very soft, and apt to be carried forward on the breast of the plough. The straw grown by this soil is thick enough, but soft and brittle, and apt to lodge in wet weather; and its grain, though more bulky than on the last soil, is thick-skinned and light. Such a soil, though deep, often rests upon retentive clay, and is easily affected by wet, though it will withstand drought for a long time. This soil has at one time been a moor, and borne heath and furze, and yields crops readily at first, though it does not continue to improve. This, like the last mentioned hungry soil, is much improved by thorough-draining, and trench-ploughing the subsoil amongst it. A soil of quite an opposite character may be found, namely, a sharp gravel upon a gravelly porous subsoil, which is admirably adapted for turnips to be raised with bone-dust, and it forms the best lair for sheep of any in cultivation, and never fails to lay thick fat on the kidneys. Both straw and grain from this soil, though not exceeding in quantity, are of fine quality. Part of the farm may be in an uncultivated state, yielding old pasture, furze, broom, or heath, presenting an unequal surface, and having patches of swamp here and there.

In judging of thin clay, hungry, and deaf soils, which are all of doubtful character, and entirely yield according to the artificial condition they are put, it will be requisite to ascertain whether they have been drained, well-wrought, or limed, and otherwise well-farmed; for if they have been so treated, and yet indicate weakness, want of stamina, or a bleached-like appearance, they have no chance of being improved; but should they have been neglected altogether, or shew evident symptoms of want of draining and manuring, and have been, moreover, ill-farmed, some good may be expected from them if well-farmed. Still their intrinsic value at the time you inspect them should be the criterion of rent. The nature of doubtful soils may receive some elucidation by examining the natural plants which grow upon them, and these have been already indicated in (538–9–40–2–9–50–2.) The uncultivated part may prove better than some under the plough; for though naturally no better, yet if brought in in a better manner, it may yield profit; because it is comparatively fresh, and will be useful by permitting the more worn-out part of the farm to remain in grass, for some years longer than the ordinary rotation, whereby it will recover its tone, and yield better crops afterwards.

If a considerable variety of soil exists on the farm, it should be observed whether they occupy different fields, which is a favourable arrangement, or whether the same field is much diversified by them, which is a troublesome one. The amount of the good and bad soils should be summed up separately, and seen which so prevail. If the bad only occupy from $\frac{1}{4}$ to $\frac{1}{3}$ of the whole, then the farm may be regarded as a good one, and its character for excellence will depend on the quality of the good soil upon it; but should there be $\frac{1}{2}$ of it bad, the proportion is too great for the good soil to do justice to itself, and, at the same time, assist the bad; and if the proportion of bad increases beyond the $\frac{1}{2}$, the value of the good falls very rapidly. Whether on good soil or bad, it is better to succeed a slovenly farmer than a tolerably good one; for the latter has sufficient skill to make the land do its utmost, with the least means of amelioration, that is, to wear it out, and it will require several years to recover itself, and there is no state of land so difficult to recover its tone as when worn out; whereas the slovenly farmer may leave the land in a dirty state, and very unpleasant to the eye, when it may not be worn out. But the fortunate chance for you would be to succeed a farmer who has brought his farm into, and has kept it, and leaves it, in full bearing. When the nature of the soil has been examined in every field, and the correct position of the farm, as pointed out above, has been ascertained, the rent it is worth is the next point to be determined; and its amount depends solely on the quantity of produce the soil can yield.

Rent.—Experience soon teaches a farmer to estimate the produce of any sort of soil; and knowing this, and the current value of grain, the gross value of produce may soon be ascertained; but what proportion of this produce should be given to the landlord for rent, is a principle which has never yet been clearly demonstrated. As to the definitions of rent propounded by political economists, they have only served to render the subject more abstruse, and less understood. The favourite theory at present seems to be, that rent is the balance remaining after deducting from the gross produce the expenses of labour, and a fair percentage for the capital employed in raising the produce. If this definition were strictly applied in practice, the tenant would pay a different rent every year; for the gross produce not only varies with the season, but with the rotation followed. The landlord would thus receive a fluctuating rent, while the tenant would derive the same profit every year; which is as much as to say, that the tenant should run no risk, let the seasons be what they may, while the landlord shall run all risks. If the definition is capable of general application, the shopkeeper should pay the larger rent the greater business he transacts. This is clearly not the principle upon which rents are regulated; for the tenant is not exempted from paying rent when his crop fails; neither does the landlord receive the largest rent when the crop is greatest.

What proportion of the produce of a farm the landlord ought to receive as rent I do not pretend to decide. The old rule gave $\frac{1}{3}$ to the landlord, $\frac{1}{3}$ to the tenant, and $\frac{1}{3}$ for labour; but this has long been departed from, and any proportions I have seen enunciated since seem as arbitrary as can be. The late Dr Coventry constructed a formula on the produce of oats, that grain being most universally grown in Scotland; others have established rules on permanent pasture, because it incurs no cost in the farming; and yet it seems odd to adopt that as a standard for the payment of labour which itself requires no labour. Still it matters not how empirically rules for calculating rent may have been discovered, provided we are furnished with a formula that will determine its amount in a satisfactory manner to landlord and tenant in all circumstances. Dr Coventry's formula was to divide the produce per acre in oats by 2, and multiply the product by $\frac{1}{10}$ of the produce. For example—

Take the produce of oats at	6	qr. per imperial acre.
Which divided by 2, is . .	3	... ...
Multiply this by $\frac{1}{10}$ of 6, . .	.6	... ...
And the product is . . .	1.8	qrs. ... ...

that is, $1\frac{8}{10}$ qrs. per acre for rent. This, when converted into money, by taking the average price for oats for crop 1843, by the fiars' price, in the Edinburgh market, at 20s. per quarter, makes the rent L.1, 16s. per acre. Carrying out the same rule to various amounts of produce, a table of ratios could be formed, which might be referred to at any time the produce of land in oats was valued. In the following table the different amounts of produce from 10 qrs. to 2 qrs. per acre are given, and to reckon the proportion of produce in decimals for rent, we have only to divide the respective numbers of the produce by 20. Thus—

Supposed produce in Oats per quarter per imperial acre.		Proportion of produce for Rent in decimals.	Supposed produce in Oats per quarter per imperial acre.		Proportion of produce for Rent in decimals.
Qrs.	Bush.		Qrs.	Bush.	
10	0	.5	5	6	.2875
9	6	.4875	5	4	.275
9	4	.475	5	2	.2625
9	2	.4625	5	0	.25
9	0	.45	4	6	.2375
8	6	.4375	4	4	.225
8	4	.425	4	2	.2125
8	2	.4125	4	0	.2
8	0	.4	3	6	.1875
7	6	.3875	3	4	.175
7	4	.375	3	2	.1625
7	2	.3625	3	0	.15
7	0	.35	2	6	.1375
6	6	.3375	2	4	.125
6	4	.325	2	2	.1125
6	2	.3125	2	0	.1
6	0	.3			

It will be observed, from this table, that while land capable of producing 10 qrs. of oats per imperial acre pays $\frac{1}{2}$ of its produce in rent, land capable of producing only 2 qrs. per acre pays only $\frac{1}{10}$ of its produce in rent. The principle upon which this great difference in the proportions is founded, is, that the cost of labour in managing land which only yields 2 qrs. per acre is as great, and perhaps fully greater than that of labouring and managing land which yields 10 qrs., whilst the returns from the respective kinds of land is as 5 : 1. Now, it seems quite fair, that if, with poor land, the tenant has due allowance afforded him, on account of its inferiority, the same consideration should be awarded to the landlord, on account of the superiority of the land he has let. Were this principle carried to an extent much beyond 10 qrs., too great a proportion of produce would seem to accrue to the landlord for rent; but the principle will, perhaps, hold good, to the extent of productiveness fully to what we are accustomed to meet with in this country.

There is one advantage which this rule is said to possess, that, should the capability of the land to produce what is estimated be correct, the mode by which the quantity to be taken for rent is apportioned will be found never to charge the land more than it can fairly be conceived to bear. Nevertheless, I conceive that the quantity apportioned for rent is too great; for example, land which will only bear 8 qrs. of oats, would be too high rented at L.3, 4s. per acre, and it is understood that Dr Coventry's valuations of rent were all too high; but should 8 qrs. of oats be considered to represent the capability of the land to bear other products of the farm in that proportion, then the rent, L.3, 4s., may not be so high as at first sight may seem. Allowing the force of this objection, if the *principle* upon which the above table is founded is correct—and I am inclined to think so—the proportion of the produce appropriated for rent may be fixed at any lower figure considered nearer the truth.

It is not to be supposed, however, that your friends will estimate the value of land by any such rule; because experience will have taught them, as it does all practical men, to come to a conclusion at once, as to the capability of every soil, *in the condition they view it;* and it is the employment of this last criterion by which they judge that stamps an intrinsic value on their judgment. They estimate the gross amount of grain which the farm will produce, and the quantity of stock it will support, in the condition they see it in, at the ordinary prices of the times, by calculation, before they ascertain the total value of the produce, and determine the rent which it can afford to pay. Though they estimate the rent of the land in its present condition in this way, they judge, besides, whether the land is capable of producing more by better farming, and may determine the rent they advise to be offered accordingly. Upon this ground, however, many rents have been offered beyond the intrinsic value of the land; and should seasons be unpropitious, prices fall, or their young friend, the new tenant, not prove a better farmer than his indifferent predecessor, the rent will soon be found by him to be too high for his capital and skill.

Circumstances having proved that fixed money-rents, when thus pitched too high, when great and permanent reduction in the price of agricultural produce operated detrimentally against the interests of the farmer, the remedy proposed against the pressure of the evil was to make the rent dependent on produce. How the remedy was to be accomplished, so as to retain a permanency of character, was not so obvious; and unless it was founded upon a permanent principle, it would be little better for the farmer's interest than when he was at the mercy of every fluctuation of prices in the market. It is clear, however, that the produce of land in a grain-rent, as in a fixed money-rent, must form the basis of every calculation; but it is as difficult to *fix* the amount of produce for rent, as a sum of money. Still, in cases of great and unlooked-for fall in price, the tenant would be safer with a grain-rent than a money one of the same value. It was never proposed, however, to pay the grain in kind, but to pay a money-rent, calculated by the average prices of the preceding year, ascertained in Scotland by the fiars. In a market of declining prices—and such has been the state of the grain markets in this kingdom since the conclusion of the long war—a grain-rent is a popular measure with tenants; but a rent entirely composed of grain is not unattended with inconvenience even to the tenant; such as with a scanty crop, when the prices are unusually high, and when the tenant has the least quantity of grain to dispose of. To remedy this inconvenience, it was proposed to pay a proportion of the rent in money, as much as the farm could not fail to pay, under the most unlooked-for and unfavourable circumstances to the farmer, and the remainder in grain; and to render the price of the grain so fixed less fluctuating, it was proposed to take the average price of all the species of grain grown on a farm, in preference to any one of them. This was the principle which Dr Coventry adopted in fixing a grain-rent, though he had confined his views to oats when he treated of a money-rent; and this is the favourite opinion now; and a rent composed of a proportion of the 3 kinds of grain most commonly grown, namely, wheat, barley, and oats, appears nearer the truth than on any other basis.

The conversion of the money-rent of 36s. per acre into grain-rent is by simple rule of three. The current prices of grain (August 1844) are 51s. per qr. for wheat, 34s. for barley, and 20s. for oats, in all 105s., and as 105 : 8, the bushels in a quarter, : : 36 : 2.743 bushels of each kind of grain, that is, 2.743 bushels of wheat, 2.743 of barley, and 2.743 of oats, in all very near 11 bushels, at the above prices of the respective grains, give 35s. 11½d. per acre as the rent, commuted into money. This is almost the same as the money-value; and shews that the prices of wheat and barley are together about the same proportion as that of oats, upon which the money-rent was alone based.

Still such an arrangement is not strictly equitable between landlord and tenant; for in abundant years, when prices are sure to fall low, the landlord would not receive his due proportion of the produce, and, on the other hand, when the crop is scanty, and prices are sure to rule high, the tenant would pay more than his proportion. To suit both parties, a minimum of rate is fixed for the sake of the landlord, that is, the value of the triple quarter, which is supposed to be, at a medium, at 105s., should never fall below, say 85s., which is the lowest rent he can receive; and for the sake of the tenant, a maximum rate is fixed, at, say 125s., which should never be exceeded, and which is the highest rent he can pay. All these expedients shew that *the* principle of a grain-rent, which should accommodate itself to existing circumstances, has not yet been discovered, and this is illustrated even in the case before us; for, a few weeks ago, wheat was 55s. and barley 30s. per quarter, collectively making, with 20s. for oats, the same value for the triple quarter as at present, namely, 105s.; but, notwithstanding this similarity in the value of the money-value at the two periods, nobody will maintain that the value of the grain was the same to the farmer, when formerly he could have obtained 55s. for his wheat, and now only 51s.* I have heard of cases where the triple quarter was fixed as the maximum rent, and no limit was placed on the minimum, an arrangement which in some years gave the landlord only 33 per cent. of the medium rent; but this arrangement was put an end to at the first opportunity of an expiring lease, and another one substituted of a more complicated character, which, in its working, has proved of no benefit to the tenant beyond that afforded by a fixed money-rent. In fact, it is quite possible to manage the terms of a grain-rent so as to render it nugatory, as respects the interest of the tenant. You thus see that the subject of grain-rents is not so easily settled between landlord and tenant as seems at first sight.

The rate of interest of money in the country should enter as an element in the calculation of rent. If the farmer is entitled to a rate of profit on the capital he invests in cultivating a farm, it is but fair the rate of profit should vary with the rate of interest. If he expects 15 per cent. on his capital when the rate of interest is 5 per cent., he should be content with 12 per cent. when the rate falls to 4 per cent.

Offering the Rent.—Having determined on the rent to be offered for the farm, by the best advice of your friends, the offer should be transmitted before the time expires specified for its reception. The common practice is for the landlord to receive offers of rent from rival candidates for the farm, and, on consideration, to choose one of the number for the tenant of his farm. If the candidates are of equal respectability in character, and possessed of sufficient capital, the one who offers the highest rent is usually chosen, and the announcement of the choice is made, on a day previously fixed by the landlord, by public advertisement or otherwise. A missive, containing the principal heads of a lease, with the rent offered engrossed in it, is signed by both landlord and tenant,

* See Quarterly Journal of Agriculture, vol. viii. p. 543-9.

and forms the basis of the agreement until a regular lease is drawn out by the landlord's man of business. This missive is as binding in law as a regular lease. The terms of the lease are learned by the offerers, on application to the landlord before the offer is given in, and should any of them seem objectionable to the offerer, he expresses his sentiments accordingly in the offer of rent, and wishes an alteration in it to that effect. On large properties, the terms of lease are usually printed, and a copy given to every person who chooses to apply for it when looking at the farm; so that no offerer may be ignorant of what the terms are before he tenders his offer.

I cannot help thinking that the landlord acts erroneously in taking in offers of *rent* from farmers desirous of taking farms. It looks very like that he is ignorant of the value of his own property, and required farmers to acquaint him of it, or is afraid to ask a rent, in case he should not ask enough. The latter is, perhaps, not the true motive for adopting so anomalous a course, as to allow his property to be valued by strangers, the probability being that he feels himself incompetent to express an opinion on the subject, not understanding country affairs; but it is evident that, if the plan is adopted with a view to ascertaining the true value of the land, the information could be obtained from some friends in the neighbourhood, who ought to know the intrinsic value of the farm much better than strangers, who, perhaps, had never visited that part of the country until they came to view the farm. In every other profession, the disposer of the goods puts a value upon them, and when he finds he cannot procure a purchaser at the price he has fixed, he lowers it; but he never demeans himself by asking the customer what he will offer for his goods—a custom only country chapmen are in the habit of practising. By much the more dignified course for a landlord to pursue, is to put a rent on the farm, and fix the conditions of lease, and invite farmers to become candidates, on the ground of selecting the *best* tenant; that is, select him from among the candidates who is known to be, or testified to be, the best farmer, and who produces evidence of possessing as much capital at the time as will not only stock but carry on the farm during the currency of the lease. Were the worthy man and good farmer thus chosen instead of the rent, I am satisfied the connection between the landlord and tenant would be more cordial and personal, and therefore more lasting, than by the present auctioneering system, which induces offerers to the belief that the farm will be given to the highest bidder. Indeed, if farms were really put up to *public* auction, and literally let to the highest bidder in public, the plan would be less objectionable, in my opinion, than the present practice of private offering, in which unfair advantage is not unfrequently taken of the knowledge obtained from the different offers to screw up that of a particular candidate whom the landlord desires for a tenant, beyond what the candidate himself considers the farm is worth. In taking farms, offerers will be found so unprincipled as to agree to any rent or condition of lease, in order to get hold of them ; and then, from the vantage-ground which possession gives them, become perfect torments to the landlord during the whole lease. Were retributive justice always experienced in this world, one should not regret seeing every proprietor who lets his farms on the system here condemned, suffer aggressions from unprincipled tenants. At all events, sympathy would be misapplied in such a case.

Lease.—The conditions of a lease may be soon adjusted, if both parties are desirous of meeting on fair terms ; but after the rent has been accepted, on the understanding that the conditions exhibited are unobjectionable, it not unfrequently happens that the offerer endeavours to negociate for other conditions, which will serve, in his estimation, to mitigate the rent he has offered, and which he feels conscious the farm cannot pay. Allowances for draining are stipulated for, and the percentage fixed on the money to be advanced by the landlord. This is a proper negociation, where the farm requires thorough-draining. The steading may require more extensive repairs than what the outgoing tenant is bound to uphold ; and even additions and alterations are suggested, such as the removal of the horse-course, and the erection of a boiler and chimney-stalk of a steam-engine for the thrashing-mill, or the construction of a dam for a water-wheel. Either of those substitutions is an improvement, and ought to be executed by the landlord. A new set of feeding hammels may be required, or the courts fitted up with turnip-troughs for young cattle, or rain-spouts put round the eaves. These, being permanent improvements, ought to be executed by the landlord. The outgoing tenant is bound to leave the fences in tenantable repair ; but, nevertheless, a new fence may have to be run across a field to make it smaller, or along the side of a wood, or public road. A farm-road may be required for access to certain fields, which cannot be reached at present but by trespassing through other fields. An embankment may be required along the side of a rivulet which occasionally overflows its banks, and damages the crops of a haugh, or the lower parts of several fields. Proper watering-pools may be required in some of the fields. These, also, are all improvements of a permanent character, and ought to be undertaken by the landlord. But in expressing his willingness to execute these improvements, the landlord may demand interest for the outlay occasioned by such works ; but instead of this, he should rather state that he intends to make them as soon as the new tenant enters into occupation. In such cases, I do not think the tenant should be called on to pay interest; for the matters demanded are only accommodations on the farm, without which it cannot be said to be in a complete state for a tenant to occupy—constituting furnishings, without which the farm is not worth the rent which the intrinsic value of the land should fetch. The case is quite different with regard to draining, because the tenant will derive *pecuniary* advantage the moment the draining is executed, and will be certain to reap them during the whole lease. All these suggestions and improvements are proper; and had every candidate, when he gave in his offer, understood they were to be adopted, they would all have been placed on the same footing; but if they only originated with a party whose offer had been accepted, it is scarcely fair towards the other offerers to make so large concessions, as otherwise they might have offered more largely, but had no reason to believe that propositions, which involved the outlay of so much money, would have favourably been received by the landlord.

It is customary, in leases, for the landlord to reserve to himself the right of opening quarries or mines, of making roads, of entering into woods through fields, and of doing anything on the farm that would benefit the estate, with the proviso of giving damage to the tenant where his crop was injured. Reservation is also made for hunting and shooting game for the landlord and his friends. The question of game has caused many heart-burnings between landlord and tenant; much of which, I am persuaded, has originated in the severe restrictions imposed on the tenants. I am sure that such tenants as are desirous to witness the sports of the field, are gratified in seeing their landlord and his friends partaking of them in a manly way. But they are displeased at seeing their crops injured at all seasons by an inordinate quantity of game, protected for no apparent cause than to gratify the unsportsmanlike desire of partaking in a *battue* once or twice in a season. Let the landlord enjoy sport as he pleases; but since he delights in killing large quantities of game in a short time with little trouble to himself and friends, it is but fair to compensate his tenants, who maintain the game for him in such numbers with their crops; and but for those crops could not have sport in the way he chooses. I have long been of opinion that tenants would make the best protectors of game upon an estate, each on his own farm; and for this service they ought to be allowed to take a shot or course a hare. The indulgence is not excessive, for they cannot enjoy it without taking out a license, and keeping a dog and paying duty for it—taxes which many farmers would not pay for the sake of becoming sportsmen; but were this privilege generally granted, not as a right, but as the basis of a mutual good understanding—the landlord giving the right to hunt and shoot to the tenant, while the tenant protected the game for the landlord—the game might devour as much corn and turnips as they please without a complaint being uttered by the tenants, whether they were sportsmen or no. If they grudged to pay for a game-certificate, and the duty on sporting-dogs, it would be their own fault.

The periods at which the rent is paid is specifically mentioned in the lease. The most favourable terms for the tenant are Candlemas and Lammas. By Candlemas, 2d February, the farmer has had time to dispose of a great part of his corn; and by Lammas he has disposed of his fat cattle, sheep, and wool. Both these terms being intermediate between Whitsunday and Martinmas, when the half-year's wages of the farm-servants, field-workers, and labourers, become due, a large accumulation of demands at one time is avoided.

The principal clauses of a lease are, perhaps, those which relate to the cropping and management of the farm. It is customary to bind the tenant to follow a particular rotation in cropping his farm. I think severe restrictions are inimical to the tenant's interests, and in many instances prove detrimental to the farm itself. For example: should a crop fail, such as of clover or turnips, in the early part of the season, the tenant is not at liberty to plough those fields, and try another species of crop; and as the ground that should have been covered by a crop becomes a receptacle for weeds, the interest of the farm suffers by such a restriction. Should a better rotation be recommended to the farmer, he cannot avail himself of it. No doubt, restrictions were necessary at a time when miscropping was the practice of the day, and they are still necessary, when a better notion of farming than to miscrop prevails, because farmers do exist who will take advantage of the soil; but there is a mode of imposing restrictions which precludes advantage being taken of the land, while, at the same time, it does not interfere with the free-agency of the farmer. All restrictions in their nature imply, that the farmer cannot manage the land, and requires a disciplinarian in the lease to teach him. The restrictions I allude to are acknowledged by all parties, as being necessary to keep the land in heart, and they are these:—that no 2 corn-crops shall follow one another; and that 1 green crop shall intervene between 2 corn-crops. With such restrictions, it is quite impossible to injure land, provided the fallow-crops are adequately manured; and it is as easy to neglect manuring fallow-crops under the severest restrictions, as under the most lax; but there is this check against a tenant in his neglect of green-crops, that they will not grow without manure. To avoid the temptation of saving money in the purchase of manure, when there is not sufficiency upon the farm, there is always a clause introduced into leases—and it is a proper one—that the land shall be well manured, and worked according to the rules of good husbandry. It is true, that, in such a clause as mentioned above, the tenant is not bound to follow any system of rotation; but no evil consequence will arise from this, because it is decidedly the tenant's own interest to follow *a* rotation; and as he must have a green crop to intervene between the corn ones, he cannot possibly have more, nor can he follow a severer rotation than the ½ of his land in corn at a time; and he will soon find that he cannot sustain the fertility of land with that proportion of corn, without purchasing a considerable quantity of extraneous manure, which he will not be inclined to do at a distance from town, and during the early part of his lease, so that he will be driven to the adoption of a favourable rotation, namely, a 5-course shift. Towards the latter end of the lease, it must be owned that the farmer's interest becomes different from that of the farm; and it is then probable he may take advantage of the land, especially if he feels assured he has to quit the farm. To check such an inclination, it becomes necessary to impose severe restrictions for the last few years of the lease; and even then, the landlord need not impose any particular rotation, because it is enough to say, there shall *not be less* than ⅔ of the farm in grass laid down in the ordinary manner, namely, after a fallow crop; and it is, of course, unnecessary to specify there shall not be more grass, for the greater extent of it cannot injure the land; and if the tenant chooses to leave ½ or more in grass, so much the better for the landlord.

I never could understand why leases are made to run for the particular term of 19 years, as that is a period which corresponds with none of the members of any rotation in existence. I think it desirable that a farm be left in the same arrangement of fields it was entered to; and for this purpose it would be better to make the lease endure such a number of years as would be a multiple of the years of the rotation which best suits the soil, than for a fixed period of 19 years for all sorts of soils and rotations. For example: A farm in the neighbourhood of a town where the 4-course shift is practised, should have a lease of 16 or 20 years; a farm of the 5-course shift, one of 20 or 25 years; a farm of 6-course shift, a lease of 18 or 24 years; and one of 8-course shift, a lease of 16 or 24 years. The advantage of this system would be, that as the same fields would be in the same crop at the end of the lease, as when it began, an easier comparison could be made whether or not the farm were in better or worse condition at the end of one lease than at that of a former; and this would lead the landlord to determine whether the land would be the better for a different course of cropping, or only required better management. In the latter case, a better farmer would be an acquisition to the estate, whilst in the former a change of rotation would effect improvement with the same tenant.

The conditions of lease are generally the same for all farms on an estate, except when a particular farm may contain Carse clay, or be a pastoral one; but if rotations are beneficial to soils, it is clear that the conditions should be made conformable to the circumstances of the individual farm. I am convinced that inattention to this matter causes many a farm to be ill-farmed, and many a farmer disappointment, or loss of capital. A weak soil cannot endure the cropping of one with a stamina, nor can a deaf soil support a stock like a sharp one. Each of these soils requires different treatment, and yet all are placed under the same category of a restrictive lease. This plan of equalizing is a simple mode of classifying the farms of an estate; but its adoption does not display much judgment, and certainly evinces no discrimination on the part of its owner. No stronger argument can be adduced for the propriety of allowing a good farmer to exercise his skill in good management, according to the circumstances of the farm, than a case of this kind, and where such liberty is granted, the farm improves, and the tenant prospers; but such a liberty cannot be granted to one tenant without extending it to all, where the conditions of lease are all alike; and thus the skill of the superior tenant is purposely cramped, for the sake of maintaining a check upon an unskilful one. Hence the strong necessity for discriminating the capabilities of farms, when they are to be let; and hence, also, the propriety of choosing the tenant, and not the rent, when a farm is sought for by different parties.

It must be understood that the remarks I have made on rent and the lease apply most justly to a farm of mixed husbandry. A Carse clay farm being entirely engaged in the raising of corn, a grain-rent seems almost indispensable for it, in order to protect both the landlord and tenant from the fluctuations which beset the corn-market. A pastoral farm has nothing to do with a grain-rent, its principle of rent being either a sum of money for the grazing, or so much a-head per annum, say 4s., for every sheep it can support. In the latter case, there being no rotation of cropping, except on the small portion of ground available to culture, the number of years of the lease may be left indefinite; but should the farm retain its stock, whether of cattle or sheep, until they are 2 or 3 years old, the practice might be converted into a principle for calculating the duration of the lease, and be made a multiple of those periods; but usually pastoral farms have not so long leases as arable, there being a mutual desire between landlords and tenants to adjust rents according to the prices of stock and wool every 7 to 14 years.

There are other clauses, in regard to cropping, besides those relating to rotations and general management, such as classing flax among the corn-crops, which is a very proper definition of its nature; and pease and beans ought to be placed in the same class, unless manured. Half-manuring on the oat-stubble in autumn, or in the drills in spring, will suffice for these legumes; but it is too much exertion for any land, even in the highest heart, to produce a crop of oats, then of beans, and then of wheat, without manure. Potatoes are sometimes prohibited being sold off a farm, unless dung be brought on in return, because they are a crop which leaves no refuse for manure; but being human food, profitable to the tenant, and raised necessarily on manured soil, I do not see the justice of this restriction. No one can raise a profitable crop of potatoes without a large application of manure; and this being the case, it seems hard to prevent its sale, especially when the culture of the crop is attended with much trouble, and its delivery, by reason of its weight, a restriction on its carriage to a distance. The hay crop is much more injurious to land than potatoes, and yet it is usually permitted to be sold. I desire to see the day when no hay shall be grown upon a farm, unless it were made of grass in a succulent state, and to no greater extent than to supply the wants of the farm, which are not great when cooked food is so easily made, and so much better for stock than any hay, unless, indeed, it is also cooked. In the neighbourhood of towns where manure can be obtained at will, the growth of hay is requisite to support the horses of the inhabitants.

There are clauses of a penal nature inserted in every lease which seem necessary for the safety of the landlord. In cases of miscropping, that is, of taking 2 corn-crops in succession, a penalty of L.5 or L.10 an acre, over and above the rent, is threatened. A large sum is required for this purpose; for if the fine were less than the advantage gained by miscropping, I am sorry to say, farmers can be found who would violate a written agreement. An obligation is laid upon the tenant to remove from his farm at the time specified in the lease without the use of the instruments of a legal dismissal; otherwise possession being considered 9 points out of 10 in favour of the possessor, it might be difficult to eject a tenant who is reluctant to leave his farm. The usual periods of the outgoing tenant leaving, and, of course, of the incoming tenant entering, a farm, in Scotland, is, at the separation of the crop from the ground, to the arable land, and to the grass-land and houses at Whitsunday. The sub-letting of a lease to another party is prevented, as well as the assignation of it to a trustee for behoof of creditors; and these are very proper clauses, otherwise the farm might be alienated from the proprietor for the whole course of a lease, for the benefit of parties with whom he has no concern.

There are 2 ways for a tenant disposing of the waygoing crop, and either way is specially provided for by clauses in the lease. The one is, when the straw and manure belong to the farm, and therefore cannot be disposed of by the outgoing tenant; in which case provision is made for sufficient accommodation in the steading for an adequate number of his people and horses to thrash the crop, and deliver it at market. In such a case the straw and dung are said to be held in *steelbow*. "By a declaration that they are steelbow, it is assumed that they are given by the landlord, and are to be returned to him, or, what is equivalent, delivered to the incoming tenant, to whom the landlord has conveyed his right to them. A sufficiency of manure, and of the materials for its formation, are thus permanently retained upon the farm."* This practice of steel-bow is a valuable boon to the incoming tenant, as it is, in effect, equivalent to his possessing as much more capital on his entry to the farm, as the straw and dung he receives are worth. The other mode of disposing of a waygoing crop, is to sell the crop, straw, and corn, by public roup, as it grows upon the ground, or "upon its foot," as it is commonly termed. The purchaser undertakes the expense of cutting down and carrying off the crop. For the convenience of purchasers who cannot afford to buy a large portion of a field, the fields are divided into lots, containing each 2 or more ridges, according to their length, the lot containing not more than from 1 to 5 acres. As oats and oat-straw are the most generally useful grain and fodder, as much of a waygoing crop is sown of oats as practicable. You will observe, from this statement, that the incoming tenant has no more chance of securing a part or whole of the crop the land he is about to occupy has raised, than any other person; and should the crop fetch higher prices than he is disposed to give, he must purchase straw elsewhere; and if he has a farm, he is obliged to purchase corn when he has no use for it, in order to secure the straw upon which it grows, and of which he is much in want. The practical effect of this system upon the incoming tenant is, that he must

* Hunter's Law of Landlord and Tenant, p. 264.—A comprehensive treatise, which should be in the possession of every landlord and tenant of a farm.

possess capital to purchase as much straw as to meet the demands and his stock in winter, and as much manure as to do justice to the land in summer; and its effect upon the farm is, that should the incoming tenant not have capital over and above what is required for the stocking of the farm, he cannot purchase a sufficient quantity of straw, or should he not be able to procure as much straw and dung as are requisite for the uses of the farm, it must suffer privation to that extent. It is true he can bring his own waygoing crop to the new farm; but unless the distance between the farms is near, it would be impracticable for himself to carry a crop, and build it in the stack-yard of the new farm, at a season when everybody is so busy as to be unable to render him any assistance. In some cases the outgoing tenant is obliged to offer $\frac{1}{2}$ of the crop in valuation to the incoming tenant, and should the latter refuse the offer, he is at liberty to sell it; and in other cases, a private agreement is made for the whole crop between the incoming and outgoing tenants, but this has no connection with the lease, of which only we are now speaking. In any case, the incoming tenant is placed, by this system, in a far worse position than by the system of steelbow, for which reason that system ought, in my opinion, to be the universal one of the country in reference to straw and dung.*

In some parts, it is customary for the tenant to pay a portion of the public burdens affecting the land. The statute-labour is a burden exclusively affecting the tenant, he having to bestow his share, along with co-tenants, of upholding the parish roads which pass through the property; and whether the labour is given in kind or by commutation in money, is the same to him, though it is most convenient for the tenant to pay the cash, and see it expended in the best way, by competent labourers. Road money and schoolmaster's salary are also sometimes paid by the tenant; and if paid by him without deduction from the rent, their amount, of course, affects the rent to that extent, when it is under consideration at the time he inspects the farm; but if only advanced by the tenant, then the receipts presented to the landlord on the rent-day are equivalent to as much rent. These are matters of accommodation between landlord and tenant, as well as many others of minor importance, which, for convenience, are inserted in the lease, such as the delivery of kain-fowls at certain seasons, the driving of coals to the mansion-house, the delivery of straw, corn, turnips, and hay in winter, for the use of the landlord's horses and cows, when he has no home-farm.†

* See Farmer's Lawyer, for customs of incoming and outgoing tenants, in different parts of Scotland, p. 334 to 347.

† It would take up too much room to give the form of a lease here. It is better to refer you

Arbitration.—There are almost always questions of minor magnitude, though, nevertheless, of importance, between the landlord and outgoing tenant, which are usually made over to the incoming tenant to see properly settled, as being the party most immediately affected by them. These consist of the state of the fences, of the gates, of the steading, and of the dwelling-houses, all which the outgoing tenant is bound to leave in tenantable repair. The most pleasant way of ascertaining the fact, whether they are left according to the terms of the lease, is by arbitration of friends mutually chosen by the outgoing and incoming tenants, with power to the arbiters to appoint an oversman, in case of difference of opinion in drawing up the decreet-arbitral. When the fences and buildings are found in a sufficient state of repair, the business is soon settled; but when otherwise, then the arbiters appoint work-people acquainted with the respective sorts of work, to inspect the state of the farm, to calculate the costs of repair, and to report their opinions either in writing, or *viva voce* before them. The decreet of the arbiters results in the outgoing tenant paying the expenses of repair to the incoming, who thus becomes obligated to leave them in a tenantable state for his successor. These are the usual subjects of arbitration; but, of course, any other subject in connection with the farm may be arbitrated by the same parties at the time.*

Stocking.—Having inspected the farm—made up your mind as to its rent—made offer of rent, and been accepted—agreed to the conditions of lease—submitted questions to arbitration between yourself and the outgoing tenant—agreed with him on several minor matters—it is now time for you to ascertain the sum you will immediately require to stock the farm. To give you a clear view of this, it will be necessary for me to enumerate the implements of husbandry you will need at successive periods, the live-stock and corn required at starting, according to the mixed husbandry and the rotation to be followed, which is a 5-course shift, on 500 acres.

To give variety to the subject, and which may actually occur, I will suppose that 50 acres of the farm are in old grass, that is, permanent pasture. A piece of old grass is reckoned by many farmers a valuable appendage to a farm; and I confess I would have great desire to possess so fine an object on my farm. There are misconceptions, however, about old grass, which should be cleared up. Some people are of opinion

to Hunter's Law of Landlord and Tenant, p. 835-9; and also to the Farmer's Lawyer, where a great deal of information on various clauses in leases, will be found in the Appendix, from p. 223 to 331.

* See Parker's Notes on the Law of Arbitration—a useful volume for a farmer's library.

that a piece of land laid down to pasture may remain so as long as they choose. This is a mistaken notion. Permanent grass, to be profitable, must be on land in high condition, and of the best quality. Such grass will feed an ox to upwards of 100 stones, and will give a full bite in autumn long after the sown grasses, though it is later of rising in spring, and less suitable for young stock of any kind than the sown grasses. Inferior land will only produce coarse permanent grass, which will not *feed* stock, and yield less profit than the land would under the plough. If your farm possesses a field of fine old pasture, I would advise you to let it remain so, even though you should have liberty to break it up; and when it requires manuring, a top-dressing of good clean earth—not full of the seeds of weeds—and quick-lime made into a compost for months, in the proportion of 5 carts of earth to 1 of lime, laid on late in autumn or winter, or early in spring, will raise a flush of grass like a renewal of old age, and eradicate every tendency to mossiness. Supposing that such a field exists on the farm, I shall make a calculation of the stocking accordingly.

As soon as a farm is taken, orders should be given for making the implements of husbandry, to be delivered at the periods mentioned below. To make the arrangements for the entry clear, I shall specify what should be done in the different months until the entry is completed.

December 1842.—The first expense incurred is the ploughing of the fallow-break, which, if done by hired labour, will cost 8s. per acre, but if done by the good-will of neighbours in a "ploughing-day," which is the custom of the country, and should be regarded by you as the earnest of a hearty welcome to a stranger amongst them, the cost will consist of ale and bread to servants, and entertainment to their masters, say 90 persons at 2s. 6d.	L.11	5	0
January 1843.—The first clearing of the courts of manure and preparing it for use, will cost, say	7	0	0
March 1843.—Attendance at hiring-markets, for the engagement of farm-servants for 1 year, from May 26. 1843, on the wages and conditions mentioned in (1977.) and (2348.), with cost of arles, , .	1	0	0
The men hired should be a steward, shepherd, cattle-man, hedger, 5 ploughmen, and a stout lad to work a horse or horses. When 2 ploughmen are engaged to live in one house, a father and a son, for example, the son receives wages in cash from L.20 to L.25, with the ordinary allowance of potatoes, say 1500 yards of drill.			
Select 11 work-horses, with short legs, thick middles, and good action, from 4 to 6 years old, at L.30 a-piece, ...	330	0	0
A brood-mare in foal, or with a foal at her foot, ...	25	0	0
A hackney, with good action, 5 or 6 years old, ...	25	0	0
Carried forward,	L.399	5	0

	L.	s.	d.
Brought forward,	L.399	5	0
6 Sets of harness, each to contain the following articles:— 2 Bridles, 2 collars, 2 cart-saddles and breeching, 3 leather back-bands, 2 pairs of long, and 1 pair of short-chains, 2 pairs of cart and 2 pairs of trace-chains, 2 cart and 1 trace belly-band, 2 iron back-bands, 1 leading-chain and belt, 1 pair of cart-ropes, long and short cart and plough reins, 2 nose-bags, at L.8,	48	0	0
1 riding saddle and bridle, and stall-collar,	5	0	0
1 Corn tun for work-stable,	3	10	0
1 Corn tun for riding-stable,	1	10	0
10 Whole-bodied single-horse carts, at L.10,	100	0	0
1 Tilt-cart, at	12	0	0
6 Corn-carts, made to mount on the single-horse cart axles, at L.3, 10s.	21	0	0
2 Single-horse cart-frames, at 30s.,	3	0	0
6 Stretchers for trace-horses, at 1s.,	0	6	0
6 Iron ploughs, at L.4, 4s.,	25	4	0
6 Slides for ploughs, at 2s. each,	0	12	0
2 Iron small ploughs, at 50s.,	4	0	0
1 Iron double mould-board plough,	3	10	0
2 Iron scufflers, at 50s.,	5	0	0
6 Pair of harrows, with master swing-tree, at 42s.,	12	12	0
1 Sledge for harrows, shod with iron,	1	0	0
6 Full sets of swing-trees, complete, at 12s.,	3	12	0
1 Extra set, do.	0	12	0
3 Long swing-trees, for drilling, at 2s. 6d.,	0	7	6
6 Feering poles, with iron points, at 1s. 3d.,	0	7	6
1 Pair of seed-harrows for grass-seeds, complete,	1	7	0
1 Finlayson's harrow or grubber,	7	0	0
2 Two-horse rollers, of metal, at L.12,	24	0	0
1 Presser roller, 2 wheeled,	6	10	0
1 Wheelbarrow,	1	0	0
1 Broadcast sowing-machine, with 3 wheels,	12	0	0
1 Turnip double-drill sowing-machine,	6	10	0
1 Bone-dust and turnip double drill sowing machine,	11	10	0
7 Graips for dung, at 3s.,	1	1	0
4 Spreading graips, at 2s.	0	8	0
7 Lime shovels, at 4s.,	1	8	0
2 Ditching shovels, at 4s. 6d.,	0	9	0
3 Ditching spades, at 4s. 6d.,	0	13	6
1 Hedge spade,	0	4	6
7 Stable forks, at 1s. 6d.,	0	10	6
2 Long do., at 2s. 3d.,	0	4	6
4 Half-long do., at 1s. 9d.,	0	7	0
4 Field and stack forks, at 1s. 9d.,	0	7	0
3 Dung-hawks, 1 with 2 prongs, 1 with 3 prongs, and 1 with 4 prongs, for cow-byre, at 4s.,	0	12	0
Carried forward,	L.726	10	0

	L.	s.	d.
Brought forward,	L.726	10	0
2 Mud hoes, at 2s. 3d.,	0	4	6
1 Grindstone,	0	10	0
3 Metal troughs for pigs, at 7s.	1	1	0
1 Hedge-knife, large,	0	7	6
1 Pruning-knife,	0	3	0
1 Breasting-knife,	0	5	0
1 Axe,	0	3	0
1 Saw,	0	6	0
1 Sledge-hammer,	0	6	0
2 Picks,	0	12	0
1 Mattock,	0	6	0
1 Iron foot-pick,	0	8	0
2 Small stone-hammers,	0	3	0
1 Iron lever and wedges,	0	15	0
2 Tar-kits,	0	5	0
2 Oil Tins or bottles,	0	5	0
1 Set of blacksmiths' tools, viz., bellows, anvils, vice, forehammer, and 2 small hammers, where a smithy is at a distance,	7	0	0
1 Joiners' bench,	0	12	0
7 Cows' bands, with chains,	0	7	0
1 Bull chain,	0	2	0
1 Imperial bushel and strike,	0	18	0
2 Forpets, or double-quart measure,	0	5	0
2 Sowing sheets,	0	2	0
2 Ruskies for carrying seed, at 5s.	0	10	0
20 Corn-sacks, at 2s.,	2	0	0
2 Stable-pails,	0	9	0
2 Horse sheets, 1 at 20s. and 1 at 10s.,	1	10	0
1 Set of Phlemes, Blood-stick, Clyster-pipe, and Drink-horn,	0	6	0
1 Garden-rake and Dutch hoe,	0	5	0
1 Pump and trough,	3	0	0
1 Bore-rod and spirit-level for draining,	1	0	0
	L.750	16	0

April 1843.—With 50 acres of old grass, the tillage land of 450 acres will be apportioned in this manner:—

90 acres of new grass.		
90 ...	2-year old grass.	
90 ...	oats.	
90 ...	60 acres	turnips and potatoes.
	30 ...	bare-fallow.
90 ...	60 ...	barley and spring wheat.
	30 ...	winter wheat.
450		

	L.	s.	d.
Carried forward,	L.750	16	0

Brought forward,		L.750 16 0
Grass-seeds 7 lb. of red clover, 5 lb. of white clover, 1½ bushels of perennial rye-grass per acre, which, over 90 acres, gives of—		
White clover, 4 cwt., at 80s., ...	L.16 0 0	
Red clover, 5 cwt. 70 lb., at 70s., ...	19 13 9	
Rye-grass, 11 qrs. 2 lb., at 25s., ...	14 1 3	
		49 15 0
Hay for horses until they go to grass at the end of May, 5 tons, at 70s., ...		17 10 0
Oats, 40 qrs. at 20s., ...		40 0 0
8 Bolls of seed-potatoes, at 10s. per boll, ...		4 0 0
Men's wages from March until May 26. 1843, 7 Men 8 weeks, at 10s. per week, ...		28 0 0
Field-workers from do. to do., ...		7 0 0
Blacksmith's work, from do. to do., ...		3 0 0
May 1843.—		
Share of the expenses of arbitration on fences and buildings, .		2 0 0
Cost of 4 stones of Swedish turnip-seed, at 8s., ...		1 12 0
... 1 ... of yellow, at 10s., ...		0 10 0
... 4½ ... of globe, at 7s., ...		1 11 6
The above quantities are calculated thus,—2½ lb. of Swedes and Hybrids, and 2 lb. of Globe per acre.		
50 Quarters of bone-dust (or other manure), at 21s., ...		52 10 0
Poultry,—geese at 2s. 6d., goslings at 1s., turkeys at 3s., young turkeys at 1s., ducks and hens at 1s., and young do. 6d., .		5 0 0
1 good 2-year old Colt or Filly, for the draught, ...		18 0 0
1 do. 1-year old do. do. do.		12 0 0
1 Well-bred Short-horn Bull, ...		25 0 0
7 Milch cows, do. at L.17 each, ...		119 0 0
20 Milch calves, part unweaned, at L.3, 10s. each, ...		70 0 0
20 1-year old steers and heifers, at L.7 each, ...		140 0 0
120 Leicester ewes and their lambs, say 160, at 52s. each, . .		312 0 0
160 Do. ewe and wether hoggs, at 36s. each, ...		288 0 0
2 Do. tups, ...		10 0 0
Servants' corn paid at May 26. 1843, in advance,—		
10 Houses this year, 1 servant in each house, 1 qr. 1 bush. barley each, = 11 qrs. 2 bush., at 34s.		19 2 3
10 Do. do., 3 bush. pease, = 3 qrs. 6 bush., at 36s. .		6 15 0
10 Do. do., 7½ qrs. oats = 75 qrs., at 20s., . . .		75 0 0
½ Year's poor-rate, ...		10 0 0
June 1843.—		
3 Scythes for mowing grass, at 10s. each, ...		1 10 0
8 Hay-rakes, at 1s. 6d. each, ...		0 12 0
1 Horse hay and stubble-rake, ...		3 15 0
1 Long ladder, 20 feet long, at 9d. per foot, ...		0 15 0
Carried forward,		L.2,065 13 9

Brought forward,	L.2,065	13	9
2 Half-long do., 12 feet long, at 9s.,	0	18	0
6 Short do., at 5s. each,	1	10	0
1 Sheep-crook,	0	3	6
1 Killing or bathing-stool,	0	3	6
2 Tubs and some bottles for holding the bath for sheep,	0	15	0
2 Pair of wool-shears,	0	6	0
1 Beam, scales, and weights, for wool,	2	10	0
1 Buisting-iron and tar-kettle for sheep,	0	3	0
2 Hangers and skewers for sheep when killed,	0	2	0
July 1843.—			
2 Potato-graips,	0	6	0
9 Baskets for potatoes, at 9d.,	0	6	9
10 Turnip-hoes for women, at 1s. 2d.,	0	11	8
10 Weed-hooks for weeding corn, at 6d.,	0	5	0
4 Rope-twisters, at 2s.,	0	8	0
September 1843.—			
6 Straw-racks for cattle in courts, at 7s.,	2	2	0
1 Turnip-rammer, 14s., and trocar, 3s. 6d., for cattle,	0	17	6
1 Turnip-cutter for sheep,	5	0	0
1 Do. for calves and cattle,	1	10	0
6 Turnip-pickers, chiselled, at 2s.,	0	12	0
6 Knives for topping and tailing turnips, at 1s. 6d.,	0	9	0
2 Hay-racks for sheep, at 30s.,	3	0	0
20 Sheep-troughs, at 6s.,	6	0	0
300 Net-stakes, at 1d. each,	1	5	0
20 Sheep-nets, at 7s. 6d. each,	7	10	0
1 Mallet for driving stakes	0	2	6
1 Hay-knife,	0	3	6
1 Chaff-cutter,	8	10	0
2 Stable-lanterns, at 3s. 6d.,	0	7	0
3 Other do., 1 for steward, 1 for shepherd, and 1 for byreman,	0	10	6
1 Horn for blowing at fodder time,	0	2	0
1 Drill-machine for sowing corn,	7	0	0
11 Qrs. of seed-wheat, at 51s.,	27	11	0
2 Baskets for pickling wheat, at 3s.,	0	6	0
May 1844.—			
2 Sows, 1 Boar, and 4 Shotts,	10	0	0
14 Calves for rearing, at 30s.,	21	0	0
Oats for 6 pair of horses 1 year, from April last year, at 34 qrs. per pair, = 204 qrs., riding-horse, do., 23 qrs., = 227 qrs., at 20s.,	227	0	0
Seed-oats, 90 acres, at 5 bush. per acre, = 56¼ qrs. at 20s.,	56	5	0
Seed-barley or wheat, 60 acres, at 3 bush. per acre, = 22½ qrs. at 34s. if barley,	38	5	0
Clover-seeds same as last year; hay-seeds got from your own hay,	35	12	9
Lime for one year,	20	0	0
Carried forward,	L.2,554	17	5

Brought forward,	L.2,544	17	5
Toll-gates, ditto,	5	0	0
1 Year's poor-rate,	20	0	0
1 Do. road-money,	5	0	0
1 Do. schoolmaster's salary,	1	0	0
1 Do. assessed taxes,	2	10	0
1 Do. insurance,	1	10	0
1 Do. for mole-catcher, at 7s. 6d. per 100 acres,	1	17	6
1 Do. blacksmith work for 6 pair of horses, keeping up all, except the mould-board of the plough, at 60s. per draught, adding for riding-horse and other jobs, 60s.,	21	0	0
Oil, grease, and tar, for one year,	2	10	0
Bath or other composition for sheep; 440 sheep at 1½ each,	2	15	0
Money-wages of 10 men-servants for the 1st year,	74	0	0
Do. of 8 women, L.6 each for the 1st year,	48	0	0
Do. of other field-workers in summer,	5	0	0
Corn for 8 men, to new corn, ½ year,	30	0	0
Oats for horses from May to harvest, 45 qrs., at 20s.,	45	0	0
To Harvest 1844.—			
Thrashing-machine, with hummeller and bruiser, 6-horse power water, complete,	140	0	0
If steam and high pressure, L.210.			
If horse, L.120.			
Dressing-machine or fanners,	7	0	0
1 Barn-steelyard and weights,	6	0	0
1 Hand-hummeller,	0	10	0
60 Corn-sacks, at 2s.,	6	0	0
1 Sack-barrow, with wheels,	1	10	0
2 Hand-barrows for lifting sacks of corn, at 14s.,	1	8	0
4 Barn weights for filling corn, at 1s. 6d.,	0	6	0
2 Oat wire-riddles, at 2s.,	0	4	0
2 Barley do. at 2s. 4d.,	0	4	8
2 Wheat do. at 3s. 3d.,	0	6	6
2 Sieves, do. at 2s. 6d.,	0	5	0
1 Slap-riddle, at 2s.,	0	2	0
1 Barn-stool,	0	3	0
1 Wooden hoe, for raking corn to riddles,	0	1	0
1 Large barn-sheet,	0	12	0
2 Chaff-sheets,	0	7	0
6 Barn brooms,	0	3	0
2 Corn shovels, at 3s.,	0	6	0
6 Sack needles, and clue of twine,	0	2	0
1 Furnace pot and grate,	1	5	0
1 Meal ark,	3	0	0
5 Bags of oat-meal at 36s. per bag,	9	0	0
28 Pairs of shearers' blankets, at 7s. 6d.,	10	10	0
Carried forward,	L.2,999	5	1

Brought forward.	L.2,999	5	1
6 Porridge kits for reapers, 3s.,	0	18	0
6 Milk tins, 2s.,	0	12	0
6 Small beer barrels, 3s.,	0	18	0
1 Gantress for large beer barrel,	0	3	0
Joiner and masons' work, for 1½ year,	10	0	0
Tenant's share of the expense of drawing out the lease,	10	0	0
Half-year's rent, paid Lammas 1844,	400	0	0
Reapers' wages upon 180 acres, 5s.,	45	0	0
Incidental expenses for 1½ year,	30	0	0
Draining,	50	0	0
	L.3,546	16	1

During this expenditure, the following sums have been received from the farm:—

2 Years clip of wool, 270 fleeces each year, = 540 at 6 lb. each = 3240 lb., at 1s. per lb.,	L.162	0	0			
150 Sheep sold at 40s. each,	300	0	0			
20 Sheep died or sturdied, sold for	10	0	0			
20 Fat cattle, at L.16 each,	320	0	0			
Pigs sold	10	0	0			
				802	0	0
Making the balance of outlay to amount to				L.2,744	16	1

The farm may be expected to yield the following returns, namely, from

30 Acres of wheat, at 33 bush. per ac.	= 124					
Deduct for seed,	11¼					
		112¾ qrs. at 51s.,		L.287	10	3
60 Acres Barley, at 42 bush. per ac.	= 315 qrs.					
Deduct for seed and servants,	45 ...					
		270	... 34s.,	459	0	0
90 Acres oats, at 48s.,	= 540 ...					
Deduct for seed, servants, and horses,	360 ...					
		180	... 20s.,	180	0	0
180		555		L.926	10	3

1 Year's cast of Sheep, wool included,	L.391	0	0			
1 ... Cattle, ...	320	0	0			
1 ... Pigs, ...	20	0	0			
1 ... Extra Stock, ...	10	0	0			
				741	0	0
				L.1,667	10	3

From which fall to be deducted the following expenses:—

Wages of 10 Men,	L.42	0	0			
... 10 Women,	70	0	0			
... Harvesters,	45	0	0			
Carried forward,	L.157	0	0	L.1,667	10	3

Brought forward,	L.157	0	0	L.1,667	10	3
Amount of Ale and Bread,	20	0	0			
... Blacksmith,	18	0	0			
... Carpenter,	12	0	0			
.. Clover seeds,	30	0	0			
... Turnip seeds,	3	13	6			
... Bone-dust and lime,	82	10	0			
Rates, taxes, and insurance,	30	0	0			
Milk calves,	20	0	0			
Petty expenses,	66	6	0			
	L.439	9	6			
Add 1 Year's rent,	800	0	0			
				1,239	9	6
Profit for maintenance,				L.428	0	9

Being 15½ per cent. per annum on the capital actually expended of L.2744 : 16 : 1, and 12 per cent. on the gross outlay of L.3546 : 16 : 1. If the prices of grain had been calculated at those of last years, at the same period, the above balance would have been reduced L.113, a sum equal to 4 per cent. on the money expended.

The sum actually passed through the hands of the new tenant in the first 1½ year after entry to the farm is L.7 : 1 : 10 per acre, and that actually laid out by him is L.5 : 9 : 9½ per acre; beside the sum required to furnish his house. The profit derived in a year for his benefit is L.428 : 0 : 9, on which he must maintain his family; and this is the entire sum he is entitled to receive every year as a return for the capital he has expended; but it is no uncommon tale to hear farmers lamenting it as a hardship, when they are unable to lay by from 10 to 15 per cent. on their capital, keeping out of view, or forgetting all the while, the maintenance they and their families have derived from the farm—they having, in fact, supported themselves ere any of the produce reached the market. The slightest consideration, however, will make it evident, that the extent of the demand a farmer can make on his farm is a per centage on the capital expended by him, equivalent to that received by every commercial or mercantile man, whatever that may be; and 15 per cent. is considered a fair return on ordinary risks. To this he is justly entitled; but in receiving it, he, as a farmer, always possesses the advantage of having his maintenance on the spot, without first having to undergo the trouble of converting his goods into money, and of then going into the market with cash in hand to purchase victuals before he can support himself and family. The agreeable immunity from such sort of trouble is worth something.

86. OF IMPROVING WASTE LAND.

> " To smoothe the rugged wilderness,
> To drain the stagnant fen, to raise the slope
> Depending road, and to make gay the face
> Of nature with th' embellishments of art."
>
> SOMERVILLE.

Having fairly taken possession of your farm, and having had time to look about you, it is more than probable that some part of the farm will be in a rude state, and not in harmonious keeping with the remainder; in which case, it should be your duty, as, no doubt, it will be your pleasure, to make its face gay, as the poet has it who has just been quoted, with the embellishments of art.

A few hints, in regard to the easiest way of making rough land available to the plough, may prove useful, as much labour may be expended on work of this kind without yielding the advantages reasonably to be expected from it. I have already described the advantages of trenching rough land with the spade instead of the plough in (1201.), and have recommended an effectual method of trenching with the spade (1202–3.), and have also stated the cost of trenching 14 inches deep, in the worst cases, to amount to L.6 : 13 : 3 the imperial acre (1204.). You will find it may be done cheaper in anyother way just to be mentioned.

When the ground consists of the site of an old plantation, it is impossible to bring it under the plough but by trenching, in the first instance, with the spade. The smallest root forms an insurmountable barrier to the plough, and the destruction to implements, injury to horses, and the time spent in removing obstructions, create more vexation than the outlay of money in trenching can inconvenience any farmer who has a sincere desire to improve land; for if he has no spare cash, he should not attempt the improvement of waste land.

In like manner, when it has been ascertained, by the sinking of pits (1203.), that the subsoil contains a great many stones, whether large or small, and though the ground may not have been the site of a plantation, or of even large shrubs, it should still be trenched, in preference to being ploughed. In ploughing *stony* ground, irrespective of other obstructions, the plough meets such opposition from large stones as to stop its progress instantaneously, and the shock not only endangers the safety of the implement, but injures the shoulders of the horses, and the animals, moreover, feel their entire frame so dreadfully shaken, as to render them quite timid when afterwards put to similar work; and even when only small stones occur, the plough cannot maintain its hold in the ground, but is easily and constantly thrown out, so that the ploughing is either imperfectly executed, or much time is lost in remedying failures.

Nay, independently of tree-roots and numerous stones, ground that has grown brushwood, such as hazel, alder, or birch, or even large broom and whin, is ploughed with much difficulty and little satisfaction. I have tried the ploughing of whinny ground intermixed with bushes of birch, and been glad to relinquish the task; it being truly painful to see the horses stopped every few minutes with sudden jerks, the ground being, after all, only irregularly laid over, and, in some places, merely scratched with the points of the coulter and sock. Employed in such work, the wind of 2 valuable horses of my own was completely destroyed. They did not work together, and blame, therefore, was not attributable to their drivers; but the keen temperament of the willing animals urged them to persevere in a work which was really beyond their strength to overcome; and as long as horses go on in work, any injury sustained by them is not observed until it produces effects beyond the reach of cure.

In all these cases, then, I would earnestly recommend, both on the score of humanity to horses, and of superiority of work, to trench the ground with the spade, instead of *attempting* to plough it. In the act of trenching, the roots of trees and brushwood, and every stone, to the depth of 14 or 16 inches, should be laid upon the surface of the trenched ground. Trenching may be executed in any season; but to allow time for after operations, it is best done in the long dry days of summer, and should be finished by early autumn. Excepting knolls of gravel, resting on a gravelly substratum, there is scarcely any waste land in this country but requires draining. In some localities, as in Kincardineshire, it is possible to trench up as many stones as will thorough-drain the ground; but the ordinary probability will be, that as many stones will be obtained as will assist the draining to a considerable extent. To ascertain the nature of the subsoil at the depth to which the drains should be sunk, namely, 30 inches, 2 or 3 drains, according to the extent of the ground, should be opened from the lowest to the highest side of the ground, and, on examination of the nature of the materials under the soil, it should be determined whether 30 inches will be a sufficient depth for the drains (909.); and, at the same time, it should be decided at what distances the drains should be placed (935–39.). On determining in this manner on the depth and numbers of drains required to render the land thoroughly dry, it will be necessary to ascertain whether as many stones have been trenched up as will supply the drains with sufficiency of materials. If so, then the flatest stones should be selected to form the soles, and a conduit built, or triangular couplings with flat stones placed upon the soles. The probability is, that the stones destined to fill the drains will require breaking with a hammer. Should the quantity of stones evidently appear inadequate to fill the drains, the bottom should be laid with drain-tiles and soles, and the stones placed carefully above them in equal proportion in each drain. The entire arrangement of the drains, in regard to the directions of the main and small drains, and their connection with each other, as also the filling in of the loose materials, should be conducted as directed in the article on draining, page 482 of volume first.

A method of trenching land has been recently introduced into Scotland, which deserves attention, both on account of its efficiency and cheapness. It consists of marking off a piece of the rough ground equal in breadth to a future ridge. A trench of about 3 feet is then marked off with a line across the ridge-breadth, and notched with the spade. The first spit of earth which removes the surface of the trench is carried off to the side of the field where the trenching will terminate. The first spit, when removed with a trenching-spade, will be from 10 to 12 inches in depth. The trench being thus cleared of its upper surface, its subsoil is stirred with *graips*, or forks, as they are called in England, having long and strong prongs, 14 or 15 inches in length. The graips are used in the same manner as the spade, but instead of the earth being lifted up whole, as with the spade, it is cut in pieces with the prongs, except in strong clay, and every stone in it brought to the surface. The earth not being lifted up and turned over, the labour required to stir the subsoil 15 inches deep is not greater than ordinary delving of the ground. In stirring the subsoil with the graip, the 2 men employed work side by side, and assist each other in breaking the ground when it rises in lumps, or in bringing the larger stones to the surface. After the trench has been gone over in this manner, the stones are laid along the side of the ground which is being trenched, and also of the last trenched ridge. Another trench is then lined off and notched, and the soil removed by the spade and placed with its surface undermost upon the subsoil which has just been stirred and freed of stones by the forks. The advantages of this mode of trenching is apparent, in its keeping, in the first place, the subsoil undermost, while, in the next place, the ground is entirely stirred to the depth of at least 24 inches, and wholly freed of stones; and this, in the last place, is a great advantage in new trenched ground, which is not easily passed over with loaded carts. The stones being laid in a row at every breadth of a ridge, are ready for the use of drains, and this trouble is included in the cost of trenching, and is a deduction from its cost; and in most strong soils as many stones will be obtained in this manner as will suffice to thorough-drain the land; but if not, sufficient will be found to form a protective covering to tiles and tile-soles. Men undertake to trench ground in this manner for L.2, 8s. the imperial acre, and this being the case, it is a much cheaper, and a far more efficient, mode of bringing in *rough* ground, than by any application of the plough. When the ground is very stony, a 2-pronged graip is employed, as meeting with less resistance. Strong clay is apt to rise in large lumps, but which, when raised, can easily be severed by a thrust or two of the graips through it. The hardest moor-band pan can be penetrated and broken with such graips. Their cost is 5s. each, and they will last for years with occasional repairs of the points of the prongs; if not made, however, of good materials, they will inevitably be fractured in work either in the handle or prongs.

Trenching is practised in Flanders, not only in rough ground, but in particular districts, in the ordinary course of rotation. "This remarkable practice," says Dr Radcliff, "is confined to the lighter soils, and is unused where the strong clay prevails. In the districts in which it is adopted, the depth of the operation varies with that of the soil; but till this shall have arrived at nearly 2 feet of mellow surface, a little is added to it each trenching, by bringing to the top a certain proportion of the under stratum, which, being exposed to the action of the atmosphere, and minutely mixed with a soil already fertilized, gradually augments the staple till the sought-for depth be acquired. In the Pays de Waes there seem to be little necessity for any farther deepening; but the repetition of the practice itself is as periodical as the recommencement of the rotation. It is performed with a spade, the iron of which is 15 inches, and the handle 2 feet in length. The labourer standing on the last formed trench, with his left hand at the bottom of the handle, and his right near the top, by the weight of his body, and without the assistance of his foot, sinks the spade about 18 inches, and, standing sideways, throws off the soil with a peculiar sleight and turn of the wrist, so as to lodge it in an oblique position in the trench, and against the preceding line of work, retiring as he casts it from the spade, and thereby effecting some little mixture of the 2 strata, though the upper surface is, at the same time, placed below the other.

"The object of this practice is not only to let a surface rest that has been 7 or 8 years employed in the production of various crops, but to bring another into action, which has not merely had the advantage of repose, but the enrichment of a considerable portion of manure, which, in a porous soil, cannot have failed to find its way to the lower stratum. To the universality of the habit for ages back, much of the fertility of the Pays de Waes is to be attributed. It is particularly observable, that in any district where trenching takes place, the quantum of manure is diminished, and the number of ploughings are less, so that, eventually, it is not so expensive a process as at first view it may appear: nevertheless, a fair proportion of manure is considered essential in the first season; for though the under stratum has not only had a long exemption from duty, as well as the annual acquisition of such parts of the manure as may have been drained through the upper soil, yet it possesses a coldness which requires a stimulant to bring it into action.

"The soil which has undergone this operation is easily worked, and the trenching seems to go forward expeditiously; indeed, in any of the light and deep soils, the labour is not severe. In the Pays de Waes it is performed by the piece, to the depth of 16 inches, for about L.1, 12s. the English acre. The cheapness of the execution is a great encouragement to the practice; but this turns upon the price of labour, which, in this district, is 15d. a-day, and chiefly upon the facility of a loose and pulverized soil. Some have sought to economize, by the use of 2 ploughs, the second working to a considerable depth; but the objection made to this by skilful farmers in the vicinity of St Nicolas was, that sufficient depth was not thereby attained, nor were the 2 strata by this operation sufficiently blended; for though, by the spade, they are made to change places, yet by the oblique manner in which the mould is placed in the trench, a certain degree of mixture of the upper and under soil is effected, which is considered of importance." *

Land so trenched and drained, and which operations may be continued during the winter and early spring, should bear, for its first crop, potatoes or turnips; and of these I would prefer turnips, because the drained land having that crop eaten upon it by sheep, will at once be put into a state of fertility. Oats seem to be the favourite crop, for the first one, with improvers of soil, ostensibly because it assists in rotting turf quicker than any other; but really because remuneration for the outlay is desiderated at once—a laudable motive only when exercised in proper time. It is out of the question to attempt winter wheat upon soil in that loosened state at so late a period of the year, even though a

* Radcliff's Report on the Agriculture of Flanders, pp. 166–9.

sufficient quantity of manure should be at hand; and the soil cannot be sufficiently pulverised by spring to ensure success in barley. In regard to oats, however, the favourite crop, they do not succeed well on trenched ground, their pabulum being situate *in* turf, which is buried in the trench. Let it be decided then, that turnips are the most advisable crop under the circumstances. Should the draining have been accomplished early in winter, in order to preserve the *surface* of the land dry, it should be feered and gathered up from the flat, fig. 133, (824.), with a very light furrow, the hint-end furrows neatly cleared out, and gaw-cuts formed in all the hollows, and across the lower head-ridge into the adjoining open ditch; in which state it will remain safe all winter until spring arrive, when it should receive a harrowing, and the drills formed and dunged, and the seed sown, as described for turnip husbandry, in section 61, p. 743 of volume second.

When waste land consists of grass free of wood, and however uneven the surface may be, the best plan is to plough it early in winter, with a neat light furrow, sufficiently narrow to make the furrow slices lie close against each other. In spring it should be sown with common oats, in whose stubble the land should be drained and levelled, where necessary, with the spade in winter, and wrought in the manner described above for turnips next season. In such a case the plough can operate effectually, and oats will rot the turf at once, and bring the surface into a proper state to be drained and levelled.

In the improvement of soil, it is desirable to have its surface as even as practicable, so that all sudden hollows and heights should be removed. No hollow should be filled up without a drain being made in it, in the first instance, to take away the water that will naturally tend towards it; and no height should be lowered without having the surface upon it replaced upon the lowered substratum.

In levelling high gathered-up crooked ridges, much precaution is requisite, because the soil which is exposed after the lowering of the ridges has taken place, will take a considerable time to be brought into a state of fertility; and this is specially the case with clay soils. There is a mode, however, of levelling such ridges which avoids this inconvenience, and is thus described by Mr James Carmichael, of Raploch farm, near Stirling; but I must first premise, that the land should be thorough-drained before an attempt is made of levelling high ridges. "In winter, ploughing the land intended to be levelled and straightened in the succeeding summer, the plough was entered in the side of the ridge, exactly in the middle between the crown and the open-furrow; and thus $\frac{1}{4}$ of the ridge was *gathered up* from the open-furrow on each ridge while the open-furrow was left as open and deep as possible. The remaining ½ of the ridge was then *cloven down*, the crown being laid completely bare, so that every ridge was divided into 2. The work thus remained high and dry during the winter. In May or June following, the plough was made to *gather up* the soil about 2 feet on each side of the crown of the ridge, upon which the subsoil was thus completely exposed for about 5 feet in breadth. This bared space on the crown was then ploughed up into drills, and removed, either by shovels or with a 2-horse levelling-box, into the deep open furrows. The plough was again employed to make drills in the crown, and the subsoil removed into the open-furrows; and the same operation was repeated as often till the open-furrows were raised apparently above the level of the crowns. This being done, part of the formerly accumulated soil on the sides of the ridges was *cloven down* by the plough upon the recently removed subsoil in the open-furrows, on the one hand, and upon the bared crowns on the other hand. The land was then cross-harrowed, and twice cross-ploughed, and harrowed alternately as deep as possible, and all inequalities removed before the ridges were re-formed straight.

" The whole process is exceedingly simple, and its advantages are perfectly obvious; for while every particle of the original surface is thus carefully preserved, the subsoil is so sparingly and equally incorporated with it, that not a doubt can remain of the beneficial result of the operation. Where the ridges have been very broad and high, it will be found necessary to remove part of the subsoil from the sides of the ridges also after the crown has been refilled, particularly when the subsoil is tilly; but this is easily performed by 2 or 3 women or boys going after the plough, and throwing the remains of till turned up into the furrows; or, should the ridges be too broad for this, the levelling-box will effect the purpose."

The labour attending this operation will depend on circumstances, such as the breadth and crookedness of the ridges, and the strength of the clay soil. In dry weather, the operation will be more expeditiously and better done than in wet, the soil being rendered light by the heat of the sun; and in changeable weather no more subsoil should be ploughed than can be removed during the course of the day.*

Besides larger inequalities of surface, there are minor ones which require emendation in the improvement of the soil, and the neglect of which renders the surface of arable land unpleasant to the sight. The inequalities I allude to, are slight hollows, low heights running

* Prize Essays of the Highland and Agricultural Society, vol. ix., p. 37-9.

across several ridges, one side of part of a ridge higher than the other, the head-ridge at the lower end of a field higher than the ends of the ridges, and such like anomalies. The only method of getting quit of these inconveniences is by the use of the *levelling-box*, figured and described below by Mr Slight. It is easily used, and its effects are satisfactory. The plough is first made to turn over the soil lightly on the height to be removed, the levelling-box, worked by a pair of horses, then follows, and takes up, from one end of the ploughed ground, as much soil as it can contain, the conductor holding by the stilts or handles. On arriving at the hollow place to be filled up, the box is made by the conductor to capsize, and get quit of its load of earth, and, when empty, levels the earth smooth with the surface, on passing over it. Of the 2 sorts of levelling-boxes, that figured below is, in my opinion, the most handy. It is easily filled by raising the handles a little, and giving the scoop a catch of the mould. It is as easily transported to its destined spot, by leaning a little weight upon the handles, which causes the box to travel on its keel. It is as easily upset, when arrived at its destined place, by raising the handles a little, when the ground catches the scoop and overturns the box, the handles striking against and resting upon the stretcher. The box may be returned to its working position on its bottom immediately, or it may move along upon its mouth, smoothing the surface till it reaches the ploughed soil, when, by a sudden jerk of the rope attached to one of the handles, it regains its working position. In all these movements the horses need not be stopped, but may proceed from one place to another without the least trouble, while the box is filling and emptied by the conductor. The other sort of box has movable handles, which are slipped off and on to gudgeons of iron, projecting from the side of the box. When the handles are let free of the gudgeons by the conductor, the box easily capsizes; but unless the horses are stopped, there is difficulty in shipping the handles quickly upon the gudgeons, and which, if not quickly done, when the box regains its working position, the edge of the scoop will again bite the ground, and capsize the box again; and will thus continue to be overturned until the handles are fastened on the gudgeons. The stopping of the horses for this purpose necessarily loses time.

[*The levelling-box or scoop.*—Fig. 600 is a pioneering implement used principally for the more expeditious removal of earth from one part of a field to another, where it is found requisite to reduce the height of that part for the purpose of raising the surface of an adjoining hollow, when the surface has been found inconveniently unequal for ploughing. The implement is one of considerable importance wherever such operations are required; for in point of economy it has been found, even in extensive engineering operations, to excel any other mode of removing earth within short limits. Its proportions and dimensions are very variable, according to the nature of the stuff to be removed,

Fig. 600.

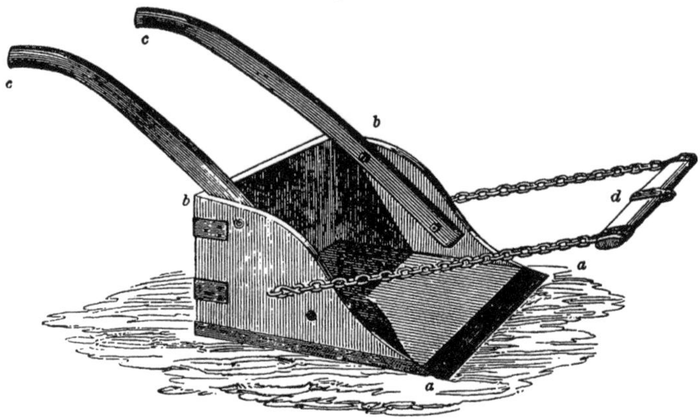

THE LEVELLING-BOX, OR SCOOP.

and the distance it may be carried, and also whether worked by one or two horses. The figure represents one of the latter kind, furnished with fixed handles and draught-chains. The dimensions of such a scoop will be 3½ feet wide at the front or cutting-edge *a a*, diminished to 3 feet 3 inches at the back *b b*, which may be 1½ feet in depth. The two sides *b a*, *b a*, having the same depth where they join the back, are curved off to nothing at the front. It is requisite for strength that it be made of hard wood; but the common willow, from its toughness and lightness, is perhaps better adapted than any other wood for this purpose, and the thickness should not be less than 1½ inch. The sole of the scoop is armed with a strong shoeing of iron covering a breadth of 4 to 5 inches of the fore-part above and below, and terminating in a sharp cutting edge. Two skeds or bolsters, 3 inches in length and 4 inches in breadth, are fixed on the lower side of the sole, thinned off forward to give facility of entrance in the soil to the cutting-edge of the scoop, and upon which it runs like a sledge when filled. All the corners are strongly bound with iron plates, and the skeds upon which it runs are covered with strong sheet or hoop iron. The handles *b c*, *b c*, of 5 feet in length, are bolted to the sides, and so fitted as to bring the extremities *c*, *c*, to a convenient distance for being held in the hands of the conductor. The draught-chains, with their stretcher *d*, are attached to an eye-bolt or a staple in either side of the scoop, and to the stretcher the common swing-tree yoke is appended either for one or two horses. In working this form of the levelling-scoop, when the conductor wishes to fill it, he allows the handles to rise so much as to allow the edge *a a* of the sole to bite the ground, and on being thus drawn forward by the horses, the box is filled to any desired extent, when, by the handles being pressed down, the front edge is relieved from the ground, and the machine travels on upon its skeds. When it has arrived at the place of discharging, the horses may be stopt or not, and on the conductor lifting up the handles, the contents from the box are discharged, and, returning it to its travelling position, the operation is repeated.

The levelling-scoop is frequently fitted up in a manner somewhat different from this, as regards the mode of working only, whereby it takes the character of a tilting machine. In this form the chief difference is in the mode of attaching the handles. A strong gudgeon is fixed in each side at the place of attachment of the chain ; and the terminations of the chains being an eye-bolt or link, it is passed upon the gudgeon. The handles, in place of being fixed to the sides, have an eye formed in their end and strengthened with iron, which passes also upon the gudgeon, and are held there by a washer and cutter. A second pair of gudgeons are strongly fixed, one upon each corner of the box at the back, in a position that will pass through the handles when they are at a proper height. The handles are here also pierced and defended with iron, so as to slide freely off and on upon these gudgeons, and their extremities brought as before to a convenient width. With these arrangements, when the conductor is guiding the scoop to the place where it is to be filled, he keeps the handles upon the two last described or hind gudgeons; and when filled and conveyed to the place of discharge, with the horses still moving on, he raises the handles till the edge of the scoop bites the ground, when he immediately spreads the handles till they leave the hind gudgeons (but retains them in his hands), and as the horses move forward the box is tilted over, turning upon the edge of the scoop as a fulcrum, and the contents are discharged. The box is now bottom upwards, but as it is furnished with two prongs rising a few inches above, and attached to the back of the box; these now take the ground, and in their turn become a fulcrum, upon which the box is righted by the still forward motion of the horses. When so righted, the conductor ships the handles upon the hind gudgeons and proceeds as before. It will be observed that the handles always remain upon the lower or principal gudgeons along with the draught-chains.—J. S.]

---

87. OF FARM BOOK-KEEPING.

"Let it be booked with the rest of this day's deeds."
II. KING HENRY IV.

It has long been alleged, that of all classes of men in business, none know the actual state of their affairs so imperfectly as farmers; because, as is also alleged, they neglect to adopt any system of book-keeping. It must be owned that these allegations against farmers, as a class, are too well-founded. It is too true, that in any given day they could not tell the exact state of their affairs; and it is equally true, that they do not practise any regular system of book-keeping.

It must be admitted, however, on the other hand, that many obstacles stand in the way of farmers adopting such a system of book-keeping as would present a correct value of their capital on any given day. It is easy for the manufacturer or mercantile man to estimate, at any time, the market value of every article in his wareroom for which he may receive or has paid cash. Not so with the farmer. There are many articles produced, and many operations performed on the farm, of the pecuniary value of which no just estimate can be made. He cannot estimate, for example, the value of every stone of straw given daily to his live-stock as fodder or as litter; he could not even weigh every stone so bestowed; and the difficulty of making a correct estimate is increased, when he is not allowed to sell straw. He cannot estimate the accumulated amount of labour, both manual and mechanical, bestowed on every acre of land, before it yields its produce, when he does not daily hire the labour of the people, nor purchase the food of the animals which perform it. He cannot ascertain the value of every cubic yard of manure produced on the farm, whether in composts or in the court-yards, when he is not allowed to dispose of it at market; nor can he make a comparative estimate of it when first taken to the dunghill, and when applied to the soil after being fermented. He cannot put a value on every ton of turnips and other green fodder consumed by stock, nor can he weigh every ton so used. He cannot put a just estimate on the value of pasture. I do not mean to say that the intrinsic weight of all these articles could not possibly be ascertained, for they could certainly all be weighed on a steelyard; but it is clear that their money value cannot be ascertained, because the farmer has no means of instituting a comparison between the value they cost him and their market value, which includes profit; for since it is by means of the produce of the farm the farmer is enabled to present a surplus produce at market, it is evident he should not estimate the produce consumed on the farm at the market value of the produce sold, the profit on which being realised in consequence of using the produce whose value is recommended to be ascertained. But besides all this, were he to attempt to ascertain the weight of every article used on the farm, in conjunction with its ordinary labour, without the employment of additional hands, he would soon find the most important operations of the field half neglected; and were he to employ additional hands, they must be paid for the special purpose, and might, perhaps, incur a greater expense in estimating the weight of articles, than the knowledge of the estimate, when attained, would be worth.

Still the frankest admission of the real difficulties which beset the farmer in estimating the actual value of the produce of his farm and of its expenses, is no sufficient excuse for his neglecting to arrange his accounts, so as to know to a certainty at any time whether he is gaining or losing on his capital. Not unfrequently the neglect is the result of sheer indolence; but in many cases I am convinced it arises solely from the parties not knowing properly how to set about keeping accounts. It was perhaps from a conviction of this latter circumstance, that has prompted several persons to publish systems of farm book-keeping; but those of them I have seen are much too complicated, and attempt to embrace many more objects than any farmer can attend to.* They, in fact, endeavour to estimate the money value of all the particulars I have enumerated above, and which is estimated as a non-essential by the farmer; and if this be a true estimate, which I reckon it to be, the attempt to estimate it in any system of book-keeping is an argument against the adoption of the system, even by those farmers who keep accurate accounts of all their money transactions, and who cannot, in truth, be charged of neglecting book-keeping.

To obviate all pretext, on your part, of urging the objection of being unacquainted with any system of book-keeping which would suit the farmer, I shall describe and illustrate one that is not theoretical, but has long been practised by farmers of acknowledged skill and success in their profession; and it has the advantage of being founded on the common mode of book-keeping taught in the schools, and followed in the counting-houses of mercantile people; so modified, however, as to suit the transactions and phraseology of the farm.

All the books required to give you a correct idea of the state of your affairs, are a *cash-book, farm-account book, corn-book, stock-book*, and *labour-account-book;* and a *ledger* is also required for open accounts, that is to say, for all transactions not concluded in ready-money. As the affairs of life are usually conducted, a ledger must be used for all credit transactions, and these are, perhaps, much more numerous than necessary, though the dealings of farmers are conducted more upon the ready-money principle than those of most people in business.

Into the *Cash-book* is entered every receipt and payment in cash connected with business. The following is an example of how it should be ruled into *Dr.* and *Cr.* columns, and of the form of the entries, which are purposely varied, to embrace almost every species of transaction connected with a farm, and more varied than they are likely to be in the time specified. The narrow column beside the money columns contains the page or folio of the ledger in which the respective accounts referred to are opened. The balance between the *Dr.* and *Cr.* column shews the amount of cash you have in hand, or which you owe. When an account is posted into the ledger, a small mark, such as a cross thus +, is made opposite it, to shew that it has been posted. This is a specimen of the *cash-book.*

* The works I more particularly allude to, are those on farm book-keeping, by the late Mr Trotter of Dreghorn, in Mid-Lothian, and by Colonel Munro of Poyntz, in Ross-shire.

CASH BOOK.

1843.			Folio of Ledger.	*Dr.* L.	s.	d.	*Cr.* L.	s.	d.
Oct.	2.	To cash on hand at this date, being balance brought forward from last year's account, .		27	13	4			
+	3.	By John Dickson, farm-steward, advanced him to meet outlays, ...	15				2	0	0
	4.	To farm, received from Robert Brown, Dalkeith, for oats sold to him in this day's market, . .		26	10	3			
+	10.	To British Linen Company, Edinburgh, withdrawn from them this day, ...	16	60	0	0			
	"	By farm, paid harvest expenses, per account, .					90	2	6
	11.	By house expenses, advanced Mrs Williams, housekeeper, to meet current outlay, . . .					10	0	0
+	13.	To British Linen Company, Edinburgh, withdrawn from them for purchase of cattle, &c.,	16	290	0	0			
	14.	By farm, paid Henry Johnstone, Musselburgh, for 6 tons rape-dust at L.5, 5s. per ton, . .					31	10	0
	"	By farm, paid for 22 short-horn bullocks, purchased this day at Jedburgh fair, ...					258	10	0
+	18.	To John Thomson, Leith, received from him for barley sold on 4th instant, ...	14	34	6	0			
	"	To farm, received from John Mackie, baker, for wheat sold to him in this day's market, . .		73	16	0			
+	19.	By British Linen Company Edinburgh, deposited with them, ...	16				100	0	0
	27.	To farm, received from Thomas Wightman, salesman, Morpeth, for sheep sold by him there,		114	15	0			
	28.	By farm, paid road-money for current year, per collector's receipt, ...					5	15	0
+	"	By British Linen Company, Edinburgh, deposited with them, ...	16				120	0	0
Nov.	1.	To farm, received from Thomas Robertson, corn-merchant, for grain sold to him this day, . .		71	8	0			
+	2.	By British Linen Company, Edinburgh, deposited with them this day, ...	16				70	0	0
+	8.	To Meadowbank Brewing Company, received from them for barley sold to their agent on the 25th ult., ...	17	72	0	0			
	11.	By farm, paid to agent of Scottish Union Insurance Company premium and duty on L.3000, insured with them on farm-stock, crop, &c., NOTE.—Of the above there is insured on the Farm house and steading, . . . L.900 0 0 Household furniture, ... 400 0 0 Thrashing-mill, steam-engine, farm-stock, and crop, ... 1,700 0 0 Together, L.3,000 0 0					6	8	9
	"	By farm, paid Public Burdens, viz.:— Poor-rate, 20 months' cess, . L.13 11 8 Schoolmaster's salary, tenant's proportion, ... 2 6 8					15	18	4
		Carried forward,		770	8	7	710	4	7

CASH-BOOK—*continued.* *Dr.* *Cr.*

Date		Particulars	Folio of Ledger.	*L.*	*s.*	*d.*	*L.*	*s.*	*d.*
1843.		Brought forward,		770	8	7	710	4	7
Nov.	22.	To British Linen Company, Edinburgh, withdrawn from them for payment of rent, servants, &c., ...	16	500	0	0			
+	,,	By farm, paid one-half year's rent of farm due at this date, ...					500	0	0
	,,	To farm, received from the landlord allowance for drains due at this time, ...		50	0	0			
	,,	By farm, paid labour-account in full for half year, from Whitsunday to Martinmas— Yearly servants—money wages, L.20 0 0 Field-workers, ... 50 0 0 Spadesmen, ... 34 10 0 L.104 10 0 Blacksmith's account, L.11 5 0 Carpenter's do. 10 14 6 21 19 6					126	9	6
				1320	8	7	1336	14	1

To save the columns of the cash-book being encumbered with a number of trifling entries, it is advisable to have a separate book, which may be carried in the pocket, to contain small payments and receipts. The payments consist of *market* and *travelling expenses*, including tolls and drink-money given to the men when they go from home with horses and carts. They consist also of *postages*, *stationery*, *discounts*, commonly called luck-pennies, *charity*, &c. The receipts are for *skins*, whether of cattle or of sheep that may have died; for *mutton* sold to work-people; *old iron* sold to the smith or founder, and such like articles. The ruling for this sort of book is simply date and money columns. The book may be named the *Petty cash-book*. Specimens of payments and receipts may prove useful; and first of *payments*.

TRAVELLING EXPENSES, POSTAGES, &c.

1844.			*L.*	*s.*	*d.*
March	10.	Coach fare to and from Edinburgh, .	1	4	6
,,	13.	Bill at M'Queen's hotel, ...	0	17	6
,,	20.	Tolls to Berwick, and share at Ordinary at the Red Lion, ...	0	4	0
April	22.	Tolls and drinks to 4 men to Berwick, in delivering grain, ...	0	6	0
June	20.	Luck-penny to Wm. Gray on a hogg,	0	1	0
July	12.	Charity to Wandering Jamie, . .	0	0	6
Aug.	7.	Sheet of postage-stamps, ...	1	0	0
Sept.	13.	1½ quire thick laid 4to, ...	0	3	9
		16 dozen envelopes, ...	0	4	0
		Sealing-wax, ...	0	1	0
Nov.	11.	3 bottles Stephens' blue writing fluid,	0	1	6
		Box of Stephens' steel pens, . .	0	4	6
		¼ 100 quills, ...	0	2	6
		½ ream of postage half-sheets, . .	1	2	4

Of petty *receipts*, this is a specimen of entries :—

SKINS, MUTTON, &c.

				L.	*s.*	*d.*
1843.						
Nov.	2.	Skin of a steer that died, . . .		0	12	0
,,	5.	Hogg sold as under—	lb.			
		James Glendinning, .	12 at 3d.	0	3	0
		Robert Aitken, . .	7 ...	0	1	9
		William Wallace, .	4½ ...	0	1	1½
		Robert Walkinshaw, .	5 ...	0	1	3
		Betty Jameson, . .	6 ...	0	1	6
		Widow Lawrie, . .	4 ...	0	1	0
		Used by the house servants,	9½			
			48			
,,	7.	Sold sheep skin to John Chirnside, .		0	1	6
,,	9.	Sold 3 cwt. old iron castings to James Carmichael, founder, for . .		0	10	0

The sums of these *petty cash* transactions should be taken every month or half year, as you feel disposed; but I would prefer taking them at the end of every year, and carrying their amounts in as many lines at once into the cash-book.

In reference to temporary proceedings, I may mention that a *Memorandum-book* will be found useful for jotting down every sale, purchase, or promise made in the way of business, not implemented immediately. It may be written in this style :—

4. October 1843. Sold 24 qrs. barley to John Thomson, Leith, at 28s. Weight 56 lb.

6. Oct. Let the cutting of the drains in the West-bank field to William Salmond, at 3d. per rood, depth 30 inches.

7. Oct. Promised to exchange 5 qrs. of spring wheat with Mr Hunter of Friartown; the carts to meet his at Tarry-toll. Inquire of him if he has any seed oats and seed potatoes to spare?

12. Oct. Try and purchase a yearling bull from Mr Hunt of Thornington for next season.

Part of the petty cash-book, because ruled in the same manner, may be usefully devoted to a *Pocket cash-book*, into which every penny passing through your hands should be entered on the spot, and every sum, whenever entered, being deducted from the sum standing in the book, or added to it, according as it has been paid away or received, the sum left will shew at a glance the cash in hand. The cash in hand should be counted *at least* once a-week, and a mark affixed to the sum in the pocket cash-book, to shew that the amount had been examined and found correct at that date. This will be found an effectual plan for preventing errors, or at least of detecting them as long as the transactions are fresh in the memory. The following example will shew at once how the pocket cash-book is kept, and the transactions in it are those, of course, which have appeared in the cash-book, with all others of the most trivial nature:—

POCKET CASH-BOOK.

1843.			L.	s.	d.
Oct.	2	Cash in hand, ...	27	13	4
,,	3	Paid to John Dickson, steward, . . .	2	0	0
			25	13	4
,,	4	By Robert Brown, Dalkeith, received for oats,	26	10	3
			52	3	7
	,,	Market expenses, ...	0	4	10
			51	18	9
,,	10	British Linen Company. Withdrawn, .	60	0	0
		Pigs. Received for 2 sold by steward, .	1	4	0
			113	2	9
,,	,,	Harvest expenses. Sundries, per note, .	90	2	6
			23	0	3
+,,	11	House expenses. Mrs Williams, . .	10	0	0
			13	0	3
,,	13	British Linen Company. Withdrawn, .	290	0	0
			303	0	3
,,	14	Farm. Paid Henry Johnstone, Musselburgh, for rape-seed, ...	31	10	0
			271	10	3
,,	,,	Farm. Paid for 22 cattle, . L.258 10 0 All expenses, . . 1 10 0	260	0	0
			11	10	3

Another separate book should be kept for entering all the *Household expenses* incurred during the year; and their gross amount should be entered at the end of the year into the cash-book. The housekeeper is the proper person to keep this book; and you can inspect it as often as you see fit, and set your initials at the date the inspection takes place.

We come now to the *ledger*, which contains every current account, specimens of a few of which are opened of different sorts of transactions, from the nature of which the entries will be easily understood. The folio of each account, 14, 15, 16, and 17, is marked at the corner of the account. The ledger, exhibiting the state of every account as it stands between you and the person in whose name the account is opened, it is clear that were all transactions conducted with ready money there would be no use for a ledger, as every transaction you then had would be closed the moment the cash which it involved was either paid or received. As much business is, however, transacted on credit, it is necessary to have a ledger to conduct them in their progress during the year, at which period they may be concluded or not according to circumstances; and, if not concluded, the balance will have to be carried forward to the account of the succeeding year. As all this involves much trouble, you should endeavour to encourage ready-money transactions as much as you can. This is a specimen of the ledger.

LEDGER. 14.

Dr. JOHN THOMSON, CORN-MERCHANT, LEITH. *Cr.*

1843.			*L.*	*s.*	*d.*	1843.			*L.*	*s.*	*d.*
Oct.	4.	To Farm,—24½ qrs. barley, at 28/,	34	6	0	Oct.	18.	By Cash, . . .	34	6	0

15.

Dr. JOHN DICKSON, FARM-STEWARD. *Cr.*

1843.			*L.*	*s.*	*d.*	1843.			*L.*	*s.*	*d.*
Oct.	3.	To Cash, . .	2	0	0						

16.

Dr. BRITISH LINEN COMPANY, BANKERS, EDINBURGH. *Cr.*

1843.			*L.*	*s.*	*d.*	1843.			*L.*	*s.*	*d.*
Oct.	19.	To Cash, . .	100	0	0	Oct.	10.	By Order, ...	60	0	0
,,	28.	,, Do. . .	120	0	0	,,	13.	,, Do. . . .	290	0	0
Nov.	2.	,, Do. . .	70	0	0	Nov.	22.	,, Do. . . .	500	0	0

17.

Dr. MEADOWBANK BREWING COMPANY, EDINBURGH. *Cr.*

1843.			*L.*	*s.*	*d.*	1843.			*L.*	*s.*	*d.*
Oct.	25.	To Farm, for 45 qrs. barley sold to their Agent,	72	0	0	Nov.	8.	By Cash, . . .	72	0	0

The next, and one of the most important books, is the *Farm-account book*. It shews on the *Dr.* column every sort of outlay on account of the farm, and on the *Cr.* side the whole incomings; and consequently by balancing this book the exact profit and loss of the farm may be ascertained. When a sale of farm produce is made *on credit*, it appears *first* in this book, as shewn in the *Cr.* column of the specimen below, on the 25th October, where the Meadowbank Brewing Company is charged for a parcel of barley. An account is then opened for this firm in the ledger, and the entry is carried to their debit, as seen at folio 17 of the ledger. A mark, such as this +, is affixed in the farm-account, to shew that the barley thus delivered has been posted; and the folio of the Company's ledger is noted in the column provided for the purpose. When payment is received from the agent of the Company, it is entered in the *Dr.* side of the cash-book, as there shewn in the entry, of date November 8th, and from thence the payment is posted to the *Cr.* side of their account in the ledger, as may be seen in that book. The payment, when received, being entered in the cash-book, it is not necessary to refer to this transaction again in the farm-account. When grain or other produce is sold for ready money, it appears, *first*, in the *Dr* side of the cash-book, as shewn at date of 4th October, and from thence is carried to the *Cr.* side of the farm-account, as shewn by the entry of oats sold to Robert Brown, corn-merchant, Dalkeith, on the 4th October; and this being a ready money transaction, does not appear in the ledger at all, and saves book-keeping, as well as confers other advantages. At the end of the agricultural year the ledger is carefully gone over, and the balance on every open account entered on a balance-sheet. To these arrears is added the cash on hand, as shewn by the cash-book, and the difference between the amount of those sums, and the corresponding amount of the preceding year, should agree exactly with the balance of the farm-account. The end of the agricultural year may happen at one period one year, and at another period in another year, according to the time the crop is safely housed, and the autumnal operations completed, and it may thus vary from the 1st October to the 1st November, but usually occurs about the middle of October.

It will be obvious to you that into the cash-book and the farm-account together, every transaction is entered that can occur, and at the same time, all transactions are so appropriately separated, that the amount of produce, and its money value which have both passed through your hands, and been duly posted, may be ascertained at any time. Such a book as this I would earnestly recommend for your adoption, and in this form :—

FARM ACCOUNT—CROP 1843. *Dr.* *Cr.*

			Folio of Ledger.	L.	s.	d.	L.	s.	d.
1843. Oct. +	4.	By John Thomson, Leith, for 24½ quarters barley, sold to him at 28s., ...	14				34	6	0
	,,	By cash received from Robert Brown, Dalkeith, for 25⅝ quarters oats sold to him this day at 21s.,					26	10	3
	10.	To cash, paid expense of cutting and carrying present crop, viz.: Wages, per note, L.57 12 6 Beer account, . 14 17 0 Bread account, 17 13 0		90	2	6			
	14.	To cash, paid Henry Johnstone, Musselburgh, for 6 tons rape-dust at L.5, 5s. per ton, . . .		31	10	0			
	,,	To cash, paid for 22 short-horn bullocks, purchased at Jedburgh fair, at L.11, 15s., . . .		258	10	0			
	18.	By cash, received from John Mackie, baker, for wheat sold to him this day, viz.: 13½ quarters white, at 48s., L.32 8 0 18 ,, red, at 46s., . 41 8 0					73	16	0
+	25.	By Meadowbank Brewing Company, for 45 quarters barley sold to their agent this day, at 32s.,	17				72	0	0
	27.	By cash, received from Thomas Wightman, salesman, Morpeth, for 85 sheep sold by him there, per his account, ...					114	15	0
	28.	To cash, paid road-money for current year, per collector's receipt, ...		5	15	0			
Nov.	1.	By cash, received from Thomas Robertson, corn-merchant, for grain sold to him this day, viz.: 23 quarters barley, at 30s., L.34 10 0 36 quarters oats, at 20s. 6d., 36 18 0					71	8	0
	11.	To cash, paid agent of Scottish Union Insurance Company premium and duty on L.3000, insured with them on farm-stock, crop, &c. . . .		6	8	9			
	,,	To cash, paid Public Burdens, viz.: Poor rate, 20 months' cess, . . L.13 11 8 Schoolmaster's salary, tenants' proportion, ... 2 6 8		15	18	4			
	22.	To cash, paid half-year's rent, due at this date,		500	0	0			
	,,	By cash, received allowance for drains, due at this time, ...					50	0	0
	,,	To cash, paid labour account in full, for half year, from Whitsunday to Martinmas, viz.: Yearly servants—money wages, L.20 0 0 Field-workers, ... 50 0 0 Spadesmen, ... 34 10 0 L.104 10 0 Blacksmith's account, L.11 5 0 Carpenter's do. 10 14 6 21 19 6		126	9	6			
				1034	14	1	442	15	3

The next set of books are those connected with the disposal and state of the corn and live-stock on the farm; and, first, in regard to the *Corn*, it is necessary to make a *Plan of the stackyard*, with all the stacks drawn in circles as they stand in the stackyard. Each circle should have inscribed within it the description of corn contained in the stack which it represents, and the name of the field in which it grew. A letter T should be inserted, after the stack has been thrashed, so as to indicate at once the stacks which have been thrashed previous to the day on which the plan is again consulted, and thus avoid the error of selecting a stack that has been thrashed. There should also be a number attached to each stack. This is a specimen of a part of the stackyard, to shew how the plan is executed; and you may consult Plate IV., containing the ground plan of the proposed steading, and the position of the stackyard.

PLAN OF THE STACKYARD.

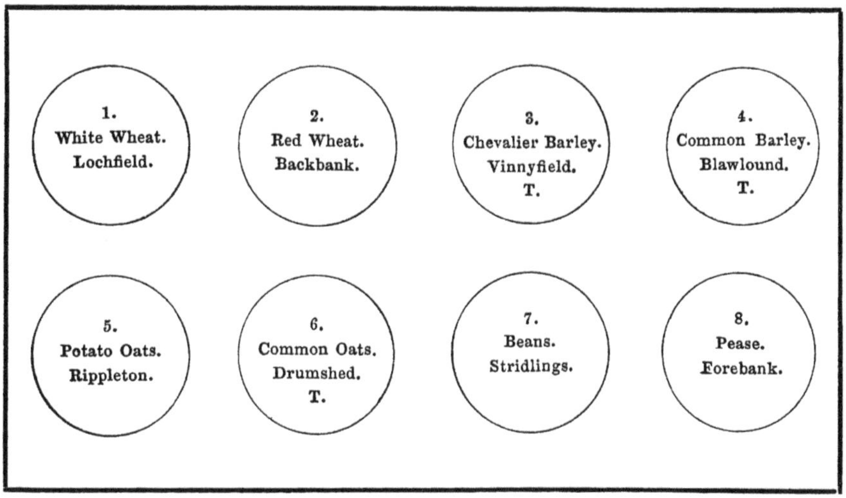

By enlarging the circle, much more information may be included within it, such as the date, when the stack was thrashed—the quantity of corn it contained—and how it was disposed of, whether sown, given to the horses or servants, or sold to millers or corn-dealers. A set of plans of stackyards of this full description for a series of years would form interesting documents for comparison of produce of one year with another.

The following is a specimen of the *Corn-account*, and *Oats* alone have been selected for the illustration, because, the account of every other sort of grain is ruled and kept in the same manner. The left hand columns shew, at a glance, the balance of the corn on hand at the date; and the right hand ones shew how the amount of produce has been disposed of, and the sums realized for it. Every bin of oats measured up in the barn is entered into this account; and whatever is delivered to a purchaser is entered here at once in the name of the person who receives it. These corn-accounts will be found records from year to year of much interest, and from which most valuable inferences may be drawn for the guidance of future operations.

CORN-ACCOUNT,—CROP 1843.

OATS.						Amount of Produce.				How disposed of.				Rate.	Value.		
						Clean.		Light.		Used.		Sold.					
1843.			Qrs.	Bs.		Qrs.	Bs.	Qrs.	Bs.	Qrs.	Bs.	Qrs.	Bs.		L.	s.	d.
Oct.	3.	To	28	4	Thrashed from No. 17. . . .	25	2	3	2								
...	4.	By	25	2	Sold to Robt. Brown, Dalkeith,							25	2	21/	26	10	3
			3	2													
...	...	By	3	2	Used by horses, ...					3	2						
...	31.	To	39	3	Thrashed from No. 22, . . .	36	0	3	3								
Nov.	1.	By	36	0	Sold to Thomas Robertson, corn-merchant, . . .							36	0	20/6	36	18	0
			3	3													
...	...	By	3	3	Used by horses, ...					3	3						
Dec.	20.	To	32	5	Thrashed from No. 21.												
...	23.	To	30	4	Do. do. No. 18.												
...	28	To	31	1	Do. do. No. 20.												
			94	2	. . . .	89	2	5	0								
		By	60	0	Paid to six yearly servants, as wages, ...					60	0						
1844.			34	2													
Jan.	3.	By	29	2	Sold to Henry Johnstone, .							29	2	19/6	28	10	4½
			5	0													
...	4.	By	5	0	Used by horses, ...					5	0						
...	16.	To	29	3	Thrashed from No. 19, . . .	27	0	2	3								
...	17.	By	24	0	Sold to James Menzies, . .							24	0	20/	24	0	0
			5	3													
...	...	By	3	0	Sold to Adam Forman, labourer,							3	0	20/	3	0	0
			2	3													
...	20.	By	2	0	Used by horses, ...					2	0						
			0	3													
...	22.	By	0	3	Used by poultry, ...					0	3						
Feb.	6.	To	29	0	Thrashed from No. 4, . . .	26	3	2	5								
...	8.	By	26	3	Sold to Jas. Montgomery, Esq.							26	3	21/	27	13	10½
			2	5													
...	9.	By	2	5	Used by horses, ...					2	5						
						203	7	16	5	76	5	143	7		146	12	6

The *Live-stock* accounts differ in form, according to the kind of stock, except cattle and sheep, the forms of whose accounts are alike. It will be observed, that the only difference of form betwixt these and the corn account consists in having no money-columns for live-stock; and the reason is, that as both cattle and sheep are disposed of to the butcher or dealer in large lots, 1 or 2 lines in a year are quite sufficient to enter their disposal, and it is not worth while ruling money-columns for a few entries; and even in the case of stock which are purchased and sold again in the course of the same season, very few entries are required; and when bought or sold, are entered at once into the cash-book. In the *live-stock accounts*, the number of heads are entered in the left-hand column, their nature being designated in words. In the right-hand columns they are subdivided according to sex and state. When any are bought or sold, their entire number is added or subtracted from the gross balance on the left hand, and according to particular sex and state they are specified in the the columns of the right hand.

LIVE-STOCK ACCOUNT.

1843.		CATTLE.	Steers.	Cows.	Calves.
May 1.	1	Bull, ...	1		
	10	Yearling steers, ...	10		
	10	Do. heifers, ...		10	
	7	Cows, ...		7	
	28		11	17	
„ 12.	3	Yearling steers bought in the Grassmarket,	3		
	31		14		
June 7.	1	Yearling heifer died, . . .		1	
	30			16	
Oct. 14.	22	Bullocks bought at Jedburgh fair,	22		
	52		36		
„ 15.	20	Weaned calves now entered in the book,			20
	72				
1844. Feb. 1.	1	Bull sold to James Wood, butcher, .	1		
	71		35		
Mar. 15.	22	Fat bullocks sold to Thos. Scott, cattle-dealer, ...	22		
	49		13		
April 2.	1	Yearling bull bought at Coldstream from Mr Hunt, Thornington, . . .	1		
	50		14	16	20

LIVE-STOCK ACCOUNT.

1844.		SHEEP.	Breeding ewes.	Feeding sheep.	Hoggs.
May 1.	120	Leicester ewes, ...	120		
	160	Do. hoggs, ...			160
	2	Do. tups, ...		2	
	282				
May 25.	1	Ewe died—sold by shepherd, . .	1		
	281		119		
June 10.	50	Barren ewes and clipt hoggs sold at Morpeth, ...	5		45
	231		114		115
„ 20.	1	Sturdied hogg sold to Wm. Gray, butcher,			1
	230				114
„ „	40	Hoggs sold to Thos. Scott, cattle-dealer,			40
	190				74
July 12.	2	A ewe (sold), and hogg died, . .	1		1
	188		113	2	73
		Account altered at this date, by addition of last crop of lambs.			
July 15.	113	Breeding ewes, ...	113		
	75	Shearlings, consisting of dinmonts, gimmers, and tups, ...		75	
	167	Hoggs, bred on farm, ... 156			
		Do., bought from shepherd, . 11			167
	355				
	70	Draft ewes and dinmonts sold to Jas. Elliot,	40	30	
	285		73	45	
	86	Gimmers selected for breeding, . .	43	43	
	371		116	88	167

Horses are so seldom disposed of, that money-columns are unnecessary to be introduced into their account, as their value can be transferred at once to the cash-book. Those sold can be omitted in the account of the following year, and those bred added to it.

LIVE-STOCK ACCOUNT.

Days.		Age.	Work.	Riding	Foal.	HORSES Name and Description.
1843.						
Oct.		8	1			Old Bob, brown.
		7	1			Grey, Rattler.
		8	1			... Mettle.
		7	1			Bay, Tailor.
		7	1			... Snip
		9	1			... Rose.
		8	1			... Brisk.
		7	1			Black, Tinker.
		7	1			... Match'em.
		8	1			Brown, Dawtie.
		8	1			... Beauty.
		3		1		Bay, Norah, out of Rose by Theseus.
		2	1			Brown colt, out of Beauty by Clydesdale.
		1			1	... filly, ... Lofty.
		2	1			Black filly, ... Mettle ... Champion.
		1			1	Black colt, ... Farmer.
		12		1		White pony.
		13		1		Bay blood brood-mare, Keerpoy, by Bumper out of a Magog mare.
		3		1		Bay colt, Sepoy, out of Keerpoy by Bustler.
		2		1		Bay filly, Pannaoola, ... Bustler.
		1		1		Bay colt, Lascar, ... King David.
					1	Bay filly, Beebee, ... Apostate.
			13	6	3	

LIVE-STOCK ACCOUNT.

SWINE. *Dr.*

DATE.		Sows.	Shotts.	Pigs.	
1843.					
Nov.	1.	2			Breeding sows.
		1			Boar.
			12		Shotts.
Dec.	27.			13	Pigs.
				3	Sold to servants.
		3	12	10	On hand.
1844.					
Jan.	3.		2	2	Sold Mrs Wilkie.
		3	10	8	On hand.
...	5.		1		Used in house.
		3	9	8	On hand.
...	17.		3		Sold to servants.
		3	6	8	On hand.
...	31.			8	Transferred to shotts.
		3	6		
			8		Transferred from pigs.
		3	14		On hand.

SWINE. *Cr.*

DATE.		Sows.	Shotts.	Pigs.	Sold or used.	*L.*	*s.*	*d.*
1843.								
Dec.	27.			1	Robert Aitken,	0	15	0
				1	James Dyot, .	0	15	0
				1	Betty Jameson,	0	15	0
1844.								
Jan.	3.		2	2	Mrs Wilkie, .	3	6	0
...	5.		1		Used in house 9 stones, .	0	0	0
...	17.		1		Robert Walkinshaw, . .	0	18	0
			1		Jas. Middlemiss,	0	18	0
			1		Widow Laurie,	0	18	0
						8	5	0

The account for *Swine* is ruled in the foregoing manner. Swine are frequently sold singly, and in small lots. The left hand columns shew at a glance the state of the swine stock of different sexes and ages, while the stock and money-columns on the right hand indicate the different states of the stock and the prices realized for them by sales to different people.

The *Farm-steward* will have to render assistance to the farmer in book-keeping, as well as in other respects. The cash and materials which pass through his hands he should mark down in his book, and report from time to time, say once a-week, to his master. An intelligent *young* man will readily adopt any simple method of keeping accounts which his master may point out; but in the case of an elderly person, who has become confirmed in a method of his own, it is better not to interfere with him, lest, as has happened in cases within my knowledge, you put him out of his own plan, without succeeding in teaching him a better. All that the steward is required to do in book-keeping is to furnish his master with an intelligent record of what has taken place. When the ploughmen receive their wages *in kind*, the steward's book is a suitable place for keeping an account of the grain, meal, or other things given to each individual ploughman from time to time. Besides this, the steward should keep memoranda of various other particulars, such as the time any mare has been served by the stallion,—the number of loads of dung driven to the dunghills of any particular field or fields,— the number of loads of turnips deposited in store,—the loads of potatoes pitted in each pit,—the date when each stack of corn was thrashed,—the quantity of corn measured up from its contents,—and many similar matters unnecessary to be enumerated. A book ruled with simple money-columns, and a column for the dates, will answer all the purpose of a memorandum, corn, and cash-book.

It is also the duty of the steward to keep an account of all the manual labour performed on the farm, whether by field-workers or day-labourers. There are various modes of keeping the time-account of field-workers and labourers, and which require very differently ruled columns, but the 2 specimens given below, 1 fortnightly, and 1 half-yearly, are as accurate and particular as such accounts require to be. They require very close ruling, and a book with a wide page.

FIELD-WORKERS' TIME-ACCOUNT. (*Fortnightly.*)

1844. May 27.	M.	T.	W.	T.	F.	S.	M.	T.	W.	T.	F.	S.	Days.	Rate.	Amount. L.	s.	d.
Mary Johnstone, .	1	½	1	¾	1	1	½	.	1	1		1	9½	10d.	0	7	11
Alison Dods, . .	.	½	1	¾	1	1	½	.	1	1	1	1	8¾	10d.	0	7	3½
Margaret Cleghorn,	1	½	1	.	1	1	.	.	1	1	1	1	8 2	10d.	0	7	1
Anne Paterson, . .	1	1	1	1	1	1	1	1	1	1	1	1	1	10d.	0	10	0
Elizabeth Gilmore,	1	½	.	.	¾	1	1	.	1	1	1	1	7¾	10d.	0	6	5½
Janet Clinkscales, .	1	.	1	1	¾	1	1	.	1		1	1	9¾	10d.	0	8	1½
Mary Brown, . .	1	½	1	1	¾	1	.	.	1		1	1	8¾	10d.	0	7	3½
Christian Lamb, .	1	½	1	1	¾	1	1	.	1	½	1	1	9¾	10d.	0	8	1½
Sophia Millar, . .	1	½	1	1	¾	1	.	.	1	½	1	.	7¾	6d.	0	3	10½
															3	6	2

In this method the vertical columns should be extended across the page, according to the number of weeks intended to be embraced between the pay-days. I give a *fortnight* as a specimen, but, of course the time may only be a week, or longer than a fortnight, to suit the convenience of parties.

The following table, besides, being suited for a *time-account*, is a most convenient form of keeping the *harvest roll of reapers*, as a great many names may be entered on a single page or sheet. It is given as a time-account, in the form to extend to half a year.

FIELD-WORKERS' TIME-ACCOUNT. (*Half-Yearly.*)

	MARY JOHNSTONE.							ALISON DODS.							MARGARET CLEGHORN.						
1844.	M.	T.	W.	T.	F.	S.	DAYS.	M.	T.	W.	T.	F.	S.	DAYS.	M.	T.	W.	T.	F.	S.	DAYS.
May 27.	1	½	1	¾	1	1	5¼	.	½	1	¾	1	1	4¼	1	½	1	.	1	1	4½
June 3.	½	.	1	1	¾	1	4¼	½	.	1	1	1	1	4½	.	.	1	1	1	1	4
... 10.	1	1	1	1	1	1	6	1	1	1	1	1	1	6	1	1	1	1	1	1	6
... 17.	1	1	1	1	1	1	6	1	1	1	1	1	1	6	1	1	1	1	1	1	6
... 24.	1	½	.	.	¾	1	3¼	1	.	1	½	1	1	4½	1	½	.	.	¾	1	3¼
July 1.	1	.	1	½	1	1	4½	1	½	.	.	¾	1	3¼	1	.	1	½	1	1	4½
... 8.	1	.	1	1	¾	1	4¾	1	.	1	1	¾	1	4¾	1	.	1	1	¾	1	4¾
... 15.	1		1	1	1	1	5	1	.	1	1	1	1	5	1		1	1	1	1	5
... 22.	1	½	1	1	¾	1	5¼	.	.	1	½	1	1	3½	1	½	1	1	¾	1	5¼
... 29.	1	.	1	½	1	1	4½	1	½	1	1	¾	1	5¼	.	.	1	.	1	.	2½
Aug. 5.	.	.	.	.	.	.		.	.	.	.	.	.		.	.	.	.	.	.	
							48¾							47							45¾

In this method there should be a separate space vertically for each person, and as many horizontal lines as there are weeks in the half year; the day of the month on which each Monday falls, being entered at the left side of the page. At the end of the half year the names should be entered on another page, the total number of days, rates, and amounts, carried out opposite to each name, and the whole summed up to shew the half years' bill of field-workers' wages.

The *Shepherd*, having the general charge of the cattle, sheep, and pigs on the farm, should also keep a book, in which he should enter the number and kind of stock bought and sold, of and to whom. He should also put down the date of the death of a sheep, or when it is killed for the use of the house,—the time when a cow, a sow, or the ewes are served,—and any memorandum of any suggestion that may occur to him to make for the benefit of the stock under his charge. The contents of his book should be communicated to his master every week.

When the whole crop of *corn* is disposed of, an *Abstract* like the following should be drawn up and inserted in a convenient part of the corn account-book. A series of such abstracts will give, at a glance, the comparative gross produce of every year, and will shew whether the farm is increasing or decreasing in productiveness. I may mention that this abstract gives the probable proportion which each kind of grain will bear to each other on a 500 acre farm under a 5-shift course, in mixed husbandry; and the gross produce is not only given, but how the quantities of grain have been disposed of, whether consumed in wages, seed, or provender. An average from a series of such abstracts would indicate pretty nearly the proportions consumed of the gross produce by each of those class of items. Taking this abstract as an example of an average one, it appears that, on such a farm, the corn consumed in wages and in seed are nearly alike, each being about $\frac{1}{14}$ of the gross produce; whereas the quantity of corn consumed in support of animals amounts to $\frac{1}{7}$ of the gross produce, or equal to both wages and seed taken together, making the whole expenditure in grain, in working the farm, amount to more than $\frac{1}{4}$ of the gross produce.

ABSTRACT OF CORN-ACCOUNT.

PRODUCE.

OATS.	Variety.	Acres.			Rate. Qrs.	Rate. Bu.		Qrs.	Bu.	Qrs.	Bu.
Butterwell field, .	Hopetoun,	34		×	7	5	=	255	0		
Bean-rig do. . .	Potato, .	37		×	8	0	=	296	0		
White-corn-lea do.	Sandy, .	30		×	7	0	=	210	0		
		—	101					——		761	0
WHEAT.											
Lang-bank, . . .	Chidham,	33		×	4	0	=	132	0		
Fore-croft, . . .	Hunter's,	27		×	4	4	=	121	4		
		—	60					——		253	4
BARLEY.											
Hinds'-acres, . .	Annat, .		42	×	6	0	=			252	0
			203							1266	4

DISPOSAL.

Acres in crop as above, . . .	203		
Produce per acre, ...	$6\frac{1}{4}$ qrs. nearly.		Qrs. Bu.
Gross produce, .	——		1266 4
		Qrs. Bu.	
Used on farm in wages, ...		88 0	
Do. Do. seed		87 4	
Do. Do. provender, ...		180 0	
			355 4
Quantity sold, ...			911 0
At the average rate per quarter of, ...			L.1 8 5
Amounting to, ...			L.1294 7 7

At the end of every agricultural year the books should be proved; and to enable you to do this strictly, an *Inventory* and *Valuation* of stock should be made, as a means of comparing one year with another. In doing this, the value of the stock should be stated as nearly as possible at what it would bring in the market at the time; but this being merely an estimate, and not an actual cash transaction, it should be entered on a page by itself, as a collateral document, and not mixed up with the balances shewn by the regular books; it not being properly a part of them, however indispensable for obtaining the real state of affairs. It is not necessary to include the value of the implements in use, as they do not vary materially in value from year to year, and should rather form an item in the statement to be kept of capital invested in the farm, along with draining, building, and other improvements, and which fall to be charged against the entire lease, and not against the particular year in which the outlay is incurred. Without such an inventory and valuation of stock the books may present a fallacious statement. For example, should the turnip crop promise to be very abundant, you may make purchases in anticipation, or retain stock which would otherwise have been disposed of from grass, and then an unfair balance would be shewn against that crop. On the other hand, a bad turnip crop may lead you to reduce the stock, and thus throw an undue amount into the credit side of the current year's account.

For the purpose of proving the books, the ledger should be carefully gone over, and the balance on every open account entered on a balance-sheet. To these arrears should be added the cash in hand shewn by the balance, if any, in the pocket cash-book, and the *difference* between the amount of both these sums, and the *corresponding sum of the previous year*—for I assume that a correct state of affairs was made up to begin with—will agree exactly with the balance of the cash-book. Should they not tally, the entries should be compared, and the additions and calculations gone over carefully; for if any entry has been erroneously calculated, or copied in the posting, or any error committed in the adding of sums together, the balance cannot tally.

It is highly probable that you consider the books I have mentioned as necessary, too numerous, and as constituting a system of book-keeping much too complicated for an ordinary practical farmer to follow, as he cannot be supposed to have so much time at his command as to write all that is requisite to maintain such a system. But very little reflection will convince you, that the books are not so numerous, nor the quantity of writing so onerous, as they appear at present, when *all* the accounts have been *separately* laid before you; but a certain number are requisite. For, you must allow that when a transaction occurs, it *ought* to be marked down somewhere; it should not be left entirely to the memory; and where can it be so *conveniently* entered as in the pocket cash-book, or petty cash-book, or memorandum-book, *all* constituting, if you choose, *one single book*, which should be always in your pocket, and, of course, may be written upon, wherever you are, whether in the field, on horseback, or in the house? The transaction *being entered there*, is safely committed for consideration, or for being transferred into other books, at the first leisure moment during the course of the current week. To pretend that *any* farmer *cannot* find leisure to *write a few lines* in 1 book, or even 2 books, during the course of a whole week, is to maintain an absurdity. The transaction, it is true, may be a trifling one, and may be undeserving of much attention, and, on that account, may lie over recorded in the pocket-book till the following week, when something else more important may occur to arrest the attention seriously towards the books; but you should be made aware, that while farmers have large transactions to perform occasionally, their ordinary business consists of a multitude of minor ones, all which, together, may not amount to a large sum of money; and if these may be neglected because they are diminutive, then, of course, by far the greatest number of transactions on a farm may be neglected. If any farmer were to venture to make such a broad statement as this with a serious face, even to his companions, he would assuredly be regarded by them as an indolent fellow that will never "do any good." But why should the farmer claim immunity from book-keeping more than men engaged in a trade or profession in which they cannot get on with any degree of success without careful book-keeping? *Their* business must be a trifling one indeed, if they have not many more entries to make in the course of one day than a farmer has in a whole week. Why, then, I repeat it, should he claim exemption from the very moderate degree of trouble required to write a few lines even every day, or, probably, only occasionally, or expect success and ease of mind in the conduct of his affairs, while indolently disinclined to bestow so little trouble? Doubtless, some farmers have prospered who were remiss in this duty, and others have failed who were most exemplary, and such instances will readily be cited in excuse by the careless and self-sufficient; but will any man with a spark of intelligence, or with a mind slightly elevated above sheer prejudice, justify the neglect of a useful and wholesome *duty* on such untenable grounds? Let me hope that you, the rising race of farmers, will emancipate our character from the reproach to which, on this ground, we have been so long and too justly exposed, from other classes of the community.

But after all it must be owned that were every farmer to keep a regular set of books, it would impose a task upon him, not so much on his time, as on his ingenuity; for it requires no little knowledge of business transactions to place every entry in its proper place; and, besides, a farmer is placed at a disadvantage in reference to recording the particulars of his business, when compared with the facilities of doing so by mercantile men, who have clerks for the purpose. To assist farmers who have not paid so much attention to the minutiæ of book-keeping as to be able to keep a regular set of books, and at the same time to encourage them to keep books of some sort, and not allow their affairs to get into confusion, I will shew them that but few books are absolutely needed to tell them whether they are gaining or losing in business.

It is clear, in the first place, that when every sum received and paid is put into the *Dr.* and *Cr.* columns of a *Cash-book*, that the balance between them will truly shew whether more money has been paid away or received. Now the most convenient plan to collect materials for the cash-book, is to keep a small *Pocket cash-book*, into which *every transaction*, whether of the most important or trifling nature, should be entered, and added or subtracted from the balance, as the case may be, and such a book will afford the readiest way of knowing every day whether the balance is for or against you. The items of transactions, when they occur, should be entered into the proper columns of the cash-book, from the pocket cash-book, once a-week.

You will observe, that transactions only which involve the payment and receipt of *cash* can be put into the cash-books; but there are many things which have nothing to do with *cash* for the time, which nevertheless must be put down somewhere, as they may involve dealings in cash afterwards. These embryo transactions should be put into a *Memorandum-book*, to be picked out and deleted from it, and entered in the cash-book whenever cash has passed from hand to hand. It would create confusion in the cash-book to enter any thing into it as a mere memorandum.

I have said before, that transactions in ready money require no ledger account; but, as you must have many transactions on credit, and you cannot know how you stand with any person without a ledger account, it is necessary to have a *Ledger* for such accounts, and which are not likely to be numerous. You must also have accounts of the quantity of *Corn* and number of *Stock*, and how each is disposed of, and it is scarcely possible to make them simpler than the forms given above.

You will observe that the prosecution of the mixed husbandry involves much more book-keeping than any other. In carse farming, corn is the only item that involves trouble; in pastoral farming, the stock alone does so; and being disposed of but seldom, and in large sums, very few entries are required in the cash-book. Farming in the neighbourhood of a town involves many retail transactions in green crops straw, and manure, which are most befitting the steward to keep a particular account of; and their results only require to be entered in the cash-book; and every such transaction being in ready money, no ledger account is needed. Dairy-farming requires much attention to stock and the products of the dairy, and the latter being supplied at times to particular customers, ledger accounts must be opened with them; but few cash entries are required, and almost none for corn, as that species of produce is mostly consumed on the farm. Mixed husbandry deals equally with stock and crop, and therefore must pay equal attention to both in book-keeping.

I adopted a plan of keeping the time of field-workers, which being simple, taking up little room, and little time to make an entry, and, at the same time, is quite lucid and correct, I recommend it, and give a specimen of its form.

A few words of explanation, in regard to the construction of this table, seems necessary. The field-workers are paid usually by the half-year: this table includes the time from Whitsunday, 27th May, to Martinmas, 22d November, 1843. The space allotted to the name of each field-worker should, therefore, be sufficient to contain the numbers and fractions of days she is likely to be employed in the field during the half-year. In regard to the meaning of the cyphers employed, each stroke signifies 1 full day, each fraction, that proportion of a day, and each dot means that the worker was not at work, on account of the farmer, on that day. The weeks are divided from one another by a long line; so there will be as many long lines as there are weeks in the half-year

The entries of each week are given in a more varied form here than they are likely to be in practice, in order the better to shew the effect of the various marks. The date of any day during the half-year can be easily ascertained, if desired, by consulting an almanac. On comparing the fortnightly table with this one, it will be observed that the days' work performed for 2 weeks opposite to the first 3 names, correspond exactly in both tables; and on comparing the half-yearly table with the same, the entries of each day's work, from May 27 to August 5 inclusive, of the same first 3 names, also correspond. On comparing this table still further with both the others, it will be seen that the space occupied by it, together with the days, and rate, and money-columns, is much smaller, and the plan simpler than either of them. The contrast, for brevity, will be still more strongly seen in its favour on comparing it with the succeeding form of labour-account.

FIELD-WORKERS' TIME-ACCOUNT.

<table>
<tr><th>NAMES.</th><th colspan="4">Weekly Time, from 27th May to November 22. 1843.</th><th>Days.</th><th>Rate.</th><th>L.</th><th>s.</th><th>d.</th></tr>
<tr><td>Mary Johnstone,</td><td>| ½ | ¾ | |</td><td>½ · | | ¾ |</td><td>| | | | | |</td><td>| | | | | |</td><td></td><td></td><td></td><td></td><td></td></tr>
<tr><td></td><td>| ½ · · ¾ |</td><td>| · | ½ | |</td><td>| · | | ¾ |</td><td>| · | | | |</td><td></td><td></td><td></td><td></td><td></td></tr>
<tr><td></td><td>| ½ | | ¾ |</td><td>| · | ½ | |</td><td>· · · · · ·</td><td></td><td>48¾</td><td>10d.</td><td>2</td><td>0</td><td>7½</td></tr>
<tr><td>Alison Dods, .</td><td>· ½ | ¾ | |</td><td>½ · | | | |</td><td>| | | | | |</td><td>| | | | | |</td><td></td><td></td><td></td><td></td><td></td></tr>
<tr><td></td><td>| · | ½ | |</td><td>| ½ · · ¾ |</td><td>| · | | ¾ |</td><td>| · | | | |</td><td></td><td></td><td></td><td></td><td></td></tr>
<tr><td></td><td>· · | ½ | |</td><td>| ½ | | ¾ |</td><td>· · · · · ·</td><td></td><td>47</td><td>10d.</td><td>1</td><td>19</td><td>2</td></tr>
<tr><td>Marg. Cleghorn,</td><td>| ½ | · | |</td><td>· · | | | |</td><td>| | | | | |</td><td>| | | | | |</td><td></td><td></td><td></td><td></td><td></td></tr>
<tr><td></td><td>| ½ · · ¾ |</td><td>| · | ½ | |</td><td>| · | | ¾ |</td><td>| · | | | |</td><td></td><td></td><td></td><td></td><td></td></tr>
<tr><td></td><td>| ½ | | ¾ |</td><td>· · | ½ | ·</td><td>· · · · · ·</td><td></td><td>45¾</td><td>10d.</td><td>1</td><td>18</td><td>1½</td></tr>
</table>

A still more full and complicated *labour-account* than any of the above is given below. It professes to enter into an exposition of the daily labour performed by each labourer; and in so far a series of such documents would form an interesting means of comparison with the seasonal labour of other years.

MONTHLY LABOUR-ACCOUNT—FOR APRIL 1844

	Names.	Monday.	Tuesday.	Wednesday.	Thursday.	Friday.	Saturday.	Days.	Rate.	Week. L.	Week. s.	Week. d.	Month. L.	Month. s.	Month. d.
1844. April 1.	John Davidson,	1, Turning dunghill,	1, Do.	1, Straw-barn.	1, Fences.	1, Do.	1, Do.	6	1/6	0	9	0			
... 8.	... ...	1, Barn.	½, Water-furrow.	1, Do.	1, Do.	1, Sowing.	1, Water-furrow.	5½	...	0	8	3			
... 15.	... ...	1, Stone-breaking.	1, Cleaning hedge.	1, Do.	1, Do.	1, Do.	1, Do.	6	...	0	9	0			
... 22.	... ...	1, Filling dung.	1, Do.	1, Do.	1, Barn,	1, Fences.	1, Do.	6	...	0	9	0	1	15	3
... 1.	Mary Johnstone, field-worker,	Corn barn.	1, Picking potatoes.	1, Thrashing-mill.	½, Gathering quicken.		1, Corn-barn.	4¼	10d.	0	3	6½			
... 8.	... ...	1. Turning dunghill.	1, Do.	1, Thrashing-mill.	1, Corn-barn.	1, Carrying seed.	½, Do.	5½	...	0	4	7			
... 15.	... ...	1, Thrashing-mill.	1, Do.	1, Corn-barn.	1, Do.	½, Gathering quicken.	1, Cutting seed-potatoes.	5½	...	0	4	7			
... 22.	... ...	1. Planting potatoes.	1, Do.	½, Cutting potatoes.	1, Planting potatoes.	...	1, Thrashing-mill.	4½	...	0	3	9	0	16	5½
... 1.	Alison Dods, do.,	¾, Corn-barn.	1, Picking potato-sets.	1, Thrashing-mill.	½, Gathering quicken.	...	1, Corn-barn.	4¼	10d.	0	3	6½			
... 8.	... ...	1. Turning dunghill.	1, Do.	1, Thrashing-mill.	1, Corn-barn.	1, Carrying seed.	½, Do.	5½	...	0	4	7			
... 15.	... ...	1, Thrashing-mill.	1, Do.	1, Corn-barn.	1, Do.	½ Gathering quicken.	1, Cutting potato-sets.	5½	...	0	4	7			
... 22.	... ...	1, Planting potatoes.	1, Do.	½, Cutting potato-sets.	1, Planting potatoes.		1, Thrashing-mill	4½	...	0	3	9	0	16	5½

This method is well adapted for a proprietor, who occupies a small home-farm, where the number of persons employed is not great. By causing his steward to render such an account monthly, he will see the monthly occupation of each person, as well as the sum required for his or her monthly payment. The tenant-farmer, at the head of his concern, will prefer the simpler and condensed method given above. In regard to detailed accounts of labour, they form interesting records of what takes place under one's eye, as reliance can then be placed on their accuracy; but when transmitted to a distance, and no check exists as to their contents, it is feared that these may be made up to suit the purposes of deception.

I present a mode of keeping stock accounts practised by a farmer well acquainted with book-keeping. It is somewhat different in form from the foregoing, but does not, in my opinion, seem to possess any advantages over them. It will be observed, however, that the mutton sold to the work-people is entered into the stock-account as the sales happen to occur, instead of being taken to the petty cash-book; and also, that the inventory of the stock in the left hand column admits the names of the stock, such as those of the cows, to be inserted at length; and, in both these respects, this form may please the fancy more than the others.

On comparing the form of the live-stock tables given by the late Sir Patrick Murray, Bart., of Ochtertyre, you will observe how confused they are in comparison with this form, or with those given above.

The remarks, however, of Sir Patrick Murray, which accompany his tables, on the benefits to country gentlemen of keeping regular accounts, are expressed in terms of so much truth and good sense, that I cannot refrain from quoting them. "If country gentlemen kept *regular* accounts of the value of all the *family supplies* derived from the lands in their occupation, and of the *works*, exclusive of *farm*-work, performed by their farm-servants and horses, it would be proved that the occupation, cultivation, and improvement of land is not so expensive a concern as it is usually conceived to be. The contrary opinion is so prevalent, that it both limits the extent of the farm operations of gentlemen who are engaged in them, and prevents others engaging in them—consequences much to be regretted, both as limiting and retarding the general agricultural improvement of the country, and as lessening the inducements of landed proprietors and others to *reside in the country*, and employing their time and capital towards its cultivation and improvement—objects which, in every point of view, are obviously of the greatest national importance in all countries."*

With the 2 following tables I shall conclude this subject:—

* Prize Essays of the Highland and Agricultural Society, vol. i. p. 258-63.

CATTLE. *Dr.*

Date.		Age.	Cows.	Steers.	Queys.	Calves.	Names and Description.
1843. Oct.	15.	7	1				Cowslip. Red and white.
		5	1				Woodbine. Red and white.
		5	1				Rosina. Red.
		5	1				Flora. White.
		4	1				Juliet. Roan.
		4	1				Alba. White.
		4	1				Silkysides. Roan.
						11	Calves.
		2		23			Steers.
		4		1			Bull. Arab. Roan.
		1			12		Queys.
		1		13			Stots.
				37			
,,	19.			1			Stot died.
			7	36	12	11	On hand.
Dec.	23.			12			Sent to Morpeth.
			7	24	12	11	On hand.
1844. April	17.			23			Sent to Newcastle.
			7	1	12	11	On hand.
						11	Transferred, as under, to stots and queys.
			7	1	12		
				5	6		Transferred, as above, from calves.
			7	6	18		On hand.
,,	30.				12		Sent to Newcastle.
			7	6	6		On hand.

CATTLE. *Cr.*

Date.		Age.	Cows.	Steers.	Queys.	Calves.	Disposal.	Price. *L.*	*s.*	*d.*
1843. Nov.	19.			1			1 Stot died.			
,,	21.						Skin sold at Berwick.	0	12	0
Dec.	23.			12			Sent to Morpeth, and sold for L.19, 10s. a-head,	234	0	0
1844. April	17.			23			Sent to Newcastle, and sold for L.12 a-head,	276	0	0
,,	30.				12		Sent to Newcastle, and sold for L.12 a-head.	144	0	0
				36	12			654	12	0

SHEEP. *Dr.* SHEEP. *Cr.*

Date.		Tups.	Ewes.	Shearlings.	Hoggs.	Lambs.	Description.
1843. Oct.	15.	2					1 shearling, 1 two-shear tup.
			90				35 aged, 25 young, 30 gimmers.
				40			Shearlings.
					120		Hoggs, ewe and wethers.
,,	2.			1			Killed for the house.
		2	90	39	120		
,,	5.				1		Died.
		2	90	39	119		
,,	30.		10	25			Sent to Newcastle.
		2	80	14	119		
Dec.	1.			1			Killed for the house.
		2	80	13	119		
,,	20.		2				Sold to James Rankine.
		2	78	13	119		On hand.

Date.		Disposal.	Weight.	Price.	Value. L.	s.	d.
			lb.				
1843. Nov.	2.	One shearling killed for the house.	59½				
,,	5.	One hogg died and sold as under,	48				
			107½				
		James Glendinning.	12	3d.	0	3	0
		Robert Aitken,	7	3d.	0	1	9
		William Wallace,	4½	3d.	0	1	1½
		Robert Walkinshaw,	5	3d.	0	1	3
		Betty Jameson,	6	3d.	0	1	6
		Widow Laurie,	4	3d.	0	1	0
		Used by house-servants,	9¼				
,,	30.		155¼				
		10 ewes, sent to Newcastle at 40s. each,			20	0	0
Dec.	1.	and 25 shearlings, sent to Newcastle at 35s. each,			43	15	0
,,	20.	One shearling killed for the house,	63				
		Two ewes sold to James Rankine,			1	12	0
			218½		65	16	7½

88. OF THE CONVENIENCES OF THE COTTAGES OF FARM-SERVANTS.

"Comfort is within,
The embers' cheerful glow,
The sound of the familiar voice,
The song that lightens toil."
SOUTHEY.

There may be no necessity of your building cottages for farm-servants on the farm you have taken; but as I have given the particulars required for constructing a good steading, as well as of the useful part of the farm-house, it may not be out of place to give also a few plans of cottages of different sizes, suitable to the families of farm-servants. The usual practice in building cottages for farm-servants is to adopt a uniform plan, upon which all are built. The practice is not founded on sound principle, nor even on expediency; because it implies that families consisting of very different numbers, should, nevertheless, be accommodated within a similar space. Instead, therefore, of a family accommodating itself to the size of the cottage, the cottage ought to be adapted to the wants of the family; and there is no way of establishing this mutual understanding between cottages and their inmates, but by building them of different capacities for accommodation, and appropriating them to families in accordance with their numbers. This is the only rational course to pursue; and in pursuance of it, it is as easy to build a given number of cottages on different plans, as on the same plan. Following out this principle, I shall give a number of plans, suited to families of different sizes, derived from various sources of practical experience, but modified, in some instances, to suit my own notions of the conveniences, comforts, and means of cleanliness which such dwellings should possess.

In order to appreciate the value of any professedly improved plan, we have only to consider the general character of the old farm-servants' houses in the country; and the remarks of Dr Gilly on the old cottages in Northumberland will apply everywhere. "The general character of the best of the old-fashioned hind's cottages in the neighbourhood is bad at the best. They have to bring every thing with them—partitions, window-frames, fixtures of all kinds, grates, and a substitute for ceiling; for they are, as I have already called them, mere sheds. They have no byre for their cows, no sties for their pigs, no pumps or wells, nothing to promote cleanliness or comfort. The average size of these sheds is about 24 feet by 16. They are dark and unwholesome. The windows do not open, and many of them are not larger than 20 inches by 16. Into this space within the shed are crowded 8, 10, and even 12 persons. How they lie down to rest, how they sleep, how they can preserve common decency, how unutterable horrors are avoided, is beyond all conception. The case is aggravated, when there is a young woman to be lodged in this confined space, who is not a member of the family, but is hired to do the field-work, for which every hind is bound to provide a female. It shocks every feeling of propriety to think, that, in a room, and within such a space as I have been describing, civilised beings should be herding together, without a decent separation of age and sex. So long as the agricultural system in this district requires the hind to find room for a fellow-servant of the other sex in his cabin, the least that morality and decency can demand is, that he should have a second apartment, where the unmarried female, and those of a tender age, should sleep apart from him and his wife." *

Every one will assent to the propriety of these remarks; and those who know what old hinds' houses are, will admit their truthfulness and pertinency. The following plans are constructed in the spirit of these remarks. Fig. 601 represents the ground-plan of a cottage, which may

Fig. 602. Fig. 601.

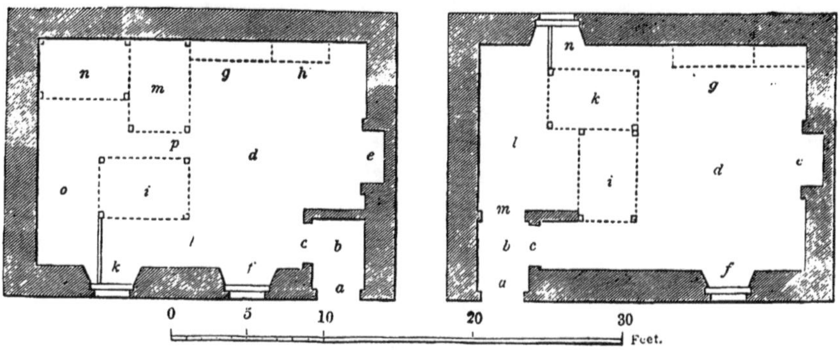

THE GROUND-PLAN OF SERVANTS' HOUSES FOR SMALL FAMILIES.

be made double, if desired, capable of accommodating a hind and his wife, and a field-worker; and if there is a girl or grown-up daughter beside, she could sleep with the field-worker. The cottage measures, inside, 22 feet in length, and 15 feet in width, and the accommodation on the floor is 15 feet by $11\frac{1}{2}$ feet: *a* is the entrance door, *b* the porch, *c* the door into the principal apartment, *d* the principal apartment, which is provided with a fire-place *e*, window *f*, plate-rack *g*, and dresser *h*; 2 box-beds *i k*, are so placed as to form partitions between the principal

* Gilly's Appeal on Behalf of the Peasantry of the Border, p. 19, 20.

apartment *d*, and store-room *l*; the bed *i* being entered by the store-room *l*, which is provided with a door *m*, and lighted by a part of the window *n*, which also lights a sort of small apartment for entering to the bed *k*.

Should the hind have a family of 3 children, beside the field-worker, the arrangement of the furniture, as shewn in fig. 602, will afford accommodation to the increased number of the inmates within the same dimensions of a cottage; *a* being the entrance door, *b* the porch, *c* the inner door, *d* the principal apartment, which has floor-room 15 feet by 12 feet; *e* the fire-place, *f* the window, *g* the plate-rack, and *h* the dresser; one bed *i*, is entered by the apartment lighted by the window *k*, and provided with a door at *l*; another bed *m* is entered from the principal apartment *d*; and a third one *n*, is situate in a separate apartment *o*—a store-room, lighted by a part of the window *k*, and entered by a passage, npon which a door is placed at *p*.*

Both these plans would accommodate a small family more comfortably than larger houses, as nothing is more uncomfortable to work-people than to be obliged to occupy a house much larger than the small quantity of furniture they usually possess can occupy. Were box-beds dispensed with, and curtained ones substituted, a thin wooden partition might be erected along and between the beds *i* and *k*, fig. 601, in the apartment *d*; and from *l* to *p*, and between the beds *m* and *n* fig. 602. It is not probable, however, that box-beds, which are really warm and comfortable within, will ever be laid aside, though of inconvenient construction for sick people. Modifications of their form have been recommended by medical men, consisting chiefly in having the back and ends to open on hinges, and the top made movable, for the sake of promoting ventilation, as well as of affording easy access to patients. Curtains suspended from movable rods, which may be drawn forward, have also been recommended to be placed in front, instead of sliding panels, to screen the people while dressing and undressing, in cases where the beds do not occupy separate apartments.

Fig. 603 is the ground-plan of a double cottage, capable of affording ample accommodation for an ordinary family. Each of the apartments, like the former, is 22 feet in length, by 15 feet in width, with a projection behind to serve as an additional apartment; *a* is the entrance door, *b* the porch, *c* the inner door, *d* the principal apartment, *e* the fire-place, *f* the window, *g* the plate-rack, and *h* the dresser; one bed *i* enters from the apartment *k*, with a door, and though dark, light may occasionally be borrowed from the window *f* close by; another bed *l* is

* See Cunningham's Designs for Farm Cottages and Steadings, Plate II.

entered from the apartment which is lighted by the window *m*, shut

Fig 603.

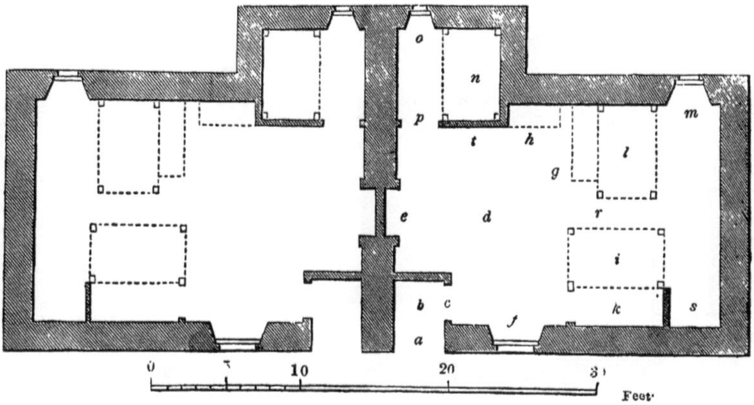

THE GROUND-PLAN OF SERVANTS' HOUSES FIT TO ACCOMMODATE AN ORDINARY FAMILY.

in by the door *r*, and serves as a store-room; the projection behind contains the bed *n*, is lighted by the window *o*, and shut in by the door *p*. By enlarging the plan of this house to 24 feet in length, an additional bed could be placed in length at *s*, lighted by the window *m*; and, in this case, in order to give the bed *l* a separate entrance, its face should be turned into the principal apartment *d*, and the plate-rack *g* removed to where the dresser *h* is, and the dresser placed at *t*. Such a house would accommodate 8 persons, and have a store-room *m* besides.

This plan accommodates a large family on one floor; and a cottage of even larger dimensions, if divided into 2 *separate* apartments, would not afford so much separate accommodation for its inmates as the arrangement of the 4 beds in fig. 603. For the accommodation of still larger families, instead of enlarging the ground floor to an inordinate extent, a second storey should be added. Fig. 604 gives the ground-plan of a double house of this description, where *a* is the entrance door; *b* the porch; *c* the door to the principal apartment *d*, which has a floor-room of 18 feet by 16 feet; *e* the fire-place; *f* the window, and the plate-rack and dresser may be placed along the wall from *c*; *g* is a store-room and milk-house, lighted by the window *h*, and shut in by the door *i*; *k* is a dark apartment projecting behind, for holding coals and other lumber, shut in by the door *l*, or it might be lighted by a window, and converted into a bed-room; and a bed could also be placed under the stair, in the store-room *g*. The stair *b* leads to the upper apartment, which is floored with boards, and extends over the entire apartment *d* and store-room *g*, and may be subdivided in many ways to contain a number of beds to accommodate a large family. The elevation of this

Fig. 604.

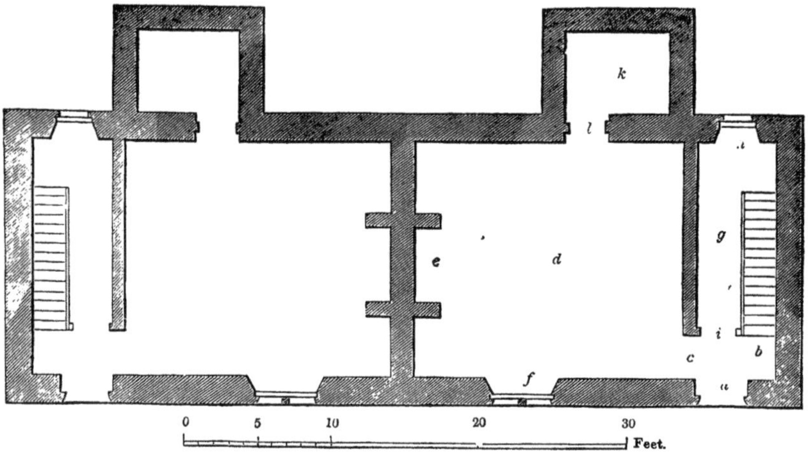

THE GROUND-PLAN OF A TWO-STOREY DOUBLE HOUSE, FIT TO ACCOMMODATE A LARGE FAMILY.

2-storey double house, fig. 605, shews how the upper apartment is lighted. The architectural appearance may be made to suit the taste of the owner. Fig. 64 is a modified plan of cottages at Howick, belonging to Earl Grey.*

Fig. 605.

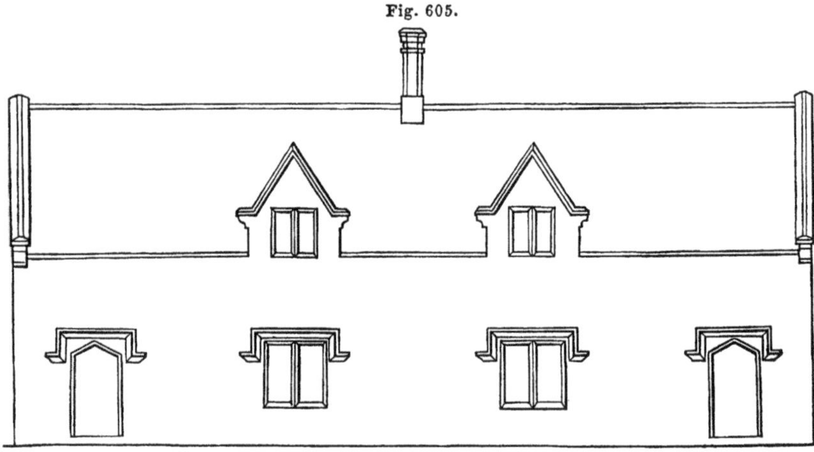

THE ELEVATION OF THE TWO-STOREY DOUBLE HOUSE.

I do not approve of ash-pits, privies, and pig-sties, being placed near dwelling-houses. It is highly proper that these accommodations should be provided for the people, but certainly not close to their dwellings. The trouble of walking a few yards to those places is well compensated in avoiding the nuisance that would be experienced in their vicinage;

* See Report of the Cottage Improvement Society of North Northumberland for 1842, p. 10 and 11.

and enforcement of obedience in their use, though situate at a few yards distance, is, I know, as easy as to any order regarding field-work. Mere proximity to such places is not alone conducive to domestic cleanliness, but the contrary. All the care bestowed by the Dutch on domestic cleanliness is not able to remove the loathsome feeling excited by the odorous exhalations of a certain kind, to be felt within the dwellings of the peasantry, and even of farmers. Those who cannot reconcile themselves to the nuisance of a water-closet scantily supplied with water within their own houses, will have no difficulty in understanding why the labouring man and his family should dislike the still more insufferable nuisance of a privy stuck against the back-wall of their cottage. Rather than use it, he and they will wander afar. On this account, I am surprised at plans being recommended and applauded, with this nuisance attached to them, flanked by an ash-pit and pig-sty.* Nor do I approve of back-doors in cottages to create unnecessary draughts. The cottage, with all its contrivances for heat, of shut doors and windows, of fires, and crowded fire-sides, is, notwithstanding, cold enough in a winter evening. Draughts produced by opposite doors is the most dangerous mode of ventilating houses for the health of the people. A cheerful fire, that always sends its smoke up the chimney, is the best ventilator a cottage can have.

As a good window is an essential article in the cottages of labourers, it may be worth while to give a description of one, made by Messrs M'Culloch and Co., Gallowgate, Glasgow, and for which they received a premium from the Highland and Agricultural Society. " This window is extremely simple in its construction, and may with safety be pronounced efficient in point of comfort and utility, while the price, it is believed, will not be higher than the cheapest description of iron windows now in use, and, for durability, will be preferable to those of any other material. The dimensions that have been recommended for the windows of ordinary cottages are, 39 inches for the height, and 24 inches for the width, within the wooden frames. The size of glass required for these frames is $7\frac{1}{4}$ by $5\frac{1}{4}$ inches. The sash is divided into 2 unequal parts, the lower part having 3 squares in height, and the upper part 2. The lower part is permanently fixed, while the upper part is constructed to turn in the vertical direction on pivots, which are situate in the line of its middle astragal; and both parts are set in a substantial wooden frame, which may either be built in while the wall is erecting, or set in afterwards in the ordinary way, with or without checked rebats, according to the taste of the proprietor. The window and its arrangements will be better understood by reference to the annexed cuts, fig. 606, shew-

* See Journal of the Royal Agricultural Society of England, vol. v. p. 237–44.

Fig. 606.

THE INSIDE ELEVATION OF AN IMPROVED WINDOW SUITED FOR COTTAGES.

Fig, 607.

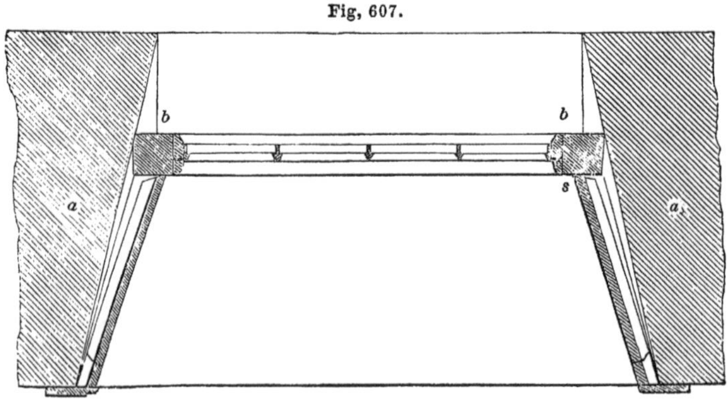

THE PLAN OF AN IMPROVED WINDOW FOR COTTAGES.

ing an inside elevation, fig. 607 a plan, and fig. 608 a vertical section, in each of which a portion of the wall is exhibited, and the same letters refer to the corresponding parts of each figure ; *a* is a portion of the surrounding wall ; *b* the wooden frame of the window ; *c* the lower sash, which is dormant ; and *d* the upper and movable sash. In fig. 608, the upper sash is represented as open for ventilation. When shut, the parts of the opening sash cover and overlap the fixed parts in such a manner as to exclude wind and water ; but when ventilation is required, the arrangement of the parts which produce this is such as to enable the housekeeper to admit air to any extent. For this purpose the notched latch *e* is jointed to a stud in the edge of the sash ; a simple iron pin or stud is also fixed in the wooden frame at *s*, and the notches of the latch being made to fall upon this stud at any reqired distance, the requisite degree of opening is secured ; and when the sash is again closed, the latch falls down parallel with, and close to, the sash. To secure the sashes when shut, the T bolt *f*, in the middle of the meeting bars, has only to be turned ¼ round, and the movable sash is held fast in close contact with the other. Fig. 606 represents the windows as finished with simple dressings, namely, plain deal shutters, facings, and sole, which, at a small expense, would give an air of neatness and comfort to the apartment, and promote a corresponding taste in the other parts of the cottage. Though the dimensions of the window here stated may be conceived sufficient for lighting an apartment of ordinary size, they can, nevertheless, be varied to suit every purpose. This may be done either by employing two such windows as above described, with a mullion of wood or of stone between them, or the single window may be enlarged by 1 or 2 squares in width, or in height, or in both directions."

Fig. 608.

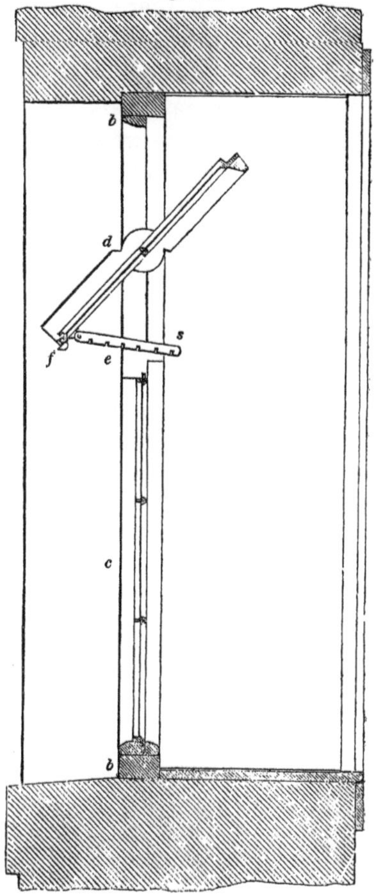

THE VERTICAL SECTION OF AN IMPROVED WINDOW FOR COTTAGES.

It is proper to mention that zinc, in the opinion of tradesmen, is too weak for window-sashes to admit of repair by an unpractised hand. Wood and lead are, for the same reasons, equally unsuitable. Malleable iron, even so thin as to impede the light but little, if the astragals are not provided with flanges for the glass to rest against, the repair must also be a work of some difficulty, and is also deemed unfit for the purpose. Cast-iron, therefore, appears to be the material least liable to objection; but astragals of cast-iron must be of considerable thickness, and such frames, consequently, could not be adapted to a very small size of glass, without materially obscuring the light. The iron sashes, as shewn above, without the wooden frames, cost 5s., and glass for such windows may be purchased at 2¾d. per square.*

The cost of erecting such cottages is inconsiderable, when their accommodation and comfortable furnishing are taken into account. Such as in figs. 601 and 602, would cost thus:—

82 Cubic yards of building,	L.14	7	0
151 Feet of hewn stone,	7	11	0
Roofing and slating,	16	16	0
Joiners' work and furnishing,	14	6	0
	L.53	0	0

They might be executed for L.48, in situations possessing extraordinary advantages. The plan affording greater accommodation in fig. 603 would cost about L.62 each; and the double cottage of fig. 604, with its elevation in fig. 605, would cost L.61, 9s. each, thus:—

Mason-work,	L.50	3	7
Joiner-work,	45	2	10
Slater-work,	17	13	4
Plaster-work,	6	13	4
Glazier work,	1	15	0
Sky-lights,	1	10	0
	L.122	18	1 †

The specification by which such cottages should be built, will be very similar to those given for the farm-house in (242.). Clumps of cottages of 2 together are, in my opinion, more picturesque objects, and more convenient for the out-door conveniences of its inmates, than a long row, though they will, no doubt, cost a little more to erect; and they

* Prize Essays of the Highland and Agricultural Society, vol. xiii., p. 548–50.

† Report of Cottage Improvement Society of North Northumberland for 1842, p. 10.

should not be built so near each other as to create a strong current of wind between them.

The following general remarks on the form and situation of hinds' houses seem judicious. " As respects the particular form and structure of a cottage, and the question whether 2 rooms on the ground-floor, or an upper storey, is to be deemed the best, I think local circumstances must always be allowed to decide. Internal space, ventilation, and dryness, are the great requisites; and if these be provided, there can be no reason for dissatisfaction on minor matters. Cottagers themselves differ much in opinion as to which form they would prefer, being influenced often by the number or ages of their children, the price of fuel, and other considerations. If the family is not large, and the circumstances are good, a second ground-floor room, though used as a bed-room, will be furnished for holiday use as a parlour, in which friends may be received to drink tea; nor will an occasional second fire be thought too costly in such a household. The case, however, is different where a number of children are to be reared with the strictest economy; here the mother will be very glad to have a room up stairs, well warmed at all times from the kitchen below. A third mistress will be able to furnish the upper room as a little parlour, no less conveniently than if placed on the ground-floor. In general, as respects hinds' cottages, I should prefer the upper storey very decidedly, because its warmth and dryness will always invite the family to use it, though unfurnished save with a bed in the corner. Not so the spare down-stairs room, which, being cold, they will allow to degenerate into a place for storing meal and potatoes. It is only when the labourer is a fixed resident, holding directly under the landowner, that he will furnish his second room on the ground-floor comfortably, and warm it by occasional fires."*

In concluding my observations on this subject, I cannot do so in better terms than those employed by Dr Gilly, in describing the condition of the Border hind even in his old-fashioned cottage. The character of the man the Reverend gentleman has drawn with truth, as I myself can testify; and I shall never cease to respect the Border hind for his sterling worth. Should it be your lot to be cast amongst them, endeavour to induce them to remain with you for years, overlooking minor faults, and remembering that an honest hind, with an honest family, all good workers, are of inestimable advantage to a farmer.

" In justice to this excellent class of persons, the hinds of the Border," observes Dr Gilly, " I must describe the interior of a cottage, when it is

* Mr Ralph Carr of Hedgeley, in Report of Cottage Improvement Society of North Northumberland, p. 15-17.

fairly 'put to rights,' after the occupant has taken possession. My description shall be that which will apply, on an average, to the one apartment of all decent agricultural labourers, where a family is not large. Take into consideration the difficulties with which he has to contend, before he can convert his shed into a tidy chamber; see what he has gathered about him, to give an air of comfort and smartness to his dwelling; look at the homely but ample provision which he has laid in for his winter supply; cast your eye over the pages of the books which fill one of his shelves; and when you have done this, you will be able to estimate the character of the meritorious class to which he belongs. We will suppose that it is in the month of December when we open his door. . . . . There are 2 beds placed within a frame-work, which take up the whole of one side of the room. In the centre of the frame-work. and between the 2 beds, is a door which opens into the space behind the beds, where many useful articles, such as pails and tubs, are stowed away; and perhaps if you look in you will see another bed on the floor in the corner. The 2 bedsteads within the frame-work are so contrived as to close in by a sliding panel. This, the whole being kept bright and clean, by rubbing and washing, affords an air of nicety and neatness to the room, which has a pleasing effect; but it is to be wished that the peasantry would substitute curtains or linen hangings for this unwholesome fashion. Confined air is an evil; and the difficulty of approaching an invalid, when surgical aid is necessary, is another evil. It may also be added, that this apparatus sometimes forms a screen, which, desirable on some occasions, is equally objectionable in others. It is hard to persuade a hind and his family to adopt a different apparatus. They plead, truly enough, that many of the cottages would be intolerably cold without such beds, in which they might box themselves up. It is also said that no decency could be observed without them. Every argument they use is an appeal for a better order of cottages, and a second room. . . . . From the panelled beds, which strike attention at first entrance, our eye rests on the dresser and shelves of pretty crockery-ware, which cover the greater part of another side of the room. . . . . Then comes the handsome clock in its tall case, and the chest of drawers, sometimes of new wainscot, sometimes of antique carved work; which many a collector could lay his hands on, and which contains decent apparel for all the family, such as they take an honest pride in wearing on all proper occasions;—and where have we a better-dressed population than among the northern peasants? The barrel of meal and the barrel of herrings occupy their places. The rack above displays some goodly flitches of bacon. White-bread loaves are seldom seen.

The girdle-cake, composed of barley and pease, and the oaten porridge, and potatoes, are the usual substitute for wheaten flour.

" Now, as to the food for our peasant's mind. One book cannot escape our attention. It is ' the big ha' Bible, once his father's pride.' Few of our hinds are without a family Bible, with notes, and with it we remark one or two smaller Bibles. The wish of good King George is realised in this part of the kingdom at least, namely, that every man in his dominions should have a Bible, and that every child should be able to read it. Within the district which has come within my own observation—and my inquiries have been directed to this point—I do not know of one single hind's family which wants a Bible, or of a child who has reached 7 years of age who cannot read, except in cases of imbecility; the Prayer-book, and some few other books of devotion, are ranged by the side of the Bible, and they all shew they have been frequently read. These spiritual treasures have given a tone to the peasant's language—they have stored his mind—they have comforted him in the hour of trial.

" In common conversation he will use provincialisms in the use and collocation of words which may grate upon refined ears; but lead them into a serious and religious train of thought and discourse, and then you will discern the majesty of biblical knowledge—then you will hear him express himself in a manner which none can do but those who are in the frequent habit of reading Scripture in the authorized version. One uniform version has served for all the Bible readers in Great Britain for more than 200 years; and the beautiful harmony arising therefrom is heard as much in the peasant's cottage as in the Queen's palace. Oh ye who doubt the full extent of the salutary effects produced by Bible reading! I have read your observations upon the uneducated use of Scripture; I have read them, with admiration of the learning and devout sentiments by which they are distinguished; but I cannot be one of your converts; I must continue to differ from you; for never have I understood the message of reconciliation more clearly, and never have I felt the force of ' Christian morals ' more powerfully, than when I have been conversing with some pious hind, whose principal reading is Scripture. . . . . One word more, on this part of my subject. The hind's cottage very seldom contains any loose productions—the vulgar ballad and the ribald jest are not to his taste.

" Such is the occupant of the miserable habitation which is usually provided for the peasantry of Northumberland and the Borders of Scotland, and, I am afraid, for the majority of agricultural labourers through the land. But a more considerate spirit is abroad; many proprietors are turning their attention to this matter; and cottages, and groups of cottages, are rising in all directions, which do honour to those who provided and built them, and secure the comfort of their inmates."*

---

89. OF THE CARE TO BE BESTOWED IN THE PRESERVATION OF IMPLEMENTS.

"Beneath thy roof secure the tackling lay."

ELTON'S HESIOD.

The farmer is often charged with neglecting his implements, by unnecessary exposure to the weather; and the charge is partially well-founded, though those who make it do not understand the cause of the apparent neglect. Implements are used both within and without doors; and those used without doors may be divided into those in use every season, and those only used occasionally. It is scarcely to be expected that implements very frequently in use, operating upon the soil, can be otherwise than constantly exposed to the weather. Fortunately, on the score of economy, the implements thus employed are of simple construction; and are, therefore, less affected by the vicissitudes of weather, as well as less costly when renewed.

The implement most frequently in use is the *plough*, and, being the chief one for operating upon the soil, is constantly exposed in the field, its situation in winter often being, as described by Thomson,

——" Where the well-used plough
Lies in the furrow, loosen'd from the frost."

Being composed of few parts, it is little liable to derangement; and being now mostly made of iron, its durability is secured. The proper treatment of this implement is, that whenever the irons, as they are termed, that is, the sock or share and coulter, become blunted, they should be sharpened, and with much ploughing in sharp gravelly soil, these may require to be sharpened every day; but besides this, those parts require laying occasionally, that is, fresh portions of iron welded to their points, to sustain their proper dimensions in length and breadth. The mould-board should be renewed before it is worn much away behind. The plough-spade should always be kept in repair, as being a useful instrument in removing the obstructing earth from the mould-board. The wedges which fasten the coulter-head in its bed being often

* Gilly's Appeal in Behalf of the Peasantry of the Border, p. 21-27.

driven out and in with stones, it has often occurred to me, that a hammer would be a useful adjunct to a plough, and might be attached somewhere in its bosom; but whether it would get leave to remain there by pilfering hands I am doubtful, though it has often excited my surprise to find the irons of ploughs so seldom meddled with in the fields, where they are so much exposed. A slide is a most useful instrument for conveying the plough upon from field to field. Ploughs are seldom painted after leaving the makers' hands; but surely they might receive a good coat of paint at least once a-year.

Harrows, being the implement most commonly in use next to the plough, are much in the field, and, of course, exposed to the weather; and though made entirely of wood, last a long time. Not being required in winter, ground not being in a state to be harrowed but when dry, they are then removed from the field, and placed in the implement-house. The tines are usually laid and made sharp once a-year, namely, just before the commencement of the oat-seed in spring, when their repair is most commonly done in a hurry. To prevent disappointment, however, at this important season, they should be laid and refixed before the harrows are set past for the winter, and then the implement will be ready for use, whenever required in spring. Harrows are never painted more than ploughs; but should be cleaned and painted when set past. A common practice, too, in conveying harrows from one field to another, is to place them in carts, which involves the separation of their 2 parts, and of the master swing-tree; whereas a sledge made on purpose contains them entire, and removes them with ease, with a single horse, from field to field.

The *roller* being only occasionally in use, in pulverising the soil, and rolling the young grass and spring crops in spring, and in pulverising the soil in summer-fallow, it is replaced in its shed whenever its services are no more wanted. It should, however, always be set past in a state of complete repair, that disappointment may not be felt at the moment it is desired to employ it; and the wood-work should be painted occasionally.

The small ploughs, such as the *ribbing* and *double mould-board plough*, being chiefly used in summer, are allowed to lie too long in the fields after their employment has ceased; and if removed before winter sets in, are placed in the implement-house dirty and worn. When no longer required, they ought to be scraped clean of earth, the irons laid and blackened, and the wood painted, before being put aside in the implement-shed.

There are no implements which receive less regard, when not in use than the whole class of *scufflers* and *grubbers*, which get leave to remain at the sides of head-ridges, and corners of turnip and potato fields, perhaps the whole winter. Many of them being made entirely of iron, do not suffer much, it is true, of deterioration from weather; but, being composed of minute parts—of tines, coulters, small wedges, and screw-bolts—these suffer from that cause, and do their work indifferently on getting loose. Instead, therefore, of being permitted to lie unregarded in the field, their worn parts should be immediately repaired, and themselves placed in the implement-shed.

All the classes of the more delicately constructed machines, such as the *grass-seed*, *drill-sowing*, and *turnip-sowing machines*, are seldom allowed to remain longer in the field than in use, but more, I fear, from being in the way of other work, than regard for their own safety; for, though removed from the field, they are almost always allowed to remain unheeded in the neighbourhood of the steading for a considerable time. Instead of this treatment, they should be immediately repaired, taken to pieces and cleaned, the journals greased, and the separate parts stowed away in the implement-house.

As to the class of small manual implements, such as *turnip-hoes*, *spreading-graips*, *dung-hawks*, *hay-knives*, *scythes*, and such like, if they are not placed in the implement-house when out of use, many of them will be lost. When scythes and hoes get worn, they should be thrown into the old-iron store, and the handles furnished with new articles.

Of all implements, *carts*, perhaps, receive the worst treatment. Though much in use in the fields, they are never left there, it is true, and are brought to the steading, but are seldom put under cover, and are subjected to every species of weather, whether to the shrinking power of the sun's rays in summer, or the rotting effects of the damps and rains of winter; and considering that carts are constructed of many parts, the wonder is they last so long with the treatment they receive. Their axles are not unfrequently neglected of grease; and as to their bodies and wheels being washed, not a mop nor a drop of water is ever thrown upon them—except when receiving ablution from rain, or in an occasional passage through a river. Carts are kept in the same state of cleanness, whether employed in clearing out a dung-court, or in bringing home a crop of potatoes for the food of man, or of turnips for that of beasts. A hole in the bottom or sides will get leave to enlarge till it lets through an article of no inconsiderable size; and a wheel-ring will be allowed to become loose, till some day it flies off altogether, to the risk of the felloes being broken to pieces. When such a mishap occurs from home, it tells strongly against the care bestowed by the steward in such matters, and places him in an awkward position with his master.

The *harness*, too, is allowed to remain in a broken state, except with such repair as a bit of string can make, until some important part gives way. The saddler should be desired to make his periodic visits at regular times, and repair whatever is amiss, even of the most trifling nature, and not delay repairs until a great extent of them are required.

The horses should not be neglected in their *shoeing*. It is not an unfrequent neglect to allow the shoes to remain on till they become loose, or the crust of the hoof grows beyond them; in either case the horses run the risk of being lamed, and if not moving upon soft earth, they would inevitably acquire thrushes and corns. The shoes should not be allowed to be worn so thin as to endanger their breaking, nor to remain so long on as to get a set in the hoof. When the shoeing is taken by the smith by the piece, he is apt to persuade the men to let the shoes hang on as long as they can; but the steward should see to the state of the horses' feet, and order them to the smithy whenever he conceives it necessary. I am aware that shoeing often interferes with field-work; but, in the busy season, the horses should be sent to the smithy in the evening, with a bunch of grass, and if the smith has shoes ready forged, which he should always have for horses he is in the habit of shoeing, not much time will be lost; but unless the steward look after this matter, the smith will be glad of the opportunity to take assistance from the man at the forehammer to serve his own purpose.

In regard to the in-door implements, the *thrashing-machine* should be cleaned out thoroughly every time a different kind of corn is to be thrashed, otherwise samples will be impure. The gudgeons are usually oiled every time the mill is in use. Whenever any part of the machine requires repair it should receive it immediately, otherwise the force to which it is subjected may render the fracture serious. All the machinery connected with the thrashing mill ought to be boarded in, in case of accidents; and such a precaution is the more necessary where women are employed, the loose state of whose dress rendering their approach to moving machinery dangerous to their safety.

Holes in *sacks* and in *barn* and *chaff-sheets* should be instantly repaired, by patching and darning; nor should a broken mesh in a riddle be overlooked, so as to render the clean-winnowing of grain unavailing.

The repairs of *slating*, *water-runs*, *drains*, or *wood-work* of the steading, usually required once a-year, are most conveniently executed in summer, when the materials soon dry, the days are long, and the steading is not in use.

The *fitting up of the implement-house* for the accommodation of the finer and smaller implements, is an arrangement of some importance, to the keeping of its floor always unencumbered, and access to particular implements, required at a time, entirely free. Wheels, loose shafts, and angular pieces of iron, are best suspended against a wall from iron pins. Articles of length, such as sowing-boxes, are best supported against a wall on brackets. Small articles of iron and other materials are best kept together on shelves. Hand-hoes, weed-hooks, and such like, are best tied together in bundles, and placed out of the way across the balks; where the scythes are also best suspended The bodies of small ploughs, grubbers, scufflers, should be placed against the foot of the walls, and kept in their position, by fastenings of cords to staples driven into the walls. If every article were put into its own place at the end of its season, confusion would be avoided, and many more articles find accommodation in the implement-house, than when every-thing is put down anywhere, without regard to order. To maintain order in the implement-house, I know, requires firmness; but enforcement of order always carries this conviction along with it, that it is easier to put a thing in its own proper place, than anywhere else, inasmuch as the place allotted for it contains and retains it in the best state and position.

I have now brought to a termination the task I had imposed upon myself in writing this work. If you will but follow the prescriptions I have given in it, for conducting the larger operations of the field, and for treating the various animals of the farm; and—not to mention the proper ploughing and manuring of the soil—as the practice of every farmer demonstrates the necessity of affording due attention to those most important, because fundamental operations;—if you finish off your fields in a manner indicating care and neatness, ploughing round their margins, and turning over the corners—if you keep your fences clean and in a state of repair—your fields free of weeds—if you give your stock abundance of fresh food at regular intervals in winter, and supply them with plenty of clean water on fresh pastures in summer—if you have the farm-roads always in a serviceable state, and every thing about the steading neat and orderly;—if you exhibit skill and taste in all these matters, and put what is called *a fine skin* on your farm, you will not fail to earn for yourself the appellation of a good and exemplary farmer; and when you have every thing about you "thus well disposed," you will find, with Hesiod of old, that profitably, as well as creditably, for you "shall glide away thy rustic year."

INDEX.

THE END.

PRINTED BY NEILL AND COMPANY, EDINBURGH.

DIRECTIONS FOR BINDING THE BOOK OF THE FARM.

The last fourteen parts are to form two volumes.

Vol. II. to end at page 728.

Vol. III. to begin with page 729.

The Book will thus, when bound, be in Three Volumes, and title-pages are given accordingly. Cancel the Title-page, Contents, and List of Plates to Vol. I. inserted in Part VII.

DIRECTIONS FOR PLACING THE PLATES.

Plates I. to XVII. to be inserted at the end of Vol. I.

... XVIII. to XXXIII. ... ... Vol. III.

The attention of the Bookbinder is particularly requested to the above directions.

EDINBURGH, 4*th September* 1844.

PRINTED BY NEILL AND COMPANY, EDINBURGH.

Gourlay Steel

Thomas Dobbie

DOUBLE HORSE CART.

THRASHING MACHINE.

Fig. 319.

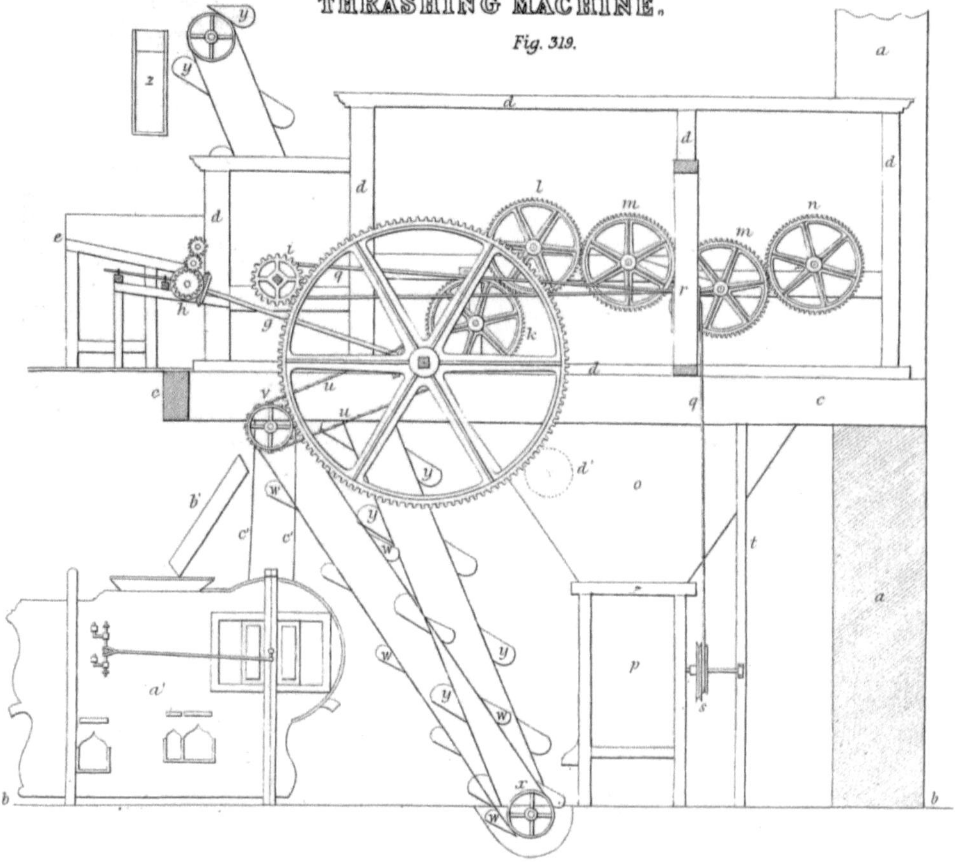

Scale of Feet

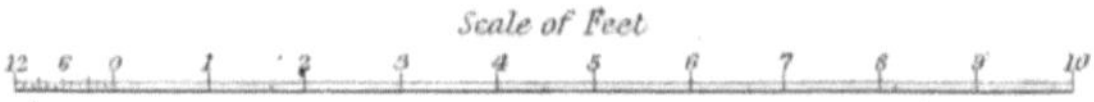

Fig. 320

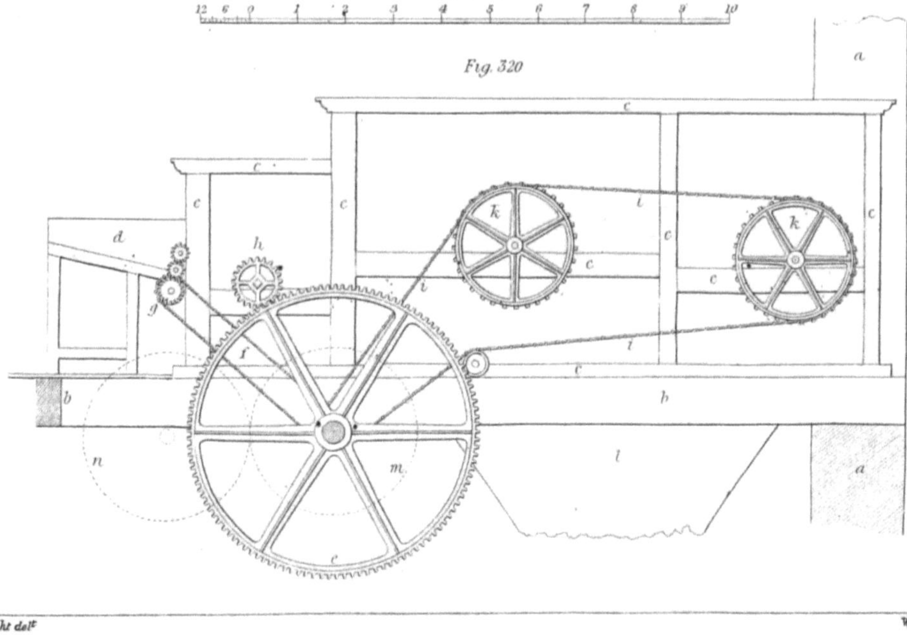

G. H. Slight del^t

W. H. Lizars sculp^t

Published by Wm Blackwood & Sons. Edinburgh & London.

THRASHING MACHINE.

Fig. 321.

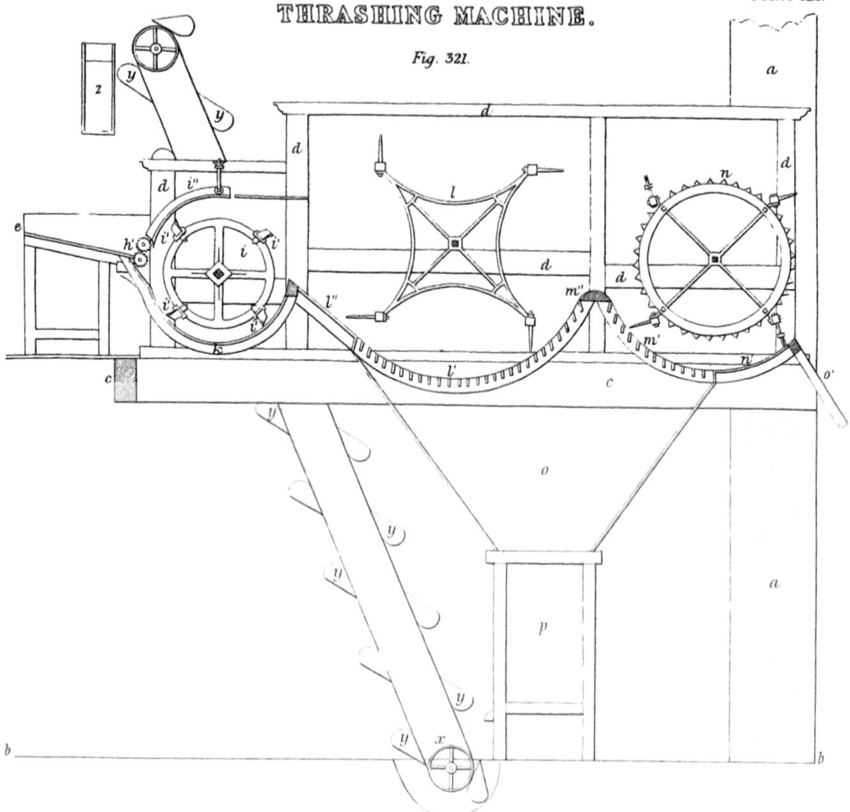

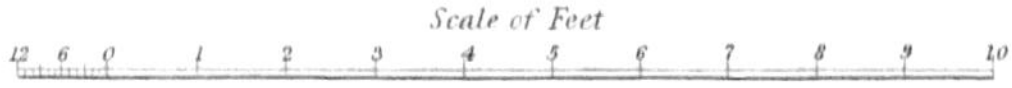

Fig. 322

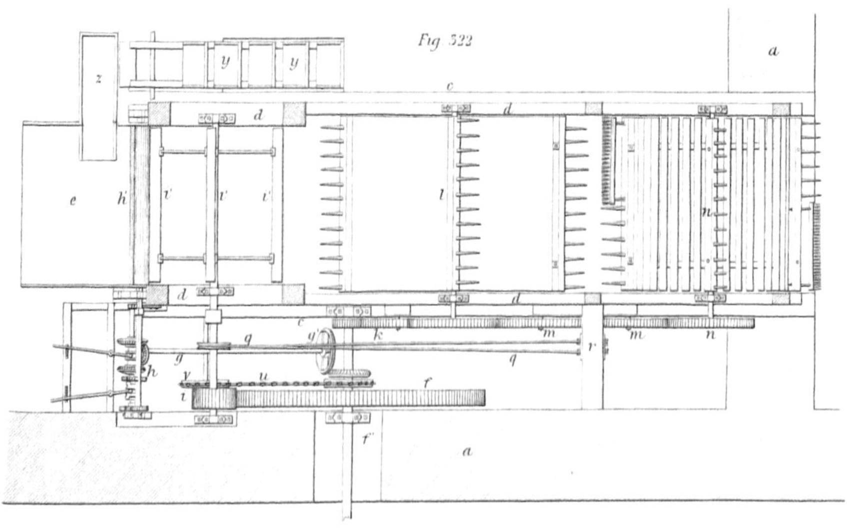

G H Slight delt W H Lizars sculpt

Published by Wm. Blackwood & Sons, Edinburgh & London.

CRANK STEAM ENGINE.

Fig 325

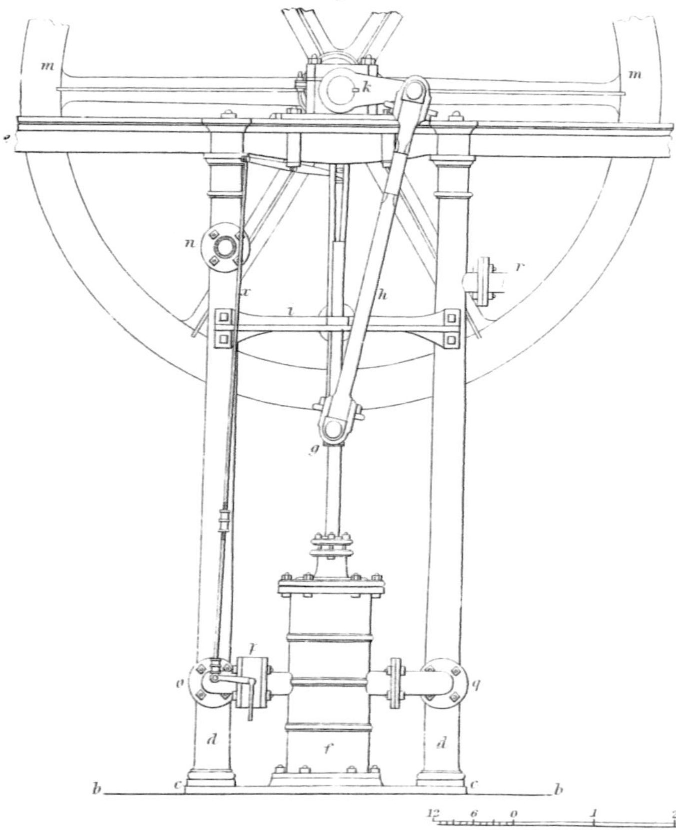

Fig 326

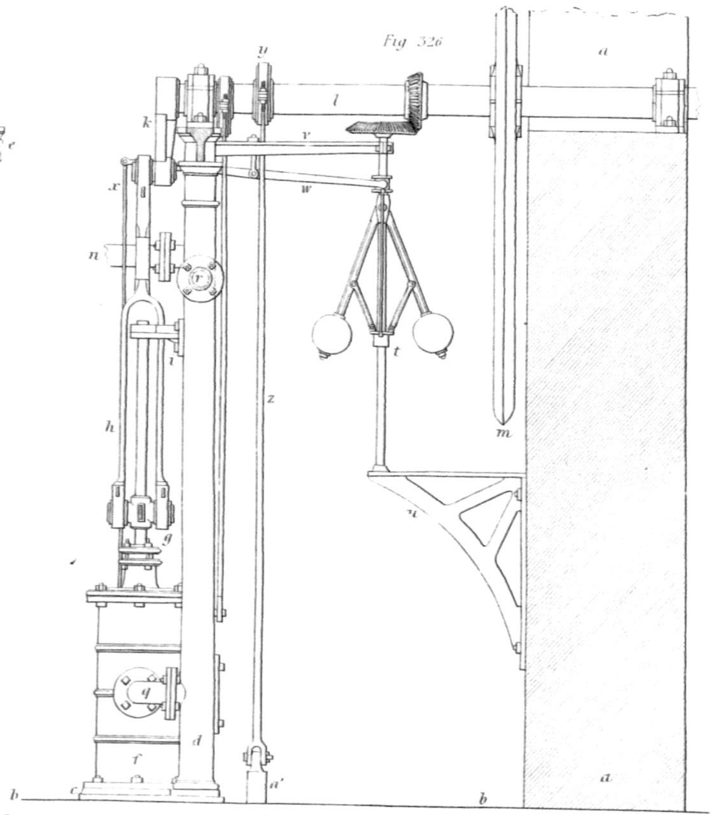

G. H. Slight delt.

BEAM STEAM ENGINE.

Fig 328.

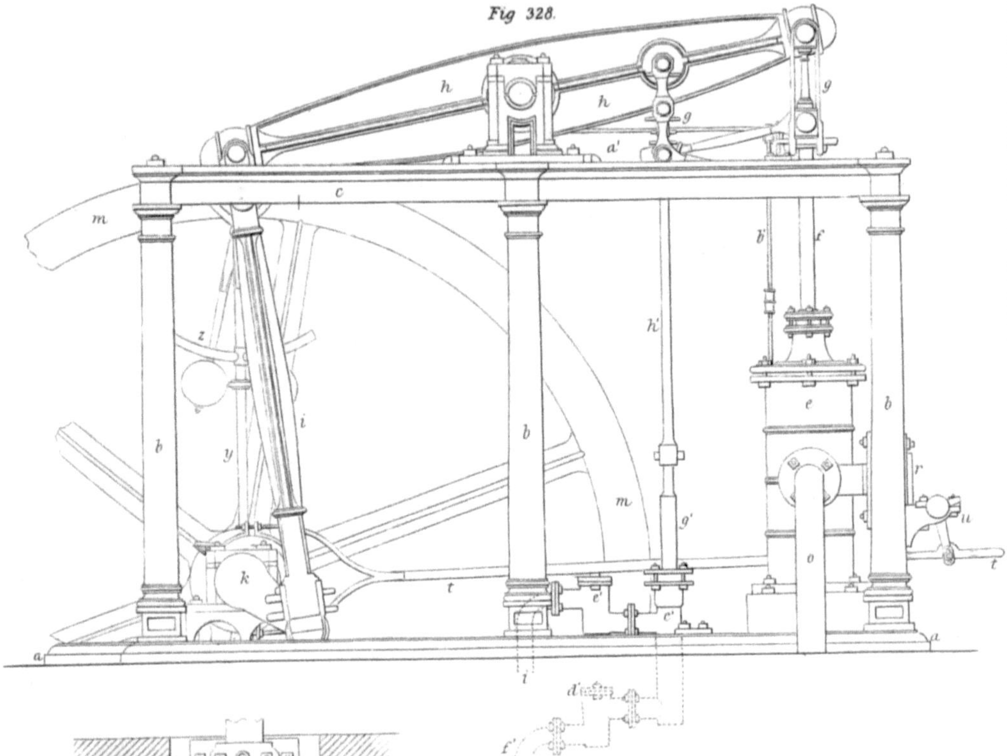

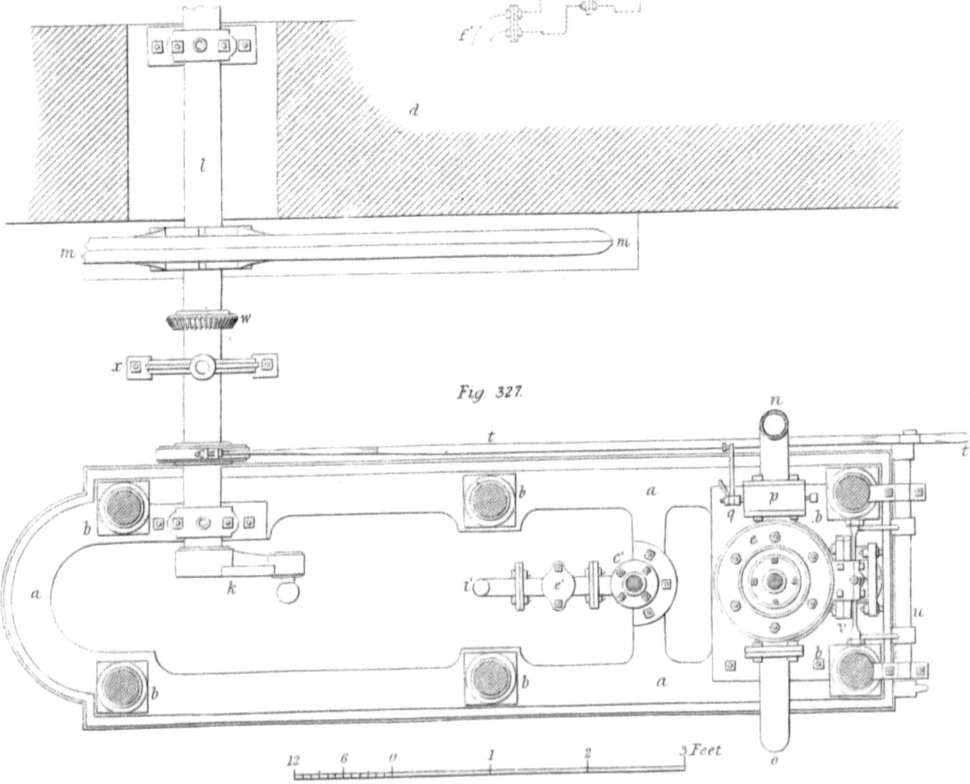

Fig 327.

G.H. Slight delt. W.H. Lizars sculpt.

Published by Wm. Blackwood & Sons, Edinburgh & London.

HORSE WHEEL.

Fig. 334

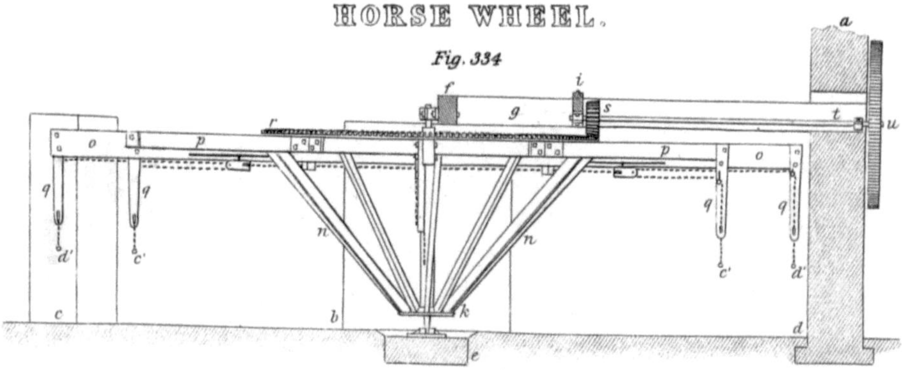

Fig. 335 a

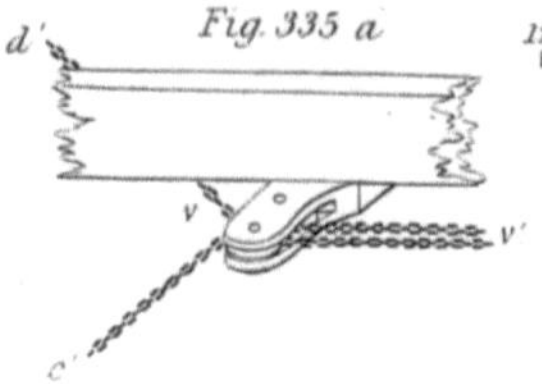

Fig 335 b

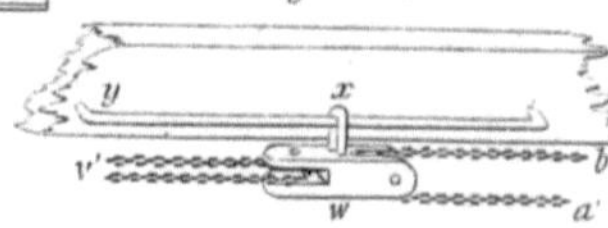

Fig. 333.

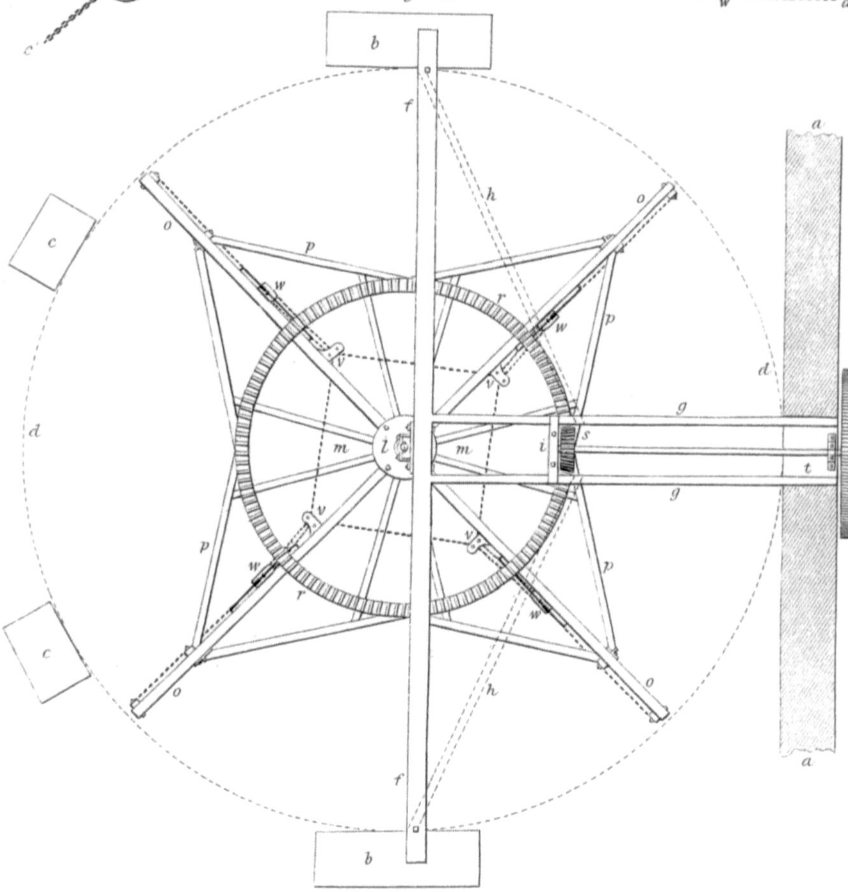

Slight del.t W.H.Lizars sculpt

Published by W.m Blackwood & Sons, Edinburgh & London.

STEAM ENGINE BOILER.

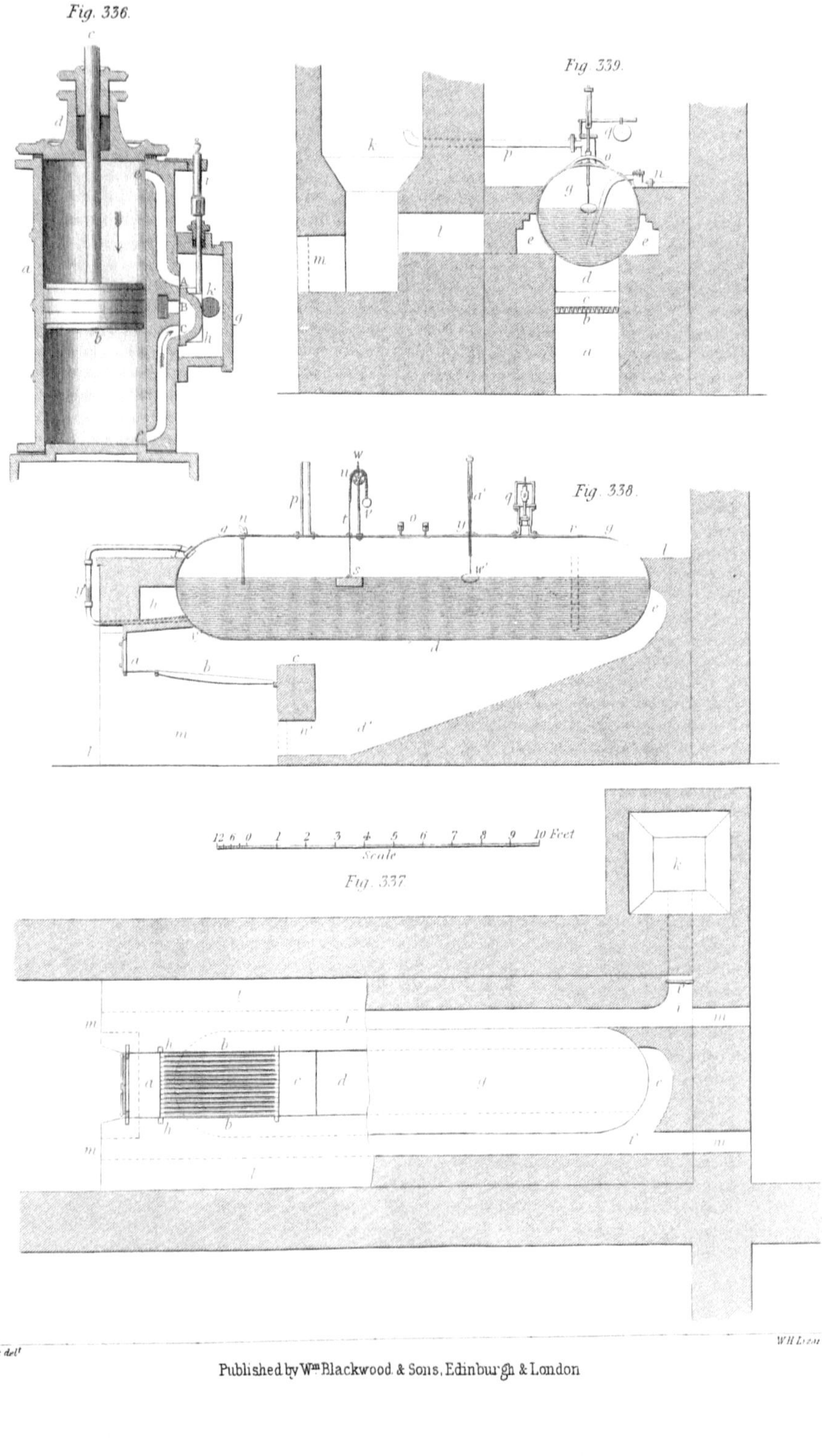

H. Slight del.t W. H. Lizars sculp.t

Published by W.m Blackwood & Sons, Edinburgh & London

WATER WHEEL.

Fig. 343.

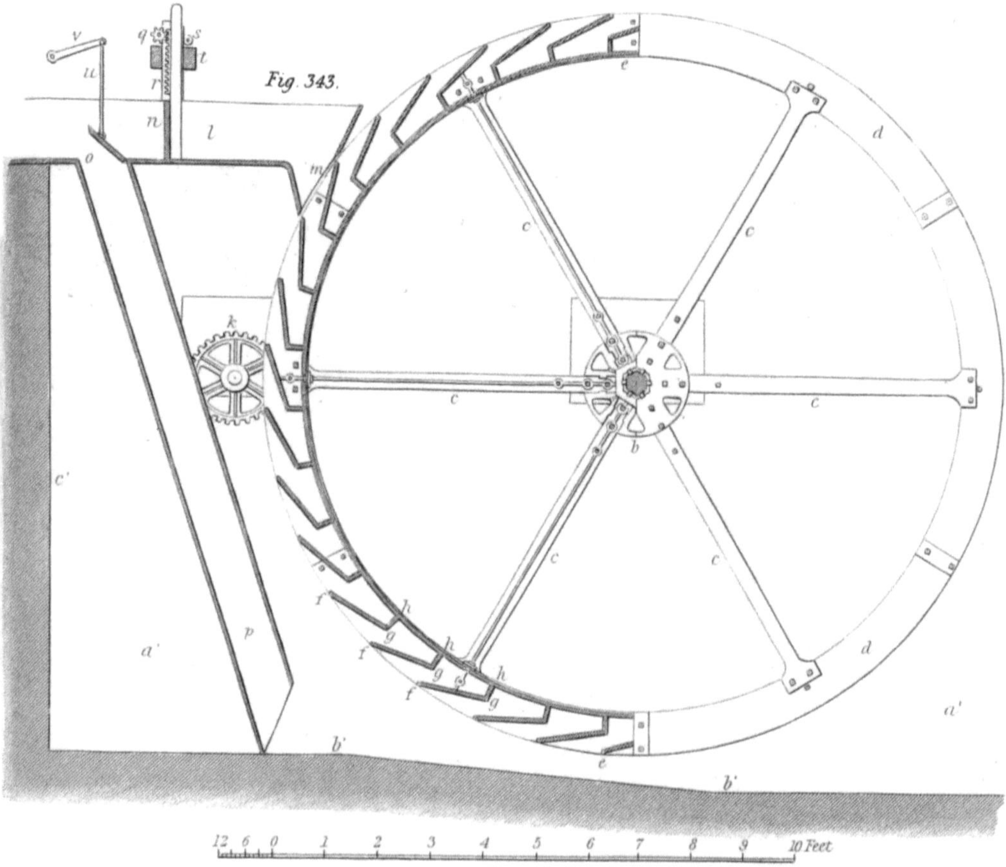

Fig. 342

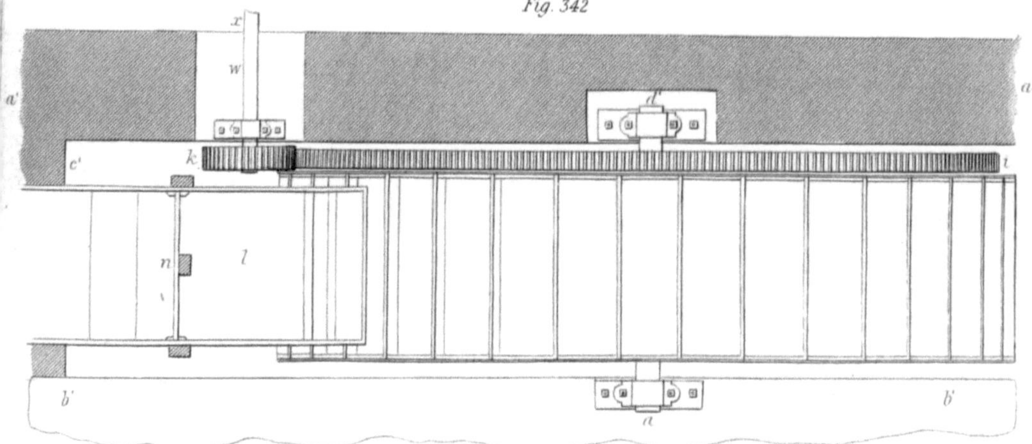

Fig. 340.

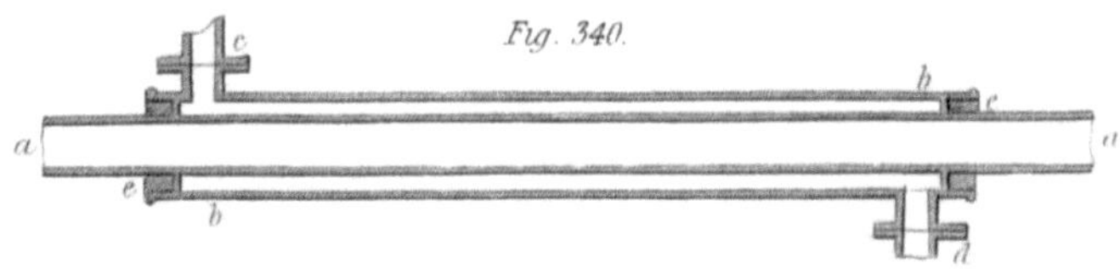

Slight del.t W.H.Lizars sculp.t

Published by W.m Blackwood & Sons, Edinburgh & London.

BROAD-CAST SOWING MACHINE.

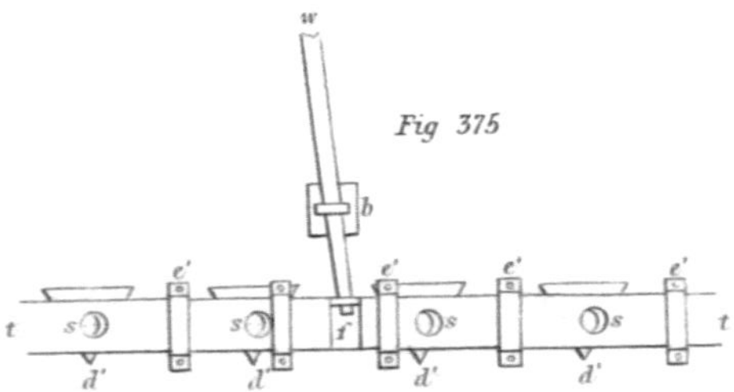

Fig 375

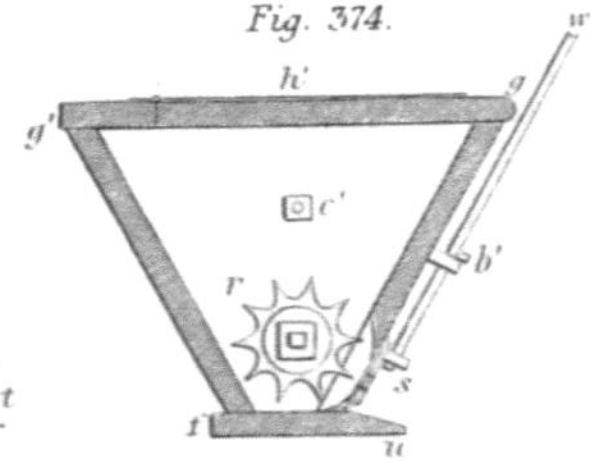

Fig. 374.

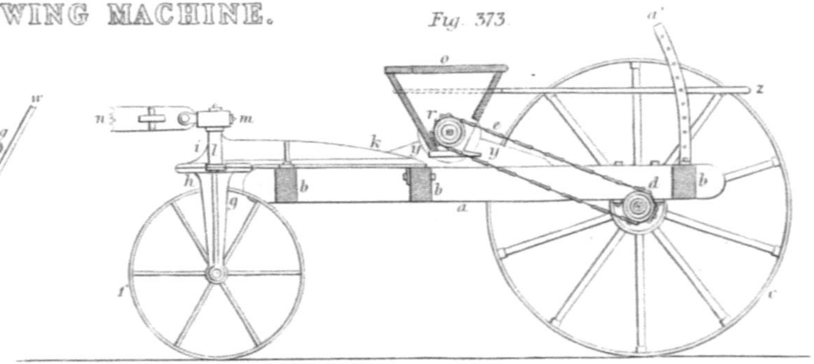

Fig. 373

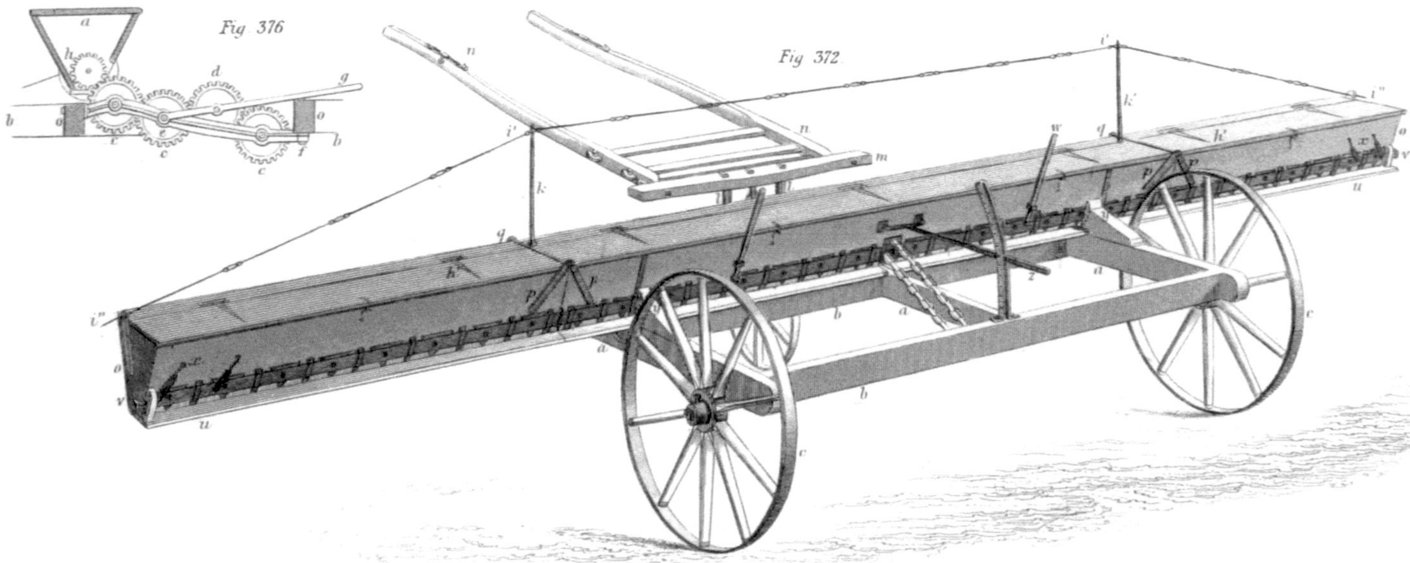

Fig 376

Fig 372

G.H. Slight delt

W.H. Lizars sculpt

Published by Wm Blackwood & Sons, Edinburgh & London

DRILL SOWING MACHINES.

Fig. 381.

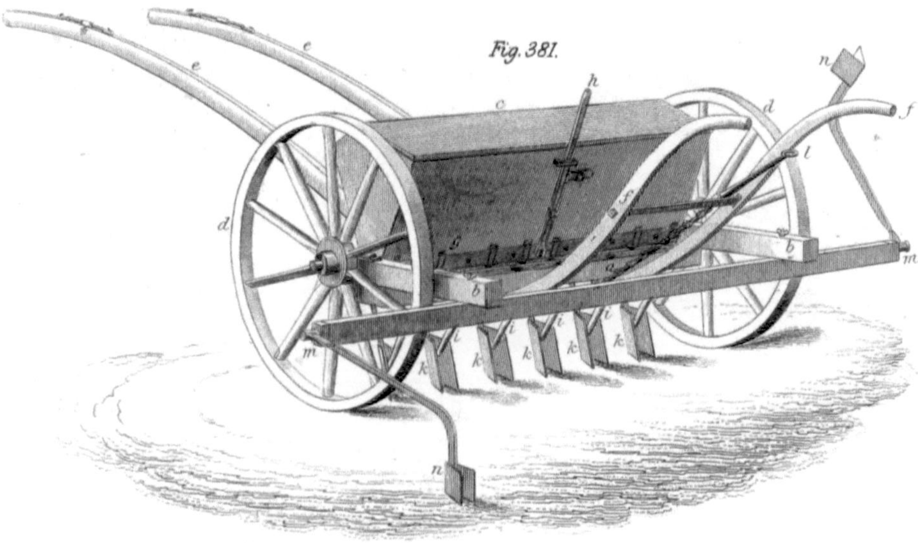

The Common Drill Sowing Machine.

Fig. 382.

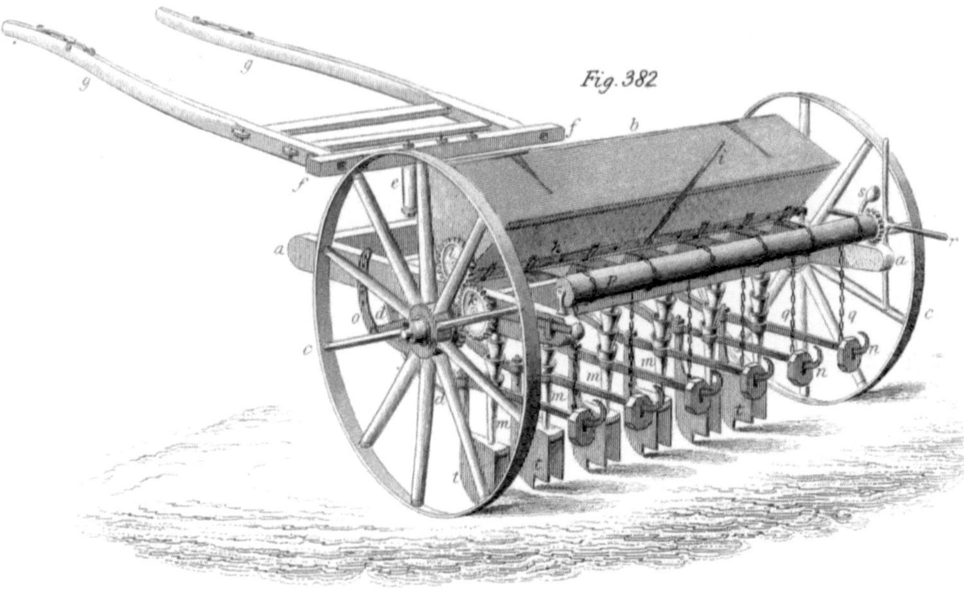

The New Lever Drill Sowing Machine.

G H Slight del.t W H. Lizars Sculp.t

Published by W.m Blackwood & Sons, Edinburgh & London.

Gourlay Steell

Thomas Dobbie

FAT WETHERS.

Published by Wm. Blackwood & Sons, Edinburgh & London.

FINLAYSON'S IMPROVED GRUBBER.

Fig. 412.

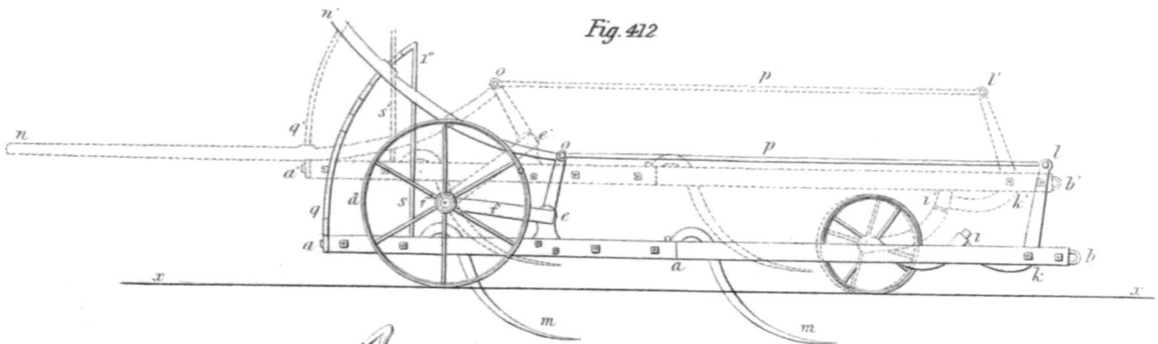

Fig. 413.

KIRKWOOD'S GRUBBER.

Fig. 414

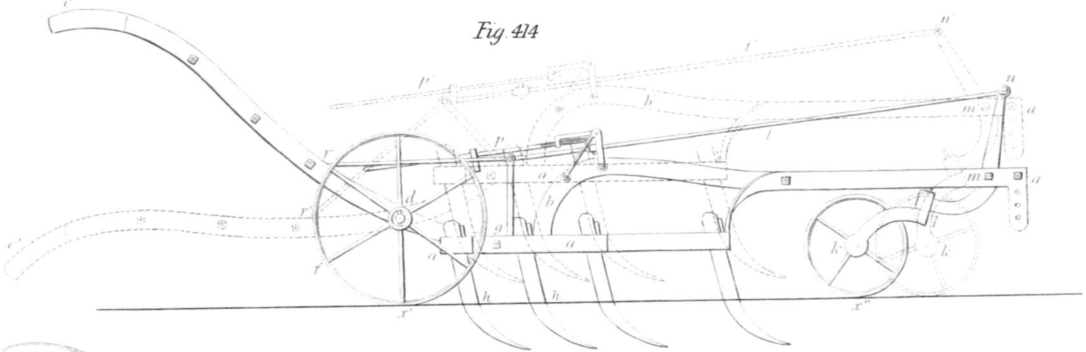

Fig. 415

G.H. Slight del.^{r}

W.H. Lizars sculp

TURNIP & BONE DUST DRILL.

Fig. 432.

HAY TEDDING MACHINE.

Fig 462

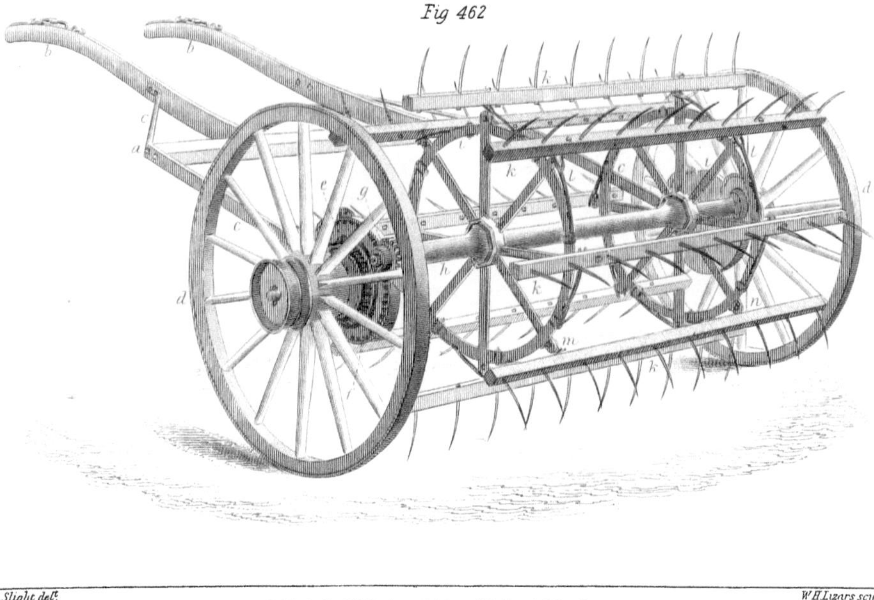

G. H. Slight delt. W. H. Lizars sculpt.

Published by Wm Blackwood & Sons, Edinburgh & London.

SHORT-HORN BULL.

THE PROPERTY OF THE AUTHOR.

John Sheriff A.R.S.A. Thos Landseer

LEICESTER TUP.

THE PROPERTY OF HIS GRACE THE DUKE OF BUCCLEUCH.